T0189562

Lecture Notes in Computer Science 11303

Commenced Publication in 1973
Founding and Former Series Editors:
Gerhard Goos, Juris Hartmanis, and Jan van Leeuwen

Editorial Board

Long Cheng · Andrew Chi Sing Leung
Seiichi Ozawa (Eds.)

Neural
Information Processing

25th International Conference, ICONIP 2018
Siem Reap, Cambodia, December 13–16, 2018
Proceedings, Part III

 Springer

Editors
Long Cheng 🆔
The Chinese Academy of Sciences
Beijing, China

Seiichi Ozawa
Kobe University
Kobe, Japan

Andrew Chi Sing Leung
City University of Hong Kong
Kowloon, Hong Kong SAR, China

ISSN 0302-9743   ISSN 1611-3349 (electronic)
Lecture Notes in Computer Science
ISBN 978-3-030-04181-6   ISBN 978-3-030-04182-3 (eBook)
https://doi.org/10.1007/978-3-030-04182-3

Library of Congress Control Number: 2018960916

LNCS Sublibrary: SL1 – Theoretical Computer Science and General Issues

This Springer imprint is published by the registered company Springer Nature Switzerland AG
The registered company address is: Gewerbestrasse 11, 6330 Cham, Switzerland

Preface

The 25th International Conference on Neural Information Processing (ICONIP 2018), the annual conference of the Asia Pacific Neural Network Society (APNNS), was held in Siem Reap, Cambodia, during December 13–16, 2018. The ICONIP conference series started in 1994 in Seoul, which has now become a well-established and high-quality conference on neural networks around the world. Siem Reap is a gateway to Angkor Wat, which is one of the most important archaeological sites in Southeast Asia, the largest religious monument in the world. All participants of ICONIP 2018 had a technically rewarding experience as well as a memorable stay in this great city.

In recent years, the neural network has been significantly advanced with the great developments in neuroscience, computer science, cognitive science, and engineering. Many novel neural information processing techniques have been proposed as the solutions to complex, networked, and information-rich intelligent systems. To disseminate new findings, ICONIP 2018 provided a high-level international forum for scientists, engineers, and educators to present the state of the art of research and applications in all fields regarding neural networks.

With the growing popularity of neural networks in recent years, we have witnessed an increase in the number of submissions and in the quality of submissions. ICONIP 2018 received 575 submissions from 51 countries and regions across six continents. Based on a rigorous peer-review process, where each submission was reviewed by at least three experts, a total of 401 high-quality papers were selected for publication in the prestigious Springer series of *Lecture Notes in Computer Science*. The selected papers cover a wide range of subjects that address the emerging topics of theoretical research, empirical studies, and applications of neural information processing techniques across different domains.

In addition to the contributed papers, the ICONIP 2018 technical program also featured three plenary talks and two invited talks delivered by world-renowned scholars: Prof. Masashi Sugiyama (University of Tokyo and RIKEN Center for Advanced Intelligence Project), Prof. Marios M. Polycarpou (University of Cyprus), Prof. Qing-Long Han (Swinburne University of Technology), Prof. Cesare Alippi (Polytechnic of Milan), and Nikola K. Kasabov (Auckland University of Technology).

We would like to extend our sincere gratitude to all members of the ICONIP 2018 Advisory Committee for their support, the APNNS Governing Board for their guidance, the International Neural Network Society and Japanese Neural Network Society for their technical co-sponsorship, and all members of the Organizing Committee for all their great effort and time in organizing such an event. We would also like to take this opportunity to thank all the Technical Program Committee members and reviewers for their professional reviews that guaranteed the high quality of the conference proceedings. Furthermore, we would like to thank the publisher, Springer, for their sponsorship and cooperation in publishing the conference proceedings in seven volumes of *Lecture Notes in Computer Science*. Finally, we would like to thank all the speakers, authors, reviewers, volunteers, and participants for their contribution and support in making ICONIP 2018 a successful event.

October 2018

Jun Wang
Long Cheng
Andrew Chi Sing Leung
Seiichi Ozawa

ICONIP 2018 Organization

General Chair

Jun Wang   City University of Hong Kong,
Hong Kong SAR, China

Advisory Chairs

Akira Hirose   University of Tokyo, Tokyo, Japan
Soo-Young Lee   Korea Advanced Institute of Science and Technology,
South Korea
Derong Liu   Institute of Automation, Chinese Academy of Sciences,
China
Nikhil R. Pal   Indian Statistics Institute, India

Program Chairs

Long Cheng   Institute of Automation, Chinese Academy of Sciences,
China
Andrew C. S. Leung   City University of Hong Kong, Hong Kong SAR,
China
Seiichi Ozawa   Kobe University, Japan

Special Sessions Chairs

Shukai Duan   Southwest University, China
Kazushi Ikeda   Nara Institute of Science and Technology, Japan
Qinglai Wei   Institute of Automation, Chinese Academy of Sciences,
China
Hiroshi Yamakawa   Dwango Co. Ltd., Japan
Zhihui Zhan   South China University of Technology, China

Tutorial Chairs

Hiroaki Gomi   NTT Communication Science Laboratories, Japan
Takashi Morie   Kyushu Institute of Technology, Japan
Kay Chen Tan   City University of Hong Kong, Hong Kong SAR,
China
Dongbin Zhao   Institute of Automation, Chinese Academy of Sciences,
China

Publicity Chairs

Zeng-Guang Hou   Institute of Automation, Chinese Academy of Sciences,
                                  China
Tingwen Huang   Texas A&M University at Qatar, Qatar
Chia-Feng Juang   National Chung-Hsing University, Taiwan
Tomohiro Shibata   Kyushu Institute of Technology, Japan

Publication Chairs

Xinyi Le   Shanghai Jiao Tong University, China
Sitian Qin   Harbin Institute of Technology Weihai, China
Zheng Yan   University Technology Sydney, Australia
Shaofu Yang   Southeast University, China

Registration Chairs

Shenshen Gu   Shanghai University, China
Qingshan Liu   Southeast University, China
Ka Chun Wong   City University of Hong Kong,
                                  Hong Kong SAR, China

Conference Secretariat

Ying Qu   Dalian University of Technology, China

Program Committee

Hussein Abbass   University of New South Wales at Canberra, Australia
Choon Ki Ahn   Korea University, South Korea
Igor Aizenberg   Texas A&M University at Texarkana, USA
Shotaro Akaho   National Institute of Advanced Industrial Science
                                  and Technology, Japan
Abdulrazak Alhababi   UNIMAS, Malaysia
Cecilio Angulo   Universitat Politècnica de Catalunya, Spain
Sabri Arik   Istanbul University, Turkey
Mubasher Baig   National University of Computer and Emerging
                                  Sciences Lahore, India
Sang-Woo Ban   Dongguk University, South Korea
Tao Ban   National Institute of Information and Communications
                                  Technology, Japan
Boris Bačić   Auckland University of Technology, New Zealand
Xu Bin   Northwestern Polytechnical University, China
David Bong   Universiti Malaysia Sarawak, Malaysia
Salim Bouzerdoum   University of Wollongong, Australia
Ivo Bukovsky   Czech Technical University, Czech Republic

Ke-Cai Cao	Nanjing University of Posts and Telecommunications, China
Elisa Capecci	Auckland University of Technology, New Zealand
Rapeeporn Chamchong	Mahasarakham University, Thailand
Jonathan Chan	King Mongkut's University of Technology Thonburi, Thailand
Rosa Chan	City University of Hong Kong, Hong Kong SAR, China
Guoqing Chao	East China Normal University, China
He Chen	Nankai University, China
Mou Chen	Nanjing University of Aeronautics and Astronautics, China
Qiong Chen	South China University of Technology, China
Wei-Neng Chen	Sun Yat-Sen University, China
Xiaofeng Chen	Chongqing Jiaotong University, China
Ziran Chen	Bohai University, China
Jian Cheng	Chinese Academy of Sciences, China
Long Cheng	Chinese Academy of Sciences, China
Wu Chengwei	Bohai University, China
Zheru Chi	The Hong Kong Polytechnic University, SAR China
Sung-Bae Cho	Yonsei University, South Korea
Heeyoul Choi	Handong Global University, South Korea
Hyunsoek Choi	Kyungpook National University, South Korea
Supannada Chotipant	King Mongkut's Institute of Technology Ladkrabang, Thailand
Fengyu Cong	Dalian University of Technology, China
Jose Alfredo Ferreira Costa	Federal University of Rio Grande do Norte, Brazil
Ruxandra Liana Costea	Polytechnic University of Bucharest, Romania
Jean-Francois Couchot	University of Franche-Comté, France
Raphaël Couturier	University of Bourgogne Franche-Comté, France
Jisheng Dai	Jiangsu University, China
Justin Dauwels	Massachusetts Institute of Technology, USA
Dehua Zhang	Chinese Academy of Sciences, China
Mingcong Deng	Tokyo University of Agriculture and Technology, Japan
Zhaohong Deng	Jiangnan University, China
Jing Dong	Chinese Academy of Sciences, China
Qiulei Dong	Chinese Academy of Sciences, China
Kenji Doya	Okinawa Institute of Science and Technology, Japan
El-Sayed El-Alfy	King Fahd University of Petroleum and Minerals, Saudi Arabia
Mark Elshaw	Nottingham Trent International College, UK
Peter Erdi	Kalamazoo College, USA
Josafath Israel Espinosa Ramos	Auckland University of Technology, New Zealand
Issam Falih	Paris 13 University, France

Bo Fan	Zhejiang University, China
Yunsheng Fan	Dalian Maritime University, China
Hao Fang	Beijing Institute of Technology, China
Jinchao Feng	Beijing University of Technology, China
Francesco Ferracuti	Università Politecnica delle Marche, Italy
Chun Che Fung	Murdoch University, Australia
Wai-Keung Fung	Robert Gordon University, UK
Tetsuo Furukawa	Kyushu Institute of Technology, Japan
Hao Gao	Nanjing University of Posts and Telecommunications, China
Yabin Gao	Harbin Institute of Technology, China
Yongsheng Gao	Griffith University, Australia
Tom Gedeon	Australian National University, Australia
Ong Sing Goh	Universiti Teknikal Malaysia Melaka, Malaysia
Iqbal Gondal	Federation University Australia, Australia
Yue-Jiao Gong	Sun Yat-sen University, China
Shenshen Gu	Shanghai University, China
Chengan Guo	Dalian University of Technology, China
Ping Guo	Beijing Normal University, China
Shanqing Guo	Shandong University, China
Xiang-Gui Guo	University of Science and Technology Beijing, China
Zhishan Guo	University of Central Florida, USA
Christophe Guyeux	University of Franche-Comte, France
Masafumi Hagiwara	Keio University, Japan
Saman Halgamuge	The University of Melbourne, Australia
Tomoki Hamagami	Yokohama National University, Japan
Cheol Han	Korea University at Sejong, South Korea
Min Han	Dalian University of Technology, China
Takako Hashimoto	Chiba University of Commerce, Japan
Toshiharu Hatanaka	Osaka University, Japan
Wei He	University of Science and Technology Beijing, China
Xing He	Southwest University, China
Xiuyu He	University of Science and Technology Beijing, China
Akira Hirose	The University of Tokyo, Japan
Daniel Ho	City University of Hong Kong, Hong Kong SAR, China
Katsuhiro Honda	Osaka Prefecture University, Japan
Hongyi Li	Bohai University, China
Kazuhiro Hotta	Meijo University, Japan
Jin Hu	Chongqing Jiaotong University, China
Jinglu Hu	Waseda University, Japan
Xiaofang Hu	Southwest University, China
Xiaolin Hu	Tsinghua University, China
He Huang	Soochow University, China
Kaizhu Huang	Xi'an Jiaotong-Liverpool University, China
Long-Ting Huang	Wuhan University of Technology, China

Panfeng Huang	Northwestern Polytechnical University, China
Tingwen Huang	Texas A&M University, USA
Hitoshi Iima	Kyoto Institute of Technology, Japan
Kazushi Ikeda	Nara Institute of Science and Technology, Japan
Hayashi Isao	Kansai University, Japan
Teijiro Isokawa	University of Hyogo, Japan
Piyasak Jeatrakul	Mae Fah Luang University, Thailand
Jin-Tsong Jeng	National Formosa University, Taiwan
Sungmoon Jeong	Kyungpook National University Hospital, South Korea
Danchi Jiang	University of Tasmania, Australia
Min Jiang	Xiamen University, China
Yizhang Jiang	Jiangnan University, China
Xuguo Jiao	Zhejiang University, China
Keisuke Kameyama	University of Tsukuba, Japan
Shunshoku Kanae	Junshin Gakuen University, Japan
Hamid Reza Karimi	Politecnico di Milano, Italy
Nikola Kasabov	Auckland University of Technology, New Zealand
Abbas Khosravi	Deakin University, Australia
Rhee Man Kil	Sungkyunkwan University, South Korea
Daeeun Kim	Yonsei University, South Korea
Sangwook Kim	Kobe University, Japan
Lai Kin	Tunku Abdul Rahman University, Malaysia
Irwin King	The Chinese University of Hong Kong, Hong Kong SAR, China
Yasuharu Koike	Tokyo Institute of Technology, Japan
Ven Jyn Kok	National University of Malaysia, Malaysia
Ghosh Kuntal	Indian Statistical Institute, India
Shuichi Kurogi	Kyushu Institute of Technology, Japan
Susumu Kuroyanagi	Nagoya Institute of Technology, Japan
James Kwok	The Hong Kong University of Science and Technology, SAR China
Edmund Lai	Auckland University of Technology, New Zealand
Kittichai Lavangnananda	King Mongkut's University of Technology Thonburi, Thailand
Xinyi Le	Shanghai Jiao Tong University, China
Minho Lee	Kyungpook National University, South Korea
Nung Kion Lee	University Malaysia Sarawak, Malaysia
Andrew C. S. Leung	City University of Hong Kong, Hong Kong SAR, China
Baoquan Li	Tianjin Polytechnic University, China
Chengdong Li	Shandong Jianzhu University, China
Chuandong Li	Southwest University, China
Dazi Li	Beijing University of Chemical Technology, China
Li Li	Tsinghua University, China
Shengquan Li	Yangzhou University, China

Ya Li	Institute of Automation, Chinese Academy of Sciences, China
Yanan Li	University of Sussex, UK
Yongming Li	Liaoning University of Technology, China
Yuankai Li	University of Science and Technology of China, China
Jie Lian	Dalian University of Technology, China
Hualou Liang	Drexel University, USA
Jinling Liang	Southeast University, China
Xiao Liang	Nankai University, China
Alan Wee-Chung Liew	Griffith University, Australia
Honghai Liu	University of Portsmouth, UK
Huaping Liu	Tsinghua University, China
Huawen Liu	University of Texas at San Antonio, USA
Jing Liu	Chinese Academy of Sciences, China
Ju Liu	Shandong University, China
Qingshan Liu	Huazhong University of Science and Technology, China
Weifeng Liu	China University of Petroleum, China
Weiqiang Liu	Nanjing University of Aeronautics and Astronautics, China
Dome Lohpetch	King Mongkut's University of Technology North Bangoko, Thailand
Hongtao Lu	Shanghai Jiao Tong University, China
Wenlian Lu	Fudan University, China
Yao Lu	Beijing Institute of Technology, China
Jinwen Ma	Peking University, China
Qianli Ma	South China University of Technology, China
Sanparith Marukatat	Thailand's National Electronics and Computer Technology Center, Thailand
Tomasz Maszczyk	Nanyang Technological University, Singapore
Basarab Matei	LIPN Paris Nord University, France
Takashi Matsubara	Kobe University, Japan
Nobuyuki Matsui	University of Hyogo, Japan
P. Meesad	King Mongkut's University of Technology North Bangkok, Thailand
Gaofeng Meng	Chinese Academy of Sciences, China
Daisuke Miyamoto	University of Tokyo, Japan
Kazuteru Miyazaki	National Institution for Academic Degrees and Quality Enhancement of Higher Education, Japan
Seiji Miyoshi	Kansai University, Japan
J. Manuel Moreno	Universitat Politècnica de Catalunya, Spain
Naoki Mori	Osaka Prefecture University, Japan
Yoshitaka Morimura	Kyoto University, Japan
Chaoxu Mu	Tianjin University, China
Kazuyuki Murase	University of Fukui, Japan
Jun Nishii	Yamaguchi University, Japan

Haruhiko Nishimura	University of Hyogo, Japan
Grozavu Nistor	Paris 13 University, France
Yamaguchi Nobuhiko	Saga University, Japan
Stavros Ntalampiras	University of Milan, Italy
Takashi Omori	Tamagawa University, Japan
Toshiaki Omori	Kobe University, Japan
Seiichi Ozawa	Kobe University, Japan
Yingnan Pan	Northeastern University, China
Yunpeng Pan	JD Research Labs, China
Lie Meng Pang	Universiti Malaysia Sarawak, Malaysia
Shaoning Pang	Unitec Institute of Technology, New Zealand
Hyeyoung Park	Kyungpook National University, South Korea
Hyung-Min Park	Sogang University, South Korea
Seong-Bae Park	Kyungpook National University, South Korea
Kitsuchart Pasupa	King Mongkut's Institute of Technology Ladkrabang, Thailand
Yong Peng	Hangzhou Dianzi University, China
Somnuk Phon-Amnuaisuk	Universiti Teknologi Brunei, Brunei
Lukas Pichl	International Christian University, Japan
Geong Sen Poh	National University of Singapore, Singapore
Mahardhika Pratama	Nanyang Technological University, Singapore
Emanuele Principi	Università Politecnica elle Marche, Italy
Dianwei Qian	North China Electric Power University, China
Jiahu Qin	University of Science and Technology of China, China
Sitian Qin	Harbin Institute of Technology at Weihai, China
Mallipeddi Rammohan	Nanyang Technological University, Singapore
Yazhou Ren	University of Science and Technology of China, China
Ko Sakai	University of Tsukuba, Japan
Shunji Satoh	The University of Electro-Communications, Japan
Gerald Schaefer	Loughborough University, UK
Sachin Sen	Unitec Institute of Technology, New Zealand
Hamid Sharifzadeh	Unitec Institute of Technology, New Zealand
Nabin Sharma	University of Technology Sydney, Australia
Yin Sheng	Huazhong University of Science and Technology, China
Jin Shi	Nanjing University, China
Yuhui Shi	Southern University of Science and Technology, China
Hayaru Shouno	The University of Electro-Communications, Japan
Ferdous Sohel	Murdoch University, Australia
Jungsuk Song	Korea Institute of Science and Technology Information, South Korea
Andreas Stafylopatis	National Technical University of Athens, Greece
Jérémie Sublime	ISEP, France
Ponnuthurai Suganthan	Nanyang Technological University, Singapore
Fuchun Sun	Tsinghua University, China
Ning Sun	Nankai University, China

Norikazu Takahashi	Okayama University, Japan
Ken Takiyama	Tokyo University of Agriculture and Technology, Japan
Tomoya Tamei	Kobe University, Japan
Hakaru Tamukoh	Kyushu Institute of Technology, Japan
Choo Jun Tan	Wawasan Open University, Malaysia
Shing Chiang Tan	Multimedia University, Malaysia
Ying Tan	Peking University, China
Gouhei Tanaka	The University of Tokyo, Japan
Ke Tang	Southern University of Science and Technology, China
Xiao-Yu Tang	Zhejiang University, China
Yang Tang	East China University of Science and Technology, China
Qing Tao	Chinese Academy of Sciences, China
Katsumi Tateno	Kyushu Institute of Technology, Japan
Keiji Tatsumi	Osaka University, Japan
Kai Meng Tay	Universiti Malaysia Sarawak, Malaysia
Chee Siong Teh	Universiti Malaysia Sarawak, Malaysia
Andrew Teoh	Yonsei University, South Korea
Arit Thammano	King Mongkut's Institute of Technology Ladkrabang, Thailand
Christos Tjortjis	International Hellenic University, Greece
Shibata Tomohiro	Kyushu Institute of Technology, Japan
Seiki Ubukata	Osaka Prefecture University, Japan
Eiji Uchino	Yamaguchi University, Japan
Wataru Uemura	Ryukoku University, Japan
Michel Verleysen	Universite catholique de Louvain, Belgium
Brijesh Verma	Central Queensland University, Australia
Hiroaki Wagatsuma	Kyushu Institute of Technology, Japan
Nobuhiko Wagatsuma	Tokyo Denki University, Japan
Feng Wan	University of Macau, SAR China
Bin Wang	University of Jinan, China
Dianhui Wang	La Trobe University, Australia
Jing Wang	Beijing University of Chemical Technology, China
Jun-Wei Wang	University of Science and Technology Beijing, China
Junmin Wang	Beijing Institute of Technology, China
Lei Wang	Beihang University, China
Lidan Wang	Southwest University, China
Lipo Wang	Nanyang Technological University, Singapore
Qiu-Feng Wang	Xi'an Jiaotong-Liverpool University, China
Sheng Wang	Henan University, China
Bunthit Watanapa	King Mongkut's University of Technology, Thailand
Saowaluk Watanapa	Thammasat University, Thailand
Qinglai Wei	Chinese Academy of Sciences, China
Wei Wei	Beijing Technology and Business University, China
Yantao Wei	Central China Normal University, China

Guanghui Wen	Southeast University, China
Zhengqi Wen	Chinese Academy of Sciences, China
Hau San Wong	City University of Hong Kong, Hong Kong SAR, China
Kevin Wong	Murdoch University, Australia
P. K. Wong	University of Macau, SAR China
Kuntpong Woraratpanya	King Mongkut's Institute of Technology Chaokuntaharn Ladkrabang, Thailand
Dongrui Wu	Huazhong University of Science and Technology, China
Si Wu	Beijing Normal University, China
Si Wu	South China University of Technology, China
Zhengguang Wu	Zhejiang University, China
Tao Xiang	Chongqing University, China
Chao Xu	Zhejiang University, China
Zenglin Xu	University of Science and Technology of China, China
Zhaowen Xu	Zhejiang University, China
Tetsuya Yagi	Osaka University, Japan
Toshiyuki Yamane	IBM, Japan
Koichiro Yamauchi	Chubu University, Japan
Xiaohui Yan	Nanjing University of Aeronautics and Astronautics, China
Zheng Yan	University of Technology Sydney, Australia
Jinfu Yang	Beijing University of Technology, China
Jun Yang	Southeast University, China
Minghao Yang	Chinese Academy of Sciences, China
Qinmin Yang	Zhejiang University, China
Shaofu Yang	Southeast University, China
Xiong Yang	Tianjin University, China
Yang Yang	Nanjing University of Posts and Telecommunications, China
Yin Yang	Hamad Bin Khalifa University, Qatar
Yiyu Yao	University of Regina, Canada
Jianqiang Yi	Chinese Academy of Sciences, China
Chengpu Yu	Beijing Institute of Technology, China
Wen Yu	CINVESTAV, Mexico
Wenwu Yu	Southeast University, China
Zhaoyuan Yu	Nanjing Normal University, China
Xiaodong Yue	Shanghai University, China
Dan Zhang	Zhejiang University, China
Jie Zhang	Newcastle University, UK
Liqing Zhang	Shanghai Jiao Tong University, China
Nian Zhang	University of the District of Columbia, USA
Tengfei Zhang	Nanjing University of Posts and Telecommunications, China
Tianzhu Zhang	Chinese Academy of Sciences, China

Ying Zhang	Shandong University, China
Zhao Zhang	Soochow University, China
Zhaoxiang Zhang	Chinese Academy of Sciences, China
Dongbin Zhao	Chinese Academy of Sciences, China
Qiangfu Zhao	University of Aizu, Japan
Zhijia Zhao	Guangzhou University, China
Jinghui Zhong	South China University of Technology, China
Qi Zhou	University of Portsmouth, UK
Xiaojun Zhou	Central South University, China
Yingjiang Zhou	Nanjing University of Posts and Telecommunications, China
Haijiang Zhu	Beijing University of Chemical Technology, China
Hu Zhu	Nanjing University of Posts and Telecommunications, China
Lei Zhu	Unitec Institute of Technology, New Zealand
Pengefei Zhu	Tianjin University, China
Yue Zhu	Nanjing University, China
Zongyu Zuo	Beihang University, China

Contents – Part III

Embedding Learning

Transfer Learning

Reinforcement Learning

Other Learning Approaches

Embedding Learning

fMRI Semantic Category Decoding Using Linguistic Encoding of Word Embeddings

Subba Reddy Oota[1](✉), Naresh Manwani[1], and Raju S. Bapi[1,2]

[1] International Institute of Information Technology, Hyderabad, India
oota.subba@students.iiit.ac.in, {naresh.manwani,raju.bapi}@iiit.ac.in
[2] School of Computer and Information Sciences, University of Hyderabad, Hyderabad, India

Abstract. The dispute of how the human brain represents conceptual knowledge has been argued in many scientific fields. Brain imaging studies have shown that the spatial patterns of neural activation in the brain are correlated with thinking about different semantic categories of words (for example, tools, animals, and buildings) or when viewing the related pictures. In this paper, we present a computational model that learns to predict the neural activation captured in functional magnetic resonance imaging (fMRI) data of test words. Unlike the models with hand-crafted features that have been used in the literature, in this paper we propose a novel approach wherein decoding models are built with features extracted from popular linguistic encodings of Word2Vec, GloVe, Meta-Embeddings in conjunction with the empirical fMRI data associated with viewing several dozen concrete nouns. We compare these models with several other models that use word features extracted from FastText, Randomly-generated features, Mitchell's 25 features. The experimental results show that the predicted fMRI images using Meta-Embeddings meet the state-of-the-art performance. Although models with features from GloVe and Word2Vec predict fMRI images similar to the state-of-the-art model, model with features from Meta-Embeddings predicts significantly better. The proposed scheme that uses popular linguistic encoding offers a simple and easy approach for semantic decoding from fMRI experiments.

Keywords: Brain decoding · Word embedding · Neural network

1 Introduction

How a human brain represents and organizes conceptual knowledge has been an open research problem that attracted researchers from various fields [1–3]. In recent studies, the topic of exploring semantic representation in the human brain has attracted the attention of researchers from both neuroscience and computational linguistic fields. Using brain imaging studies Neuroscientists have shown that distinct spatial/temporal patterns of fMRI activity are associated with different stimuli such as face or scrambled face [4], semantic categories of pictures,

© Springer Nature Switzerland AG 2018
L. Cheng et al. (Eds.): ICONIP 2018, LNCS 11303, pp. 3–15, 2018.
https://doi.org/10.1007/978-3-030-04182-3_1

including tools, animals, and buildings, playing a movie, etc. [5–10]. These experimental results postulate how the brain encodes meaning of words and knowledge of objects, including theories that meanings are encoded in the sensory-motor cortical areas [11–13]. Such findings would also facilitate making predictions about breakdown in the function and their spatial location in different neurological disorders. Theoretical and empirical studies have been conducted to explore categorization of animate and inanimate objects and the brain representation of these semantic differences [14,15]. Linguists have identified different semantic meanings corresponding to individual verbs as well as the types of a noun that can fill those semantic meanings, for example, WordNet [16], VerbNet [17], and BabelNet [18]. Tom Mitchell's group at CMU pioneered studies that demonstrated common semantic representation for various nouns in terms of shared brain activation patterns across subjects [19]. In [20], presented the idea of detecting the cognitive state of a human subject based on the fMRI data by exploring different classification techniques.

The key aspect that lies at the heart of many of the fMRI decoding studies is the establishment of an associative mapping of the linguistic representation of nouns or verbs and the corresponding brain activation patterns elicited when subjects viewed these lexical items. Mitchell's team designed a computational model to predict the brain responses using hand-crafted word vectors as input to map the correlation between word embeddings and brain activity involved in viewing the words [19]. Since, Mitchell's 25 dimensional (dim) vector that uses fixed set of contextual dim (such as see, hear, eat etc.) will face word sense disambiguity and our high dimensional word vectors would have better basis for sense disambiguation as they use co-occurrence frequencies from large corpora. For example, the lexical item "Bank" has multiple semantic senses, such as the "bank of a river" or a "financial institution" based on the context. In fact, this forms motivation for our proposal of word embeddings in place of fixed context vectors. In recent times, linguistic representation of lexical items in computational linguistics is largely through a dense, low-dimensional and continuous vector called word-embedding [21,22]. Common word embeddings are generated from large text corpora such as Wikipedia and statistics concerning the co-occurrence of words is estimated to build such embeddings [23,24]. Some of the most popular word embedding models are Word2Vec [23], GloVe [25] and Meta-Embeddings [26]. The recent popular approach FastText [24] is a fast and effective method to learn word representations and can be utilized for text classification. Since FastText embeddings are trained for understanding morphological variations and most of the syntactic analogies are morphology-based, FastText embeddings do significantly better on the syntactic analogies than on semantic tasks [24]. [23] introduced continuous Skip-gram model in Word2Vec that is an efficient method for learning high-quality distributed vector representations that capture a large number of precise syntactic and semantic word relationships. Global Vectors for word representations (GloVe) [25] model combines the benefits of the Word2Vec skip-gram model when it comes to word analogy tasks, with the benefits of matrix factorization methods that can exploit global

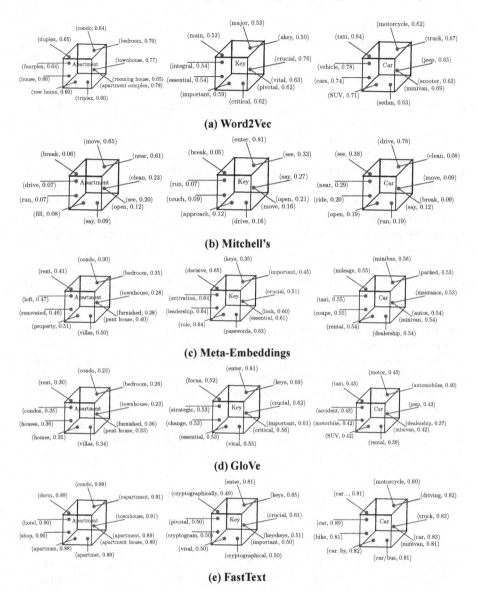

Fig. 1. Top 10 features for the words "*Apartment*" (left), "*Key*" (center) and "*Car*" (right) generated from the five word embedding methods.

statistical information. In [26], the idea of Meta-Embeddings has been proposed and has two benefits compared to individual embedding sets: enhancement of performance and improved coverage of the vocabulary.

Recently, the success of deep learning based word representations has raised the question whether these models might be able to make association between brain activations and language. In [27], authors proposed a model that combines the experience based word representation model with the dependency based word2vec features. The resulting model yielded better accuracies. However, this paper does not discuss which are most predicted voxels in various brain regions for different word embedding models and also does not give results on brain activations corresponding to multiple senses of a word. A recently published article [28] that gives a strong, independent support for our proposed approach of using word embedding representations for brain decoding models in place of carefully hand-crafted feature vectors. This paper aims at building a brain decoding system in which words and sentences are decoded from corresponding brain images. However, our approach addresses, in addition, the encoding problem where we try to build a system which learns associative mapping encoding words into corresponding fMRI images. Also, this paper uses ridge regression whereas we used Multi-layer feed forward neural network to learn the non-linear associative mapping between semantic features and brain activation responses.

In this paper, we propose a method to study the correlation between brain activity involved in viewing a word and corresponding word embedding (such as Word2Vec, GloVe, Meta-Embeddings, FastText and Mitchell's 25 [19]). To the best of our knowledge, this is the first time a comparative study is made of various existing, popular word embeddings for decoding brain activation. We propose a three-layer neural network architecture in which the input is a word embedding vector and the target output is the fMRI image depicting brain activation corresponding to the input word in line with the state-of-the-art approaches [19].

The structure of the paper is as follows. In Sect. 2, we discuss the motivation towards using word embeddings. Section 3 describes the approach we are using to build the model, while Sect. 4 presents comparative results of various models along with the statistical significance of the results. In Sect. 5, we give the conclusions and future work.

2   Motivation for Using Word-Embeddings

The word embeddings like Word2Vec, Glove etc., are known to capture the semantics of words based on the context as well as the co-occurrence of different words. We use these as features to capture the associative relationship between the meaning encoded in word embedding and the observed brain activation. So, Whenever the brain looks at a word, we assume that it tries to relate the word with some object/action, its properties, and other words with similar meaning. We consider the following example.

We observe the top 10 similar words for **Apartment**, **Key**, and **Car**. We obtain these similar words using different word embeddings which are given in Fig. 1. In Word2Vec, the similar words are semantically similar to Apartment, key and Car. On the other hand, GloVe and Meta-Embeddings give not only semantically similar words but also related words like {*rental, parked, accident, insurance, etc.*} for **Car**, {*role, decisive, passwords, activation, leadership, etc.*} for **Key** and {*furnished, rent, renovated, etc.*} for **Apartment**. These related words have the higher probability in Meta-Embeddings approach compared to those obtained with GloVe Embedding. These word embeddings are generated using just the text data without considering any brain activity specific features.

Table 1. Top 10 features for the word "**Celery**" generated from the six methods

(1)	(2)	(3)	(4)	(5)	(6)	(7)
broccoli 0.71	eat 0.35	carrots 0.16	carrots 0.24	eat 0.19	eat 0.837	cabbage 0.74
bellpeppers 0.69	taste 0.24	onions 0.16	cabbage 0.33	taste 0.18	taste 0.346	carrots 0.74
parsley 0.69	fill 0.051	parsley 0.18	cauliflower 0.35	fill 0.012	fill 0.315	onions 0.73
cilantro 0.68	see 0.063	broccoli 0.20	onion 0.35	see 0.07	see 0.243	spinach 0.73
cabbage 0.68	clean 0.054	garlic 0.20	parsley 0.38	clean 0.018	clean 0.115	garlic 0.72
cauliflower 0.67	open 0.042	cabbage 0.21	broccoli 0.38	open 0.08	open 0.060	tomato 0.70
tomato 0.67	smell 0.189	carrot 0.21	garlic 0.38	smell 0.026	smell 0.059	potatoes 0.70
lettuce 0.67	touch 0.061	spinach 0.22	potatoes 0.40	touch 0.019	touch 0.029	parsnips 0.69
cherry 0.66	say 0.094	cauliflower 0.22	turnips 0.40	say 0.092	say 0.016	sweetroot 0.69
Brussels 0.66	hear 0.021	asparagus 0.23	lettuce 0.41	hear 0.032	hear 0.000	lemongrass 0.69

(1) Word2Vec(Top 10), (2) Word2Vec similarity (with Mitchell's 25 words), (3) GloVe(Top 10) (4) Meta-Embeddings (Top 10), (5) Meta-Embeddings similarity (with Mitchell's 25 words) (Top 10) (6) Mitchell's 25 (Top 10), (7) FastText (Top 10)

On the other hand, Mitchell's feature vectors would be, by design, related to stimulus-modality-specific brain regions, as the learning model associates sensory features that have large weights with dominant evoked responses in related sensory cortical areas. The word embedding methods (Word2Vec, GloVe, and Meta-Embeddings) encode the meaning in terms of co-occurrence frequencies of other words in the corpus and thus may not relate to various modules of the brain the way Mitchell's hand-crafted features are designed.

It is interesting to understand the closeness of various word embeddings with Michell's 25. Table 1 describes similar words for "celery" based on various word embeddings as well as Mitchell's 25. As word embeddings and Mitchell's operate on different dimensions, we checked if they have similar underlying semantics. We estimated similarity scores for embedding vector for celery with embedding vectors for various feature words used in Mitchell's. Table 1 shows that the resulting score vector is quite similar, pointing out that the underlying similarity of semantics between vector-based encoding and Mitchell's. In this way, even these methods seem to capture the meaning in a way similar to Mitchell's scheme and perhaps might learn to elicit appropriate brain activation.

3 Proposed Approach

In this paper, we use a 3-layer neural network architecture as shown in Fig. 2 to build a trainable computational model that predicts the neural activation for any given stimulus word (**w**). Given a random stimulus word (**w**), we provide semantic features associated with (**w**) as input (generated from one of the six different methods, namely, Word2Vec, GloVe, Meta-Embeddings, FastText, Randomly-generated, and Mitchell's 25 [19]). The second step involves hidden layer representation and is accomplished via N hidden neurons in the hidden layer. Hidden neurons are fully connected to the input layer and the connection weights are learned through an adaptation process. The third step predicts the neural fMRI activation at every voxel location in the brain as a weighted sum of neural activations contributed by each of the hidden layer neurons. More precisely, the predicted activation z_v at voxel v in the brain for word w is given by

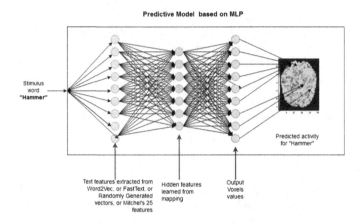

Fig. 2. 3-layer neural network architecture for decoding fMRI brain activation

$$z_v = \sum_{j=1}^{N} c_{vj} f(net_j) + c_{j0} \tag{1}$$

$$f(net_j) = tanh(\sum_{i=1}^{M} c_{ij} x_i + c_{i0}) \tag{2}$$

where, $f(net_j)$ is the value of the jth hidden neuron for word w, N is the number of hidden neurons present in the model, and c_{vj} is a learned coefficient that specifies the degree to which the jth intermediate semantic feature activates a voxel in the output layer.

4   Experimental Results and Observations

In this section, we describe the details of experiments conducted and the use of various word embeddings and observations thereof. We first describe the datasets used for our study.

4.1   FMRI Dataset Description

We used CMU fMRI data[1] of nine healthy subjects. These nine healthy subjects viewed 60 different word-picture pairs six times each. The 60 arbitrary stimuli included five items from each of the 12 semantic categories (animals, body parts, building parts, buildings, furniture, clothing, insects, kitchen items, tools, vegetables, vehicles, other man-made items). For each stimulus, we computed a mean fMRI image over its six repetitions and the mean of all 60 of these stimuli was then subtracted to get the final representation image.

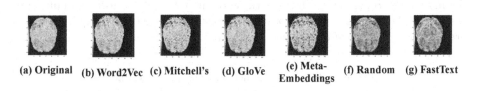

(a) Original   (b) Word2Vec   (c) Mitchell's   (d) GloVe   (e) Meta-Embeddings   (f) Random   (g) FastText

Fig. 3. Predicting fMRI images for given stimulus word "Bell"

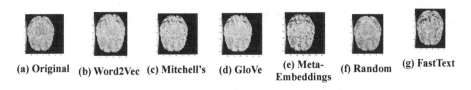

(a) Original   (b) Word2Vec   (c) Mitchell's   (d) GloVe   (e) Meta-Embeddings   (f) Random   (g) FastText

Fig. 4. Predicting fMRI images for given stimulus word "Arm"

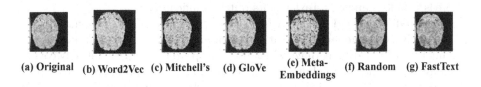

(a) Original   (b) Word2Vec   (c) Mitchell's   (d) GloVe   (e) Meta-Embeddings   (f) Random   (g) FastText

Fig. 5. Predicting fMRI images for given stimulus word "Bee"

---

[1] Available at http://www.cs.cmu.edu/~fmri/science2008/data.html.

4.2   Architecture Used and Training Strategy

The 3-layer neural network had 100 nodes in the hidden layer. Table 2 describes the other parameter settings for proposed model. At the input layer, we use semantic features of stimulus word. These semantic features could be any one of Word2Vec, GloVe, Meta-Embeddings, FastText, Randomly generated vectors, or Mitchell's 25 features. The reason behind using random features is to set a baseline control study. We trained separate computational models for each of the 9 participants using all the four input encoding methods. Each trained model was evaluated by means of a "leave-one-out" cross-validation approach in which the model was repeatedly trained with 59 of the 60 available word stimuli and associated fMRI images. Each trained model was then tested by requiring it to predict the fMRI image for the one "held-out" word.

Table 2. 3-layer neural network parameter setting

Parameters	Values
Hidden layer size	100
Optimizer	Adam
Activation	Tanh
Momentum, learning rate	0.9, 0.001

4.3   Statistical Analysis of Predicted fMRI Images

Figures 3, 4 and 5 compare the ground truth fMRI image and the corresponding predicted fMRI images using all the six methods for the words "bell", "arm" and "bee". It can be observed from the Figs. 3, 4 and 5 that the predicted fMRI images corresponding to Word2Vec, GloVe, and Mitchell's features look visually similar to the actual fMRI image obtained during the empirical experiment, whereas Random and FastText results differ significantly.

The predicted fMRI images when Meta-Embeddings are used have more robust activation compared to that of the original fMRI images. From this we can infer that Meta-Embeddings which use multiple data sources, not only covers semantically similar words but also gets closer to how the brain seems to represent. However, the activation regions seem largely similar in all the approaches except that of approaches using Random and FastText embeddings.

We use the rescaled mean squared error (R^2) as a metric to measure the error between predicted and target fMRI brain images. Kruskal-Wallis rank test was used for comparing mean ranks across the six methods in nine subjects. The one-way ANOVA test confirmed that there was a statistically significant difference between Meta-Embeddings, FastText, and Randomly generated vectors. Table 3 shows mean ranks of nine subjects when using six methods. From Table 3, we can observe that Meta-Embeddings, GloVe, and Mitchell's features are not statistically significantly different from one another in all the nine subjects. This

Table 3. Statistical significance (One-way ANOVA test) among the six methods reported individually per subject

#Subject	(1)	(2)	(3)	(4)	(5)	(6)	F statistic	p-value
Subj-1	0.2867	0.5562	0.5587	0.5561	−0.05600	−0.0078	17.6136	1.332e−15*
Subj-2	0.2963	0.3169	0.3194	0.3064	−0.0600	−0.0089	16.1014	2.620e−14*
Subj-3	0.2963	0.2924	0.2972	0.2911	−0.0600	−0.0089	13.1922	8.552e−12*
Subj-4	0.4327	0.4273	0.4319	0.4253	0.3208	0.3435	7.6373	7.840e−07*
Subj-5	0.1918	0.1800	0.1883	0.1805	−0.2231	−0.5236	29.7585	1.110e−16*
Subj-6	−0.8066	−0.8213	−0.8008	−0.7797	−1.2333	−1.4631	1.4862	0.1935
Subj-7	0.2015	0.1896	0.1961	0.1924	−0.1820	−0.1564	13.5018	3.677e−08*
Subj-8	0.2270	0.2200	0.2280	0.2213	−0.1469	−0.1710	29.7879	1.110e−16*
Subj-9	0.1816	0.1751	0.1778	0.1735	−0.3220	−0.2670	15.0325	5.497e−09*

(1) Word2vec, (2) Mitchell's 25, (3) Glove, (4) Meta-Embeddings, (5) Random, (6) FastText
*$p < 0.05$

Table 4. *Post-hoc* multiple comparison of the six embedding schemes

Post-hoc	Subjects (significance)								
	Subj-1	Subj-2	Subj-3	Subj-4	Subj-5	Subj-6	Subj-7	Subj-8	Subj-9
(1) vs (2)	0.001**	0.8995	0.8995	0.8995	0.8995	0.8995	0.89947	0.89947	0.89947
(1) vs (3)	0.001**	0.8995	0.8995	0.8995	0.8995	0.8995	0.89947	0.89947	0.89947
(1) vs (4)	0.001**	0.8995	0.8995	0.8995	0.8995	0.8995	0.89947	0.89947	0.89947
(1) vs (5)	0.001**	0.001**	0.001**	0.001**	0.001**	0.3799	0.001**	0.001**	0.001**
(1) vs (6)	0.001**	0.001**	0.001**	0.0087**	0.001**	0.7800	0.001**	0.001**	0.001**
(2) vs (3)	0.8995	0.8995	0.8995	0.8995	0.8995	0.8995	0.8995	0.8995	0.8995
(2) vs (4)	0.8995	0.8995	0.8995	0.8995	0.8995	0.8995	0.8995	0.8995	0.8995
(2) vs (5)	0.4777	0.001**	0.001**	0.001**	0.001**	0.4069	0.0010**	0.001**	0.001**
(2) vs (6)	0.7998	0.001**	0.001**	0.0172*	0.001**	0.8050	0.001**	0.001**	0.001**
(4) vs (3)	0.8995	0.8995	0.8995	0.8995	0.8995	0.8995	0.8995	0.8995	0.8995
(4) vs (5)	0.4789	0.001**	0.001**	0.001**	0.001**	0.3325	0.001**	0.001**	0.001**
(4) vs (6)	0.8009	0.001**	0.001**	0.0219*	0.001**	0.7343	0.001**	0.001**	0.001**

(1) Word2Vec features, (2) Mitchell's 25 features , (3) Glove features, (4) Meta-Embeddings features
(5) Randomly-generated features, (6) FastText features, **$p < 0.01$, *$p < 0.05$

leads us to conclude that all these methods have similar performance. Word2Vec approach is statistically significantly different as compared to Mitchell's approach only in the case of subject-1 (see Table 4). The *post-hoc* Scheffe's test results in Table 4 show that R^2 values of Meta-Embeddings, GloVe, Mitchell's and Word2Vec differ significantly from those of the FastText vectors at $p = 0.001$ and Random vectors at $p = 0.001$. No significant differences were observed between mean ranks of the Meta-Embeddings, GloVe, Word2Vec and Mitchell's 25 features.

4.4   Mapping Semantics onto the Brain

To evaluate our computation model, we examine the fMRI signatures for the features used in six methods shown in Fig. 6 for subject-2. These input features represent the model's learned decomposition of neural representations into their component semantic features and depict substantial activities in different regions of the brain. From Fig. 6, we observe that predicted activations in multiple cortical regions using Meta-Embeddings approach seems similar to the state-of-the-art Mitchell's method. Some of the semantic features such as "riding", "see", "say" and "fear" associated with the word "Bicycle" used in Mitchell's method lead to activations in the corresponding brain regions such as the "Premotor Area", "Occipital lobe/visual cortex", "Superior temporal gyrus/auditory cortex" and "Insula". In Meta-Embeddings, features like "riding", "spoke", "surly" associated with the word "Bicycle" predicted similar activations as that of the Mitchell's method. However, the models using embedding methods such as Word2Vec and GloVe predicted activations only in the "Occipital lobe/visual cortex" and "Superior temporal gyrus/auditory cortex". Whereas the model using Randomly generated features failed to predict activations in the corresponding brain regions.

4.5   Statistical Analysis Across Subjects

Kruskal-Wallis rank test was used for comparing median ranks across Word2Vec, GloVe, Meta-Embeddings, Mitchell's 25, Random and FastText methods performed across all subjects. The one-way ANOVA test confirmed that there was statistically significant difference between average error of predicted fMRI image when using Word2Vec, GloVe, Meta-Embeddings, Mitchell's 25, Random and FastText methods ($p = 0.001$) with a median rank of 0.2190 for Word2Vec, 0.2477 for GloVe, 0.2434 for Meta-Embeddings, 0.2459 for Mitchell's, -0.1281 for Random and -0.04139 for FastText. A *post hoc* Scheffe's test showed that average error of predicted fMRI image for Random and FastText methods differed significantly from those of the other four methods: Word2Vec, GloVe, Meta-Embeddings and Mitchell's 25 at $p = 0.0010053$. No significant differences were

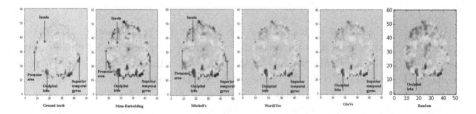

Fig. 6. Predicted fMRI image for the word "Bicycle". One representative horizontal slice (Taken at $z = 19$) for each method is displayed. From left to right: ground truth, Meta-Embeddings, Mitchell's, Word2Vec, GloVe and Randomly-generated features used to learn different decoding models.

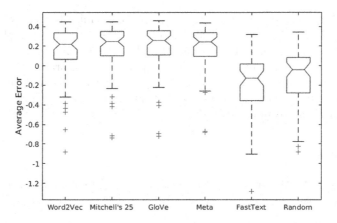

Fig. 7. Box-plot for average error of all subjects for six methods. Horizontal lines represent median ranks, notches represent 95% confidence interval. Median rank of Random and FastText methods are significantly less than that of other four methods ($p = 0.0010053$).

observed between median ranks of the other four word embedding methods. From Fig. 7, we can observe that the average error for the models using embeddings Word2Vec, GloVe, Meta-Embeddings and features from Mitchell's 25 is similar and is significantly different from the average errors of the models using FastText and Random features.

5   Conclusion

This study employs the existing popular word embeddings such as Word2Vec, GloVe, Meta-Embeddings and FastText to scrutinize the semantic representations in brain activity as measured by fMRI. One of the main observations from our study was that while Mitchell's hand-crafted features were designed to cover multi-modal activity of the brain covering several brain regions, the corpus-based word embedding models are based on word co-occurrence based statistics and thus lack the multi-modal context embedded in Mitchell's feature vector. Such general word embedding encoding schemes tend to give a strong within-category coverage for the input words and try to project this across different brain regions through associative mapping learned in the 3-layer neural network. Thus the current study can be considered a feasibility study of using generic word embedding schemes for brain decoding rather than painstakingly assembling hand-crafted features. Experimental results reveal that the R^2 error between Mitchell's approach and the other schemes such as Word2Vec, GloVe and Meta-Embeddings is small and the statistical significance of the results also points out that both the approaches are similar in their final outcome. In future, we would like to include both image based and word based features generated using pretrained embeddings from different (multi-modal or multi-view) genres so that such feature vectors will also have an opportunity to learn mapping to multi-modal sensory and association regions of the brain. This might give us more insights into the mapping process of multi-modal representations to brain response and eventually improve the decoding accuracy of brain activation with such predictive solutions. The source code is publicly available at https://github.com/subbareddy248/BrainDecoding so that researchers and developers can work on this exciting problem collectively.

References

1. Caramazza, A., Mahon, B.Z.: The organization of conceptual knowledge: the evidence from category-specific semantic deficits. Trends Cogn. Sci. **7**(8), 354–361 (2003)
2. Mahon, B.Z., Caramazza, A.: What drives the organization of object knowledge in the brain? Trends Cogn. Sci. **15**(3), 97–103 (2011)
3. Tong, F., Pratte, M.S.: Decoding patterns of human brain activity. Annu. Rev. Psychol. **63**, 483–509 (2012)
4. Clark, V.P., Maisog, J.M., Haxby, J.V.: fMRI study of face perception and memory using random stimulus sequences. J. Neurophysiol. **79**(6), 3257–3265 (1998)
5. Carlson, T.A., Schrater, P., He, S.: Patterns of activity in the categorical representations of objects. J. Cogn. Neurosci. **15**(5), 704–717 (2003)
6. Howell, D.C.: Statistical Methods for Psychology. Cengage Learning, Belmont (2012)
7. Haxby, J.V., Gobbini, I.M., Furey, M.L., Ishai, A., Schouten, J.L., Pietrini, P.: Distributed and overlapping representations of faces and objects in ventral temporal cortex. Science **293**(5539), 2425–2430 (2001)
8. Ishai, A., Ungerleider, L.G., Martin, A., Schouten, J.L., Haxby, J.V.: Distributed representation of objects in the human ventral visual pathway. Proc. Natl. Acad. Sci. **96**(16), 9379–9384 (1999)
9. Cox, D.D., Savoy, R.L.: Functional magnetic resonance imaging (fMRI) "brain reading": detecting and classifying distributed patterns of fMRI activity in human visual cortex. Neuroimage **19**(2), 261–270 (2003)
10. Polyn, S.M., Natu, V.S., Cohen, J.D., Norman, K.A.: Category-specific cortical activity precedes retrieval during memory search. Science **310**(5756), 1963–1966 (2005)
11. Caramazza, A., Shelton, J.R.: Domain-specific knowledge systems in the brain: the animate-inanimate distinction. J. Cogn. Neurosci. **10**(1), 1–34 (1998)
12. Crutch, S.J., Warrington, E.K.: Spatial coding of semantic information: knowledge of country and city names depends on their geographical proximity. Brain **126**(8), 1821–1829 (2003)
13. Samson, D., Pillon, A.: Orthographic neighborhood and concreteness effects in the lexical decision task. Brain Lang. **91**(2), 252–264 (2004)
14. Cree, G.S., McRae, K.: Analyzing the factors underlying the structure and computation of the meaning of chipmunk, cherry, chisel, cheese, and cello (and many other such concrete nouns). J. Exp. Psychol. Gen. **132**(2), 163 (2003)
15. Mahon, B.Z., Caramazza, A.: The orchestration of the sensory-motor systems: clues from neuropsychology. Cogn. Neuropsychol. **22**(3–4), 480–494 (2005)
16. Miller, G.A., Beckwith, R., Fellbaum, C., Gross, D., Miller, K.J.: Introduction to wordnet: an on-line lexical database. Int. J. Lexicogr. **3**(4), 235–244 (1990)

17. Kipper-Schuler, K.: VerbNet: a broad-coverage, comprehensive verb lexicon. Ph.D. thesis, University of Pennsylvania (2005)
18. Navigli, R., Ponzetto, P.S.: BabelNet: building a very large multilingual semantic network. In: Proceedings of the 48th Annual Meeting of the Association for Computational Linguistics, pp. 216–225 (2010)
19. Mitchell, T.M., et al.: Predicting human brain activity associated with the meanings of nouns. Science **320**(5880), 1191–1195 (2008)
20. Singh, V., Miyapuram, K.P., Bapi, R.S.: Detection of cognitive states from fMRI data using machine learning techniques. In: Proceedings of the 20th International Joint Conference on Artificial Intelligence, pp. 587–592 (2007)
21. Hinton, G.E., Mcclelland, J.L., Rumelhart, D.E.: Distributed representations, parallel distributed processing: explorations in the microstructure of cognition, Volume 1: Foundations (1986)
22. Turney, P.D., Pantel, P.: From frequency to meaning: vector space models of semantics. J. Artif. Intell. Res. **37**, 141–188 (2010)
23. Mikolov, T., Sutskever, I., Chen, K., Corrado, G., Dean, J.: Distributed representations of words and phrases and their compositionality. In: Advances in Neural Information Processing Systems, pp. 3111–3119 (2013)
24. Bojanowski, P., Grave, E., Joulin, A., Mikolov, T.: Enriching word vectors with subword information. arXiv preprint arXiv:1607.04606 (2016)
25. Pennington, J., Socher, R., Manning, C.: GloVe: global vectors for word representation. In: Proceedings of the 2014 Conference on Empirical Methods in Natural Language Processing (EMNLP), pp. 1532–1543 (2014)
26. Yin, W., Schütze, H.: Learning meta-embeddings by using ensembles of embedding sets. arXiv preprint arXiv:1508.04257 (2015)
27. Abnar, S., Ahmed, R., Mijnheer, M., Zuidema, W.: Experiential, distributional and dependency-based word embeddings have complementary roles in decoding brain activity. In: Proceedings of the 8th Workshop on Cognitive Modeling and Computational Linguistics (CMCL 2018), pp. 57–66 (2018)
28. Pereira, F., et al.: Toward a universal decoder of linguistic meaning from brain activation. Nat. Commun. **9**(1), 963 (2018)

Named Entity Disambiguation via Probabilistic Graphical Model with Embedding Features

Weixin Zeng[1], Jiuyang Tang[1,2], Xiang Zhao[1,2(✉)], Bin Ge[1,2], and Weidong Xiao[1,2]

[1] College of System Engineering, National University of Defense Technology, Changsha, China
{zengweixin13,jytang,xiangzhao,gebin,wdxiao}@nudt.edu.cn
[2] Collaborative Innovation Center of Geospatial Technology, Wuhan, China

Abstract. Named entity disambiguation (NED) is the task of linking ambiguous mentions in text to their corresponding entities in a given knowledge base, such as Wikipedia. State-of-the-art NED solutions harness neural networks to generate abstract representations, i.e., embeddings, of mentions and entities, based on which the disambiguation process can be achieved by finding entity with the most similar representation to mention. Nevertheless, the coherence among mentions, and their corresponding entities, is yet neglected. To fill this gap, in this work, we put forward intra, an approach effectively integrating embedding features into a collective disambiguation framework, i.e., probabilistic graphical model. Markov Chain Monte Carlo sampling and SampleRank algorithm are implemented for model parameters learning and inference. We evaluate intra on existing dataset against several state-of-the-art NED systems, which validates the effectiveness of our proposed method.

Keywords: Named entity disambiguation
Probabilistic graphical model

1 Introduction

Named entity disambiguation (NED), also named as entity linking (EL), is the task of determining true meanings for mentions in text, which is crucial to many text processing related tasks, such as knowledge extraction, knowledge fusion and sentiment analysis. The specific disambiguation process can be observed from Fig. 1. Note that *entities* are unique identifiers of objects in the world, such as people, organizations, and locations (e.g., *Ronan Keating*), while *mentions* are surface forms of entities, which can appear in various forms and contain a certain degree of ambiguity, such as abbreviations and nick names (e.g., *Ronan*).

Example 1. *In Fig. 1, there is a piece of feedback for the movie Lady Bird. Nevertheless, it is hard for a computer to make sense of this review due to the*

© Springer Nature Switzerland AG 2018
L. Cheng et al. (Eds.): ICONIP 2018, LNCS 11303, pp. 16–27, 2018.
https://doi.org/10.1007/978-3-030-04182-3_2

ambiguity of important words (mentions), namely, **Ronan, Lucas** *and* **Timothée,** *in which case NED can be harnessed to resolve the ambiguous mentions.*

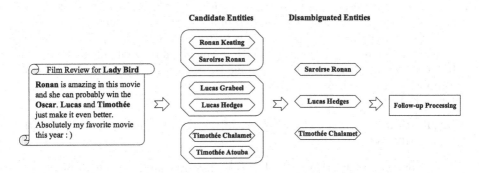

Fig. 1. An example of entity linking.

Concretely, NED is comprised of two main stages—candidate entities generation and candidate entities ranking. While the former targets at generating possible entities for mentions, also named as candidate entities, the latter ranks candidate entities and outputs the true entity. Referring to Example 1, the first stage generates candidate entities for mentions, e.g., *Ronan Keating* and *Saroirse Ronan* for mention Ronan. Then comes the crucial part, which is to devise strategies to rank the candidates and obtain the true corresponding entity, such as *Saroirse Ronan* for Ronan.

With reference to the strategy for ranking candidate entities, which determines the accuracy of the whole NED process, current methods can roughly be divided into three categories, individual [4], collective [8,11], and neural network based NED methods [6,12]. Individual NED methods rank candidate entities solely based on the similarity/co-occurrence between mentions and candidates, which fail to capture the interactions between mentions in the same document. Considering that mentions in the same document should conform to a certain topic and their corresponding entities ought to be somewhat related, collective NED methods establish relations between candidate entities and construct the mention-entity graph, on which graph-based algorithm or probabilistic graphical model (PGM) based methods are applied to output the results. Nevertheless, these methods all neglect the contribution made by latent features contained in texts of both mentions and entities, such as abstract representations, which have been proven effective by neural network based NED methods. That being said, most of neural network based approaches focus on designing complicated neural network structures to generate representations and capture the similarity between mentions and candidate entities, whereas interactions among candidate entities have been overlooked.

In a nutshell, the deficiency of current methods is two-fold. On one hand, the coherence among candidate entities is neglected in most approaches; on another, latent features in mention/entity texts are also not fully utilized.

In this work, we propose to integrate latent features captured by deep neural network into a collective framework, i.e., PGM, so as to achieve effective and efficient entity linking. Specifically, in PGM, factors are established between mentions and entities, or among entities, which can reflect the strength of relationships via the values calculated in accordance to the pre-defined feature functions, namely, popularity, name similarity, text similarity and entity relatedness. As for model inference and learning, we implement Markov Chain Monte Carlo (MCMC) [14] sampling and SampleRank [1] algorithms. The effectiveness of our proposed approach is verified via empirical evaluations, along with feature ablation test.

Contributions. The main contributions of this article can be summarized into three ingredients:

- We propose an effective named entity disambiguation method based on probabilistic graphical model with embedding features, intra, to facilitate downstream applications.
- We are among the first to integrate latent features generated from neural networks, i.e., embeddings, into a collective linking framework, i.e., PGM, so as to achieve superior linking performance.
- intra is evaluated on a real-life dataset, and the comparative results against other state-of-the-art approaches verify the effectiveness of intra.

2   Related Work

Early works on NED, which follow the independent methodology, tend to design a set of useful features to calculate similarities between mentions and entities and rank candidate entities merely according to the semantic matching score. Although methods of this kind [4,9] can achieve good experimental results, semantic coherences among candidate entities are neglected. Considering the deficiencies in independent NED approaches, collective NED methods [8,11] are put forward. This line of works assume that mentions in the same document are semantically coherent, which should fit in the textual topic of the whole document. Therefore, the resulting entities are also expected to have high relatedness and the problem is converted to finding matching pairs maximizing the coherence.

Over recent years, neural networks have also been applied to NED. Recurrent neural network (RNN) [6,12], convolutional neural network (CNN) [6] and attention mechanism [12] are harnessed to extract more effective latent features to model mention and entity representations. The representations are then utilized for similarity and relatedness computation so as to determine the most possible candidate entities. Nevertheless, merely few works [12] integrate the similarity score generated by neural networks into the collective disambiguation framework, especially PGM. As is pointed out in [7], PGMs can fully express the relations between mentions and entities or among entities via pre-defined features, hence serving as a competitive collective liking framework. However, in [7], the contribution made by latent features has been neglected, which are utilized in our work and attain superior performance.

3 Methodology

3.1 Model Representation

PGMs are probabilistic models for which graphs are harnessed to express the conditional dependence structures between variables, among which the Markov network, an undirected graphical model, is composed of random variables which have the Markov property (memoryless property of a stochastic process). In Markov network, a potential (also called a factor in factor graph) is associated with each complete subgraph.

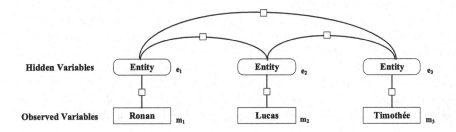

Ronan is amazing in this movie and she can probably win the **Oscar**. **Lucas** and **Timothée** just make it even better. Absolutely my favorite movie this year :)

Fig. 2. Factor graph of named entity disambiguation.

A Markov network can represent the joint probability distribution over the variables in the graph. The full joint distribution is obtained as follows:

$$P(\boldsymbol{X}) = \frac{1}{Z} \prod_i \phi_i(\boldsymbol{X}) \qquad (1)$$

where $\boldsymbol{X} = (x_1, x_2, ..., x_n)$ represent all the variables and C denote all the cliques in the graph. A clique i is a subset of all the nodes in graph, in which the nodes are connected. The potentials are denoted by ϕ and $Z = \sum_x \prod_i \phi_i(\boldsymbol{X})$ is the normalization constant (partition function), where x refers to a specific assignment. Note that the product of potentials can be converted to summation by adopting log linear model, where $\phi_i(\boldsymbol{X})$ can be rewritten as $e^{\ln \phi_i(\boldsymbol{X})}$ or $e^{\varepsilon_i(\boldsymbol{X})}$, and $\ln \phi_i(\boldsymbol{X}) = \varepsilon_i(\boldsymbol{X}) = w_i f_i(\boldsymbol{X})$. Consequently, potentials can be further represented by a set of features with associated weights, $P(\boldsymbol{X}) = \exp[\sum_i w_i f_i(\boldsymbol{X})]/Z$, where f_i represents the i-th feature and w_i is its corresponding weight, which will be learned during the model learning process.

With reference to the entity linking problem, we adopt the factor graph, which is a special form of Markov network and an undirected bipartite graph connecting variables and factors, to model the problem. Each factor represents a function over the variables it is connected to. In this factor graph, mentions $m_1, m_2, ..., m_n$ and the corresponding entities $e_1, e_2, ..., e_n$ are variables. The former is already known, while the latter needs to be determined. Therefore, the joint probability is represented as:

$$P(\boldsymbol{Y}|\boldsymbol{X};w) = \frac{1}{Z(\boldsymbol{X})}\prod_{\phi_i}e^{\phi_i} = \frac{1}{Z(\boldsymbol{X})}\prod_{\phi_i}e^{w_i f_i(\boldsymbol{X},\boldsymbol{Y})} \tag{2}$$

where $\boldsymbol{Y}$ and $\boldsymbol{X}$ represent hidden and observed variables respectively. w denotes the parameters that need to be learned. ϕ is called factor here, which connects certain observed and random variables. Similar to previous definition, the values attached to factors can be represented by a set of features with associated weights, hence $\phi_i = w_i * f_i(\boldsymbol{X}, \boldsymbol{Y})$. Additionally, $Z(\boldsymbol{X}) = \sum_{\boldsymbol{y}}\prod_i \phi_i(\boldsymbol{X})$, where $\boldsymbol{y}$ refer to the assignments of $\boldsymbol{Y}$.

We further explain the factor graph representation by extending Example 1. As can be seen in Fig. 2, there are three observed variables, namely, mentions Ronan, Lucas and Timothée. The corresponding entities they refer to are random variables. The values for entity variables are chosen from candidate entities, and each assignment is called a *State*, such as *Saroirse Ronan* for Ronan, *Lucas Hedges* for Lucas and *Timothée Chalamet* for Timothée. The *State* that maximizes the probability over the whole graph is considered as the final result.

3.2   Features

The features are introduced in this subsection, which are essential for calculating the probability over the whole graph, as presented in Eq. 2. Each factor is denoted as a white square in Fig. 2, the value of which is determined via feature functions.

Popularity. Popularity, represented as $f_1(m, e)$, is established between mentions and entities. It denotes the frequency that an entity is referred given the name of a mention, divided by the total frequencies of all possible candidate entities, which reflects the possibility that a candidate entity is the true entity on the basis of prior knowledge. Specifically, we obtain the popularity feature score by harnessing the data provided in [7].

Name Similarity. This feature is also defined over mentions and entities, which represents the similarity between a mention and a candidate entity solely based on their names. Previous work [7] directly uses edit distance to calculate the string similarity, which might not be appropriate since both mentions and entities can appear in various forms, including random capital characters or misspellings. As a result, in our work, we jointly embed words and entities to the same low-dimensional vector space, so as to obtain the abstract, but more appropriate representations. The specific process is detailed in next section. Given mention m, candidate entity e and joint word and entity embeddings, the name similarity $f_2(m, e)$ is defined as the cosine similarity between $embed(m)$ and $embed(e)$.

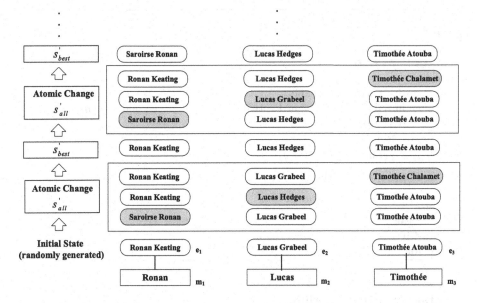

Fig. 3. MCMC sampling process.

Note that if mention is composed of more than one word, $embed(m)$ is the averaged embedding of all the word embeddings.

Text Similarity. Defined over mentions and candidate entities, this feature captures the similarity between the context of a mention and the description text of an entity, which can also benefit the disambiguation process. Instead of representing a piece of text as a sparse bag-of-words vector constructed from all of its tokens, we obtain a continuous low-dimensional text representation by adding up and averaging word embeddings in the context. Suppose a piece of text T consists of t words $w_1, w_2, ..., w_t$, the representation of this text is $Embed(T) = \sum_i^t embed(w_i)/t$. After obtaining the text representation of mention context T_m and entity description T_e, the text similarity $f_3(m, e)$ can be attained via cosine similarity between $Embed(T_m)$ and $Embed(T_e)$. Note that the entity description is retrieved from the first paragraph of its corresponding Wikipedia page.

Entity Relatedness. This feature aims at capturing the coherence among entities. For each pair of assigned entities e_i and e_j, a factor is added and the feature value is determined in accordance to the Topic-Specific PageRank values. Concretely, Topic-Specific PageRank is obtained by applying random walk process on the Wikipedia Graph [3] and we adopt the values provided by [7].

Formally, suppose $tspr(e_i, e_j)$ and $tspr(e_j, e_i)$ are the topic-specific pagerank values between entity e_i and entity e_j. While the start point for the former is entity e_i, the latter initiates from entity e_j. The value of entity relatedness $f_4(e_i, e_j)$ can be calculated via summing up $tspr(e_i, e_j)$ and $tspr(e_j, e_i)$. Note that if $e_i = e_j$, $f_4(e_i, e_j) = 1$.

Algorithm 1. MCMC for Inference.

Input:

mentions $\boldsymbol{m} = m_1, m_2, ..., m_n$, candidate entities for each mention Can_m, feature values f_1, f_2, f_3, f_4

Output:

state s with maximal probability

1: random initialization of assigned entities $\boldsymbol{e}_0 = e_1^0, e_2^0, ..., e_n^0$ and $s = (\boldsymbol{m}, \boldsymbol{e}_0)$
2: **for** $u \in [1, n]$ **do**
3:     $s'_{all} \leftarrow Atom(s)$
4:     $s'_{best} \leftarrow \arg\max_{s'' \in s'_{all}} (P(s''))$
5:     **if** $P(s'_{best}) > P(s)$ **then**
6:         $s \leftarrow s'_{best}$
7:     **else**
8:         break
9:     **end if**
10: **end for**

3.3 Inference and Learning

In this subsection, we will mainly introduce how to infer the results with the factor graph model and how to learn the parameters in the model.

The simplest way to determine a set of entities $e_1, e_2, ..., e_n$ for given mentions $m_1, m_2, ..., m_n$ is to numerate all the possible assignments and choose the one that maximizes the overall probability over the graph. Nevertheless, exhaustion is time-consuming with the increase of candidate entities. As a consequence, we resort to Markov Chain Monte Carlo (MCMC) sampling procedure [1] to iteratively infer the results. Indeed, there are many other algorithms that might improve the outcome and we will leave it as future study.

Concretely, as described in Algorithm 1, the first step is to randomly generate an initial state, where the entities for mentions are chosen stochastically. Then we perform *atomic change*, represented as $Atom()$, of the state, during which we merely change the entity assignment for one mention, and accordingly a total of s'_{all} states can be obtained via *atomic change* of state s. That is, $Atom()$ is composed of n actions, where n is the number of mentions in a given state s. In the m-th action ($0 < m \leq n$), we change the entity assignment for the m-th mention by randomly picking an entity from the candidate entities, while keeping the entity assignments for other mentions the same as state s, and hence, a new state is generated with only one entity assignment changed. Consequently, $Atom(s)$ generates n new states, denoted by s'_{all}.

Among s'_{all}, only the state with highest probability over the graph, s'_{best}, is kept, which is further compared with the previous state s based on the probability over graph $P()$. If the probability of the optimal atomic change state s'_{best} is larger than that of previous state, s'_{best} is regarded as the new s; otherwise it will break from the original loop. The final state s contains the eventual assignment

---

Algorithm 2. SampleRank with MCMC for Learning.

Input:
  $q : \mathbb{S} \to \mathbb{S}$: MCMC transition step
  $\chi : \mathbb{S} \to \mathbb{R}$: corrected mapped ratio (compared to the ground truth)
  $\mathbb{D}$: training set

Output:
  $\frac{1}{T} \sum_{t=1}^{T} \boldsymbol{w}_t$

1: $\boldsymbol{w}_0 \leftarrow \boldsymbol{0}$
2: **for** $\sigma \in \mathbb{D}$ **do**
3:     random initialization of assigned entities $\boldsymbol{e}_0 = e_1^0, e_2^0, ..., e_n^0$ and $s_1 = (\boldsymbol{m}, \boldsymbol{e}_0)$
4:     **for** $t \in [1, \#samples]$ **do**
5:         perform a MCMC step:
             $s_{t+1} \leftarrow s_{t,best}' = q(s_t)$
6:         $s^+ = \arg\max_{s^* \in \{s_t, s_{t+1}\}} \chi(s^*)$
           $s^- = \arg\min_{s^* \in \{s_t, s_{t+1}\}} \chi(s^*)$
           $\boldsymbol{\nabla} = \boldsymbol{p}(s^+) - \boldsymbol{p}(s^-)$
7:         **if** $\boldsymbol{w}_t' \boldsymbol{\nabla} < \chi(s^+) - \chi(s^-)$ and
           $\chi(s^+) \neq \chi(s^-)$ **then**
8:             $\boldsymbol{w}_{t+1} = \boldsymbol{w}_t + \eta \boldsymbol{\nabla}$
9:         **end if**
10:        **if** $(\neg accept(s_{t+1}, s_t, \boldsymbol{w}_t)$ **then**
11:            $s_{t+1} \leftarrow s_t$
12:        **end if**
13:    **end for**
14: **end for**

---

of entities, which are considered as the results yielded by the MCMC inference process. The procedure is also illustrated in Fig. 3.

As for learning the parameters $\boldsymbol{w} = [w_1, w_2, w_3, w_4]$, the weights for four various features, we adopt SampleRank algorithm [14], which is integrated in a MCMC sampling process, to achieve efficient parameter optimization. The algorithm updates the parameters by comparing pairs of states and an object function for comparison, χ, is defined to calculate the score of a state when compared with the ground-truth assignments in terms of the corrected linked mentions. Concretely, $\chi(s) = \#correct/\#all$, where $\#correct$ and $\#all$ represent the number of correctly linked mentions and all mentions respectively. The specific training procedure is depicted in Algorithm 2, where $\mathbb{S}$ and $\mathbb{R}$ denote the collection of all states and real numbers correspondingly. Plus, the output is the averaged value of all weights over time step T. Line 2–5 are MCMC steps, which are detailed in Algorithm 1, but here the new state is accepted first so as to update the parameters, which is further processed according to line 10 and 11. The specific update process is elaborated in line 6–9, where $\boldsymbol{p}$ refers to a vector recording the values of different features with respect to a given state s, $\boldsymbol{\nabla}$ is the vector recording the value differences over each feature between two states, w' is the transposition of vector w, and η is the learning rate.

Notably, the core of our inference algorithm is, SampleRank [14], which aims to minimize margin violations between arbitrary configuration pairs. Since this is not the main contribution of this paper, we leave the details behind the rationale to the original article.

4   Experiment

Candidate Entities Generation. Despite the fact that entity ranking method is crucial to the overall NED performance, its previous step, candidate entities generation, determines the upper bound of the linking accuracy. Concretely, chances are that in the candidate generation step, the candidate entities generated for some mentions do not contain the true entity, thus leading to wrong linking result in spite of following steps. Consequently, in our work, we adopt the dataset provided by [11], which includes candidate entities for each mention.

Joint Entity and Word Embedding. For calculating name similarity and text similarity features, the embeddings of words and entities are needed, which are obtained via a joint training process. Embeddings are n-dimensional vectors of concepts (words/entities) which describe the similarities between these concepts using cosine similarity. We extend traditional skip-gram model [14], which generates word representations that can help predict context words given a specific word, to joint embedding model, which integrates entities and regards entities as special form of words. Concretely, we use python package Gensim for embedding training, and the training material is obtained from Wikipedia, where anchor texts in the documents are replaced by entity identifiers.

Experimental Settings. The dataset we use for evaluation is AIDA-CoNLL [8], which is composed of three parts, AIDA-train for training, AIDA-A for held-out, and AIDA-B for testing. The corresponding number of documents are 946, 216 and 231 respectively. There are 34,956 mentions in total, in which 7,136 are NIL, meaning that there are no corresponding entities.

Following previous work, the micro-F1 score is harnessed as evaluation metric. F1 score is the harmonic value of *precision* and *recall*. While *precision* takes into consideration all mentions that are linked by the system and computes the correctness, *recall*, on the other hand, reflects the fraction of correctly linked mentions over all the mentions that should be linked. Mirco denotes that the value is averaged over the aggregation of all mentions (across all texts).

We compare intra with four other entity linking systems. DoSeR [15] integrates word and entity embeddings into a collective framework, on which sophisticated graph algorithm is implemented to achieve effective disambiguation results. WAT [13] is an improved version of TagMe [2], which harnesses graph-based and vote-based algorithms to approximate the coherence among entities. NERFGUN [7] is the most similar to our work, whereas the latent features generated by neural networks are not taken into consideration. Babelfy [10] is a graph-based entity linking system which uses the BabelNet semantic network for disambiguation, and it supports multi-lingual entity disambiguation task. Additionally, there are two more solutions [5,12] with good performance, whereas [12] did not report the results on AIDA dataset, and the results reported in [5] were generated by using a previous version of GERBIL, which has been pointed out by [7] that the results might be subject to considerable changes with the update of GERBIL. They did not offer public API's neither, and we do not consider them in our experiment for fair comparison.

With regard to training details of intra, the training dataset is AIDA-train, which is iterated five times to output the optimal results. The learning rate η is set to 0.01, the choice of which is further discussed below.

Result and Discussion. We first compare the result of intra with other state-of-the-art systems, then explore the contribution made by each feature.

As for the Micro-F1 results over AIDA-B, it can be easily observed that intra attains the best result over all methods. Babelfy is originally targeted at multi lingual entity linking, thus might not be carefully tuned for NED over a specific language. WAT and NERFGUN improve the result by 5% and 6% respectively, which can be attributed to the well-designed collective linking algorithms and hand-crafted features. With the introduction of word and entity embeddings, as well as latent features in text, DoSeR further enhances F1 score by 6% in comparison to NERFGUN. Nevertheless, it is over matched by intra with 3%, which might be justified that the factor graph can better integrate deep latent features with other features and yield more promising linking results (Table 1).

Table 1. Micro-F1 score over AIDA-B.

Method	Babelfy	WAT	NERFGUN	DoSeR	intra
Micro-F1	0.66	0.71	0.72	0.78	**0.81**

We then explore the contribution made by each feature by means of removing one feature at a time. As presented in Table 2, all four features are significant to the overall Micro-F1 score, among which popularity and text similarity are comparatively more essential, since the former is crucial in most situations where the entity referred by mention is exactly the most popular entity, and text similarity can help eliminate irrelevant candidate entities by taking into consideration the context information.

When it comes to the choice of hyper-parameter η, it should be resulted from experiments on the held-out dataset, i.e., AIDA-A. Therefore, we report

Table 2. Micro-F1 score after removing certain feature.

Feature	intra	intra-f1	intra-f2	intra-f3	intra-f4
Micro-F1	0.81	0.52	0.72	0.59	0.71

Table 3. Micro-F1 score on AIDA-A with different choices of η.

η	0.001	0.01	0.03	0.1
Micro-F1	0.74	0.77	0.73	0.69

the corresponding results, which reveals that smaller learning rate $\eta = 0.001$ leads to a slower learning process, despite nearly equally promising results. In contrast, larger η makes the convergence harder to achieve and also gives rise to inferior outcome. Consequently, we set $\eta = 0.01$ and apply it on the test set (Table 3).

5   Conclusions

In this work, aimed at developing an efficient NED framework, we propose intra, an approach based on PGM integrated with embedding features generated by neural networks. The NED problem is first represented as a factor graph, and four features are defined for the calculation of joint probability over the graph. Then on the basis of the probabilistic graph, MCMC sampling and SampleRank algorithms are implemented for model parameter learning and inference. We finally evaluate intra on existing dataset against several state-of-the-art NED systems, which validates the effectiveness of our proposed method.

Acknowledgments. This work was partially supported by NSFC under grants Nos. 61872446, 71690233 and 71331008.

References

1. Andrieu, C., de Freitas, N., Doucet, A., Jordan, M.I.: An introduction to MCMC for machine learning. Mach. Learn. **50**(1–2), 5–43 (2003)
2. Ferragina, P., Scaiella, U.: Fast and accurate annotation of short texts with Wikipedia pages. IEEE Softw. **29**(1), 70–75 (2012)
3. Moro, A., Raganato, A., Navigli, R.: Entity linking meets word sense disambiguation: a unified approach. TACL **2**, 231–244 (2014)
4. Dredze, M., McNamee, P., Rao, D., Gerber, A., Finin, T.: Entity disambiguation for knowledge base population. In: COLING 2010, Beijing, China, pp. 277–285 (2010)
5. Ganea, O., Ganea, M., Lucchi, A., Eickhoff, C., Hofmann, T.: Probabilistic bag of hyperlinks model for entity linking. In: WWW 2016, Montreal, Canada, pp. 927–938 (2016)
6. Gupta, N., Singh, S., Roth, D.: Entity linking via joint encoding of types, descriptions, and context. In: EMNLP 2017, Copenhagen, Denmark, pp. 2681–2690 (2017)
7. Hakimov, S., ter Horst, H., Jebbara, S., Hartung, M., Cimiano, P.: Combining textual and graph-based features for named entity disambiguation using undirected probabilistic graphical models. In: Blomqvist, E., Ciancarini, P., Poggi, F., Vitali, F. (eds.) EKAW 2016. LNCS (LNAI), vol. 10024, pp. 288–302. Springer, Cham (2016). https://doi.org/10.1007/978-3-319-49004-5_19

8. Hoffart, J., et al.: Robust disambiguation of named entities in text. In: EMNLP 2011, Edinburgh, UK, pp. 782–792 (2011)

9. Mihalcea, R., Csomai, A.: Wikify!: linking documents to encyclopedic knowledge. In: CIKM 2007, Lisbon, Portugal, pp. 233–242 (2007)

10. Moro, A., Cecconi, F., Navigli, R.: Multilingual word sense disambiguation and entity linking for everybody. In: ISWC 2014, Riva del Garda, Italy, pp. 25–28 (2014)

11. Pershina, M., He, Y., Grishman, R.: Personalized page rank for named entity disambiguation. In: NAACL HLT 2015, Denver, Colorado, pp. 238–243 (2015)

12. Phan, M.C., Sun, A., Tay, Y., Han, J., Li, C.: NeuPL: attention-based semantic matching and pair-linking for entity disambiguation. In: CIKM 2017, Singapore, pp. 1667–1676 (2017)

13. Piccinno, F., Ferragina, P.: From TagME to WAT: a new entity annotator. In: ERD 2014, Queensland, Australia, pp. 55–62 (2014)

14. Wick, M.L., Rohanimanesh, K., Bellare, K., Culotta, A., McCallum, A.: SampleRank: training factor graphs with atomic gradients. In: ICML 2011, Washington, pp. 777–784 (2011)

15. Zwicklbauer, S., Seifert, C., Granitzer, M.: Robust and collective entity disambiguation through semantic embeddings. In: SIGIR 2016, Pisa, Italy, pp. 425–434 (2016)

Category-Embodied Knowledge Embedding

Maoyuan Zhang[1], Qi Wang[1(✉)], Zhou Xu[2,3], Jianping Zhu[1], Shuyuan Sun[1],
and Yang Wen[1]

[1] School of Computer, Central China Normal University,
Wuhan, People's Republic of China
nlpwq@mails.ccnu.edu.cn
[2] School of Computer Science, Wuhan University, Wuhan, People's Republic of China
[3] Department of Computing,
The Hong Kong Polytechnic University, Kowloon, Hong Kong

Abstract. Knowledge graph (KG) embedding, which transforms both
the entities and relations into continuous low-dimensional continuous
vector space, has attracted considerable research. A large amount of models have been proposed for knowledge graph embedding. However, most
previous approaches only regard the knowledge graph as a set of triples,
ignoring the categories of the entities. In this paper, we take advantages
of category information by modelling the category-specific embedding.
Specially, we see the interaction between the category embedding and KG
embedding as a closed loop, in which the category embedding and KG
embedding are promoted mutually. Triples along with their categories
are represented in a unified framework, in which way the embedding of
triples are category-aware. We evaluate our model on multiple real-world
KGs, and it show impressive improvements on link prediction and triple
classification compared with other baselines.

Keywords: Distributed representation
Knowledge graph representation

1 Introduction

Freebase [1], DBpedia [9] and NELL [6] are the common Knowledge Graphs
(KGs), which have become crucial resources to store structured facts and benefit
many intelligent applications, such as named entity recognition [13], web search
[12] and question answering [2]. Commonly, a typical KG stores the structured
information as multi-relational data and formalizes it as triple fact of the form
of (head entity, relation, tail entity). Each entity is represented as a node in
the KG, and the relation is the edge connecting the head entity and tail entity.
However, the entity and relation in the KG usually be represented as discrete
symbols according to the original knowledge representation, which makes the
KG hard to be used [20].

© Springer Nature Switzerland AG 2018
L. Cheng et al. (Eds.): ICONIP 2018, LNCS 11303, pp. 28–37, 2018.
https://doi.org/10.1007/978-3-030-04182-3_3

In recent years, the KG embedding, i.e., projecting either the entries or the relations into a continuous vector space with lower dimensions while preserving the intrinsic structure of the original graph has aroused great research interest. As the key branch of embedding methods, the translation-based methods, such as TransE [3], TransG [15], TransR [10], describe the knowledge as a translation operation from head entity to tail entity by defining a relation-specific score function $f_r(h,t)$ to measure the plausibility of the triple in the latent space.

The previous translation-based methods are effective in representing structured data. However, most methods merely focus on the structure information, and the semantic information located in the category of entities is largely ignored. In this paper, we jointly learn the embedding of semantic category and KG embedding, which incorporates the category information into the structure-based representation learning of KG.

There are also works that incorporate the category information into KG embedding models. SSE [8], the most relevant model to our work, employs two manifold learning algorithm for representation learning, based on the semantic smoothness assumption. The main restriction of SSE is that it supposes that each entity happens to belong to one category, which is inconsistent with the actual KG. Besides, TKRL [17] is a type-embodied representation learning model, which handles the multiple category labels by projecting the entity with type-specific matrix and modelling relation as translation operation between the projected entities. However, if suffers from the relative high space complexity because it assigns a specific projection matrix for each category.

To take advantages of the category information, we propose a unified framework for jointly learn the embeddings of KG and entities' categories on it. Inspired by the work [7], we regard the category embedding and KG embedding as a closed loop, where each category embedding is a multivariate Gaussian distribution, and the embedding of each entity is generated by a Gaussian mixture distribution. On the one hand, the KG embedding is used to guide the category embedding, that is because it preserves a good structure in low dimensional space [5]. On the other hand, the category embedding feedback is employed as a category-aware constraint with intuitive that entities in the same category should be close to others in the low-dimension space.

We summarize our contribution as follows: (1) For all we know, we are the first to introduce the category embedding to enhance the KG embedding. (2) The experimental results show impressive improvement on link prediction and triple classification compared with other baselines.

2   Related Work

As the intriguing work of embedding the entities and relationships of multi-relational in low-dimensional vector space, **TransE** [3] employed the score function $f_r(h,t) = \|h + r - t\|$ to denote the consistency of triple (h,r,t). TransE can well model the one-to-one relations. However, if there exist more complicated relations [20], TransE can not work well. To address this issue, previous studies have proposed some variants of TransE, such as TransH [18] and

TransR [18]. Both TransR and TransH are designed to model the unique representation of entities in distinct relations by projection methods. TransG was a generative model that used clustering method (Chinese Restaurant Process) to update entity vectors. As a matter of fact, a variety of category information can be incorporated to further improve the performance of embedding task [20]. A straightforward method to employ category information, as investigated in [11], was to construct the triples such as $(e, Belongsto, c_e)$ and incorporate them into ordinary training examples. **SSE** [8] employed two manifold learning algorithm for finding the built-in geometric structure of the embedding space, and enforced entities of the same category to be closer in the embedding space, which is similar to our idea that embeddings of entities within the same semantic category tend to be similar. TKRL [17] was also a translation-based model in which entities were projected by type-specific matrices. Given a fact (h, r, t), it first projected h and t using type-specific projection matrices, and then applied a translation-based framework to model relationships. The category of entity was also used for controlling the occurrence possibility of head and tail positions for specific relations. For instance, negative examples of violations of entity category constraints were excluded from training set [19], or built with relatively low probabilities [17].

3 Methology

In this section, we introduce our approach, CEKE (Category-Embodied Knowledge Graph Embedding), for jointly learning the representation of the given knowledge graph based on the structure information and semantic category information.

3.1 Preliminaries

Knowledge Graph which is a set of triples in form of (h, r, t), $h, r \in \mathcal{E}$, and $r \in \mathcal{R}$ where $\mathcal{E}$ is the collection of entity and $\mathcal{R}$ is the collection of relation. Given a knowledge graph $G = (\mathcal{E}, \mathcal{R})$, the problem of knowledge graph embedding aims to transform both the entity e and the relation r into continuous low-dimensional space [20].

As the intriguing work of embedding the entities and relationships of multi-relational in low-dimensional vector space, TransE [3] introduced an exemplification which explained the relationship r as a translation operating on the embedding of the head entity h and tail entity t, and used the score function $f_r(e_i, e_j) = \|\mathbf{e_i} + r - \mathbf{e_j}\|$ to denote the plausibility of (e_i, r, e_j). A lower score indicates the better explanation of the triple based on the structure-based information. In order to obtain the embeddings, a margin-based ranking loss, i.e.,

$$O_1 = \sum_{(e_i, r, e_j) \in G} \sum_{(e_i', r, e_j') \in G'} \left[\gamma + f_r(e_i, e_j) - f_r(e_i', e_j') \right]_+ \quad (1)$$

is minimized.

In this paper, we also try to jointly learn the semantic category embeddings. Inspired by [7], each semantic category is assumed to be a Gaussian component, which is characterized by a mean vector indicating the semantic category center and a covariance matrix indicating the member entities' spread. Suppose there are K categories on the graph G. The mixing coefficient $z_i \in \{1, 2, ...K\}$ denotes category assignment of each entity.

Definition 1. Semantic Category Embedding: The embedding of semantic category k $(k = 1, 2, ..., K)$ is a multivariate Gaussian distribution $N(\psi_k, \Sigma_k)$, where $\psi_k \in \mathbb{R}^d$ is the mean vector, and $\Sigma_k \in \mathbb{R}^{d*d}$ is the covariance matrix.

3.2 Semantic Category Embedding

In order to make the embedding space embody the category information, this paper we explicitly learn the semantic category embedding, and based on the similar idea to assume that entities within the same category should have similar embeddings and lie close to the class center [8].

Although there are several methods to constrain the embedding space by leveraging the semantic category information, they don't have explicit notation of semantic category embedding because the main goal is KG embedding. In this paper, we propose a jointly learning framework to integrate the KG embedding and semantic category embedding in one single objective function based on Gaussian mixture model. More specifically, the semantic category embedding is defined as a multivariate Gaussian distribution, and the embedding of each entity is generated by a Gaussian mixture distribution from a category $z_i = k$, i.e., $p(e_i|z_i = k; e_i, \psi_k, \Sigma_k)$ is also a multivariate Gaussian distribution, we have:

$$p(e_i|z_i = k; e_i, \psi_k, \Sigma_k) = N(e_i|\psi_k, \Sigma_k) \tag{2}$$

For all the entities in $\mathcal{E}$, the likelihood is defined as follows:

$$\prod_{e=1}^{|\mathcal{E}|} \sum_{k=1}^{K} p(z_i = k) p(e_i|z_i = k; e_i, \psi_k, \Sigma_k) \tag{3}$$

where $p(z_i = k)$ is the probability of entity e_i belonging to category k, and the mixing coefficients $p(z_i = k)$ is denoted as π_{ik} for simplicity.

We have the objective function:

$$O_2 = \sum_{e=1}^{|\mathcal{E}|} \log \sum_{k=1}^{K} \pi_{ik} N(e_i|\psi_k, \Sigma_k), \ s.t. \sum_{k=1}^{K} \pi_{ik} = 1 \tag{4}$$

It is worth mentioning that we can learn the representation of the semantic category by optimizing π_{ik} and (ψ_k, Σ_k).

3.3 Joint Embedding

Our approach consists of two main components, i.e., the knowledge graph embedding and semantic category embedding. We integrated O_2 into the margin-based ranking loss (i.e. Eq. (1)) adopted in previous structure-based KG embedding models, and proposed the CEKE model. To utilize the semantic category information, we integrate them by jointly optimizing the unification objective function:

$$O = \sum_{(e_i,r,e_j)\in G} \sum_{(e_i',r,e_j')\in G'} \left[\gamma + f_r(e_i,e_j) - f_r(e_i',e_j')\right]_+ - \lambda_1 O_2 \qquad (5)$$

where the first term in O makes the embedding space compatible with the observed triples, and the second term further enforces the embedding space to be semantically related. Hyperparameter λ_1 is defined to make a trade-off between the two cases.

We see the interaction between these two embedding tasks as a closed loop. The knowledge embedding is initialized by TransE, and then the category embedding is to be optimized. That is, two entities in the same category tend to have similar embedding and get closer to the category center ψ_k. Suppose that we now have known the mixed category indicator $\pi_{ik}'s$ and the category embedding (ψ_k, Σ_k). Then we can optimize the embeddings of entities and relations with the known categories embeddings. Based on the closed loop, the community assignment Π and community embedding (ψ, Σ) can be further optimized by the updated parameters e.

3.4 Learning Model

To jointly learn the category embedding and knowledge graph embedding, the ultimate objection function for CEKE is defined as follows

$$O\left(\mathbf{e},\mathbf{r},\Pi,\Psi,\Sigma\right) = O_1(\mathbf{e},\mathbf{r}) - \lambda_1 O_2(\mathbf{e},\Pi,\Psi,\Sigma) \qquad (6)$$

Our final optimization problem becomes:

$$(\mathbf{e}^*,\mathbf{r}^*,\Pi^*,\Psi^*,\Sigma^*) \longleftarrow \arg\min_{\forall k diag(\Sigma_k)>0} O\left(\mathbf{e},\mathbf{r},\Pi,\Psi,\Sigma\right) \qquad (7)$$

where we introduce a constraint of $diag(\Sigma_k) > 0$ for each of $k \in \{1, 2, ..., K\}$ in order to avoid the singularity issue of optimizing O [7].

4 Inference

As the ultimate objective is composed of two embedding task, we firstly iteratively optimize (Φ, Ψ, Σ) with a constrained minimization given $(\mathbf{e}, \mathbf{r})$, and then optimize $(\mathbf{e}, \mathbf{r})$ with an unconstrained minimization given (Φ, Ψ, Σ). The iterative update strategy is adopted.

Fix $(\mathbf{e}, \mathbf{r})$, optimize (Π, Ψ, Σ). In this case, the optimization problem in Eq. 7 is simplified as the negative log-likelihood of GMM, and we can infer (Π, Ψ, Σ) via exception maximization (EM) algorithm. The constraint is satisfied by randomly resetting $\Sigma_k > 0$ and $\psi_k \in \mathbb{R}^d$ whenever a $diag(\Sigma_k)$ starts to have zero, which is suggested by [7]. We have the closed-form solution as

$$\pi_{ik} = \frac{N_k}{|\mathcal{E}|} \tag{8}$$

$$\psi_k = \frac{1}{N_k} \sum_{i=1}^{|\mathcal{E}|} \gamma_{ik} \mathbf{e_i} \tag{9}$$

$$\Sigma_k = \frac{1}{N_k} \sum_{i=1}^{|\mathcal{E}|} \gamma_{ik} (\mathbf{e_i} - \psi_k)(\mathbf{e_i} - \psi_k)^T \tag{10}$$

where $\gamma_{ik} = \frac{\pi_{ik} N(\mathbf{e_i}|\psi_k, \Sigma_k)}{\sum_{k'=1}^{K} \pi_{ik'} N(\mathbf{e_i}|\psi_{k'}, \Sigma_{k'})}$ and $N_k = \sum_{i=1}^{|\mathcal{E}|} \gamma_{ik}$.

Fix (Π, Ψ, Σ), optimize $(\mathbf{e}, \mathbf{r})$. In this paper, we use a stochastic gradient descent (SGD) algorithm to optimize the subject embedding according to Eq. 7. Noticing that it is inconvenient to compute the gradient of e because of the summation within the logarithm therm of O_2. Therefore, the upper bound of O_2 is incorporated. We have:

$$O_2' = \sum_{s=1}^{|\mathcal{E}|} \sum_{k=1}^{K} \pi_{ik} \log N(\mathbf{e_i}|\psi_k, \Sigma_k), \; s.t. \sum_{k=1}^{K} \pi_{ik} = 1, \tag{11}$$

and we redefine the objective function $O'(\mathbf{e}, \mathbf{r}, \Pi, \Psi, \Sigma) = O_1 - \lambda_1 O_2'$, and $O' \leqslant O$. The derivatives are calculated using the back-propagation algorithm.

5   Experiments and Analysis

In this section, we conduct an empirical study to evaluate the performance of our proposed model CEKE and other baselines on two benchmark tasks including link prediction [3] and triple classification [14].

Data Sets. Our experiments are evaluated by four public datasets. As for the first three datasets[1], LOCATION, SPORT, and NELL186 are the subsets of NELL, and are created by the author of Ref. [8]. The last one FB15K [4] generated from Freebase is commonly used in previous methods. We set K as the number of label of entity in each data set. Table 1 lists the statistics of datasets.

Baseline Methods. We employ the TransE [3], TKRL [17] as baselines, we also compare our model with SSE [8] which achieved great improvement on

---

[1] http://www.aclweb.org/anthology/P/P15/.

Table 1. Statistics of datasets.

Dataset	#Ent	#Rel	#Cat	#Train	#Valid	#Test
LOCATION	380	8	5	430	144	144
SPORT	1,520	8	4	2,296	765	765
NELL186	14,463	18	35	31,134	5,000	5,000
FB15K	14,951	1,345	30	483,142	50,000	59,071

the same datasets by incorporating manifold regularization, denoted as (-LE/-LLE), and the setting proposed in Ref. [11], which employed the entity category information in a more direct way. That is, we creat a triple $(e, BelongsTo, c_e)$ for every existing triple, where c_e is the category label of entity e, and the setting is named TrasnE-Cat. As a footnote, since all our experiments are performed on the same datasets, we directly make a comparison between our model and the baseline methods in Ref. [8].

5.1 Link Prediction

Link prediction is a subtask of knowledge graph completion which focuses on completing the triple (e_i, r, e_j) when e_i or e_j is missing.

Evaluation Setting. Similar to TransE and its variants, we use three measures as our evaluation metrics: for each test triple (e_i, r, e_j), the head entity e_i (or the tail entity e_j) is replaced with every entity e $\in \mathcal{E}$ existing in the knowledge graph. Then, we can get the distance values of the candidate triples. We ascendingly sort these triples based on their scores and then obtain the original triple. Rather than requiring the best answer, link prediction pays much more attention to ranking a set of candidate: (1) Mean Rank (denoted as Mean): the average rank of valid entities; (2) Median Rank (denoted as Median): the median of ranks; 3) Hist@10: the proportion of testing triple ranked in top 10 predictions.

Table 2. Experimental results on link prediction.

Datasets	LOCATION			SPORT			NELL186			FB15K		
Metrics	Mean	Median	Hist@10	Mean	Median	Hist@10	Mean	Median	Hist@10	Mean	Median	Hist@10
TransE	30.94	10.70	50.56	362.66	62.90	43.86	426.98	28.00	34.29	193.67	12.00	47.97
TransE-Cat	28.48	**8.90**	52.43	320.30	86.40	37.46	309.01	27.50	34.92	193.04	12.00	47.48
TransE-LE	28.59	**8.90**	53.06	183.10	23.20	45.83	245.80	24.00	36.64	191.83	11.00	49.72
TransE-LLE	28.03	9.20	52.36	231.67	52.40	43.18	241.83	29.00	35.15	190.36	11.00	48.61
TKRL	27.06	10.1	52.07	199.25	37.44	46.12	238.21	26.75	38.00	**184.00**	-	**69.4**
CEKE	**25.85**	9.09	**55.20**	**180.83**	**21.50**	**46.57**	**224.09**	**22.05**	**40.17**	186.41	**10.30**	55.66

Results. Table 2 lists the evaluation results on the test set. From the results, we observe that: (1) CEKE consistently outperforms all baselines on all the datasets with all the metrics. It indicates that the semantic category information do improve the performance on link prediction. Although using the same category in the graph, SSE's performance is inferior to our model, which shows the strength of our joint learning framework.

5.2 Triple Classification

The purpose of the triple classification is to determine the correctness or incorrectness of a given triple (e_i, r, e_j). The baseline setting is same with link prediction. We test our model on SPORT, LOCATION and NELL186.

Evaluation Setting. The setting for triple classification is simple: for each triple (e_i, r, e_j), if the energy function is lower than the certain threshold δ_r, then the triple will be classified as positive. Otherwise as negative. The common metrics on test sets include micro-averaged accuracy with regard to each triple (denoted as Micro), macro-averaged accuracy in the light of each relation (denoted as Macro) and mean average precision (denoted as MAP) which also concerns the relation-specific class.

Implementation Details. Since the three datasets have not released negative triples in previous works, we follow the strategy used in Ref. [16] for constructing negative samples. The thresholds δ_r are determined based on the validation set by maximizing the indicator Micro-ACC.

Table 3. Experimental results on triple classification.

Datasets	LOCATION			SPORT			NELL186		
Metrics	Micro	Macro	MAP	Micro	Macro	MAP	Micro	Macro	MAP
TransE	86.11	81.66	89.09	75.52	73.78	75.46	89.87	84.86	95.40
TransE-Cat	82.5	77.81	88.20	75.09	74.23	78.82	92.65	87.16	96.56
TransE-LE	86.39	81.50	89.23	79.88	77.34	81.86	92.71	87.94	**97.01**
TransE-LLE	87.01	83.03	89.53	80.29	**77.71**	82.99	**94.93**	94.97	96.84
TKRL	87.33	82.06	89.16	77.98	76.45	80.92	94.01	91.22	95.67
CEKE	**87.69**	**84.78**	**90.23**	**81.02**	76.93	**84.02**	93.31	**95.80**	96.30

Results. Evaluation results on the test sets are reported in Table 3. We can draw the following conclusions from observations: (1) The proposed model CEKE achieves superior performance compared with all the baseline methods on triple classification task. We speculate the reason for improvements is that representations of entities within one category are usually more similar, and entities maybe share common attributes, which helps the classification task. (2) With the consideration of semantic category information the accuracy has been improved, which further demonstrates the effectiveness and flexibility of CEKE.

6   Conclusion

In this paper, we propose the knowledge graph embedding model CEKE which is able to incorporate the category information. By learning the KG embedding and category embedding in a unified framework, our model can learn the embeddings that are aware of semantic category information. The results of our empirical study demonstrate that CEKE obtains substantial improvements compared with TransE and baseline methods.

Acknowledgments. The authors would like to acknowledge the support provided by the Research Planning Project of National Language Committee (No. YB135-40) and the Humanity and Social Science Youth Foundation of Ministry of Education of China (No. 15YJC870029).

References

1. Kurt, B., Colin, E., Praveen, P., Tim, S., Jamie, T.: Freebase: a collaboratively created graph database for structuring human knowledge. In: SIGMOD Conference 2008, pp. 1247–1250 (2008)
2. Antoine, B., Sumit, C., Jason, W.: Question answering with subgraph embeddings. In: Computer Science, pp. 615–620 (2014)
3. Antoine, B., Nicolas, U., Alberto, G., Jason, W., Yakhnenko, O.: Translating embeddings for modeling multi-relational Data. In: Advances in Neural Information Processing Systems 2013, pp. 2787–2795 (2013)
4. Antoine, B., Jason, W., Ronan, C., Yoshua, B.: Learning structured embeddings of knowledge bases. In: AAAI Conference on Artificial Intelligence 2011, pp. 301–306 (2011)
5. Cao, S., Lu, W., Xu, Q.: GraRep: learning graph representations with global structural information. In: ACM International on Conference on Information and Knowledge Management 2015, pp. 891–900 (2015)
6. Carlson, A., Betteridge, J., Kisiel, B., Settles, B., Mitchell, T.M.: Toward an architecture for never-ending language learning. In: AAAI Conference on Artificial Intelligence 2010, pp. 1306–1313 (2010)
7. Cavallari, S., Zheng, V., Cai, H., Chang, C., Cambria, E.: Learning community embedding with community detection and node embedding on graphs. In: ACM on Conference on Information and Knowledge Management 2017, pp. 377–386 (2017)
8. Guo, S., Wang, Q., Wang, L., Wang, L., Guo, L.: SSE: semantically smooth embedding for knowledge graphs. IEEE Trans. Knowl. Data Eng. **29**(4), 884–897 (2017)
9. Lehmann, J.: DBpedia: a large-scale, multilingual knowledge base extracted from wikipedia. Semant. Web **6**(2), 167–195 (2015)
10. Lin, L., Liu, Z., Zhu, X.: Learning entity and relation embeddings for knowledge graph completion. In: AAAI Conference on Artificial Intelligence 2015, pp. 2181–2187 (2015)
11. Maximilian, N., Volker, T., Hans, P.: Factorizing YAGO: scalable machine learning for linked data. In: International World Wide Web Conference 2012, pp. 271–280 (2012)
12. Szumlanski, S., Gomez, F.: Automatically acquiring a semantic network of related concepts. In: ACM on Conference on Information and Knowledge Management 2010, pp. 19–28 (2010)

13. Tanev, H.: A WordNet-based approach to named entities recognition. In: The Workshop on Building and Using Semantic Networks, Association for Computational Linguistics 2002, pp. 1–7 (2002)
14. Wang, Z., Zhang, J., Feng, J., Chen, Z.: Knowledge graph embedding by translating on hyperplanes. In: Association for the Advancement of Artificial Intelligence 2014, pp. 1112–1119 (2014)
15. Xiao, H., Huang, M., Hao, Y., Zhu, X.: TransG: a generative mixture model for knowledge graph embedding. In: Computer Science, pp. 2316–2325 (2015)
16. Xie, R., Liu, Z., Jia, J., Luan, H., Sun, M.: Representation learning of knowledge graphs with entity descriptions. In: International Joint Conference on Artificial Intelligence 2016, pp. 2659–2665 (2016)
17. Xie, R., Liu, Z., Sun, M.: Representation learning of knowledge graphs with hierarchical types. In: International Joint Conference on Artificial Intelligence 2016, pp. 2965–2971 (2016)
18. Wang, Z., Zhang, J., Feng, J., Chen, Z.: Knowledge graph embedding by translating on hyperplanes. In: AAAI-Association for the Advancement of Artificial Intelligence 2014, pp. 1112–1119 (2014)
19. Krompaß, D., Baier, S., Tresp, V.: Type-constrained representation learning in knowledge graphs. In: Arenas, M., et al. (eds.) ISWC 2015. LNCS, vol. 9366, pp. 640–655. Springer, Cham (2015). https://doi.org/10.1007/978-3-319-25007-6_37
20. Wang, Q., Mao, Z., Wang, B., Guo, L.: Knowledge graph embedding: a survey of approaches and applications. IEEE Trans. Knowl. Data Eng. **29**(12), 2724–2743 (2017)

Unsupervised Ensemble Learning Based on Graph Embedding for Image Clustering

Xiaohui Luo, Li Zhang$^{(\boxtimes)}$, Fanzhang Li, and Chengxiang Hu

School of Computer Science and Technology and Joint International Research
Laboratory of Machine Learning and Neuromorphic Computing,
Provincial Key Laboratory for Computer Information Processing Technology,
Soochow University, Suzhou 215006, Jiangsu, China
{xhluo,cxhu}@stu.suda.edu.cn, {zhangliml,lfzh}@suda.edu.cn

Abstract. Manifold learning has attracted more and more attention in machine learning for past decades. Unsupervised Large Graph Embedding (ULGE), which performs well on the large-scale data, has been proposed for manifold learning. To improve the clustering performance, a novel Unsupervised Ensemble Learning based on Graph Embedding (UEL-GE) is explored, which takes ULGE to get low-dimensional embeddings of the given data and uses the K-means method to obtain the clustering results. Furthermore, the multiple clusterings are corrected by using the bestMap method. Finally, the corrected clusterings are combined to generate the final clustering. Extensive experiments on several data sets are conducted to show the efficiency and effectiveness of the proposed ensemble learning method.

Keywords: Ensemble learning · Image clustering
Dimension reduction · Manifold learning

1 Introduction

Image clustering plays a key role in fields of image browsing, image retrieval, image annotation, as well as image indexing [1–5]. As we know that images are high-dimensional data. High-dimensional data may contain complex information, however, most of which are useless and redundant. To obtain useful knowledge, it is very time-consuming and laborious to deal with high-dimensional data directly. The higher the data dimension is, the greater the required resources will grow exponentially, which is prone to the curse of dimensionality [6]. Therefore, in order to efficiently acquire potential valuable information in high-dimensional data, it is necessary to reduce the dimensionality of data. Dimensionality reduction can keep most of the information in the data while embedding high-dimensional data into a low-dimensional space [7,8]. A large number of experiments have shown that dimensionality reduction can effectively eliminate redundant features, and learning tasks, such as visualization, can be easily carried out in the low-dimensional space [9].

© Springer Nature Switzerland AG 2018
L. Cheng et al. (Eds.): ICONIP 2018, LNCS 11303, pp. 38–47, 2018.
https://doi.org/10.1007/978-3-030-04182-3_4

Over the past few decades, with the development of manifold learning [10], many successful methods have been proposed, such as Laplacian Eigenmaps (LE) [11], Local Linear Embedding (LLE) [7], Isometric Feature Mapping (ISOMAP) [8], Locality Preserving Projections (LPP) [12], Spectral Regression (SR) [13], and Unsupervised Large Graph Embedding (ULGE) [14]. These methods have received extensive attention and successful applications related to machine learning, data mining, as well as computer vision.

ULGE, a novel manifold learning method, adopts an anchor-based strategy [15] and a parameter-free neighbor assignment strategy [16] to construct the similarity matrix, which is symmetric, positive semi-definite and doubly stochastic, with rank p (where p denotes the reduced dimension). ULGE uses a random sampling method to select anchors. The random sampling method is simple and has a low computational complexity. ULGE achieves significant performance on multiple data sets. However, it is difficult for the random sampling method to guarantee the representation of anchors. When anchors are representative, the clustering accuracy after dimensionality reduction will be improved. On the contrary, when anchors cannot guarantee the representation, the clustering accuracy will be correspondingly low and the experimental results are unstable. There is also a great fluctuation on the clustering accuracy.

To address this issue, this paper proposes an efficient method, called Unsupervised Ensemble Learning based on Graph Ebedding (UEL-GE) which combines multiple ULGEs. Ensemble learning [17] is an active field in machine learning, which can improve the clustering performance by building and combining multiple classifiers. The diversity of UEL-GE can be guaranteed by randomly generating anchors and adopting K-means clustering. Comparative experiments on five data sets show that the clustering accuracy of UEL-GE has improved significantly compared with ULGE.

The rest of this paper is organized as follows: In Sect. 2, the framework of the proposed method UEL-GE is investigated. In Sect. 3, a series of experiments are carried out to show the efficiency and effectiveness of the proposed method. The paper ends with final remark in Sect. 4.

2   Unsupervised Ensemble Learning Based on Graph Embedding

The framework of the proposed method UEL-GE is shown in Fig. 1. Generally, UEL-GE has two steps. In Step 1, multiple clusterings are generated by applying ULGEs and K-means. Since the labels of obtained clusterings are messy, we use the *bestMap* method [18] to correct the labels and combine the corrected clustering to obtain the final clustering results. In the following, we first introduce ULGE, and then show the diversity and the combination rule in UEL-GE.

2.1   ULGE

Given a sample matrix $\mathbf{X} = [\mathbf{x}_1, \mathbf{x}_2, \ldots, \mathbf{x}_n]^T \in \mathbb{R}^{n \times d}$ with the d-dimensional sample $\mathbf{x}_i \in \mathbb{R}^d$ and the sample number n, the goal of ULGE is to reduce the

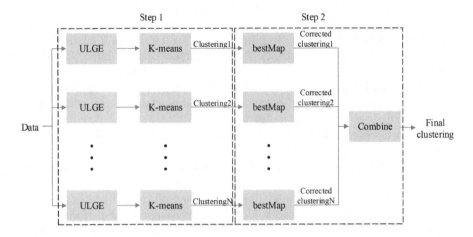

Fig. 1. Framework of UEL-GE

dimension of large-scale data [14]. First, ULGE adopts an anchor-based strategy to seek m anchors, where $m \ll n$. Anchor generation is the most important step of the anchor-based strategy. Generally, ULGE randomly selects m anchors from the given sample set. Let $\mathbf{U} \in \mathbb{R}^{m \times d}$ represent the matrix of whole anchors. Second, the distances $h(\mathbf{x}_i, \mathbf{u}_j)$ between a sample and an anchor are calculated, where $\mathbf{x}_i$ is the i-th sample and $\mathbf{u}_j$ is the j-th anchor. In order to keep the calculation simple, ULGE uses the square of the Euclidean distance to define $h(\mathbf{x}_i, \mathbf{u}_j) = \|\mathbf{x}_i - \mathbf{u}_j\|_2^2$. Then, a parameter-free neighbor assignment strategy is adopted to construct a similarity matrix [14]. The strategy can be cast into the following problem:

$$\min_{\mathbf{Z}} \sum_{j=1}^{m} h(\mathbf{x}_i, \mathbf{u}_j) z_{ij} + \gamma \sum_{j=1}^{m} z_{ij}^2 \quad s.t. \ \mathbf{z}_i^T \mathbf{1} = 1, z_{ij} \geq 0 \quad (1)$$

where γ is a parameter and $\mathbf{Z} \in \mathbb{R}^{n \times m}$ is the similarity matrix between samples and anchors, z_{ij} represents the similarity between the i-th sample and the j-th anchor. We resort the distances $h(\mathbf{x}_i, \mathbf{u}_1), \ldots, h(\mathbf{x}_i, \mathbf{u}_m)$ in ascending order and assign them to d_j^i $(j = 1, 2, \ldots, m)$. In other words, $d_1^i \leq d_2^i \leq \cdots \leq d_m^i$. According to [16], the value of γ can be set as

$$\gamma = \frac{k}{2} d_{k+1}^i - \frac{1}{2} \sum_{j=1}^{k} d_j^i \quad (2)$$

where k is k-nearest anchors for the sample.

When $\mathbf{u}_j$ is the k-nearest anchors of $\mathbf{x}_i$, the similarity z_{ij} is obtained by (1) and (2). When $\mathbf{u}_j$ is not the k-nearest anchors of $\mathbf{x}_i$, $z_{ij} = 0$. Thus, the similarity matrix $\mathbf{Z}$ can be denoted as

$$z_{ij} = \begin{cases} \frac{d_{k+1}^i - h(\mathbf{x}_i, \mathbf{u}_j)}{\sum_{l=1}^k (d_{k+1}^i - d_l^i)}, & \text{if } \mathbf{u}_j \in \Gamma(i) \\ 0, & \text{otherwise} \end{cases} \qquad (3)$$

where $\Gamma(i)$ is the set of k-nearest anchors for the sample $\mathbf{x}_i$.

According to $\mathbf{Z}$, the symmetric, positive semi-definite, doubly stochastic similarity matrix $\mathbf{A}$ can be obtained by

$$\mathbf{A} = \mathbf{Z}\mathbf{\Delta}^{-1}\mathbf{Z}^T \qquad (4)$$

where $\mathbf{\Delta} \in \mathbb{R}^{m \times m}$ is a diagonal matrix with $\Delta_{jj} = \sum_{i=1}^n z_{ij}$.

In order to make the rank of $\mathbf{A}$ be equal to p, we should perform eigenvalue decomposition on $\mathbf{A}$ [19]. Let $\mathbf{A}^* = \mathbf{F}_p \mathbf{\Lambda}_p \mathbf{F}_p^T$, where $\mathbf{\Lambda}_p$ is the diagonal matrix consisting of the first p largest eigenvalues of the matrix $\mathbf{A}$, and $\mathbf{F}_p$ consists of the eigenvectors corresponding to the first p largest eigenvalues.

Finally, the projection matrix $\mathbf{W}$ can be calculated as follows.

$$\min_{\mathbf{W}} \|\mathbf{X}\mathbf{W} - \mathbf{F}_p\|_F^2 + \alpha\|\mathbf{W}\|_F^2 \qquad (5)$$

where $\alpha \geq 0$ is the regularization parameter, $\mathbf{W}$ is the projection matrix and $\|\cdot\|_F$ is the Frobenius norm of a matrix.

2.2   Diversity in UEL-GE

In most clustering problems, ensemble learning has more powerful prediction ability than a single method. In practice, it has been confirmed that a good ensemble learning must meet the individual learners to be "good" and "different", "good" requires individual learners to maintain a high degree of correctness whereas "different" requires individual learners satisfy diversity. Here, the correctness is determined by ULGE and K-means, and experiment results in [14] verified that ULGE satisfies the correctness of ensemble learning. In addition, the performance of K-means is obvious to all.

In what follows, we discuss the diversity in UEL-GE. Diversity is obtained in the learning process by using four common schemes [20]: data sample perturbation, input attribute perturbation, output representation perturbation, and algorithm parameter perturbation. Here, the data sample perturbation method is used to enhance the diversity. ULGE randomly generate anchors, which can lead to differences in the resulting projection matrix. Furthermore, K-means is unstable which also provides diversity.

Therefore, the correctness and diversity in the proposed ensemble learning can be guaranteed.

2.3   Combination Rule in UEL-GE

Under the premise of ensuring correctness and diversity, ensemble learning generates multiple sets of individual learners, and then uses appropriate methods to combine individual learners to achieve a more significant effect than a single learner.

According to the requirement of ensemble learning, N individual learners that meet the correctness and the diversity are guaranteed. ULGE can get the projection matrix $\mathbf{W}$, and then the low-dimensional embedding $\mathbf{Y}$ can be calculated by $\mathbf{Y} = \mathbf{XW}$. The K-means method is used to cluster the low-dimensional embedding $\mathbf{Y}$ to obtain the clustering $c(\mathbf{Y})$. Then, each sample can be partitioned to a corresponding category. However, the tags clustered by the K-means method are messy, so it is necessary to correct the clustering results. The correction method, $bestMap$ was proposed for dealing with this situation [18,21]. Namely,

$$cc_t(\mathbf{Y}) = bestMap(\mathbf{rand\_c}, c_t(\mathbf{Y}))(t = 1, 2, \ldots, N) \qquad (6)$$

where $\mathbf{rand\_c}$ is a reference clustering label which is randomly selected from the N clustering labels $c_t(\mathbf{Y})$ $(t = 1, 2, \ldots, N)$, and $cc_t(\mathbf{Y})$ is the corrected clustering results. Because the reference clustering label is randomly selected, it increases the diversity of ensemble learning. The $bestMap(\mathbf{rand\_c}, c_t(\mathbf{Y}))$ function maps the clustering results $c_t(\mathbf{Y})$ to the data labels $\mathbf{rand\_c}$ [18].

Thus, we obtain the N corrected clusterings in low-dimensional subspaces. Then, a vote scheme is adopted to combine these N corrected clusterings to a final clustering. The combination strategy of the ensemble algorithm is defined as follows:

$$finalC(\mathbf{y}_i) = \arg\max_{q \in L} \sum_{t=1}^{N} I(cc_t(\mathbf{y}_i) = q) \qquad (7)$$

where $L = \{1, 2, \ldots, cnum\}$, $cnum$ is the number of classes of samples $\mathbf{X}$ and $cc_t(\mathbf{y}_i) \in L$, $\mathbf{y}_i$ is the i-th sample of low dimensional embedding $\mathbf{Y}$, and $finalC(\mathbf{y}_i)$ represents the final prediction value of the sample $\mathbf{y}_i$. Moreover, $I(\cdot)$ is an indicator function, when "·" is true, the value is 1 whereas the value is 0 when "·" is false.

3   Experimental Results

In this section, the advantage of UEL-GE is validated on five data sets. In addition, we perform K-means with the original data as the baseline and compare UEL-GE with other unsupervised algorithms, such as Principal Component Analysis (PCA) [22], LPP, SR, and ULGE.

3.1   Data Sets

We conduct experiments on five different data sets, i.e., USPS [23], DBRHD [24], COIL100 [25], ORL [26], and UMIST [27]. Both USPS and DBRHD are handwritten digit data sets. COIL100 is an object clustering and recognition data set, which contains 100 different objects. ORL and UMIST are face databases. ORL is made up of 40 distinct faces and each face has 10 different images. UMIST consists of 575 images of 20 people. The detail description of the data sets is listed in Table 1.

Table 1. Detail description of data sets

Data sets	Samples	Features	Classes
USPS	9298	256	10
DBRHD	10992	16	10
COIL100	7200	1024	100
ORL	400	1024	40
UMIST	575	1024	20

3.2 Parameter Setting

Following [14], we set the parameters in the compared methods. The reduced dimension of all the methods is set as the number of classes in these data sets. The neighbor sample value for both LPP and SR, and neighbor anchor value for both ULGE and UEL-GE are set to 5. Both LPP and SR use the same Gaussian kernel, where the parameter of bandwidth is set as 1. The regularization parameter α in SR, ULGE, and UEL-GE is 0.01. The number of anchors m in ULGE and UEL-GE takes 50% of the sample data set. N is set as 16.

3.3 Evaluation Metric

In order to compare the performance of these methods, the K-means method is used to cluster the low-dimensional embedding. The clustering performance is measured by the clustering accuracy (ACC) and the normalized mutual information (NMI). In experiments, all the algorithms are performed ten times, and the K-means clustering is also performed ten times after each dimension reduction. We report the mean ACC and NMI results for all compared methods.

All the experiments are implemented in MATLAB R2014a. The concerned experiments are running on a 3.19 GHz Intel(R) Core (TM) i5-6500 CPU computer with 4GB memory and Windows 10.

3.4 Results Analysis

Tables 2 and 3 report the ACC and NMI of different methods on five data sets, respectively. In the tables, the best method is highlighted in bold. From the tables we can easily see that UEL-GE performs better on most data sets. The clustering performance of UEL-GE is obviously superior to other comparative methods. Compared with other methods, ULGE has only a weak advantage, and is not even better than other algorithms on some datasets. When multiple ULGEs are combined, the performance has been improved a lot.

3.5 Parameter Analysis

There are several parameters involved in our method. We mainly analyze two important parameters, the anchor number m and the number of integrated

Table 2. Accuracy on five data sets (%)

Data sets	USPS	DBRHD	COIL100	ORL	UMIST
Baseline	65.7±0.71	70.2±1.44	48.9±0.57	55.3±0.96	42.2±0.55
PCA	64.6±1.12	68.6±2.08	49.4±0.53	56.1±0.94	42.9±0.74
LPP	65.8±1.11	67.8±1.38	41.4±0.75	32.4±0.17	30.2±0.50
SR	65.5±1.48	73.8±1.69	52.3±0.88	**61.0±0.75**	54.8±1.51
ULGE	65.5±1.30	74.2±1.94	55.5±0.93	55.6±2.37	51.4±2.63
UEL-GE	**68.8±3.33**	**77.9±3.54**	**59.9±0.79**	59.9±2.18	**57.1±4.44**

Table 3. Normalized mutual information on five data sets (%)

Data sets	USPS	DBRHD	COIL100	ORL	UMIST
Baseline	61.0±0.28	68.1±0.28	77.2±0.13	77.3±0.49	65.3±0.34
PCA	59.6±0.37	67.8±0.52	77.3±0.12	77.4±0.39	65.7±0.46
LPP	63.3±0.48	66.3±0.84	74.4±0.29	56.9±0.21	48.6±0.27
SR	70.3±0.66	70.6±0.28	80.6±0.46	79.8±0.26	**77.3±0.97**
ULGE	70.3±0.96	71.4±0.69	82.6±0.31	77.4±1.38	72.9±1.76
UEL-GE	**70.9±1.36**	**72.4±1.51**	**84.1±0.32**	**80.1±0.97**	76.4±2.82

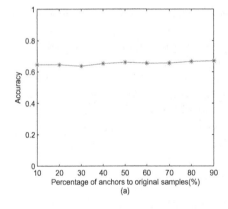 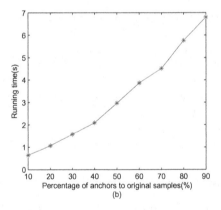

(a)   (b)

Fig. 2. Performance indexes vs. the percentage of anchors to original samples, (a) Accuracy, and (b) Running time.

individual learners N. Experiments on parameter analysis are constructed on the USPS data set.

We assume that the anchor number m is the percentage of the sample data set. We use ULGE to select parameter m. Figure 2(a) shows that the more anchors we select, the better performance of ULGE we achieve. The curve is on the rise, and when the anchor number takes 50%, the curve has a local maximum value.

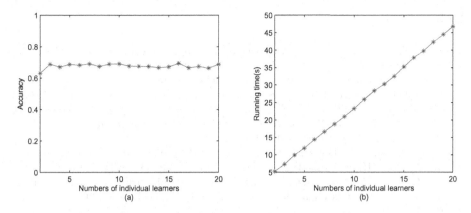

Fig. 3. Performance indexes vs. the numbers of individual learners, (a) Accuracy, and (b) Running time.

However, Fig. 2(b) shows that the increase in the anchors will result in an increase in running time. So, the anchor number takes 50% of the sample data set.

The number of individual learners N is determined by UEL-GE. From Fig. 3 we can see, the change under different N is not obvious. But with the increment of N, the performance of UEL-GE increases a little and the running time increases a lot. In our experiments, the selection of parameter N is 16.

4   Conclusions

In this paper, we propose UEL-GE which combines multiple ULGEs. Compared with a single method, UEL-GE uses individual learners and its performance is much better than a single method. A series of comparative experiments are carried out to verify the performance of the proposed method. One of our future work is to investigate another anchor generation method to replace the random selection method. In addition, the ensemble learning in this paper is simple, in the future, we will use an effective ensemble learning to improve the performance of algorithm.

Acknowledgments. This work was supported in part by the National Natural Science Foundation of China under Grant Nos. 61373093, 61402310, 61672364 and 61672365, by the Soochow Scholar Project of Soochow University, by the Six Talent Peak Project of Jiangsu Province of China and by the Graduate Innovation and Practice Program of colleges and universities in Jiangsu Province.

References

1. Yang, Y., Xu, D., Nie, F.P.: Image clustering using local discriminant models and global integration. IEEE Trans. Image Process. **19**(10), 2761–2773 (2010)
2. Hancer, E., Ozturk, C., Karaboga, D.: Artificial bee colony based image clustering method. In: 2012 IEEE Congress on Evolutionary Computation, pp. 1–5. IEEE (2012)
3. Zhao, Z.L., Liu, B., Li, W.: Image clustering based on extreme K-means algorithm. IEIT J. Adapt. Dyn. Comput. **1**, 12–16 (2012)
4. Hong, S., Choi, J., Feyereisl, J.: Joint image clustering and labeling by matrix factorization. IEEE Trans. Patt. Anal. Mach. Intell. **38**(7), 1411–1424 (2016)
5. Liu, Q.S., Sun, Y.B., Wang, C.T.: Elastic net hypergraph learning for image clustering and semi-supervised classification. IEEE Trans. Image Process. **26**(1), 452–463 (2017)
6. Keogh, E., Mueen, A.: Curse of Dimensionality. Springer, US (2011)
7. Roweis, S.T., Saul, L.K.: Nonlinear dimensionality reduction by locally linear embedding. Science **290**(5500), 2323–2326 (2000)
8. Tenenbaum, J.B., Silva, V.D., Langford, J.C.: A global geometric framework for nonlinear dimensionality reduction. Science **290**(5500), 2319–2323 (2000)
9. Choi, J.Y., Bae, S.H., Qiu, X.: High performance dimension reduction and visualization for large high-dimensional data analysis. In: Proceedings of the 10th IEEE/ACM International Conference on Cluster, Cloud and Grid Computing, pp. 331–340. IEEE Computer Society (2010)
10. Seung, H.S., Lee, D.D.: The manifold ways of perception. Science **290**(5500), 2268–2269 (2000)
11. Belkin, M., Niyogi, P.: Laplacian eigenmaps and spectral techniques for embedding and clustering. In: Proceedings of the 14th International Conference on Neural Information Processing Systems: Natural and Synthetic, pp. 585–591. MIT Press (2002)
12. He, X.F., Niyogi, P.: Locality preserving projections. Adv. Neural Inf. Process. Syst. **16**(1), 186–197 (2003)
13. Cai, D., He, X.F., Han, J.: Spectral regression: a unified subspace learning framework for content-based image retrieval. In: Proceedings of the 15th ACM International Conference on Multimedia, Augsburg, Germany, pp. 403–412 (2007)
14. Nie, F.P., Zhu, W., Li, X.L., Unsupervised large graph embedding. In: Proceedings of the 31st AAAI Conference on Artificial Intelligence (AAAI), San Francisco, USA, pp. 2422–2428 (2017)
15. Liu, W., He, J., Chang, S.F.: Large graph construction for scalable semi-supervised learning. In: Proceedings of the 27th International Conference on Machine Learning, pp. 679–686 (2010)
16. Nie, F.P., Wang, X., Jordan, M.I.: The constrained laplacian rank algorithm for graph-based clustering. In: Thirtieth AAAI Conference on Artificial Intelligence, pp. 1969–1976. AAAI Press (2016)
17. Zhou, Z.H.: When semi-supervised learning meets ensemble learning. Front. Electr. Electr. Eng. China **6**(1), 6–16 (2011)
18. Lovcsz, L., Plummer, M.D.: Matching theory. In: Annals of Discrete Mathematics, no. 29 (1986)
19. Eckart, C., Young, G.: The approximation of one matrix by another of lower rank. Psychometrika **1**(3), 211–218 (1936)

20. Zhou, Z.H.: Machine Learning. Tsinghua University Press, Beijing (2016). (in Chinese)
21. Chen, X.L., Cai, D.: Large scale spectral clustering with landmark-based representation, In: AAAI Conference on Artificial Intelligence, pp. 313–318 (2011)
22. Wold, S., Esbensen, K., Geladi, P.: Principal component analysis. Chemom. Intell. Lab. Syst. $2(1)$, 37–52 (1987)
23. Lecun, Y., Boser, B., Denker, J.S.: Backpropagation applied to handwritten zip code recognition. Neural Comput. $1(4)$, 541–551 (1989)
24. Alimoglu, F., Alpaydin, E.: Methods of combining multiple classifiers based on different representations for pen-based handwritten digit recognition. In: Proceedings of the 5th Turkish Artificial Intelligence and Artificial Neural Networks Symposium (TAINN), Istanbul, Turkey (1996)
25. Nayar, S.: Columbia Object Image Library (1996)
26. Samaria, F.S., Harter, A.C.: Parameterisation of a stochastic model for human face identification. In: 1994 IEEE Applications of Computer Vision, pp. 138–142. IEEE (1994)
27. Graham, D.B., Allinson, N.M.: Characterising virtual eigensignatures for general purpose face recognition. Face Recog. Form Theory Appl. $163(2)$, 446–456 (1998)

Potential Probability of Negative Triples in Knowledge Graph Embedding

Shengyue Luo$^{(\boxtimes)}$ and Wei Fang

Department of Computer Science,
Beijing University of Posts and Telecommunications, Beijing, China
{luo_shengyue, fang_wei}@bupt.edu.cn

Abstract. In this paper, we propose the concept of triples' potential probability. Typically, knowledge graph only contains positive triples. Most of knowledge representation methods treat the replaced triples, which replace the head/tail entities or relations with other entities or relations randomly, as negative triples. Actually, not all triples are absolutely negative triples after substitution. It could be a positive triple essentially, but has not been discovered yet. Considering the problems arising from the above situation, we propose the potential probability to solve it. First, we utilize the co-occurrence of relations and paths in the knowledge graph to find potentially correct probabilities of some negative triples. Then we add these triples with potential probabilities to the training model. Finally, we take the experiments on two translation-based models, TransE and TransH, using four public datasets. Experimental results show that our method greatly enhances the performance of the target embedding models.

Keywords: Knowledge graph · Potential probability · Knowledge embedding
Relation path

1 Introduction

Knowledge Graph (KG) maps the facts of the objective world into pieces of knowledge [1]. It applies to intelligent Q&A [2], personalized recommendation, social networks and other fields. Knowledge in large-scale public knowledge graphs such as FreeBase [3], WordNet [4], ConceptNet [5] etc. is represented by triple facts such as (cake, IsA, dessert). Knowledge graph embedding technique can transform the knowledge in KG into a low-dimension vector while maintaining the original structure and semantic information. Converting discrete facts into continuous vector representations is beneficial to increase the flexibility of knowledge graph applications [6]. Besides, since most of machine learning algorithms can handle the low-dimension vector as input easily, knowledge graph embedding can reduce the difficulty of subsequent tasks, like feature engineering or similarity calculation [7] etc.

Because the public dataset only contains golden triples, common translation-based models generally construct negative triples by replacing the head/tail entity or relation randomly at first. In the above example, head entity and tail entity are cake and dessert, respectively. Then they try to distinguish positive and negative triples through a margin-based ranking loss function. However, there are always many missing facts

© Springer Nature Switzerland AG 2018
L. Cheng et al. (Eds.): ICONIP 2018, LNCS 11303, pp. 48–58, 2018.
https://doi.org/10.1007/978-3-030-04182-3_5

actually, even a knowledge graph with millions of triples. A so-called negative triple we randomly generate may be a positive triple essentially.

A knowledge graph always contains relation path information which is convenient to acquire and use. It is known that multiple-steps paths between entities can express their sematic relationships to some extent. PTransE [8] is an algorithm that introduces related relation path information to triples. Inspired by relation path, we propose the concept of potential probability by utilizing the characteristic of co-occurrence in knowledge graph data in this work. More specifically, as shown in Fig. 1, if the path < bornIn (r_1), at Location (r_2) > and the relation nationality (r_e) have a high probability of co-occurrence, we assume that there is a relation r_e with high potential probability between e_1 and e_2.

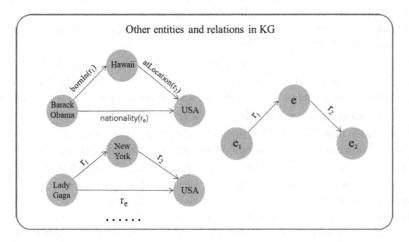

Fig. 1. Example of co-occurrence in KG.

In experiments, we evaluate the performance of our method on the link prediction task, using four public datasets: FB15k, FB15k-237, WN18 and WN18RR. The knowledge representation algorithm TransE [9] and TransH [10] which convert the triples into low dimension vectors are our baselines. Experiments based on them show that our method can greatly improve the performance of some of the most commonly used translation-based models. The main contributions of this work are concluded as follows:

- We propose a method that can be applied to common translation-based models. It can calculate the potentially correct probability of some negative triples to improve the effect of knowledge graph embedding.
- We only use the original data of the knowledge graph and results of each iteration, without any external resources, which reduce the actual use-cost of the model.
- In experiment, our approach has significant improvements compared with the baseline.

2   Related Work

In a sense, knowledge graph embedding is a sub-domain of network embedding. Because it contains semantic information, knowledge graph embedding requires a more targeted model than general network embedding [11]. After several years' development, knowledge embedding technique has lots of great models, which is mainly divided into two categories, translation-based model and semantic-based matching model.

The translation-based model generally uses a distance-based scoring function to evaluate the correctness of a triple. TransE [9] is a classic translation-based model that uses a simple algorithm to transform knowledge into the low dimension vectors. More details about its score function will be discussed in Sect. 3.1. Because this algorithm is too uncomplicated, it doesn't do well in complex relationships (one-to-many, many-to-one and many-to-many). To solve this problem, TransH [10] projects the head and tail entities onto the hyperplane of the corresponding relation. Its score function is: $f_r(\mathbf{h}, \mathbf{t}) = \|\mathbf{h}_\perp + \mathbf{r} - \mathbf{t}_\perp\|_{1/2}^2$ where $\mathbf{h}_\perp = \mathbf{h} - \mathbf{w}_r^{\mathrm{T}}\mathbf{h}\mathbf{w}_r$ and $\mathbf{t}_\perp = \mathbf{t} - \mathbf{w}_r^{\mathrm{T}}\mathbf{t}\mathbf{w}_r$. Both of the above algorithms process the knowledge in the same semantic space. TransR/CTransR [12] projects the head and tail entities into another semantic space corresponding to its relation. Its score function is: $f_r(\mathbf{h}, \mathbf{t}) = \|\mathbf{M}_r\mathbf{h} + \mathbf{r} - \mathbf{M}_r\mathbf{t}\|_{1/2}^2$. TransD [13], which computes faster than TransR/CTransR, not only considers the relations, but also the influence of different types of entities on the projection matrix. Its score function is: $f_r(\mathbf{h}, \mathbf{t}) = -\|\mathbf{h}_\perp + \mathbf{r} - \mathbf{t}_\perp\|_2^2$ where $\mathbf{h}_\perp = \mathbf{M}_{rh}\mathbf{h}$ and $\mathbf{t}_\perp = \mathbf{M}_{rt}\mathbf{t}$. TranSparse [14] proposes two models, TranSparse (share) and TranSparse (separate), which focus on solving the problems of heterogeneity and imbalance in KGs, respectively.

The sematic-based matching model uses a similarity-based score function to evaluate the correctness of a fact. It maps entities and relations into an implicit semantic space for similarity measures. Semantic Matching Energy (SME) defines a score function of neural network to capture correlations between entities and relations by using matrix operations. The score function is $f_r(\mathbf{h}, \mathbf{t}) = \mathbf{g}_{left}^{\mathrm{T}}\mathbf{g}_{right}$. It has two semantic matching energy functions including a linear form $\mathbf{g}_\eta = \mathbf{M}_{\eta 1}\mathbf{e}_\eta + \mathbf{M}_{\eta 2}\mathbf{r} + \mathbf{b}_\eta$ [15] and a bilinear form $\mathbf{g}_\eta = (\mathbf{M}_{\eta 1}\mathbf{e}_\eta) \otimes (\mathbf{M}_{\eta 2}\mathbf{r}) + \mathbf{b}_\eta$ [16] where $\eta = \{left, right\}$, $\otimes$ is the Hadamard product. Single Layer Model (SLM) is a baseline of NTN model [17], which designs a nonlinear neural network to represent the score function $f_r(\mathbf{h}, \mathbf{t}) = \mathbf{u}_r^T f(\mathbf{M}_{r1}\mathbf{h} + \mathbf{M}_{r2}\mathbf{t} + \mathbf{b}_r)$. Neural Tensor Network (NTN) is an extension of SLM. It takes into account the second-order correlations of the nonlinear neural network. The score function is: $f_r(\mathbf{h}, \mathbf{t}) = \mathbf{u}_r^T f(\mathbf{h}^{\mathrm{T}}\mathbf{M}_r\mathbf{t} + \mathbf{M}_{r1}\mathbf{h} + \mathbf{M}_{r2}\mathbf{t} + \mathbf{b}_r)$ where $f()$ is the *tanh* operation, $\mathbf{M}_r \in \mathbb{R}^{d \times d \times k}$ is a three-way tensor, $\mathbf{M}_{r1}, \mathbf{M}_{r2} \in \mathbb{R}^{k \times d}$ are weight matrixes and $\mathbf{b}_r$ is the bias.

Except the above models, there are others that introduce relation path information in knowledge embedding technology. PTransE [8], which is based on TransE model, has a more complete understanding of knowledge by introducing the relation path information. CKRL [18] assume the existing of noise triples in KGs, and we can apply the noise detection by various approaches including considering relation path information.

3 Our Method

We first define the notations used in this paper. Entities and relations are denoted by E and R, respectively. Our goal is to encode both entities and relations in $\mathbf{R}^k$. Triple facts are represented as (h, r, t), where h denotes the head entity, r denotes a relation and t denotes the tail entity. We use S to represent the positive triple set (triples in KG), use S' to represent original negative triples (triplets generated by replacing the head/tail entity or relation randomly). And G represents the generated triples with potential probabilities, where $G \subseteq S'$.

Figure 2 is a step description of our method, which explains the flow of the entire algorithm briefly.

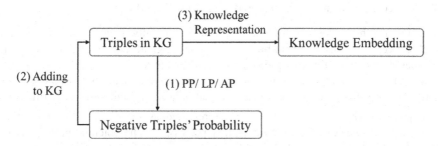

Fig. 2. Flowchart of our improved knowledge embedding algorithm.

3.1 Training

The key idea of most current translation-based models is that the vector of the relation r can be regarded as a translation between the head entity vector and the tail entity vector when the triple (h, r, t) holds. For example, the score function of classic model TransE [9] is:

$$f(h, r, t) = \|\mathbf{h} + \mathbf{r} - \mathbf{t}\|_{1/2}, \tag{1}$$

where $\mathbf{h}$, $\mathbf{r}$, $\mathbf{t}$ are vectors of head entity, relation and tail entity, respectively. Using the score function can measure the correctness of a triple to some extent. Other models have some differences in calculating the score of triple facts. We hope that positive triples' score will close to 0, and negative triples' score will be as large as possible through the algorithm.

Objective Formalization. Due to the introduction of the concept of potential probability for some negative triples, we improve the margin-based ranking loss function as follows:

$$L = \sum_{(h,r,t) \in S} \sum_{(h',r,t') \in S'} \max(0, (1 - \delta)\lambda + f(h, r, t) - f(h', r, t')), \tag{2}$$

where f is the score function of a definite model. $f(h, t, r)$ is the score of positive triple and $f(h', t', r')$ is that of negative triple. δ is potential probability if the negative triple has, otherwise it is 0. $\lambda > 0$ is the hyper-parameter of margin. The less differences between positive and negative triples, the higher potential probability is.

The negative triple set was generated by using the following rule:

$$S' = \{(h', r, t)|h' \in E\} \cup \{(h, r, t')|t' \in E\}, \quad (h, r, t) \in S \tag{3}$$

Note that the triple generated by the above method is called original negative triple. It might belong to G or not. Obviously, it is not in S, which indicates that it is not positive definitely.

3.2   Generating Negative Triples with Potential Probability

To distinguish the triples which could be positive and really negative, we introduce a novel concept, **potential probability** for some original negative triple facts. It describes the likelihood of a negative triple to be positive.

If a relation has a high frequency of co-occurrence with a path, we assume that the relation is more likely to exist between the path's head and tail entities. Formally, a given path $p = (r_1, r_2, \ldots, r_m)$ can be represented as $E_h \xrightarrow{r_1} \ldots \ldots \xrightarrow{r_m} E_t$, where $h \in E_h$, $t \in E_t$, since there are multiple entity pairs (h, t). The potential probability of (h, r_e, t) will be calculated as follows:

$$PP(h, t, r_e) = \frac{P(r_e * p_i)}{P(p_i)}, \tag{4}$$

where $P(p_i)$ represents the probability of path p_i in KG, and operation $*$ means r_e and p_i co-occurrence. The potential probability of a triple might be very large (near to 1) sometimes, which makes it pretty close to a positive triple. Actually, differences should be existed between them, so we try to solve this problem in following two ways.

Linear Smoothing. The first method we proposed is using two different linear functions to decrease the large potential probabilities (We assume that small potential probabilities are less likely to cause model errors). Linear smoothing function can be written as follows:

$$LP(h, t, r_e) = \begin{cases} PP_{(h,t,r_e)}, & x < \omega \\ aPP_{(h,t,r_e)} + (1-a)\omega, & x \geq \omega \end{cases} \tag{5}$$

where a and ω are hyper-parameters. To some extent, the potential probabilities which are larger than ω can be reduced through this function.

Adaptive Transformation. We expect that the scores of the positive triples will increase, and the negative triples will become smaller after each step's training. It can help us distinguish the triples effectively. So, the score of positive and negative triples will be updated in each iteration. In the same way, we consider that the potential probability of negative triples also need to be updated. Taking the TransE model as an example, for a positive triple (h_i, r_i, t_i), we assume that it generates a negative triple (h_i, r_i, t_j) by replacing the tail entity randomly. Replacing the head entity also works. The change of r_i in a triple before and after an iteration is represented as follows:

$$\Delta_{(r_i,t)} = \|\mathbf{h}_i - \mathbf{t}\|_{1/2} - \|\mathbf{h}'_i - \mathbf{t}'\|_{1/2}, \quad \mathbf{t} = \{\mathbf{t}_i, \mathbf{t}_j\} \tag{6}$$

where $\mathbf{h}_i$, $\mathbf{t}_i$, $\mathbf{t}_j$ and $\mathbf{r}_i$ are vectors of entities and relation before one iteration. $\mathbf{h}'_i, \mathbf{t}'_i, \mathbf{t}'_j$, and $\mathbf{r}'_i$ are vectors of entities and relation after the iteration.

If the changes of relation r_i in positive and negative triples are very similar, we consider that the potential probability of this negative triple should increase, otherwise decrease. The potential probability of negative triples will be updated as follows:

$$AP_{(h,t,r_i)} = \left(e^{-\left|\Delta_{(r_i,t_i)} - \Delta_{(r_i,t_j)}\right|} + \varepsilon \right) \bullet PP_{(h,t,r_i)} \tag{7}$$

where ε is a hyper-parameter.

3.3 Optimization and Implementation Details

We use stochastic gradient descent (SGD) to optimize our model. In practice, we enforce the constraints as follows:

$$\|\mathbf{h}\|_2 \leq 1, \quad \|\mathbf{r}\|_2 \leq 1, \quad \|\mathbf{t}\|_2 \leq 1. \quad \forall h, r, t \tag{8}$$

Taking into account the diversity of relations, we add the inverse relation during training process like [8]. We believe that if there is a relation r from e_1 to e_2, a relation r^{-1} should exist from e_2 to e_1. In addition, for any triple in the KG, we consider all paths which length equals to 2 between the head entity and the tail entity as initial paths. But a path could be particularly long in KG, which makes the calculation speed of the algorithm decrease, as well as the memory of consumption become larger. Consequently, we restrict the length of path to at most 2-steps, for it is impractical to list all paths with various lengths.

4 Experiment and Analysis

4.1 Datasets

For evaluation, we use 4 public knowledge graph datasets: FB15k, FB15k-237, WN18, WN18RR. FB15k is a subset of FreeBase that contains a large amount of objective world facts. The FB15k-237 [19] is a subset of FB15k, which removes a large number of redundant relations. Triples in FB15k-237 are reduced by more than a half ultimately.

WN18 is a subset of WordNet, which stores many sematic relations between English words. WN18RR [20] is a subset of WN18. It only deletes 7 relations of WN18, but it greatly increases the difficulty of knowledge embedding. The details of datasets are in Table 1.

4.2 Experimental Setting

We implement experiments on three strategies by using TransE and TransH as baselines: the potential probability without modification (PP), the potential probability using the linear function (LP) for smoothing, as well as adaptive transformation with each iteration (AP). Obviously, it is not difficult to apply our method to other translation-based models.

Since the amount of data in FB15k and FB15k-237 is very large, only the relations that contain the path with a potential probability greater than N are selected. We test situations of $N = \{0.5, 0.6, 0.7, 0.8, 0.9\}$. Considering the calculation speed and experimental accuracy, we choose 0.8 for FB15k and 0.5 for FB15k-237 finally. This setting is inversely proportional to the data size of FB15K and FB15K237, which is consistent to our expectation. We use SGD to train our model with the learning rate α set among $\{0.001, 0.003, 0.01, 0.03, 0.1\}$. We select margin λ among $\{0.1, 1, 2, 3, 5\}$, hyper-parameter ε among $\{0.001, 0.003, 0.01, 0.03, 0.1\}$, a among $\{0.8, 0.85, 0.9, 0.95\}$, ω among $\{0.5, 0.6, 0.7, 0.8, 0.9\}$ and L1 or L2 distances. To ensure the relative fairness, we set training epoch to 1000, batch-size to 100, embedding dimension k to 50 and use bern rules [10] for the replacement of head/tail entity in all datasets.

Table 1. Statistics of datasets.

Dataset	#Relation	#Entity	#Train	#Valid	#Test
FB15k	1,345	14,951	483,142	50,000	59,071
FB15k-237	237	14,541	272,115	17,535	5,000
WN18	18	40,943	141,442	5,000	5,000
WN18RR	11	40,943	86,835	3,034	3,134

Since the data size of the four public datasets is not very large, the smaller learning rate can effectively avoid overfitting. The best configuration obtained by valid set are: $\alpha = 0.001$, $\lambda = 2$, $a = 0.95$, $\omega = 0.8$, $\varepsilon = 0.03$ and taking L1 as dissimilarity on FB15k; $\alpha = 0.001$, $\lambda = 3$, $a = 0.95$, $\omega = 0.8$, $\varepsilon = 0.01$ and taking L1 as dissimilarity on FB15k-237; $\alpha = 0.001$, $\lambda = 5$, $a = 0.8$, $\omega = 0.5$, $\varepsilon = 0.01$ and taking L1 as dissimilarity on WN18; $\alpha = 0.001$, $\lambda = 5$, $a = 0.95$, $\omega = 0.8$, $\varepsilon = 0.001$ and taking L1 as dissimilarity on WN18RR.

Table 2. Evaluation results on link prediction with TransE.

Dataset	FB15k				FB15 k-237			
Metric	Mean rank		Hits@10		Mean rank		Hits@10	
	Row	Filt.	Row	Filt.	Row	Filt.	Row	Filt.
TransE [9]	243	125	34.9	47.1	–	–	–	–
TransE	213	104	44.32	62.06	369	239	29.22	40.85
TransE (PP)	203	107	50.5	**68.38**	368	**234**	**32.24**	**43.96**
TransE (LP)	198	**102**	50.51	68.36	372	237	32.08	43.88
TransE (AP)	**197**	**102**	**50.67**	68.21	373	239	31.97	43.72
Dataset	WN18				WN18RR			
Metric	Mean rank		Hits@10		Mean rank		Hits@10	
	Row	Filt.	Row	Filt.	Row	Filt.	Row	Filt.
TransE [9]	263	251	75.4	89.2	–	–	–	–
TransE	231	218	78.17	90.88	3601	3587	45.05	47.86
TransE (PP)	211	200	80.63	**93.65**	2988	2974	46	48.95
TransE (LP)	**191**	**179**	80.59	93.41	2988	2974	**46.2**	**49.49**
TransE (AP)	196	184	**80.7**	93.53	**2899**	**2885**	45.93	49.11

Table 3. Evaluation results on link prediction with TransH.

Dataset	FB15 k				FB15k-237			
Metric	Mean rank		Hits@10		Mean rank		Hits@10	
	Row	*Filt.*	*Row*	*Filt.*	*Row*	*Filt.*	*Row*	*Filt.*
TransH [10]	212	87	45.7	64.4	–	–	–	–
TransH	183	**62**	44.53	62.3	357	195	29.11	41.07
TransH (PP)	**181**	62	**46.52**	**64.51**	355	195	29.6	41.42
TransH (LP)	183	63	46.11	64.32	**353**	**193**	29.55	41.69
TransH (AP)	183	63	46.46	64.18	357	196	**29.78**	**41.8**
Dataset	WN18				WN18RR			
Metric	Mean rank		Hits@10		Mean rank		Hits@10	
	Row	Filt.	Row	Filt.	Row	Filt.	Row	Filt.
TransH [10]	400	388	73	82.3	–	–	–	–
TransH	238	226	76.78	89.71	3472	3459	42.21	44.85
TransH (PP)	227	215	77.2	89.74	2978	2904	**43.96**	**45.9**
TransH (LP)	246	234	**77.43**	89.76	**2967**	**2741**	43.3	45.75
TransH (AP)	**221**	**209**	77.25	**89.87**	3014	2933	43.84	45.01

4.3   Link Prediction

Evaluation Protocol. Following the same protocol of [9], we use two measures as evaluation criteria: (1) Mean Rank of positive triples, and (2) Hits@10 which refers to the proportion of positive triples in the top ten. We use two evaluation settings, "Row" and "Filter". "Row" does not delete the generated negative triples appearing in the training set, valid set or test set. And it still regards them as negative triples. Another evaluation setting filters out all these triples before ranking.

Result. The results of link prediction experiment are shown in Tables 2 and 3. The first line in the tables is the results of the baseline models which copy from [9, 10], respectively. And the second line is the best results we got after trying two baseline models ourselves. From the experimental results, we obverse that: (1) Our three ways, which introduce the potential probability, significantly outperform in mean rank and Hits@10 on four datasets compared with the baselines of TransE and TransH. Experimental results show that the potential probabilities of some negative triples provide a great supplement for knowledge embedding in KGs. (2) Inspired by TransE, comparative experiments of mean rank (left) and Hits@10 (right) on FB15k data are

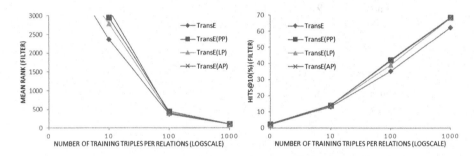

Fig. 3. Effect of knowledge embedding with different number of training triples on FB15k.

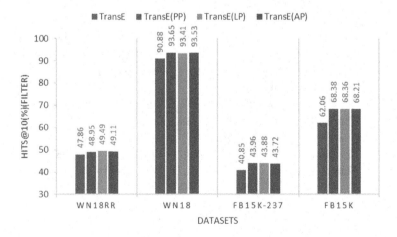

Fig. 4. Results of Hits@10 (Filter) on 4 datasets.

presented in Fig. 3. In the picture, the effect of TransE is better than (or similar with) other methods under the same circumstances, when the number of data is small. But with the increase of data, our method become better and better than TransE. This situation is more apparent in Fig. 4. The number of data in dataset gradually increases from left to right, with the increasingly obvious differences between our three methods and TransE in Hits@10.

5   Conclusion and Future Work

In this paper, we propose a concept of potential probability of some negative triples, which aims to calculate the probably positive probability of some negative triples. To measure the value objectively, we apply two algorithms, linear smoothing and adaptive transformation. In addition, we only use the internal structure data in KGs to simplify the calculation and method usage scenario. Experimental results indicate that adding potential probability to some negative triples during the training process helps to increase the effect of knowledge embedding.

We will explore the following research directions in the future: (1) Internal structure data in KG still contain lots of information, which are worth paying more attention. We will explore the implementation of translating calculated path information to potential probability dynamically during the training process. It makes the potential probability of a negative triple change dynamically, too. (2) External information can provide a lot of relation and entity details [21]. DKRL [22], ETRL [23] and IKRL [24] combine the entities' textural or image information to improve the knowledge embedding results. We will attempt to combine the external information to enrich the description of entities and relations, which makes the calculation of the triple's potential probability more consistent with the objective facts.

References

1. Singhal, A.: Official Google Blog: Introducing the Knowledge Graph: things, not strings, pp. 1–8. Official Google Blog (2012)
2. Liu, Z., Li, K., Qu, D.: Knowledge graph based question routing for community question answering. In: Liu, D., Xie, S., Li, Y., Zhao, D., El-Alfy, El-Sayed M. (eds.) ICONIP 2017. LNCS, vol. 10638, pp. 721–730. Springer, Cham (2017). https://doi.org/10.1007/978-3-319-70139-4_73
3. Fellbaum, C.: WordNet: An Electronic Lexical Database, vol. 71, p. 423. MIT Press, Cambridge (1998)
4. Bollacker, K., Evans, C., Paritosh, P., Sturge, T., Taylor, J.: Freebase: a collaboratively created graph database for structuring human knowledge. In: SIGMOD Conference, pp. 1247–1250 (2008)
5. Speer, R., Havasi, C.: Representing general relational knowledge in ConceptNet 5 (2012)
6. Jia, Y., Wang, Y., Jin, X., Lin, H., Cheng, X.: Knowledge graph embedding: a locally and temporally adaptive translation-based approach. ACM Trans. Web 12(2), 1–33 (2017)

7. Guo, S., Wang, Q., Wang, B., Wang, L., Guo, L.: Semantically smooth knowledge graph embedding. In: Meeting of the Association for Computational Linguistics and the, International Joint Conference on Natural Language Processing, pp. 84–94 (2015)
8. Lin, Y., Liu, Z., Luan, H., Sun, M., Rao, S., Liu, S.: Modeling relation paths for representation learning of knowledge bases. In: Proceedings of the 2015 Conference on Empirical Methods in Natural Language Processing, pp. 705–714, Lisbon, Portugal (2015)
9. Bordes, A., Usunier, N., Garcia-Duran, A., Weston, J., Yakhnenko, O.: Translating embeddings for modeling multi-relational data. In: International Conference on Neural Information Processing Systems, pp. 2787–2795. Curran Associates Inc. (2013)
10. Wang, Z., Zhang, J., Feng, J., Chen, Z.: Knowledge graph embedding by translating on hyperplanes. In: Proceedings of the Twenty-Eighth AAAI Conference on Artificial Intelligence, pp. 1112–1119. AAAI Press (2014)
11. Wang, Q., Mao, Z., Wang, B., Guo, L.: Knowledge graph embedding: a survey of approaches and applications. IEEE Trans. Knowl. Data Eng. 29(12), 2724–2743 (2017)
12. Lin, Y., Liu, Z., Zhu, X., Zhu, X., Zhu, X.: Learning entity and relation embeddings for knowledge graph completion. In: Proceedings of the Twenty-Ninth AAAI Conference on Artificial Intelligence, vol. 108, pp. 2181–2187. AAAI Press (2015)
13. Ji, G., He, S., Xu, L., Liu, K., Zhao, J.: Knowledge graph embedding via dynamic mapping matrix. In: Meeting of the Association for Computational Linguistics and the International Joint Conference on Natural Language Processing, pp. 687–696 (2015)
14. Ji, G., Liu, K., He, S., Zhao, J.: Knowledge graph completion with adaptive sparse transfer matrix. In: Proceedings of the Thirtieth AAAI Conference on Artificial Intelligence, pp. 985–991. AAAI Press (2016)
15. Bordes, A., Glorot, X., Weston, J.: Joint learning of words and meaning representations for open-text semantic parsing. In: Proceedings of International Conference on Artificial Intelligence & Statistics, pp. 127–135 (2012)
16. Bordes, A., Glorot, X., Weston, J., Bengio, Y.: A semantic matching energy function for learning with multi-relational data. Mach. Learn. 94(2), 233–259 (2014)
17. Socher, R., Chen, D., Manning, C.D., Ng, A.Y.: Reasoning with neural tensor networks for knowledge base completion. In: International Conference on Neural Information Processing Systems, pp. 926–934. Curran Associates Inc. (2013)
18. Xie, R., Liu, Z., Sun, M.: Does William Shakespeare really write Hamlet? knowledge representation learning with confidence. In: Thirty-Second AAAI Conference on Artificial Intelligence (2018)
19. Toutanova, K., Chen, D.: Observed versus latent features for knowledge base and text inference. In: Proceedings of the 3rd Workshop on Continuous Vector Space Models and their Compositionality, pp. 57–66 (2015)
20. Dettmers, T., Minervini, P., Stenetorp, P., Riedel, S.: Convolutional 2D knowledge graph embeddings (2017)
21. Wang, Z., Zhang, J., Feng, J. Chen, Z.: Knowledge Graph and Text Jointly Embedding. In: EMNLP, pp. 1591–1601 (2014)
22. Xie, R., Liu, Z., Jia, J., Luan, H., Sun, M.: Representation learning of knowledge graphs with entity descriptions. In: Proceedings of the Thirtieth AAAI Conference on Artificial Intelligence, pp. 2659–2665. AAAI Press (2016)
23. Ouyang, X., Yang, Y., He, L., Chen, Q., Zhang, J.: Representation learning with entity topics for knowledge graphs. In: Li, G., Ge, Y., Zhang, Z., Jin, Z., Blumenstein, M. (eds.) KSEM 2017. LNCS (LNAI), vol. 10412, pp. 534–542. Springer, Cham (2017). https://doi.org/10.1007/978-3-319-63558-3_45
24. Xie, R., Liu, Z., Luan, H. Sun, M.: Image-embodied knowledge representation learning. In: IJCAI, pp. 3140–3146 (2017)

Event Causality Identification by Modeling Events and Relation Embedding

Zhenyu Yang, Wei Liu[⊠], and Zongtian Liu

School of Computer Engineering and Science,
Shanghai University, Shanghai 200444, China
18201789699@163.com, {liuw,ztliu}@shu.edu.cn

Abstract. Events and event relations contain high-level semantic information behind texts. In this paper, we mainly discuss event causality relation identification. Traditional approaches of causality relation identification rely on the recognition of casual relationship connectives or manual features of causality relationships, and these methods have disadvantage of low recognition coverage and being lack of adaptive. To solve this problem, we propose a novel model based on modeling event and event relation. We use word sequence around event trigger as input data and use event based Siamese Bi-LSTM network to model events by encoding the event representations into a fixed size vectors, and then these events representations are applied in relation embedding training and prediction. Experimental results show that the proposed method can achieve better effect on CEC 2.0 corpus.

Keywords: Siamese network · Event relation · LSTM · CEC

1 Introduction

Natural language organized texts express higher-level semantic information through events. Recognizing these events and the relationships between these events can help computers easily understand the precise meaning of texts and lay a solid foundation for the reasoning and modeling of event ontology.

We define an event as a thing happens in a certain period of time and place, in which some actors participate and show some features of action, also accompany with the changing of internal status [1]. An event trigger is the word that most exactly expresses the occurrence of an event. For example: in the sentence "the earthquake happened yesterday caused 21 wounded". "wounded" is a trigger of event. Event trigger is the most significant signal of event in texts.

Event can be formalized as a 6-tuple $e = (A, O, T, P, S, L)$. We call elements in 6-tuple event elements, and represent action, object, time, place, status, language expression respectively. In natural language processing, we mainly focus on participants, objects, time, and location of an event. These elements present as word in natural language and contains important information of events.

Causality relation is a kind of common and important relation between events. If an event e_1 happened, the another event e_2 happens with the probability above the threshold of causality, there is a causality relation between e_1 and e_2. Causality relation

© Springer Nature Switzerland AG 2018
L. Cheng et al. (Eds.): ICONIP 2018, LNCS 11303, pp. 59–68, 2018.
https://doi.org/10.1007/978-3-030-04182-3_6

can be divided into explicit causality and implicit causality. Explicit causality denotes those relations exist connectives exactly express the relation between events. Implicit causality denotes those relations lack exact connectives and need to be speculated by the contexts. In addition, there're three relations between events beside causality relation, which include composition relation, follow relation and concurrency relation. If an event e can be decomposed to several sub-events e_i with smaller granularity, there exists composition relation between e and e_i. If in a certain length of time, the occurrence of event e_1 follows the occurrence of the event e_2 at above specified threshold, there exists a follow relation between e_1 and e_2. If there are event e_1 and event e_2 occur simultaneously in a certain period of time, there is a concurrency relation between e_1 and e_2.

Current researches on causality relation identification are mostly based on the feature selection, pattern matching and rule reasoning. These approaches of causality relation identification can't realize the context and identify the implicit causality relation in texts.

In recent years, deep learning (DL) within the machine learning field has shown that it can be successfully applied to reduce the data dimension by extracting deep features of data and use those features to present better results than traditional machine learning methods. Although there are preliminary applications of DL in many natural language processing (NLP) tasks. There are few researches on causality relation identification based on DL. Therefore, we propose a new method based on Siamese network. Firstly we use Bi-LSTM network to capture the semantic information in events and generate event representations which cover event elements and event triggers. Then we use the element-wise difference between events to predict the causality relation. The experimental results show that our proposed model has achieved better performance in causality relation identification. In addition, event representations generated by our proposed model also achieve satisfactory results in the task of event classification.

The remained of this paper is organized as follows: we describe the related works in Sect. 2. Our proposed model is described in Sect. 3. Section 4 presents our experimental results. Finally, we conclude in Sect. 5.

2 Related Work

2.1 Siamese Network

Siamese network is a special type of neural network architecture which is widely applied in calculating the similarity of pair of inputs like texts or pictures [2–4]. Siamese network proposed by Chopra consists of two identical neural networks with shared parameters and the last layers of two networks are then fed to a contrastive loss function which calculates the similarity between two inputs. Chopra's work illustrates the method for learning complex similarity metrics with a face verification application. Recently, Siamese Network is also applied in NLP. Kenter [5] presented the Siamese CBOW model based on Siamese Network. Siamese CBOW handles the task of sentence representation by training word embedding directly, and then trains a sentence embedding by predicting from its surrounding sentence representations. Muller [6] proposed their Manhattan LSTM (MaLSTM) for assessing the semantic similarity metric between sentences. The work demonstrates that a simple LSTM is capable of modeling complex semantics if the representations are explicitly guided.

2.2   Causality Relation Identification

Broadly speaking, causality relation identification refers to the method of knowing whether an event causes another. By analyzing the verbs that express causality relation in French, Garcia [7] proposed a COATIS system to extract the explicit causality relation in French. Khoo [8] proposed an automatic method for identifying causality relation in Wall Street Journal text using linguistic clues and pattern-matching. Girju [9] searched for causal verbs through the Internet and WordNet to establish the Lexico-syntactic model, which enables automatic recognition of causality relations for specific events.

However these methods based on pattern-matching are domain-specific and require a lot of artificial markings. Therefore, recent studies have used methods based on machine learning and statistical probabilities to identify causality relation.

For example, Marco [10] adopted the Naive Bayesian to identify explicit causality relation by analyzing the probabilities of words between adjacent sentences. Inui [11] used support vector machine (SVM) to identify explicit causality relation in corpus by using the specific language components between the indicator and the sentence. Zhong [12] proposed a method based on cascaded model to identify explicit causality relation.

Although methods above work well, they are limited to the identification of explicit causality relation. In fact, there're a lot of implicit causality relations in texts. Therefore, there are also researchers who have studied the identification of implicit causality relation.

Fu [13] casted the causality relation identification as event sequence labeling and proposed dual-layers CRFs model to label the causal relation of event sequence. Yang [14] proposed correlation degree RCE to describe the probabilities between events and set threshold as a binary prediction to predict an event pair as causality or not.

The researches of causality relation identification above are mostly based on the feature selection, pattern matching and rule reasoning. Some scholars pay attention to the causality connectives rather than the relation between semantic information of events. In this paper, we propose a method to generate event representations based on event trigger and event elements. Event representations are used to predict the causality relation between events.

3   Proposed Model

3.1   Structure of Proposed Model

Researchers in the field of Knowledge Graph (KG) embed knowledge graph components (entities and relations) in continuous vector space while preserving properties of the original data, such as TransE [15], TransH [16] and TransD [17]. In TransE, relations are represented as translation embedding in vector space, if a triplet (subject, relation, object) exists in KG, we want that object should be close to subject + relation, while subject + relation should be far away from object if the triplet doesn't exist. Once the model has learned an embedding vector for each entity and relation, predictions will be performed by using the same translation approach in embedding space. For example, the prediction for a given subject-relation is generated by searching for the nearest neighbor entity of subject + relation in vector space.

In the field of event-oriented knowledge representation, events and event relations can be considered as special entities and relations. If we use certain methods to represent events and event relations in continuous vector space, we can also predict the relation type between events.

Based on the ideas above, this paper proposes our proposed model based on Siamese Architecture shown in Fig. 1. There are two networks Bi-LSTM$_a$ and Bi-LSTM$_b$ which each processes one of the events in a given pair and they share parameters. We use Bi-directional long short time memory (Bi-LSTM) networks to obtain event representations. Then event representations generated by Siamese LSTM Network are used to train relation embedding.

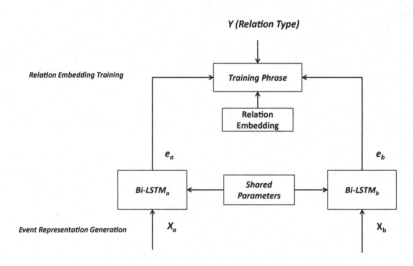

Fig. 1. The training process of proposed model

3.2 Event Representation Generation

Word embedding is the collective name for a set of language modeling and feature learning techniques in natural language processing where words or phrases from the vocabulary are mapped to vectors of real numbers. Word embedding proposed by Mikolov [18, 19] can be trained to capture semantic and syntactic relationships between words, by mapping related words to vectors that lie close in the embedding vector space. In summary, word embedding provides us an efficient method to represent word in vector space. In this paper, pre-trained word embedding is used to convert words into dense vectors.

In order to represent event, we introduce a sequence model Recurrent Neural Network (RNN). RNN is a powerful model for learning features from sequential data. RNN model is suitable for our inputs which are sequences of words, and since neural networks receive fixed size vectors or matrixes as input, words are converted into word embedding before used as inputs. Bi-directional RNN (Bi-RNN) uses a finite sequence to model sequence based on past and future contexts. This is done by concatenating the hidden states of two RNN, one processing the sequence from left to right, the other one from right to left. We can update the hidden state of each timestamp t as following:

$$h_{ft} = \sigma\big(W_f h_{t-1} + U_f x_t + b_f\big) \tag{1}$$

$$h_{bt} = \sigma(W_b h_{t+1} + U_b x_t + b_b) \tag{2}$$

$$h_t = h_{ft} \oplus h_{bt} \tag{3}$$

In formulas above, h_{ft} is the hidden state of timestamp t along the forward direction (from left to right), h_{bt} is the hidden state of timestamp t along the backward direction (from right to left), h_t is the hidden state at timestamp t and $\oplus$ denotes the concatenating operation between two vectors.

Although RNNs present acceptable performance in sequences processing, the optimization of the weight matrixes is difficult because its backpropagated gradients vanish over long sequences. LSTM networks are introduced to avoid the long-term dependency problem. Like RNNs, LSTM sequentially updates a hidden-state representation.

In this paper, we use Bi-RNNs with LSTM cell which is called Bi-LSTM and introduced above to learn event representation. The learning process is shown in Fig. 2.

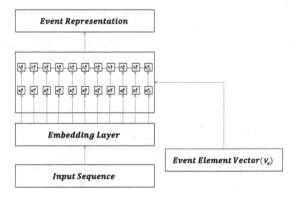

Fig. 2. The training process of proposed model

In this paper, word sequences with fixed length are used as input to represent events. Word sequences contain five words behind event triggers, event trigger word and five words after event triggers in texts. We use "<pad>" to represent paddings in word sequences which make length of input sequence equal. In CEC, we find that the average distance between event triggers and event element(such as time, place and object) are 3.4 and 96% of event elements can be covered when the length of word sequence is eleven. So we set the length of word sequence as eleven. Firstly, word sequence is converted to dense word embedding by embedding layer, and then input into Bi-LSTM model. After the processing of Bi-LSTM model, we finally get hidden set $H = \{h_0, h_1,.. h_{10}\}$. The event representation e can be obtained by following formula. Where f_t represents the feature of event trigger and f_e represents the feature of event elements. We use hidden state h_5 which is the hidden state of event trigger to represent feature of event trigger f_t. When $\alpha = 0$ event representation e excludes the feature of event elements.

$$e = (1 - \alpha) * f_t + \alpha * f_e \tag{4}$$

In addition to the feature of event trigger, our event representation also focuses on feature of event elements. One-hot vector v_e is used to denote whether the word in timestamp i of input sequence is event elements. Feature of event elements can be obtained as following:

$$f_e = \frac{1}{\sum_{i=0}^{L} v_{e_i}} \sum_{i=0}^{L} v_{e_i} * h_i \tag{5}$$

Where $v_{ei} \in \{0,1\}$ is the value in i-dimension of v_e, h_i is the hidden state in timestamp i generated by Bi-RNN model which is discussed in the above, L is the length of input sequence.

3.3 Training Relation Embedding

Given a training set S of triplets (e1, e2, r) composed of two events e1, e2 and a relation $r \in R$, our model learns the representations of events and relations. The basic idea in this step is minimize the Dist(e1, e2, r) for each training example. Dist(e1, e2, r) is calculated as following:

$$Dist(e1, e2, r) = ||e1 + r - e2|| \tag{6}$$

In the task of relation identification, we introduce the loss function as following, where c is a constant, r_{pos} is the relation between e_1 and e_2 r_{neg} represent the negative relation between e_1 and e_2. The second item on the right of the equation is the training example, while the third item is the corrupted example we generated in order to make $e_1 - e_2$ be away from the corrupted event relation.

$$loss = c - Dist(e1, e2, r_{pos}) + Dist(e1, e2, r_{neg}) \tag{7}$$

r_{neg} is calculated as following:

$$r_{neg} = \frac{1}{N-1} \sum_{r \in R - \{r_{pos}\}} r \qquad (8)$$

4 Experimental Result

4.1 Experiment Dataset

Our experimental dataset is CEC 2.0. CEC 2.0 is an event-based Chinese natural language corpus developed by the Semantic Intelligence Laboratory of Shanghai University. It has collected 333 newspaper reports about earthquakes, fires, traffic accidents, terrorist attacks and food poisoning. We labeled event triggers, participants, objects, times, places and relationships between events by using a semi-automatic method. Statistics of events and relationships labeled exactly is shown in Table 1.

Table 1. Statistics of event types and event relation

Event type	Amount	Event relation type	Amount
Perception	264	Follow relation	702
StateChange	996	Causality relation	806
Emergency	667	Concurrency relation	504
Statement	859	Overall	2008
Action	1121		
Operation	1245		
Movement	469		
Overall	5621		

4.2 Event Causality Identification

We compare our proposed model's results with other models shown in Table 2. Yang [14] defined causal correction degree (RCE) to predict whether causality exists between events. Zhong [12] proposed a cascaded model based on the bootstrapping algorithm to identify causality relation. Girju's method [9] is based on pattern-matching. From the results, we find absolute increment when α increases, and the highest F-Measure is 83.82%. At the same time, we also notice that the performance decline when $\alpha > 0.2$ and proposed model ($\alpha = 0.5$) even achieves worse result than proposed model ($\alpha = 0$). The result demonstrates that the feature of event elements really work in the event representations and enrich the semantic information of the event. However, if the model focuses on event elements excessively, important information will be ignored. Compared with other models, Proposed model ($\alpha = 0.2$) has shown slight improvement in F-Measure. The proposed model's ability to capture the semantic information of the event is likely to be one of the reasons of improvement in performance.

Table 2. Performance Comparison of all models in causality relation identification

Method	Precision (%)	Recall (%)	F-Measure (%)
Yang's method [14]	62.20	58.00	59.90
Zhong's method [12]	**85.39**	77.53	81.27
Girju's method [9]	73.91	**88.69**	80.63
Proposed model (α = 0)	79.01	80.34	79.67
Proposed model (α = 0.1)	82.07	81.16	81.61
Proposed model (α = 0.2)	83.01	84.65	**83.82**
Proposed model (α = 0.3)	82.51	81.62	82.07
Proposed model (α = 0.4)	82.63	79.29	80.93
Proposed model (α = 0.5)	77.04	79.89	78.44

4.3   Event Recognition

In this paper, we apply Bi-LSTM network in proposed model to learn event representation which can represent the content of events. To evaluate the practicality of our representations of events generated in our proposed model, we applied the Bi-LSTM network trained for the task of event relation identification into the task of event classification. We use SVM classifier to classify the events in CEC.

   We also compare our proposed model's results with other models proposed for the task of event classification shown in Table 3. Fu et al. [20] proposed classifier based on SVM and dependency parsing. Zhao et al. [21] proposed a classifier based on maxium entropy with defined features. Our proposed model exactly capture context information of events, and the event embeddings perform well in the task of event classification.

Table 3. Performance Comparison with related works in event classification

Method	Precision (%)	Recall (%)	F-Measure (%)
Event representation generated by proposed model + SVM classifier	**81.10**	**81.16**	**81.01**
Fu's method [20]	71.60	67.20	69.30
Zhao's method [21]	57.14	64.22	60.48

5   Discussion and Conclusion

This paper presented a novel method for event causality relation identification based on modeling events and relations on dense vector space. We use word sequence around event triggers as input and learn event embedding by Siamese Bi-LSTM network in relation identification task. The Bi-LSTM learns the features of event trigger and event elements. Experimental results show that our method achieves good performance and the best F-Measure of the causality relation arrives at 83.82%. Furthermore, we applied Bi-LSTM network trained in relation identification to generate event representations and use them in event classification task. The results show that event representations perform very well and our proposed model really capture important context information of events.

In future work, we will improve the performance and scalability of proposed model, meanwhile we will try to apply the approach in proposed model in event reasoning and find out more semantic information behind events and relations and dig out more event knowledge for event-based natural language processing.

References

1. Liu, Z.T., et al.: Research on event-oriented ontology model. Comput. Sci. **36**(11), 189–192 (2009)
2. Chopra, S., Hadsell, R., Lecun, Y.: Learning a similarity metric discriminatively, with application to face verification. In: 2005 IEEE Computer Society Conference on Computer Vision and Pattern Recognition (CVPR 2005), pp. 539–546. IEEE, Piscataway (2005)
3. Norouzi, M., Fleet, D.J., Salakhutdinov, R.R.: Hamming distance metric learning. In: Advances in Neural Information Processing Systems, pp. 1061–1069. Curran Associates, New York (2012)
4. Baraldi L., Grana C., Cucchiara R.: A deep siamese network for scene detection in broadcast videos. In: Proceedings of the 23rd ACM International Conference on Multimedia, pp. 1199–1202. ACM, New York (2015)
5. Kenter, T., Borisov, A., de Rijke, M.: Siamese CBOW: optimizing word embeddings for sentence representations (2016). arXiv preprint: arXiv:1606.04640
6. Mueller, J., Thyagarajan, A.: Siamese recurrent architectures for learning sentence similarity. In: Thirtieth AAAI Conference on Artificial Intelligence, pp. 2786–2792. AAAI, Menlo Park (2016)
7. Garcia, D.: COATIS, an NLP system to locate expressions of actions connected by causality links. In: Plaza, E., Benjamins, R. (eds.) EKAW 1997. LNCS, vol. 1319, pp. 347–352. Springer, Heidelberg (1997). https://doi.org/10.1007/BFb0026799
8. Khoo, C.S.G., Kornfilt, J., Oddy, R.N., Myaeng, S.H.: Automatic extraction of cause-effect information from newspaper text without knowledge-based inferencing. Literary Linguist. Comput. **13**(4), 177–186 (1998)
9. Girju, R.: Automatic detection of causal relations for question answering. In: Proceedings of the ACL 2003 Workshop on Multilingual Summarization and Question Answering, vol. 12, pp. 76–83. ACL, Stroudsburg (2003)
10. Marcu, D., Echihabi, A.: An unsupervised approach to recognizing discourse relations. In: Proceedings of the 40th Annual Meeting on Association for Computational Linguistics, pp. 368–375. ACL, Stroudsburg (2002)
11. Inui, T., Inui, K., Matsumoto, Y.: What kinds and amounts of causal knowledge can be acquired from text by using connective markers as clues? In: Grieser, G., Tanaka, Y., Yamamoto, A. (eds.) DS 2003. LNCS (LNAI), vol. 2843, pp. 180–193. Springer, Heidelberg (2003). https://doi.org/10.1007/978-3-540-39644-4_16
12. Zhong, J., Long, Y., Tian, S.: Causal relation extraction of uyghur emergency events based on cascaded model. Zidonghua Xuebao/Acta Autom. Sin. **40**(4), 771–779 (2014)
13. Fu, J., Liu, Z., Liu, W.: Using dual-layer CRFs for event causal relation extraction. IEICE Electron. Express **8**(5), 306–310 (2011)
14. Yang, J., Liu, Z., Liu, W.: Identify causality relationships based on semantic event. J. Chin. Comput. Syst. **36**(3), 433–437 (2016)

15. Bordes, A., Usunier, N., Garcia-Duran, A., Weston, J., Yakhnenko, O.: Translating embeddings for modeling multi-relational data. In: International Conference on Neural Information Processing Systems, pp. 2787–2795. Curran Associates, New York (2013)

16. Wang, Z., Zhang, J., Feng J.: Knowledge graph embedding by translating on hyperplanes. In: AAAI - Association for the Advancement of Artificial Intelligence, pp. 1112–1119. AAAI, Menlo Park (2014)

17. Ji, G., He, S., Xu, L.: Knowledge graph embedding via dynamic mapping matrix. In: Meeting of the Association for Computational Linguistics and the International Joint Conference on Natural Language Processing, pp. 687–696. ACL, Stroudsburg (2015)

18. Le, Q.V., Mikolov, T.: Distributed representations of sentences and documents. In: Proceedings of the 31st International Conference on Machine Learning, pp. 1188–1196. JMLR (2014)

19. Mikolov, T., Chen, K., Corrado, G., Dean, J.: Efficient estimation of word representations in vector space (2013). arXiv preprint: arXiv:1301.3781

20. Fu, J., Liu, Z., Zhong, Z.: Chinese event extraction based on feature weighting. Inf. Technol. J. 9(1), 184–187 (2010)

21. Zhao, Y., Qin, B., Che, W., Liu, T.: Research on Chinese event extraction. J. Chin. Inf. Process. 22(1), 3–8 (2008)

Topic-Bigram Enhanced Word Embedding Model

Qi Yang, Ruixuan Li$^{(\boxtimes)}$, Yuhua Li, and Qilei Liu

School of Computer Science and Technology,
Huazhong University of Science and Technology,
Wuhan 430074, China
{ayang7,rxli,idcliyuhua,resol1992}@hust.edu.cn

Abstract. In this paper, we propose a novel model which exploits the topic relevance to enhance the word embedding learning. We attempt to leverage the hidden topic-bigram model to build topic relevance matrices, then learn the Topic-Bigram Word Embedding (TBWE) by aggregating the context as well as corresponding topic-bigram information. The topic relevance weights are updated with word embeddings simultaneously during the training process. To verify the validity and accuracy of the model, we conduct experiments on word analogy task and word similarity task. The results show that the TBWE model can achieve the better performance in both two tasks.

Keywords: Topic-bigram · Semantic enhance
Word embedding learning

1 Introduction

Natural Language Processing (NLP) tasks have always been a hot research topic in artificial intelligence. In order to alleviate the issues of dimension disaster and semantic gap appearing in traditional language models, where each word in the vocabulary is represented as a long vector with only one non-zero element, Xu et al. [23] first utilized neural networks to deal with NLP tasks by using word embedding. Conceptually, word embedding involves a mathematical embedding from "one-hot" representation per word to a continuous vector space with a much lower dimension. Since then, neural network models have been widely applied to obtain word representations, and the representative works includes NNLM (Neural Network Language Model) [2], the Hierarchical Neural Language Model [16], Recurrent Neural Network [13], and the Word2Vec model [12]. The principle is that words sharing common contexts in the corpus should be located in close proximity in the continuous embedding space, so that the syntactic and semantic information can be considered simultaneously.

Owing to the advance in simplicity and robustness, Word2Vec model has been subsequently expanded. Some works tried to integrate more auxiliary information for word embedding learning. As a results, more domain-specific word representations can be learnt. Recent studies also considered learning topical word

© Springer Nature Switzerland AG 2018
L. Cheng et al. (Eds.): ICONIP 2018, LNCS 11303, pp. 69–81, 2018.
https://doi.org/10.1007/978-3-030-04182-3_7

embedding based on both context and their topics [10], which was expressively used for contextual word embeddings and document embeddings. However, they still ignore the topic correlation and influence of word position. In this paper, we propose a novel word embedding learning model to take the advantage of latent semantic information, considering both the position information and the topical correlation. We incorporate the Markovian dependency between topics of a sequence data into the word embeddings, namely **Topic-Bigram Enhanced Word Embedding** (TBWE). The basic idea of TBWE is to make the use of correlation information between topics, so that context-word pairs can be modelled by both the contextual information and topic relevance weights in the sequence data.

Specifically, we employ the Topic-Bigram model [1] to obtain the topic relevance weights. Assuming a first-order Markovian dependency, the probability of a topic associated with each observation depends on the previous one. In this way, given a sequence $w_d = [w_{d,1}.w_{d,2}...w_{d,N_d}]$ for trace d where $w_{d,j}$ denotes j-th token, the latent topic sequence $z_d = [z_{d,1}.z_{d,2}...z_{d,N_d}]$ is associated with w_d, and $z_{d,j} \in \{1, ..., K\}$. Multinomial parameters $\vartheta_d^{h,k}$, the mixing coefficient of the topic sequence $h.k$ for the trace d, can be estimated to help learn better word representation vectors. We design two TBWE models, **TBWE-1** and **TBWE-2**, based on CBOW and Skip-gram model [14] respectively, as shown in Fig. 1, where topic relevance matrix $\Theta_i^{z_t,z_{t+1}}$ indicates the dependency weight between topic z_t and z_{t+1}, and i is the distance between them.

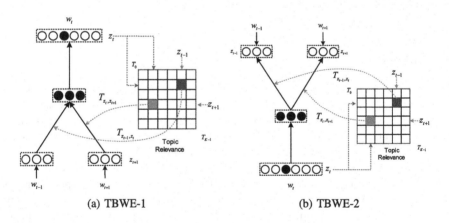

(a) TBWE-1   (b) TBWE-2

Fig. 1. TBWE model, topic relevance indices the dependency weights when generating the sequence data, and we incorporate the normalized weights into (a) CBOW, (b) Skip-gram, respectively.

As topics can capture underlying semantic information, we regard each topic as the semantic tag, so that the topic relevance weights encode their inherent semantic relationships in sequence data. For each word-context pair, semantic relationships can be weighted by topic relevance weights. We propose the TBWE model to learn the word embeddings with position-dependent topic relevance weights using the Stochastic Gradient Ascent (SGA) algorithm.

The main contribution of this work is to extend the Word2Vec [12] model to incorporate topic dependency information into basic word embedding representation, considering both the contextual words and correlation between their underlying themes. The experiments conducted on word analogy task and word similarity task, demonstrate that the performance of the proposed model is improved comparing to baseline models.

2   Related Work

2.1   Neural Language Models

Neural language models use continuous representations of words to make their predictions. Bengio et al. [2] proposed the feed-forward neural network language model (NNLM) by using the concatenating vectors to represent the previous text for modelling N-gram conditional probability. Then Mikolov et al. [13] proposed recurrent neural network language model (RNNLM), using more comprehensive contextual information to predict the word iteratively, not only the N-gram.

However word embeddings is just a by-product in above training models, rather than the purpose. The C&W model proposed by Ronan et al. [3], which trains word embeddings as the target firstly, scores the N-gram phrases directly according to the occurrence frequency. Then, Mikolov et al. [12] proposed the Word2Vec model, including two model architectures CBOW (Continuous Bag-of-Words) and Skip-gram model, to improves the efficiency significantly by removing the hidden layer and transforming neural network structure to log linear structure. But the computational complexity still increases with the corpus size, where the time complexity of conditional probability is $O(|V|)$. Mikolov then introduced two optimization techniques for more efficient learning, namely Hierarchical Softmax and Negative Sampling [14]. Among the neural language models and their variations, the Word2Vec model is the most popular one. Our work is also the extension that incorporates the topic relevance information into the training process of Word2Vec model.

2.2   Integrating Auxiliary Knowledge

Some researchers also consider incorporating different natural language factors into word embeddings for domain-specific analysis. Yu et al. [24] combined the prior knowledge of the semantic relations between words in the dictionary (such as synonyms, etc.) to train the word embeddings. The Sentiment-Specific Word Embedding model (SSWE) [21] uses both the syntactic contexts of words and the sentiment polarity of sentences to learn the sentiment-specific word embedding based on C&W model. Hu [6] proposed the SG ++ model to help twitter sentiment analysis by exploiting the emotion information and negation factor of words based on the Skip-gram model. Meanwhile, many researches take the

POS into account. Levy et al. [8] integrated the syntactic dependency into the Skip-gram model. Liu et al. [9] averaged the word embeddings of context words in CBOW model using POS relevance weights of each word-context pair. All the above works only considered the syntax information, but the semantic knowledge is still ignored.

2.3   Topic and Semantic Language Models

Topic models which learn the latent topic distributions for documents, have been widely used in data mining, sentiment analysis and recommendation. As a combination with the word embedding, Ren et al. [18,19] proposed the Topic and Sentiment-enriched Word Embedding model to learn topic-enriched word embeddings which considered the polysemy phenomenon of sentiment-baring words, and further used for improving twitter sentiment classification. Compared to using the target word only to predict context words in Skip-Gram, Liu et al. [10] integrated topics into basic word embeddings representation so that the topical word embeddings can model different meanings of a word with different context. Joint Topic-Semantic model [22] even puts topics of documents into training for social recommendation.

Motivated by the achievement of TWE [10] and PWE [9], this paper proposes a novel model TBWE which exploits topic dependency as the auxiliary semantic knowledge to model each word-context pair in the sequence data. The basic idea of TBWE is that, the topic is treated as tag associated with each word, while word embedding is dependent on the contextual words and surrounding topics, which means that the topic dependency of the context is also considered.

3   Our Models

3.1   CBOW and Skip-Gram

Before describing the main framework in detail, we first review the CBOW and Skip-gram model [14], which are designed for learning word embedding more efficient.

The CBOW model utilizes the average value of contextual word vector as input representation, without considering the order information, shown in Eq. 1.

$$I\left(c\right) = \frac{1}{2k-1} \sum_{w_j \in c} v\left(w_j\right) \tag{1}$$

where c indicates the context of word w, $v\left(w_j\right)$ is the corresponding representation vector of w_j, k is context window size and $-k \leq j \leq k$, $j \neq 0$. CBOW model utilizes the softmax regression to estimate the word prediction probability, shown in Eq. 2.

$$P\left(w|c\right) = \frac{exp\left(v\left(w\right)^T \cdot I\left(c\right)\right)}{\sum_{w' \in V} exp\left(v\left(w'\right)^T \cdot I\left(c\right)\right)} \tag{2}$$

Given the corpus D with T tokens, the objective of CBOW is to maximize the log probability:

$$\mathbb{Q}(D) = \frac{1}{T} \sum_{(w,c)\in D} \log\left(P\left(w|c\right)\right) \tag{3}$$

Contrast to CBOW model, Skip-gram only selects one contextual word representation as input to predict the target word every time, and the objective is to maximize the log probability:

$$\sum_{(w,c)\in D} \sum_{w_j \in c} \log\left(P\left(w_j|w\right)\right)$$

therein,

$$P\left(w_j|w\right) = \frac{exp\left(v\left(w_j\right)^T \cdot v\left(w\right)\right)}{\sum_{w_{j'}\in V} exp\left(v\left(w_{j'}\right)^T \cdot v\left(w\right)\right)} \tag{4}$$

3.2   TBWE-1

Since topics represent the ultimate factors underlying a token appearance in the sequence, the correlation between topics can better model the semantic themes evolution, which contributes to learning more meaningful word embeddings.

TBWE-1 extends the CBOW model by incorporating topic relevance weights to learn word embedding, shown in Fig. 1(a). In order to utilize the weight matrix filled with topic relevance values, TBWE-1 adopts the weighted sum operation to calculate the input representation for target words during training. Incorporate the weight matrix into Eq. 1:

$$I\left(c\right)_{tbwe} = \frac{1}{2k-1} \sum_{w_j \in c} \left(\Theta_i^{z_t,z_j} \cdot v\left(w_j\right)\right) \tag{5}$$

As introduced in Sect. 1, the $\Theta_i^{z_t,z_j}$ represents the semantic relevance weight between topic z_t and z_j with distance i , which can be obtained by the Topic-Bigram model [1]. However, the multinomial parameters $\vartheta_d^{h,k}$ of Topic-Bigram model is associated with each trace d, so we aggregate over all topic-bigrams in the corpus, i.e., $\Theta_1^{h,k} = \sum_{d\in C} tf \cdot \vartheta_d^{h,k}$, used as the weighting matrix after normalizing.

Considering that topic-bigram model mainly estimates the relevance parameters between successive words, so the position distance of the corresponding matrix is 1. To simplify this process, weighting matrix with different distance will adopt the same values with $\Theta_1^{h,k}$ when initialization.

3.3   TBWE-2

Similar to TBWE-1, TBWE-2 model incorporates the topic relevance weights when estimating word prediction probability. Specifically, the conditional probability $P\left(w_j|w\right)$ integrates the corresponding topic relevance weights and position distance between w and w_j, shown in Eq. 6.

$$P\left(w_j|w, z_w, z_{w_j}\right) = \frac{exp\left(v\left(w_j\right)^T \cdot v\left(w\right) \cdot \Theta_i^{z_w, z_{w_j}}\right)}{\sum_{w_j' \in V} exp\left(v\left(w_j'\right)^T \cdot v\left(w\right) \cdot \Theta_i^{z_w, z_{w_j}}\right)} \tag{6}$$

where $\left(z_w, z_{w_j}\right)$ are the topic tags of word pair (w, w_j), and the subscript i, the position distance, determines the weighting matrix used.

We adopt the negative sampling optimization to further improve the training speed of TBWE-2. And the conditional probability integrates corresponding topic relevance weight like:

$$p\left(u|w, z_w, z_{w_j}\right) = \begin{cases} \sigma\left(v(w)^T \theta^u \Theta_i^{z_t, z_{t+1}}\right), & L^u\left(w\right) = 1 \\ 1 - \sigma\left(v(w)^T \theta^u \Theta_i^{z_t, z_{t+1}}\right), & L^u\left(w\right) = 0 \end{cases} \tag{7}$$

where σ is the binomial logic regression, θ^u is the auxiliary vector, and $L^u\left(w\right)$ indicates whether $u = w_j$ or not. The other notations used here are the same defined with [14].

3.4 Optimization and Parameter Estimation

In TBWE model, we first use the Topic-Bigram model [1] to obtain topic assignments for each training token, and further estimate the topic relevance weights with position distance being 1 to initialize the weighting matrix. For the consideration of simplifying model, weighting matrices with different distances are the same before training, the value of each item is corresponding to topic-pair dependency weight, and then updated jointly with the word representations during learning.

For efficiency purpose, we adopt the negative sampling technique during training in both TBWE-1 and TBWE-2 model. Learning TBWE-1 model follows the similar optimization scheme as PWE model used in [9], and the key issue for training TBWE-1 model is to calculate the derivatives of the topic relevance weighting matrices, where the partial derivative with respect to the topic relevance weight can be computed as:

$$\frac{\partial \mathbb{Q}\left(D\right)}{\partial \Theta_i^{z_t, z_{t+j}}} = \frac{\partial \mathbb{Q}\left(D\right)}{\partial I(c)_{tbwe}} \cdot v\left(w_j\right) \tag{8}$$

As for the update of topic relevance weights in TBWE-2, we propose to apply the stochastic gradient ascent algorithm (SGA) for parameters learning, and gradients are calculated using the back-propagation algorithm, the corresponding partial derivatives of the objective function are:

$$\nabla = \frac{\partial \mathbb{L}\left(D\right)}{\partial \Theta_i^{z_w, z_{\tilde{w}}}} = [L^w(u) - \theta^u \Theta_i^{z_w, z_{\tilde{w}}}] \cdot \theta^u \cdot v(\tilde{w})^T \tag{9}$$

then, we could update the $\Theta_i^{z_w, z_{\tilde{w}}}$ by:

$$\Theta_i^{z_w, z_{\tilde{w}}} = \Theta_i^{z_w, z_{\tilde{w}}} + \eta \cdot \nabla \tag{10}$$

where $u \in \{w\} \cup NEG^{\tilde{w}}\left(w\right)$ and η is the learning rate. The update process of other two key parameters $\theta^u, v\left(\tilde{w}\right)$ are similar to above. More details are described in Algorithm 1.

Algorithm 1. Negative Sampling of TWBE-2

Input: word w and context c, topic relevance Θ_1
Output: word representations $(v(\widetilde{w}), \widetilde{w} \in c)$
1: Determine the distance i of $(w, \widetilde{w})$
2: Initialize $\theta^u, v(\widetilde{w}), \Theta_i$
3: **for** $\widetilde{w} \in c$ **do**
4:     sampling $NEG^{\widetilde{w}}(w)$
5:     $e_1 \leftarrow 0, e_2 \leftarrow 0$
6:     **for** $u \in \{w\} \cup NEG^{\widetilde{w}}(w)$ **do**
7:         $q \leftarrow \sigma \left(v(\widetilde{w})^T \cdot \theta^u \cdot \Theta_i^{z_u, z_{\widetilde{w}}} \right)$
8:         $g \leftarrow \eta \left[L^w(u) \right] - q$
9:         $e_1 \leftarrow e_1 + g \cdot \theta^u \cdot \Theta_i^{z_u, z_{\widetilde{w}}}$
10:         $e_2 \leftarrow e_1 + g \cdot \theta^u \cdot v(\widetilde{w})^T$
11:         $\theta^u \leftarrow \theta^u + g \cdot v(\widetilde{w}) \cdot \Theta_i^{z_u, z_{\widetilde{w}}}$
12:     **end for**
13:     $v(\widetilde{w}) \leftarrow +e_1$
14:     $\Theta_i^{z_u, z_{\widetilde{w}}} \leftarrow +e_2$
15: **end for**

---

4 Experiments

4.1 Experimental Setup

Experimental Corpus: In this paper, we select the English Wikipedia Corpus[1] to learn topic-bigram word embedding which includes about 1.28M articles and 26.5M tokens in total. In experiments, we test our proposed models in two tasks, word analogy task and word similarity task, for performance evaluation. And we use the analogical reasoning dataset SYN introduced by Mikolov [12] in word analogy task which contains approximately 9 K semantic and 10.5 K syntactic analogy questions. The word similarity task is tested on five evaluation sets: MC [15], RG [20], WordSim-353 [4], SimLex-999 [5], SCWS [7] and RW [11], the five datasets will be introduced in detail in following subsection.

Experiment Setting: We first learn topic assignments with Topic-Bigram model for each token in the corpus, and topic number is set to $T = 20$ and $T = 50$, to compare the effect of topic number, while iterative number being $I = 500$. As for word embedding, we set the dimensions to 300, learning rate is set to 0.025, and then test the effect of TBWE model with different parameters, that is, the negative sample number and context window size. We use the accuracy as a standard to evaluate the performance of word analogy task, and compute the Spearman rank correlation coefficient between similarity scores in the word similarity task for comparison.

We don't conduct comparative experiments with TWE models, which using the word representations to learning the topic representations and word

---

[1] http://www.psych.ualberta.ca/~westburylab/downloads/westburylab.wikicorp. download.html.

representations are same with baseline Skip-gram model, thus it is quite different with our goal of integrating topical information into word representations.

4.2 Qualitative Analysis

In order to demonstrate the characteristics of each TBWE model, we first do the qualitative analysis on a relatively smaller dataset and manually select the most similar words of target words, *batman*, *florida*, *turing*, and *dancing*, those have been analysed in dependency-based word embeddings [8]. As a comparison, we also find similar words obtained using Skip-gram model with window size is 5, where the topic number in TBWE models is 20. The results are shown in Table 1.

Table 1 shows that the proposed TBWE-2 model results in similar sets with Skip-gram models while totally different results are obtained in TBWE-1 model for target word *batman*. And *dracula* ranks beyond *hollywood* and *comics* in TBWE-1, the most likely reason is the similarity of bats image. For target words *florida* and *turing*, our results are trying to find words with more semantic

Table 1. Target words and the Top-5 nearest neighbour words introduced by different embeddings

Target word	Skip-gram	TBWE-1	TBWE-2
batman	superman	superman	superman
	miniseries	adventures	smallville
	smallville	comedy	bytb
	starring	animated	animated
	bytb[a]	dracula	miniseries
florida	oklahoma	texas	lakeland
	nebraska	arizona	minnesota
	fresno	alabama	jacksonville
	lakeland	michigan	alabama
	lauderdale	virginia	pasadena
turing	unsolvable	computation	entscheidungsproblem
	undecidable	brasenose	hodges
	hodges	planing	undecidable
	alonzo	wayback	halting
	halting	algorithm	alonzo
dancing	dance	dance	dance
	dances	blues	dances
	folk	bluegrass	folk
	bluegrass	folk	reggae
	tunes	jazz	tunes

[a] bytb stands for Batman Yesterday, Today and Beyond

similarity. It's evident that TBWE relfect more *functional* aspect to describe what the target w is, similar distinction to [8]. This observation appears with *dancing* as well, capturing words representing different *dance style* types, not only associated. The qualitative analysis yields that the TBWE models can better model the semantic relationships, as expected.

4.3   Syntactic Word Analogy Task

Given a word-pair $\langle A, B \rangle$ and a single word C, the analogy task [12] would try to find the word D whose relationship with C is similar to $\langle A, B \rangle$. For example, given the word-pair $\langle Beijing, China \rangle$ and word $Japan$, word $Tokyo$ is the correctest answer. In word embedding space, those words with the maximum cosine similarity to $v(B) - v(A) + v(C)$ are selected out being the correct answers set, represented as $Top_N(\widetilde{v})$. Only when the $Top_N(\widetilde{v})$ set exactly includes the answer word in evaluation set can the question be regarded as answered correctly.

We test the word analogy task on SYN dataset, and experiment with 4 training conditions, which are contexts with window size being 5 and 10, the number of negative sampling being 10 and 20, and two different TBWE models that TBWE-1a and TBWE-2a with 20 topics, TBWE-1b and TBWE-2b with 50 topics. The experimental results in Table 2 indicate that TBWE-2 model with 20 topics outperforms all other models. And the better performance of TBWE models prove that topic dependency information is beneficial for learning word embeddings. As for comparison with PWE, which incorporates Part-of-Speech relevance weights for learning word embeddings, TBWE models achieve better performance because they can capture more contextual semantic information in sequence data based on Topic-Bigram models, while syntactic information can also been described by topic clusters to some extent.

Of the two TBWE models with different topic number, it is unexpected that TBWE model with less topic number achieves the better performance, the most possible reason is that too many topics would weaken the intensity of the dependency. Interestingly, when increasing the size of context window from 5 to 10, the performance improvement of TBWE-1 model is more obvious than the other models, in which case, TBWE-1 model can make more effective use of contextual semantic information. However, the advantages of negative sampling are not so significant.

Table 2. The Accuracy of different models in analogy task

Parameters		Models						
Window	Sampling	CBOW	Skip	PWE	TBWE-1a	TBWE-2a	TBWE-1b	TBWE-2b
5	10	0.499	0.555	0.530	0.537	**0.574**	0.518	0.558
10	10	0.525	0.577	0.577	0.560	**0.587**	0.544	0.573
5	20	0.511	0.548	0.519	0.526	**0.575**	0.527	0.563
10	20	0.536	0.575	0.557	0.555	**0.585**	0.550	0.571

4.4 Word Similarity Task

We test the word similarity task on five evaluation sets: MC [15], RG [20], WordSim-353 ([4]), SimLex-999 [5], SCWS ([7]) and RW [11], which contain 30, 65, 353, 999, 1,762, and 2,034 pairs of words respectively. Furthermore, the words in WordSim-353, MC, RG are mostly frequent words, while SimLex-999, SCWS and RW have much more rare words and unknown words (i.e., unseen words in the training corpus) than the first three sets [17].

Following the previous works, word similarity task is another classic task to evaluate the performance of word embedding method. In this paper, we measure the similarity of two words w_i and w_j by the Spearman's rank correlation coefficient ρ^2. Firstly, we test the models with different parameters on WordSim-353 dataset, and all the experimental results are given in Table 3. Then, in the further comparison of same settings on different datasets, we set the context window size and the number of negative samples to be 10, the final results are shown in Table 4. The proposed TBWE model outperforms other baseline in word similarity tasks, including PWE and basic Word2Vec models. This further proves that integrating the topic dependency weights can better model the sequential context patterns, and the relevance weights between topics are useful for word embeddings.

Through comparison and analysis of the results, it can be found out that the performance is not always improved along with the increase of contextual

Table 3. Performance of leveraging topic dependency on the word similarity task

Parameters		Models						
Window	Sampling	CBOW	Skip	PWE	TBWE-1a	TBWE-2a	TBWE-1b	TBWE-2b
5	10	0.560	0.588	0.606	0.613	**0.673**	0.570	0.639
10	10	0.597	0.572	0.632	0.641	0.657	0.598	**0.661**
5	20	0.565	0.630	0.609	0.634	0.639	0.569	**0.645**
10	20	0.610	0.617	0.626	0.607	**0.659**	0.589	0.657

Table 4. Performance on different datasets, both the context window size and number of negative sampling are set to 10

Datasets	Models						
	CBOW	Skip	PWE	TBWE-1a	TBWE-2a	TBWE-1b	TBWE-2b
RG	0.609	0.521	0.582	0.604	**0.632**	0.582	0.628
RW	0.314	0.373	0.355	0.345	0.366	0.333	**0.374**
MC	0.619	0.671	0.614	0.648	**0.693**	0.558	0.649
SCWS	0.608	0.629	0.646	0.639	**0.647**	0.620	0.645
SimLex-999	0.257	0.323	0.318	0.321	**0.337**	0.295	0.327

---

[2] https://en.wikipedia.org/wiki/Spearman%27s_rank_correlation_coefficient.

window size when keeping other parameters constant, and same scenario also occurs when increasing the number of negative samples. However, the TBWE-2 model still achieves the best performance with different topic numbers. The reason for the better performance may because that TBWE-2 model, which is based on Skip-gram model, can absorb the dependency information better, so we believe the TBWE-2 model can achieve better performance given more dependent data for learning.

4.5 Compliexity Analysis

Compared with the typical CBOW and Skip-gram model, the TBWE model does not modify the basic architecture for word embedding learning, only integrates the needed weighting matrix of topic relevance to incorporate contextual semantic dependency into training explicitly. The number of additional parameters is $win \times |K| \times |K|$, where win indicates the window size of context, and K is the number of topics. As a contrast, the total parameters in TBWE model is less than in TWE model $(K + V) \times e$. In CBOW and Skip-gram model, the model parameters are $O\left(e \times V\right)$, where the e is the vector dimension and V is the vocabulary size. Considering that V is far bigger than K, TBWE model can still guarantee the efficiency of model training.

In computational complexity, we continue to adopt the same optimization strategy with PWE model during training in TBWE model, our models consider both contextual topic correlation weights and position information for calculating the conditional probability, rather than the Part-of-Speech relevance weights. Compared with typical embedding models, TBWE models require additional parameters to record the contextual semantic information, but the computational complexity does not increase too much relatively. Note that we don't take the overhead of Topic-Bigram model into account.

5   Conclusion and Future Work

In this paper, we proposed a topic-bigram enhanced word embedding model, which learns word representation with the auxiliary knowledge about topic dependency weights. Topic relevance value in the weighting matrices is incorporated into word-context prediction process during the training. And we evaluate our TBWE model on two typical tasks including syntactic word analogy and word similarity tasks. The experimental results show that our models outperform baseline models.

In the further research, we are going to consider to combine the syntactic and semantic knowledge together to learn more informative word embeddings, incorporating POS relevance weights into TBWE model in a more efficient way. And we want to evaluate the model in different topic numbers and data size.

Acknowledgments. This work is supported by the National Key Research and Development Program of China under grants 2016QY01W0202 and 2016YFB0800402, National Natural Science Foundation of China under grants 61572221, U1401258, 61433006, 61502185 and 61772219, Major Projects of the National Social Science Foundation under grant 16ZDA092, Science and Technology Support Program of Hubei Province under grant 2015AAA013, Science and Technology Program of Guangdong Province under grant 2014B010111007 and Guangxi High level innovation Team in Higher Education Institutions—Innovation Team of ASEAN Digital Cloud Big Data Security and Mining Technology.

References

1. Barbieri, N., Manco, G., Ritacco, E., Carnuccio, M., Bevacqua, A.: Probabilistic topic models for sequence data. Mach. Learn. **93**(1), 5–29 (2013)
2. Bengio, Y., Ducharme, R., Vincent, P., Janvin, C.: A neural probabilistic language model. J. Mach. Learn. Res. **3**, 1137–1155 (2003)
3. Collobert, R., Weston, J.: A unified architecture for natural language processing: deep neural networks with multitask learning. Machine Learning. In: Proceedings of the 25th International Conference (ICML 2008), vol. 307, pp. 160–167. ACM, Helsinki, Finland (2008)
4. Finkelstein, L., et al.: Placing search in context: the concept revisited. In: Proceedings of the 10th International World Wide Web Conference, WWW 2001, pp. 406–414. ACM, Hong Kong, China (2001)
5. Hill, F., Reichart, R., Korhonen, A.: Simlex-999: evaluating semantic models with (genuine) similarity estimation. Comput. Linguist. **41**(4), 665–695 (2015)
6. Hu, Q., Pei, Y., Chen, Q., He, L.: SG++: word representation with sentiment and negation for twitter sentiment classification. In: Proceedings of the 39th International conference on Research and Development in Information Retrieval, SIGIR 2016, pp. 997–1000. ACM, Pisa, Italy (2016)
7. Huang, E.H., Socher, R., Manning, C.D., Ng, A.Y.: Improving word representations via global context and multiple word prototypes. In: Proceedings of the 50th Annual Meeting of the Association for Computational Linguistics, ACL 2012, vol. 1, pp. 873–882. The Association for Computer Linguistics, Jeju Island, Korea (2012)
8. Levy, O., Goldberg, Y.: Dependency-based word embeddings. In: Proceedings of the 52nd Annual Meeting of the Association for Computational Linguistics, ACL 2014, vol. 2, pp. 302–308. The Association for Computer Linguistics, Baltimore (2014)
9. Liu, Q., Ling, Z., Jiang, H., Hu, Y.: Part-of-speech relevance weights for learning word embeddings. CoRR abs/1603.07695 (2016)
10. Liu, Y., Liu, Z., Chua, T., Sun, M.: Topical word embeddings. In: Proceedings of the 29th AAAI Conference on Artificial Intelligence, pp. 2418–2424. AAAI Press, Austin, Texas, USA (2015)
11. Luong, T., Socher, R., Manning, C.D.: Better word representations with recursive neural networks for morphology. In: Proceedings of the 17th Conference on Computational Natural Language Learning, CoNLL 2013, pp. 104–113. ACL, Sofia, Bulgaria (2013)
12. Mikolov, T., Chen, K., Corrado, G., Dean, J.: Efficient estimation of word representations in vector space. CoRR abs/1301.3781 (2013)

13. Mikolov, T., Karafiát, M., Burget, L., Cernocký, J., Khudanpur, S.: Recurrent neural network based language model. In: Proceedings of the 11th Annual Conference of the International Speech Communication Association, pp. 1045–1048. ISCA, Makuhari, Chiba, Japan (2010)

14. Mikolov, T., Sutskever, I., Chen, K., Corrado, G.S., Dean, J.: Distributed representations of words and phrases and their compositionality. In: Proceedings of the 27th Annual Conference on Neural Information Processing Systems, vol. 26, pp. 3111–3119. Advances in Neural Information Processing Systems, Lake Tahoe, Nevada, USA (2013)

15. Miller, G.A., Charles, W.G.: Contextual correlates of semantic similarity. Lang. Cogn. Process. **6**(1), 1–28 (1991)

16. Morin, F., Bengio, Y.: Hierarchical probabilistic neural network language model. In: Proceedings of the 10th International Workshop on Artificial Intelligence and Statistics, AISTATS 2005. Society for Artificial Intelligence and Statistics, Bridgetown, Barbados (2005)

17. Qiu, S., Cui, Q., Bian, J., Gao, B., Liu, T.: Co-learning of word representations and morpheme representations. In: Proceedings of the 25th International Conference on Computational Linguistics, COLING 2014, pp. 141–150, Dublin, Ireland (2014)

18. Ren, Y., Wang, R., Ji, D.: A topic-enhanced word embedding for twitter sentiment classification. Inf. Sci. **369**, 188–198 (2016)

19. Ren, Y., Zhang, Y., Zhang, M., Ji, D.: Improving twitter sentiment classification using topic-enriched multi-prototype word embeddings. In: Proceedings of the 30th AAAI Conference on Artificial Intelligence, pp. 3038–3044. AAAI Press, Phoenix, Arizona, USA (2016)

20. Rubenstein, H., Goodenough, J.B.: Contextual correlates of synonymy. Commun. ACM **8**(10), 627–633 (1965)

21. Tang, D., Wei, F., Qin, B., Yang, N., Liu, T., Zhou, M.: Sentiment embeddings with applications to sentiment analysis. IEEE Trans. Knowl. Data Eng. **28**(2), 496–509 (2016)

22. Wang, H., Wang, J., Zhao, M., Cao, J., Guo, M.: Joint topic-semantic-aware social recommendation for online voting. In: Proceedings of the 2017 ACM on Conference on Information and Knowledge Management, CIKM 2017, pp. 347–356. ACM, Singapore (2017)

23. Xu, W., Rudnicky, A.: Can artificial neural networks learn language models? In: 6th International Conference on Spoken Language Processing, ICSLP 2000/INTER-SPEECH 2000, pp. 202–205. ISCA, Beijing, China (2000)

24. Yu, M., Dredze, M.: Improving lexical embeddings with semantic knowledge. In: Proceedings of the 52nd Annual Meeting of the Association for Computational Linguistics, ACL 2014, vol. 2, pp. 545–550. The Association for Computer Linguistics, Baltimore (2014)

Hybridized Character-Word Embedding
for Korean Traditional Document Translation

Hosang Yu, Gil-Jin Jang, and Minho Lee[✉]

School of Electronics Engineering, Kyungpook National University,
1370 Sankyuk-Dong, Puk-Gu, Taegu 702-701, South Korea
youhs4554@gmail.com, gjang@knu.ac.kr, mholee@gmail.com

Abstract. Translating traditional documents is quite laborious and time consuming for human translators owing to the voluminous nature and a complexity
of grammatical patterns. In recent times, a neural network-based machine
translation architecture such as sequence-to-sequence (seq2seq) model showed
superior performance in translation. However, it suffers out-of-vocabulary
(OOV) issue when dealing with very complex and vocabulary languages such as
Chinese characters, resulting in performance degradation. To cope with the
OOV issue, we propose a new method by combining word embedding and
character embedding to supplement loss from unknown words with character
embedding. Experimental results show that the proposed method is efficient to
translate old Korean archives (Hanja) to modern Korean documents (Hangul).

Keywords: Neural machine translation · Deep learning
Natural language processing · Seq2seq · Character-word embedding

1 Introduction

Understanding traditional documents is very important and meaningful in historical and
cultural context. Now, various national and private institutions are investing enormous
budgets to train the analysis specialists. However, it takes lots of time and budget to
build an expert group who can translate old archives into modern language. Moreover,
most of the old records in Korea were written in Chinese which were introduced to
Korea about 2000 years ago and changed into various forms through era. Even if the
experts are trained, since various languages of Northeast Asia are mixed with different
languages such as Bengali, Sanskrit, Mongolian, and Koranic, it is even difficult for the
experts to understand them completely. Unfortunately, the recent decline of interests in
traditional record researches and the aging of researchers has become a serious problem, and the analysis of traditional records will become more difficult in future.

A neural network-based machine translation model (Neural Machine Translation,
NMT) can be a solution to overcome those problems. It is composed of two recurrent
neural networks (RNN) that serve as encoder and decoder. It is called as Sequence-To-
Sequence (seq2seq) models that show the state-of-the-art performance in machine
translation. It can learn the mapping relationships between input and output languages
through end-to-end learning.

© Springer Nature Switzerland AG 2018
L. Cheng et al. (Eds.): ICONIP 2018, LNCS 11303, pp. 82–89, 2018.
https://doi.org/10.1007/978-3-030-04182-3_8

However, this approach also has problem in dealing with languages with a very large number of word units such as Chinese. For example, while the size of the trainable word embedding matrix is limited, the number of words in Chinese is so numerous. So, there is a high probability with many unknown words. Since all of these unrecognized words are treated as unknown in the seq2seq model, they may have the same meaning even if they are completely different meaning. This phenomenon makes lowers translation performance as well as learning difficulty. This is called the out-of-vocabulary (OOV) problem (Fig. 1).

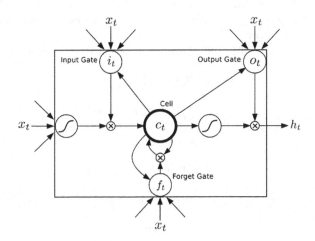

Fig. 1. LSTM unit

There have been various studies to solve OOV problem in natural language processing. Santos and Guimaraes (2015) got the state-of-the-art results in Portuguese and Spanish corpus by applying character-level embedding to named entity recognition problems along with word embedding [1]. Kim et al. (2016) showed a positive result by constructing a neural language model using only character embedding [2]. Ma et al. (2016) utilized several embedding techniques, including character trigrams, to learn pre-learned label embedding in named entity recognition [3].

As word is a combination of individual characters, character embedding naturally can cope with OOV issues. In general, the text consists of a combination of not only words but also letters, and the meaning of the words corresponds to the composition of the characters (e.g. Chinese). So, character-level approach is natural choice to avoid additional word segmentation issue (Chen et al. 2015a) [4]. Therefore, applying deep-learning techniques to the complex languages tend to favor character embedding over word embedding (Zheng et al. 2013) [5].

In this study, to cope with OOV issues, we propose a new approach by combining word embedding and character embedding to supplement loss from unknown words with character embedding. Our method can improve the translation performance in the existing seq2seq based model. To this end, we implement a NMT system based on the structure of 'seq2seq-with-attention' model which is trained using traditional literature of the Choson Dynasty corpora.

2   Related Works

2.1   Long Short-Term Memory (LSTM)

Recurrent neural network (RNN) has feedback loops to store memory which is important for sequential data. LSTM [6], a popular memory unit for RNN, enables training with very long sequential data by applying gating mechanism, to reduce long-term dependency problem.

Gating mechanism used in LSTM can be represented as follows:

$$i_t = \sigma(W_{xi}x_t + W_{hi}h_{t-1} + b_i) \tag{1}$$

$$f_t = \sigma\left(W_{xf}x_t + W_{hf}h_{t-1} + b_f\right) \tag{2}$$

$$o_t = \sigma(W_{xo}x_t + W_{ho}h_{t-1} + b_o) \tag{3}$$

$$g_t = tanh(W_{xc}x_t + W_{hc}h_{t-1} + b_c) \tag{4}$$

$$c_t = f_t \circ c_{t-1} + i_t \circ g_t \tag{5}$$

$$h_t = o_t \circ tanh(c_t) \tag{6}$$

where σ denotes the sigmoid function $(1 + e^{-x})^{-1}$ which squashes real-value within a range [0, 1], '$\circ$' means Hadamard product. When an input x_t is given at time t, input gate i_t, forget gate f_t, output gate o_t and cell candidate (input modulation gate) g_t are calculated. If the current input has more important information compared to the previous input, then it would be close to 1 and f_t would be close to 0 that makes LSTM forget previous data and updates based on the current input. By incorporating this mechanism, LSTM selectively stores information from long sequential data and prevents long term dependency problem.

2.2   Sequence-to-Sequence Model

The basic seq2seq model consists of two recurrent neural networks that serve as encoder and decoder. The encoder transforms the input sentence into a thought vector, and the decoder generates the sentence from the thought vector by refining the prominent element in the semantic expression of the input sentence [7].

The Seq2seq model uses a memory element such as a long short-term memory (LSTM) or a gated recurrent unit (GRU) as a unit cell of the recurrent neural network to solve the long-term dependence problem of a long sequence inputs. This memory element learns how to manage internal information of a memory element to process a given sequential input [8]. The training of seq2seq model is done by gradient descent to minimize time averaged cross-entropy loss between the actual and generated words probability distribution. In this process, the learning of the gate parameters involved in the internal information management of each memory element is performed.

3   Proposed Model

When translating Chinese characters, the vanilla seq2seq model has limitations in two aspects. The first limitation is that the seq2seq model only understands the sentence in the forward direction. (1) The Chinese sentences can have a completely different meaning depending on the arrangement of the words, or (2) the meaning of the word does not change, but the word arrangement is modified to emphasize the meaning. If the sentence to be translated has a problem that corresponds to (1), there is no problem in using the basic model. However, in case of (2), the seq2seq model is likely to output a completely different translation result. Therefore, the proposed model constructs an encoder of bi-directional recurrent neural network structure so that the sentence can be understood in the forward and backward direction.

The second limitation is that the decoder of the basic seq 2seq model generates a sentence using only one thought vector. There is a limitation in that the encoder can refine the semantic expression of a sentence and compress it into a vector of a limited size. Therefore, if the input sentence is long, the translation performance can be adversely affected [9]. Therefore, the proposed model solves those limitations by applying attention mechanism [10].

We also combine word embedding with character embedding. The word embedding consists of the proper nouns extracted from the translation corpora, and the character embedding consists of other characters.

3.1   Hybrid of Word and Character Embedding

Unlike English, Chinese is a language written in characters rather than words. Most proper nouns in Chinese are only meaningful when the characters are combined. So extracting proper nouns is an important task to handle very complex and huge vocabulary languages such as Chinese characters. Therefore, constructing a word dictionary by distinguishing proper nouns not only reduces unnecessary character embedding processes, but also helps learning by reducing the input sequence length.

For Chinese, we extracted the proper nouns by filtering based on parentheses which are one of the notations for specifying the proper nouns in the translation corpora we have.

For Korean, after eliminating both Chinese characters and special characters, word segmentation was performed using a morpheme analyzer, and a dictionary was constructed. The Korean morphological analyzer used in the experiment provides a POS tagging function for proper noun. This function was used to extract proper nouns.

In this paper, we propose a new method by combining word embedding and character embedding, which learns both proper nouns and non-proper noun word embedding. This method is effective in reducing language redundancy by separating proper nouns from a group of characters. In addition, we cope with the OOV problem by supplementing the meaning of unknown proper nouns with character embedding.

3.2   Attention Mechanism

Attention mechanism provides a method that allows the decoder to peek at the hidden states computed in the encoder. In other words, attention makes it possible to treat them as a dynamic memory of the input information. By doing, attention mechanism can improve the performance for longer source sentences [10].

Attention mechanism can be summarized as three equations as follows:

$$\alpha_{ts} = \frac{\exp\left(score\left(h_t, \overline{h_s}\right)\right)}{\sum_{s'=1}^{S} \exp\left(score\left(h_t, \overline{h_{s'}}\right)\right)} \tag{7}$$

$$c_t = \sum_s \alpha_{ts} \overline{h_s} \tag{8}$$

$$a_t = f(c_t, h_t) = \tanh(W_c[c_t; h_t]) \tag{9}$$

We compare each target and source hidden states (h_t and h_s) and normalize 0 to 1 using Softmax. This normalized value is "attention weight" representing the relevance between each source and target hidden state as shown in Eq. (7). And then, we get weighted average of source states. This value is "context vector" c_t in Eq. (8). Lastly, we concatenate context vector and target hidden state, and apply projection and pass to tanh activation. This value is "Attention Vector" a_t containing information of current attention decisions in Eq. (9).

4   Experiments

The experiment verifies the validity of the proposed method by comparing with only character embedding and combining word embedding and character embedding after proper noun extraction. Comparisons for experiments are made under the same conditions.

4.1   Dataset

The training data are collected from http://db.itkc.or.kr. The parallel corpus consists of a sentence composed of Chinese characters and a sentence translated into modern Korean. The total number of samples is 413,916 but only 223,387 samples with the token length of 100 or less are selected for experiment. The tokenization process is performed to construct the word dictionary of Chinese and Korean. The Chinese characters are basically separated by one character, but words inside the parentheses signifying the proper nouns are treated as a single word. In case of Korean, word segmentation is performed using the KKMA morpheme analyzer [11] provided in KoNLPy Python package.

After performing the tokenization process, a word dictionary for the token found in the training data is constructed. The word dictionary additionally includes a <PAD> token for padding, a <UNK> token for a word that does not exist in the words dictionary, and an <EOS> token to indicate the end of the sentence. Each word in the words dictionary is converted into an index in the words dictionary of the word.

4.2 Implementation Details

The encoder-decoder has two layers of 1,024 units of LSTM each, and the encoder uses a bi-directional RNN structure. In order to train alignment between two different languages, the attention mechanism is applied to the entire input sentence. The initial state of the encoder is initialized to zero and all initial weights are set to a uniform distribution of [−0.01, 0.01]. We set the batch size as 64, vocabulary size as 40K, the dimension of the embedding vector as 300, and the dimension of the attention unit as 512. In addition, to prevent learning deterioration due to the imbalance of the length of sentences in the batch, buckets are constructed by grouping the data according to the sentence length, and learning is performed separately for each bucket. The translation model is learned by the Adam optimizer with the initial learning rate of 1e−4.

4.3 Results and Discussion

Table 1 shows translation scores of the proposed method compared to character embedding method only. In case of using only the character embedding, the embedding matrix for unit Chinese character is trained without the proper noun ex-traction process for the Chinese characters. BLEU [12], ROUGE-L [13], and METEOR [14] scores are used for quantitative performance evaluation of translation performance. Based on these results, we claim that our method helps improve the translation performance.

Table 1. Translation scores on the Choson dynasty documents.

Condition	BLEU-4	ROUGE-L	METEOR
Only character embedding	0.1732	0.3894	0.2113
Combine character & word embedding (proposed)	0.2561	0.5582	0.3031

5 Conclusion

We separated proper nouns from whole group of characters through proper noun extraction process, which can reduce the language redundancy as well as length of sequences. By virtue of combining the character embedding and the word embedding, we can complement meaning loss in each embedding method to cope with the OOV problem to some extent.

However, learning embedding matrices to complement each other by combining character-word embedding is not a fundamental solution to the OOV problem. It may not be covered by both of embedding methods when the embedded matrix does not have enough size. Hence, hierarchical embedding with a stroke-character-word structure is required. In other words, it would be useful approach for considering the stroke-level embedding and word-character level embedding to complement a meaning loss, which will be investigate in near future.

Acknowledgments. This work was partly supported by Institute for Information & communications Technology Promotion (IITP) grant funded by the Korea government (MSIT) (2016-0-00564, Development of Intelligent Interaction Technology Based on Context Awareness and Human Intention Understanding) and the National Research Foundation of Korea (NRF) grant funded by the Korea government (MSIP) (No. NRF-2017M3C1B6071400).

References

1. dos Santos, C.N., Guimaraes, V.: Boosting named entity recognition with neural character embeddings. arXiv preprint arXiv:1505.05008 (2015)
2. Kim, Y., Jernite, Y., Sontag D., Rush, A.M.: Character-aware neural language models. In: AAAI, pp. 2741–2749 (2016)
3. Ma, Y., Cambria, E., Gao, S.: Label embedding for zero-shot fine-grained named entity typing. In: Proceedings of COLING 2016, The 26th International Conference on Computational Linguistics: Technical Papers, pp. 171–180 (2016)
4. Chen, X., Xu, L., Liu, Z., Sun, M., Luan, H.-B.: Joint learning of character and word embeddings. In: IJCAI, pp. 1236–1242 (2015)
5. Zheng, X., Chen, H., Xu, T.: Deep learning for Chinese word segmentation and POS tagging. In: Proceedings of the 2013 Conference on Empirical Methods in Natural Language Processing, pp. 647–657 (2013)
6. Hochreiter, S., Schmidhuber, J.: Long short-term memory. Neural Comput. **9**(8), 1735–1780 (1997)
7. Cho, K., et al.: Learning phrase representations using RNN encoder-decoder for statistical machine translation. arXiv preprint arXiv:1406.1078 (2014)
8. Graves, A.: Supervised sequence labelling with recurrent neural networks (2012). http://books.google.com/books
9. Cho, K., Van Merriënboer, B., Bahdanau, D., Bengio, Y.: On the properties of neural machine translation: encoder-decoder approaches. arXiv preprint arXiv:1409.1259 (2014)
10. Bahdanau, D., Cho, K., Bengio, Y.: Neural machine translation by jointly learning to align and translate. arXiv preprint arXiv:1409.0473 (2014)
11. Lee, D.-J., Yeon, J.-H., Hwang, I.-B., Lee, S.-G.: KKMA: a tool for utilizing Sejong corpus based on relational database. J. KIISE Comput. Pract. Lett. **16**(11), 1046–1050 (2010)
12. Papineni, K., Roukos, S., Ward, T., Zhu, W.-J.: BLEU: a method for automatic evaluation of machine translation. In: Proceedings of the 40th Annual Meeting on Association for Computational Linguistics, pp. 311–318 (2002)
13. Lin, C.-Y.: Rouge: a package for automatic evaluation of summaries. In: Proceeding of Workshop on Text Summarization Branches Out (2004)
14. Banerjee, S., Lavie, A.: METEOR: an automatic metric for MT evaluation with improved correlation with human judgments. In: Proceedings of the ACL Workshop on Intrinsic and Extrinsic Evaluation Measures for Machine Translation and/or Summarization, pp. 65–72 (2005)
15. Sundermeyer, M., Alkhouli, T., Wuebker, J., Ney, H.: Translation modeling with bidirectional recurrent neural networks. In: Proceedings of the 2014 Conference on Empirical Methods in Natural Language Processing (EMNLP), pp. 14–25 (2014)
16. Wu, Y., et al.: Google's neural machine translation system: bridging the gap between human and machine translation. arXiv preprint arXiv:1609.08144 (2016)

17. Auli, M., Galley, M., Quirk, C., Zweig, G.: Joint language and translation modeling with recurrent neural networks. In: Proceedings of the 2013 Conference on Empirical Methods in Natural Language Processing, pp. 1044–1054 (2013)
18. Cambria, E., White, B.: Jumping NLP curves: a review of natural language processing research. IEEE Comput. Intell. Mag. **9**(2), 48–57 (2014)
19. LeCun, Y., Bengio, Y., Hinton, G.: Deep learning. Nature **521**(7553), 436–444 (2015)
20. Chung, J., Gulcehre, C., Cho, K., Bengio, Y.: Empirical evaluation of gated recurrent neural networks on sequence modeling. arXiv preprint arXiv:1412.3555 (2014)
21. Sutskever, I., Vinyals, O., Le, Q.V.: Sequence to sequence learning with neural networks. In: Advances in Neural Information Processing Systems, pp. 3104–3112 (2014)
22. Sutskever, I., Martens, J., Hinton, G.E.: Generating text with recurrent neural networks. In: Proceedings of the 28th International Conference on Machine Learning (ICML-11), pp. 1017–1024 (2011)
23. Mikolov, T., Kombrink, S., Burget, L., Černocký, J., Khudanpur, S.: Extensions of recurrent neural network language model. In: Acoustics, Speech and Signal Processing (ICASSP), pp. 5528–5531 (2011)
24. Mikolov, T., Karafiát, M., Burget, L., Černocký, J., Khudanpur, S.: Recurrent neural network based language model. In: Eleventh Annual Conference of the International Speech Communication Association, pp. 1045–1048 (2010)
25. Mikolov, T., Sutskever, I., Chen, K., Corrado, G.S., Dean, J.: Distributed representations of words and phrases and their compositionality. In: Advances in Neural Information Processing Systems, pp. 3111–3119 (2013)
26. Goldberg, Y.: A primer on neural network models for natural language processing. J. Artif. Intell. Res. **57**, 345–420 (2016)
27. Mikolov, T., Chen, K., Corrado, G., Dean, J.: Efficient estimation of word representations in vector space. arXiv preprint arXiv:1301.3781 (2013)
28. Turney, P.D., Pantel, P.: From frequency to meaning: vector space models of semantics. J. Artif. Intell. Res. **37**, 141–188 (2010)
29. Elman, J.L.: Distributed representations, simple recurrent networks, and grammatical structure. Mach. Learn. **7**(2–3), 195–225 (1991)

Word Embedding Based on Low-Rank Doubly Stochastic Matrix Decomposition

Denis Sedov[1]([⊠]) and Zhirong Yang[1,2]

[1] Department of Computer Science, Aalto University, Espoo, Finland
denis.sedov@aalto.fi
[2] Department of Computer Science, Norwegian University of Science
and Technology, Trondheim, Norway
zhirong.yang@ntnu.no

Abstract. Word embedding, which encodes words into vectors, is an important starting point in natural language processing and commonly used in many text-based machine learning tasks. However, in most current word embedding approaches, the similarity in embedding space is not optimized in the learning. In this paper we propose a novel neighbor embedding method which directly learns an embedding simplex where the similarities between the mapped words are optimal in terms of minimal discrepancy to the input neighborhoods. Our method is built upon two-step random walks between words via topics and thus able to better reveal the topics among the words. Experiment results indicate that our method, compared with another existing word embedding approach, is more favorable for various queries.

Keywords: Nonnegative matrix factorization · Word embedding
Cluster analysis · Doubly stochastic

1 Introduction

In recent years machine learning (ML) that involves text data has found many real-world applications [6,11,13]. Each data item in these applications is a sequence of words and other tokens. Originally each word is represented by its ID. However, this is not suitable for machine learning, where most common ML algorithms admit vectors as their input. One-hot encoding is inefficient when the vocabulary is large. Therefore word embedding which finds a low-dimensional vectorial representation of words is a fundamental starting point.

A good word embedding method should respect the relations among the words. It is commonly to learn an embedding vector space where the neighborhoods of the words are approximately preserved. Two typical approaches include *Word2Vec* [8] which maximizes the likelihood of each word given their neighbors (or in the reversed way) and *GloVe* which minimizes a weighted squared loss between the input and output pairwise relations. Some variants of Word2Vec and GloVe have been proposed subsequently [3,5,12].

© Springer Nature Switzerland AG 2018
L. Cheng et al. (Eds.): ICONIP 2018, LNCS 11303, pp. 90–100, 2018.
https://doi.org/10.1007/978-3-030-04182-3_9

However, embeddings learned by the above approaches may not provide optimal similarities between the words. After the word vectors are obtained, their pairwise similarities require external measures such as cosine similarity, which can be suboptimal because the learning objective involves non-normalized word vectors. Moreover, the negative sampling trick in Word2Vec provides only an approximating surrogate. Theoretically it remains unknown whether the ad hoc choice of negative distribution guarantees that the original CBOW or Skip-Gram objectives are optimized or not.

In this paper we present a new nonnegative matrix factorization (NMF) method and apply it to learn vectorial representation of words. Our method factorizes the doubly stochastically constrained approximating matrix. In this way we directly optimize over the normalized word vectors and provide their optimal similarities in the embedding space in terms of least approximation discrepancy. Unlike Word2Vec, our method does not require extra stochastic approximation tricks or assumptions on negative distributions. We test our method on two popularly used text data sets and compare it with the Word2Vec results. Our results indicate that the proposed method is often more favorable for various k-nearest-neighbor queries.

The remaining of the paper is organized as follows. In Sect. 2 we review the word embedding problem and two existing embedding methods. Next we present our new NMF method and show how to apply it to learn probabilistic representation of words in Sect. 3. Our optimization algorithm is presented in Sect. 4. Experimental setting and results are presented in Sect. 5. Then in Sect. 6 we conclude the work and discuss some future directions.

2   Brief Review of Previous Word Embedding Methods

A text corpus can be treated as a sequence of words and some other tokens such as punctuations. Originally each word is represented by their id in the vocabulary. Because many modern machine learning methods admit vectors as input, conventionally the word ids are converted into their one-hot encodings. That is, the i-th word in the N-sized vocabulary is represented by an N-dimensional vector with the i-entry is 1 and the others are zeros. Obviously, such one-hot encoding is inefficient when N is large. A low-dimensional (r-dim with $r \ll N$) vector encoding, called word embedding, is needed for more efficient learning tasks.

Word embeddings should respect the proximity of words in the original sequence. A common requirement is that if two words often appear nearby, their mapped points in the embedding space should be close. On the other hand, if two words seldom co-occur in the same neighborhood, they should be placed distantly in the embedding.

One way to implement the above requirement is to maximize the likelihood of a language model. For example, the Word2Vec Skip-Gram method finds the vectors $\{w_t\}_{t=1}^N$ of the words which maximizes

$$\mathcal{L}(\{w_t\}_{t=1}^N) = \frac{1}{T} \sum_{t=1}^T \sum_{j \in \mathcal{N}(t)} \log P\left(\text{word}_j | \text{word}_t\right), \tag{1}$$

where $\mathcal{N}(t)$ is the neighborhood of location t and the conditional likelihood is defined as

$$P\left(\text{word}_j | \text{word}_t\right) = \frac{\exp\left(w_j^T w_t\right)}{\sum_{i=1}^N \exp\left(w_i^T w_t\right)} \tag{2}$$

Another approach is to approximately preserve the probability that a word appears in the neighborhood of another word. For example, GloVe implements the approximation by minimizing a weighted squared loss [9], assuming log-normal noise in the observed neighboring frequencies.

3 Low-Rank Doubly Stochastic Matrix Decomposition

Although Word2Vec and GloVe are widely used, they do not provide a metric in the embedding space for retrieval. Cosine similarity as a conventional choice in natural language processing is often used to calculate, for example, k-nearest neighbors of a query in the embedding space. However, the cosine similarities between words are not optimized during the embedding learning. Therefore the retrieval based on such an external metric may not respect the original data distribution.

We observe that the mismatch arises mainly because the word vectors are not normalized in the learning objective, but they are normalized in the metric for retrieval. To overcome this problem we propose to use a new learning objective which explicitly involves the normalized word vectors. First, we employ the doubly stochasticity constraint to normalize the similarities in the embedding space, which enforces that each row or column of the output similarity matrix has unitary sum. This means each word in the embedding space has equal total similarity and denoises the imbalanced effect in the input space. Second, we find a low-rank nonnegative matrix which factorizes the doubly stochastic matrix, which significantly reduces the dimensionality of word vectors. Our optimization is based on multiplicative updates which are widely used in nonnegative matrix factorization (see Sect. 4).

Let W be the word embedding matrix (rows as word vector codes). It has been shown that the above factorization problem can be reformulated as follows (see Theorem 1 in [14]):

$$\underset{W \geq 0}{\text{minimize}} \; \mathcal{J}(W) = D(S \| \widehat{S}) \tag{3}$$

$$\text{subject to } \widehat{S}_{ij} = \sum_{k=1}^r \frac{W_{ik} W_{jk}}{\sum_v W_{vk}}, \tag{4}$$

$$\sum_{k=1}^r W_{ik} = 1, \; i = 1 \ldots, N. \tag{5}$$

Here $D()$ is an information divergence measuring the discrepancy between between the input and output proximities S and $\widehat{S}$. We adopt the Kullback-Leibler divergence

$$D(S||\widehat{S}) = \sum_{i=1}^{N}\sum_{j=1}^{N} \left(S_{ij} \log \frac{S_{ij}}{\widehat{S}_{ij}} - S_{ij} + \widehat{S}_{ij} \right) \qquad (6)$$

because it accounts for Poisson noise and thus better for sparsity in S.

The doubly stochastic similarity matrix $\widehat{S}$ in embedding space provides a probabilistic interpretation. Let $W_{ik} = P(\text{topic}_k|\text{word}_i)$, the probability of assigning the ith data object to the kth topic. Without preference to any particular word, we impose a uniform prior $P(\text{word}_j) = 1/N$ over the words. With this prior, we can compute by the Bayes' formula

$$P(\text{word}_j|\text{topic}_k) = \frac{P(\text{topic}_k|\text{word}_j)P(\text{word}_j)}{\sum_{v=1}^{N} P(\text{topic}_k|\text{word}_v)P(\text{word}_v)} \qquad (7)$$

$$= \frac{P(\text{topic}_k|\text{word}_j)}{\sum_{v=1}^{N} P(\text{topic}_k|\text{word}_v)}. \qquad (8)$$

Then we can see that

$$\widehat{S}_{ij} = \sum_{k=1}^{r} \frac{W_{ik}W_{jk}}{\sum_{v=1}^{N} W_{vk}} \qquad (9)$$

$$= \sum_{k=1}^{r} \frac{P(\text{topic}_k|\text{word}_j)}{\sum_{v=1}^{N} P(\text{topic}_k|\text{word}_v)} P(\text{topic}_k|\text{word}_i) \qquad (10)$$

$$= \sum_{k=1}^{r} P(\text{word}_j|\text{topic}_k)P(\text{topic}_k|\text{word}_i) \qquad (11)$$

$$= P(\text{word}_j|\text{word}_i). \qquad (12)$$

That is, if we define a bipartite graph with the words and topics as graph nodes, $\widehat{S}_{ij}$ is the probability that the ith word node reaches the jth word node via a topic node (see Fig. 1). It is easy to verify that $\widehat{S}_{ij} = \widehat{S}_{ji}$. Therefore the output similarity matrix $\widehat{S}$ is doubly stochastic.

4   Optimization

We implement the optimization in Eqs. 3 to 5 by multiplicative updates. To minimize an objective $\mathcal{J}$ over a nonnegative matrix W, we first calculate the gradient and separate it into two nonnegative parts ($\nabla_{ik}^{+} \geq 0$ and $\nabla_{ik}^{-} \geq 0$):

$$\nabla_{ik} \overset{\text{def}}{=} \frac{\partial \mathcal{J}}{\partial W_{ik}} = \nabla_{ik}^{+} - \nabla_{ik}^{-}. \qquad (13)$$

topics

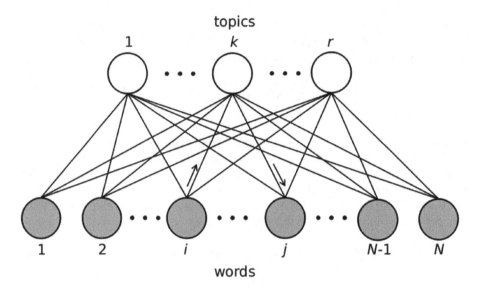

words

Fig. 1. Word-Topic bipartite graph for N words and r topics ($r < N$). The arrows show a Word-Topic-Word random walk path, which starts at the ith word node and ends at the jth word node via the kth topic node.

Usually the separation can easily be identified from the gradient. Then the algorithm iteratively applies a multiplicative update rule

$$W_{ik} \leftarrow W_{ik} \frac{\nabla_{ik}^-}{\nabla_{ik}^+} \tag{14}$$

until convergence. Such algorithms have several attractive properties, as they naturally maintain the positivity of W and do not require extra effort to tune learning step size. For a variety of NMF problems, such multiplicative updates monotonically decrease $\mathcal{J}$ after each iteration and therefore W can converge to a stationary point [15].

We cannot directly apply the above multiplicative fixed-point algorithm to the proposed learning objective because there are probability constraints on the W rows. Projecting the W rows to the probability simplex after each iteration would often lead to poor local minima in practice.

Instead, we employ a relaxing strategy [18] to handle the probability constraint. We first introduce Lagrangian multipliers $\{\lambda_i\}_{i=1}^N$ for the constraints:

$$\mathcal{L}(W, \lambda) = \mathcal{J}(W) + \sum_i \lambda_i \left(\sum_{k=1}^r W_{ik} - 1 \right). \tag{15}$$

This suggests a preliminary multiplicative update rule for W:

$$W_{ik}' = W_{ik} \frac{\nabla_{ik}^- - \lambda_i}{\nabla_{ik}^+}, \tag{16}$$

where

$$\frac{\partial \mathcal{J}}{\partial W} = \underbrace{\left[\left(W^T Z W \right)_{kk} s_k^{-2} \right]}_{\nabla_{ik}^+} - \underbrace{\left[2 \left(Z W \right)_{ik} s_k^{-1} \right]}_{\nabla_{ik}^-}, \tag{17}$$

with $Z_{ij} = S_{ij}/\widehat{S}_{ij}$ and $s_k = \sum_{v=1}^{N} W_{vk}$. Imposing $\sum_k W'_{ik} = 1$ and isolating λ_i, we obtain

$$\lambda_i = \frac{b_i - 1}{a_i}, \tag{18}$$

where

$$a_i = \sum_{l=1}^{r} \frac{W_{il}}{\nabla_{il}^+}, \text{ and, } b_i = \sum_{l=1}^{r} W_{il} \frac{\nabla_{il}^-}{\nabla_{il}^+}. \tag{19}$$

Putting this λ back in Eq. 16, we obtain

$$W_{ik} \leftarrow W_{ik} \frac{\nabla_{ik}^- a_i + 1 - b_i}{\nabla_{ik}^+ a_i}. \tag{20}$$

To maintain the positivity of W, we add b_i to both the numerator and denominator, which does not change the fixed point and gives the ultimate update rule:

$$W_{ik} \leftarrow W_{ik} \frac{\nabla_{ik}^- a_i + 1}{\nabla_{ik}^+ a_i + b_i}. \tag{21}$$

The above calculation steps are summarized in Algorithm 1. In implementation, one does not need to construct the whole matrix $\widehat{S}$. The ratio $Z_{ij} = S_{ij}/\widehat{S}_{ij}$ only requires calculation on the non-zero entries of S.

The above algorithm obeys a monotonicity guarantee provided by the following theorem.

Theorem 1. *Denote W^{new} the updated matrix after each iteration of Algorithm 1. It holds that $\mathcal{L}(W^{new}, \lambda) \leq \mathcal{L}(W, \lambda)$ with $\lambda_i = (b_i - 1)/a_i$.*

The proof follows the Majorization-Minimization procedure [4,15,16] and is a direct corollary of Theorem 2 in [14]. The theorem shows that Algorithm 1 jointly minimizes the approximation error and drives the rows of W towards the probability simplex. The Lagrangian multipliers are adaptively and automatically selected by the algorithm, without extra human tuning effort. The quantities b_i are the row sums of the unconstrained multiplicative learning result, while the quantities a_i balance between the gradient learning force and the probability simplex attraction. Besides convenience, we find that this relaxation strategy works more robustly than the brute-force projection after each iteration.

Algorithm 1. Optimization algorithm of our method

Input: input similarity matrix S, number of topics r, positive initial guess of W.
Output: word embedding matrix W (rows as word vectors).
repeat

$$\widehat{S}_{ij} = \sum_{k=1}^{r} \frac{W_{ik}W_{jk}}{\sum_v W_{vk}}$$

$$Z_{ij} = S_{ij}/\widehat{S}_{ij}$$

$$s_k = \sum_{v=1}^{N} W_{vk}$$

$$\nabla_{ik}^{-} = 2\left(ZW\right)_{ik} s_k^{-1}$$

$$\nabla_{ik}^{+} = \left(W^T ZW\right)_{kk} s_k^{-2}$$

$$a_i = \sum_{l=1}^{r} \frac{W_{il}}{\nabla_{il}^{+}}, \quad b_i = \sum_{l=1}^{r} W_{il}\frac{\nabla_{il}^{-}}{\nabla_{il}^{+}}$$

$$W_{ik} \leftarrow W_{ik}\frac{\nabla_{ik}^{-}a_i + 1}{\nabla_{ik}^{+}a_i + b_i}$$

until W converges under the given tolerance

---

5   Experiments

To compare the performance of our method with Word2Vec, we train word embeddings on two publicly available datasets and then construct k-nearest neighbor tables for specific word queries. Finally, we perform qualitative analysis on these tables and show that our method is better at capturing semantic relation between the words. The codes used in the experiments are available online[1].

Both training datasets used during the experiments represent a collection of English Wikipedia articles. The first dataset is `WikiText-2` [7], which consists of 2.5M tokens with 33K words in the vocabulary. We also include `text8` dataset[2], which is almost 7 times larger than `WikiText-2` and consists of 17M tokens and 254K words in the vocabulary. During the experiments, we used only top 20K most frequent words for both datasets.

For fair comparison, we followed the same default setting in the original version of Word2Vec[3]. Both methods are trained on the same set of vocabulary words, the word embedding dimension is set to 200 and the size of word neighborhood equals to 8.

The results for `WikiText-2` dataset are presented in Table 1. We can see that sometimes word neighbors produced by Word2Vec are not close semantically to the query words, whereas our method was able to produce much better results. For example, for the word "asteroid" the proposed method produces words like

---

[1] https://users.aalto.fi/~sedovd1/Matrix_decomp_WE/.

[2] http://mattmahoney.net/dc/textdata.html.

[3] https://github.com/tmikolov/word2vec.

Table 1. Seven nearest neighbors for `WikiText-2` dataset using: (top) our method and (bottom) Word2Vec.

Word	Neighbors
camera	footage, shots, shooting, screen, shoot, setting, showing
zoo	gorillas, bars, spiders, exhibit, Pattycake, sharing, lowland
literature	literary, poets, languages, language, writings, tradition, references
moon	observations, observation, Venus, solar, transit, measurements, atmosphere
coin	coins, dollar, dollars, Mint, purchase, fund, costs
leather	silk, cloth, wrapped, manufactured, mud, synthetic, mills
cold	warm, heat, exposed, hot, winter, falling, dry
spring	winter, summer, kept, fall, brief, Over, arrival
queen	ruler, mentions, kings, supreme, throne, kingdom, succession
asteroid	planets, probe, orbit, spacecraft, NASA, Solar, orbits
Word	Neighbors
camera	reggae, synthesizers, backup, retro, boots, carriage, bouncing
zoo	griffin, Avis, Reader, headpiece, earthworks, Sleat, Owl
literature	Kannada, writings, Fu, tradition, poets, Vaishnava, historical
moon	skeletal, reactivity, equilibrium, sodium, spectral, infrared, triangular
coin	dollar, convention, solution, price, potential, program, annular
leather	twigs, pipes, longitudinal, bags, triangular, tapered, gum
cold	curved, snow, reaches, beak, rough, bars, loop
spring	autumn, 1900s, 1940s, 1880s, 1870s, 2000s, 1800s
queen	Blanche, Wentworth, diplomat, bodyguard, mathematician, relates, prince
asteroid	molecule, insect, orientation, isotope, triangle, undirected, flash

"planets", "orbit" and "spacecraft", which are all related to cosmos, whereas Word2Vec yields words like "molecule", "insect" and "orientation" that share very little in common. Moreover, it shows that the performance of Word2Vec can be quite poor for small datasets.

Table 2 shows the results for `text8` dataset. We can see that the increase in text corpus size helps to obtain more meaningful embeddings. However, Word2Vec tends to produce rather rare and specific neighbors for the query words, whereas our method produces more common words. For example in case of Word2Vec, the closest neighbors to the word "dracula" are "stoker" and "bram", which constitute the name of the author, who wrote the corresponding novel, as well as "lugosi" and "bela", which are related to the name of the actor portraying Dracula. On the other hand, by using our method, the close neighbors consist of the words "frankenstein" and "godzilla". These three words constitute

Table 2. Seven nearest neighbors for `text8` dataset using: (top) our method and (bottom) Word2Vec.

Word	Neighbors
green	blue, red, white, yellow, black, color, brown
airport	railway, rail, downtown, airlines, traffic, train, metropolitan
celebrity	interviews, credits, kids, favorite, talent, joy, charity
microsoft	windows, operating, apple, mac, os, dos, macintosh
ancient	greek, middle, historical, pre, latin, medieval, tradition
byzantine	emperors, dynasty, ottoman, conquered, constantinople, conquest, rulers
roman	empire, church, catholic, holy, eastern, christian, ancient
dinosaur	dinosaurs, prehistoric, fossils, habitat, insect, specimen, elephants
dracula	frankenstein, vampire, noir, adaptations, godzilla, cyberpunk, horror
godzilla	sequel, monsters, monster, adventure, anime, horror, robot
Word	Neighbors
green	shade, lantern, purple, onion, violet, herring, panther
airport	heathrow, ferry, destinations, monorail, airline, flights, hub
celebrity	britney, quiz, vh, listings, portrayals, futurama, syndicated
microsoft	novell, xp, excel, hypercard, borland, netscape, macromedia
ancient	hellenistic, etruscan, sumerian, vedic, mycenaean, phoenician, hellenic
byzantine	achaemenid, seleucid, assyrian, justinian, hittite, heraclius, frankish
roman	byzantine, frankish, aztec, claudian, gaius, aurelius, seleucid
dinosaur	mammal, reptiles, lizard, dodo, zebra, bipedal, skeleton
dracula	stoker, bram, vampire, lugosi, bela, poirot, remake
godzilla	lugosi, toho, miniseries, bela, remake, akira, highlander

the group of the iconic horror movie monsters and are well associated with each other.

6   Conclusion

We have proposed a new word embedding method which is based on low-rank decomposition of doubly stochastic similarity matrix in the embedding space. Unlike previous approaches, our method provides not only the low-dimensional word vectors but also their pairwise similarity metric for subsequent applications such as retrieval. The resulting similarities are explicitly optimized in terms of least Kullback-Leibler divergence to the input similarity matrix. We have proposed an optimization algorithm based on multiplicative updates for minimizing the presented cost function. Experiment results have shown that our method works better for two selected text corpora compared to the state-of-the-art word embedding method in terms of providing more meaningful k-nearest neighbors in the embedding space.

There are several future directions. In this work we have used a batch-mode optimization algorithm, which could be replaced by using distributed and stochastic learning techniques, for example co-distillation [1], towards a more scalable and more efficient method. We could also incorporate Bayesian treatment of the embedding vectors, for example, using Dirichlet priors [10,14] and automatic rank determination [17]. Moreover, the discrepancy between input and output proximity matrices could be replaced by other learnable information divergence [2]. In addition, our methods is ready to be applied in other domains, for example, finding the embedding vectors of k-mers in DNA sequences.

Acknowledgment. The work is supported by Finnish Academy (grant numbers 307929 and 314177) and the Telenor-NTNU AI Lab project.

References

1. Anil, R., Pereyra, G., Passos, A., Ormandi, R., Dahl, G.E., Hinton, G.E.: Large scale distributed neural network training through online distillation. In: International Conference on Learning Representations (2018)
2. Dikmen, O., Yang, Z., Oja, E.: Learning the information divergence. IEEE Trans. Pattern Anal. Mach. Intell. **37**(7), 1442–1454 (2015)
3. Dingwall, N., Potts, C.: Mittens: an extension of glove for learning domain-specialized representations. In: NAACL-HLT, pp. 212–217 (2018)
4. Hunter, D., Lange, K.: A tutorial on MM algorithms. Am. Stat. **58**(1), 30–37 (2004)
5. Ling, W., Dyer, C., Black, A., Trancoso, I.: Two/too simple adaptations of Word2Vec for syntax problems. In: NAACL-HLT, pp. 1299–1304 (2015)
6. Manning, C., Raghavan, P., Schütze, H.: Introduction to Information Retrieval. Cambridge University Press, New York (2008)
7. Merity, S., Xiong, C., Bradbury, J., Socher, R.: Pointer sentinel mixture models. In: International Conference on Learning Representations (2017)
8. Mikolov, T., Sutskever, I., Chen, K., Corrado, G., Dean, J.: Distributed representations of words and phrases and their compositionality. In: Advances in Neural Information Processing Systems, pp. 3111–3119 (2013)
9. Pennington, J., Socher, R., Manning, C.: Glove: global vectors for word representation. In: Empirical Methods in Natural Language Processing, pp. 1532–1543 (2014)
10. Sinkkonen, J., Aukia, J., Kaski, S.: Component Models for Large Networks. CoRR abs/0803.1628 (2008)
11. Stamatatos, E., Kokkinakis, G., Fakotakis, N.: Automatic text categorization in terms of genre and author. Comput. Linguist. **26**(4), 471–495 (2000)
12. Stergiou, S., Straznickas, Z., Wu, R., Tsioutsiouliklis, K.: Distributed negative sampling for word embeddings. In: Proceedings of the Thirty-First AAAI Conference on Artificial Intelligence, pp. 2569–2575 (2017)
13. Turian, J., Ratinov, L., Bengio, Y.: Word representations: a simple and general method for semi-supervised learning. In: Proceedings of the 48th Annual Meeting of the Association for Computational Linguistics, pp. 384–394 (2010)

14. Yang, Z., Corander, J., Oja, E.: Low-rank doubly stochastic matrix decomposition for cluster analysis. J. Mach. Learn. Res. **17**(187), 1–25 (2016)
15. Yang, Z., Oja, E.: Unified development of multiplicative algorithms for linear and quadratic nonnegative matrix factorization. IEEE Trans. Neural Netw. **22**(12), 1878–1891 (2011)
16. Yang, Z., Peltonen, J., Kaski, S.: Majorization-minimization for manifold embedding. In: International Conference on Artificial Intelligence and Statistics, pp. 1088–1097 (2015)
17. Yang, Z., Zhu, Z., Oja, E.: Automatic rank determination in projective nonnegative matrix factorization. In: Vigneron, V., Zarzoso, V., Moreau, E., Gribonval, R., Vincent, E. (eds.) LVA/ICA 2010. LNCS, vol. 6365, pp. 514–521. Springer, Heidelberg (2010). https://doi.org/10.1007/978-3-642-15995-4_64
18. Zhu, Z., Yang, Z., Oja, E.: Multiplicative updates for learning with stochastic matrices. In: Kämäräinen, J.-K., Koskela, M. (eds.) SCIA 2013. LNCS, vol. 7944, pp. 143–152. Springer, Heidelberg (2013). https://doi.org/10.1007/978-3-642-38886-6_14

Meta-path Based Heterogeneous Graph Embedding for Music Recommendation

Qianqi Fang[1,2], Ling Liu[2(✉)], Junliang Yu[2], and Junhao Wen[2]

[1] Ping An Trust Co., Ltd., Shenzhen 518048, China
[2] School of Big Data and Software Engineering, Chongqing University,
Chongqing 400044, China
gigifang@foxmail.com, {liuling,yu.jl,jhwen}@cqu.edu.cn

Abstract. The prosperous online music streaming industry makes personalized music recommendation a topic worthy of extensive study. Traditional music recommendation techniques which are based on conventional collaborative filtering or acoustic content features usually sufffer from data sparsity or time-consuming computation problems, respectively. In fact, online music services not only generate listening history for each user but also accumulate a large amount of heterogeneous data including performers, tags, ownerships and so on. Capturing underlying user preference from the heterogeneous data to enhance music recommendation is transparently promising, because on one hand these data can mitigate the sparsity of listening history while incorporating them into recommendation model is computationally affordable. To this end, in this paper we propose a novel music recommendation approach. It first models the music system as a heterogeneous music graph. Then, to make full use of the heterogeneous data, carefully designed meta-paths are used to dig up the information lying in the graph. Finally, we learn user preferences through a combination of Bayesian Personalized Ranking model and heterogeneous embedding representation learning. Extensive experimental analysis on real-world public dataset validates that the proposed approach outperforms the baselines, especially on cold start users.

Keywords: Music recommendation · Heterogeneous graph
Meta-path · Embedding learning

1 Introduction

With the rapid development of online music platforms, the digital music industry now accounts for more than 50% of the global music revenue, 59% of which is streaming in contrast to downloads [13]. To meet the great commercial demand, the novel online personalized music recommendation service is urgently to be developed. Traditional recommendation systems which are usually based on listening history or acoustic content features are often limited by the data sparsity and time-consuming computation problems, leading to a degradation in recommendation quality. Actually, online music services not only generate listening

© Springer Nature Switzerland AG 2018
L. Cheng et al. (Eds.): ICONIP 2018, LNCS 11303, pp. 101–113, 2018.
https://doi.org/10.1007/978-3-030-04182-3_10

history for each user but also accumulate a large amount of heterogeneous data including performers, tags, and ownerships, which are shown in Fig. 1. According to the recent study [12], people tend to express themselves and make claims about their identities through listening preferences, which are often reflected by the attributes of songs like tags. Consequently, capturing underlying user preference from the heterogeneous data to enhance music recommendation is transparently promising. Besides, there are two advantages to make full use of these heterogeneous data. One hand, they can act as the auxiliary information to complement the listening history, which could mitigate the data sparsity problem. On the other hand, compared with extracting acoustic content features from a large scale of audio sources, it is computationally affordable to incorporate these metadata into the recommendation model.

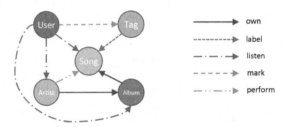

Fig. 1. Heterogenous music graph

Existing studies [7,18,19] based on heterogeneous graphs encode different kinds of information in the music system as well. However, they either only pay attention to the explicit links without discovering latent interactions or are time-consuming (e.g. attempting to exhaust all patterns in a complex graph). In fact, if we explore the heterogeneous graph with empirically designed patterns and then capture both the explicit and implicit information with techniques like embedding representation learning, the recommendation is more likely to be improved.

To this end, in this paper, we present a novel music recommendation approach which learns user preference via meta-path based heterogeneous graph embedding. Methodologically, the proposed approach first models the music recommender system as a heterogeneous music graph. To uncover the underlying information hiding in the interactions among different types of entities and edges, carefully designed meta-paths [16], which could characterize composite relations between users and songs, guide the random walks to explore the graph by generating heterogeneous node sequences which are called *preference corpus*. Inspired by the success of network embedding [6], we finally learn user preferences through a combination of Bayesian Personalized Ranking model and heterogeneous embedding representation learning [5]. In this way, we fuse both the listening history and metadata into our recommendation model. To summarize, our main contributions are listed as follows:

- We formulate the user-song interactions and metadata as a heterogeneous music graph, which benefits to the learning for user preferences.
- To fully capture user preferences, we conduct meta-path based random walks to explore the heterogeneous music graph, and then learn the fine-grained user preferences through heterogeneous embedding learning.
- We rigorously conduct experiments to validate the effectiveness of the proposed approach in music recommendation.

2 Related Work

Most existing music recommenders can be grouped into two categories: content-based approach, collaborative filtering (CF) based approach.

Content-based approaches typically focus on extracting and analyzing the music audio content features like Mel Frequency Cepstral Coefficients (MFCC), using these intrinsical features to model the similarity between new tracks and tracks in users' listening history [2,4,9,15]. These approaches enable the system to generate high quality recommendation even in the case that the listening logs are deficient. However, they are usually computationally expensive and can not capture the preference of each user.

CF is the most common technique which not only used in the domain of music recommendation, but also for other types of recommender systems. In contrast to content-based methods, CF-based methods pay attention to user listening history to generate personalized recommendations for individuals. Most previous CF-based models are extensions of matrix factorization [1,8,17,22–24], which have been proven to be effective. However, these approaches may suffer from the data sparsity problem, which are inadequate for the scenario that recommending songs to new users. To tackle the problem, some work that combines content and CF-based approaches named hybrid approaches [14,20].

Although CF-based approaches have been proven to be effective, most of them only studied the user-song interactions. In practice, loads of heterogeneous data can be incorporated to further improve the recommendation performance, including the performers, ownerships, and tags of songs. In recent literatures [7,19], the heterogeneous data are taken into consideration, but these work is limited by high computational cost and overlook of implicit information. Motivated by the success of word/network embedding techniques [6,10], a few music recommender systems adopt representation learning to music recommendation [3,18]. In these work, the play sequences and metadata are exploited to model users' preferences and similarities of songs through embeddings. In particular, Cheng et al. [3] develop a two-stage model which combines matrix factorization and play sequence embedding. However, we argue that the loosely coupled design of this method may cause a loss of information and therefore it results in a suboptimal recommendation quality.

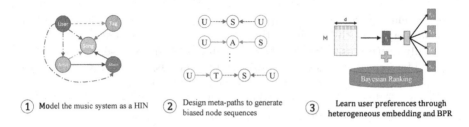

① **M**odel the music system as a HIN   ② Design meta-paths to generate biased node sequences   ③ Learn user preferences through heterogeneous embedding and BPR

Fig. 2. Overview of our proposed approach - HMR

3   Preliminaries

Before the introduction of the proposed model, we first give some illustrations about the later mentioned concepts.

Definition 1. *Heterogenous Graph*: A graph $\mathbf{G} = (V, E, C)$ in which each node v and each link e are associated with their mapping functions $\phi(v) : V \rightarrow C_V$ and $\phi(e) : E \rightarrow C_E$, respectively. denote The entity sets and relation types are denoted by C_V and C_E, where $|C_V| + |C_E| > 2$.

The heterogeneous music graph used in our model (i.e. Fig. 1) is presented with Users (U), Songs (S), Albums (Al), Artists (A) and Tags (T) as nodes, wherein edges involve 'listen' (U-S), 'own' (Al-S and A-S), 'perform' (A-S), 'mark' (U-T), and 'label' (T-S) relations. By considering a heterogeneous music graph as input, we formalize the problem of music recommendation over a heterogeneous music graph as follows.

Given a heterogeneous music graph $\mathbf{G}$, the task of music recommendation is to learn a ranking function for each user u. Formally, the ranking function is defined as follows.

$$f : (u, \mathbf{G}) \rightarrow Ranked\_list :$$
$$r_1(m) \succeq ...r_i(p) \succeq r_{i+1}(q)... \tag{1}$$

where $r_i(p) \succeq r_{i+1}(q)$ encodes that user u prefer song p rather than song q.

4   Proposed Model

In this section, the **H**eterogenous music graph based **M**usic **R**ecommendation method (**HMR**) is presented. The proposed model is composed of two main parts: (1) generating node sequences over the heterogenous music graph in which listening history and metadata of songs are encoded; and (2) learning user preferences through heterogeneous embedding. The overview of our proposed method is illustrated in Fig. 2.

4.1 Generating Node Sequences over Heterogenous Music Graph

As the heterogeneous music graph $\mathbf{G}$ is usually formed in a large scale, how to reduce the computational cost while comprehensively exploiting $\mathbf{G}$ is the most concerned challenge. Besides the solutions with a holistic view that have a high cost, a sensible alternative is meta-path [16], which has been adopted to lots of heterogeneous mining scenarios like academic network analysis.

Table 1. Meta-paths designed for preference extraction.

Path	Schema	Description
P_1	$U \xrightarrow{l} S$	User listens to a song
P_2	$U \xrightarrow{l} A \xrightarrow{pe} S$	User listens to a song performed by an artist
P_3	$U \xrightarrow{l} Al \xrightarrow{o} S$	User listens to a song from an album
P_4	$U \xrightarrow{m} T \xrightarrow{la} S$	User marks a tag to a song
P_5	$U \xrightarrow{l} A \xrightarrow{o} Al \xrightarrow{pe} S$	User listens to a song from an album performed by an artist

* $\xrightarrow{l}$ denotes the 'listen' relation, $\xrightarrow{pe}$ denotes the 'perform' relation, $\xrightarrow{o}$ denotes the 'own' relation, $\xrightarrow{m}$ denotes the 'mark' relation and $\xrightarrow{la}$ denotes the 'label' relation.

Specifically, a meta-path is denoted in the form of $V_1 \xrightarrow{R_1} V_2 \xrightarrow{R_2} \cdots \xrightarrow{R_{n-1}} V_n$, wherein $R = R_1 \circ R_2 \cdots \circ R_n$ that defines a composite relation between its start type V_1 and end type V_l [16]. For example, the path $U \xrightarrow{l} A \xrightarrow{pe} S$ characterizes a composite relation that a user listens to a song peformed by an artist. According to the schema of $\mathbf{G}$ shown in Fig. 1, we carefully design 5 types of meta-paths presented in Table 1, which are reasonable and define complex relations between users and songs with different types of metadata involved. Actually, more meta-paths can be designed when more entities and relations are included. Here we just focus on the schema of Fig. 1 for simplicity.

The designed meta-paths demonstrate the strategy of information mining at an abstract level. In practice, we explore the heterogeneous music graph by conducting random walks to generate concrete node sequences under the guidance of meta-paths. It should be noted that, compared to the songs which have been played for few times by a user, the frequently listened songs are more likely to reflect the user's preference. The random walks therefore should pay more attention to the frequently listened songs or other entities which are repeatedly clicked. Here we show how meta-paths guide the random walks to generate biased node sequences called *preference corpus*.

Given a meta-path schema $\mathcal{P} = V_1 \xrightarrow{R_1} V_2 \xrightarrow{R_2} \cdots \xrightarrow{R_{n-1}} V_n$, the transition probability of random walks at step k is defined as follows:

$$
p(v^{k+1}|v^k, \mathcal{P}) = \begin{cases} \frac{\phi_{v^k}(v^{k+1})}{\sum_1^m \phi_{v^k}(v^m)} & (v^k, v^{k+1}) \in \{\xrightarrow{l}, \xrightarrow{o}, \xrightarrow{pe}, \xrightarrow{m}\}, \\ 1 & (v^k, v^{k+1}) \in \{\xleftarrow{pe}, \xleftarrow{o}\}, \\ \frac{1}{|N_{v^{k+1}}|} & (v^k, v^{k+1}) \in \{\xleftarrow{l}, \xleftarrow{la}, \xleftarrow{m}\}, \\ 0 & (v^{k+1}, v^k) \notin E. \end{cases} \tag{2}
$$

As can be seen from Eq. (2), at each step of the random walk, the next node type is decided by the pre-defined meta-path $\mathcal{P}$. When $(v^k, v^{k+1}) \in \{\xrightarrow{l}, \xrightarrow{o}, \xrightarrow{pe}, \xrightarrow{m}\}$, the frequently played songs, albums, artists or marked tags are more likely to be chosen. Here we define $\phi_{v^k}(v^m)$ as the visited times of v_k's neighborhood. It should be noted that meta-paths are generally used in a symmetric way. In this paper, they are used likewise, which means $V_{n+1} = V_{n-1}$. If $k + 1 > n$, the meta-path will be expanded recursively (e.g. $U \xrightarrow{l} S \xleftarrow{l} U$). Consequently, when $(v^k, v^{k+1}) \in \{\xleftarrow{l}, \xleftarrow{la}, \xleftarrow{m}\}$, we uniformly select successor user node. Besides, we consider that each song is only included in one album and each album and each song only belong to one artist. So when the relation is $\xleftarrow{in}$ or $\xleftarrow{o}$, the transition probability should be 1.

Guiding by the meta-paths, the explicit interactions are captured. Furthermore, with the proceeding of the random walks, entities which are not explicitly connected in $\mathbf{G}$ are also linked by meta-paths. In other words, the latent interactions are also perceived through information propagation in random walks, which would be helpful in discovering users' implicit preferences.

4.2 Learning User Preferences Through Heterogeneous Embedding

After the *preference corpus* has been collected, the next step is to extract the encoded user preferences. Inspired by the success of network embedding [6], we come up with the idea that user preferences can be learned through embedding representation learning. As the *preference corpus* consists of heterogeneous data, we feed them to the heterogenous Skip-Gram proposed by [5] for learning node representations $\mathbf{Y} \in \mathbb{R}^{|V| \times d}$. Formally, the objective function of heterogenous Skip-Gram is defined as:

$$
\arg\max_{\mathbf{Y}} \sum_{v \in V} \sum_{v_t^m \in N(v^k)} \log p(v_t^m|v^k; \mathbf{Y}), \tag{3}
$$

where $N(v^k)$ is the neighbors of v^k within the context w and $p(v_t^m|v^k; \mathbf{Y})$ as the heterogenous softmax function is defined with the formula that $p(v_t^m|v^k; \mathbf{Y}) = \frac{e^{y_{v_t^m} \cdot y_{v^k}}}{\sum_{v \in V_t} e^{y_v \cdot y_{v^k}}}$, where y_v is the corresponding row of $\mathbf{Y}$ with index of v, representing the learned embedding of node v, and V_t is the node set with type t in $\mathbf{G}$. To reveal the information hiding in the sequence, the heterogenous Skip-Gram maximizes the likelihood of nodes co-occurrence within a context. But when computing the normalization factor, it only draws upon nodes in the same type set instead of computing all nodes, which is distinct from the common Skip-Gram model. Besides, in order to accelerate the learning, we further adopt negative sampling [10] to avoid the complexity of computing the normalization factors.

After the representation learning, we could recommend songs to users through similarity computation between embeddings. However, this simple way leads to a low performance in our validation experiments when compared with other baselines. So we consider that a ranking model should be jointly learned to enhance the recommendation performance.

In practice, recommended songs are usually presented as an ordered list, songs with higher rankings in the list are more likely to be noticed. Let P_u and N_u denote the positive (listened) and negative (not listened) song sets of user u, and $\mathbf{Z} \in \mathbb{R}^{m \times d}$ and $\mathbf{Q} \in \mathbb{R}^{n \times d}$ denote the user and song latent matrix, respectively. As an widely used "one-class collaborative filtering" model, Bayesian Personalized Ranking (BPR) [11] aims to model the preference-order for each user by maximizing the following posterior probability on $\mathbf{\Theta} \equiv (\mathbf{Z}, \mathbf{Q})$:

$$\prod_{u \in U} \mathcal{P}(x_{ui} \succeq x_{uj} | \mathbf{\Theta}) \mathcal{P}(\mathbf{\Theta}), i \in P_u, j \in N_u, \tag{4}$$

where $x_{u\cdot}$ denotes the score of user u on one of the candidate items, $\mathcal{P}(x_{ui} \succeq x_{uk} | \mathbf{\Theta})$ is defined as $\sigma(x_{ui} - x_{uk})$ and the predicted preference score $x_{ui} = \mathbf{Z}_u \cdot \mathbf{Q}_i$.

Algorithm 1. HMR

Input: Heterogenous music graphs $\mathbf{G}$, meta-paths $\mathcal{P}$, #walks per user H, walk length l, embedding dimension d, window size w, listening history D

Output: Recommended song list $\mathbf{L}$ for each user.

1 Initialize node embeddings $\mathbf{Y}$ and latent matrix $\mathbf{Z}$ and $\mathbf{Q}$;
2 **for** *path p in $\mathcal{P}$* **do**
3   **for** *user i in V_u* **do**
4   **for** $j = 1 \rightarrow H$ **do**
5   seq += MetaPathRandomWalk($\mathbf{G}$,p,i,l);

6 **while** *not converged* **do**
7      $\mathbf{Y}(\mathbf{Z}) \leftarrow$ HeterogeneousSkipGram($\mathbf{Y}$,w,seq);
8      $\mathbf{Z}, \mathbf{Q} \leftarrow$ BayesianPersonalizedRanking($\mathbf{Z}, \mathbf{Q}, D$);

9 **return** $\mathbf{L}$

To combine BPR and heterogeneous embedding learning, a practicable idea is to share the user embeddings across two models. That means, $\mathbf{Z}$ bridges these two models. On one hand, it is the user latent matrix in BPR and on the other hand is also the user embeddings in heterogeneous Skip-Gram. By taking negative log-form of posterior probability, the objective function of the conjunct model is presented as follows,

$$\mathcal{L} = -\sum_u \sum_{i \in P_u} \sum_{j \in N_u} \ln(\sigma(x_{ui} - x_{uj})) + \frac{\lambda_\Theta}{2}(||\mathbf{Z}||_F^2 + ||\mathbf{Q}||_F^2)$$

$$- \lambda_G \arg\max_Y \sum_{v \in V, k=0}^{|V|} \sum_{v_t^m \in N(v^k)} \log p(v_t^m|v^k; \mathbf{Y}, \mathbf{Z}), \tag{5}$$

where λ_G is introduced to control the impact of the heterogeneous data and λ_Θ controls the magnitude of $\mathbf{Z}$ and $\mathbf{Q}$. By doing so, the user preferences lying in the listening history and heterogeneous music graph are both integrated into $\mathbf{Z}$.

To find the possible optimal parameters, we perform stochastic gradient descent on $\mathbf{Z}$ and $\mathbf{Q}$. The final model then can generate recommendation for user u by computing $\mathbf{Q} \cdot \mathbf{Z}_u^T$. To summarize, the overall process of the proposed model is presented in Algorithm 1. To the best of our knowledge, we are the first to adopt meta-path based heterogeneous graph embedding to music recommendation.

5   Experimental Results

In this section, we aim to answer the following three questions by experiments: (1) Can HMR outperform the baselines? (2) Can HMR improve the recommendation quality for cold start users. (3) How heterogeneous data impact the performance?

5.1   Experimental Settings

Datasets. A real-world datasets, Xiami [19] are used in our experiments, which includes 4,266 users, 11,592 artists, 30,913 albums, 66,823 songs,12,528 tags and 1,020,543 listening records. We first order all listening records for each user by timestamp and extract 80% of the data as the training set, from which we randomly select 10% as the validation set. For the rest 10% of the data, we consider that the users and songs that appear in the training and validation sets to obtain the test set. We pick the best parameters of compared methods according to their best average performance on the validation set. Then we conducted the experiments for 10 times and presented their average performances.

Comparison Methods. In order to demonstrate the superiority of our approach, we compare HMR with the following methods for personalized item ranking.

- **MostPop (MP):** This method generates a non-personalized recommendation for each user based on how many times songs are listened.
- **BPR:** This is the classical bayesian personalized ranking model [11] based on pairwise assumption for item ranking.
- **IPF:** This method [21] is based on random walk and fuses users' long and term preferences.

- **Song2vec:** This model [3] adopts word embedding techniques to model play sequence, and embeds the song similarity into matrix factorization to boost the latent feature learning.

Parameter Settings. For all model-based methods, the coefficient λ of penalty terms is set as 0.01 and the dimension of latent features d is specified as 20. For HMR, the number of walk started with each user is $H = 20$, the length of each walk is $l = 20$, the dimension of embedding is $d = 20$, the window size is $w = 7$, and $\lambda_G = 0.5$. Other settings are picked according to the performance on validation set. Notice that all the designed meta-paths evenly generate the preference corpus.

Evaluation Metrics. Three common metrics, *Precision@K*, *Recall@K* and *MAP@K* (Mean average precision) are used to measure the recommendation quality.

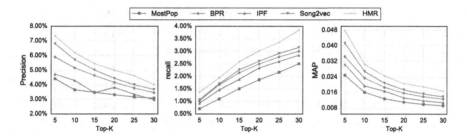

Fig. 3. Performance comparison with counterparts

5.2   Recommendation Performance

As is shown in Fig. 3, the findings can be summarized as follows.

1. In all cases, our model HMR outperforms other opponents by large margins, showing significant improvement.
2. The embedding learning based models have advantages over conventional models. The performances of HMR and Song2vec are both better than that of others, which shows the effectiveness of embedding learning in extracting implicit interactions lying in the auxiliary information.
3. In comparison to BPR, HMR shows an evident improvement. As BPR is the basic component of HMR, the result confirms that combining the heterogeneous data through embedding learning can lead to a better recommendation performance.
4. HMR also shows a superiority to Song2vec. We infer that is because HMR can extract more latent user preference by modeling the system as a heterogeneous music graph. Moreover, we also argue that modeling the play sequence, as

Song2vec does, works excessively on details, which is prone to be interferered by noises to a certain extent. Unlike Song2vec, HMR pays more attention to metadata by means of meta-paths while conducting biased random walks to avoid noises. In this way, the reliable metadata and listening history are both utilized.

5.3   Performance on Cold-Start Users

The key challenge for recommender systems is to recommend items to new (cold-start) users who have few consumption records. Music recommender systems also suffer from this problem. In this part, we want to check if HMR can improve the recommendation performance on cold-start users. To this end, we test the performance of baselines on users who have less than 20 listening records.

Table 2. Performance on cold-start users

Dataset	Metric	MP	BPR	IPF	Song2vec	HMR	Improv
Xiami	Prec@10	0.388%	0.836%	0.716%	0.865%	**0.948%**	9.595%
	Prec@20	0.284%	0.582%	0.492%	0.597%	**0.701%**	17.755%
	Rec@10	1.706%	2.905%	2.333%	3.617%	**4.172%**	15.344%
	Rec@20	2.174%	3.936%	2.970%	4.423%	**5.004%**	13.135%
	MAP@10	0.00477	0.01346	0.01142	0.02257	**0.02835**	25.609%
	MAP@20	0.00517	0.01437	0.01201	0.02310	**0.03062**	32.554%

According to Table 2, we could clearly observe that the proposed model HMR outperforms other baselines in terms of the performance on cold-start users. In particular, HMR shows great advantages on the ranking related metric - MAP@K, which indicates that the generated recommendation list provides higher ranks for songs in the test set. As expected, Song2vec also shows a decent performance, which helps corroborate that introducing auxiliary information is promising. Based on the result, HMR is proven to be effective on recommending in the scenario where few listening logs are recorded.

5.4   Impact of Parameter λ_G

In HMR, a hyper-parameter λ_G is introduced to control the impact of the heterogeneous data. We investigate the sensitivity of λ_G by testing the recommendation performance on *Precision@10*, *Recall@10* and *MAP@10* under a range of $[0, 1]$ with a step 0.1.

As can be observed from Fig. 4, specifically, HMR reaches the peak on *Precision@10* when $\lambda_G = 0.6$ and gets the best performance on *Recall@10* and *MAP@10* when $\lambda_G = 0.7$. It seems that a smaller λ_G may result a degradation in recommendation quality, which is consistent with our expectation that incorporating heterogeneous data can strongly boost the model. However, large λ_G

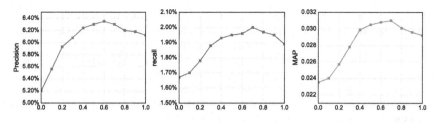

Fig. 4. Impact of λ_G

values may also turn the model down because of the overweight of the side information. Overall, the curves are basically smooth, which shows the robustness of incorporating the heterogeneous data.

6 Conclusion

This paper aims to utilize the heterogeneous data as auxiliary information to enhance recommendation. Inspired by the recent advances of network embedding, we propose a novel music recommendation approach called HMR. It first generates sequences of nodes under the guidance of the specifically designed meta-paths in which types of heterogeneous data are maximally encoded to explore the graph. To learn the user preferences, HMR then combines representation learning and bayesian personalized ranking learning. Experiments on real-word datasets show that HMR significantly improves the quality of music recommendation.

Acknowledgement. This research is supported by the Graduate Scientific Research and Innovation Foundation of Chongqing (cys17035), the National Natural Science Foundation of China (61472021).

References

1. Benzi, K., Kalofolias, V., Bresson, X., Vandergheynst, P.: Song recommendation with non-negative matrix factorization and graph total variation. In: 2016 IEEE International Conference on Acoustics, Speech and Signal Processing (ICASSP), pp. 2439–2443 (2016)
2. Cheng, R., Tang, B.: A music recommendation system based on acoustic features and user personalities. In: Cao, H., Li, J., Wang, R. (eds.) PAKDD 2016. LNCS (LNAI), vol. 9794, pp. 203–213. Springer, Cham (2016). https://doi.org/10.1007/978-3-319-42996-0_17
3. Cheng, Z., Shen, J., Zhu, L., Kankanhalli, M., Nie, L.: Exploiting music play sequence for music recommendation. In: Twenty-Sixth International Joint Conference on Artificial Intelligence, pp. 3654–3660 (2017)
4. Dieleman, S., Schrauwen, B.: Deep content-based music recommendation. In: International Conference on Neural Information Processing Systems, pp. 2643–2651 (2013)

5. Dong, Y., Chawla, N.V., Swami, A.: metapath2vec: Scalable representation learning for heterogeneous networks. In: Proceedings of the 23rd ACM SIGKDD International Conference on Knowledge Discovery and Data Mining, pp. 135–144 (2017)
6. Goyal, P., Ferrara, E.: Graph embedding techniques, applications, and performance: a survey. Knowl. Based Syst. **151**, 78–94 (2018)
7. Guo, C., Liu, X.: Automatic feature generation on heterogeneous graph for music recommendation. In: International ACM SIGIR Conference on Research and Development in Information Retrieval, pp. 807–810 (2015)
8. Hu, Y., Koren, Y., Volinsky, C.: Collaborative filtering for implicit feedback datasets. In: 2008 Eighth IEEE International Conference on Data Mining, pp. 263–272 (2008)
9. Jannach, D., Lerche, L., Kamehkhosh, I.: Beyond "hitting the hits": generating coherent music playlist continuations with the right tracks. In: Proceedings of the Ninth ACM Conference on Recommender Systems, pp. 187–194. ACM (2015)
10. Mikolov, T., Sutskever, I., Chen, K., Corrado, G.S., Dean, J.: Distributed representations of words and phrases and their compositionality. In: Neural Information Processing Systems, pp. 3111–3119 (2013)
11. Rendle, S., Freudenthaler, C., Gantner, Z., Schmidt-Thieme, L.: BPR: Bayesian personalized ranking from implicit feedback. In: Uncertainty in Artificial Intelligence, pp. 452–461 (2009)
12. Rentfrow, P.J., Mcdonald, J.A., Oldmeadow, J.A.: You are what you listen to: young people's stereotypes about music fans. Group Process. Intergroup Relat. **12**(3), 329–344 (2009)
13. Schedl, M., Knees, P., Gouyon, F.: New paths in music recommender systems research. In: The Eleventh ACM Conference, pp. 392–393. ACM (2017)
14. Schedl, M., Schnitzer, D.: Hybrid retrieval approaches to geospatial music recommendation. In: Proceedings of the 36th International ACM SIGIR Conference on Research and Development in Information Retrieval, pp. 793–796 (2013)
15. Soleymani, M., Aljanaki, A., Wiering, F., Veltkamp, R.C.: Content-based music recommendation using underlying music preference structure. In: IEEE International Conference on Multimedia and Expo, pp. 1–6 (2015)
16. Sun, Y., Han, J., Yan, X., Yu, P.S., Wu, T.: Pathsim: meta path-based top-k similarity search in heterogeneous information networks. Proc. VLDB Endow. **4**, 992–1003 (2011)
17. Vall, A., Skowron, M., Knees, P., Schedl, M.: Improving music recommendations with a weighted factorization of the tagging activity. In: ISMIR, pp. 65–71 (2015)
18. Wang, D., Deng, S., Zhang, X., Xu, G.: Learning music embedding with metadata for context aware recommendation. In: Proceedings of the 2016 ACM on International Conference on Multimedia Retrieval, pp. 249–253 (2016)
19. Wang, D., Xu, G., Deng, S.: Music recommendation via heterogeneous information graph embedding. In: International Joint Conference on Neural Networks, pp. 596–603 (2017)
20. Wang, X., Wang, Y.: Improving content-based and hybrid music recommendation using deep learning. In: Proceedings of the 22nd ACM International Conference on Multimedia, pp. 627–636 (2014)
21. Xiang, L., Yuan, Q., Zhao, S., Chen, L., Zhang, X., Yang, Q., Sun, J.: Temporal recommendation on graphs via long-and short-term preference fusion. In: Proceedings of the 16th ACM SIGKDD International Conference on Knowledge Discovery and Data Mining, pp. 723–732 (2010)

22. Yu, J., Gao, M., Rong, W., Song, Y., Fang, Q., Xiong, Q.: Make users and preferred items closer: recommendation via distance metric learning. In: Liu, D., Xie, S., Li, Y., Zhao, D., El-Alfy, E.-S.M. (eds.) ICONIP 2017. LNCS, vol. 10638, pp. 297–305. Springer, Cham (2017). https://doi.org/10.1007/978-3-319-70139-4_30
23. Yu, J., Gao, M., Rong, W., Song, Y., Xiong, Q.: A social recommender based on factorization and distance metric learning. IEEE Access 5, 21557–21566 (2017)
24. Yu, J., Gao, M., Song, Y., Zhao, Z., Rong, W., Xiong, Q.: Connecting factorization and distance metric learning for social recommendations. In: Li, G., Ge, Y., Zhang, Z., Jin, Z., Blumenstein, M. (eds.) KSEM 2017. LNCS (LNAI), vol. 10412, pp. 389–396. Springer, Cham (2017). https://doi.org/10.1007/978-3-319-63558-3_33

Knowledge Graph Embedding via Entities' Type Mapping Matrix

Md Mostafizur Rahman[1,2]([⊠]) and Atsuhiro Takasu[1,2]

[1] National Institute of Informatics, Tokyo, Japan
[2] Sokendai (The Graduate University for Advanced Studies), Tokyo, Japan
{rahman,takasu}@nii.ac.jp

Abstract. Knowledge graph (KG) is the most popular method for presenting knowledge in search engines and other natural-language processing (NLP) applications. However, KG remains incomplete, inconsistent, and not completely accurate. To deal with the challenges of KGs, many state-of-the-art models, such as TransE, TransH, and TransR, have been proposed. TransE and TransH use one semantic space for entities and relations, whereas TransR uses two different semantic spaces in its embedding model. An issue is that these proposed models ignore the category-specific projection of entities. For example, the entity *"Washington"* could belong to the person or location category depending on its context or relationships. An entity may therefore involve multiple types or aspects. Considering all entities in just one semantic space is therefore not a logical approach to building an effective model. In this paper, we propose TransET, which maps each entity based on its type. We can then apply any other existing translation-distance-based embedding models such as TransE or TransR. We evaluated our model using two tasks that involve link prediction and triple classification. Our model achieved a significant and consistent improvement over other state-of-the-art models.

Keywords: Knowledge graph · Translation based model
Link prediction

1 Introduction

Large knowledge graph (KG) implementations such as DBpedia [8], YAGO [15], and Freebase [2] incorporate very large numbers of entities and attributes to keep pace with the rate of information generation in the Web, smart systems, and social life. KGs are used in many AI-based tasks such as inferring new knowledge, question answering, relations in social-network applications, and item recommendation. Evolving with new facts while organizing and maintaining existing knowledge is an increasingly difficult task, which is why KGs are far from complete and can be very sparsely populated.

Although existing KGs contain billions of entities and relations, they still have gaps and may contain incorrect facts. "Link prediction" means predicting

© Springer Nature Switzerland AG 2018
L. Cheng et al. (Eds.): ICONIP 2018, LNCS 11303, pp. 114–125, 2018.
https://doi.org/10.1007/978-3-030-04182-3_11

relations/facts between entities based on existing triples in a KG. Traditionally, KGs represent the relations/facts between their various entities as triples. A triple can be represented as (h, r, t), where h and t are entities in the real world and r is a relation between h and t. For example, consider the triple (*Tokyo, CapitalOf, Japan*), where *Tokyo, Japan* are the head and tail entities, respectively, and *CapitalOf* is the relation. The purpose of link prediction is to detect unknown pairs of head and tail entities that are correlated via some relation.

The concept of "embedding" has also been widely used for representing words and texts [1,11], with many embedding models having been proposed for KG completion. Most of these models fall into one of three categories: bilinear models, neural-network-based models, and translation-distance-based models.

The translation-distance-based models have gained popularity both for their simplicity and their effectiveness, where they have achieved state-of-the-art performance. Bordes et al. [5] proposed TransE, which is the simplest and smartest way of predicting the links in a KG. TransE was inspired by Mikolov's skip-gram model [10,11]. It learns vector embeddings for entities and relations, with relations being represented as translations in the embedding space. The basic principle is that $\boldsymbol{h} + \boldsymbol{r} \approx \boldsymbol{t}$, where (h, r, t) holds. Here, $\boldsymbol{h}, \boldsymbol{r}, \boldsymbol{t}$ are each embeddings of h, r, t, respectively. To solve the one-to-many/many-to-one/many-to-many issues in TransE, TransH [17] has been proposed. It involves a principle stating that entity representations will differ based on various relations. Similarly, TransR [9] assumes that each relation has its own embedding space. However, TransR proposes using separate spaces for entities and relations.

In these proposed models, the entities' type has been completely ignored. In the real world, entities can be categorized in terms of several types, such as person, movie, or organization. We can often assume that entities of the same type should share strong similarities and that, in their relation, their type also plays an important role. As an example, the *HasNationality* relation requires a person-type head entity and a location/country-type tail entity. On the other hand, the *CapitalOf* relation requires location-type entities for both head and tail. We can imagine the existence of two different triples for these two relations: (*Washington, HasNationality, U.S.*) and (*Washington, CityOf, U.S.*). Here, the entity "Washington" plays two completely different roles in these two relations, based on their type. In our model, we explicitly define the role of entity type in a relation. We propose a model, TransET, where, for each relation r, entities are mapped based on both type and relationship.

In this paper, our contributions are as follows: (1) we propose a model, where we explicitly consider entity types and mapping matrices are designed by considering entity types; (2) TransET can be easily combined with other state-of-the-art models (e.g., TransE, TransR) to produce more accurate predictions; (3) we prepare a new dataset collected from Freebase and make it available publicly to facilitate similar lines of research. We use this new dataset to compare our model with existing models in experiments that demonstrate TransET's superior performance.

2   Related Work

In its present state, KG technology is far from fully mature, although link prediction is an effective approach to completing a KG. Various models have been proposed to address the link-prediction issue. The models proposed to date differ in terms of their scoring function.

First, we describe the notation used in this paper. A triple is denoted by (h, r, t), where h is the head entity, r is the relation, and t is the tail entity. The bold letters $\boldsymbol{h}$, $\boldsymbol{r}$, and $\boldsymbol{t}$ denote embeddings of h, r, and t, respectively, in an embedding space $\mathbb{R}^n$. $f_r(\boldsymbol{h}, \boldsymbol{t})$ is the scoring function of the model under consideration.

2.1   Unstructured Model (UM)

UM [3,4] is the preliminary image of TransE, considering only entities as embeddings. Because UM ignores relations, its scoring function is a simplification of that used in TransE. The scoring function is given as:

$$f_r(\boldsymbol{h}, \boldsymbol{t}) = ||\boldsymbol{h} - \boldsymbol{t}||_{l_{1/2}},\qquad(1)$$

where $\boldsymbol{h}$ and $\boldsymbol{t}$ are the embeddings of head and tail, respectively.

2.2   Structure Embedding (SE)

Bordes proposed the SE model [6], which introduces two different matrices to project separately the head and tail entities for each relation. Its scoring function is defined as follows:

$$f_r(\boldsymbol{h}, \boldsymbol{t}) = ||\boldsymbol{M}_{rh}\boldsymbol{h} - \boldsymbol{M}_{rt}\boldsymbol{t}||_{l_{1/2}},\qquad(2)$$

where $\boldsymbol{M}_{rh}$ and $\boldsymbol{M}_{rt}$ are the projection matrices for the head and tail, respectively.

2.3   TransE, TransH, and TransR/CTransR

TransE [5] learns embedding as $\boldsymbol{h} + \boldsymbol{r} \approx \boldsymbol{t}$ where (h, r, t) holds. Therefore, $(\boldsymbol{h} + \boldsymbol{r})$ is very close to $\boldsymbol{t}$. TransE is the most popular translation-distance-based embedding model and is both very simple and fast. The scoring function of TransE is:

$$f(\boldsymbol{h}, \boldsymbol{t}) = ||\boldsymbol{h} + \boldsymbol{r} - \boldsymbol{t}||_{l_{1/2}},\qquad(3)$$

which is low if (h, r, t) holds and is high otherwise.

Many researchers [9,17] have claimed that TransE has problems in representing one-to-many, many-to-one, and many-to-many relations, with a number of models being proposed to address these issues.

The first such effort was TransH [17], which represents relations by hyperplanes. This model projects entities on the hyperplane corresponding to a relation. A single entity can have different representations on different hyperplanes. TransH models the relation r as $\boldsymbol{r}$ on a hyperplane with the normal vector $\boldsymbol{w_r}$. Given a triple (h, r, t), the entity representations $\boldsymbol{h}$ and $\boldsymbol{t}$ are projected on the hyperplane of $\boldsymbol{w_r}$ with the restriction that $\|\boldsymbol{w_r}\| = 1$. The calculation is expressed as:

$$\boldsymbol{h}_\perp = \boldsymbol{h} - \boldsymbol{w_r}^\top \boldsymbol{h} \boldsymbol{w_r},$$
$$\boldsymbol{t}_\perp = \boldsymbol{t} - \boldsymbol{w_r}^\top \boldsymbol{t} \boldsymbol{w_r}. \tag{4}$$

The scoring function is very similar to TransE:

$$f_r(\boldsymbol{h}, \boldsymbol{t}) = \|\boldsymbol{h}_\perp + \boldsymbol{r} - \boldsymbol{t}_\perp\|_{l_{1/2}}. \tag{5}$$

TransR [9] also addressed the flaws of TransE, but in a slightly different way than did TransH. TransR considers separate spaces for entities and relations, but the main principle is that entities and relations are completely different types of objects, implying that they should not occupy the same vector space. Given a triple (h, r, t), TransR projects the entity representations $\boldsymbol{h}$ and $\boldsymbol{t}$ into the space specific to a relation r. That is:

$$\boldsymbol{h}_r = \boldsymbol{h} \boldsymbol{M}_r, \quad \boldsymbol{t}_r = \boldsymbol{t} \boldsymbol{M}_r, \tag{6}$$

where $(\boldsymbol{h}, \boldsymbol{t}) \in \mathbb{R}^n$, $\boldsymbol{r} \in \mathbb{R}^m$, and $\boldsymbol{M}_r \in \mathbb{R}^{n \times m}$ represents the projection matrix from the entity space to the relation space for relation r. The scoring function is:

$$f_r(\boldsymbol{h}, \boldsymbol{t}) = \|\boldsymbol{h}_r + \boldsymbol{r} - \boldsymbol{t}_r\|_{l_{1/2}}. \tag{7}$$

CTransR is an extension of TransR proposed by the same authors. In this model, entity pairs for a relation are clustered into different groups, and the pairs in the same group share the same unique relation vector.

2.4 Other Models

For the link-prediction and triple-classification tasks, bilinear and neural-network-based models are also popular. RESCAL [12,13] is a bilinear model, with each relation being represented by an n-by-n matrix in an embedding space $\mathbb{R}^n$ and the scores for the triples being calculated by a bilinear mapping. Another bilinear model, proposed by Trouillon [16], uses complex numbers instead of real numbers and takes the conjugate of the embedding of the tail entity before calculating the bilinear mapping.

The SLM model [14], proposed by Socher, concatenates head and tail entities as an input layer to the nonlinear hidden neural layer and has the scoring function:

$$f_r(\boldsymbol{h}, \boldsymbol{t}) = \boldsymbol{u}_r^\top f(\boldsymbol{T}_{r1} \boldsymbol{h} + \boldsymbol{T}_{r2} \boldsymbol{t} + \boldsymbol{b}_r), \tag{8}$$

where $\boldsymbol{T}_{r1}$ and $\boldsymbol{T}_{r2}$ are weighting matrices and $f(\cdot)$ is the *tanh* operation.

The NTN model [14] is an extension of the SLM model. It considers second-order correlations as inputs to nonlinear hidden neural networks. Its scoring function is:

$$f_r(\boldsymbol{h}, \boldsymbol{t}) = \boldsymbol{u}_r^\top f(\boldsymbol{h}^\top \boldsymbol{T}_r \boldsymbol{t} + \boldsymbol{T}_{rh} \boldsymbol{h} + \boldsymbol{T}_{rt} \boldsymbol{t} + \boldsymbol{b}_r), \tag{9}$$

where $\boldsymbol{T}_r$ represents a three-way tensor, $\boldsymbol{T}_{rh}$ and $\boldsymbol{T}_{rt}$ denote weighting matrices, $\boldsymbol{b}_r$ is the bias, and $f(\cdot)$ is the *tanh* operation. To date, NTN has proved computationally expensive and has scalability issues.

Krompaß [7] proposed latent-variable models that consider relation and entity type constraints. However, such models add more redundancy than do translation-distance-based models. For this reason, they can have an overfitting problem. Our model considers only entity type constraints, aiming to retain simple model estimation. It exploits linear mapping and involves less parameter overhead than the latent-variable models. Although neural-network-based models also tend to encounter overfitting, the standard advantage of such models is that they can capture many kinds of relations.

3   Our Method

Translation-distance-based embedding models mostly follow TransE. Both TransE and TransH assume embeddings of entities and relations within the same space $\mathbb{R}^n$. However, relations and entities are completely different objects and it may not be appropriate to represent them in a common semantic space. Although TransH extends modeling flexibility by employing relation hyperplanes, it does not fully address the restrictions of this assumption. In contrast, the entities in TransR are mapped to vectors in different relational spaces, according to their relations. However, none of these models consider the significance of entity type. Therefore, they cannot judge the exact role of each entity, based on its relation. In our model, we deliberately include the type information. The entity's type can be incorporated easily by introducing an entity type-mapping matrix. For the relations, the entity type information plays a significant role. For example, the "*CapitalOf*" relation would imply that both the head and the tail would be location type entities. Using type information in a translation model can improve the efficiency of any such model.

Because an entity may belong to several subcategories, considering all the subcategories when modeling the type will make the model complex, thereby increasing the time and space complexity significantly. To address this issue in our model, we consider only the basic or most prominent type for an entity. As an example, a "Person-type" entity could be subcategorized as an "Actor" or a "Player." In our model, we overlook the subcategories and retain only the most prominent class of the entity.

Our proposed TransET model can easily be combined with other state-of-the-art translation-distance-based models. For this purpose, we introduce an entity-type-based projection matrix $\boldsymbol{M}_p \in \mathbb{R}^{n \times n}$ for each type of entity. For a given triple (h, r, t), the projection matrix $\boldsymbol{M}_p$ can be different, if h and t have a different type. This applies for both head and tail entities:

$$h_p = hM_p, \quad t_p = tM_p, \tag{10}$$

where $(h, t) \in \mathbb{R}^n$. In our model, we enforce the constraints $||h||_2 \leq 1$, $||t||_2 \leq 1$, $||hM_p||_2 \leq 1$, and $||tM_p||_2 \leq 1$. It is not mandatory to have the same dimensionality for entity embeddings and entities' type embeddings. However, in our experiments to learn vectors and matrices, we keep the same dimensionality. The scoring function for each specific relation r (where $r \in \mathbb{R}^n$) is correspondingly defined as:

$$f_r(h, t) = ||h_p + r - t_p||_{l_{1/2}}. \tag{11}$$

Equation 3 gives the scoring function of TransE. If we incorporate TransE into the entities' type-mapping matrix model, denoted by $TransET_{TransE}$, then the model's scoring function will be given by Eq. 11. TransE has the same vector representation for each entity. In TransET, the head and tail entities are mapped according to their type.

In TransR, entities are mapped separately to a relation space. Equations 6 and 7 denote the embeddings and the scoring function of the TransR method. If we incorporate TransR into the entities' type-mapping matrix model ($TransET_{TransR}$), then the embeddings of head and tail entities become:

$$h_p^r = (hM_p)M_r, \quad t_p^r = (tM_p)M_r \tag{12}$$

and the scoring function is defined as:

$$f_r(h, t) = ||h_p^r + r - t_p^r||_{l_{1/2}}. \tag{13}$$

The TransR mapping matrix M_r is the same for both head and tail entities. The key difference between our proposed model ($TransET_{TransR}$) and the traditional TransR is that, before projecting to a relation space, entities are mapped via the entities' type-mapping matrix. In the relation space, entities therefore have vector representations based on their type and a specific relation property.

In the above discussion, we have shown how to combine TransET with TransE and TransR. In the same way, we could combine TransET with other translation-distance-based models.

4   Training

We define a margin-based loss function as the objective for training:

$$L = \sum_{(h,r,t)\in S} \sum_{(h',r,t')\in S'} max(0, f_r(h, t) + \gamma - f_r(h', t')). \tag{14}$$

Here, γ is the margin, S is the set of correct triples, and S' is the set of incorrect triples. Existing KGs should only contain correct triples. An S' is constructed by replacing a head or a tail entity in an existing triple.

We use a stochastic gradient descent (SGD) method to minimize L. TransR initializes the embeddings of entities and relations obtained from TransE. To avoid overfitting of $TransET_{TransR}$, we also initialize entity and relation embeddings with the results of $TransET_{TransE}$.

5   Experiment

Our proposed model was evaluated via two tasks: link prediction [5] and triple classification [14]. This section discusses the experimental procedures for our model.

Table 1. Datasets.

Dataset	#E	#R	#Train	#Valid	#Test
FB10K	10,056	141	69,210	5,447	6,479
FB13	75,043	13	316,232	5,908	23,733

5.1   Datasets

In this study, we used data from Freebase [2], which is a very popular KG and is used for various NLP experiments. Our model explicitly requires entity type information. We therefore prepared an "FB10K" dataset from Freebase, which contained the head and tail entities' type information, using the following procedure. We retrieved the relations, with their heads and tails, which are real-world entities. We annotated the basic types for those head and tail entities by using human judgment on the relations. The basic type information for real-world entities had been collected from the schema.org[1] vocabulary. According to the definitions in schema.org, there are 10 basic types of real-world entity. Although one entity may involve various subcategories/subtypes, we focused only on basic entity types, aiming to keep the model simple. Freebase provides the hierarchical type information about entities via their type/instance field. Each entity may involve multiple hierarchical types. For example, entity "David Schwimmer" (Freebase ID: /m/016tbr) has 22 hierarchical types, including /people/person, /award/award_winner, /celebrities/celebrity, and /film/director.

It would be possible to consider all hierarchical types for each entity when building the model. Unfortunately, this would raise the time and space complexity enormously, as the previous example illustrates. Because a KG may contain millions of entities and billions of relational facts, considering all hierarchical types is simply not realistic. The datasets used in this paper can be obtained from https://github.com/Rahman29/Conference.

In this paper, the "FB13" dataset has been employed to evaluate the triple classification task. This dataset contains negative triples, which is helpful for this particular task. Moreover, it has only 13 relations and we annotated the relations with respect to basic types according to the schema.org vocabulary. Table 1 shows the statistics for the datasets used in this paper.

---

[1]   http://schema.org.

5.2   Link Prediction

For a triple (h, r, t), the link-prediction task predicts the missing h or t, given the relation and the other entity. The results were evaluated by ranking the predicted head or tail entity, as calculated by the scoring function $f_r(h, t)$ for test triples.

Table 2. Evaluation results for link prediction.

Dataset	FB10K			
Metric	Mean Rank		Hits@10	
	Raw	Filter	Raw	Filter
$TransE(unif)$	534	351	47.05	61.90
$TransE(bern)$	529	346	46.35	61.75
$TransR(unif)$	402	278	54.96	70.68
$TransR(bern)$	390	248	57.30	72.82
$TransET_{TransE}(unif)$	401	196	50.56(**+3.51**)	63.41(**+1.51**)
$TransET_{TransE}(bern)$	374	152	51.40(**+5.05**)	64.43(**+2.68**)
$TransET_{TransR}(unif)$	304	112	56.20(**+1.24**)	73.30(**+2.62**)
$TransET_{TransR}(bern)$	296	110	60.40(**+3.1**)	74.50(**+1.68**)

For our experiments, we adopted the same protocol as that used in TransE. For each testing triple (h, r, t), we corrupted it by replacing the tail t or head h with every entity e in the KG or the current dictionary and calculated a probabilistic score for the corrupted triple (h, r, e) or (e, r, t), respectively, in terms of the scoring function $f_r(h, e)$. It's the "Raw" setting protocol. Because we have corrupted the triples randomly, this same triple may already exist in the actual KG and would be considered correct. During the ranking, it is logically possible that such triples may appear before the original triple. To eliminate this issue, we intentionally remove those corrupt triples that are created by replacing h or t randomly but that already exist in the KG before computing the rank of each testing triple. They may exist in any of the training, valid, or testing sets. This revised setting protocol is called the "Filter" setting. In addition, we employ the same two sampling methods, "bern" and "unif," that were used in the previous studies.

Two evaluation metrics were used: average Mean Rank and Hits@10 (the proportion of testing triples whose rank did not exceed those of the top 10 predictions). For both settings, a lower Mean Rank and a higher Hits@10 imply a better performance.

We used the same dataset to compare our models with the baseline models. To obtain the best settings for our models, we tuned five parameters. We selected the margin γ from the set $\{0.5, 1, 2\}$, the dimensionality of entity and relation vectors from $\{50, 100\}$, the learning rate α from $\{0.001, 0.0001, 0.0005\}$, the number of training triples in each mini-batch from $\{20, 50, 300, 1440\}$, and

Table 3. Relation-specific head or tail prediction examples (Filter setting).

Models	TransE		TransR		$TransET_{TransE}$		$TransET_{TransR}$	
Relations	Hits@10		Hits@10		Hits@10		Hits@10	
	head	tail	head	tail	head	tail	head	tail
film_in_this_genre	**100**	10.52	**100**	11.0	**100**	**13.30**	**100**	13.0
olympics_participated_in	40.62	**100**	44.90	**100**	45.10	**100**	**47.28**	**100**
directed_by	94.20	95.65	96.27	96.27	94.0	95.65	**97.50**	**97.50**
films_production_designed	86.67	93.33	86.81	94.0	87.32	99.49	**93.80**	**100**
film_release_region	94.74	**100**	**96.49**	**100**	95.71	**100**	96.48	**100**
country	96.96	96.96	**100**	98.59	97.0	96.98	96.90	**100**
award_winner	**100**	50.10	**100**	58.38	**100**	51.67	**100**	**53.56**
/film/film/language	7.70	91.52	15.66	**97.0**	10.0	91.51	**15.90**	96.83

the dissimilarity measure in the embedding scoring function from $\{L_1, L_2\}$. The optimal configurations were $\gamma = 1$, $d = 50$, $\alpha = 0.0005$, $B = 50$, and using L_1 as the dissimilarity function for $TransET_{TransE}$ and $TransET_{TransR}$. We also had to find optimal parameter settings for TransE and TransR. For TransE, they were $\gamma = 1$, $d = 50$, $\alpha = 0.001$, $B = 50$, and using L_1 as the dissimilarity function. For TransR, they were $\gamma = 1$, $d = 50$, $k = 50$ (dimensionality for relation vectors), $\alpha = 0.001$, $B = 1440$, and using L_1 as the dissimilarity function.

Table 2 presents the results for link prediction. The values in parentheses are the improvements over the respective base model. Our models outperformed all other methods for both "bern" and "unif", using the experimental dataset with both Raw and Filter settings. The results appear to show that the "bern" sampling method performs slightly better than the "unif" method. TransE is the method that improved most significantly using this dataset.

For the Raw setting, $TransET_{TransE}$ and $TransET_{TransR}$ achieved 51.40% and 60.4% of Hits@10, respectively, for "bern," which are 5.05%, and 3.1% higher than those for TransE and TransR, respectively. Our models showed even better performance for the Filter setting. $TransET_{TransE}$ and $TransET_{TransR}$ achieved 64.43% and 74.5% of Hits@10, respectively, for "bern," which are 2.68% and 1.68% higher than those for TransE and TransR, respectively.

Table 3 reports the Hit@10 (head/tail prediction) results for some relations. The improvement shown by $TransET_{TransE}$ and $TransET_{TransR}$ compared with TransE and TransR for these relations is a very promising.

As noted, the entities' type plays a crucial role with respect to its relations. It is therefore logical that incorporating type information could be utilized to achieve better performance in the link-prediction task. We believe that the projection of entities based on the entities' type-mapping matrix would improve the performance of the proposed models. In these models, we are using entities' type information in addition to the entities themselves, enabling the projected vectors to exhibit more semantic information than the vectors in TransE and TransR models.

5.3   Triple Classification

The triple classification task checks whether a given triple (h, r, t) is correct or incorrect. When it was first introduced in the NTN model, it acted as a binary classification. Because the task requires negative samples, we employed the "FB13" dataset, which is a benchmark dataset from Freebase involving 13 relations. In this dataset, we incorporate entity type information.

To implement this task, we set a relation-specific threshold σ_r. For a triple (h, r, t), if the dissimilarity score (computed by the scoring function f_r) is below the σ_r threshold, then the predicted triple is positive. Otherwise, the prediction is negative. The value for σ_r is determined in accordance with the classification accuracy.

Table 4. Triple classification for the FB13 dataset.

Model	Accuracy (%)
$TransE(unif)$	70.9
$TransE(bern)$	81.5
$TransR(unif)$	74.7
$TransR(bern)$	82.5
$TransET_{TransE}(unif)$	74.1 (+3.1)
$TransET_{TransE}(bern)$	83.3 (+1.8)
$TransET_{TransR}(unif)$	78.9 (+4.2)
$TransET_{TransR}(bern)$	**86.8** (+4.3)

Table 4 shows the evaluation results for triple classification. The parameters and evaluation results for TransE and TransR were obtained directly from the original papers. We see from the table that $TransET_{TransE}$ is more accurate than $TransE$, and $TransET_{TransR}$ is more accurate than $TransR$. These results imply that incorporating type information can improve accuracy.

6   Conclusion

In this paper, we have proposed an embedding model based on entities' type projection matrices. We consider only the basic entity-category label/type to project the type of the entities. The strength of this model is that it can be combined easily with other translation-distance-based models to improve accuracy without making the models more complex. The TransET model is conceptually simple and can demonstrate highly competitive results for link prediction and triple classification. The underlying idea of TransET was applied to the most popular translation-distance-based models (TransE and TransR) with the experimental results showing better performances than for the basic TransE and

TransR models. TransET can improve the results of other translation-distance-based models that are based on TransE. We can therefore conclude that the entities' type-based diversity in a KG is an important factor and that the entities' type-mapping matrix is suitable for modeling KGs.

We exploited the negative-sampling method during the training phase using the same approach as in TransE. We observed that same-type entities tend to converge and form clusters, which can lead to errors in some cases. In the near future, we aim to develop a data-sampling algorithm to address this problem. We also intend to perform experiments on a wider variety of datasets.

Acknowledgments. This work was supported by a JSPS Grant-in-Aid for Scientific Research (B) (15H02789, 15H02703).

References

1. Bengio, Y., Ducharme, R., Vincent, P., Jauvin, C.: A neural probabilistic language model. J. Mach. Learn. Res. **3**, 1137–1155 (2003)
2. Bollacker, K., Evans, C., Paritosh, P., Sturge, T., Taylor, J.: Freebase: a collaboratively created graph database for structuring human knowledge. In: Proceedings of the 2008 ACM SIGMOD International Conference on Management of Data, pp. 1247–1250. ACM (2008)
3. Bordes, A., Glorot, X., Weston, J., Bengio, Y.: Joint learning of words and meaning representations for open-text semantic parsing. In: Artificial Intelligence and Statistics, pp. 127–135 (2012)
4. Bordes, A., Glorot, X., Weston, J., Bengio, Y.: A semantic matching energy function for learning with multi-relational data. Mach. Learn. **94**(2), 233–259 (2014)
5. Bordes, A., Usunier, N., Garcia-Duran, A., Weston, J., Yakhnenko, O.: Translating embeddings for modeling multi-relational data. In: Advances in Neural Information Processing Systems, pp. 2787–2795 (2013)
6. Bordes, A., Weston, J., Collobert, R., Bengio, Y., et al.: Learning structured embeddings of knowledge bases. In: AAAI, vol. 6, p. 6 (2011)
7. Krompaß, D., Baier, S., Tresp, V.: Type-constrained representation learning in knowledge graphs. In: Arenas, M., et al. (eds.) ISWC 2015. LNCS, vol. 9366, pp. 640–655. Springer, Cham (2015). https://doi.org/10.1007/978-3-319-25007-6_37
8. Lehmann, J., et al.: Dbpedia-a large-scale, multilingual knowledge base extracted from Wikipedia. Semant. Web **6**(2), 167–195 (2015)
9. Lin, Y., Liu, Z., Sun, M., Liu, Y., Zhu, X.: Learning entity and relation embeddings for knowledge graph completion. In: AAAI, vol. 15, pp. 2181–2187 (2015)
10. Mikolov, T., Chen, K., Corrado, G., Dean, J.: Efficient estimation of word representations in vector space. arXiv preprint arXiv:1301.3781 (2013)
11. Mikolov, T., Sutskever, I., Chen, K., Corrado, G.S., Dean, J.: Distributed representations of words and phrases and their compositionality. In: Advances in Neural Information Processing Systems, pp. 3111–3119 (2013)
12. Nickel, M., Tresp, V., Kriegel, H.P.: A three-way model for collective learning on multi-relational data. In: ICML, vol. 11, pp. 809–816 (2011)
13. Nickel, M., Tresp, V., Kriegel, H.P.: Factorizing YAGO: scalable machine learning for linked data. In: Proceedings of the 21st International Conference on World Wide Web, pp. 271–280. ACM (2012)

14. Socher, R., Chen, D., Manning, C.D., Ng, A.: Reasoning with neural tensor networks for knowledge base completion. In: Advances in Neural Information Processing Systems, pp. 926–934 (2013)
15. Suchanek, F.M., Kasneci, G., Weikum, G.: YAGO: a core of semantic knowledge. In: Proceedings of the 16th International Conference on World Wide Web, pp. 697–706. ACM (2007)
16. Trouillon, T., Dance, C.R., Gaussier, É., Welbl, J., Riedel, S., Bouchard, G.: Knowledge graph completion via complex tensor factorization. J. Mach. Learn. Res. **18**(1), 4735–4772 (2017)
17. Wang, Z., Zhang, J., Feng, J., Chen, Z.: Knowledge graph embedding by translating on hyperplanes. In: AAAI, vol. 14, pp. 1112–1119 (2014)

A Sentence Similarity Model Based on Word Embeddings and Dependency Syntax-Tree

Wenfeng Liu[1,2], Peiyu Liu[1,3(✉)], Jing Yi[1,4], Yuzhen Yang[2],
Weitong Liu[1,3], and Nana Li[1,3]

[1] School of Information Science and Engineering, Shandong Normal University,
Jinan 250014, China
285219011@qq.com, liupycn@163.com, 2980475748@qq.com,
1328271168@qq.com, 1647137750@qq.com
[2] School of Computer, Heze University, Heze 274015, China
331271247@qq.com
[3] Shandong Provincial Key Laboratory for Distributed Computer Software
Novel Technology, Jinan 250014, China
[4] School of Computer Science and Technology, Shandong Jianzhu University,
Jinan 250101, China

Abstract. How to effectively measure the similarity between two sentences is a challenging task in natural language processing. In this paper, we propose a sentence similarity comparison method that combines word embeddings and syntactic structure. First of all, by generating the corresponding syntactic tree, we synthetically analyze the two sentences and block them according to the syntactic components. Secondly, we prune the syntactic tree, remove the stop words and perform morphological restoration. Then, some important operations will be performed, such as passive flipping, negative flipping, and so on. Finally, the similarity of two sentence pairs is calculated by weighting the block embeddings of the syntactic tree. Experiments show the effectiveness of this method.

Keywords: Word embeddings · Dependency syntax tree · Sentence similarity
Syntactic structure

1 Introduction

Measuring the similarity between sentences is a basic task in natural language processing tasks. It is widely used in various tasks such as information retrieval, text clustering, text classification, machine translation, question answering system. The effectiveness of these tasks or applications depends to a large extent on the accuracy of the sentence's similarity. Therefore, studying the similarity between sentences is a crucial basic work in natural language processing.

The importance of different syntactic elements in the sentence is different. As we all know, the composition of syntactic components in English sentences mainly consists of subject, predicate, object, predicate, attributive, adverbial, object complement and appositive. Different natural language processing tasks or applications focus on

© Springer Nature Switzerland AG 2018
L. Cheng et al. (Eds.): ICONIP 2018, LNCS 11303, pp. 126–137, 2018.
https://doi.org/10.1007/978-3-030-04182-3_12

different syntactic components, such as the news headlines lay more emphasis on the subject, predicate and object.

Syntactic structure is a legal representation of a meaningful sentence, and only the expression that is consistent with the syntactic structure of the sentence makes sense. Words are the least textual representation and represent a meaningful and linguistic component in a sentence. The vectorized representation of words provides a basis for the similarity of words, sentences, or larger-sized textual representations. As the basic unit of text semantic information, words have drawn much attention from researchers. Since traditional one-hot representation is not effective in calculating word similarity, for example, "PC" and "computer" do not have any similarity, while they denote the same meaning in semantics. And how to integrate the contextual semantic information of words into their representation has always been a research focal point. Until 2013, due to limitations of hardware technology, researchers have used statistical machine learning methods to solve the problem of text representation. The distributed hypothesis was proposed by Harris [1], and was further expanded by Firth [2]. With the development of hardware technology and optimization algorithms, the distribution hypothesis has been widely concerned.

In recent years, with the extensive application of deep learning in natural language processing, many new models of distributed representations have been emerged. The current word vectors are mainly based on Word2Vec, GloVe, FastText [3] or their improved versions, which have achieved very good representation in many natural language processing tasks. Our approach exploits the improved syntactic structure and embeddings of sentences. Furthermore, it is also an unsupervised method, and does not need any labelled sentence similarity data. Compared with the current mainstream methods, our method has achieved the best results on data set of SemEval-2015 task 2.

2   Related Work

2.1   The Dependency Syntax Tree

With the integration and evolution of statistical natural language processing, the dependency syntax analysis has been deeply discussed and wide range of applications. The deterministic dependency that based data-driven syntactic analysis is one of the main methods in statistical dependency analysis, in which one word to be analyzed is taken one after another in a particular direction. Each input produces a single analysis until the sequence. Each operation acquires a unique syntactical expression from a defined sequence of analysis actions, sometimes with backtracking and patching. The dependency analysis strategy is a syntax-driven and list-based deterministic approach [4]. While the dependency parsing approach proposed by Yamada and Matsumoto [5] and Nivre and Nilsson [6] can be regarded as a data-driven and stack-based deterministic approach. In subsequent studies, Andor et al. [7] developed the SyntaxNet system that based on arc-transformed syntactic analysis and achieved very good results in processing of the English text. Therefore, the dependency syntax analysis based on arc-transformed has become the popular method of constructing the dependency syntax tree. And it is also the basis of our approach.

2.2 Word Embeddings

Words are the smallest and meaningful language units. In recent years, the most widely used approach is the distributed representation that based on neural networks [8], which is also referred to as word vectors or word embeddings. This method uses neural network technology to model the relationship between the context and the target word. Due to the flexibility of neural networks, the biggest advantage of this approach is that it can represent the complex contexts. Xu and Alex [9] first tried to solve the bi-gram language model by neural network. Bengio and Senecal [10] proposed a method to accelerate training of a neural probabilistic language model. They have obtained the word vectors while learning the language model. Mnih and Hinton [11] proposed a Log-Bilinear language model (LBL) based on NNLM. Its structure is a log bilinear structure, while the NNLM model structure is a neural network structure.

The vector-based inverse language model (ivLBL) [12] has actually abandoned the log bilinear structure although the name inherits the log bilinear language model (LBL) structure. And the recurrent neural network based language model (RNNLM) proposed by Mikolov [13] directly models $P(w_i \mid w_1, w_2, ..., w_{i-1})$. The C&W model proposed by Collobert and Weston [14] is the first to directly generate word vectors. Mikolov *et al.* [15] put forward the CBOW (continuous Bag-of-Words) model and skip-gram model. Their purpose of the two models is to obtain word embeddings in a more efficient way, and the two models have been widely used. The literature [16] presents a comparison between several multi-word term aggregation methods of distributional context vectors applied to the task of semantic similarity and relatedness in the biomedical domain. Pengjin *et al.* [17] successfully applied the improved bag-of-words model to text classification. Its core idea is to train a different word vector for each type of text to improve the classification effect.

3 Sentence-Level Similarity Model

In recent years a number of sentence similarity calculation models have been put forward. For example, the similarity between sentence pairs in is obtained by calculating the semantic similarity between words in two sentences [18]. The formula is as follows:

$$SIM(A, B) = \frac{1}{2}\left(\frac{\sum_{i=1}^{m} a_i}{m} + \frac{\sum_{j=1}^{n} b_j}{n}\right) \tag{1}$$

where, m and n denote the number of words in sentences A and B respectively. i denotes the i-th word in sentence A, j denotes the j-th word in sentence B. a_i represents the highest value of similarity between the i-th word in sentence A and all words in sentence B. b_j denotes the highest value of similarity between the i-th word in sentence B and all words in sentence A. This idea does not consider the syntax structure of the sentence.

Exploiting the syntactic structure of sentences is also a common method to compare the similarities of theirs. The first step is to construct the syntactic tree of sentence pairs and use the similarity of sentence structure components for calculating the similarity of theirs. Based on the skeleton dependency tree, Lévy [19] discussed the sentence similarity in which the predicate serves as the root of the tree, and words that depend on the predicate serve as the leaf node. Literature [20] proposed a semantic-based method to calculate the similarity of sentences. This method calculates the similarity between two sentences by calculating the effective collocation of two sentences. The formula of sentence-pairs similarity is as follows:

$$SIM(Sen1, Sen2) = \frac{W_1 + W_2}{\max\{PairCount1, PairCount2\}} \tag{2}$$

W_1 denotes the effective matching weight of the words between *Sen1* and *Sen2*, W_2 represents the effective matching weight of the words between *Sen2* and *Sen1*. *PairCount1* is the number of valid matches of the *Sen1* to *Sen2*, and *PairCount2* is that of *Sen2* to *Sen1*. This method only makes use of syntactic match of words, so it can not consider syntactic information of sentences as a whole.

4   The Similarity Model Based on Word Embeddings and Dependency Syntax-Tree

Word embeddings are distributed representations of words and have certain semantic information. A meaningful sentence must conform to the syntactic structure of the corresponding language. Therefore, it is of great significance to integrate the syntactic structure when comparing the similarity of sentence pairs. The similarity of sentence pairs is based on the assumption that the input sentence pairs should conform to the grammatical relationship of the language, and we can perform various calculations on the word embeddings more naturally. Figure 1 is the framework of the model.

4.1   The Proposed Approach

The model proposed in this paper is mainly divided into the following steps.

Step 1: Constructing the Dependency Syntax Tree

We perform arc-based dependency syntax and construct the dependency syntax tree of a sentence. This corresponds to the Sect. 4.2, as shown in Figs. 2 and 3.

Step 2: Blocking the Sentences

According to the syntactic dependency, the sentences are divided into different syntactic blocks, such as *subject block, predicate block, object block, etc.*

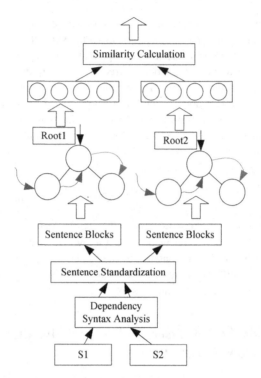

Fig. 1. The framework of our model

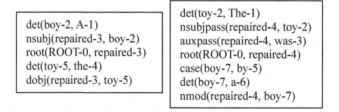

Fig. 2. The dependencies of two sentences

Step 3: Sentence Standardization

If the predicate block is a passive tense, the sentence should be performed passive rollover. That is, the *subject block* and *the object block* are exchanged. At the same time, the "*be*" or "*been*" verb in *the predicate block* will be deleted. The *step 2* and *step 3* correspond to the Sect. 4.3.

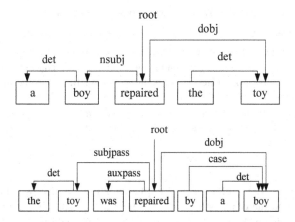

Fig. 3. Expanded dependency syntax tree

Step 4: Block Embeddings

The stop words (such as "*by*" and "*the*") in the syntactic blocks will be deleted. Using the word embeddings of multiple words in the syntactic blocks, we construct the block embeddings of the syntactic blocks, calculate the alignment probability of the blocks between the sentence pairs.

Step 5: Sentence Embeddings

Based on the block embeddings of the syntactic blocks, we construct the sentence embeddings.

Step 6: Similarity Calculation

Used the euclidean distance calculation formula, it is easy to calculate the similarity of two sentences.

4.2   Arc-Based Transformation Syntax Tree

Syntax-Tree model, which is composed of a *stack*, a *buffer* and a set of *arcs*, is a transition-based dependency parser which is incrementally constructed [21]. It handles words from left to right. These unprocessed words of sentence are put in *buffer*, part-of-speech (POS) of theirs are put in a list. The processed words are pushed into *stack*, and *arcs* describe the dependencies between words.

At the beginning, all the words of a sentence are placed in *the buffer*. At each step, the parser can only do one of the three operations: *op_shift*, *op_left_arc* and *op_right_arc* (as shown in Table 1).

Given an input sentence $[w_1, ..., w_n]$, we define a set of *states* S ($s*$ denotes the start state). However, the most important is to define the decision function $D(s)$. According to the state information (word, POS on top of the stack or in the buffer), $D(s)$ will decide to run one of the three operations (*op_shift*, *op_left_arc*, *op_rig-ht_arc*).

Table 1. Three operations of the parser

Operation	Description
op_shift	Pushes the next_word(w_2) of the current_word(w_1) onto the top of *the stack*
op_left_arc	Pops two words(w_1 w_2) on the top from the *Stack*, attaches the second(w_1) to the first(w_2), creates an arc pointing to the left, pushes the first word back onto *the stack*
op_right_arc	Pops the top two words(w_1 w_2) from the *stack*, attaches the second(w_1) to the first(w_2), creates creates a right arc, and pushes the second(w_1) back onto *the stack*

In order to improve the efficiency of implementation, we use two stacks, one is for handling words and the other is to deal with their part of speech. And two buffers are needed. Two words on the top of the stack, two in buffer and their part of speech are considered which operation will be executed. According to corresponding operation, the two stacks and two corresponding buffers are performed the same processing. We use the following feature combination templates,

{*stack_word1, stack_word2, buffer_word1, buffer_word2, stack_pos1, stack_pos2, buffer_pos1, buffer_pos2*},

Each combination of features corresponds to one of the three operations.

Finally we obtain the dependency tree of the words and the dependency tree of the part of speech about a sentence. According to the obtained syntax trees, different syntax blocks have different weights. In the syntax block, the closer the block center word is, the higher the weight.

Figure 2 is the dependency syntax trees that generated on two sentences "A boy repaired the toy" and "The toy was repaired by a boy". The numbers indicate the order in which the words appear in the sentence.

4.3   Syntactic Block and Passive Transformation

According to the dependencies of the words in the sentence, such as "*boy*" and "*repaired*", the dependency tag is "*nsubj*", which is dependent on subject and predicate. Any word that points to "boy" belongs to the subject block. The division of other syntactic blocks is similar as that of subject block (such as predicate block, object block, etc.), as shown in Fig. 4. If the predicate block is a passive tense, and the "*be*" or "*been*" is followed by the passive tense of the verb, then the sentence should be passively flipped. Doing so can maximize the elimination of the difference in syntactic structure. The "*be*" or "*been*" in the predicate will be deleted, simultaneously, the subject block and the object block are exchanged. The label on the arc changes accordingly. For example, the label "*subjpass*" should be changed to "*nsubj*".

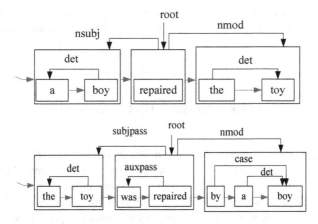

Fig. 4. The blocked dependency syntax tree

4.4   The Embeddings of Syntactic Block

The dependency relations between the syntactic blocks indicate the dependencies between these syntactic blocks. For these tags, the alignment model is used to calculate the alignment probabilities between the sentence pairs. According to different syntactical components, they are spliced according to the sequence of *subject*, *predicate*, and *object*. The embeddings of the syntactic block are represented by the distance-weighted summation of word embeddings in the block. The closer the distance from the core words in the block, the greater the weight.

Different words in a syntactic block have different degrees of importance. According to the dependency relationship formed in the syntactic tree block, we allocate different weights based on the distance from the center word and use weighted summation to obtain the embeddings of the syntactic block. In this way, the dimension of the syntactic block is the same as that of word. But the two different sets of embeddings are distributed in different vector spaces. The syntactic block embeddings are spliced into the sentence embeddings (as is shown in Fig. 5).

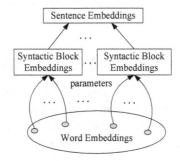

Fig. 5. The construction diagram of sentence vector

The similarity of embeddings in different vector spaces does not make any sense, although their dimensions are the same. The formula for the syntax block vector is shown in Eq. (3).

$$b_i = \sum_{w_j \in c} \lambda_j * [e(w_j)] \tag{3}$$

where, $\sum_{w_j \in c} \lambda_j = 1$, b_i represents i-th syntax block of the sentence, $e(w_j)$ denotes the embeddings of w_j .

For example, the sentence "There are a few beautiful flowers", after word segmentation and lemmatization, it is composed of five words, "there", "be", "a few", "beautiful" and "flower". The corresponding word embeddings are e1, e2, e3, e4 and e5. The key word of the sentence is "flower". So when the phrase is a word vector, the word embeddings of "flower" is more important than others words. In short, sentence-level vectors are not simply cumulative. Therefore, according to different tasks, syntactic structure and syntactic components are used to learn more powerful vectors.

4.5   Splicing Syntactic Blocks

In order to reduce the size of the dictionary, lemmatization will be performed. For example, different tenses of the verbs should be converted into the roots of theirs. For "*repaired*", "*repairs*", "*repairing*", all of them should be transformed to "*repair*". If the sentence is negative, we add a penalty factor to the syntactic block that contained negative words. The sentence vector is constructed as shown in Eq. (4).

$$S = [\alpha * \lambda_1 * block_1; \beta * \lambda_2 * block_2; \gamma * \lambda_3 * block_2] \tag{4}$$

block₁, *block₂* and *block₁* represent *the subject module, the predicate module*, and *the object module*, respectively. α, β and γ are normalized regulators. λ_1, λ_2 and λ_3 are negative adjustment factors, which take different parameter values according to the number of negative words in the negative word dictionary (see Eq. (5)).

$$\lambda_i = \begin{cases} 1 & \text{even number of negative words} \\ -1 & \text{others} \end{cases} \tag{5}$$

where, $i = 1, 2, 3$.

5   Experiments and Results Analysis

The data set for experiments has 200 data pairs that selected from the Task 2 of SemEval-2015. The basic word embeddings used in this paper are trained by the Skip-gram model on Wikipedia datasets. The deep learning framework is MXNet[1] developed by amazon's deep learning group. The dimension of the word embeddings is 300.

---

[1] http://mxnet.incubator.apache.org/.

The baselines of our experiments are as follows:

(1) Word embeddings-based Summation (**WE+SU**), this method directly averages the word embeddings of words that appear in sentence;
(2) Combined word embeddings and word similarity (**WE+SI**);
(3) Combined word embeddings and skeleton dependency tree (**WE+SD**);
(4) Combined word embeddings and semantic dependency tree (**WE+SE**);

The method of this paper is to combine word embeddings and arc-based dependency syntax tree for sentence comparison (**WE+ST**).

The difference between the model and human judgment is treated as the evaluation value. In addition, the absolute value of human judgment or corresponding model is less than or equal to 1. The larger the difference is, the less similar. The larger the difference between them, the more dissimilar they are. The smaller the difference, the more similar they are (as is shown in Eq. (6)).

The similarity of i-th sentence pairs is calculated by the following formula:

$$\text{Score}_i = 1 - |M_i - N_i| \tag{6}$$

where M_i represents the similarity value of i-th sentence pairs that judged by human, N_i is that of model's. Under a certain model, the overall similarity value with n sentence pairs is calculated by Eq. (7).

$$\text{SIM} = \frac{1}{n}\sum_{i=1}^{n}\text{Score}_i \tag{7}$$

where n represents the number of sentence pairs that evaluated.

It's easy to see from Tables 2 and 3 that the WE+SU model is the worst, because the model is simple and crude. And it does not consider syntactic structure or word order. Therefore, it weakens the expression ability of sentence-level vectors. Obviously the method of skeleton structure dependency (WE+SD) is better than the three methods (WE+SU, WE+SE, and WE+SI) which do not consider the syntactic structure. Our method (WE+ST) works best because it not only considers the syntactic structure information of the sentence, but also utilizes the different syntactic blocks. All of the blocks splice into sentence embeddings based on syntactic blocks that can compute sentence similarity of different lengths effectively. In particular, for the third and fifth test cases in Table 2, since our model has considered negative sentences and negative emotional words, it is very close to the result of human label.

Table 2. Comparison of experimental results of various models

Models	SIM
WE+SU	0.55
WE+SI	0.62
WE+SE	0.66
WE+SD	0.73
WE+ST	**0.85**

Table 3. Test cases

	Manual scoring	WE +SU	WE +SI	WE +SE	WE +SD	WE +ST
A boy repaired the toy The toy was repaired by a boy	1.00	0.91	0.93	0.93	0.97	**1.00**
Turkish riot police tear gas Taksim Square protest Turkish riot police enter Taksim Square	0.85	0.92	0.72	0.73	0.76	**0.81**
Egyptian police fire tear gas at protesters in Cairo Police don't fire tear gas at protesters in Cairo	−1.00	0.93	0.92	0.90	0.87	**−0.96**
Flooding in southern China kills 37 Blast in northern Syria kills 6	0.30	0.88	0.81	0.74	0.72	**0.45**
China yuan weakens to 6.1559 against USD China yuan strengthens to 6.1767 against USD	−1.00	0.95	0.92	0.93	0.89	**−0.98**

6 Conclusion

This paper proposes a new sentence similarity model that based on word embeddings and dependency syntax structure. We study the role of syntactic blocks in semantic representations of sentence-level, and explore the dependencies of sentence blocks as well as the passive translation of sentence clauses, and so on. In order to guarantee the order of words in sentence to a certain extent, our model splices the constructed syntactic block embeddings into sentence embeddings. Finally, the similarity values are calculated by calculating the euclidean distances of the sentence pairs. The performance of the model is measured by comparison with human judgment. And our model has achieved the best results. However, this model does not consider all the syntactic components, and the different syntactic blocks used in different applications are not the same. The main task of the next step is to further consider the syntactic elements in the syntactic structure as much as possible, and construct similarity calculation models of sentence pairs on different natural language processing tasks.

Acknowledgments. This work was supported by the national natural science foundation of China (61373148, 61502151), Shandong social science planning project (17CHLJ18, 17CHLJ33, 17CHLJ30), the natural science foundation of Shandong province (ZR2014FL010) and Shandong province department of education (J15LN34).

References

1. Harris, Z.S.: Distributional structure. Word **10**(2–3), 146–162 (1954)
2. Firth, J.: A synopsis of linguistic theory 1930–1955. Stud. Linguist. Anal. Oxf. Philol. Soc. **41**(4), 1–32 (1957)

3. Bojanowski, P., Grave, E., Joulin, A., Mikolov, T.: Enriching word vectors with subword information. Trans. Assoc. Comput. Linguist. **5**, 135–146 (2017)
4. Covington, M.A.: A fundamental algorithm for dependency parsing. In: 39th Annual ACM Southeast Conference, pp. 95–102. ACM Press, Pisa (2001)
5. Yamada, H., Matsumoto, Y.: Statistical dependency analysis with support vector machines. In: 8th International Workshop on Parsing Technologies, pp. 195–206. ACL Press, Nancy (2003)
6. Nivre, J., Nilsson, J.: Three algorithms for deterministic dependency parsing. Comput. Linguist. **34**(4), 513–553 (2003)
7. Andor, D., et al.: Globally Normalized transition-based neural networks. In: 54th Annual Meeting of the Association for Computational Linguistics, pp. 2442–2452. ACL Press, Berlin (2016)
8. Tian, J., Zhang, T., Qin, A., Shang, Z., Tang, Y.Y.: Learning the distribution preserving semantic subspace for clustering. IEEE Trans. Image Process. **26**(12), 5950–5965 (2017)
9. Xu, W., Alex, R.: Can artificial neural networks learn language models? In: 6th International Conference on Spoken Language Processing, pp. 202–205. China Military Friendship Publish, Beijing (2000)
10. Bengio, Y., Senecal, J.S.: Adaptive importance sampling to accelerate training of a neural probabilistic language model. IEEE Trans. Neural Netw. **19**(4), 713–722 (2008)
11. Mnih, A., Hinton, G.: Three new graphical models for statistical language modelling. In: 24th International Conference on Machine Learning, pp. 641–648. ACM Press, Corvallis (2007)
12. Mnih, A., Kavukcuoglu, K.: Learning word embeddings efficiently with noise-contrastive estimation. Adv. Neural. Inf. Process. Syst. **2013**, 2265–2273 (2013)
13. Mikolov, T.: Statistical language models based on neural networks. Technical report, Google Mountain View (2012)
14. Collobert, R., Weston, J.: A unified architecture for natural language processing: deep neural networks with multitask learning. In: 25th International Conference on Machine Learning, Helsinki, Finland, pp. 160–167 (2008)
15. Mikolov, T., Chen, K., Corrado, G., Dean, J.: Efficient estimation of word representations in vector space. In: International Conference on Learning Representations, pp. 1–12. Hans Publisher, Scottsdale (2013)
16. Henry, S., Cuffy, C., Mcinnes, B.T.: Vector representations of multi-word terms for semantic relatedness. J. Biomed. Inform. **77**, 111–119 (2018)
17. Jin, P., Zhang, Y., Chen, X., Xia, Y.: Bag-of-embeddings for text classification, In: 25th International Joint Conference on Artificial Intelligence, pp. 2824–2830. AAAI Press, New York (2016)
18. Deng, H., Zhu, X., Li, Q.: sentence similarity calculation based on syntactic structure and modifier. Comput. Eng. **43**(9), 240–244 (2017)
19. Lévy, B.: Robustness and efficiency of geometric programs the Predicate Construction Kit (PCK). Comput. Aided Des. **72**(1), 3–12 (2016)
20. Bin, L.I., Liu, T., Bing, Q., Sheng, L.I.: Chinese sentence similarity computing based on semantic dependency relationship analysis. Appl. Res. Comput. **12**, 15–17 (2003)
21. Liu, W., Liu, P., Yang, Y., Gao, Y., Yi, J.: An attention-based syntax-tree and tree-LSTM model for sentence summarization. Int. J. Perform. Eng. **13**(5), 775–782 (2017)

Transfer Learning

Semi-coupled Transform Learning

Jyoti Maggu$^{(\boxtimes)}$ and Angshul Majumdar

Indraprastha Institute of Information Technology, New Delhi 110020, India
{jyotim, angshul}@iiitd.ac.in

Abstract. This work introduces semi-coupled transform learning. Given training data in two domains (source and target), it learns a transform in each of the domains such that the corresponding coefficients are (linearly) mapped from the source to the target. Since the mapping is in one direction (source to target) but not the other way round, we call it 'semi-coupled'. Our work is the analysis equivalent of (semi) coupled dictionary learning. The proposed technique has been applied in two problems. The first being image super-resolution and the second, cross lingual document retrieval. In both the cases, our proposed transform learning based formulation excels considerably over existing techniques.

Keywords: Dictionary learning · Deep learning · Reconstruction

1 Introduction

There are many problems in image processing and computer vision which can be recast in the framework for transfer learning. For example consider the example of single image super-resolution; the objective is to create a high resolution image from a low-resolution one. There are many signal processing (sparsity) based techniques to solve this problem [1–3]. However in recent times, dictionary learning based approaches are preferred owing to their improved performance. For each of the domains (high resolution and low resolution) two dictionaries are learnt, such that the coefficients of low resolution dictionary can be linearly mapped onto the high resolution dictionary [4–7]. After the training phase, when a new low resolution image is input, it learns the corresponding coefficients from the learnt dictionary; the coefficients in turn are mapped to that of the high resolution version by the learnt linear map. From the thus formed high resolution coefficients, the corresponding high resolution image is synthesized.

This formulation falls under the purview of coupled dictionary learning. Similar approaches have been applied to other problems, e.g. photo sketch synthesis [5, 6], RGB to Depth image matching [8], pose varying face matching [9] and visible (VIS) to near infra-red (NIR) face matching [9]. In photo sketch matching the problem is to match a person's sketch to that of a digital photo – this is usually used in law enforcement. Similar problems exists in the other domains as well – e.g. matching between visible image collected during daylight and that of NIR image collected during night [9].

L. Cheng et al. (Eds.): ICONIP 2018, LNCS 11303, pp. 141–150, 2018.
https://doi.org/10.1007/978-3-030-04182-3_13

Although one finds most applications of coupled dictionary learning in vision problems, it has been used in computational linguistics as well [10]. There in the problem is cross lingual document retrieval. The query is in one language (source) and the problem is to find the documents from the other (target) language.

Dictionary learning is a synthesis approach, i.e. it learns a basis (dictionary) that can regenerate/synthesize the data from the learned coefficients. This work proposes an analysis version of coupled dictionary learning. The analysis version of dictionary learning is new and is dubbed as transform learning [11–14]/analysis sparse coding [15]. It learns an analysis basis/transform so as to generate the coefficients when operated on the data.

Transform learning enjoys certain theoretical advantages over dictionary learning [11, 12]. For the same number of basis (i.e. equal sized dictionary and transform), transform learning can capture significantly more variability in the data (by its ability to represent more number of sub-spaces) compared to its synthesis counterpart – dictionary learning. In machine learning this boils down to the problem of over-fitting and generalizability. Given the fixed volume of data, the number of dictionary atoms required (to capture the variability) will be larger than the number of basis in the transform. This means that a dictionary will have more parameters to learn compared to a transform. Which in turn means, that the dictionary learning approach would overfit with limited amount of training data; the issue will be less pronounced in the learned transform.

Transform learning is a new formulation. So far it has seen limited application in solving inverse problems [11–14]. Some recent studies [15–17] have applied analysis sparse coding for feature generation in machine learning problems. Owing to the relative nascency of transform learning we will discuss it in the next (literature review) section; coupled dictionary learning will also be discussed. The proposed formulation of coupled analysis sparse coding will be described in Sect. 3. The experimental results will be in Sect. 4. The conclusions of this work and further direction of research will be discussed in Sect. 5.

2 Literature Review

2.1 Transform Learning

Transform learning is the analysis version of dictionary learning; it analyses the data by learning a transform/basis to produce coefficients. Mathematically this is expressed as,

$$TX = Z \tag{1}$$

Here T is the transform, X is the data and Z the corresponding coefficients.
The following transform learning formulation was proposed [11, 12] –

$$\min_{T,Z} \|TX - Z\|_F^2 + \lambda \left(\|T\|_F^2 - \log \det T \right) + \mu \|Z\|_1 \tag{2}$$

The factor $-\log \det T$ imposes a full rank on the learned transform; this prevents the degenerate solution ($T = 0, Z = 0$). The additional penalty $\|T\|_F^2$ is to balance scale; without this $-\log \det T$ can keep on increasing producing degenerate results in the other extreme.

In [11, 12], an alternating minimization approach was proposed to solve the transform learning problem. This is given by –

$$Z \leftarrow \min_Z \|TX - Z\|_F^2 + \mu\|Z\|_1 \tag{3a}$$

$$T \leftarrow \min_T \|TX - Z\|_F^2 + \lambda\left(\varepsilon\|T\|_F^2 - \log \det T\right) \tag{3b}$$

Updating the coefficients (3a) is straightforward. It can be updated via one step of soft thresholding. This is expressed as,

$$Z \leftarrow signum(TX) \cdot \max(0, abs(TX) - \mu) \tag{4}$$

Here $\odot$ indicates element-wise product.

In the initial paper on transform learning [11], a non-linear conjugate gradient based technique was proposed to solve the transform update. In the more refined version [12], with some linear algebraic tricks they were able to show that a closed form update exists for the transform.

$$XX^T + \lambda\varepsilon I = LL^T \tag{5a}$$

$$L^{-1}XZ^T = USV^T \tag{5b}$$

$$T = 0.5R\left(S + (S^2 + 2\lambda I)^{1/2}\right)Q^T L^{-1} \tag{5c}$$

The first step is to compute the Cholesky decomposition; the decomposition exists since $XX^T + \lambda\varepsilon I$ is symmetric positive definite. The next step is to compute the full SVD. The final step is the update step. One must notice that L^{-1} is easy to compute since it is a lower triangular matrix. The proof for convergence of such an update algorithm can be found in [13].

There are only a handful of papers on this topic. Theoretical aspects of transform learning are discussed in [11–13]. In [14] it is used to solve inverse problems. Exactly the same formulation has been dubbed as 'analysis sparse coding' when applied to feature generation [15].

2.2 Coupled Dictionary Learning

The idea of coupled dictionary learning was proposed in [4–9]. Let there be two domains – 1 and 2. X_1 and X_2 are the training data for the two domains. Coupled dictionary learning trains two dictionaries D_1 and D_2 (along with their coefficients

$Z_1$and Z_2) and linear coupling maps from domain1 to 2 – M_{12} and from 2 to 1 – M_{21}. Mathematically this is expressed as,

$$\min_{D_1,Z_1,D_2,Z_2,M_{12},M_{21}} \|X_1 - D_1 Z_1\|_F^2 + \|X_2 - D_2 Z_2\|_F^2$$
$$\mu\left(\|Z_2 - M_{21}Z_1\|_F^2 + \|Z_1 - M_{12}Z_2\|_F^2\right) + \eta\left(\|Z_1\|_0 + \|Z_2\|_0\right) \qquad (6)$$

Here we have abused the notations slightly; the l_0-norm is defined on the vectorised version of the Z's. Solving (6) may apparently be a daunting task. However when segregated into separate sub-problems, they have well known solutions such as [18].

During testing, say the signal is available in domain 1 and the corresponding signal in domain 2 needs to be generated; such problems can arise in photo sketch synthesis and image super-resolution. The learnt dictionary in domain 1 is used to generate the coefficients.

$$\min_{z_1^{test}}\left\|x_1^{test} - D_1 z_1^{test}\right\|_F^2 + \eta\left\|z_1^{test}\right\|_1 \qquad (7)$$

The generated coefficients of domain 1 are now transformed to domain 2 by the learnt linear map: $\hat{z}_2^{test} = M_{21}z_1^{test}$. For classification problems usually a classifier is trained with the coefficients Z_1 and Z_2. During testing (e.g. RGB to NIR matching), $\hat{z}_2^{test}$ is run through the classifier for domain 2. For synthesis problems (e.g. super-resolution) the high resolution image is synthesized by $\hat{x}_2^{test} = D_2\hat{z}_2^{test}$.

3 Proposed Semi-coupled Transform Learning

Today dictionary learning is a popular representation learning tool. A little analysis shows that for a synthesis dictionary of size $m \times n$, with sparsity (number of non-zero elements in Z) k, the number of sub-spaces is ^nC_k for k-dimensional sub-spaces. For analysis transform learning of size $p \times d$, with co-sparsity l the number of sub-spaces is ^pC_l for sub-spaces of dimension $d - l$. If we assume equal redundancy, i.e. $p = n = 2d$, and equal dimensionality of the sub-space, i.e. $k = d - l$, the number of analysis sub-spaces will be n where as the number of synthesis sub-spaces are $k\log_2(n/k)$ (via Stirling's approximation); usually $n \gg k\log_2(n/k)$. For example with $n = 700$, $l = 300$ *and $k = 50$*, the number analysis sub-spaces are 700 whereas the number of synthesis sub-spaces are only 191.

This analysis means that for a transform and dictionary of same dimensions, an analysis transform is able to capture significantly more variability in the data compared to a synthesis dictionary. In other words, for a fixed training set a smaller sized transform need to be learned compared to a dictionary. From the machine learning perspective, given the limited training data, learning fewer parameters for the transform has less chance of over-fitting than learning a larger number of synthesis dictionary atoms. Hence, for limited training data, as is the case with most annotated document retrieval problems, transform learning can be assumed to yield better generalizability (and hence better results) compared to dictionary learning. Hence, we propose to base our work on coupled transform learning.

This work proposes a semi-coupled formulation; we learn a single directional map (from source to target). Suppose there are domains – 1 and 2. Say X_1 and X_2 are the corresponding training data. Semi-coupled analysis sparse coding learns two transforms T_1 and T_2 (one for each domain) and their corresponding features Z_1 and Z_2, so that the features from one of the domains can be linearly mapped (M) into the other. Semi-coupling is practical. For example, in photo sketch identification, one needs to find the digital photograph from a photo sketch – not the other way round. If there is a need for bi-directionality, e.g. in RGB and NIR matching, we can always learn two semi-coupled transforms from one domain to the other.

Mathematically our formulation is expressed as,

$$\min_{T_1,T_2,Z_1,Z_2,M} \|T_1X_1 - Z_1\|_F^2 + \|T_2X_2 - Z_2\|_F^2 + \mu\|Z_2 - MZ_1\|_F^2$$
$$+ \lambda\left(\varepsilon\|T_1\|_F^2 + \varepsilon\|T_2\|_F^2 - \log \det T_1 - \log \det T_2\right) + \eta\left(\|Z_1\|_1 + \|Z_2\|_1\right) \quad (8)$$

The alternating minimization approach is used for solving (8). It can be segregated into the following sub-problems.

$$P1: \min_{T_1}\|T_1X_1 - Z_1\|_F^2 + \lambda\left(\varepsilon\|T_1\|_F^2 - \log \det T_1\right)$$

$$P2: \min_{T_2}\|T_2X_2 - Z_2\|_F^2 + \lambda\left(\varepsilon\|T_2\|_F^2 - \log \det T_2\right)$$

$$P3: \min_{Z_1}\|T_1X_1 - Z_1\|_F^2 + \mu\|Z_2 - MZ_1\|_F^2 + \eta\|Z_1\|_1$$
$$\equiv \min_{Z_1}\left\|\begin{pmatrix} T_1X_1 \\ \sqrt{\mu}Z_2 \end{pmatrix} - \begin{pmatrix} I \\ \sqrt{\mu}M \end{pmatrix}Z_1\right\|_F^2 + \mu\|Z_1\|_1$$

$$P4: \min_{Z_2}\|T_2X_2 - Z_2\|_F^2 + \mu\|Z_2 - MZ_1\|_F^2 + \eta\|Z_2\|_1$$
$$\equiv \min_{Z_2}\left\|\begin{pmatrix} T_2X_2 \\ \sqrt{\mu}MZ_1 \end{pmatrix} - \begin{pmatrix} I \\ \sqrt{\mu}I \end{pmatrix}Z_2\right\|_F^2 + \eta\|Z_2\|_1$$

$$P5: \min_{M}\|Z_2 - MZ_1\|_F^2$$

Sub-problems P1 and P2 are standard transform updates. We already know how to update them (5a–5c). Sub-problem P3 and P4 are standard updates for sparse transform coefficients (4); they just require one step of soft thresholding. Updating the map is easy since P5 is a simple least square problem. This concludes the training phase.

In fully coupled transform learning, one would have to learn another linear map from domain 2 to domain 1. As mentioned at the onset this is not required in most cases. Even if it is required we can learn another semi-coupled transform from 2 to 1. But learning a fully coupled transform in a single problem means one more variable (linear map) to solve. Given the limited training data, solving more variables/parameters would lead to over-fitting. Hence we consciously avoid such a formulation.

During testing, it can be applicable to two kinds of problems. In the first, one can carry out analysis in the feature domain. For such, the coefficients in domain 2 are used to learn a classifier. During testing the sample is given in domain 1. From which the corresponding feature is generated by sparse coding –

$$z_1^{test} \leftarrow signum(T_1 x_1^{test}) \cdot \max\left(0, abs(T_1 x_1^{test}) - \mu\right) \tag{9}$$

From the features of domain 1, the target domain features are generated by $\hat{z}_2^{test} = M z_1^{test}$. These features are input to the learnt classifier for final results.

There can be a second possibility, where the analysis is carried out not on the transform features (z-domain), but on the samples itself (x-domain). In such a case instead of stopping at $\hat{z}_2^{test}$, one needs to synthesize the corresponding sample. This is done by solving the inverse problem $T_2 \hat{x}_2^{test} = \hat{z}_2^{test}$. Once $\hat{x}_2^{test}$, one can carry out further analysis in the sample domain.

4   Experimental Results

4.1   Image Super-Resolution

For image super-resolution, we train on the CIFAR-100 dataset. These are 32×32 images (HR – high resolution). Our interest is in 4 (2×2)-fold super-resolution. During training, we blur and down-sample the CIFAR images to 16×16 (LR – low resolution). We follow a patch-based technique. The LR images form the source and the HR the corresponding targets. From the LR images we extract 8×8 patches and their corresponding 16×16 patches from the target HR images. On this our proposed semi-coupled transform learning (SCTL) formulation is run.

The training is carried out on the 50K training images of CIFAR-100. The remaining 10K test images are used for validation. The tuned parameter values we obtained are $\lambda = 0.1$, $\varepsilon = 1$, $\mu = 0.5$ and $\gamma = 0.05$.

We have compared our method with the coupled dictionary learning (CDL) formulation [6]. The authors of [6] have compared with other super-resolution techniques [3–5] and have shown to supersede them; they also improve the baseline bicubic interpolation technique. Therefore it is enough to show that our proposed technique (SCTL) yields better results than CDL [6].

We have carried out experiments both on greyscale and RGB images. For RGB, the image was converted to YCbCr space. The super-resolution technique was only applied on the illuminance channel. For the others, simple bicubic interpolation is done. The results are shown in the following Fig. 1. If one concentrates on the sharp edges (for example Lena's nose), one can see that CDL images are blurred compared to our proposed method.

Owing to limitations in space, we cannot show results on the other test images. But the PSNR values for others (including Lena) are shown in Table 1. The results

Fig. 1. Original (left). CDL [6] (mid). Proposed (right)

establish the superiority of our proposed method over CDL [6] and hence over [3–5] (since [6] showed improvement upon them). In all cases we improve upon the state-of-the-art by more than 2 dB – this is a significantly large improvement. To put it in context, [6] improves upon the prior works [3–5] by 1 to 1.5 dB. Here we improve upon the best known [6] by more than 2 dB in every case.

Table 1. PSNR for Super-resolution

Image name		Lena	Barbara	Pepper	Cameraman
Color	CDL [6]	30.79	28.21	29.76	27.86
	Proposed	33.03	30.28	31.81	30.14
Grayscale	CDL [6]	31.27	28.98	30.46	28.70
	Proposed	34.55	31.17	32.68	30.85

One notices that the performance (of both algorithms) for the grayscale image is always better than the color counterpart. This is because in color imaging, only the illuminance channel is properly super-resolved; the other channels are simply extrapolated using bicubic interpolation.

4.2   Cross Lingual Document Retrieval

In this work we follow the exact evaluation protocol outlined in [19]. We test all algorithms on the Europarl data set of documents in English and Spanish, and a set of Wikipedia articles in English and Spanish that contain inter language links between them (i.e., articles that the Wikipedia community have identified as comparable across languages). For the Europarl data set, we use 52,685 documents as training, 11,933 documents as a development set, and 18,415 documents as a final testset. Documents are defined as speeches by a single speaker, as in [20]. For the Wikipedia set, we use 43,380 training documents, 8,675 development documents, and 8,675 final test.

For both corpora, the terms are extracted by word breaking all documents, removing the top 50 most frequent terms and keeping the next 20,000 most frequent terms. No stemming or folding is applied. We assess performance by testing each document in English against all possible documents in Spanish, and vice versa. We measure the Top-1 accuracy (i.e., whether the true comparable is the closest in the test set), and the Mean Reciprocal Rank (MRR) of the true comparable, and report the average performance over the two retrieval directions. Ties are counted as errors.

We have compared our method against Oriented Principal Component Analysis (OPCA) and Coupled Probabilistic Latent Semantic Analysis (CPLSA) – two best performing methods proposed in [19]; and coupled dictionary learning (CDL) [21]. The dimensions for the projections are given in the respective papers. For our problem $\lambda = 0.1$, $\varepsilon = 1$, $\mu = 1$ and $\eta = 0.05$ is used for both semi coupled analysis sparse coding and symmetrically coupled analysis sparse coding. The number of projections used is 300.

The final results are shown in Tables 2 and 3.

Table 2. Comparable document retrieval on Europarl

Algorithm	Accuracy	MRR
OPCA	0.9742	0.9806
CPLSA	0.9716	0.9782
CDL	0.9812	0.9839
Proposed	0.9954	0.9896

Table 3. Comparable document retrieval on Wikipedia

Algorithm	Accuracy	MRR
OPCA	0.7255	0.7734
CPLSA	0.4579	0.5130
CDL	0.7279	0.7742
Proposed	0.7868	0.8002

For these experiments, we use the unpaired t-test with Bonferroni correction to determine the smallest set of algorithms that have statistically significantly better accuracy than the rest. The p-value threshold for significance is chosen to be 0.05.

For the Europarl there is no statistically significant difference between OPCA and CPLSA. CDL is significantly better than them. Our proposed techniques are even better than CDL. There is no statistically significant different between our two algorithms.

For the Wikipedia dataset, OPCA and CDL are statistically similar; both of them are significantly better than CPLSA. Our proposed coupled analysis sparse coding techniques show significant improvement over OPCA and CDL. Even for this dataset there is no statistically significant difference between semi-coupled analysis sparse coding and the symmetrically coupled counterpart.

5 Conclusion

Transform learning is a recently proposed analysis representation learning technique. So far it has been used for solving inverse problems in signal and image processing. Only a handful of short papers have used it for simple feature extraction. This is the first paper that solves a domain adaptation/transfer learning type of complex machine learning problem based on the transform learning approach.

We have showcased our result for two tasks. The first one is a synthesis problem where the task is to super-resolve from a low resolution image. Previously coupled dictionary based techniques have shown significant success in this problem. Our proposed transform learning based formulation improves upon the state-of-the-art.

The second problem is an analysis problem where the task is cross lingual document retrieval. In this task, we have shown that the proposed method surpasses the previous state-of-the-art.

In future, we would like to extend our work to other image processing problems (photo sketch synthesis, cross pose recognition etc.) as well as to non-image domains. Superficially we would be interested in solving cross domain multi-media information retrieval problems.

In very recent times (papers have been accepted but yet to be published) there have been studies on coupled autoencoders [22] and coupled analysis/transform learning [23, 24]. The later studies are related to our proposal. However, they have been used for other tasks that are different from the aim of this paper.

References

1. Kim, K.I., Kwon, Y.: Single-image super-resolution using sparse regression and natural image prior. IEEE Trans. Pattern Anal. Mach. Intell. **32**(6), 1127–1133 (2010)
2. Yang, J., Wright, J., Huang, T.S., Ma, Y.: Image super-resolution via sparse representation. IEEE Trans. Image Process. **19**(11), 2861–2873 (2010)
3. Han, X.-H., Chen, Y.-W.: Sparse representation for image super-resolution. In: Chen, Y.-W., C. Jain, L. (eds.) Subspace Methods for Pattern Recognition in Intelligent Environment. SCI, vol. 552, pp. 123–150. Springer, Heidelberg (2014). https://doi.org/10.1007/978-3-642-54851-2_6
4. Yang, J., Wang, Z., Lin, Z., Cohen, S., Huang, T.: Coupled dictionary training for image super-resolution. IEEE Trans. Image Process. **21**(8), 3467–3478 (2012)
5. Wang, S., Zhang, L., Liang, Y., Pan, Q.: Semi-coupled dictionary learning with applications to image super-resolution and photo-sketch synthesis. In: IEEE Conference on Computer Vision and Pattern Recognition (CVPR), pp. 2216–2223 (2012)
6. Huang, D.A., Frank Wang, Y.C.: Coupled dictionary and feature space learning with applications to cross-domain image synthesis and recognition. In: Proceedings of the IEEE International Conference on Computer Vision, pp. 2496–2503 (2013)

7. Gu, S., Zuo, W., Xie, Q., Meng, D., Feng, X., Zhang, L.: Convolutional sparse coding for image super-resolution. In: Proceedings of the IEEE International Conference on Computer Vision, pp. 1823–1831 (2015)
8. Das, N., Mandal, D., Biswas, S.: Simultaneous semi-coupled dictionary learning for matching RGBD data. In: Proceedings of the IEEE Conference on Computer Vision and Pattern Recognition Workshops, pp. 243–251 (2016)
9. Mudunuri, S.P., Biswas, S.: A coupled discriminative dictionary and transformation learning approach with applications to cross domain matching. Pattern Recogn. Lett. **71**, 38–44 (2016)
10. Mehrotra, R., Agrawal, R., Haider, S.A.: Dictionary based sparse representation for domain adaptation. In: Proceedings of the 21st ACM International Conference on Information and Knowledge Management, pp. 2395–2398 (2012)
11. Ravishankar, S., Bresler, Y.: Learning sparsifying transforms. IEEE Trans. Signal Process. **61**(5), 1072–1086 (2013)
12. Ravishankar, S., Wen, B., Bresler, Y.: Online sparsifying transform learning-Part I: algorithms. J. Sel. Top. Signal Process. **9**, 625–636 (2015)
13. Ravishankar, S., Bresler, Y.: Online sparsifying transform learning-Part II: convergence analysis. IEEE J. Sel. Top. Signal Process. **9**(4), 637–646 (2015)
14. Ravishankar, S., Bresler, Y.: Efficient blind compressed sensing using sparsifying transforms with convergence guarantees and application to MRI. SIAM J. Imaging Sci. **8**(4), 2519–2557 (2015)
15. Shekhar, S., Patel, V.M., Chellappa, R.: Analysis sparse coding models for image-based classification. In: IEEE International Conference on Image Processing (ICIP), pp. 5207–5211 (2014)
16. Maggu, J., Majumdar, A.: Alternate formulation for transform learning. In: Proceedings of the Tenth Indian Conference on Computer Vision, Graphics and Image Processing, p. 50. ACM (2016)
17. Maggu, J., Majumdar, A.: Robust transform learning. In: IEEE ICASSP, pp. 1467–1471 (2017)
18. Aharon, M., Elad, M., Bruckstein, A.: K-SVD: an algorithm for designing overcomplete dictionaries for sparse representation. IEEE Trans. Signal Process. **54**(11), 4311 (2006)
19. Platt, J.C., Toutanova, K.: Association for computational linguistics. In: Conference on Empirical Methods in Natural Language Processing, pp. 51–261 (2011)
20. Mimno, D., Wallach, H.M., Naradowsky, J., Smith, D.A., McCallum, A.: Polylingual topic models. In: Proceedings of the Conference on Empirical Methods in Natural Language Processing, pp. 880–889 (2009)
21. Mehrotra, R., Chu, D., Haider, S.A., Kakadiaris, I.A.: Towards Learning Coupled Representations for Cross-Lingual Information Retrieval
22. Gupta, K., Bhowmick, B., Majumdar, A.: Motion blur removal via coupled autoencoder. In: IEEE International Conference on Image Processing (ICIP), pp. 480–484 (2017)
23. Gupta, K., Bhowmick, B., Majumdar, A.: Coupled analysis dictionary learning to inductively learn inversion: application to real-time reconstruction of biomedical signals. In: IEEE IJCNN 2018 (accepted)
24. Nagpal, S., Singh, M., Singh, R., Vatsa, M., Noore, A., Majumdar, A.: Face sketch matching via coupled deep transform learning, vol. 206, pp. 5429–5438 (2017)

A Refined Spatial Transformer Network

Chang Shu[1,2], Xi Chen[2], Chong Yu[2,3], and Hua Han[2,4,5(✉)]

[1] University of Chinese Academy of Sciences, Beijing, China
[2] Institute of Automation, Chinese Academy of Sciences, Beijing, China
{shuchang2015,xi.chen,han.hua}@ia.ac.cn
[3] The Faculty of Mathematics and Statistics, Hubei University, Wuhan, China
yuchong19921021@163.com
[4] CAS Center for Excellence in Brain Science and Intelligence Technology,
Beijing, China
[5] School of Future Technology, University of Chinese Academy of Sciences,
Beijing, China

Abstract. Spatial invariance to geometrically distorted data is of great importance in the vision and learning communities. Spatial transformer network (STN) can solve this problem in a computationally efficient manner. STN is a differentiable module which can be inserted in a standard CNN architecture to achieve spatial transformation of data. STN and its variants can handle global deformation well, but lack the ability to deal with local spatial variation. Hence how to achieve a better manner of spatial transformation within a neural network becomes a pressing matter of the moment. To address this issue, we design a module to estimate the difference between the ground truth and STN output. The difference is measured in the form of motion field. The motion field is utilized to refine the spatial transformation predicted by STN. Experimental results reveal that our method outperforms the state-of-the-art methods in the cluttered MNIST handwritten digits classification task and planar image alignment task.

Keywords: Spatial invariance · Geometrical distortion
Spatial transformer networks · Motion field
Refined spatial transformer network

1 Introduction

Deep learning has achieved great success in the field of computer vision, and has pushed state-of-the-art results forward. Deep learning encounters a lot of opportunities but also faces many challenges at the same time. One of the challenges is how to make neural networks spatially invariant.

In order to reduce the influence of geometric distortion, one attempt is to design spatially invariant representation. Handcrafted features like SIFT [17], SURF [1], BRISK [14] or features extracted by CNN [6,11,21,22] may obtain spatial invariance in some degree. However, they may not cover critical features

© Springer Nature Switzerland AG 2018
L. Cheng et al. (Eds.): ICONIP 2018, LNCS 11303, pp. 151–161, 2018.
https://doi.org/10.1007/978-3-030-04182-3_14

we actually need. What's more, their emphases are different making it difficult to integrate them in an unified framework. Deep learning incorporates max-pooling layer to get spatial invariance by merely outputting the maximum from a sub-region. Receptive field of max-pooling layer is relatively small, neural networks won't achieve spatial invariance unless equipped with plenty of it. However, an excess of max-pooling may lose some crucial details of intermediate feature maps.

Data augmentation can increase networks' tolerance on geometric distortion through doing spatial transformation to input data. This approach trades off a sharp increase in the amount of training data for limited spatial invariance.

These methods don't get to the root of the problem. Spatial Transformer Network (STN) [10] uses sampling to warp image, as shown in the Fig. 1. This approach is differentiable making it possible to be integrated into a neural network. STN proposes a brand new way to offer spatial invariance by achieving spatial transformation within a neural network. In this case, the network will be able to transform input image to desired pose and shape to avoid negative effects caused by spatial position variation and geometric distortion.

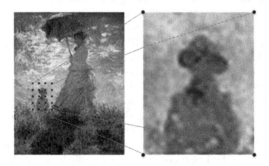

Fig. 1. An illustration of zoom-in transformation achieved by sampling. Blue lines indicate mappings. Black points represent sampling grid. (Color figure online)

The transformation estimated by a STN or its variants is parametric. In practice, it is unlikely to eliminate all the spatial variations of the input data with parametric transformations. To address this issue, we design a module to estimate the difference between the ground truth and STN output. The difference is measured in the form of motion field. The motion field is then used to refine the spatial transformation predicted by STN. The motion field is non-parametric making it more appropriate to represent nonlinear deformation in real scenes. In order to control the smoothness and non-linearity of the motion field, bending energy penalty and smoothness penalty are introduced as regular terms. Finally, we verify our method's effectiveness on the tasks of cluttered MNIST handwritten digits classification and planar image alignment, our method outperforms state-of-the-art methods.

Our main contributions of this work can be summarized as follows:

- We propose a novel neural network architecture which can achieve spatial transformation of data.
- We use bending energy penalty and smoothness penalty to control smoothness and non-linearity of estimated motion field.
- We experimentally demonstrate that refined-STN's spatial transformation ability is much more powerful than STN and its variants.

2 Related Work

2.1 Spatial Transformer Networks

In the field of image classification where deep learning is widely used, insertion of STN will eliminate partial deformation which would do harm to classification accuracy. And a state-of-the-art performance has been achieved on the CUB bird dataset [23]. Furthermore, many neural networks [2,7,24,25] apply STN to tackle the task of image alignment, since STN can offer an end-to-end training manner to networks. STN adopts affine transformation, projective transformation and thin plate spline transformation for warping. These transformations focus on global deformation of an image, but lack the ability to deal with local distortion of details. Inverse Compositional Spatial Transformer Network (IC-STN) [15] makes improvements on the basis of STN with help of inverse compositional Lucas and Kanada (LK) theory. IC-STN recurrently generates linear transformations using the same module, those linear transformations will work together to warp input image. The composition of a series of linear transformations is still linear, therefore IC-STN still suffers from inadequate spatial transformation capacity.

2.2 Parametric Warping

The parametric transformation used by a STN or its variants can be mathematically represented by this function:

$$\phi(p) = c + \alpha^T p + \sum_i w_i f_i(p) \tag{1}$$

Where p is a certain position index of an image, $\phi(p)$ is its motion we need to estimate. $c + \alpha^T p$ represents a linear transformation, where c is a bias term and α is a weighting coefficient. $\sum_i w_i f_i(p)$ approximates a nonlinear transformation by a linear combination of a series of basis functions, where f_i is a nonlinear basis function and w_i is its weighting coefficient. A basis function is a particular basis for a function space. Every continuous function in the function space can be represented as a linear combination of basis functions. Different forms of basis function f_i derive different image deformation methods. When f_i is an "U function", which is $(p - p_i)^2 \log (p - p_i)^2$, the corresponding approach is called thin plate spline [3]. When f_i is a "B function", which is $C_n^i (1 - p)^{n-i} p^i$, the corresponding approach is called B spline [8]. Different function spaces spanned by different basis functions are applied to approximate nonlinear deformation function. However, in practice, the deformation within an image is beyond fitting capability of these parametric transformations.

3   Refined Spatial Transformer Network

3.1   Proposed Method

To address above issues, we design a module to approximate the difference between the ground truth and STN output. The difference is measured in the form of motion field. This motion field is used to refine the spatial transformation predicted by STN. The structure of our proposed network is shown in Fig. 2.

Fig. 2. The pipeline of our refined-STN method. We get a base transformation from base transformation estimator (BTE) and a motion field from refiner. The motion field is then used to refine the base transformation. The refined transformation is applied to get the warped image.

Refined-STN can be roughly divided into 4 modules: base transformation estimator, converter, sampler and refiner. The first three modules mimic a standard STN. Base transformation estimator takes source image as input and outputs base transformation parameters. Converter turns input base transformation parameters into corresponding motion field, then sampler uses it to warp source image. Transformed source image is sent into refiner. Refiner produces a motion field which is added to the base transformation to get refined spatial transformation. The refined spatial transformation is utilized to warp the source image to get the final result. These 4 modules will be explained in detail in the following subsections.

Base Transformation Estimator. Base transformation estimator is a regression network which takes source image as input, and outputs base transformation parameters. Here the base transformation we use is affine transformation. Information is extracted from input image by convolution and max-pooling operation. Through fully connected layer it turns into 6 parameters of affine transformation.

Converter. Converter converts obtained affine transformation parameters into corresponding motion field by using the following equation:

$$
\begin{bmatrix} u \\ v \end{bmatrix} = \begin{bmatrix} a-1 & b & c \\ d & e-1 & f \end{bmatrix} \begin{bmatrix} x \\ y \\ 1 \end{bmatrix}
\tag{2}
$$

Where, $a \sim f$ are 6 parameters of an affine transformation, (x, y) is a coordinate in the image, (u, v) is the corresponding motion.

Sampler. Sampler does transformation to the input images according to the input motion field, it applies sampling to achieve image deformation like STN [10]. We adopt bilinear interpolation to carry out sampling:

$$T(\mathbf{p}) = \sum_{i=1}^{4} I(\widehat{\mathbf{p}}_i) \, |(1,1) - |\widehat{\mathbf{p}}_i - \mathbf{p} - \mathbf{w}||_2^2 \tag{3}$$

Where T is the output image, I is the input image, $\mathbf{p}$ and $\widehat{\mathbf{p}}_i$ are integer coordinates on the image, and $\mathbf{w}$ is the motion of $\mathbf{p}$, $\widehat{\mathbf{p}}_i$ is a 4-pixel neighbor (top-left, top-right, bottom-left, bottom-right) of $\mathbf{p}+\mathbf{w}$.

We achieve bilinear sampling by backward mapping to avoid boundary effect [15]. Rather than mapping from the input image to the output image, we do the opposite. Every pixel in the output image will be iteratively mapped into the input image to find its value.

Refiner. Refiner takes transformed source image as input and outputs motion field for refined transformation. Its structure is analogous to U-net [19], which includes contraction operation and expansion operation to capture critic information and skip-connection to make the most of feature maps from previous layers. What differs from the U-net is that deconvolutional layers fill in for up-sampling layers. In addition, we introduce bending energy penalty [9] and smoothness penalty respectively from approximation theory and optical flow theory [4,16]. We add them as regularization terms in the last layer of this module to control the output motion field $\mathbf{w}$:

$$\ell = \alpha \, |\Delta \mathbf{w}|_2^2 + \beta \, |\nabla \mathbf{w}|_2^2 \tag{4}$$

The first term is the bending energy penalty which penalizes only nonlinear transformation since it gets zero for any linear transformation. The second term is the smoothness penalty aiming at making motion field smoother.

They will be incorporated in the loss function that refined-STN optimizes.

3.2 Mathematical Representation of Refined-STN Pipeline

Our method can be mathematically formulated as follows. The base transformation estimator (BTE) regresses to an base transformation (affine transformation) for each input image.

$$\begin{bmatrix} a\ b\ c \\ d\ e\ f \end{bmatrix} = BTE(\mathbf{I_{in}}) \tag{5}$$

The affine transformation (the first term in the right-hand side of Eq. (6)) is then turned into corresponding motion field (the left-hand side of Eq. (6)) by converter. The second term in the right-hand side of Eq. (6) contains all the coordinates in the output image.

$$\begin{bmatrix} u_1\ u_2\ \cdots \\ v_1\ v_2\ \cdots \end{bmatrix} = \begin{bmatrix} a-1 & b & c \\ d & e-1 & f \end{bmatrix} \begin{bmatrix} x_1\ x_2\ \cdots \\ y_1\ y_2\ \cdots \\ 1\ \ 1\ \ \cdots \end{bmatrix} \tag{6}$$

Sampler transforms the input image according to the motion field.

$$\mathbf{I_B} = sampler(\mathbf{I_{in}}, \begin{bmatrix} u_1 \ u_2 \ \cdots \\ v_1 \ v_2 \ \cdots \end{bmatrix}) \tag{7}$$

Transformed image ($\mathbf{I_B}$) is then brought into refiner to regress to another motion field.

$$\begin{bmatrix} u_1^* \ u_2^* \ \cdots \\ v_1^* \ v_2^* \ \cdots \end{bmatrix} = refiner(\mathbf{I_B}) \tag{8}$$

Newly generated motion field is added to the original motion field to get refined motion field. Finally, the refined motion field is used to transform the input image into final position and shape.

$$\mathbf{I_R} = sampler(\mathbf{I_{in}}, \begin{bmatrix} u_1 + u_1^* \ u_2 + u_2^* \ \cdots \\ v_1 + v_1^* \ v_2 + v_2^* \ \cdots \end{bmatrix}) \tag{9}$$

The process from $\mathbf{I_{in}}$ to $\mathbf{I_B}$ mimics the pipeline of an affine-STN. After base transformation, most geometric distortion is eliminated, but there is still a big gap compared with what we expect. So refiner is designed to capture the gap and fill it. At the cluttered MNIST classification experiment, the accuracy is improved by 2.7% after refining process. At the planar face alignment experiment, the refiner reduces the end-point error by 1.4181. Hence experiments demonstrate that our design greatly improves the original STN's ability to conduct spatial transformation.

4   Experiments

In this section, we will describe implement details of compared methods and our proposed method. MNIST handwritten digits classification and planar image alignment are two classical experiments to test spatial transformation ability of a neural network. We will report experimental results of these two experiments in following subsections.

4.1   Implemental Details

The structure of STN is set as: $[conv(20, (5, 5)) \rightarrow P] \times 3 \rightarrow dense(50) \rightarrow dense(6)$. Where $conv(20, (5, 5))$ denotes a 20-filter 5×5 convolutional layer, P denotes a 2×2 max-pooling layer, D denotes a dropout layer whose rate is 0.5, and $dense(50)$ denotes a 50-unit fully connected layer.

The U-net we use in the refiner is set as follows: The filter sizes of all the convolutional layers are all 3×3. The filter number of each layer begins with 32, will double after each downsampling layer and halve after each deconvolutional layer. 4 downsampling layers and 4 deconvolutional layers are included, they all have a stride of 2. A ReLU nonlinearity is used after each convolutional layer.

The U-net output is brought to a bottleneck layer in the form of $[conv(2, (1, 1))]$ to get a standard motion field. And α and β in (4) are set to 0.01 and 1. We choose

Adam [12] as optimization method, and default parameters are used. Learning rate is 10^{-4}, and it will be reduced by factor of 10 when valid loss stops decreasing. We train the network for 10 K iterations with a batch size of 256.

For classification task, the images are padded with zero turning the image size to 64×64 pixels. In addition, a simple CNN as back-end to get classification results. Its structure is: $[conv(32, (3,3)) \rightarrow P \rightarrow D] \times 3 \rightarrow dense(256) \rightarrow dense(10) \rightarrow softmax$. The training loss is cross entropy.

For planar image alignment task, transformed source image will be directly output as alignment result, therefore the networks are trained in an end-to-end fashion. The ℓ_2 error between alignment result and target image is used as training loss.

4.2 Cluttered MNIST Classification

To prove that refined-STN can offer better spatial invariance within a classification network, we test methods on the cluttered MNIST handwriting digits database[1]. This database is modified from the classical MNIST handwriting digits database [13], it is cluttered with noise. The data set contains 50k image pairs for training, 10k image pairs for validation and 10k image pairs for testing. The image size is 60×60 pixels.

Table 1. Comparison of experimental performances in terms of cross entropy and accuracy.

Method	Cross entropy	Accuracy
CNN	0.1560	0.9508
Affine-STN [10]	0.0954	0.9703
TPS-STN [10]	0.1152	0.9772
C-STN-2 [15]	0.0714	0.9770
C-STN-4 [15]	0.0585	0.9817
IC-STN-2 [15]	0.0813	0.9749
IC-STN-4 [15]	0.0771	0.9766
refined-FTN	**0.0540**	**0.9910**

Performances of all these methods are illustrated in Table 1. Compared with the CNN method which has no spatial transformation ability, the rest of methods with spatial transformation ability get much higher accuracy. TPS-STN achieves higher accuracy than Affine-STN for thin plate spline supports nonlinear transformation. C-STN and IC-STN get higher accuracy than other STN based methods proves that combination of multiple linear transformations can truly improve spatial transformation ability. Our proposed refined-STN method

---

[1] https://s3.amazonaws.com/lasagne/recipes/datasets/mnist_cluttered_60x60_6distortions.npz.

achieves best result, because it can support more complicated transformation. Our method can be simply regraded as a combination of affine-STN and refiner. So we can see that the accuracy is improved by 2.7% after refining process.

4.3 Planar Face Alignment

In order to exhibit our method's powerful ability to transform an image, we apply it in the alignment of the images before and after warping. We use images from a human face database published by [18], and the image size is 64×64 pixels. We warp them with random linear transformation and elastic deformation, as shown in the first two rows in Fig. 3. The parameters setting of the elastic deformation

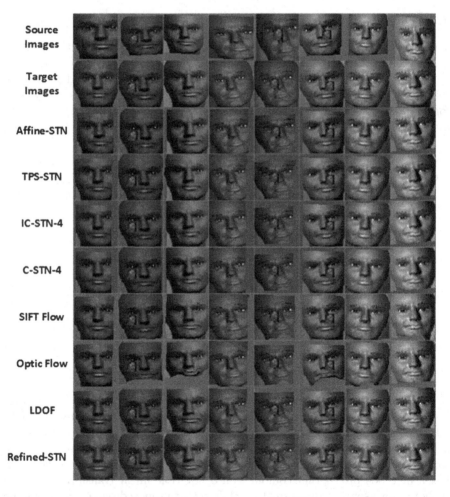

Fig. 3. An illustration of performance on human face alignment. Source images are perturbed from target images. Methods aim to deform source images to align with target images.

follows [20]. Training set contains 300k image pairs and testing set contains 60k image pairs, and 20 percent of training set is split for validation.

The same with previous experiment, Affine-STN, TPS-STN, C-STN and IC-STN are used as baselines in this experiment. Moreover, three optical flow based alignment methods are also introduced as baselines: (1) **Optic Flow** [4] is a classical variational optical flow method which turns optical flow equation into a linear system by using variational approach. (2) **SIFT Flow** [16] is derived from optical flow methods using SIFT feature rather than brightness to evaluate alignment accuracy. (3) **LDOF** [5] improves on **Optic Flow**, it also turns optical flow equation into a linear system by using variational approach, additionally it offers better initial estimation when solving the linear system. We use endpoint error (EPE) to measure alignment accuracy, which is the Euclidean distance between two images, averaged over all pixels. In order to avoid interference of background, we only evaluate accuracy on the central part of each image with the size of 40×40 pixels. Performances of all these methods are shown in the Fig. 3. Besides we give quantized evaluation in the Table 2.

Table 2. Comparison of experimental performances in terms of EPE and time.

Method	EPE	Time
Affine-STN [10]	6.9781	**0.0019s**
TPS-STN [10]	6.4329	0.0031s
IC-STN-4 [15]	6.3330	0.0032s
C-STN-4 [15]	6.3193	0.0031s
SIFT Flow [16]	8.1266	0.5783s
Optic Flow [4]	6.1780	0.0994s
LDOF [5]	5.9467	0.2572s
refined-STN	**5.5600**	0.0034s

In terms of speed, neural network based methods are several orders of magnitude faster than other methods. Optic flow method, SIFT flow method and LDOF method all adopt an iterative scheme to deal with large displacement; nevertheless, deep learning methods directly get transformation parameters through convolution, consequently they take much less time. Although refined-STN has more complicated structure, it gets similar speed with TPS-STN, C-STN and IC-STN. Affine-STN is faster than other STN based methods, but it sacrifices a lot of accuracy.

In terms of accuracy, we can see from Table 2 that SIFT flow gets largest error since this discrete-matching based method can't achieve sub-pixel accuracy. Compared to STN based methods, optical flow based methods achieve better accuracy since they support nonlinear transformation by estimating optical flow field. We can see from the row 8, column 3 in the Fig. 3 that when encountering large displacement, optic flow method collapses. LDOF method can deal with

large displacement since it offers good initial estimation of optical flow field
before solving optical flow equation. Our refined-STN method outperforms the
rest of methods, because refiner makes our result more detailed (we can see
from Fig. 3 that refined-STN get smoother lip than others, showing that it can
handle local spatial variation well). Our method can be simply regraded as a
combination of affine-STN and refiner. So we can see that the refiner reduces
the end-point error by 1.4181.

5   Conclusion

In this paper, we design a module in convolutional neural network to refine
output of a standard STN to achieve better spatial transformation. Experiments demonstrate that our design greatly improves the original STN's ability
to conduct spatial transformation. Our proposed refined-STN method achieves
superior performance in the experiments of cluttered MNIST handwritten digits
classification and planar image alignment.

In the field of classification, stretching and distortion are common factors
that influence neural networks' classification efficacy. Addition of a layer with
spatial transformation ability is direct and effective way to deal with this issue.

Before refined-STN, STN based methods can't exceed optical flow based
methods, because their spatial transformation ability is limited. However,
refined-STN method outperforms optical flow based methods in terms of both
accuracy and speed. And refined-STN achieves similar speed and much better
accuracy compared with other STN based methods.

Acknowledgments. This paper is supported by National Science Foundation of
China (No. 61673381, 61201050, 61701497), Scientific Instrument Developing Project
of Chinese Academy of Sciences (No. YZ201671), Bureau of International Cooperation,
CAS (No. 153D31KYSB20170059), and Special Program of Beijing Municipal Science
& Technology Commission (No. Z161100000216146).

References

1. Bay, H., Ess, A., Tuytelaars, T., Van Gool, L.: Speeded-up robust features. Comput. Vis. Image Underst. **110**(3), 404–417 (2008)
2. Bhagavatula, C., Zhu, C., Luu, K., Savvides, M.: Faster than real-time facial alignment: a 3d spatial transformer network approach in unconstrained poses. In: The
   IEEE International Conference on Computer Vision, vol. 2, p. 7 (2017)
3. Bookstein, F.L.: Principal warps: thin-plate splines and the decomposition of defor-
   mations. IEEE Trans. Pattern Anal. Mach. Intell. **11**(6), 567–585 (1989)
4. Brox, T., Bruhn, A., Papenberg, N., Weickert, J.: High accuracy optical flow estimation based on a theory for warping. In: Pajdla, T., Matas, J. (eds.) ECCV 2004.
   LNCS, vol. 3024, pp. 25–36. Springer, Heidelberg (2004). https://doi.org/10.1007/
   978-3-540-24673-2_3
5. Brox, T., Malik, J.: Large displacement optical flow: descriptor matching in variational motion estimation. IEEE Trans. Pattern Anal. Mach. Intell. **33**(3), 500
   (2011)

6. Bruna, J., Mallat, S.: Invariant scattering convolution networks. IEEE Trans. Pattern Anal. Mach. Intell. **35**(8), 1872–1886 (2013)
7. Chang, C.H., Chou, C.N., Chang, E.Y.: CLKN: Cascaded Lucas-Kanade Networks for image alignment. In: The IEEE Conference on Computer Vision and Pattern Recognition (2017)
8. De Boor, C.: A Practical Guide to Splines. Springer, Heidelberg (1978)
9. Galway, L.: Spline Models for Observational Data. Siam, Philadelphia (1990)
10. Jaderberg, M., Simonyan, K., Zisserman, A., et al.: Spatial transformer networks. In: Advances in Neural Information Processing Systems. pp. 2017–2025 (2015)
11. Kanazawa, A., Sharma, A., Jacobs, D.: Locally scale-invariant convolutional neural networks. Comput. Sci. (2014)
12. Kingma, D., Ba, J.: Adam: a method for stochastic optimization. Comput. Sci. (2014)
13. Lecun, Y., Cortes, C.: The MNIST database of handwritten digits (2010)
14. Leutenegger, S., Chli, M., Siegwart, R.Y.: BRISK: binary robust invariant scalable keypoints. In: The International Conference on Computer Vision, pp. 2548–2555 (2011)
15. Lin, C., Lucey, S.: Inverse compositional spatial transformer networks. In: The IEEE Conference on Computer Vision and Pattern Recognition, pp. 2252–2260 (2017)
16. Liu, C., Yuen, J., Torralba, A.: Sift flow: dense correspondence across scenes and its applications. IEEE Trans. Pattern Anal. Mach. Intell. **33**(5), 978–994 (2011)
17. Lowe, D.G.: Distinctive image features from scale-invariant keypoints. Int. J. Comput. Vis. **60**(2), 91–110 (2004)
18. Peng, Y., Ganesh, A., Wright, J., Xu, W., Ma, Y.: RASL: robust alignment by sparse and low-rank decomposition for linearly correlated images. IEEE Trans. Pattern Anal. Mach. Intell. **34**(11), 2233–2246 (2012)
19. Ronneberger, O., Fischer, P., Brox, T.: U-Net: convolutional networks for biomedical image segmentation. In: Navab, N., Hornegger, J., Wells, W.M., Frangi, A.F. (eds.) MICCAI 2015. LNCS, vol. 9351, pp. 234–241. Springer, Cham (2015). https://doi.org/10.1007/978-3-319-24574-4_28
20. Simard, P.Y., Steinkraus, D., Platt, J.C., et al.: Best practices for convolutional neural networks applied to visual document analysis. In: International Conference on Document Analysis and Recognition, vol. 3, pp. 958–962 (2003)
21. Sohn, K., Lee, H.: Learning invariant representations with local transformations. In: International Conference on Machine Learning, pp. 1339–1346 (2012)
22. Stollenga, M.F., Masci, J., Gomez, F., Schmidhuber, J.: Deep networks with internal selective attention through feedback connections. In: Advances in Neural Information Processing Systems, pp. 3545–3553 (2014)
23. Wah, C., Branson, S., Welinder, P., Perona, P., Belongie, S.: The Caltech-UCSD birds-200-2011 dataset (2011)
24. Wu, W., Kan, M., Liu, X., Yang, Y., Shan, S., Chen, X.: Recursive spatial transformer (rest) for alignment-free face recognition. In: The IEEE International Conference on Computer Vision, pp. 3792–3800 (2017)
25. Zhang, H., He, X.: Deep free-form deformation network for object-mask registration. In: The IEEE Conference on Computer Vision and Pattern Recognition, pp. 4261–4269 (2017)

Convolutional Transform Learning

Jyoti Maggu[1(✉)], Emilie Chouzenoux[2,3], Giovanni Chierchia[2],
and Angshul Majumdar[1]

[1] Indraprastha Institute of Information Technology of Delhi, Okhla Industrial Estate,
Delhi, India
{jyotim,angshul}@iiitd.ac.in
[2] LIGM, UMR CNRS 8049, Univ. Paris Est Marne-la-Vallée,
Champs-sur-Marne, France
emilie.chouzenoux@univ-mlv.fr, giovanni.chierchia@esiee.fr
[3] CVN, INRIA Saclay, CentraleSupélec, Univ. Paris Saclay, Gif sur Yvette, France

Abstract. This work proposes a new representation learning technique
called convolutional transform learning. In standard transform learning,
a dense basis is learned that analyses the image to generate the representation from the image. Here, we learn a set of independent convolutional
filters that operate on the images to produce representations (one corresponding to each filter). The major advantage of our proposed approach
is that it is completely unsupervised; unlike CNNs where labeled images
are required for training. Moreover, it relies on a well-sounded minimization technique with established convergence guarantees. We have
compared the proposed method with dictionary learning and transform
learning on standard image classification datasets. Results show that our
method improves over the rest by a considerable margin.

Keywords: Representation learning · Transform learning
Convolutive models · Image classification · Alternating optimization
Proximal approaches

1 Introduction

Learning representations from the data has always been an interesting problem
for the machine learning community. A model is trained from the data to represent it in some other domain, and the learned coefficients in the other domain are
used as features for solving tasks such as classification and reconstruction. There
has been extensive research on learning good representations from data using
well-known techniques like auto-encoders [1–3], convolutional neural networks
(CNN) [4,5], dictionary learning [6–12], and more recently transform learning
[13–20].

The key idea behind CNN is to reduce drastically the number of connections
to be learned by assuming that only a few learnt convolutional filters are enough

This work was supported by the CNRS-CEFIPRA project under grant NextGenBP
PRC2017.

L. Cheng et al. (Eds.): ICONIP 2018, LNCS 11303, pp. 162–174, 2018.
https://doi.org/10.1007/978-3-030-04182-3_15

to analyse the entire image. This automatically leads to improved generalization performance, and to a reduction of over-fitting effects. Nowadays, the success of CNNs have become so pervasive that in top tier conferences more than half of the papers are based on it. However there are some stark shortcomings. First, CNNs cannot be learned without supervision since they are based on backpropagation. Getting large volumes of labeled data is a challenge in many application fields outside digital imaging, e.g. medical imaging and remote sensing. Secondly, there is no guarantee that the learned filters are mutually different; CNN just initializes them randomly and depends on the non-convergence of backpropagation algorithm to maintain the mutual difference.

In dictionary learning, a dictionary is learned from the data such that it can synthesize the data from the learned coefficients [6,7]. Inspired by the success of CNN models, there has been recently an increased interest for convolutional dictionary learning models, where the sought dictionary is expressed as convolutive operators associated to kernels with various sizes and shapes [10–12]. The field is still nascent and the performance of such techniques have yet to reach those of CNNs.

Transform learning can be viewed as the analysis equivalent of dictionary learning, where a basis (transform) is learned such that it analyzes the data to generate the coefficients [13–15]. Such formulation has been mainly used for the solution of inverse problems arising in image and signal processing; there are only a handful of studies that use it for machine learning tasks. In [8], transform learning (dubbed as analysis sparse coding) was used for unsupervised feature extraction. A later work [9] imposed discriminative penalties on it. In [17], a kernelized version of transform learning has been proposed. Deep versions of transform learning are also getting developed [18,19].

A possible issue with the dictionary learning formulation is its synthesis nature; in neural network terms, this would correspond to a feed-backward neural network. On the other hand, transform learning based techniques are interpretable as a feed-forward neural network. Motivated by this observation, and by the promising results obtained by convolutional models (either based on CNNs or dictionary learning), we introduce in this work a novel transform learning strategy, called convolutional transform learning. To understand our proposal, one needs to rethink transform learning as a neural network. Instead of looking at a transform as a basis, one can think of it as connections from the input (data) to the representation (coefficient). With this interpretation, a conceptual extension to a convolutional formulation is then natural. The learning of filters/weights will be unsupervised which is an additional advantage in contrast to CNN. A spectral barrier penalty will be employed, in order to promote the diversity of the learned filters, expecting improved performance in terms of analysis metrics. Our learning procedure will be sufficiently versatile so that the proposed formulation can easily be extended to any machine learning problem.

In the following sections, we describe the proposed formulation, the associated optimization algorithm, we present experimental results on several image-based datasets, and finally we draw conclusions.

2 Background

In this section, we recall the concepts of dictionary and transform learning.

2.1 Dictionary Learning

Dictionary learning is a popular approach to learn a representation from the data in an unsupervised fashion. From the given input data S, a dictionary/basis D and coefficients/features X are learned in such a way that the data S can be reproduced from the learned dictionary and coefficients. Mathematically, this is represented as

$$S = DX. \tag{1}$$

For learning the sparse representations (D, X), the most popular technique is probably K-SVD [7], which aims to solve the following problem:

$$\underset{D,X}{\text{minimize}} \, \|S - DX\|_F^2 \quad \text{such that} \quad \|X\|_0 \leqslant \tau, \tag{2}$$

with $\tau > 0$ the desired level of sparsity. Other techniques, based on more sophisticated priors can also be used [9–12].

2.2 Transform Learning

Dictionary learning can be seen as the task of inferring a synthesis transform from the data. The dual task of inferring an *analysis* transform from the data is called transform learning. Mathematically, this concept is expressed as $ST \approx X$, where T is the analysis transform, S is the data, and X the corresponding coefficients. For instance, in [13], the following formulation was proposed to estimate the matrices T and X:

$$\underset{T,X}{\text{minimize}} \, \|ST - X\|_F^2 + \lambda(\|T\|_F^2 - \log \det T) + \beta\|X\|_1, \tag{3}$$

with $\lambda > 0$ and $\beta > 0$. Hereabove, the $-\log \det$ term imposes a full rank on the learned transform; this prevents the degenerate solution $T = 0, X = 0$. The additional penalty $\|T\|_F^2$ is to balance scale; without this the $-\log \det$ term can keep on increasing and producing degenerate results in the other extreme. Both of these additional constraints promote the good conditioning of the learned transform. Finally, the term $\|X\|_1$ imposes a sparsity constraint on the learned coefficients.

Transform learning model is expected to be more general than dictionary learning in its notion of compressibility. It also leads to a faster learning scheme as the sparse coding step is simply one step of thresholding as contrast to dictionary learning, where the sparse coding step typically involves the inversion of a linear system. Our proposal is to extend the above formulation to the case when matrix T encodes a convolutive structure, mimicking one layer of CNN.

3 Proposed Approach

We now introduce our formulation of convolutional transform learning in Sect. 3.1, the associated optimization algorithm in Sect. 3.2, and the mathematical derivations in Sects. 3.3–3.4.

3.1 Convolutional Transform Learning

Let us consider a dataset $\left\{s^{(k)}\right\}_{1\leqslant k\leqslant K}$ with K entries in $\mathbb{R}^N$. Our convolutional transform learning formulation relies on the key assumption that matrix T gathers a set of M kernels $t_1,\ldots,t_M$ with M entries, i.e.

$$T = [t_1 \mid \cdots \mid t_M] \in \mathbb{R}^{M\times M}. \tag{4}$$

The proposed model then reads:

$$(\forall k \in \{1,\cdots,K\}) \quad S^{(k)}T \approx X_k. \tag{5}$$

Hereabove, $\left(S^{(k)}\right)_{1\leqslant k\leqslant K} \in \mathbb{R}^{N\times M}$ are Toeplitz matrices associated to $(s^{(k)})_{1\leqslant k\leqslant K}$ such that:

$$(\forall k \in \{1,\ldots,K\})\, S^{(k)}T = \left[S^{(k)}t_1 \mid \cdots \mid S^{(k)}t_M\right]$$
$$= \left[t_1 * s^{(k)} \mid \cdots \mid t_M * s^{(k)}\right] \tag{6}$$

where $*$ is a discrete convolution operator with suitable padding, and

$$(\forall k \in \{1,\ldots,K\}) \quad X_k = \left[x_1^{(k)} \mid \cdots \mid x_M^{(k)}\right], \tag{7}$$

contains the coefficients associated to each entry $k \in \{1,\ldots,K\}$ of the dataset.

Let us denote:

$$X = [X_1^\top \mid \cdots \mid X_K^\top]^\top \in \mathbb{R}^{NK\times M}. \tag{8}$$

The goal is then to estimate (T,X) from $\left\{s^{(k)}\right\}_{1\leqslant k\leqslant K}$. To this aim, we propose to solve the following optimization problem generalizing (3) to our convolutional learning framework:

$$\underset{T\in\mathbb{R}^{M\times M},X\in\mathbb{R}^{NK\times M}}{\text{minimize}} F(T,X) \tag{9}$$

where the objective function F is defined, for every $T \in \mathbb{R}^{M\times M}$ and every $X \in \mathbb{R}^{NK\times M}$ as:

$$F(T, X) = \frac{1}{2} \sum_{m=1}^{M} \sum_{k=1}^{K} \|t_m * s^{(k)} - x_m^{(k)}\|_2^2$$

$$+ \sum_{m=1}^{M} \sum_{k=1}^{K} \left(\beta \|x_m^{(k)}\|_1 + \iota_{[0,+\infty[}(x_m^{(k)}) \right)$$

$$+ \mu \|T\|_F^2 - \lambda \log \det T \tag{10}$$

$$= \frac{1}{2} \sum_{k=1}^{K} \|S^{(k)} T - X_k\|_F^2 + \mu \|T\|_F^2$$

$$- \lambda \log \det T + \beta \|X\|_1 + \iota_{[0,+\infty[^{NK \times M}}(X). \tag{11}$$

Hereabove, function $\iota_{[0,+\infty[}$ denotes the indicator function of the positive orthant, equals to 0 for nonnegative entries, $+\infty$ elsewhere. Moreover, $(\lambda, \mu, \beta) \in]0, +\infty[^3$ are regularization parameters.

3.2   Optimization Algorithm

The resolution of Problem (9) requires an efficient algorithm for dealing with nonsmooth functions and hard constraints. In the optimization literature, proximal algorithms constitute one of the most efficient approaches to tackle such problems [22–24]. The key tool in those methods is the proximity operator [25,26] of a proper, lower semi-continuous, convex function $\psi : \mathbb{R}^N \mapsto]-\infty, +\infty]$ defined as:[1]

$$(\forall \widetilde{x} \in \mathbb{R}^N) \quad \operatorname{prox}_\psi(\widetilde{x}) = \arg \min_{x \in \mathbb{R}^N} \psi(x) + \frac{1}{2} \|x - \widetilde{x}\|^2. \tag{12}$$

Problem (9) fits nicely into the framework provided by the alternating proximal algorithm from [24,27]. For any initialization $T^{[0]} \in \mathbb{R}^{M \times M}$ and $X^{[0]} \in \mathbb{R}^{NK \times M}$, its iterations are as follows:

$$\begin{array}{l} \text{For} \quad n = 0, 1, \ldots \\ \left\lfloor \begin{array}{l} T^{[n+1]} = \operatorname{prox}_{\gamma_1 F(\cdot, X^{[n]})} \left(T^{[n]} \right) \\ X^{[n+1]} = \operatorname{prox}_{\gamma_2 F(T^{[n+1]}, \cdot)} \left(X^{[n]} \right) \end{array} \right. \end{array} \tag{13}$$

where γ_1 and γ_2 are some positive constants. The convergence of sequence $(T^{(n)}, X^{(n)})_{n \in \mathbb{N}}$ to a minimizer of F is guaranteed, as a consequence of the convergence properties of the proximal regularization of Gauss-Seidel method algorithm established in [24]. In the remaining of this section, we show that the updates on both variables T and X have closed form expressions, and thus can be computed with high precision in an efficient manner.

---

[1] See also http://proximity-operator.net/.

3.3   Update of T

Let $n \in \mathbb{N}$. Then, by definition,

$$T^{[n+1]} = \text{prox}_{\gamma_1 F(\cdot, X^{[n]})}\left(T^{[n]}\right) \tag{14}$$

$$= \text{argmin}_{T \in \mathbb{R}^{M \times M}} \frac{1}{2} \sum_{k=1}^{K} \|S^{(k)}T - X_k^{[n]}\|_F^2$$

$$+ \mu\|T\|_F^2 - \lambda \log \det T + \frac{1}{2\gamma_1}\|T - T^{[n]}\|_F^2. \tag{15}$$

Using [28], we deduce that:

$$T^{[n+1]} = \frac{1}{2}\Lambda^{-1/2}V\left(\Sigma + (\Sigma^2 + 2\lambda I_M)^{1/2}\right)U^\top, \tag{16}$$

with

$$\Lambda = \sum_{k=1}^{K}(S^{(k)})^\top S^{(k)} + \gamma_1^{-1}I_M + 2\mu I_M, \tag{17}$$

the singular value decomposition:

$$U\Sigma V^\top = \left(\sum_{k=1}^{K}(X_k^{[n]})^\top S^{(k)} + \gamma_1^{-1}T^{[n]}\right)\Lambda^{-1/2}, \tag{18}$$

and I_M the identity matrix of $\mathbb{R}^M$.

Remark for Rectangular T: Let us emphasize that our approach, and the above update can easily be generalized to the case when matrix T is rectangular, that is $T \in \mathbb{R}^{M_1 \times M_2}$ with non necessarily equality between M_1 and M_2. Then, the penalization term on T should be replaced by:

$$(\forall T \in \mathbb{R}^{M_1 \times M_2}) \quad R(T) = \begin{cases} \mu\|T\|_F^2 - \lambda\sum_{m=1}^{M}\log(\lambda_m) & \text{if } T \in \mathcal{S}_M^{++}, \\ +\infty & \text{otherwise,} \end{cases} \tag{19}$$

with $M = \min(M_1, M_2)$, $(\lambda_m)_{1 \leqslant m \leqslant M}$ are the singular values of T and $\mathcal{S}_M^{++}$ indicates the set of matrices $T \in \mathbb{R}^{M_1 \times M_2}$ with strictly positive singular values (i.e. T has rank equals to M). The gradient of (19) on its definition domain reads:

$$(\forall T \in \mathcal{S}_M^{++}) \quad \nabla R(T) = 2\mu T - \lambda T^\dagger, \tag{20}$$

with $(\cdot)^\dagger$ the pseudo-inverse operation (equivalent to inverse, when $M_1 = M_2 = M$). Using Proposition 24.68 from [21], we can determine the new update for variable T in our algorithm: Let $n \in \mathbb{N}$. Then:

$$T^{[n+1]} = \text{prox}_{\gamma_1 F(\cdot, X^{[n]})}\left(T^{[n]}\right) \tag{21}$$

$$= \text{argmin}_{T \in \mathbb{R}^{M_1 \times M_2}} \frac{1}{2} \sum_{k=1}^{K} \|S^{(k)}T - X_k^{[n]}\|_F^2 + \mu\|T\|_F^2 + \lambda R(T)$$

$$+ \frac{1}{2\gamma_1}\|T - T^{[n]}\|_F^2 \tag{22}$$

$$= \frac{1}{2}\Lambda^{-1}U\text{Diag}\left(\left[\sigma_1 + (\sigma_1^2 + 2\lambda)^{1/2}, \ldots, \sigma_M + (\sigma_M^2 + 2\lambda)^{1/2}, 0, \ldots, 0\right]\right)V^\top \tag{23}$$

with

$$\Lambda^\top\Lambda = \sum_{k=1}^{K}(S^{(k)})^\top S^{(k)} + \gamma_1^{-1}I_{M_1} + 2\mu I_{M_1}, \tag{24}$$

and the singular value decomposition:

$$U\Sigma V^\top = \left(\sum_{k=1}^{K}(X_k^{[n]})^\top S^{(k)} + \gamma_1^{-1}T^{[n]}\right)\Lambda^{-1}, \tag{25}$$

with $U \in \mathbb{R}^{M_1 \times M_1}$, $V \in \mathbb{R}^{M_2 \times M_2}$ orthogonal matrices and

$$\Sigma = \text{Diag}\left([\sigma_1, \ldots, \sigma_M, 0, \ldots, 0]\right).$$

The impact of log-det term in (10) is straight-forward. Such penalty allows to ensure that the kernels are diverse enough to capture good correlations and hence generate good features. Changing the penalty parameter associated to the log-det term has an important impact on the learned kernels. When the kernel size equals the number of its elements (i.e., square case), then a full rank property is enforced on T, and in the limit case when μ tends to infinity, the operator T is such that $T^{-1} = \frac{2\mu}{\lambda}T$.

3.4   Update of X

Let $n \in \mathbb{N}$. Then, using the definition of the proximity operator,

$$X^{[n+1]} = \text{prox}_{\gamma_2 F(T^{[n+1]}, \cdot)}\left(X^{[n]}\right) \tag{26}$$

$$= \text{argmin}_{X \in \mathbb{R}^{KN \times M}} \frac{1}{2}\sum_{k=1}^{K}\|S^{(k)}T^{[n+1]} - X_k\|_F^2$$

$$+ \beta\|X\|_1 + \iota_{[0,+\infty[^{KN \times M}}(X) + \frac{1}{2\gamma_2}\|X - X^{[n]}\|_F^2. \tag{27}$$

By relying on the useful properties of the proximity operator listed in [26], we obtain that, for every $k \in \{1, \ldots, K\}$,

$$X_k^{[n+1]} = \max\left(S_{\frac{\gamma_2\beta}{\gamma_2+1}}\left(\frac{X_k^{[n]} + \gamma_2 S^{(k)}T^{[n+1]}}{\gamma_2+1}\right), 0\right) \tag{28}$$

where $\mathcal{S}_\theta$ denotes the soft thresholding operator with parameter $\theta \geqslant 0$, i.e.:

$$(\forall u \in \mathbb{R}) \quad \mathcal{S}_\theta(u) = \begin{cases} u + \theta & \text{if} \quad u < -\theta \\ 0 & \text{if} \quad u \in [-\theta, \theta] \\ u - \theta & \text{if} \quad u > \theta. \end{cases} \quad (29)$$

4 Numerical Results

To assess the performance of the proposed approach, we considered the following datasets of small-to-medium size, on which we performed feature extraction.

YALE [29]. The Yale dataset contains 165 images of 15 individuals, downscaled to 32-by-32 pixels. There are 11 images per subject, one per different facial expression or configuration. For our experiments, we shuffled all the samples, and took 70% for training and 30% for testing. Moreover, we generated different train/test splits: YALE-2,...,YALE-8. In a YALE-p dataset, p images per subject are kept in train set, and $11 - p$ images are kept in test set. So doing, train set contains $15p$ images and test set contains $15(11 - p)$ images.

E-YALE-B [30]. The Extended Yale B database contains 2432 images with 38 subjects under 64 illumination conditions. Each image is cropped to 192-by-168 pixels, and downscaled to 48-by-42 pixels. For our experiments, we shuffled all the samples, took 70% for training and 30% for testing.

AR-Face [31]. This database contains more than 4000 images of 126 different subjects (70 male and 56 female). The images have various facial expressions, the lighting varies, and some of the images are partially occluded by sunglasses and scarves. For our experiments, we selected 2600 images of 100 individuals (50 males and 50 females), that is 26 different images for each subject. Train set contains 2000 images and 600 images are kept in test set. Each image has 540 features.

4.1 Classification Accuracy

We compared the proposed feature extraction approach (ConvTL – convolutional transform learning) with transform learning (TL) [8] and dictionary learning (DL) [7]. Since our method is unsupervised, it is only fair to compare with other unsupervised representation learning tools. As these are all unsupervised learning methods, we evaluated their performance by feeding the extracted features to a supervised classifier and then computing the classification accuracy. We also performed the classification directly on raw images (Raw). For the classification task, we used two popular techniques: nearest neighbor (NN) and support vector machine (SVM). Our algorithm was ran until convergence (typically 10 iterations are sufficient), with parameters $\gamma_1 = \gamma_2 = 1$. For every tested method, the hyperparameters were cross-validated. The results are reported in Table 1.

We found that the proposed method (ConvTL) yields better results than regular transform learning (TL) for all the considered datasets and classifiers, while being better than dictionary learning (DL) on all the datasets when using nearest neighbor classifier, and on YALE, E-YALE-B, YALE-2, YALE-6, YALE-7, and YALE-8 when using SVM classifier.

To complete our analysis, we also compared to a convolutional neural network (CNN) trained on raw images through a standard supervised classification procedure. We used a custom CNN composed of the following layers: $Conv[64 \times 3 \times 3]$ $\rightarrow ReLU \rightarrow Pool[2 \times 2] \rightarrow BNorm \rightarrow Conv[128 \times 3 \times 3] \rightarrow ReLU \rightarrow Pool[2 \times 2]$ $\rightarrow BNorm \rightarrow Dropout \rightarrow FC[256] \rightarrow ReLU \rightarrow FC[classes] \rightarrow Softmax$.

According to the results reported in Table 1, the proposed ConvTL compares favorably with the CNN. This may be related to the fact that CNNs are known to require large training sets in order to achieve breakthrough performance, whereas the considered datasets are small.

Another important observation is that in most of our experiments on downsampled data, we have observed that SVM outperforms KNN. Intuitively, when we have a limited set of points in many dimensions, SVM tends to be very good because it should be able to find the linear separation that should exist. Moreover, SVM is expected to be robust to outliers since it only uses the most relevant points to find the linear separation (support vectors). In general, if we have large dataset in a low dimensional space then KNN is probably a suitable choice. If we have few points in the dataset, lying in a high dimensional space, then a linear SVM is probably better.

Our classification accuracy is comparable to the one obtained with CNN. It should however be emphasized that the upvote for the proposed methodology is its unsupervised way of learning convolved features in contrast to CNN, where convolved features are learned in a supervised manner.

The learned features by the proposed method are general enough to be used for other image processing tasks by making small changes in the formulation.

4.2   Computational Time

The proposed method is tested on small size images which are downsampled from the original full size images. While the DL and TL methods take one to ten seconds for learning representations, the proposed approach takes around one minute. The difference in terms of computational time is simply related to the fact that, in case of TL and DL, the transform requires a matrix-vector product while in the proposed approach, convolution and deconvolution operations are needed.

4.3   Analysis of the Learned Kernels

A given number M_2 of kernels with $M_1 = M_2^2$ coefficients is learned to ideally represent the dataset. Each kernel t_m is convolved with the image s to generate a different feature vector x_m. The intra-kernel diversity is taken care by the penalties in the proposed formulation.

Table 1. Classification accuracy on benchmark datasets.

	DATASET	Raw	TL	DL	ConvTL
NN classifier	YALE	58.00	68.00	54.00	**70.00**
	E-YALE-B	71.03	72.28	71.72	**84.00**
	AR-Faces	55.00	53.50	54.50	**56.00**
	YALE-2	43.40	49.63	43.70	**51.85**
	YALE-3	49.40	48.33	47.50	**55.83**
	YALE-4	52.38	50.48	44.76	**54.28**
	YALE-5	51.11	53.33	44.44	**54.44**
	YALE-6	53.33	50.67	50.67	**57.33**
	YALE-7	60.20	61.67	53.33	**66.67**
	YALE-8	63.60	57.78	57.78	**71.11**
SVM classifier	YALE	68.00	78.00	80.00	**88.00**
	E-YALE-B	93.24	94.21	95.58	**97.38**
	AR-Faces	87.33	84.33	**97.67**	88.87
	YALE-2	58.52	51.11	58.52	**62.22**
	YALE-3	62.50	60.83	**66.67**	64.17
	YALE-4	60.95	53.33	**64.76**	64.52
	YALE-5	66.67	57.78	**68.89**	66.67
	YALE-6	73.33	61.33	81.33	**82.67**
	YALE-7	80.00	66.67	78.33	**83.33**
	YALE-8	80.00	71.11	80.00	**84.44**
CNN classifier	YALE	84.00	-	-	-
	E-YALE-B	98.60	-	-	-
	AR-Faces	95.50	-	-	-
	YALE-2	62.96	-	-	-
	YALE-3	64.17	-	-	-
	YALE-4	67.60	-	-	-
	YALE-5	74.44	-	-	-
	YALE-6	76.00	-	-	-
	YALE-7	81.67	-	-	-
	YALE-8	82.22	-	-	-

Figure 1 shows the kernels learned on YALE dataset, for different sizes $M_2 \in \{3, 5, 7, 9\}$. One can observe that the proposed algorithm is capable of learning nontrivial and nonidentical kernels, thanks to the regularization on T present in

(9). In particular, the results reported in Table 1 were obtained by fixing $M_2 = 5$, which corresponds to a good trade-off between model accuracy and complexity.

Since $M_1 > M_2$ here, the estimated T is rectangular and over-complete. The retrieved kernels are distinct from each other, as soon as $\mu > 0$. In contrast, if we had considered a large number of small size kernels (i.e., rectangular case with $M_1 < M_2$), T would have been under-complete and the number of distinct kernels would be equals to the smallest dimension of T, that is M_1; the others being some linear combination of each other. The results for this scenario are not presented here due to the lack of space.

Note that the initialization of T plays no role in the learning process, since the optimization problem in (9) is convex.

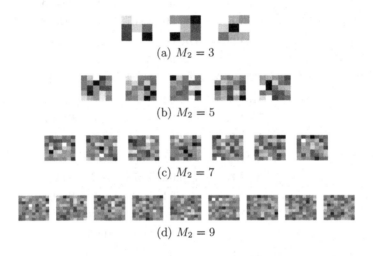

(a) $M_2 = 3$

(b) $M_2 = 5$

(c) $M_2 = 7$

(d) $M_2 = 9$

Fig. 1. Kernels learned on YALE dataset.

5   Conclusion

This paper introduces a novel representation learning technique, named convolutional transform learning. Comparison was performed with the off-the-shelf dictionary learning and transform learning formulations on image classification tasks. In the future, we plan to compare with several other representation learning techniques, namely autoencoder and its convolutional version, restricted Boltzmann machine and its convolutional version discriminative variants of dictionary and transform learning.

Our current formulation relies on an efficient alternating optimization technique with sounded theoretical guarantees. When applied to large scale problems, the approach can nonetheless be quite time consuming, so that, in the future we plan to parallelize portions of the algorithm with the aim to improve its computational efficiency. This will allow us to compare with deeper versions of the aforesaid techniques on larger datasets. The next possible extension to the proposed method could be making it multilayered architecture involving various size and number of kernels in each layer. One can expect the multilayer formulation to scale well with large number of full size images.

References

1. Vincent, P., Larochelle, H., Lajoie, I., Bengio, Y., Manzagol, P.-A.: Stacked denoising autoencoders: learning useful representations in a deep network with a local denoising criterion. J. Mach. Learn. Res. **11**, 3371–3408 (2010)
2. Kingma, D.P., Welling, M.: Auto-encoding variational bayes. In: Proceedings of ICLR, Banff, Canada, April 2014
3. Makhzani, A., Shlens, J., Jaitly, N., Goodfellow, I.: Adversarial autoencoders. In: Proceedings of ICLR, San Juan, Puerto Rico, May 2016
4. LeCun, Y., et al.: Backpropagation applied to handwritten zip code recognition. J. Neural Comput. **1**(4), 541–551 (1989)
5. Lecun, Y., Bottou, L., Bengio, Y., Haffner, P.: Gradient-based learning applied to document recognition. Proc. IEEE **86**, 2278–2324 (1998)
6. Lee, D.D., Seung, H.S.: Learning the parts of objects by non-negative matrix factorization. J. Nature **401**(6755), 788–791 (1999)
7. Aharon, M., Elad, M., Bruckstein, A.: K-SVD: an algorithm for designing overcomplete dictionaries for sparse representation. J. IEEE Trans. Signal Process. **54**(11), 4311–4322 (2006)
8. Shekhar, M., Patel, S., Chellappa, R.: Analysis sparse coding models for image-based classification. In: Proceedings of ICIP, pp. 5207–5211, Paris, France (2014)
9. Guo, J., Guo, Y., Kong, X., Zhang, M., He, R.: Discriminative analysis dictionary learning. In: Proceedings of AAAI, pp. 1617–1623, Phoenix, AZ, USA (2016)
10. Huang, F., Anandkumar, A.: Convolutional dictionary learning through tensor factorization. In: NIPS Workshop: Feature Extraction, pp. 116–129, Montreal, Canada (2015)
11. Garcia-Cardona, C., Wohlberg, B.: Convolutional dictionary learning, Preprint arXiv:1709.02893 (2017)
12. Papyan, V., Romano, Y., Sulam, J., Elad, M.: Convolutional dictionary learning via local processing, Preprint arXiv:1705.03239 (2017)
13. Ravishankar, S., Bresler, Y.: Learning sparsifying transforms. J. IEEE Trans. Signal Process. **61**(5), 1072–1086 (2013)
14. Ravishankar, S., Wen, B., Bresler, Y.: Online sparsifying transform learning - Part I. J. IEEE J. Sel. Topics Signal Process. **9**(4), 625–636 (2015)
15. Ravishankar, S., Bresler, Y.: Online sparsifying transform learning - Part II. J. IEEE J. Sel. Topics Signal Process. **9**(4), 637–646 (2015)
16. Chabiron, O., Malgouyres, F., Tourneret, J.Y., Dobigeon, N.: Toward fast transform learning. Int. J. Comput. Vis. **114**, 195 (2015)
17. Maggu, J., Majumdar, A.: Kernel transform learning. J. Pattern Recognit. Lett. **98**, 117–122 (2017)
18. Maggu, J., Majumdar, A.: Greedy deep transform learning. In: Proceedings of ICIP, Beijing, China (2017)
19. Maggu, J., Majumdar, A.: Unsupervised deep transform learning. In: Proceedings of ICASSP, Calgary, Canada (2018)

20. Fagot, D., Fevotte, C., Wendt, H.: Nonnegative Matrix Factorization with Transform Learning, Preprint arXiv:1705.04193, December 2017
21. Bauschke, H.H., Combettes, P.L.: Convex Analysis and Monotone Operator Theory in Hilbert Spaces, 2nd edn. Springer, New York (2017). https://doi.org/10.1007/978-3-319-48311-5
22. Chouzenoux, E., Pesquet, J.C., Repetti, A.: A block coordinate variable metric forward-backward algorithm. J. Global Optim. **66**(3), 457–485 (2016)
23. Bolte, J., Sabach, S., Teboulle, M.: Proximal alternating linearized minimization for nonconvex and nonsmooth problems. J. Math. Program. **146**(1–2), 459–494 (2014)
24. Attouch, H., Bolte, J., Svaiter, B.F.: Convergence of descent methods for semialgebraic and tame problems: proximal algorithms, forward-backward splitting, and regularized Gauss-Seidel methods. J. Math. Program. **137**(1), 91–129 (2013)
25. Moreau, J.J.: Proximité et dualité dans un espace hilbertien. J. Bull. Soc. Math. France **93**, 273–299 (1965)
26. Combettes, P.L., Pesquet, J.C.: Proximal splitting methods in signal processing. In: Fixed-Point Algorithms for Inverse Problems in Science and Engineering, pp. 185–212. Springer, New York (2010). https://doi.org/10.1007/978-1-4419-9569-8_10
27. Bolte, J., Combettes, P.L., Pesquet, J.C.: Alternating proximal algorithm for blind image recovery. In: Proceedings of ICIP, pp. 1673–1676, Hong Kong, China (2010)
28. Chouzenoux, E., Benfenati, A., Pesquet, J.C.: A proximal approach for a class of matrix optimization problems, Tech. Rep. (2017). http://arxiv.org/abs/1801.07452
29. Bellhumer, P.N., Hespanha, J., Kriegman, D.: Eigenfaces vs. fisherfaces: recognition using class specific linear projection. J. IEEE Trans. Pattern Anal. Mach. Intell. **17**(7), 711–720 (1997)
30. Lee, K.C., Ho, J., Kriegman, D.: Acquiring linear subspaces for face recognition under variable lighting. J. IEEE Trans. Pattern Anal. Mach. Intell. **27**(5), 684–698 (2005)
31. Martinez, M., Benavente, R.: The AR face database, Tech. Rep., CVC 24 (1998)

Delving into Diversity in Substitute Ensembles and Transferability of Adversarial Examples

Jie Hang, KeJi Han, and Yun Li[✉]

School of Computer Science and Technology,
Nanjing University of Posts and Telecommunications, Nanjing, China
liyun@njupt.edu.cn

Abstract. Deep learning (DL) models, e.g., state-of-the-art convolutional neural networks (CNNs), have been widely applied into security-sensitivity tasks, such as facial recognition, automated driving, etc. Then their vulnerability analysis is an emergent topic, especially for black-box attacks, where adversaries do not know the model internal architectures or training parameters. In this paper, two types of ensemble-based black-box attack strategies, *iterative cascade ensemble strategy* and *stack parallel ensemble strategy*, are proposed to explore the vulnerability of DL system and potential factors that contribute to the high-efficiency attacks are examined. Moreover, two pairwise and non-pairwise diversity measures are adopted to explore the relationship between the diversity in substitutes ensembles and transferability of crafted adversarial examples. Experimental results show that proposed ensemble adversarial attack strategies can successfully attack the DL system with ensemble adversarial training defense mechanism and the greater the diversity in substitute ensembles enables stronger transferability.

Keywords: Black-box attack · Vulnerability
Ensemble adversarial attack · Diversity · Transferability

1 Introduction

Deep learning models are often vulnerable to adversarial examples: malicious inputs modified to yield erroneous model outputs, while appearing unmodified to human observers at inference phase [1–4]. Potential attacks include confusing vehicle behavior in automated driving or having malicious content like malware identified as legitimate. Yet, all existing adversarial example attacks require explicit knowledge of the model internals or its training data (white-box). However, to search for adversarial examples of a real world system, such knowledge may not be available. In this situation, the target model is a *black-box* to the attacker. Therefore, it is quite difficult to extract information about the decision boundary of target models, which is usually a pre-requisite to design input perturbations that result in erroneous predictions. However, previous works have shown that *transferability* exists between different models, i.e., the adversarial examples can transfer from one model to another [1, 5–8]. Such a property can be leveraged to perform black-box attacks. In other words, the attacker can query the target system, and establish a *substitute model* based on the query results [9]. Then the attacker can

© Springer Nature Switzerland AG 2018
L. Cheng et al. (Eds.): ICONIP 2018, LNCS 11303, pp. 175–187, 2018.
https://doi.org/10.1007/978-3-030-04182-3_16

generate the adversarial examples for the substitute model, and these adversarial examples may transfer to disorder the target system. For example, an adversary who seeks to penetrate a computer network rarely has access to the specifications of the deployed intrusion detection system, however they can observe its outputs for any chosen inputs [10]. These observed input-output pairs will be used to produce synthetic datasets, and to train a substitute model approximating the target system. Therefore, the adversarial examples generated by substitutes are more likely to transfer to confuse the target system.

However, conventional attack strategies notoriously only consider to train a single substitute to craft adversarial examples with a weak transfer capability in black-box attack scenario, which is easily defended by existed defense mechanism [11–13]. Papernot et al. [14] have proposed ensemble adversarial training technique, which is an extension of adversarial training [1, 15], to increase robustness of DL models against black-box attacks. Thus, new attack strategies should be designed to explore the vulnerability of DL models with ensemble adversarial training.

In this paper, we propose two types of ensemble-based black-box attack strategies, *iterative cascade ensemble strategy* and *stack parallel ensemble strategy*, to implement more powerful black-box attacks against DL models and demonstrate that the ensemble adversarial training does not significantly increase the robustness and security of DL models. Besides, potential factors that contribute to the effective attacks against DL models are examined from three perspectives: the transferability of substitutes, the diversity of substitutes, and the number of substitutes. Ensemble adversarial black-box attack strategies and strategy analysis will be emphatically introduced in Sect. 2. The comparison experiment results on real world data sets and feasibility exploration are reported in Sect. 3 and paper concludes in Sect. 4.

2   Ensemble-Based Black-Box Attack Strategy

Before introducing the attack strategies, we will briefly introduce the architecture of substitutes and transferable adversarial examples generation algorithms used in this paper. For the input $x \in R^D$, the composition of functions modeled by the substitute can be formalized as [16]:

$$F(x) = softmax(f_n(\theta_n, f_{n-1}(\theta_{n-1}, \ldots f_2(\theta_2, f_1(\theta_1, x))))) \tag{1}$$

where each function f_i for $i \in 1 \ldots n$ is modeled by a layer of neurons, each layer is parameterized by a weight vector θ_i impacting each neuron's activation. The output of the last layer is computed by using the softmax function, which ensures that the output vector $F(x)$ satisfies $0 \leq F(x)_i \leq 1$, and $F(x)_1 + \ldots + F(x)_c = 1$, where c is the number of classes.

Transferable adversarial examples are generated by substitute through carefully introducing human indistinguishable perturbations to the original examples, then these generated adversarial examples $x^* \in R^D$ can transfer to confuse target model O, i.e., $O(x^*) \neq O(x)$. Currently proposed adversarial examples generation algorithms mainly include gradient-based (e.g., FGSM [1], I-FGSM [2], R + FGSM [14], etc.) and optimization-based (e.g., Carlini L_∞ Attack [17]), and specific details are described below:

Fast Gradient Sign Method (FGSM) is a single-substitute attack method. It finds the adversarial perturbation that yields the highest increase of the loss function under L_∞-norm. The update equation is

$$x^* = x + \alpha \cdot sign\left(\nabla_x loss\left(1_y, F(x)\right)\right) \tag{2}$$

where α controls the magnitude of adversarial perturbation, 1_y is the one-hot encoding of the ground truth label of y. I-FGSM is a straightforward way to extend the FGSM by using a better iterative optimization strategy and R + FGSM significantly increases the power of the FGSM by adding gaussian noise to inputs before computing the gradient.

Carlini L_∞ Attack is a stronger single-substitute attack method proposed recently. It finds the adversarial perturbation r by using an auxiliary ω as

$$r = \frac{1}{2}\left(\tanh(\omega) + 1\right) - x \tag{3}$$

Then the loss function optimizes the auxiliary variable ω_n

$$\min_\omega \left\| \frac{1}{2}(\tanh(\omega) + 1) \right\| + c \cdot f\left(\frac{1}{2}(\tanh(\omega) + 1)\right) \tag{4}$$

The function $f(\cdot)$ is defined as

$$f(x) = \max\left(Z(x)_{1_y}, -\max\{Z(x)_i : i \neq 1_y\}, -\kappa\right) \tag{5}$$

where $Z(x)_i$ is the logits output for class i, and κ controls the confidence gap between the adversarial class and true class.

Yet, these single-substitute attack algorithms achieve unsatisfactory attack performance in black-box attack scenario. Then, we attempt to ensemble multiple pre-trained substitutes to produce adversarial examples with more powerful transferability in the form of iterative cascade ensemble and stack parallel ensemble, as illustrated in Fig. 1.

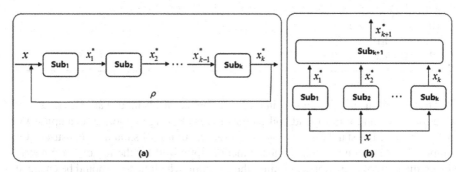

Fig. 1. Illustration of Iterative Cascade Ensemble Strategy (a) and Stack Parallel Ensemble Strategy (b).

2.1   Iterative Cascade Ensemble Strategy

Iterative cascade ensemble strategy employs a cascade structure, as shown in Fig. 1(a), where each substitute of cascade will receive adversarial examples $x_j^*(j \in [0, k])$ generated by its preceding substitute, and output its counterparts to the next substitute. During each iteration, the output of the k-th substitute x_k^* will be used as the input to the first substitute. Output results obtained from the k-th substitute after ρ iterations are final adversarial examples. Before implementing the iterative cascade ensemble strategy, the adversary first requires to train k heterogeneous substitute models with various synthetic datasets, which are constructed by observed input-output pairs and their augmentation with Jacobian-based technique [9]. In order to obtain more effective adversarial examples, each substitute is trained based on various architectures of deep neural networks. Afterwards, FGSM or Carlini L_∞ Attack is adopted as a classic attack algorithm for each substitute to craft adversarial examples. Finally, the adversaries can cascade multiple pre-trained substitutes and iteratively maximize each loss of substitute to obtain the final adversarial examples. The iterative cascade attack procedure is outlined in Algorithm 1.

Algorithm 1: Iterative Cascade Ensemble Strategy for Generating Transferable Adversarial Examples —— Iter_Casc

---

Input: normal example x, the ground true label y, a substitute F, the number of substitutes k, perturbation amplitude α, gaussian noise amplitude ε, the maximum iterative epochs ρ

Output: Transferable adversarial example x_k^*

1:    Initialize the value of $\varepsilon, \alpha, k, \rho$
2:    $x_0^* = x + \varepsilon \cdot sign(N(0^D, 1^D))$
3:    **while** $\rho > 0$ **do**
4:    $\quad$ **for** $j = 1$ to k **do**
5:    $\quad\quad$ $loss\left(1_y, F_j(x_{j-1}^*)\right) = -\sum_{t=1}^{c}\left(1_{y_t} \cdot \log F_j(x_{j-1}^*)_t\right)$
6:    $\quad\quad$ $grads = \nabla_{x_{j-1}^*} loss(1_y, F_j(x_{j-1}^*))$
7:    $\quad\quad$ $x_j^* = x_{j-1}^* + \alpha \cdot sign(grads)$ //$L_\infty (L_1$ or L_2) distance metric
8:    $\quad$ **end for**
9:    $\quad$ $x_0^* = x_k^*$
10:   $\quad$ $\rho = \rho - 1$
11:   **end while**
12:   **return** x_k^*

The algorithm first requires to initialize the value of all input variables $\varepsilon, \alpha, k, \rho$ (where $\varepsilon = \alpha/2$ and $k = \rho$), and add gaussian noise to original normal examples. For each substitute, the standard cross entropy loss function [18] should be constructed to compute gradient to maximize the loss function optimized for the L_∞ distance metric. The gradient of loss function determines the direction which feature should be changed. During each iteration, generated adversarial example x_k^* will be assigned to x_0^* as input of the first substitute. Until the loop iteration ends, the final transferable adversarial examples are obtained from the output of k-th substitute.

2.2   Stack Parallel Ensemble Strategy

Stack parallel ensemble strategy employs a parallel structure, as shown in the Fig. 1(b), where each substitute of parallel will receive the original legitimate example x, and output result x_j^* ($j \in [1, k]$) will be combined with a linear way as new input of the $k + 1$ substitute. Output results obtained from the $k + 1$ substitute are final adversarial examples. Before implementing the parallel ensemble strategy, the adversary first requires to train $k + 1$ heterogeneous substitute models with various synthetic datasets, which are constructed by observed input-output pairs and their augmentation with Jacobian-based technique [9]. In order to achieve more effective adversarial examples, each substitute is still trained based on various architectures of deep neural networks. Afterwards, FGSM or Carlini L_∞ Attack is adopted as a classic attack algorithm for each substitute to craft adversarial examples. Finally, the adversary can parallel multiple pre-trained substitutes and maximize each loss of substitute to obtain adversarial examples. The stack parallel attack procedure is outlined in Algorithm 2.

Algorithm 2: Stack Parallel Ensemble Strategy for Generating Transferable Adversarial Examples —— Stack_Paral

---

Input: normal example x, the ground true label y, a substitute F, the number of substitutes $k+1$, perturbation amplitude α, gaussian noise amplitude ε
Output: Transferable adversarial example x_{k+1}^*

1:   Initialize the value of ε, α, k and $x_{mid}^* = 0^D$
2:   $x_0^* = x + \varepsilon \cdot sign(N(0^D, 1^D))$
3:   **for** $j = 1$ to k **do**
4:        $loss(1_y, F_j(x_0^*)) = -\sum_{t=1}^{c}(1_{y_t} \cdot \log F_j(x_0^*)_t)$
5:        $grads = \nabla_{x_0^*} loss(1_y \cdot F_j(x_0^*))$
6:        $x_j^* = x_0^* + \alpha \cdot sign(grads)$ //$L_\infty (L_1$ or $L_2)$ distance metric
7:        $x_{mid}^* = x_{mid}^* + x_j^*$ // sum and save output results of each substitute
8:   **end for**
9:   $x_{mid}^* = x_{mid}^* / k$ // linear combination
10: $loss(1_y, F_{k+1}(x_{mid}^*)) = -\sum_{t=1}^{c}(1_{y_t} \cdot \log F_{k+1}(x_{mid}^*)_t)$
11: $grads = \nabla_{x_{mid}^*} loss(1_y, F_{k+1}(x_{mid}^*))$
12: $x_{k+1}^* = x_{mid}^* + \alpha \cdot sign(grads)$
13: **return** x_{k+1}^*

The algorithm still requires to initialize the value of all input variables ε, α, k, x_{mid}^* (where $\varepsilon = \alpha/2, x_{mid}^* = 0^D$), add gaussian noise to original legitimate examples and compute gradient of constructed loss function. The gradient of loss function determines the direction which feature should be changed. For the top-k substitutes, generated adversarial example x_j^* $(j \in [1, k])$ will be combined with a linear way and save to x_{mid}^* as new input of the $k + 1$ substitute. The final transferable adversarial examples are achieved from the output of $k + 1$ substitute.

2.3 Strategy Analysis

Empirical evidence has shown that adversarial examples appear in wide regions, spanning a contiguous subspace of high dimensionality and a large portion of this space is shared between different models, thus enabling transferability [1, 7, 19]. Ian Goodfellow et al. first proposed Gradient Aligned Adversarial Subspace (GAAS) [7] method to find multiple independent orthogonal adversarial directions to directly evaluate the dimensionality of the adversarial subspace. The dimensionality of adversarial subspaces is relevant to the transferability problem: the higher the dimensionality, the more likely the subspaces of substitute and target model will intersect significantly. As proposed in [7], the decision boundaries learned by both the substitute and target model must be extremely close to each another in adversarial direction. Adversarial direction is defined by x and $x^* : d_{adv} = (x^* - x)/x^* - x_2$, where adversarial example x^* (blue dot) is generated from test example (brown dot) x to be misclassified by substitute $F(x)$: $\mathrm{argmin}_{\varepsilon > 0} F(x^* : x + \varepsilon \cdot d_{adv}) \neq F(x)$, as shown in Fig. 2(a). That is, the cross-boundary distance (the red double-ended arrows) in adversarial direction between the decision boundaries of substitute and target model must be very short. In other words, the shorter the distance, the stronger transferability.

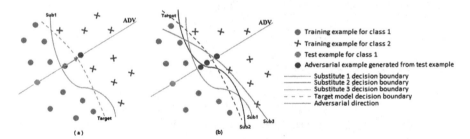

Fig. 2. Illustration of a binary misclassification procedure in the adversarial direction over a 2D input domain. (Color figure online)

Actually, it is difficult to guarantee the trained substitute accurately approximating the target black-box model and the adversarial direction is also not unique, which lead to the weak transferability of crafted adversarial examples. However, if adversarial examples remain adversarial for multiple substitutes, it is more likely to transfer to disorder the target model, as shown in Fig. 2(b). From the Fig. 2(b), we can observe that an adversarial example (blue dot) generated by our proposed ensemble-based black-box attack strategies crossing the decision boundaries of k (e.g. $k = 3$) substitutes, has a greater probability to cross the decision boundary of target model. This fully illustrates the ensemble-based black-box attack strategies effectively shorten the cross-boundary distance and improve the transferability of generated adversarial examples.

3 Experiments

All experiments[1] use Tensorflow[2] framework and cleverhans library[3]. To demonstrate the effectiveness and feasibility of the proposed ensemble-based black-box attack strategy, we empirically compare the conventional single-substitute attack algorithms described previously, e.g., FGSM, I-FGSM, R + FGSM and Carlini L_∞ attack, and expose the potential factors that contribute to the high-efficiency attacks.

3.1 Setup

Four benchmark datasets for two tasks, i.e., digit recognition and traffic sign recognition, are used in experiments. Details about datasets are listed in Table 1. The target classifier as black-box model in this work are trained with training data of each dataset. For each dataset, few unused test examples, as query inputs, are used to query target classifier and produce synthetic datasets augmented by observed input-output pairs. Then, diverse convolutional neural network architectures, as shown in Table 2, are selected to train substitutes with various synthetic datasets for ensemble to implement black-box attack tasks.

Table 1. Summary of 4 benchmark datasets

Name	Training data	Test data	Features	Labels	Task
MNIST	50000	10000	$28 \times 28 \times 1$	10	Digit recognition
USPS	7291	2007	$16 \times 16 \times 1$	10	Digit recognition
GTSRB	39209	12630	$32 \times 32 \times 3$	43	Traffic sign recognize
BelgiumTSC	4575	2534	$32 \times 32 \times 3$	62	Traffic sign recognize

Table 2. Neural network architectures used in this work for substitute and target model training. Conv: convolution layer, FC: fully connected layer, Relu: activation function

Target Model	Substitute1	Substitute2	Substitute3	Substitute4	Substitute5
Conv(128,3,3)+Relu Conv(64,3,3)+Relu Dropout(0.25) FC(128)+Relu Dropout(0.5) FC+Softmax	Conv(64,8,8)+Relu Conv(128,6,6)+Relu Conv(128,5,5)+Relu Dropout(0.5) FC+Softmax	FC(300)+Relu Dropout(0.5) FC(300)+Relu Dropout(0.5) FC(300)+Relu Dropout(0.5) FC(300)+Relu Dropout(0.5) FC+Softmax	Conv(32,3,3)+Relu Conv(32,3,3)+Relu Conv(64,3,3)+Relu FC(200)+Relu Dropout(0.5) FC+Softmax	Conv(32,3,3)+Relu Conv(32,3,3)+Relu Conv(64,3,3)+Relu Conv(128,3,3)+Relu Conv(128,3,3)+Relu Drop(0.2) FC(512)+Relu Dropout(0.5) FC+Softmax	Conv(64,3,3)+Relu Conv(64,3,3)+Relu Conv(128,3,3)+Relu Conv(128,3,3)+Relu FC(256)+Relu FC(256)+Relu FC+Softmax

---

[1] Codes is available at https://github.com/HangJie720/Ensemble_Adversarial_Attack.

[2] https://www.tensorflow.org/?hl=zh-cn.

[3] https://github.com/tensorflow/cleverhans.

Two diverse measurements, *Success rate* and *Transfer rate*, are redefined to evaluate the vulnerability of DL models according to Eqs. 6. and 7.

$$\frac{1}{nk} \sum_{j=1}^{k} \sum_{i=1}^{n} \mathbb{I}\left(F_j\left(x_i^*\right) \neq F_j(x_i)\right) \qquad (6)$$

$$\frac{1}{n} \sum_{i=1}^{n} \mathbb{I}\left(O\left(x_i^*\right) \neq O(x_i)\right) \qquad (7)$$

where $\mathbb{I}(\cdot) = 1$ represents generated adversarial example is misclassified, and 0, otherwise. These two metrics are used to measure the error rate of substitute and target model respectively.

3.2   Results

This section first quantitatively analyzes the vulnerability of DL models under success rate and transfer rate measurement. Afterwards, we empirically compare the conventional single-substitute attack algorithms based on FGSM and Carlini L_∞ attack for different datasets. Finally, possible factors that contribute to the higher transfer rate are explored from two aspects, the diversity of substitutes and the number of substitutes k.

Figure 3 demonstrates that deep learning models are extremely susceptible to adversarial examples generated by proposed ensemble-based black-box attack strategies under different perturbation amplitude α.

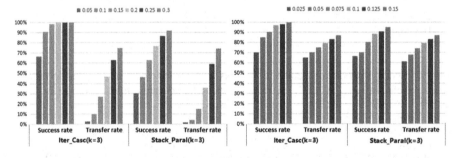

Fig. 3. Success rate and Transfer rate of adversarial examples generated by ensemble-based black-box attack strategies under different perturbation amplitude on MNIST and GTSRB.

The transferability of adversarial examples generated by each substitute and cascading or paralleling any k substitutes (e.g. $k = 3, 5$) are illustrated in Table 3 and Fig. 4. Experiments demonstrate that the adversarial examples crafted by *iterative cascade ensemble strategy* achieve higher transfer rate than *stack parallel ensemble strategy* dramatically. Both obtain superior attack performance to other single-substitute attack algorithms. We also can observe that optimization-based algorithm (e.g. Carlini L_∞ attack) provided for each substitute to iterative cascade ensemble obtain greater transferability than gradient-based algorithm (e.g. FGSM). Figure 5 demonstrates that our proposed ensemble-based black-box attack strategies are still aggressive to target classifier trained with ensemble adversarial training defense mechanism.

Table 3. Transfer rate of adversarial examples generated by single-substitute, iterative cascade ensemble strategy and stack parallel ensemble strategy based on **FGSM** and **Carlini L_∞ attack** for different datasets.

FGSM ($\alpha = 0.3$)	MNIST	USPS	GTSRB	BelgiumTSC
Sub1	32.47%	30.44%	59.32%	58.22%
Sub2	37.00%	38.20%	55.27%	49.64%
Sub3	18.57%	25.42%	50.23%	50.12%
Sub4	19.04%	27.62%	45.55%	43.65%
Sub5	16.61%	25.29%	40.29%	49.23%
Iter_Casc (k = 3)	**58.01%**	**53.23%**	**65.89%**	**64.68%**
Stack_Paral (k = 3)	**50.00%**	**48.27%**	**61.36%**	**60.00%**
Carlini L_∞ attack ($\kappa = 0 \kappa = 0$)	MNIST	USPS	GTSRB	BelgiumTSC
Sub1	12.50%	10.50%	12.50%	28.20%
Sub2	12.50%	12.00%	20.50%	19.65%
Sub3	1.50%	2.50%	9.50%	2.10%
Sub4	0.50%	1.50%	5.55%	3.55%
Sub5	1.00%	0.50%	8.50%	1.20%
Iter_Casc (k = 3)	**94.50%**	**90.00%**	**100.00%**	**100.00%**
Stack_Paral (k = 3)	**17.50%**	**20.00%**	**30.50%**	**35.00%**

Moreover, possible factors that contribute to the higher transfer rate are explored from two perspectives: the diversity of substitutes and the number of substitutes k.

(1) The number of substitutes k. The experiment results are shown in Fig. 6. for our proposed ensemble-based black-box attack strategies, which indicates that the larger the value of k, the higher transfer rate of generated adversarial examples.

(2) The diversity of substitutes. Two averaged pairwise measures [20] (the Q statistics, the correlation coefficient ρ) and two non-pairwise measures [20] (The entropy measure E, the Kohavi-Wolpert variance KW) are selected to analyze the relationship between the diversity of substitute and transferability of generated adversarial examples. Experimental results are listed in Table 4, where I, II, III and IV represent the four strategies to generate the substitutes, such as, the substitutes are same, trained with different training sets, trained with different architectures and trained with different training sets and architectures respectively. Comparative experimental results demonstrate that the greater the diversity of substitutes, the stronger the transferability of adversarial examples. Thus, all same substitutes used in I-FGSM obtain the lowest transfer rate, as shown in Fig. 4.

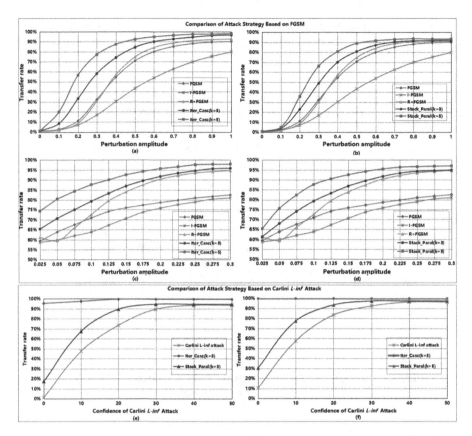

Fig. 4. Transfer rate of adversarial examples crafted by disparate attack strategies on two major classification tasks. Ensemble strategies compared with single-substitute attack algorithms based on FGSM under differ perturbation amplitude α are shown in Fig. (a)–(d). Ensemble strategies compared with single-substitute attack algorithms based on Carlini L_∞ Attack under different confidence κ are shown in Fig. (e) and (f).

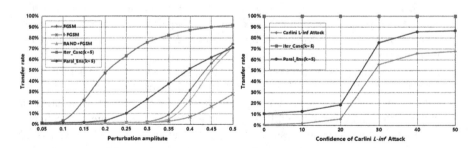

Fig. 5. Weakly defense performance of target classifier trained with ensemble adversarial training defense mechanism.

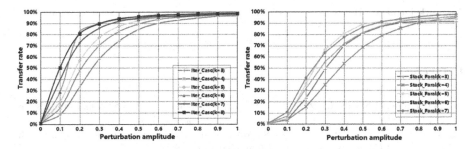

Fig. 6. Transfer rate of adversarial examples crafted by iterative cascade ensemble strategy and stack parallel ensemble strategy with different number of substitutes k.

Table 4. The relationship of diversity in substitute cascade/parallel ensembles and transferability of generated adversarial examples. ($\uparrow$) represents the measure value of diversity is increased, ($\downarrow$) represents the measure value of diversity is decreased.

MNIST	Transfer Rate		Diversity Measure Value			
	Iter_Casc (k = 3)	Stack_Paral (k = 3)	$Q(\downarrow)$	$\rho(\downarrow)$	$Ent(\uparrow)$	$KW(\uparrow)$
I	16.89%	10.89%	1.0000	1.0000	0.0000	0.0000
II	20.35%	18.52%	0.8900	0.7343	0.4900	0.1089
III	40.23%	34.53%	0.6432	0.5321	0.6235	0.2336
IV	58.01%	50.00%	0.3411	0.2300	0.7800	0.3345
GTSRB	Iter_Casc (k = 3)	Stack_Paral (k = 3)	$Q(\downarrow)$	$\rho(\downarrow)$	$Ent(\uparrow)$	$KW(\uparrow)$
I	70.44%	66.24%	1.0000	1.0000	0.0000	0.0000
II	79.26%	72.81%	0.7100	0.6911	0.5300	0.2033
III	88.12%	80.36%	0.5302	0.3510	0.7122	0.3010
IV	95.89%	93.80%	0.2201	0.1800	0.8201	0.4700

4 Conclusion

In this paper, we propose two types of ensemble-based black-box attack strategies, *iterative cascade ensemble strategy* and *stack parallel ensemble strategy*, to explore the vulnerability of deep learning system. Experimental results show that our proposed ensemble adversarial attack strategies can successfully attack the deep learning system trained with ensemble adversarial training defense mechanism. The adversarial examples generated by *iterative cascade ensemble strategy* achieve better transferability than *stack parallel ensemble strategy* dramatically. Both obtain superior attack performance to other single-substitute attack algorithms. We also can observe that the diversity in substitute ensembles is an important factor to influence the transferability of generated adversarial examples.

Acknowledgement. This work was partially supported by Natural Science Foundation of China (No. 61603197, 61772284, 41571389).

References

1. Goodfellow, I., Shlens, J., Szegedy, C.: Explaining and harnessing adversarial examples. In: 3rd International Conference on Learning Representations (2015)
2. Kurakin, K., Goodfellow, J., Bengio, S.: Adversarial examples in the physical world. In: 5th International Conference on Learning Representations (2017)
3. Biggio, B., et al.: Evasion attacks against machine learning at test time. In: Blockeel, H., Kersting, K., Nijssen, S., Železný, F. (eds.) ECML PKDD 2013. LNCS (LNAI), vol. 8190, pp. 387–402. Springer, Heidelberg (2013). https://doi.org/10.1007/978-3-642-40994-3_25
4. Papernot, N., McDaniel, P., Jha, S., Fredrikson, M., Berkay Celik, Z., Swami, A.: The limitations of deep learning in adversarial settings. In: 1st IEEE European Symposium on Security and Privacy (EuroS&P), pp: 372–387. IEEE Press, New York (2016)
5. Szegedy, C., et al.: Intriguing properties of neural networks. In: 2nd International Conference on Learning Representations (2014)
6. Papernot, N., McDaniel, P.: Transferability in machine learning: from phenomena to black-box attacks using adversarial samples. arXiv preprint arXiv:1605.07277 (2016)
7. Tramèr, F., Papernot, N., Goodfellow, I., Boneh, D., McDaniel, P.: The space of transferable adversarial examples. arXiv preprint arXiv:1704.03453 (2017)
8. Liu, Y., Chen, X., Liu, C., Song, D.: Delving into transferable adversarial examples and black-box attacks. In: 5th International Conference on Learning Representations (2017)
9. Papernot, N., McDaniel, P., Goodfellow, I., Jha, S., Berkay Celik, Z., Swami, A.: Practical black-box attacks against machine learning. In: ASIACCS, pp: 506–519. ACM, New York (2017)
10. Papernot, N., Mcdaniel, P., Sinha, A., Wellman, M.: Towards the science of security and privacy in machine learning. arXiv preprint arXiv:1611.03814 (2016)
11. Meng, D.Y., Chen, H.: MagNet: a two-pronged defense against adversarial examples. In: ACM SIGSAC Conference on Computer and Communications Security, pp. 135–147. ACM, New York (2017)
12. Wang, Q.L., Guo, W.B., Zhang, K.X., Ororbia Ii, A. G., Xing, X., Liu, X.: Adversary resistant deep neural networks with an application to malware detection. In: 23rd ACM SIGKDD International Conference on Knowledge Discovery and Data Mining, pp: 1145–1153. ACM, New York (2017)
13. Bhagoji, A, J., Cullina, D., Mittal, P.: Dimensionality Reduction as a Defense against Evasion Attacks on Machine Learning Classifiers. arXiv preprint arXiv:1704.02654 (2017)
14. Tramèr, F., Kurakin, A., Papernot, N., Boneh, D., McDaniel, P.: Ensemble adversarial training: attacks and defenses. In: 6th International Conference on Learning Representations (2018)
15. Kurakin, A., Goodfellow, I., Bengio, S.: Adversarial machine learning at scale. In: 5th International Conference on Learning Representations (2017)
16. Papernot, N., McDaniel, P., Wu, X.: Distillation as a defense to adversarial perturbations against deep neural networks. In: 2016 IEEE Symposium on Security and Privacy (SP), vol. 00, pp. 582–597. IEEE, USA (2016)

17. Carlini, N., Wagner, D.: Towards evaluating the robustness of neural networks. In: 38th IEEE Symposium on Security and Privacy, pp. 39–57. IEEE, USA (2017)
18. Boer, P.T.D., Kroese, D.P., Mannor, S., Rubinstein, R.Y.: A tutorial on the cross-entropy method. Ann. Oper. Res. **134**(1), 19–67 (2005)
19. Grosse, K., Papernot, N., Manoharan, P., Backes, M., McDaniel, P.: Adversarial perturbations against deep neural networks for malware classification. arXiv preprint arXiv:1606.04435 (2016)
20. Ludmila, I.K., Whiker, C.J.: Measures of diversity in classifier ensembles and their relationship with the ensemble accuracy. Mach. Learn. **2**(51), 181–207 (2003)

Transfer Learning Using Progressive Neural Networks and NMT for Classification Tasks in NLP

Ravi Shankar Devanapalli[✉] and V. Susheela Devi

Indian Institute of Science, CV Raman Road, Bengaluru 560012, Karnataka, India
dravishankar7@gmail.com, susheela@iisc.ac.in
https://www.iisc.ac.in/

Abstract. Recently neural networks are obtaining state of the art results on many NLP tasks like sentiment classification, machine translation, etc. However one of the drawbacks of these techniques is that they need large amounts of training data. Even though there is a lot of data being generated everyday, not all tasks have large amounts of data. One possible solution when data is not sufficient is using transfer learning techniques. In this paper, we explored methods of transfer learning (or sharing the parameters) between different tasks so that the performance on the low data resource tasks is improved. We have first tried to replicate the prior results of transfer learning in semantically related tasks. When we have semantically different tasks, we tried using Progressive Neural Networks. We also experimented on sharing the encoder from neural machine translator to classification tasks.

Keywords: Progressive neural networks · Transfer learning
Neural machine translator encoder

1 Introduction

Transfer learning is a paradigm in machine learning where the knowledge gained by solving a problem related to a task or domain is applied to solve a different but related problem. There will be a source domain (which usually has large training data) and a target domain (which usually has small training data). Using the knowledge learned from the source domain, transfer learning aims to improve the performance on the target domain.

Transfer Learning techniques have been effectively used in fields like image processing [8] and were able to achieve good results. However, in NLP, Transfer Learning has been loosely applied and conclusions are not consistent. A previous work [1] showed that transfer learning is helpful when the tasks are semantically similar. But most of the time we may not find a source task with large amount of data which is semantically similar to the target task. We experimented with different architectures to address this difficulty. We used progressive neural networks and NMT and succeeded to some extent. Researchers [4] used Progressive

© Springer Nature Switzerland AG 2018
L. Cheng et al. (Eds.): ICONIP 2018, LNCS 11303, pp. 188–197, 2018.
https://doi.org/10.1007/978-3-030-04182-3_17

Neural Networks (PNN) on games with reinforcement learning and achieved near state of the art performance. Their main idea is to use columns of architectures trained on different datasets and use them along with a new column on the target dataset. This also prevents catastrophic forgetting while training the target domain.

In machine translation, sequence to sequence [5] learning using neural networks have given state of the art results. In sequence to sequence learning, the general architecture consists of an encoder and a decoder. The source language sentence is given as input to encoder, which gives an encoded vector. Then that encoded vector is given as input to the decoder which then decodes into another language. We used the encoder of neural machine translator, which is trained on English to Vietnamese translation for classification tasks.

The rest of the paper is organized as follows. Section 2 discusses the related work. Section 3 describes the datasets we used for the experiments. Direct transfer learning, progressive neural networks and neural machine translator are discussed in Sects. 4, 5 and 6. Conclusions and future work are described in Sects. 7 and 8.

2 Related Work

Transfer learning is being successfully applied to a variety of problems in computer vision across domains and applications [8]. In NLP, researchers [1] showed that neural network based transfer learning helped when the source and target tasks are semantically related. They have used INIT and MULT methods while transferring the parameters. Researchers [2] have also shown that neural network based transfer learning improves the performance of models for sequence tagging and shown that even problems across domains and applications can benefit (though may not be significantly) from transfer learning. Researchers [3] proposed transfer learning schemes for personalized language modelling using LSTM's. They have also considered low computational resources like mobile and showed that transfer learning helps in faster training in such cases.

Researchers [4] have proposed Progressive Neural Networks to leverage transfer learning and avoid catastrophic forgetting. They are immune to forgetting and can leverage prior knowledge via lateral connections to previously learned features. We used this architecture in NLP classification tasks and tried to improve the performance on target datasets.

Sequence-to-sequence (seq2seq) have enjoyed great success in a variety of tasks such as machine translation models, speech recognition, text summarization, etc. NMT system uses encoder and decoder architecture for translating from one language to another. We have trained a NMT on English to Vietnam language and then used its encoder for classification tasks.

3   Datasets

The following datasets have been used by us for our experiments:

IMBD: A large dataset by IMDb for binary sentiment classification (positive vs. negative) - 25k sentences [9]

Movie Review (MR): A small dataset by Rotten Tomatoes for binary sentiment classification $\sim$ 10k sentences[1].

Question Classification (QC): A small dataset for 6-way question classification (e.g., location, time, and number) $\sim$ 5000 questions[2].

SNLI: A large dataset for sentence entailment recognition. The classification objectives are entailment, contradiction, and neutral $\sim$ 500k pairs [7].

SICK: A small dataset with exactly the same classification objective as SNLI $\sim$ 10k pairs[3].

MSRP: A small dataset for paraphrase detection.The objective is binary classification: judging whether two sentences have the same meaning $\sim$ 5000 pairs[4].

Quora dataset: It contains duplicate questions pairs with labels indicating whether the pair of questions request the same information $\sim$ 400k question pairs[5].

IWSLT English-Vietnamese corpus: We used this dataset to train NMT $\sim$ 133k pairs[6].

4   Direct Transfer Learning

A typical transfer learning scenario is by initializing the weights of a target task by the weights trained on the source task and fine tuning them. First we tried to replicate the results of the previous work [1] using INIT method. We used the CNN and LSTM architectures as shown in Figs. 1 and 2 respectively. We used a CNN for each sentence and after max pooling we concatenated the two vectors and used two hidden layers and an output layer. We used filter widths of 4, 5, 6 and 7 in CNN and 1024 units in hidden layer 1 and 2. Similarly, We used an LSTM for each sentence and the last hidden states are concatenated and two hidden layers and one output layer were used. We used 256 hidden units in LSTM cell and 1024 units in hidden layer 1 and 2. For datasets with one input sentence (IMDB and QC), we used only one CNN or LSTM. We have trained the model on IMDB and then transferred the parameters to MR and QC

---

[1] http://www.cs.cornell.edu/people/pabo/movie-review-data/

[2] http://cogcomp.cs.illinois.edu/Data/QA/QC/

[3] http://alt.qcri.org/semeval2014/task1/

[4] https://www.microsoft.com/en-us/download/details.aspx?id=52398

[5] https://data.quora.com/First-Quora-Dataset-Release-Question-Pairs

[6] https://nlp.stanford.edu/projects/nmt/

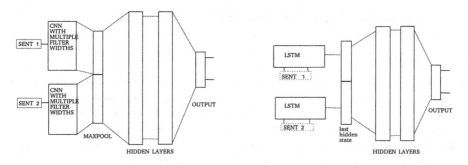

Fig. 1. CNN architecture   **Fig. 2.** LSTM architecture

datasets. Similarly we have trained the model on SNLI and then transferred the parameters to SICK and MSRP. We also trained the model on Quora and then transferred the parameters to SICK and MSRP. Results are shown in Table 1.

When we transferred the parameters from IMDB to MR, the accuracy is improved by 1.65% (CNN) and 1.06% (LSTM). From IMDB to QC, there is not much change in accuracy. The reason for this is that IMDB and MR are semantically similar datasets whereas IMDB and QC are semantically different.

Table 1. Accuracies obtained with transfer of parameters from IMDB to MR and QC datasets and from SNLI and Quora to SICK and MSRP datasets.

Dataset	Paper[1] (without Transfer)	Paper[1] (with Transfer)	Without Transfer (CNN)	With Transfer (CNN)	Without Transfer (LSTM)	With Transfer (LSTM)
MR	75.1	80.9	77.55	79.20	78.16	79.22
QC	90.8	90.40	89.45	89.33	89.40	89.86
SNLI to SICK	70.9	77.6	70.45	76.98	76.39	80.79
Quora to SICK	-	-	70.45	73.89	76.39	75.08
SNLI to MSRP	69	68.8	69.29	68.17	69.39	68.78
Quora to MSRP	-	-	69.29	69.68	69.39	69.21

Transfer of parameters from SNLI to SICK appears to be successful with 6.53% (CNN) and 4.40% (LSTM) increase in accuracy. We can observe that there is decrease of 1.12% (CNN) and 0.61% (LSTM) from SNLI to MSRP. SNLI and SICK are semantically similar datasets, whereas SNLI and MSRP are not. This suggests that transfer learning is more prone to semantics. From Quora to SICK, the accuracy has increased by 3.44% in CNN architecture and decreased by 1.31% in LSTM architecture, while there is not much change from Quora to MSRP in both architectures. Quora is not related to SICK or MSRP. This implies that there may be positive or negative transfer between unrelated datasets depending on the type of architecture used.

5   Progressive Neural Networks

From the above experiments, it is evident that initializing and fine tuning are helpful when the source and target tasks are semantically similar. In general it is not always possible to get a source which is semantically similar to our target dataset. In that case if we train a model on some dataset and transfer the parameters to the model on target dataset, we may or may not get satisfactory performance. So we tried to address this problem using Progressive neural networks (PNN). Progressive networks retain a pool of pretrained models throughout training. In this architecture, there are columns of networks. Each column consists of a neural network. There can be many columns and each column is initialized with a pretrained neural network and the last column is randomly initialized. There are lateral connections between the hidden layers of each column to the hidden layer of the last column. The pretrained weights are freezed while the lateral weights are trainable. In normal transfer learning, initializing and fine tuning may help the target task but it will forget the source task functionality. Since the pretrained weights are freezed, the source tasks functionality is retained in these networks. Progressive networks are a step in the direction of continual learning.

Researchers [4] applied progressive networks on reinforcement learning tasks (Atari and 3D maze games). We have applied progressive networks to NLP tasks. The idea is that since in this architecture, the target task is initialized with both pretrained weights and randomly, if we do not have a semantically similar source task, then we can still use the pretrained weights of a source task to initialize a column with those weights. Since there are randomly initialized layers, the target task will use whatever parameters are required from the source task and learn the rest from updating the randomly initialized weights. The idea is that there will not be any negative transfer and in the worst case it will give the same results without any degradation. Also to initialize the weights from the source task, the architecture of the target task and the source task has to be the same. But in progressive networks, even if the source architecture is different, we can still use the pretrained weights.

5.1   Implementation

We used two architectures, CNN and LSTM. Figure 3 shows the PNN with CNN architecture. For source dataset, we used filter widths of 4, 5, 6 and 7 in CNN and 1024 units in hidden layer 1 and 2. For target dataset we used filter widths of 4, 5, 6 and 7 in CNN and 512 units in hidden layer 1 and 2. The dimensions of lateral weights are 800×1024 (concatenated maxpool vector of source to hidden layer 1 of target), 1024×512 (hidden layer 1 of source to hidden layer 2 of target) and $1024 \times 2/3/6$ (number of labels in target dataset). PNN with LSTM is also similar, instead of maxpooling from CNN, we used the last hidden state of LSTM. For source dataset, we used 256 hidden units in LSTM cell and 1024 units in hidden layer 1 and 2. For target dataset we used 256 hidden units in LSTM cell and 512 units in hidden layer 1 and 2. The dimensions of lateral weights are 512×1024 (concatenated maxpool vector of source to hidden layer 1 of target), 1024×512 (hidden layer 1 of source to hidden layer 2 of target) and $1024 \times 2/3/6$ (number of labels in target dataset). Table 2 shows the results of PNN with CNN and LSTM.

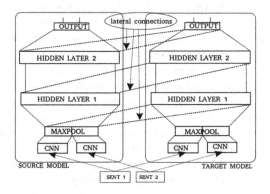

Fig. 3. Architecture of PNN with CNN.

Table 2. Accuracies using different methods of sharing the parameters using CNN (left) and LSTM (right).

CNN	Original	Direct Transfer	PNN	Modified PNN	LSTM	Original	Direct Transfer	PNN	Modified PNN
IMDB to MR	77.55	79.20	78.30	79.30	IMDB to MR	78.16	79.22	77.90	79.07
IMDB to QC	89.45	89.33	90.10	89.88	IMDB to QC	89.40	89.86	89.60	89.60
SNLI to SICK	70.45	76.98	75.61	76.22	SNLI to SICK	76.39	80.79	79.66	81.60
Quora to SICK	70.45	73.89	73.10	74.76	Quora to SICK	76.39	75.08	75.88	76.54
SNLI to MSRP	69.29	68.17	70.02	70.10	SNLI to MSRP	69.39	68.78	68.98	69.51
Quora to MSRP	69.29	69.68	69.27	70.14	Quora to MSRP	69.39	69.21	69.10	69.91

From the results, we can observe that if the tasks are not semantically similar, direct transfer learning may degrade the performance, but the progressive networks showed slightly improved or same performance but did not degrade. From the results we can observe that when the tasks are semantically similar, PNN performed worse than the direct transfer. This means that, when the tasks are semantically similar it is more advantageous to initialize and fine tune. So we did a slight modification to the progressive networks. Instead of freezing the parameters from the source tasks, we made them as trainable. This modification gave the best results of all. The idea is that if the tasks are semantically similar then initializing and fine tuning the source task parameters actually helps. So instead of freezing those parameters, we let them train.

6   Neural Machine Translator

Sequence-to-sequence (seq2seq) models have enjoyed great success in a variety of tasks such as machine translation [5], speech recognition, text summarization, etc. NMT system first reads the source sentence using an encoder to build a "encoded" vector, a sequence of numbers that represents the sentence meaning; a decoder, then, processes the encoded vector to emit a translation. The encoded vector thus contains sufficient lexical and semantic information to fully reconstruct a sentence in another language.

We have experimented with encoder of a neural machine translator (NMT) to the classification tasks. A neural machine translator [5] with 2-layer LSTMs of 512 units with bidirectional encoder, embedding dimension of 512 and Luong attention is used to train on IWSLT English-Vietnamese corpus (133k examples). A Bleu score of 24.0 was obtained on the test data. Later, the trained model was used to extract a 512 dimensional encoded vector of a sentence. The encoded vector of dimension 512 is given as input to the feed forward neural network with two hidden layers and one output layer. The encoded vector is taken as the last hidden state of the LSTM. The number of units in hidden layer 1 are 1024 for each sentence and there are 1024 units in hidden layer 2. Results are shown in Figs. 5, 6, 7, 8, 9 and 10.

6.1   PNN with NMT

In this section we exploited the Progressive neural networks by using a different architecture altogether. We have combined the encoded vector from NMT and CNN/LSTM architecture. Figure 4 shows the architecture with CNN. We used filter widths of 4, 5, 6 and 7 in CNN and 1024 units in hidden layer 1 and 2. We used the same NMT encoder as in previous section. The dimension of encoded vector is 512 and so the dimension of the lateral weights is 1024×1024. Similarly with LSTM we used the last hidden state instead of max pooling from CNN. We used 256 hidden units in LSTM cell and 512 units in hidden layer 1 and 2. We initialized the CNN/LSTM part randomly and also with parameters from a model trained on source. Results are shown in Figs. 5, 6, 7, 8, 9 and 10.

From the results in Figs. 5, 6, 7, 8, 9 and 10, we can observe that on SICK (for CNN) and MSRP datasets the accuracies are on par with the direct models. However, on QC and MR datasets, the performance is worse than the original one. From this we can say that the encoded vector is capturing meaning of the sentence to a certain extent.

We can observe that there is a major increase in accuracy for MR and QC datasets with NMT and CNN/LSTM compared to that of only encoded vector from NMT. NMT encoded vector with direct transfer performed better or on par with direct transfer alone on all datasets for both CNN and LSTM (except from SNLI to SICK with LSTM). We can observe that combining encoded vector of NMT with other models gives good performance compared to individual models alone. Thus we can say that encoded vector from NMT is giving some complementary information to the CNN or LSTM architectures.

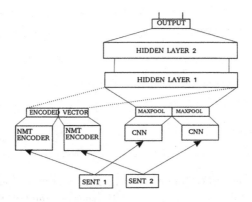

Fig. 4. PNN architecture with NMT encoder and CNN.

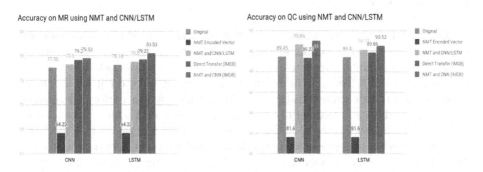

Fig. 5. Accuracies on MR dataset.   **Fig. 6.** Accuracies on QC dataset.

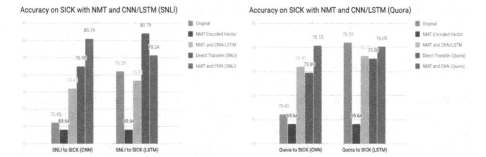

Fig. 7. Accuracies on SCIK dataset with **Fig. 8.** Accuracies on SCIK dataset with transfer from SNLI.   transfer from Quora.

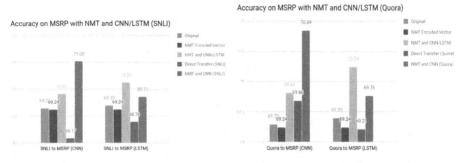

Fig. 9. Accuracies on MSRP dataset with transfer from SNLI.

Fig. 10. Accuracies on MSRP dataset with transfer from Quroa.

7 Conclusions

We performed various experiments on different methods of sharing the parameters. Initializing and fine tuning are helpful when the source and target datasets are semantically similar. Next we used progressive neural networks for the smaller datasets. Instead of freezing the source parameters, it is better if we let them train. We used the encoded vector of NMT encoder, which performed reasonably on some datasets. Instead of using the encoded vector alone, if we used it in progressive network, then the performance is improved. Encoded vector from NMT and direct transfer from source performed better than direct transfer alone. Thus we can say that encoded vector from NMT is giving some complementary information to the CNN or LSTM architectures.

Even though initializing and fine tuning are popular methods of sharing the parameters, it may not help in all cases. Progressive neural networks gave improved performance even when the datasets are not semantically similar. So we can conclude that it is better to use progressive neural networks instead of initializing and fine tuning the parameters.

8 Future Work

We can combine different architectures in progressive neural networks. So this work can be further extended by experimenting with various architectures. We have trained NMT on a smaller dataset. We can train it on a large corpus and experiment with it. Also Google has used a Multilingual Neural Machine Translation System [6] which is trained on more than one language to language pair. Since it is trained on many language pairs, it might capture the meaning of a sentence to a greater extent than a NMT trained on a one-to-one language. We can experiment with the encoder of that model on various tasks.

References

1. Mou, L., et al.: How transferable are neural networks in NLP applications? In: Proceedings of the 2016 Conference on Empirical Methods in Natural Language Processing (EMNLP), pp. 478–489 (2016)
2. Yang, Z., Salakhutdinov, R., Cohen, W.W.: Transfer learning for sequence tagging with hierarchical recurrent networks. In: ICLR 2017 (2017)
3. Yoon, S., Yun, H., Kim, Y., Park, G., Jung, K.: Efficient Transfer Learning Schemes for Personalized Language Modeling using Recurrent Neural Network CoRR 2017, volume: abs/1701.03578
4. Rusu, A.A., et al.: Progressive Neural Networks CoRR 2016, volume: abs/1606.04671
5. Luong, M.-T., Brevdo, E., Zhao, R.: Neural Machine Translation (seq2seq) Tutorial (2017). https://github.com/tensorflow/nmt
6. Johnson, M., et al.: Google's Multilingual Neural Machine Translation System: Enabling Zero-Shot Translation. CoRR 2016, volume: abs/1611.04558
7. Bowman, S.R., Angeli, G., Potts, C., Manning, C.D.: A large annotated corpus for learning natural language inference. In: Proceedings of the 2015 Conference on Empirical Methods in Natural Language Processing (EMNLP). Association for Computational Linguistics (2015)
8. Oquab, M., Bottou, L., Laptev, I., Sivic, J.: Learning and transferring mid-level image representations using convolutional neural networks. In: The IEEE Conference on Computer Vision and Pattern Recognition (CVPR), June 2014
9. Maas, A.L., Daly, R.E., Pham, P.T., Huang, D., Ng, A.Y., Potts, C.: Learning word vectors for sentiment analysis. In: Proceedings of the 49th Annual Meeting of the Association for Computational Linguistics: Human Language Technologies, June 2011

Deep Transfer Learning via Minimum Enclosing Balls

Zhilong Deng[1], Fan Liu[2](✉), Jiangjiang Zhao[1], Qiang Wei[1], Shaoning Pang[3], and Yue Leng[4]

[1] Algorithm Group, Research and Development Department,
Online Services of China Mobile, Beijing, China
{dengzhilong,zhaojiangjiang,weiqiang}@cmos.chinamobile.com
[2] Artificial Intelligence Lab, Meituan-Dianping Group, Beijing, China
liufan16@meituan.com
[3] Department of Computing, Unitec Institute of Technology, Private Bag 92025,
Auckland 1142, New Zealand
ppang@unitec.ac.nz
[4] Algorithm Group, Artificial Intelligence Lab,
Emotibot Technologies Limited, Beijing, China
yueleng@emotibot.com
http://online.10086.cn
http://www.unitec.ac.nz
http://www.emotibot.com

Abstract. Training of deep learning algorithms such as CNN, LSTM, or GRU often requires large amount of data. However in real world applications, the amount of data especially labelled data is limited. To address this challenge, we study Deep Transfer Learning (DTL) in the context of Multitasking Learning (MTL) to extract sharable knowledge from tasks and use it for related tasks. In this paper, we use Minimum Closed Ball (MEB) as a flexible knowledge representation method to map shared domain knowledge from primary task to secondary task in multitasking learning. The experiments provide both analytic and empirical results to show the effectiveness and robustness of the proposed MEB-based deep transfer learning.

Keywords: Multi-task learning · Deep transfer learning
Learner-independent multi-task learning · Minimum enclosing ball

1 Introduction

For machine learning from real world applications, we often encounter lack of data issue. For example, we have a classification task with sufficient annotation data in one domain, but there is a related task in another domain that does not have enough annotation data. The data distribution between the two tasks is different. For deep learning (DL), an implementation on a real world application needs often large amount of labelled data for training. In actual projects,

L. Cheng et al. (Eds.): ICONIP 2018, LNCS 11303, pp. 198–207, 2018.
https://doi.org/10.1007/978-3-030-04182-3_18

marking high-quality data requires a large number of knowledgeable labelers, so obtaining a sufficient number of annotation instances is extremely difficult, time consuming, and expensive. In this sense, we consider in this work a Multi-Task Learning (MTL) framework, where transfer learning can be conducted to enhance the deep learning effectiveness with existing other source of data. We call this machine learning mechanism as deep transfer learning (DTL).

MTL is inspired by the simultaneous representation of multiple tasks in daily human activities. For example, if one knows how to identify a dog from an image, then it would be easy to learn how to identify a cat from the image, because the process of identifying the shape of the animal in the image is similar. Alternatively, if one can play basketball or play netball, he can easily master another. The goal of MTL is to discover the similarities between related tasks by mimicking the mechanism of human brain. By sharing knowledge among related tasks, MTL has achieved significant improvements over single-task learning (STL) in many practical scenarios such as pattern recognition and financial forecasting.

Transfer learning (TL) is the core of MTL. In the traditional MTL study, the task correlation evaluation and TL process are realized by constructing a specific learner. The learning process of TL here is not transparent, because the TL method is adapted to a specific type of learner, so the learned shareable knowledge is not available for a new type of learner. In the context of MTL, we apply Minimum Enclosing Ball (MEB), a learner-independent and adaptive approach to perform deep transfer learning. In the experiment, we present both practical applications and experimental analytic results to demonstrate adaptability and efficiency, taking into account MEB between task correlation and correlation interpretation for multi-task machine learning. Unlike previous TL methods for a specified learner, MEB incorporates learner independencies into MTL, thus empowering deep learners for MTL (i.e., deep transfer learning).

2   Deep Learners

Traditional machine learning very much relies on steps of feature extraction. They transform training data, and augment it with additional features, in order to improve learning effectiveness of machine learning algorithms. Deep learning, inspired by information processing and communication patterns of biological neural systems, changes this regulation by employing a cascade of multiple layers of nonlinear processing units for feature extraction and transformation. In this work, we three popular deep classifiers including CNN, LSTM and GRU as the learners for deep transfer learning.

CNN [14] is one of the most widely used and most effective feed-forward neural network classifiers in the fields of computer vision, natural language processing, and speech recognition. The CNN model is expected to minimize the dependence of the model on artificial features and automatically identify pattern [6]. They can recognize extreme transformed patterns (such as handwritten characters) and are robust to scaling or rotation operations.

LSTM [5] was proposed by Sepp Hochreiter and Juergen Schmidhuber to solve the gradient explosion problem in RNN. LSTM unit is composed of a cell, an input gate, an output gate and a forget gate. The cell is responsible for "remembering" values over arbitrary time intervals. Each gate is considered as a "conventional" artificial neuron, as in a multilayer perceptron (or feedforward) neural network. The activation function is used to compute a weighted sum in order to control the flow of values that goes through the connections of the LSTM. Thus, it is suitable for back-propagation.

GRU [3] was introduced by Kyunghyun Cho et al. as a gating mechanism in recurrent neural networks. GRU is a simple variant of LSTM that shares many of its properties. GRU has fewer parameters than LSTM. It's performance is often better than that of LSTM, and running speed is approximately three times faster than LSTM.

Deep transfer learning is firstly promoted in image processing. In image processing, low-level neural network layer extraction features are not strongly associated with specific tasks, and these features can be shared among different tasks. Donahue et al. [4] suggest that high-level layers are also transferable in general visual recognition, Yosinski et al. [15] further investigated the transferability of neural layers in different levels of abstraction. Mou et al. [9] conduct systematic case studies and provided an illuminating picture on the transferability of neural networks in NLP. They experimented in two different scenarios: (1) knowledge transfer on tasks with the same semantics but different data sets; (2) knowledge transfer between tasks with the same network topology but different semantics, so that network structure parameters can be shared.

The same as traditional classifiers, the above deep learners can easily handle the training and test data in the same distribution. To adopt knowledge/data from other domain, it would be desirable to develop a learner learning, otherwise we have to re-collect new data to rebuild model as the distribution changes. The proposed DTL gives a learner independent TL solution to mitigate this gap.

3   MTL Framework

Let T^0 be a primary task with training data $D^0 = [X^0, Y^0]$ and T^k be a secondary task with training data $D^k = [X^k, Y^k]$. Theoretically, a primary task may have more than one correlate task, so $k = 1, \ldots, m$. In this paper, we assume that there is only one related task for a primary task, so $k = 1$. The relatedness R^{0k} of T^0 and T^k is defined on the training set of two tasks and the hypotheses for the related tasks as,

$$R^{0k} = f_R(\mathcal{L}(D^0), \mathcal{L}(D^k), D^0, D^k), \tag{1}$$

where f_R can be any static or dynamic relatedness measure between two correlate tasks, such as Hamming distance or coefficient of correlation. $\mathcal{L}$ can be any type of deep learner that can learn functional relatedness rather than physical relatedness as a learning system for MTPR, for example in ηMTL [12], $\mathcal{L}$ is specified as an ANN learner.

In Eq.(1), task relatedness is evaluated on $\mathcal{L}$ and shared knowledge is specified to $\mathcal{L}$. By performing task relatedness assessment on a specific deep learner, it becomes more efficient to model and share knowledge using an integrated and unified process by the learning system. On the other hand, such a learner-dependent approach also has disadvantages that are difficult to overcome. The shared knowledge can only be used for specific methods and the similarity measure process cannot be decoupled from DTL.

For DTL, we propose a similarity measure to improve the ability of arbitrary deep classifiers/learner for MTL problems, which will in turn enable a learner-independent DTL procedure. In this model, the relatedness measure is defined as,

$$R^{0k} = f_R(D^0, D^k).$$ (2)

We expect to seek a physical relatedness criterion by eliminating the influence of $\mathcal{L}$ to measure the "correlation" of tasks, where the correlation between two tasks as a set of samples that can improve performance for both two tasks. Please find the detailed interpretation of the task correlation and task relatedness in [8,10,11].

4   MEB-based Transfer Learning

MEB is considered as the knowledge representation for a learner-flexible DTL, because of the following observations. First, the size of the MEB itself is flexible, and regardless of the size of the data set, all sample points are enclosed in the smallest ball by the MEB. Second, MEB is used as a knowledge representation method, which is independent of the specific learner, the enclosed data is not associated with the model, and the shared knowledge can be applied to any model.

We use an MTL which is assembled with DTL learning, which is different from traditional TL-based MTL. This approach combines a learner-independent DTL module with a traditional STL approach. This method is similar to combining batteries of different brands as a whole. We can replace a run out battery with any other brand battery. If we can only replace with the same brand of battery, then the advantage of compatibility and independence will be lost.

Motivated by this, we propose a learner independent MEB-based MTL, $\mathcal{L}_i(T^1) + TL(T^1, T^2) + \mathcal{L}_j(T^2, TL)$, where the DTL is independent to $\mathcal{L}_i$, and MTL can be compatible with any type of learner.

We extract shared knowledge by using MEB as knowledge carrier and transfer the knowledge between two correlated tasks. That is, DTL via MEBs neither dependent on $\mathcal{L}_i$ nor dependent on the MEB, so any type of deep learner can be used for MTL by adopting $TL(T^1, T^2)$ from the primary task.

Given dataset D^0 and D^k from two correlated tasks T^0 and T^k respectively, for any subset $d^0 \subset D^0$ in one class, according to [1,7,13], a subspace can be spanned by modelling a minimum enclosing ball,

$$B^0_{c,(1+\epsilon)r} = MEB(d^0_i)$$ (3)

where $B^0_{c,(1+\epsilon)r}$ is able to determine if the MEB contains the new input instance.

To verify the utility of $B^0_{c,(1+\epsilon)r}$ for T^k, we cast the MEB into T^k data space, and we have

$$B^{0 \to k}_{c^{0 \to k}, r^{0 \to k}} = CAST(B^0_{c,(1+\epsilon)r}, D^0, D^k) \qquad (4)$$

where $B^{0 \to k}_{c,(1+\epsilon)r}$ is the resulting MEB casting $B^0_{c,(1+\epsilon)r}$ in T^k data space, and the $CAST$ function is implemented by calculating the casting MEB center $c^{0 \to k}$ and the casting MEB radius $r^{0 \to k}$, respectively.

$$c^{0 \to k} = (c^0 - c^k)\frac{r^k_{max}}{r^0_{max}}, \qquad (5)$$

and

$$r^{0 \to k} = \frac{r^k_{max}}{r^0_{max}}r^0. \qquad (6)$$

where r^0_{max} is the radius of MEB over D^0, and r^k_{max} is the radius of MEB over D^k.

The obtained $B^{0 \to k}_{c,(1+\epsilon)r}$ is expected to cast a subset S^k instances in D^k. $B^0_{c,(1+\epsilon)r}$ is judged as a sharable data space by T^k, if all instances of S^k belong to one class in T^k. The instances enclosed by $B^0_{c,(1+\epsilon)r}$ are the correlation data of T^0 to T^k. In this way, given $\forall d^0 \subset D^0$, the entire sharable feature space is obtained as a merge of all MEBs that satisfy the correlation definition and the smoothness assumption [2] as: given two instances located in a high-density region, if one is enclosed in a sharable MEB, so for the other instance,

$$B^*_x = \{b^0_i\} \cup \{x\}$$
$$Subject \ to \ b^0_i \in one \ of \ D^1 \ class, \ and \ b^{0 \to k}_i \in one \ of \ D^k \ class$$
$$d(c, x_j) > r, d(c, x_i) < r, \ and \ d(x_i, x_j) < \theta.$$

where θ is the distance threshold indicating the data distribution density.

5   Experimental Results

We have performed experiments on UCI benchmark datasets and assess the proposed approach as a case study in previous study [11]. In this paper, the industry dataset that we used for MTL experiments is collected from the customer online services database of China Mobile. This dataset records the reason of call-in for customers from Henan and Yunnan province of China. We consider the classification of Henan and Yunan data as two distinct tasks in that a different data collection criterion is applied in these two provinces. However, these two tasks apparently are related to each other as both are about the same online customer service. In the experiment, we employed the proposed DTL to extract learning knowledge from one province dataset and transferred to the leaning of other province dataset. The information of two datasets is provided in Table 1.

Table 1. China Mobile customer online services database for Henan and Yunnan provinces

Datasets	#Classes	#Train Samples	#Test Samples
Henan	9	17891	4847
Yunnan	45	17214	4675

Table 2. Results of classification for with and without DTL in between Henan and Yunnan provinces datasets. Final classification improvement of the tasks based on three different deep learning, i.e., CNN, LSTM and GRU for the datasets. The two values in each cell are the value without DTL and the value increment by the proposed DTL, respectively.

(a) Task 2 (Yunnan) → Task 1 (Henan)

Classifier	Precision	Recall	F1
CNN	$87.56\% - 1.06\%$	$87.13\% + 5.05\%$	$87.35 + \mathbf{1.89\%}$
LSTM	$82.43\% + 2.73\%$	$88.41\% - 0.46\%$	$85.32 + \mathbf{1.22\%}$
GRU	$86.46\% - 0.54\%$	$88.75\% + 3.06\%$	$87.59 + \mathbf{1.18\%}$

(b) Task 1 (Henan) → Task 2 (Yunnan)

Classifier	Precision	Recall	F1
CNN	$70.31\% + 1.97\%$	$60\% + 2.68\%$	$64.75\% + \mathbf{2.39\%}$
LSTM	$58.18\% + 4.34\%$	$58.39\% + 2.41\%$	$58.29\% + \mathbf{3.36\%}$
GRU	$60.93\% + 5.37\%$	$64.17\% + 1.13\%$	$62.51\% + \mathbf{3.26\%}$

5.1 With Versus Without TL

We conduct a series of experiments to test the proposed DTL by comparing the performance with that of simple deep learning (i.e, without TL). We perform DTL from Task 1 (Henan) to Task 2 (Yunnan), and from Task 2 (Yunnan) back to Task 1 (Henan), respectively. We calculate and record the difference of precision, recall, and F1 score. Table 2 shows experiment result on two calling reason datasets. We can draw the conclusion that although MTL with DTL does always outperform that without DTL, the proposed DTL brings positive improvement on all tasks regardless of the type of deep classifier for MTL.

5.2 Independent Versus Dependent TL

We performed also a number of experiments for comparing the proposed DLT with other classifier dependent TL approaches [9,15]. The comparison of the calling reason datasets between with and without learner independent DTL for the three evaluated criteria (precision, recall and F1 score) is conducted both from Task 1 to Task 2 and Task 2 back to Task 1. The independent DTL (i.e., MEB-based) approach obtained better results than the dependent DTL. The results of this comparison are shown in Table 3.

Table 3. Compared results of deep transfer learning between independent and dependent learner approaches.

(a) Task 2 (Yunnan) → Task 1 (Henan)

Classifier	Evaluation	Dependence	Independence
CNN	Precision	87.69%	86.5%
	Recall	89.54%	92.18%
	F1 Score	88.6%	**89.24%**
LSTM	Precision	84.4%	85.15%
	Recall	87.66%	87.95%
	F1 Score	86%	**86.53%**
GRU	Precision	86.79%	85.92%
	Recall	89.3%	91.81%
	F1 Score	88.03%	**88.77%**

(b) Task 1 (Henan) → Task 2 (Yunnan)

Classifier	Evaluation	Dependence	Independence
CNN	Precision	68.8%	72.28%
	Recall	64.86%	62.68%
	F1 Score	66.77%	**67.14%**
LSTM	Precision	61.21%	62.52%
	Recall	59.52%	60.80%
	F1 Score	60.30%	**61.65%**
GRU	Precision	63.43%	66.30%
	Recall	65.72%	65.30%
	F1 Score	64.55%	**65.77%**

5.3 Neural Network Structure

In the experiment, we used three different neural network structures, i.e., CNN, LSTM and GRU. For CNN, a multiple-layers network structure is employed in order to capture the global semantic information of text. Consider CNN can only obtain local features. We reduced in all experiments the over-fitting problem of the model and speeded up the training of the model by adding batch normalization and drop out. Figure 1 gives the different neural network structures used in our experiments.

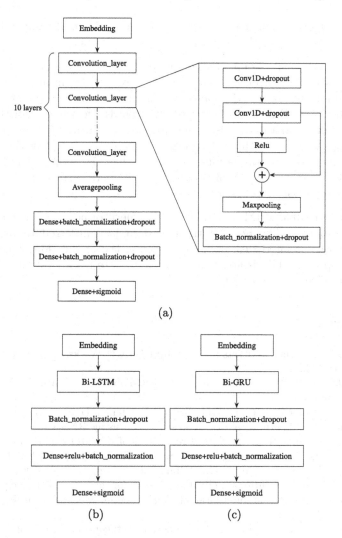

Fig. 1. Illustration of neural network structure for experiments, (a) Convolution, (b) LSTM, and (c) GRU.

6   Conclusion and Future Work

This paper focuses on the DTL scenario, where the learning system learns the shareable knowledge from one task and uses it to improve the final outcome of a different but related task. The proposed deep transfer learning method has the following advantages and disadvantages.

Independence. Deep learner independence is the main advantage of the proposed DTL. We use three well studied deep learner (e.g., CNN, LSTM, and

GRU) to verify it in call reason datasets. The experimental results show that our method can improve the accuracy for all three learners.

Validity. Experiments conducted on industry datasets, and calling reason datasets show that our approach can effectively extract potentially useful shared information for improved performance in related areas. The deep transfer learning method via MEB achieves better results than the learning algorithm without DTL. Furthermore, we also compare an algorithm with deep learner-independence with one without deep learner-independence to test the technical stability. The experimental results has shown that the proposed independent deep transfer learning approach obtain a better performance on F1 score in various disciplines of deep learner.

Efficiency. MEB-based DTL method has the merit of no need for re-collection new data progress. The need of training data is transferred from one task domain to other related task domain. However, the application of DL to two unrelated topics in MTL is attracting more and more research interest and will become a new research hotspot.

References

1. Badoiu, M.: Optimal core sets for balls. In: DIMACS Workshop on Computational Geometry (2002)
2. Chapelle, O., Scholkopf, B., Zien, A.: Semi-supervised Learning. MIT Press, Cambridge (2006)
3. Cho, K., van Merrienboer, B., Gülçehre, Ç., Bahdanau, D., Bougares, F., Schwenk, H., Bengio, Y.: Learning phrase representations using RNN encoder-decoder for statistical machine translation. In: Proceedings of the 2014 Conference on Empirical Methods in Natural Language Processing (EMNLP), pp. 1724–1734 (2014)
4. Donahue, J., Jia, Y., Vinyals, O., Hoffman, J., Zhang, N., Tzeng, E., Darrell, T.: Decaf: A deep convolutional activation feature for generic visual recognition. In: Proceedings of the 31st International Conference on Machine Learning (ICML), pp. 647–655 (2014)
5. Hochreiter, S., Schmidhuber, J.: Learning to forget: continual prediction with lstm. Neural Comput. **12**(10), 2451–2471 (2000)
6. Jure, Z., Yann, L.: Stereo matching by training a convolutional neural network to compare image patches. J. Mach. Learn. Res. **17**, 65:1–65:32 (2016)
7. Kumar, P., Mitchell, J.S.B., Yildirim, E.A.: Approximate minimum enclosing balls in high dimensions using core-sets. J. Exp. Algorithmics (JEA) **8**, 1–1 (2003)
8. Liu, F.: Minimum Enclosing Ball-Based Learner Independent Knowledge Transfer for Correlated Multi-task Learning. Master's thesis, School of Computing and Mathematical Sciences, Auckland University of Technology, September 2010
9. Mou, L., Meng, Z., Yan, R., Li, G., Xu, Y., Zhang, L., Jin, Z.: How transferable are neural networks in NLP applications? In: Proceedings of the 2016 Conference on Empirical Methods in Natural Language Processing (EMNLP), pp. 479–489 (2016)
10. Pang, S., Liu, F., Kadobayashi, Y., Ban, T., Inoue, D.: Training minimum enclosing balls for cross tasks knowledge transfer. In: ICONIP, pp. 375–382, November 2012

11. Pang, S., Liu, F., Kadobayashi, Y., Ban, T., Inoue, D.: A learner-independent knowledge transfer approach to multi-task learning. Cogn. Comput. **6**(3), 304–320 (2014)
12. Silver, D.L., Mercer, R.E.: Selective functional transfer: Inductive bias from related tasks. In: IASTED International Conference on Artificial Intelligence and Soft Computing (ASC2001), pp. 182–189 (2002)
13. Tsang, I.W., Kwok, J.T., Cheung, P.M.: Core vector machines: fast SVM training on very large data sets. J. Mach. Learn. Res. **6**, 363–392 (2005)
14. LeCun, Y., Bottou, L., Bengio, Y., Haffner, P.: Gradient-based learning applied to document recognition. Proc. IEEE **86**, 2278–2324 (1998)
15. Yosinski, J., Clune, J., Bengio, Y., Lipson, H.: How transferable are features in deep neural networks? In: NIPS, pp. 3320–3328, December 2014

Cross-domain Recommendation with Probabilistic Knowledge Transfer

Qian Zhang, Dianshuang Wu, Jie Lu$^{(\boxtimes)}$, and Guangquan Zhang

Decision Systems and e-Service Intelligence Laboratory,
Center for Artificial Intelligence, Faculty of Engineering and Information Technology,
University of Technology Sydney, Sydney, Australia
{qian.zhang-1,dianshuang.wu,jie.lu,guangquan.zhang}@uts.edu.au

Abstract. Recommender systems have drawn great attention from both academic and practical area. One challenging and common problem in many recommendation methods is data sparsity, due to the limited number of observed user interaction with the products/services. To alleviate the data sparsity problem, cross-domain recommendation methods are developed to share group-level knowledge in several domains so that recommendation in the domain with scarce data can benefit from domains with relatively abundant data. However, divergence exists in the data of similar domains so that the extracted group-level knowledge is not always suitable to be applied in the target domain, thus recommendation accuracy in the target domain is impaired. In this paper, we propose a cross-domain recommendation method with probabilistic knowledge transfer. The proposed method maintain two sets of group-level knowledge, profiling both domain-shared and domain-specific characteristics of the data. In this way users' mixed preferences can be profiled comprehensively thus improves the performance of the cross-domain recommender systems. Experiments are conducted on five real-world datasets in three categories: movies, books and music. The results for nine cross-domain recommendation tasks show that our proposed method has improved the accuracy compared with five benchmarks.

Keywords: Recommender systems
Cross-domain recommender systems · Knowledge transfer
Probabilistic model

1 Introduction

Recommender systems are rapidly developed and widely used in e-commerce and online shopping website [7]. These systems aim to provide recommendations to users to help them choose products or services they need in the era of information explosion. One basic and challenging issue is the data sparsity problem, which greatly impairs the performance of recommender systems, leading to poor user experience and their unsatisfactory [11]. Cross-domain recommender systems are developed to deal with this problem in the fierce market competition [1]. One user

© Springer Nature Switzerland AG 2018
L. Cheng et al. (Eds.): ICONIP 2018, LNCS 11303, pp. 208–219, 2018.
https://doi.org/10.1007/978-3-030-04182-3_19

may not have enough data in one domain, but have more data in another domain. The abundance of data in another domain can assist the recommendation in a specific target domain. By taking the advantages of data in multiple domains, cross-domain recommender systems can exploit the relatively dense data in the source domain to assist recommendation with scarce data in the target domain.

Cross-domain recommender systems can be clustered into three groups: cross-domain recommender systems with side information, with partially overlapping entities and with non-overlapping entity. Due to the privacy issue, users are always de-identified and the correspondence between users are not available [13]. In this paper, we focus on the most commonly happened scenario: cross-domain with non-overlapping entities. Some methods are developed to handle this problem by transferring knowledge from group-level. Users and items are clustered into groups and knowledge is shared through group-level rating patterns. For example, codebook transfer (CBT) clusters users and items into groups and extracts group-level knowledge as a "codebook" [4]. Later, a probabilistic model named rating matrix generative model (RMGM) is extended from CBT, relaxing the hard group membership to soft membership [5]. These two methods cannot ensure that the information on the two groups from two different domains is consistent, and the effectiveness of knowledge transfer is not guaranteed.

Although cross-domain recommender systems have gained lots of attention and efforts from academia, they still suffer the "negative transfer" problem [6, 8]. The main reason is that data collected from two correlated domains are probably from two related but different distributions. Most existing methods on cross-domain recommendation ignore the domain shifts and extract group-level knowledge directly without considering domain-specific characteristics. The group-level knowledge is not suited to the target domain, thus degrades the performance of the cross-domain recommender system.

In this paper, we investigate how to improve the performance of cross-domain recommender systems by exploring domain-specific characteristics in each domain, which is crucial when divergence exists in the data. The proposed method assumes that ratings are generated from two sets of group-level knowledge. One is shared across multiple domains, i.e. the domain-shared knowledge, while the other is domain-specific knowledge which is different for each domain. Users' mixed preferences in a target domain can be profiled with the help of common features extracted from data in other domains and reserve some unique features of the data from the target domain. In this way, group-level knowledge is shared while domain-specific knowledge remained. Probabilistic model is a suitable and powerful way to generate two sets of rating patterns from two different priors, and then generate the ratings. Thus, we propose a probabilistic method for knowledge transfer in cross-domain recommendation (ProbKT). The main contributions of this paper are as follows:

(1) A cross-domain recommendation method ProbKT is developed to enable transferring domain-share knowledge while remaining domain-specific characteristics at the same time.

(2) ProbKT has advantages in effectively transferring knowledge in multiple domains with similar data where divergence may exist.

(3) The proposed method ProbKT is evaluated on five real-world datasets with nine cross-domain recommendation tasks comparing with five other non-transfer or cross-domain recommendation methods. The results show that our proposed method outperforms other recommendation methods in sparse data.

The rest of the paper is organized as follows. Section 2 gives some preliminary and a formal description of the problem. Section 3 describes our method using probabilistic model to enable cross-domain recommendation in multiple domains. In Sect. 4, we present our experiments on five real-world datasets containing three data categories. Finally, in Sect. 5, conclusion is provided with some future directions of this research.

2 Preliminary and Problem Formation

In this section, cross-domain recommendation by tri-factorization is briefly introduced. The problem targeted in this paper is also formally formulated.

2.1 Cross-domain Recommendation by Tri-Factorization

Matrix factorization projects both users and items onto the same latent space so that they are comparable, and through their inner products reconstructs the rating matrix [3]. Similarly, the rating matrix $\boldsymbol{R} \in \mathbb{R}^{M \times N}$ (bold letters represent matrixes) can be factorized into three matrixes (suppose there are M users and N items). Users and items are clustered into several latent groups and in the middle is the group-level rating pattern: $\boldsymbol{R} = \boldsymbol{U}\boldsymbol{S}\boldsymbol{V}^T$, where $\boldsymbol{U} \in \mathbb{R}^{M \times K}$ is user group membership matrix, representing users clustered into K groups, $\boldsymbol{V} \in \mathbb{R}^{N \times L}$ is item group membership matrix, representing items clustered into L groups and $\boldsymbol{S} \in \mathbb{R}^{K \times L}$ is the group rating pattern matrix, i.e. the group-level knowledge.

Say rating matrixes in D domains are available, denoted as $\boldsymbol{R}^d$. The assumption of the cross-domain recommender systems is that the group-level knowledge can be shared if these domains are similar. Thus for the D rating matrixes are reconstructed as:

$$\hat{\boldsymbol{R}}^d = \boldsymbol{U}^d \boldsymbol{S} (\boldsymbol{V}^d)^T \tag{1}$$

2.2 Problem Formulation

In our problem setting, there is no correspondence on the users/items across the domains and users/items are treated as completely different. We assume that on both the source and target domains the data are explicit ratings. The problem is formally defined as:

Given D rating matrixes $\boldsymbol{R}^D = \{\boldsymbol{R}^1, ..., \boldsymbol{R}^d, ..., \boldsymbol{R}^D\}$, $\boldsymbol{R}^d \in \mathbb{R}^{M^d \times N^d}$, our goal is to develop a cross-domain recommendation method to assist the recommendation task of predicting the rating using knowledge in one target domain $\boldsymbol{R}^t$

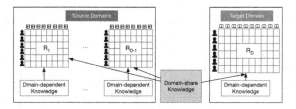

Fig. 1. An example to explain how knowledge is shared in the ProbKT method.

from all the other rating matrixes in $\boldsymbol{R}^D$, where for each source domain data $\mathcal{U}^d \cap \mathcal{U}^t = \emptyset$ and $\mathcal{I}^d \cap \mathcal{I}^t = \emptyset$.

3 Cross-domain Recommendation with Probabilistic Knowledge Transfer

In this section, our proposed ProbKT method is to learn a joint probabilistic model using data in multiple domains.

3.1 The Method Description

As we have seen in the tri-factorization model, the group-level knowledge can be shared cross domains. Since users and items have no intersections, the group-level knowledge may not be totally the same, especially when divergence exists in data between two domains. If we force the group-level knowledge to be the same in two domains, it is likely that ratings predicted are not accurate in the target domain since the extracted knowledge from the source domain is not effectively adapted. We assume that only partial group-level knowledge can be shared between two domains. For a matrix factorization model, it is not easy to solve the optimization problem with constraints on part of the group-level matrix. As shown in Fig. 1, the domain-shared knowledge contributes to data in each domain. Except that, each domain has its own domain-dependent knowledge that contribute to its corresponding domain. In this example shown in Fig. 1, knowledge is extracted from several datasets in movie domain and applied to the targeted book domain. The cross-domain recommender systems is to recommend books to users assisted by data in movie domain. Here, a probabilistic model fits our assumption and is suitable to solve the knowledge transfer issue for this cross-domain recommendation scenario.

Given user-item rating matrixes in D domains $\boldsymbol{R} = \{\boldsymbol{R}^1, ..., \boldsymbol{R}^d, ..., \boldsymbol{R}^D\}$, where $\boldsymbol{R}^d \in \mathbb{R}^{M^d \times N^d}$ in the domain d, M^d is the number of users and N^d is the number of items. In each domain, users and items are from K user groups and L item groups. In these groups, K^s user groups are shared groups between these domains while K^e user groups are specific groups for each domain ($K^s + K^e = K$). Similar notations L^s and L^e go for item groups. One rating R_{ij}^d represents the rating provided by user i on item j. For this rating, it is associated with two

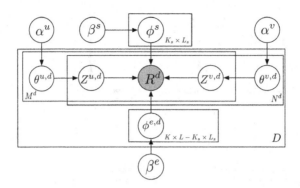

Fig. 2. Graphical model representation of the ProbKT method.

latent variables: Z_{ij}^u and Z_{ij}^v, which represent the user-group and item-group of this rating. There are two group-level rating patterns: ϕ^s representing domain-shared group-level knowledge and ϕ^e representing domain-specific group-level knowledge.

The graphical model representation is in Fig. 2. The generative process is as follows:

(1) For domain-shared user-item joint groups (k^s, l^s)
    Generate $\phi_{k^s,l^s}^s \sim Dir(\beta^s)$
(2) For each domain d
    (a) For domain-specific user-item joint groups (k^e, l^e)
        Generate $\phi_{k^e,l^e}^e \sim Dir(\beta^e)$
    (b) For each user $i = 1, ..., M^d$
        Generate $\theta_i^{u,d} \sim Dir(\alpha^u)$
    (c) For each item $j = 1, ..., N^d$
        Generate $\theta_j^{v,d} \sim Dir(\alpha^v)$
    (d) For the rating R_{ij}^d
        (i) Generate a user group $Z_{ij}^u \sim Multi(\theta_i^{u,d})$
        (ii) Generate an item group $Z_{ij}^v \sim Multi(\theta_j^{v,d})$
        (iii) For user groups $Z_{ij}^u = 1, ..., K^s$ and item groups $Z_{ij}^v = 1, ..., L^s$
            Generate a rating $R_{ij}^d \sim Multi(\phi_{Z_{ij}^u,Z_{ij}^v}^s)$;
            For user groups $Z_{ij}^u = 1, ..., K^s$ and item groups $Z_{ij}^v = L^s + 1, ..., L$,
            user groups $Z_{ij}^u = K^s + 1, ..., K$ and item groups $Z_{ij}^v = 1, ..., L$
            Generate a rating $R_{ij}^d \sim Multi(\phi_{Z_{ij}^u,Z_{ij}^v}^{e,d})$

Here, α^u, α^v, β^e and β^s are hyper-parameters of the Dirichlet priors. Through this model, the domain-shared knowledge ϕ^s and domain-independent knowledge ϕ^e compose the group-level knowledge ϕ in the target domain. The group-level rating matrix, together with the user group membership matrix and item group membership matrix are calculated by: $S_{kl} = \sum_{r=1}^{R_0} r\phi_{klr}$, $U_{ik}^t = \theta_{ik}^u$, and $V_{jl}^t = \theta_{jl}^v$. In our experiment settings, ratings are all from $\{1, 2, 3, 4, 5\}$, thus $R_0 = 5$.

3.2 The Method Learning

For our probabilistic method, we have latent variable set $\Omega = \{\theta^{u,1:\sum_{d=1}^{D} M^d},$
$\theta^{v,1:\sum_{d=1}^{D} N^d}, Z^{u,1:\sum_{d=1}^{D} M^d N^d}, Z^{v,1:\sum_{d=1}^{D} M^d N^d}, \phi^{e,1:D(KL-K_sL_s)}, \phi^{s,1:KL}\}$
and parameter set $\Theta = \{\alpha^u, \alpha^v, \beta^s, \beta^e\}$. We use Jensen's inequality to acquire
the lower bound of the log-likelihood of our proposed probabilistic method:

$$\log P(\boldsymbol{R}^d; \Theta) \geq \mathbb{E}_Q[\log P(\boldsymbol{R}^d, \Omega; \Theta)] - \mathbb{E}_Q[\log Q(\Omega)] \tag{2}$$

$Q(\Omega)$ is the approximating distribution governed by a set of variational
parameters $\Omega = \{\gamma^{u,1:\sum_{d=1}^{D} M^d}, \gamma^{v,1:\sum_{d=1}^{D} N^d}, \xi^{u,1:\sum_{d=1}^{D} M^d N^d}, \xi^{v,1:\sum_{d=1}^{D} M^d N^d},$
$\eta^{e,1:D(KL-K_sL_s)}, \eta^{s,1:KL}\}$. The distance between the lower bound and the true
log-likelihood is the KullbackLeibler (KL) divergence. The KL divergence is zero
if the distribution $Q(\Omega)$ is equal to the true posterior. Distribution $Q(\Omega)$ should
be restricted that it is tractable while at the same time allowing it to provide a
good approximation to the true posterior distribution. Usually, mean field theory
[10] is used and the distribution $Q(\Omega)$ is:

$$Q(\Omega) = \prod_{k=1}^{K_s} \prod_{l=1}^{L_s} Dir(\eta_{k,l}^s) \prod_{d=1}^{D} \{\prod_{i=1}^{M^d} Dir(\gamma_i^{u,d}) \prod_{j=1}^{N^d} Dir(\gamma_j^{v,d}) \prod_{k=K_s+1}^{K} \prod_{l=1}^{L}$$

$$Dir(\eta_{k,l}^{e,d}) \prod_{k=1}^{K_s} \prod_{l=L_s+1}^{L} Dir(\eta_{k,l}^{e,d}) \prod_{i=1}^{M^d} \prod_{j=1}^{N^d} [Multi(\xi_{i,j}^{u,d}) Multi(\xi_{i,j}^{v,d})]\} \tag{3}$$

The optimization of minimizing the KL-divergence can be done using the
following update equations:

$$\gamma_{i,k}^{u,d} = \alpha_k^u + \sum_{j=1}^{N^d} I_{i,j}^d \xi_{i,j,k}^{u,d} \tag{4}$$

$$\gamma_{j,l}^{v,d} = \alpha_l^v + \sum_{i=1}^{M^d} I_{i,j}^d \xi_{i,j,l}^{v,d} \tag{5}$$

In each domain $d \in \{1, ..., D\}$, for $k \in \{1, ..., K_s\}$,

$$\xi_{i,j,k}^{u,d} = exp\{\psi(\gamma_{i,k}^{u,d}) - \psi(\hat{\gamma}_i^{u,d}) + \sum_{l=1}^{L_s} \xi_{i,j,l}^{v,d}(\psi(\eta_{k,l,R_{i,j}^d}^s) - \psi(\hat{\eta}_{k,l,r}^s)) +$$

$$\sum_{l=L_s+1}^{L} \xi_{i,j,l}^{v,d}(\psi(\eta_{k,l,R_{i,j}^d}^{e,d}) - \psi(\hat{\eta}_{k,l,r}^{e,d}))\} \tag{6}$$

where $\hat{\gamma}_i^{u,d} = \sum_{k=1}^{K} \gamma_{i,k}^{u,d}, \hat{\eta}_{k,l,r}^s = \sum_{r=1}^{R_0} \eta_{k,l,r}^s$ and $\hat{\eta}_{k,l,r}^{e,d} = \sum_{r=1}^{R_0} \eta_{k,l,r}^{e,d}$.
For $k \in \{K_s + 1, ..., K\}$,

$$\xi_{i,j,k}^{u,d} = exp\{\psi(\gamma_{i,k}^{u,d}) - \psi(\hat{\gamma}_i^{u,d}) + \sum_{l=1}^{L} \xi_{i,j,l}^{v,d}(\psi(\eta_{k,l,R_{i,j}^d}^{e,d}) - \psi(\hat{\eta}_{k,l,r}^{e,d}))\} \tag{7}$$

$$\eta_{k,l,r}^s = \beta_r^s + \sum_{d=1}^{D} \sum_{i=1}^{M_d} \sum_{j=1}^{N_d} I_{i,j,r}^d \xi_{i,j,k}^{u,d} \xi_{i,j,l}^{v,d} \tag{8}$$

$$\eta_{k,l,r}^{e,d} = \beta_r^e + \sum_{i=1}^{M_d} \sum_{j=1}^{N_d} I_{i,j,r}^d \xi_{i,j,k}^{u,d} \xi_{i,j,l}^{v,d} \tag{9}$$

4 Experiments

In this section, the proposed method ProbKT is evaluated. First, we introduce the datasets and the used evaluation metrics in Sect. 4.1, followed by experimental settings and the baseline methods in Sect. 4.2. The results of the experiments are presented in Sect. 4.3.

4.1 Datasets and Evaluation Metrics

To test our proposed method, the source domain data and the target domain data are chosen where they are similar but still have divergence between them. The five real-world datasets we used are: EachMovie[1], Movielens1M[2], LibraryThing[3], Amazon Book[4] and YahooMusic[5]. Each of these datasets is publicly available. Numerous experiments are conducted on those in single-domain recommendation methods, but experiments of those datasets on cross-domain recommendation methods are deficient. For AmazonBooks, we removed all users who had given exactly the same rating for every book, as these data are not effective for constructing a recommender system [12]. EachMovie and LibraryThing were normalized to the range of $\{1, 2, 3, 4, 5\}$ before conducting experiments. Refer to [12] about details of the five datasets.

For all the datasets, we filtered out items that are rated less than 10 times and users that who have rated less than 20 items. In our experiment setting, the source domain dataset is more dense than the target domain data. 500 users and 1000 items are randomly chosen for both the source domain and the target domain. But for the target domain, 300 users are randomly selected to be new customers, who are given only 5 observed ratings, and the left are put in the test set. The details of the chosen subsets are listed in Table 1. Three categories are in our chosen datasets, and our recommendation tasks are all the combinations of the three categories.

Evaluation metrics are root mean square error (RMSE) and mean absolute error (MAE): $RMSE = \sqrt{\sum_{u,v,X_{uv} \in Z} \frac{(\hat{X}_{uv} - X_{uv})^2}{|Z|}}$ and $MAE = \sum_{u,v,X_{uv} \in Z} \frac{|\hat{X}_{uv} - X_{uv}|}{|Z|}$, where Z is the test set, and $|Z|$ is the test ratings number.

---

[1] http://www.cs.cmu.edu/~lebanon/IR-lab/data.html#intro.
[2] http://grouplens.org/datasets/movielens/1m/.
[3] https://www.librarything.com.
[4] http://jmcauley.ucsd.edu/data/amazon/.
[5] https://webscope.sandbox.yahoo.com/catalog.php?datatype=r.

Table 1. Description of subsets in five real-word datasets

Data type	Data source	Domain	Sparsity	Average
Movie	EachMovie	source	96.00%	4.32
	Movielens1M	target	98.50%	2.91
book	LibraryThing	source	87.43%	3.97
	AmazonBook	target	97.87%	3.13
music	YahooMusic_1	source	95.70%	4.14
	YahooMusic_2	target	97.27%	2.66

4.2 Baselines and Experimental Settings

We use three non-transfer learning methods and two cross-domain methods for comparison. The non-transfer learning methods were: Pearson's correlation coefficient (PCC) [2], single value decomposition (SVD) [3] and FMM [9]. The cross-domain methods were: RMGM [5] and CBT [4]. These two cross-domain recommendation methods are all developed without fully considering the domain-shift widely existed in data of two domains. PCC is a user-based CF recommendation method, and the neighborhoods was set to 50. The latent feature number in SVD, FMM, CBT and RMGM was fixed at 40. In SVD, the learning rate was set to 0.003 and the regularization factor was set to 0.015. To be fair, the number of user groups and item groups in our proposed method are both set to be the same at latent feature number in other methods. The number of domain-shared user and item groups is set to be half of the total number of user/item groups. For hyper-parameters, α^s and α^t are set to be 0.1 while β^e and β^s are set to be 0.2. Twenty random initializations in our experiments are set and the averaged results and standard deviations are reported.

4.3 Results

The experiment results of our proposed ProbKT compared with the other five baselines on two accuracy metrics are presented in Tables 2, 3 and 4. Overall, ProbKT has the best performance in all the nine tasks. These results indicate that ProbKT can extract common knowledge to share between the source and target domain, which can help increase the recommendation accuracy in the target domain.

Compared with non-transfer learning recommendation methods like PCC, FMM and SVD, experiment results suggest that transfer learning in ProbKT is effective since it can improve the recommendation. The other two cross-domain recommendation methods CBT and RMGM, sometimes fail to improve the performance of recommender systems in the target domain. According to the results, these two methods are largely dependent on their basic method, FMM. If FMM can make accurate recommendation in the target domain, it is likely these two methods can transfer knowledge from the source domain to the target domain.

Table 2. Cross-domain recommendation results on the movie target domain

Methods		Source data	MAE	RMSE
non-trans	PCC	-	1.2123	1.5722
	FMM	-	1.3529±0.0025	1.6751±0.0019
	SVD	-	1.0949±0.0049	1.3489±0.0035
CDRS	CBT	movie	1.3772±0.0375	1.7152±0.0332
		book	1.1640±0.0209	1.4429±0.0323
		music	1.2094±0.0122	1.5177±0.0216
	RMGM	movie	1.3098±0.0188	1.6327±0.0189
		book	1.1692±0.0063	1.4547±0.0075
		music	1.2371±0.0135	1.5409±0.0152
	ProbKT	movie	**0.9980±0.0016**	**1.2152±0.0016**
		book	**0.9937±0.0017**	**1.2144±0.0021**
		music	**0.9991±0.0027**	**1.2222±0.0027**

Otherwise, their performance is not as good as the non-transfer learning recommendation method SVD. ProbKT, on the other hand, does not have this
concern, due to that the probabilistic model is more flexible and able to fit to
different datasets.

Compared with the other two cross-domain recommendation methods CBT
and RMGM, ProbKT also has better performance. The core part of CBT is to
directly apply extracted group-level knowledge from the source domain to the
target domain without adaptation or adjustment. RMGM replaced the hard-
membership of user/item groups in CBT to soft-membership and relaxes the constraints, which enhances its effectiveness of transferring knowledge. But these two
methods cannot properly deal with the divergence existed between two domains.
Through the probabilistic method we have proposed in modeling the group-
level knowledge in two parts: domain-dependent part and domain-independent
part. The domain-independent part is the knowledge that shared between two
domains. The domain-dependent part is able to capture the characteristics in
each domain and allows the method to be more flexible.

4.4 Parameter Analysis and Complexity Analysis

We analyzed how the parameters K and L affect the performance of ProbKT.
Due to the space limitation, only the result of movie as source domain and target
domain is presented. To analyze K and L, grid search is used with evaluation
metrics of both MAE and RMSE as shown in Fig. 3. The result of analysis shows
that the larger of K and L, the better of the performance of ProbKT. This fits to
the intuitive fact that more groups of user and items are, the more delicate the

Table 3. Cross-domain recommendation results on the book target domain

Methods		Source data	MAE	RMSE
non-trans	PCC	-	1.1802	1.4907
	FMM	-	1.0274±0.0075	1.2631±0.0082
	SVD	-	1.2117±0.0133	1.5283±0.0166
CDRS	CBT	movie	1.2001±0.0142	1.5228±0.0127
		book	1.0775±0.0076	1.3489±0.0155
		music	1.1024±0.0039	1.4003±0.0091
	RMGM	movie	1.0235±0.0077	1.2691±0.0108
		book	1.0197±0.0064	1.2623±0.0102
		music	1.0228±0.0043	1.2644±0.0094
	ProbKT	movie	**0.9757±0.0044**	**1.1811±0.0049**
		book	**0.9838±0.0044**	**1.1858±0.0050**
		music	**0.9822±0.0054**	**1.1889±0.0070**

Table 4. Cross-domain recommendation results on the music target domain

Methods		Source data	MAE	RMSE
non-trans	PCC	-	1.4843	1.8539
	FMM	-	1.3040±0.0060	1.5787±0.0053
	SVD	-	1.4778±0.0134	1.7195±0.0181
CDRS	CBT	movie	1.7532±0.0208	2.0958±0.0223
		book	1.5843±0.0122	1.8599±0.0182
		music	1.6251±0.0329	1.9246±0.0504
	RMGM	movie	1.3369±0.0148	1.6216±0.0149
		book	1.3670±0.0162	1.6470±0.0218
		music	1.3413±0.0197	1.6261±0.0228
	ProbKT	movie	**1.3053±0.0069**	**1.5030±0.0058**
		book	**1.3050±0.0083**	**1.5037±0.0069**
		music	**1.3114±0.0077**	**1.5072±0.0081**

probabilistic method can model the users and items. However, the complexity of the method will significantly increase with the increase of K and L. The time consumed by different K is shown in Table 5. For simplicity, the setting of L is set to be the same as K. The time consumption shown in Table 5 is for 20 iterations of the proposed method. For fair comparison with other baselines, we choose 40 for both K and L for ProbKT in our comparison experiments. The total complexity of ProbKT is $O(n)$.

Table 5. Time consumption with different settings of K

K	MAE	RMSE	time(s)
$K = 10$	1.0464±0.0072	1.2563±0.0076	469.15
$K = 20$	1.0282±0.0032	1.2358±0.0035	566.04
$K = 30$	1.0106±0.0031	1.2198±0.0026	704.02
$K = 40$	0.9981±0.0023	1.2153±0.0024	848.16
$K = 50$	0.9922±0.0025	1.2168±0.0027	1045.00
$K = 60$	0.9891±0.0026	1.2190±0.0020	1239.54
$K = 70$	0.9861±0.0017	1.2199±0.0017	1573.37
$K = 80$	0.9854±0.0017	1.2207±0.0015	2228.76
$K = 90$	0.9838±0.0016	1.2206±0.0013	2663.10
$K = 100$	0.9829±0.0017	1.2209±0.0009	4691.15

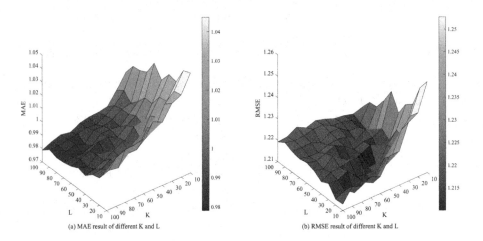

(a) MAE result of different K and L   (b) RMSE result of different K and L

Fig. 3. Results with different settings on parameter K and L.

5   Conclusion

In this paper, we develop a probabilistic method named ProbKT for cross-domain recommendation in multiple domains. Unlike previously developed cross-domain recommendation methods sharing group-level knowledge, our proposed ProbKT method takes both domain-share and domain-specific knowledge into consideration. Through generating group-level knowledge from two different priors, ProbKT relaxes the constrains of previous cross-domain recommendation methods sharing group-level knowledge. In this way, ProbKT is more flexible and is able to deal with data from multiple domains with divergence existed. With the virtue of knowledge extracted from multiple source domains, ProbKT alleviates the data sparsity problem and increases the prediction accuracy in cross-domain recommendation. Experiments on five real-world datasets with nine cross-domain recommendation tasks demonstrate that our method ProbKT achieves the best performance compared with five baselines including both non-transfer learning and cross-domain recommendation methods. In the future, we will try to develop methods that can deal with heterogeneous data in this problem setting. Also, a Bayesian deep learning will be established as a more delicate method involving various user behaviors.

Acknowledgment. This work was supported by the Australian Research Council (ARC) under Discovery Grant [DP170101632].

References

1. Cremonesi, P., Quadrana, M.: Cross-domain recommendations without overlapping data: myth or reality? In: Proceedings of the 8th ACM Conference on Recommender Systems, pp. 297–300. ACM (2014)
2. Deshpande, M., Karypis, G.: Item-based top-n recommendation algorithms. ACM Trans. Inf. Syst. **22**(1), 143–177 (2004)
3. Koren, Y., Bell, R., Volinsky, C.: Matrix factorization techniques for recommender systems. Computer **42**(8), 30–37 (2009)
4. Li, B., Yang, Q., Xue, X.: Can movies and books collaborate? Cross-domain collaborative filtering for sparsity reduction. IJCAI **9**, 2052–2057 (2009)
5. Li, B., Yang, Q., Xue, X.: Transfer learning for collaborative filtering via a rating-matrix generative model. In: Proceedings of the 26th Annual International Conference on Machine Learning, pp. 617–624. ACM (2009)
6. Lu, J., Behbood, V., Hao, P., Zuo, H., Xue, S., Zhang, G.: Transfer learning using computational intelligence: a survey. Knowl. Based Syst. **80**, 14–23 (2015)
7. Lu, J., Wu, D., Mao, M., Wang, W., Zhang, G.: Recommender system application developments: a survey. Decis. Support Syst. **74**, 12–32 (2015)
8. Lu, J., Xuan, J., Zhang, G., Luo, X.: Structural property-aware multilayer network embedding for latent factor analysis. Pattern Recogn. **76**, 228–241 (2018)
9. Si, L., Jin, R.: Flexible mixture model for collaborative filtering. In: Proceedings of the 20th International Conference on Machine Learning, pp. 704–711 (2003)
10. Xing, E.P., Jordan, M.I., Russell, S.: A generalized mean field algorithm for variational inference in exponential families. In: Proceedings of the 19th Conference on Uncertainty in Artificial Intelligence, pp. 583–591. Morgan Kaufmann Publishers Inc. (2002)
11. Xu, J., Yao, Y., Tong, H., Tao, X., Lu, J.: Rapare: a generic strategy for cold-start rating prediction problem. IEEE Trans. Knowl. Data Eng. **29**(6), 1296–1309 (2017)
12. Zhang, Q., Wu, D., Lu, J., Liu, F., Zhang, G.: A cross-domain recommender system with consistent information transfer. Decis. Support Syst. **104**, 49–63 (2017)
13. Zhao, L., Pan, S.J., Yang, Q.: A unified framework of active transfer learning for cross-system recommendation. Artif. Intell. **245**, 38–55 (2017)

Transfer Learning with Active Queries for Relational Data Modeling Across Multiple Information Networks

Ke-Jia Chen$^{(\boxtimes)}$, Kai Zhang, Xi-Lin Jiang, and Yunyun Wang

Jiangsu Key Laboratory of Big Data Security & Intelligent Processing, Nanjing University of Posts and Telecommunications, Nanjing 210023, Jiangsu, China
chenkj@njupt.edu.cn

Abstract. This paper studies the relationship prediction problem in multi-network scenarios, aiming to overcome the network sparsity challenge where the labeled data (connected node pairs) are much less than the unlabeled data (unconnected node pairs). The TAQIL framework is proposed by using transfer learning to get knowledge from the related source networks and then use active learning to query the labels of the most informative instances from the oracle in the target network. A new query function is also proposed in order to better use the parameters output by the transfer learning method. The alternate use of transfer learning and active learning allows adaptive transfer of knowledge across multiple networks to mitigate *cold start* and meantime improve the prediction accuracy with active queries in the target network. The experimental results on both non-network datasets and network datasets demonstrate the significant improvement in prediction accuracy compared with several benchmark methods and related state-of-art methods.

Keywords: Collective link prediction · Transfer learning
Active learning · Heterogeneous information networks

1 Introduction

Many realistic prediction tasks are essentially relational data modeling in information networks [5]. For example, personalized recommendation involves predicting the potential preference relationship between users and items (e.g. books, films, products, advertisements) based on the observed relationships in the form of users' past clicks or ratings on items. However, the relational data in real information networks are often sparse. In the recommendation system, the majority of users may only rate a few items. In the online social network, the existing social relationships could take only a small part of all the possible relationships. Moreover, it is more difficult to infer relationships accurately especially when the information network is new. Therefore, the network sparsity problem (or *new network* problem) brings great challenges to relational data modeling and relationship prediction.

© Springer Nature Switzerland AG 2018
L. Cheng et al. (Eds.): ICONIP 2018, LNCS 11303, pp. 220–229, 2018.
https://doi.org/10.1007/978-3-030-04182-3_20

If the nodes pairs in information networks are treated as instances and the relationships between the nodes are treated as labels, the above task is similar to the problem of learning from limited amount of the labeled data, where there exists relatively insufficient labeled data (i.e. connected node pairs) but plenty of unlabeled data (i.e. unconnected node pairs).

Transfer learning and active learning are two effective approaches to overcome this challenge. In transfer learning [11], a high quality classifier can be trained from data in the target domain by utilizing data from the related source domains. On the basis of transferred data, active learning [13] can be used to query labels for the most informative unlabeled data from an oracle in target domain, thereby further improving the classification performance. The technique of combining transfer learning and active learning has been widely used in [7–10,12,14,16,17].

In practice, there may exist multiple information networks with similar types. So the prediction tasks in multiple networks are strongly correlated [4]. In this paper, we propose a method named TAQIL (Transfer Learning with Active Queries in Inferring Links) which combines transfer learning and active learning to model relational data in multiple information networks. It aims to accelerate the learning process, mitigate the network sparsity challenge and accurately infer relationships in the target network.

In TAQIL, the classifier is initially trained using plenty of examples in the source network and limited examples in the target network based on TrAdaBoost [3] strategy. It is then used to select instances to label (based on a proposed query function) by querying in the target network. Once labeled, the new examples are added into the training set and the TrAdaBoost process is iterated. In this paper, the training set and the test set are constructed using instances based on node pairs in the network and the instances are labeled according to the relationship between nodes.

The rest of this paper is organized as follows. Some related work are reviewed in Sect. 2. Section 3 firstly gives the notations and problem formulation and then describes the details of the proposed framework. The experimental results are presented and analyzed in Sect. 4. Section 5 concludes the paper.

2   Related Work

In recent years, there have been increasing interests in combing transfer learning and active learning. The former can improve the model by using plenty of labeled data in the source domain with limited labeled data in the target domain; The latter can query the labels actively in the target domain to adapt the model.

Some of the existing methods perform transfer learning and active learning alternately. The framework AcTraK in [14] builds a classifier in the source domain to predict labels for the target domain, and queries the oracle only if the prediction is of low confidence. In [12], a domain separator is built to distinguish target domain data from source domain data, and further used to query labels of those target domain examples that are not similar to examples from the source domain. The method in [10] trains two individual classifiers from the data in the source and target domains respectively, and then uses the Query By Committee (QBC) strategy to actively select instances from the target domain. The transfer process in [16] is achieved under the assumption that the changes in all marginal and conditional distributions are smooth and then combined with active learning methods.

Other methods integrate transfer and active learning into a unified framework. The method of Kale and Liu [9] presents a framework to combine the agnostic active learning algorithm and transfer learning, aiming to improve the performance of an active learner in the target domain with labeled data from the source domain. The framework in [2] integrates transfer and active learning into a single convex optimization problem. It computes the weights of source domain data and selects the samples from the target domain data simultaneously. Kale et al. [8] later propose a hierarchical active transfer learning method HATL, which exploits cluster structure shared between different data domains to perform transfer learning. Huang and Chen [7] jointly perform transfer learning and active learning by querying the most valuable information from the source domain when the oracle is unavailable in the target domain.

In multiple network link prediction problem, the links in the target network are usually sparse, so data from auxiliary networks can be transferred to help the prediction in the target network [1,18]. But it is rare to use transfer learning and active learning together to solve this problem. The most related work to this paper is TranFG (transfer-based factor graph) proposed by Tang et al. [15]. It classifies the type of social relationships by learning across heterogeneous networks and use active learning to further improve the performance of predictor. Another related work recently published in [19] is an active transfer learning method for recommendation system, where active learning is used to construct entity correspondences across systems.

3   The Proposed Method

This section introduces the proposed method TAQIL (Transfer Learning with Active Queries for Inferring Links). We define notation and problem formulation in Sect. 3.1 and describe TAQIL in Sect. 3.2.

3.1   Problem Formulation

We first define the notations used in the paper (Table 1).

To simplify the explanation, the paper currently studies the transfer prediction model between two information networks: a source network and a target network, but the framework can also be generalized to multiple networks. Given a source network G_s with abundantly relationship data and a target network G_t with limited relationship data, our objective is to learn a function $\phi : (G_t|G_s) \to R_t$ which can infer the potential relationships in G_t by leveraging the supervised data from G_s. Here, $|E_s^r| >> |E_t^r|$ and $|E_t^u| >> |E_t^r|$. In other words, the number of relationships in the source network is more larger than

Table 1. Symbol definition

G	An information network
E^r	A set of node pairs with relationship in G
E^u	A set of node pairs without relationship in G
x	The feature vector for a given node pair
X	The set of x in G
L	The labeled dataset in G, $L = \{x_i, f(x_i)\}$, $x_i \in X$ $(i = 1, ..., n)$, n is the number of examples
$f(x)$	A Boolean function denoting the label of x
ϕ	A prediction function for E^u in G

that in the target network, and the target network is sparse with an extreme case of $|E_t^r| = 0$.

3.2 The TAQIL Framework

The proposed TAQIL framework aims to solve the cold-start problem in inferring links in G_t, where the relative sufficient number of labeled examples L_s in G_s are used to train a classification model and this model is further improved with active queries from the target network G_t. This process can be repeated for several times.

Figure 1 illustrates the TAQIL framework across two reviewer networks: Epinions and Slashdot. In our problem setting, the labeled data is insufficient in the target network (Slashdot) but sufficient in the source network (Epinions). In the Slashdot network, the oracle is available, so the unlabeled data (unconnected node pairs) can be iteratively selected to query their labels (like or dislike). Our objective is to leverage the labeled data in the Epinions network to help infer the new relationships in the Slashdot network with least queries.

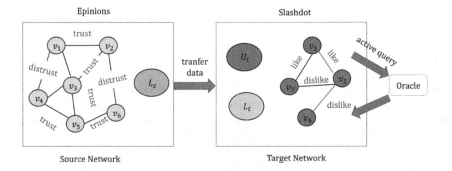

Fig. 1. Example of TAQIL framework across two information networks

In TAQIL, the TrAdaBoost transfer model [3] is used as the preliminary attempt because in the case of only a small amount of target network data L_t, the algorithm can leverage the old source network data L_s to construct a high-quality classification model for G_t. For active learning, we consider a variant known as pool-based active learning and use uncertainty sampling strategy. The traditional query function chooses the instances nearest to the classification plane as the most informative instances (Eq. 1).

$$min_{x^t \in U^t} |w^t x^t| \tag{1}$$

In order to reduce the average weighted training loss on the diff-distribution data, TrAdaBoost outputs the hypothesis h_k and its weight β_k after N iteration $(k = \lceil N/2 \rceil, ..., N)$. Therefore, a new query function is defined in Eq. 2 which leverages the output of TrAdaBoost. Here, $h_k(x_i^t) = P(f_t(x) = 1|x_i^t)$ denotes the possibility of the connection between a pair of nodes which is represented by x_i^t. The smaller the value of Eq. 2, the more informative the instance x_i^t. In other words, the instance selected by the query function from U_t is not the instance nearest to one classification plane, but the instance nearest to $\lfloor N/2 \rfloor$ classification planes.

$$min_{x_i^t \in U_t} |\sum_{k=\lceil N/2 \rceil}^{N} ln(1/\beta_k)(h_k(x_i^t) - 0.5)| \tag{2}$$

The TAQIL framework is summarized in Algorithm 1. Notice that L_s and L_t includes all the connected node pairs as positive examples and some of the unconnected node pairs as negative examples. The negative examples currently are random sampled without using semi-supervised technique. *Learner* is the basic learning model and Q is the maximum number of examples labeled by the oracle.

$$\phi(x_i^t) = \begin{cases} 1, & \sum_{k=\lceil N/2 \rceil}^{N} ln(1/\beta_k)h_k(x_i^t) \geqslant 0.5\sum_{k=\lceil N/2 \rceil}^{N} ln(1/\beta_k) \\ 0, & \text{otherwise.} \end{cases} \tag{3}$$

4   Experiment

4.1   Settings

We evaluate our proposed framework first on non-network datasets (the *Mushroom* dataset from the UCI machine learning repository[1] and two text datasets *20Newsgroup*[2] and *Reuters-21578*[3]) and eventually on three real information network datasets *Epinions, Slashdot* and *MobileU*[4].

---

[1] http://www.ics.uci.edu/mlearn/MLRepository.html.
[2] http://people.csail.mit.edu/jrennie/20Newsgroups/.
[3] http://www.daviddlewis.com/resources/testcollections/.
[4] http://arnetminer.org/socialtieacross/.

Algorithm 1. Description of TAQIL

Input:
   L_s, L_t, U_t, Learner
Output:
   The actively transfer predictor: ϕ
1: Call TrAdaBoost(L_s, L_t, Learner, N)
2: Get h_k and β_k, $k = \lceil N/2 \rceil$,...,N
3: **for** i=1,...,Q **do**
4:     Call function Query(h_k, β_t, T_u)
5:     Select x_q^t from U_t using Eq. 2
6:     Query the label y_q^t of x_q^t from an oracle
7:     $L_t = L_t \cup (x_q^t, y_q^t)$, $U_t = U_t/(x_q^t, y_q^t)$
8:     Call TrAdaBoost(L_s, L_t, Learner, N) again
9:     Update h_k and β_k
10: **end for**
11: Calculate ϕ using Eq. 3

The settings in non-network datasets are consistent with the experimental settings in [3]. For network datasets, we use Epinions as the source network, and Slashdot and MobileU as the target network respectively.

All tasks, the datasets they used and experimental settings are listed in Table 2. The name of *edible-poisonous* indicates that all the positive instances are from the category *edible*, while negative ones from *poisonous*. The tasks 2–4 are named in the same way. In the last two tasks, positive instances are from the positive relationships (trust or like) and negative instances are from the negative relationships (distrust or dislike). Six commonly used structural features (common neighbors, Jaccard coefficient, in-degree, out-degree, total-degree and Sorensen Index) [6] are calculated to describe node pairs. In experimental settings, each target data is split into three subsets: 50% of the data is used for testing, 49% of the data constitutes unlabeled data U_t and 1% of the data constitutes labeled data L_t. The maximum number of the active queries is set to 150 and the number of iteration N is set to 100. All experiment results are the average values using 10-fold cross validation.

Table 2. Datasets and experiment settings

Task		Datasets	Settings	
			$|L_t \cup U_t|$	L_s
1:	edible-poisonous	Mushroom	4,608	3,156
2:	rec-sci	20Newsgroup	1,186	1,190
3:	rec-talk	20Newsgroup	1,137	1,137
4:	orgs-people	Reuters-21578	1,016	1,046
5:	Eponions2Slashdot	Epinions & Slashdot	1,020	370
6:	Epinions2MobileU	Epinions & MobileU	1,020	314

4.2   Comparison Methods

The following methods are compared with our method.

- *Active* is a classifier model based on active learning. It is trained with data only from L_t (i.e. labeled data in the target domain) and uses uncertainty sampling strategy.
- *Active-s* has the same settings with *Active* except that it is trained with data from both L_t and L_s (i.e. labeled data in the source domain).
- *TAQIL* is our method proposed for inferring new relationships but can also be used to classify the instances in non-network datasets.
- *TAQIL-b* is a baseline method which has the same settings with *TAQIL* except that it uses random sampling rather than uncertainty sampling in active queries.
- *Actrack* [14] is a related active transfer learning method, which selects an essential instance to construct an inductive model from the target dataset, and then a transfer classifier is trained with labeled data from L_t and L_s.
- *HATL* [8] is a state-of-art active transfer learning method, which exploits cluster structure shared between different data domains to perform transfer learning and to generate effective label queries during active learning.
- *TranFG* [15] is used to compare with our method in relationship prediction tasks in information networks. It incorporates social theories into a factor graph model by using transfer learning and active learning. Finally, It can predict the types of relationships in the target network.

In the first four methods, we use Logistic Regression as the base classifier. In Actrack and HATL, we use SVM as the base classifier due to its superior performance.

4.3   Experimental Results

The performance curves with increasing queries of six methods (TAQIL, TAQIL-b, HATL, Active-s, AcTrack and Active) in non-network datasets are plotted and compared in Fig. 2. The performance curves with increasing queries of five methods (TAQIL, TAQIL-b, TranFG, Active-s and Active) in network datasets are plotted and compared in Fig. 3. The detailed accuracy values in all tasks of comparative methods with 150 label queries are listed in Table 3.

Figure 2 shows that the proposed method TAQIL outperforms the competing methods throughout the entire label query budget. As expected, Active leads to the worst performance on most datasets due to the limited training data. The performance of Active-s is much higher than Active because the former uses the source domain data in the training set. When comparing TAQIL with TAQIL-b, the proposed method is always superior to its baseline method. It verifies that with the new query function TAQIL can select the most informative unlabeled instance from U_t each time. The accuracy value of HATL and TAQIL is close in Fig. 1, but the complexity of HATL is higher because it needs to get hierarchical cluster tree over U_t and L_s before the transfer process. The performance of

Table 3. The accuracy value of all methods with 150 label queries

Methods	Task1	Task2	Task3	Task4	Task5	Task6
Active	0.8123	0.8383	0.7431	0.8062	0.7643	0.6836
Active-s	0.8863	0.9189	0.8994	0.8258	0.7948	0.7923
TAQIL	**0.9504**	**0.9583**	**0.9522**	**0.8711**	0.8896	**0.8636**
TAQIL-b	0.9304	0.9475	0.9403	0.8375	0.8649	0.7857
Actrack	0.8808	0.8908	0.8508	0.8295	—	—
HALT	0.9203	0.9478	0.9512	0.8604	—	—
TranFG	—	—	—	—	**0.8908**	0.8108

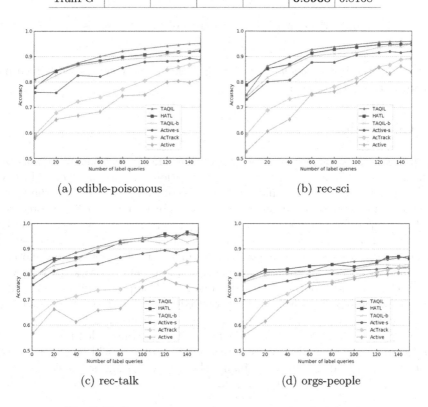

(a) edible-poisonous   (b) rec-sci

(c) rec-talk   (d) orgs-people

Fig. 2. Performance comparison on non-network datasets

AcTrack is quite lower than TAQIL. The possible reason is that Actrack only queries the experts for the class label in case of mislabeling but TAQIL uses TrAdaBoost to iteratively optimize the weights and uses multiple classification planes to select the most informative example to be labeled.

Figure 3 shows similar observation that TAQIL achieves the better performance than Active, Active-s and TAQIL-b. It even outperforms the state-of-art method TranFG in most label query number settings. The experiment verifies

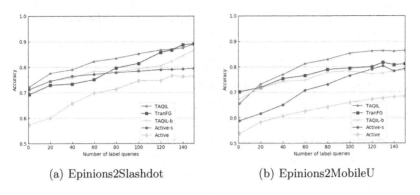

(a) Epinions2Slashdot   (b) Epinions2MobileU

Fig. 3. Performance comparison on real network datasets.

that our method is beneficial for inferring links in the new network which has not much labeled instances.

5   Conclusion

In this paper, we propose a practical active transfer learning framework for inferring links in multiple networks, where relationships are insufficient in both source and target networks, and labels can be actively queried from the target domain. The framework transfers supervised information from the source network to help infer relationships in the target network. A new query function is defined to actively query the label of the most informative example which is the nearest one to $\lfloor N/2 \rfloor$ hyper planes output by the TrAdaBoost method. Experiments on 6 tasks from non-network datasets and network datasets validate the effectiveness of the proposed framework. Comparing with other related active transfer learning methods, the proposed method performs better not only in multiple network settings but also in other classification tasks.

In the future, we plan to extend the framework for transfer learning with multiple source networks. Also, other active query strategies will be experimented under the proposed framework. Finally, the active learning strategies for handling temporal drift will be studied since information networks often evolve over time.

Acknowledgments. This research was supported by the National Natural Science Foundation of China (No. 61571238 and No. 61603197).

References

1. Cao, B., Liu, N.N., Yang, Q.: Transfer learning for collective link prediction in multiple heterogenous domains. In: Proceedings of International Conference on Machine Learning, pp. 159–166. Citeseer (2010)
2. Chattopadhyay, R., Fan, W., Davidson, I., Panchanathan, S., Ye, J.: Joint transfer and batch-mode active learning. In: Proceedings of International Conference on Machine Learning, pp. 253–261 (2013)
3. Dai, W., Yang, Q., Xue, G.R., Yu, Y.: Boosting for transfer learning. In: Proceedings of International Conference on Machine Learning, pp. 193–200. ACM (2007)
4. Dong, Y., Zhang, J., Tang, J., Chawla, N.V., Wang, B.: CoupledLP: link prediction in coupled networks. In: Proceedings of ACM SIGKDD International Conference on Knowledge Discovery and Data Mining, pp. 199–208. ACM (2015)
5. Getoor, L., Diehl, C.P.: Link mining: a survey. SIGKDD Explor. 7(2), 3–12 (2005)
6. Hasan, M.A., Zaki, M.J.: A survey of link prediction in social networks. In: Aggarwal, C. (ed.) Social Network Data Analytics, pp. 243–275. Springer, Boston (2011). https://doi.org/10.1007/978-1-4419-8462-3_9
7. Huang, S.J., Chen, S.: Transfer learning with active queries from source domain. In: Proceedings of IJCAI, pp. 1592–1598 (2016)
8. Kale, D., Ghazvininejad, M., Ramakrishna, A., He, J., Liu, Y.: Hierarchical active transfer learning. In: Proceedings of SIAM International Conference on Data Mining, pp. 514–522. SIAM (2015)
9. Kale, D., Liu, Y.: Accelerating active learning with transfer learning. In: Proceedings of IEEE International Conference on Data Mining, pp. 1085–1090. IEEE (2013)
10. Li, S., Xue, Y., Wang, Z., Zhou, G.: Active learning for cross-domain sentiment classification. In: IJCAI, pp. 2127–2133 (2013)
11. Pan, S.J., Yang, Q.: A survey on transfer learning. IEEE Trans. Knowl. Data Eng. 22(10), 1345–1359 (2010)
12. Saha, A., Rai, P., Daumé, H., Venkatasubramanian, S., DuVall, S.L.: Active supervised domain adaptation. In: Gunopulos, D., Hofmann, T., Malerba, D., Vazirgiannis, M. (eds.) ECML PKDD 2011. LNCS (LNAI), vol. 6913, pp. 97–112. Springer, Heidelberg (2011). https://doi.org/10.1007/978-3-642-23808-6_7
13. Settles, B.: Active learning literature survey. Univ. Wisconsin-Madison, Madison, WI. Technical report, CS Technical report 1648 (2009)
14. Shi, X., Fan, W., Ren, J.: Actively transfer domain knowledge. In: Daelemans, W., Goethals, B., Morik, K. (eds.) ECML PKDD 2008. LNCS (LNAI), vol. 5212, pp. 342–357. Springer, Heidelberg (2008). https://doi.org/10.1007/978-3-540-87481-2_23
15. Tang, J., Lou, T., Kleinberg, J., Wu, S.: Transfer learning to infer social ties across heterogeneous networks. ACM Trans. Inf. Syst. (TOIS) 34(2), 7 (2016)
16. Wang, X., Huang, T.K., Schneider, J.: Active transfer learning under model shift. In: Proceedings of IEEE International Conference on Machine Learning, pp. 1305–1313 (2014)
17. Yang, L., Hanneke, S., Carbonell, J.: A theory of transfer learning with applications to active learning. Mach. Learn. 90(2), 161–189 (2013)
18. Zhang, J., Kong, X., Philip, S.Y.: Predicting social links for new users across aligned heterogeneous social networks. In: Proceedings of IEEE International Conference on Data Mining, pp. 1289–1294. IEEE (2013)
19. Zhao, L., Pan, S.J., Yang, Q.: A unified framework of active transfer learning for cross-system recommendation. Artif. Intell. 245, 38–55 (2017)

Semi-supervised Transfer Metric Learning with Relative Constraints

Rakesh Kumar Sanodiya[1](✉), Sriparna Saha[1], Jimson Mathew[1],
and Prateek Bangwal[2]

[1] Indian Institute of Technology Patna, Patna, India
rakesh.pcs16@iitp.ac.in
[2] University of Petroleum and Energy Studies, Dehradun, India

Abstract. Distance metric learning is one of the most important aspects behind the performance of numerous algorithms under the data mining paradigm. In this article, we propose a new method for transfer metric learning under semi-supervised setting, using the concept of relative distance constraints to exploit more information from the unlabeled data present in the target task. We need an appropriate distance function for extracting useful information from unlabeled data. For this purpose, we use the concept of pairwise relative distance constraints. With the help of few labeled data, we obtain the pair-wise similarities in the form of inequality and equality constraints. We use the concept of Bregman projection to satisfy such constraints to the initial distance matrix that is composed of both labeled and unlabeled data, and then construct the appropriate K-nearest neighbor graph using this matrix, which provides better results regardless of the dimension of the data.

Keywords: Metric learning · Transfer learning
Semi-supervised learning · Relative distance comparisons

1 Introduction

A number of data mining algorithms such as classification (K-NN) and clustering (K-means) perform well if an appropriate distance matrix is provided to them [1]. So it is very vital to have a good distance metric for satisfactory results of any classification/clustering algorithms. Metric learning is being one of the most researched topics in recent years and many researchers have come up with different methods. Generally, these methods can be divided into the following categories: (a) Supervised metric learning, which learns the metric from the labeled data or known task; classification is a prime example of the supervised metric learning, (b) Unsupervised metric learning, which uses the information available within the data for example clustering and (c) Semi-supervised metric learning can easily be understood as the combination of supervised and unsupervised learning. This metric takes care of both labeled information as well as the geometric information from the unlabeled data. A number of machine learning

© Springer Nature Switzerland AG 2018
L. Cheng et al. (Eds.): ICONIP 2018, LNCS 11303, pp. 230–241, 2018.
https://doi.org/10.1007/978-3-030-04182-3_21

algorithms such as supervised learning expect the labeled data to train their respective models but in reality, the amount of labeled data present in the real-world scenario is very less but the available unlabeled data is more. It is very time-consuming, costly and tedious work to label the data manually for such supervised model. Thus, the model which is learned does not provide proper results. Therefore, semi-supervised metric learning is preferred over the rest of two learning paradigms as it helps us to overcome the problem related to the deficiency of labeled data.

In this paper, we consider circumstances similar to semi-supervised metric learning [2] where very few labeled data is available. In multi-task learning and transfer learning scenario, one task may have very few labeled data for learning but at the same time, it is also possible that there are some other related tasks which have sufficient labeled data. The semi-Supervised learning framework accomplishes the unlabeled information, on the other side, transfer learning and multi-task learning [3] try to mitigate the scarcity of the labeled data by considering some related tasks and to help in improving the learning performance. In other words, when a person tries to learn to ride bicycle and tricycle then, the learning experience of riding bicycle can be used for learning tricycle.

In both transfer metric learning and multi-task learning, one task helps another task in improving the learning performance, but both are different in terms of having different objective functions and problem setting. There are basically two types of tasks in the transfer learning (a) source task, (b) target task. It is accepted that there is sufficient information in the source tasks but in the target task, there is not enough data present. In transfer metric learning (TML) [4] the metric and the task covariance between the source task and target task have been learned and brought together within a convex formulation. The main aim of the transfer metric learning is to learn the metric of target task with the assistance of source tasks without enhancing the execution of the given source tasks. This is the primary distinction between transfer learning and multi-task learning. However, the multi-task learning objective is to enhance the performance of every task at the same time [4]. Based on this method, regularized distance metric learning (RDML) [5] based transfer metric learning is proposed which is the extension of the transfer learning.

The semi-supervised extension of TML named STML improves the performance by considering the manifold assumption and clustered data. STML uses the initial distance matrix generated using Gaussian kernel function to set up a K-nearest neighbor graph $G = (\nu, \varepsilon)$ with local scaling [6], vertex set denoted as $\nu = \{1, ..., n_m\}$ related to the labeled and unlabeled data points, and the edges represented in set ε as $\varepsilon \subseteq \nu \times \nu$ denoting the relationship between the data points [4]. But, this Gaussian kernel function can not quantify appropriate distances between the data points if the available data is having high dimensions. Therefore, this will affect the performance of this STML algorithm.

In this work, we formulate a distance function as discussed in the paper [7] which considers relative distance comparisons in the form of constraints to express pairwise similarity and dissimilarity between all the data points [8].

This constraint set C is composed of equality C_{eq} and inequality C_{neq} constraints. Our distance function also considers non-linear transformation. In order to find out an appropriate distance function, we satisfy all such constraints onto the initial Gaussian kernel matrix using the Bregman projection [9,10]. Thus, learned matrix is used for generating the more appropriate K-NN graph which improves the performance of STML algorithm.

The major contributions of this paper are as follows:

- This paper proposed a transfer metric learning method under the semi-supervised setting with the help of the relative distance constraints.
- For generating the equality and inequality relative constraints only few labeled data is required, which makes our approach practically useful in real scenarios.
- We used Gaussian kernel to generate initial matrix then satisfying such generated constraints onto this matrix using the Bregman projection, which help us to generate a more appropriate distance matrix. Later, this adjusted matrix is used for generating more appropriate K-NN graph.
- Because the K-NN graph we have is more appropriate therefore the performance of existing STML improves.
- To evaluate the performance of our proposed approach, experimented results are presented on two data sets: Hand written letter dataset and USPS digit dataset.

2   Background and Related Work

Let us assume that we have a training dataset (labeled) as $\{(x_i, y_i)\}_{i=1}^{n}$ where the ith data point is $x_i \in \mathbb{R}^d$ and the respective class label is given as $y_i \in \{1, ..., C\}$. Then, Regularized distance metric learning (RDML) [5] can be formulated as follows:

$$\min_{\Sigma} \frac{2}{n(n-1)} \sum_{j,k} g\left(y_{j,k}\left[1 - \|x_j - x_k\|_{\Sigma}^2\right]\right) + \frac{\lambda}{2}\|\Sigma\|_F^2 \tag{1}$$

$$\text{s.t. } \Sigma \succeq 0$$

Where λ is a regularization parameter that maintains regularization term and empirical loss. And $\|\Sigma\|_F$ is the Frobenius norm of a matrix Σ.
And

$$y_{j,k} = \begin{cases} 1 & if y_j = y_k \\ -1 & otherwise; \end{cases}$$

$\Sigma \succeq 0$ indicates it is a PSD (positive semidefinite matrix) matrix, and $\|x_j - x_k\|_{\Sigma}^2 = (x_j - x_k)^T \Sigma (x_j - x_k)$, $g(c) = max(0, \alpha - c)$ indicates hinge loss. Here the value of α must be between 0 to 1. In the semi-supervised extension of transfer metric learning (STML) setting, let $m - 1$ source tasks and one T_m target task are given then the optimization problem can be formulated as follows:

$$\min_{\Sigma,\omega_m,\omega,\Omega} \frac{2}{n_m(n_m - 1)} \sum_{j,k} g\left(y_{j,k}^m \left[1 - \left\|x_j^m - x_k^m\right\|_{\Sigma_m}^2\right]\right) + \frac{\lambda_1}{2}\|\Sigma_m\|_F^2$$

$$+ \frac{\lambda_2}{2}tr(\tilde{\Sigma}\Omega^{-1}\tilde{\Sigma}^T) + \lambda_3 tr(\Sigma_m X_m L X_m^T) + \frac{\lambda_2}{2}tr(\tilde{\Sigma}\Omega^{-1}\tilde{\Sigma}^T) + \lambda_3 tr(\Sigma_m X_m L X_m^T)$$

s.t. $\Sigma \succeq 0$

$$\Omega = \begin{pmatrix} \frac{1-\omega}{m-1} & \omega_m \\ \omega_m^T & \omega \end{pmatrix}, \tilde{\Sigma} = (\tilde{\Sigma}_s, vec(\Sigma_m))$$

$$\omega(1 - \omega) \geq (m - 1)\omega_m^T\omega_m$$

$$(2)$$

It can be proved that the Eq. 3 follows the convex formulation [11]. The optimization procedure for STML can be done by dividing it into two subproblems. We first optimize with respect to Σ_m while keeping ω_m and ω fix, and then optimize ω_m and ω while keeping the Σ_m fixed.

Keeping ω_m and ω fixed and only concerning the Σ_m:

$$\min_{\Sigma_m} \frac{2}{n_m(n_m - 1)} \sum_{j<k} g(y_{j,k}^m[1 - \|x_j^m - x_k^m\|_{\Sigma_m}^2]) + \frac{\lambda_1'}{2}\|\Sigma_m\|_F^2 - \frac{\lambda_2'}{2}tr(\Sigma_m M)$$

$$+ \lambda_3 tr(\Sigma_m X_m L X_m^T)$$

s.t. $\Sigma_m \succeq 0$

$$(3)$$

Keeping Σ_m fixed and concerning the ω_m and ω:

$$\min_{\omega_m,\omega,f,t,h_j,r_j} -t$$

$$\text{s.t.} \frac{1-\omega}{m-1} \geq t\lambda_1$$

$$\mathbf{f} = \mathbf{U}^{\mathbf{T}}(\omega_m - t\psi_{12})$$

$$\sum_{j=1}^{m-1} h_j \leq \omega - t\psi_{22}$$

$$(4)$$

$$r_j = \frac{1-\omega}{m-1} - t\lambda_j, \forall j$$

$$\frac{f_j^2}{r_j} \leq h_j, \forall j$$

$$\omega(1 - \omega) \geq (m - 1)\omega_m^T\omega_m$$

Initially, it is assumed that all source tasks and target tasks are unrelated to each other. Therefore, we keep the value of $\omega = \frac{1}{m}$ and ω_m as zero vector. The problem

(4) is a second-order cone programming problem (SOCP) [12] and we can solve it very easily if the number of the task is very less.

3   Proposed Work

Our proposed method (STMLR) is mixed of semi-supervised transfer metric learning and kernel based relative distance metric learning. The architecture of our proposed method is shown in Fig. 1. Initially, we have two or more independent source tasks and one target task, target task is having enough unlabeled data but very less labeled data. Either of RDML or ITML can be used to learn and obtain the prior metric of the source tasks. These learned source task matrix (prior metric) will help in learning the target task matrix. First, we consider target task. Then, we use the Gaussian kernel discussed in Sect. 3.1 to obtain our initial kernel matrix K_0 of the target task. Since this initial kernel matrix does not quantify the appropriate distances between the data points to generate the graph, we have to find out an appropriate distance function to measure the distances between data points. Therefore, we use the relative distance constraints generated in Sect. 3.2 onto the initial kernel matrix K_0 to learn the appropriate distance function. We use inner product between any data points in Kernel space discussed in Sect. 3.3 to determine the distance between them. We seek for a kernel matrix K after projecting the initial kernel matrix K_0 onto this pairwise constraint matrix. The log-det Bregman divergence discussed in Sect. 3.4 helps to keep the distance between the initial kernel matrix K_0 and kernel matrix K as minimum as possible. After satisfying all the constraints using the Bregman Projection, we get final kernel matrix K shown in Fig. 2. Now this final kernel matrix K of the target task is used by the k-nearest neighbor algorithm to generate an appropriate graph W. We determine graph Laplacian matrix L of this graph by subtracting the graph matrix w from the diagonal matrix of the graph, i.e., $L = D - W$. This appropriate graph Laplacian matrix L is given to fourth term $\lambda_3 tr(\Sigma_m X_m L X_m^T)$ of STML Eq. 3.

3.1   Similarity Between Data Points Using Gaussian Kernel

Let us say we have l_m number of labeled data points, denoted as $\left\{(x_j^m, y_j^m)\right\}_{j=1}^{l_m}$, and u_m unlabeled data points, denoted as $\left\{x_j^m\right\}_{j=l_m+1}^{l_m+u_m}$ (i.e., a total of $n_m = l_m + u_m$ data points are given to us), for the target task T_m. In order to generate similarity graph, we use the Gaussian kernel function to find out the similarity between the data points. Because of its smoothing property, the choice of Gaussian kernel is so obvious: Let us say we have given a positive definite kernel R, then there exists its relative space of functions, P. The kernel helps us to determine the characteristics of the functions in the space P [13]. It turns out that if R is a Gaussian kernel, the functions in the space P are very smooth. So, a learned function (e.g, a regression function, principal components in RKHS as in kernel PCA) is very smooth.

Let us consider a similarity graph $G = (\nu, \varepsilon)$ with vertex set $\nu = 1, ..., n_m$ analogous to both the labeled data points as well as the unlabeled data points and the edge set $\varepsilon \subseteq \nu \times \nu$ defining the existing relationships between data points. Now the weight of each edge between any pair of data points, x_i^m and x_j^m, is calculated by using the following Gaussian kernel function.

$$w_{ij} = \begin{cases} exp(\frac{-\|x_i^m - x_j^m\|_2^2}{\sigma_i \sigma_j}) & if\, x_i^m \in N_k(x_j^m) \;\; or \;\; x_j^m \in N_k(x_i^m) \\ 0 & otherwise \end{cases} \qquad (5)$$

where $\|.\|_2$ is the notation for the two-norm of a vector, $N_k(x_i^m)$ denotes the neighborhood set of K-nearest neighbors of x_i^m, and $\sigma_i(\sigma_j)$ is the distance between $x_i^m(x_j^m)$ and its Kth nearest neighbor.

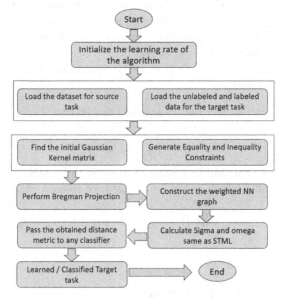

(a) Flow chart of proposed architecture

Fig. 1. Showing flow chart of our proposed architecture

3.2   Relative Distance Constraints C_{eq} and C_{neq}

The constraints available in the constraint set C report knowledge about the distances between the data points present in the dataset D. This reported knowledge is given by using the distance function φ. The motive of this distance function is not to calculate accurate distances between any data points. But it expresses the information in the form $\varphi(p, q) < \varphi(p, r)$ for some $p, q, r \in D$. In particular, constraints we have, can be viewed in two categories, i.e., C can be divided into two sets, inequality set C_{neq} and equality set C_{eq}. The set C_{neq} has constraints

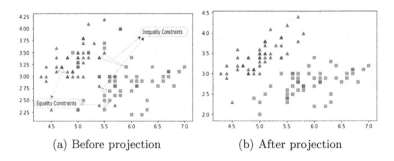

(a) Before projection   (b) After projection

Fig. 2. 2 dimensional representation of data before or after Bregman projection

where a point out of three points is an outlier. This means that the distance of one point (outlier) to other two data points is much larger than that of two other points. The set c_{eq} has constraint where no item is an outlier, i.e., the distances of all three items are then approximately the same. Both the constraints C_{eq} and C_{neq} were also used in the paper [7].

For the given three data points p, q, r, the constraints in C_{neq} can be defined as the following inequalities

$$(p \leftarrow q|r) : \mu\varphi(p,q) \leq \varphi(q,r) \tag{6}$$

and

$$(q \leftarrow p|r) : \mu\varphi(q,p) \leq \varphi(q,r) \tag{7}$$

Here, $\mu > 1$ is some constant parameter. Similarly, the constraints in C_{eq} for the data points p, q and r which are in equal distance to each other can be defined as follows:

$$\varphi(p,q) = \varphi(q,r) = \varphi(p,r) \tag{8}$$

3.3 Development to a Kernel Space

The constraints available in set C are extended to a kernel Hilbert space with the function $\Theta : D \rightarrow \mathbb{R}^m$. In this space, the inner product between any two points p and q can be defined by using the symmetric matrix, K, that is, $K_{pq} = \Theta(p)^T \Theta(q)$. We also assume that in this space this kernel K uses some unknown distance function, φ. Thus, the inner product between any two points, p and q, by using some unknown distance function, φ, in Hilbert Space is defined as follows:

$$\varphi(p,q) = \|\Theta(p) - \Theta(q)\|^2 = K_{pp} - 2K_{pq} + K_{qq} \tag{9}$$

Now, we consider the Eqs. (6), (7) and (8), the inequality constraints in the kernel space. Let v_p be a vector with zeros and value 1 at position p. Matrix form for the Eq. (9) can be expressed as:

$$K_{pp} - 2K_{pq} + K_{qq} = (v_p - v_q)^T K(v_p - v_q) = tr(K(v_p - v_q)(v_p - v_q)^T) \tag{10}$$

where tr(T) denotes the trace of matrix T and we know the fact that $K = K^T$, using the above equation we can modify Eq. (6) as:

$$\mu tr(K(v_p - v_q)(v_p - v_q)^T) - tr(K(v_p - v_r)(v_p - v_r)^T) \leq 0$$
$$tr(K\mu(v_p - v_q)(v_p - v_q)^T - K(v_p - v_r)(v_p - v_r)^T) \leq 0$$
$$tr(K(\mu(v_p - v_q)(v_p - v_q)^T - (v_p - v_q)(v_p - v_r)^T)) \leq 0$$
$$tr(KC_{p\leftarrow q|r}) \leq 0$$

where $C_{(p\leftarrow q|r)} = \mu(v_p - v_q)(v_p - v_q)^T - (v_p - v_r)(v_p - v_r)^T$ matrix representation of corresponding constraint. For $C_{q\leftarrow p|r}$ for Eq. (7) can be easily formed similarly.

Using similar technique we can represent the constraints in the set C_{eq}. Since the constraint C_{eq} implies all the points are equidistant, we can write the first equation for the constraint $(p, q, r) \in C_{eq}$ as

$$tr(KC_{p\leftrightarrow q,r}) = 0 \qquad (11)$$

3.4 The Bregman Projection and It's Log Determinant Divergence in Kernel Learning

Initially, we construct initial kernel matrix K_0 by using the Gaussian kernel discussed in Sect. 3.1, this matrix is projected onto the convex set of constraints. After performing this, we get positive semi-definite matrix, K. Our main objective is to minimize the distance between K and K_0 by using the Bregman divergence. The Bregman divergence between K and K_0 is as follows:

$$D_\Theta(K, K_0) = \Theta(K) - \Theta(K_0) - tr(\Delta_\Theta(K_0)^T(K - K_0)) \qquad (12)$$

where θ is a strictly-convex real-valued function, and $\Delta_\Theta(K_0)$ is gradient calculated at K_0. If we set $\Theta(K) = -\log det(k)$, i.e., the log-determinant (logdet) matrix divergence will be:

$$D_{ld}(K, K_0) = tr(K, K_0^{-1}) - \log det(KK_0^{-1}) - n \qquad (13)$$

4 Experimental Results

In this section, we first study semi-supervised transfer metric learning (STML) and semi-supervised transfer metric learning with relative constraints (STMLR) empirically, and compare their performances. We used source code offered by [4] to implement our STMLR algorithm and also used CVX toolbox to solve the optimization problem. All the source task metrics are learned using the RDML algorithm. It has already been proved in the paper [4] that existing STML outplayed the ITML, RDML algorithm in terms of performance. Now, We evaluate our proposed approach on two popular datasets, handwritten letter classification[1] and USPS digit classification[2] to show the performance over the existing (STML).

---

[1] https://archive.ics.uci.edu/ml/datasets/letter+recognition.
[2] https://archive.ics.uci.edu/ml/support/Optical+Recognition+of+Handwritten+Digits.

4.1 Handwritten Letter Classification

We generate six binary classification tasks by considering data points corresponding to the tasks from the handwritten letter classification dataset. Each task can have approximately 1,000 positive and 1,000 negative data points. Each data point of the task has 16 features to represent an image of the letter. To perform the experiment, we consider the following tasks, namely, h/n, m/n, g/y, a/o, c/e and f/t. While performing experiments, one of them becomes target task and others become source tasks. We keep 0.5 as learning rate and constant value for gamma priors (alpha and beta), for the classification purpose we used the KNN classifier where k is initialized as $K = 3$. In order to see the effect of varying the size of training data on the performance of both the algorithms,

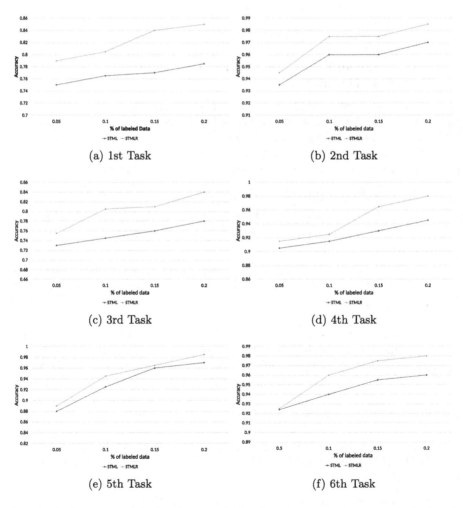

(a) 1st Task

(b) 2nd Task

(c) 3rd Task

(d) 4th Task

(e) 5th Task

(f) 6th Task

Fig. 3. Performance on the handwritten letter classification with one target task and other tasks as source tasks in STML and STMLR

we vary the percentage of training data from 5% to 20%. The results shown in Fig. 3 show that our approach performs better for every task in comparison to the existing STML.

4.2   USPS Digit Classification

The USPS digit dataset contains 7,291 examples with each of 256 features. From this dataset also, we have considered four classification tasks namely, 3/4, 4/5, 3/5 and 5/6. We use the same experiment setting as used for handwritten letter classification dataset. Here also, results shown in Fig. 4 clearly illustrate that our approach performs better in comparison to STML.

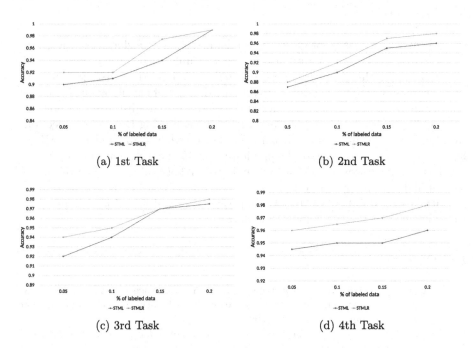

Fig. 4. Performance on the USPS Digit classification with one target task and other tasks as source tasks in STML and STMLR

4.3   Discussion of Results

For both the datasets we compared our algorithm, STMLR with the existing algorithm, STML, for the 5%, 10%, 15% and 20% of labeled data.

- For the Handwritten letter dataset we have achieved better accuracy than the STML as shown in Fig. 3. For example In 1st task for 5%, 10%, 15%, 20% labeled data we achieved 79%, 80.5%, 84%, 85% accuracy respectively, which is better when compared to STML. There are five more similar tasks and for each case STMLR performed better.

– Also, For the USPS digit dataset we have achieved better accuracy than the STML as shown in Fig. 4. For example In 1st task for 5%, 10%, 15%, 20% labeled data we achieved 92%, 92%, 97.5%, 99% accuracy respectively, which is better when compared to STML. There are three more similar tasks and for each case STMLR performed better.

We have also conducted t-test [12] at 0.05 significance level to check whether results obtained are statistically significant or happened by chance. Results obtained using t-test clearly show that the obtained performance improvements are statistically significant.

5   Conclusion

In this paper, we have proposed a semi-supervised transfer metric-learning method to exploit more information from the unlabeled data present in the target task. For this purpose, we have found out appropriate distance function to measure an accurate distance between any data points available in the higher dimension of the target task. To measure this appropriate distance function, we used relative distance constraints and satisfied them using Bregman projection. Results show that our approach performs better when compared to STML in higher dimensional data also. The previously satisfied constraints can get unsatisfied as the projection of the constraints is not orthogonal. Therefore, in future, we will find out an appropriate subset of constraints by using some evolutionary algorithm (like NSGA-III, AMOSA) while satisfying multiple-objective functions which ensure good partitioning of the data.

References

1. Xing, E.P., Jordan, M.I., Russell, S.J., Ng, A.Y.: Distance metric learning with application to clustering with side-information. In: Advances in Neural Information Processing Systems, pp. 521–528 (2003)
2. Bilenko, M., Basu, S., Mooney, R.J.: Integrating constraints and metric learning in semi-supervised clustering. In: Proceedings of the Twenty-First International Conference on Machine Learning, p. 11. ACM (2004)
3. Caruana, R.: Multitask learning. Mach. Learn. **28**(1), 41–75 (1997)
4. Zhang, Y., Yeung, D.Y.: Transfer metric learning with semi-supervised extension. ACM Trans. Intell. Syst. Technol. **3**(3), 54 (2012)
5. Jin, R., Wang, S., Zhou, Y.: Regularized distance metric learning: theory and algorithm. In: Advances in Neural Information Processing Systems, pp. 862–870 (2009)
6. Zelnik-Manor, L., Perona, P.: Self-tuning spectral clustering. In: Advances in Neural Information Processing Systems, pp. 1601–1608 (2005)
7. Amid, E., Gionis, A., Ukkonen, A.: A kernel-learning approach to semi-supervised clustering with relative distance comparisons. In: Appice, A., Rodrigues, P.P., Santos Costa, V., Soares, C., Gama, J., Jorge, A. (eds.) ECML PKDD 2015, Part I. LNCS (LNAI), vol. 9284, pp. 219–234. Springer, Cham (2015). https://doi.org/10.1007/978-3-319-23528-8_14

8. Pei, Y., Fern, X.Z., Rosales, R., Tjahja, T.V.: Discriminative clustering with relative constraints (2014). arXiv preprint: arXiv:1501.00037
9. Kulis, B., Sustik, M.A., Dhillon, I.S.: Low-rank kernel learning with Bregman matrix divergences. J. Mach. Learn. Res. **10**, 341–376 (2009)
10. Bregman, L.M.: The relaxation method of finding the common point of convex sets and its application to the solution of problems in convex programming. USSR Comput. Math. Math Phys. **7**(3), 200–217 (1967)
11. Argyriou, A., Evgeniou, T., Pontil, M.: Convex multi-task feature learning. Mach. Learn. **73**(3), 243–272 (2008)
12. Lobo, M.S., Vandenberghe, L., Boyd, S., Lebret, H.: Applications of second-order cone programming. Linear Algebr. Appl. **284**(1–3), 193–228 (1998)
13. Chung, M.K.: Gaussian kernel smoothing. LNCS, pp. 1–10 (2012)

Fuzzy Domain Adaptation Using Unlabeled Target Data

Hua Zuo$^{(\boxtimes)}$, Guangquan Zhang, and Jie Lu

DeSI Lab, Centre for Artificial Intelligence, FEIT, University of Technology
Sydney, 81 Broadway, Ultimo, NSW 2007, Australia
{hua.zuo,guangquan.zhang,jie.lu}@uts.edu.au

Abstract. Transfer learning has been emerging recently and gaining more attention because of its ability to deal with "small labeled data" issue in new markets and for new products. It addresses the problem of leveraging knowledge acquired from previous domain (a source domain with a large amount of labeled data) to improve the accuracy of tasks in the current domain (a target domain with little labeled data). Fuzzy rule-based transfer learning methods are developed due to the ability to dealing with the uncertainty in domain adaptation scenarios. Although some effort is made to develop the fuzzy methods, they only apply the knowledge of the labeled data in the target domain to assist the model's construction. This work develops a new method that explores and utilizes the information contained in the unlabeled target data to improve the performance of the new constructed model. The experiments on both synthetic datasets and real-world datasets illustrate the effectiveness of our method, and also give the application scope of applying it.

Keywords: Domain adaptation · Transfer learning · Machine learning
Fuzzy rules · Regression

1 Introduction

Machine learning [1] has gained a great achievement in many areas, such as finance, military, entertainment, and so on. And many machine learning methods are developed to handle the practical situations [2]. Although these methods work well in some cases, there is a big obstacle that impeded the further development of the traditional machine learning methods. This obstacle comes from an assumption that the model only works well in the condition that the training data and testing data have the same statistical characteristics, i.e. the same feature space and distributions. But in the data-shortage and rapid-changing environments, the constructed model always has a poor performance, and building a new is impossible due to the insufficient labeled data.

Human own the ability of applying the knowledge acquired previously to solve the current task. For example, recognizing an apple will be helpful for identifying a new fruit, such as a peach or a pear. Because of this ability, human could continuously accumulated knowledge and adapt to the new and challenging environment. Transfer learning has been emerging recently and becoming more and more popular due to its ability of knowledge transfer.

© Springer Nature Switzerland AG 2018
L. Cheng et al. (Eds.): ICONIP 2018, LNCS 11303, pp. 242–250, 2018.
https://doi.org/10.1007/978-3-030-04182-3_22

There are increased attention focusing on transfer learning [3], and more methods are developed to handle the real cases in artificial intelligence to expand the application of transfer learning [4, 5]. Some well-known examples of transfer learning include prediction, image recognition, recommend systems, and natural language processing. Since the development of transfer learning is based on machine learning, many notable machine leaning models [6–8] are applied as the basic learning in the transfer leaning methods. Additionally, researchers in deep learning exploring the transfer ability of deep models [9]. For more information about transfer learning, there are some well-written survey papers that summarize the current transfer learning methods and give the clear categories to review them [10, 11].

Although transfer learning exhibits an upward trend, there is still a huge gap between existing work and domain adaptation tasks. For instance, most of the current transfer learning focus on solving the classification problems, but there is little work on the regression tasks in the domain adaptation problems. Additionally, the ignorance of the uncertainty phenomenon in the transfer learning problems weak the application scope of the current works. Since only little labeled or no labeled data available in the target domain, the insufficient information lead to the uncertainty in the learning and model's construction process. However, the application of fuzzy systems in the transfer learning problems has shown a good results and light the way of handling the uncertainty issues.

We have done some work at solving the domain adaptation problems in regression tasks using fuzzy rule-based models [12, 13]. We have proposed three algorithms that deal with three different fuzzy transfer learning cases separately. In the first case, we consider the discrepancy of the distributions in the feature space, which will lead to the different conditions of fuzzy rules in the source and target domain. An algorithm of changing the input space through mappings is presented to solve the distribution gap between two domains. In the second case, other than the feature distributions, consider the difference in the output space, and an algorithm of changing the linear functions is proposed to adjusting the output space to make it fitted with the target data. The third algorithm combines the first two, modifies both the conditions and conclusion of the fuzzy rules to make them compatible with the target domain. All the work in these papers use the labeled target data to lead the construction of transformation mappings, and ignore the data without labels. Here, we propose a new method that explores and utilizes the knowledge contained in the unlabeled target data to improve the performance of the constructed target model.

The structure of this work is as follows. Section 2 presents some basic definitions in transfer learning, and the learning model applied in our method, Takagi-Sugeno fuzzy model. In Sect. 3, we propose a new method, which uses both labeled and unlabeled target data. In Sect. 4, synthetic and real-world datasets are used to analyse the performance of our method and test its effectiveness in dealing with practical situations. The final section concludes the paper and outlines future work.

2 Preliminary

Some definitions are introduced first to given the readers the basic knowledge of transfer learning. And then, a fuzzy rule-based system, Takagi-Sugeno fuzzy model, is formulated.

2.1 Transfer Learning

Definition 1 (Domain) [3]: A domain is denoted by $D = \{F, P(X)\}$, where F is a feature space, and $P(X)$, $X = \{x_1, \cdots, x_n\}$, are the probability distributions of the instances.

Definition 2 (Task) [3]: A task is denoted by $T = \{Y, f(\cdot)\}$, where $Y \in R$ is the output, and $f(\cdot)$ is an objective predictive function.

Definition 3 (Transfer Learning) [3]: Given a source domain D_s, a learning task T_s, a target domain D_t, and a learning task T_t, transfer learning aims to improve learning of the target predictive function $f_t(\cdot)$ in D_t using the knowledge in D_s and T_s where $D_s \neq D_t$ or $T_s \neq T_t$.

Transfer learning uses the knowledge obtained from previous domains (source domain) to help build the model for dealing with the tasks in the current domain (target domain).

2.2 Fuzzy Rule-Based Model

The basic learning model used here is the Takagi-Sugeno fuzzy model, which consists of c rules as follows:

$$\text{If } x \text{ is, } A_i(x, v_i) \text{ then } y_i \text{ is } L_i(x, a_i) \quad i = 1, \ldots, c \tag{1}$$

where x is the input, and y_i is the output of applying the corresponding rule. v_i is the centers of the prototype (cluster), and a_i determines the linear functions in the conclusions of the fuzzy rules. Thus, the output of the fuzzy system is y with the following representation.

$$y = \sum_{i=1}^{c} A_i(x, v_i) L_i(x, a_i) \tag{2}$$

The construction of the Takagi-Sugeno fuzzy model involves a learning process based on a given labeled datasets. First, the data are divided based on Fuzzy C-means algorithm, which could help cluster the data and find out the centers of the clusters v_i. Based on the v_i, the coefficients of a_i are computed through an optimization procedure.

3   Methodology

In our previous papers, we proposed the methods of changing the input and output spaces of the source domain to fit the current tasks. The labeled target data are used to guide the construction of the mappings that connects the domains. But the unlabeled target data are not used to help the construction of target model. In the transfer learning scenarios, there are a large amount of target data without labels that also contains much information of target domain. Therefore, how to utilize the unlabeled target data is a critical step in enhancing the performance of transfer learning between domains.

In this work, the knowledge contained in the unlabeled target data will be explored and applied to improve the performance of the constructed model for the target domain. The following steps outline this fuzzy rule-based domain adaptation method, which utilizes target data with and without labels for solving the regression tasks in target domain.

Step 1: Train a fuzzy model (fuzzy rules) for the source domain.

In the source domain, a mass of labeled data are available. Suppose the dataset in the source domain is denoted as $D = \left\{ \left(x_1^s, y_1^s \right), \cdots, \left(x_{N_s}^s, y_{N_s}^s \right) \right\}$, where $x_k^s \in R^n$ $(k = 1, \cdots, N_s)$ is an n-dimensional variable, $y_k^s \in R$ is a continuous variable, and N_s gives the number of labeled data. Based on the dataset D, a supervised learning process is executed to train the source model and gain a set of fuzzy rules.

Referring to the numbers of fuzzy rules in the source and target domains, we want to claim that the rules of source domain could be modified and transferred to solve the target tasks as if the number of fuzzy rules in the source domain is greater than in the target domain. Since the Takagi-Sugeno fuzzy model uses nonlinearly weighted linear functions to fit a curve. Each cluster indicates a separate area in the input space, and the corresponding linear function represents the action applied in that area. More clusters, or fuzzy rules, represents more precise of the partition and action described in the output space. Thus, it is reasonable to set an adequate number of fuzzy rules when building a Takagi-Sugeno fuzzy model to get good performance.

We consider two cases here to indicate the relationship of the numbers of fuzzy rules in two domains. If the source domain has no less fuzzy rules' number than the target domain, then the fuzzy rules of source domain could be modified and used to handle the regression tasks in the target domain. If the source domain has less fuzzy rules' number than the target domain, we could adopt the strategy of retaining the source model with rules no less than that in the target domain. Therefore, this also can be regarded as a criteria of selecting an appropriate domain from multiple domains for the target domain.

Therefore, determining the fuzzy rules' number is quite important when building a Takagi-Sugeno fuzzy model. Although we claim that using more rules to construct a Takagi-Sugeno fuzzy model is reasonable, we still need the prior knowledge to estimate the number of fuzzy rules for a specific domain or dataset. Here, we apply the IGMM model [14] to find out the data's structure and provide a guide to determine the number of trained rules. IGMM implements the process of mixing Gaussian distributions to fit the data distribution, which detects the data structure in the data-based learning process.

After analysing the results from IGMM, a model M^s is trained based on the source dataset D, and a set of fuzzy rules are obtained with formulation as follows:

$$\text{if } x_k^s \text{ is } A_i\left(x_k^s, v_i^s\right), \text{ then } y_k^s \text{ is } L_i\left(x_k^s, a_i^s\right) \quad i = 1, \cdots, c \tag{3}$$

There are c fuzzy rules, and each rule is governed by the center of cluster v_i^s, and the linear function a_i^s.

Step 2: Modify the fuzzy rules of source domain to fit the target tasks.

The target dataset H consists of two subsets: H_L with labeled data, and H_U with unlabeled data. $H = \{H_L, H_U\} = \left\{\left\{\left(x_1^t, y_1^t\right), \cdots, \left(x_{N_{t1}}^t, y_{N_{t1}}^t\right)\right\}, \left\{x_{N_{t1}+1}^t, \cdots, x_{N_t}^t\right\}\right\}$, where $x_k^t \in R^n$ $(k = 1, \cdots, N_t)$ is the n-dimensional input variable, $y_k^t \in R$ corresponds to the labels for the data in H_L. The numbers of data in H_L and H_U are N_{t1} and $N_t - N_{t1}$ respectively, and satisfy $N_{t1} \ll N_t$, $N_{t1} \ll N_s$.

Since the distributions of $\left\{x_1^s, \cdots, x_{N_s}^s\right\}$ and $\left\{x_1^t, \cdots, x_{N_t}^t\right\}$ are different, the model trained on D could not be used to do the prediction tasks in H.

We adopt the approach of changing the input space by constructing a mapping between the input variables between two domains [13]. Different with our previous method, which only applies the labeled target data to train the mapping, this work uses the information contained in unlabeled target data to enhance the accuracy of the target model.

After the transformation of the mapping in the input space, the fuzzy rules in M^s are changed, and target model M^t is obtained:

$$\text{if } x_k^t \text{ is } A_i\left(\mathbf{\Phi}\left(x_k^t\right), \mathbf{\Phi}\left(v_i^s\right)\right), \text{ then } y_k^t \text{ is } L_i\left(\left(\mathbf{\Phi}(x_k^t), a_i^s\right)\right) i = 1, \cdots, c \tag{4}$$

where $\mathbf{\Phi}$ is the transformation mapping for the input space. $\mathbf{\Phi} = [\Phi_1 \cdots \Phi_n]$ indicates that the mapping for each input variable is built separately, and the structures of them are the same, which is constructed of network with one hidden layer. The detailed structure of the mappings could refers to our previous paper [13].

To optimize the parameters in $\mathbf{\Phi}$, target data with and without labels are used to train and modify the existing fuzzy rules. The optimized cost function is:

$$S = \sqrt{\frac{1}{N_{t1}} \sum_{k=1}^{N_{t1}} \left(\sum_{i=1}^{c} \frac{A_i\left(\mathbf{\Phi}(x_k^t), \mathbf{\Phi}(v_i^s)\right)}{\sum_{j=1}^{c} A_j\left(\mathbf{\Phi}(x_k^t), \mathbf{\Phi}(v_j^s)\right)} L_i\left(\mathbf{\Phi}(x_k^t), a_i^s\right) - y_k^t\right)^2}$$
$$+ \lambda_1 \sqrt{\frac{1}{N_{t1} * h} \sum_{k=1}^{N_{t1}} \sum_{l=1}^{h} \left(y_k^t - y_k^t(l)\right)^2 * \exp\left(-\left\|x_k^t - x_k^t(l)\right\|\right)} + \frac{\lambda_2}{2} w^T w \tag{5}$$

There are three items in the cost function in (5). The first item focuses on the labeled target data, which guides the learning process in a supervised way. The second item intend to utilize the unlabeled data to optimize the parameters of the mappings. In the regression problems, it is reasonable to assume that the closer data in the input space have similar outputs. With this assumption, the outputs of unlabeled target data are estimated and considered to be approximate to the output of the nearest instance with label. Thus, the h-nearest data $\left\{x_k^t(1)\cdots x_k^t(h)\right\}$ in $\boldsymbol{H}_U$ are found for each labeled target data x_k^t, and the corresponding outputs for $\left\{x_k^t(1)\cdots x_k^t(h)\right\}$ are expected to be similar with the output of x_k^t. $\exp(-\left\|x_k^t - x_k^t(l)\right\|$ defines the degree of the closeness to make sure that the closer data have more approximate outputs. The third item controls the complexity of the constructed prediction model.

4 Experiments

The synthetic datasets are applied first to validate our proposed method, and indicate the application scope of it. Secondly, our method is used to solve some real-world domain adaptation problems.

To evaluate this fuzzy method, the performance indexes are given in advance. The RMSE is chosen to test the model. In addition, the generalization ability of the constructed model is also important, so five-fold cross validation is used in the models' construction process. To keep consistence, when testing the performance of a model in solving the target tasks, the target dataset $\boldsymbol{H}_U$ is used to estimate the ability of the model in fitting target data.

4.1 Experiments on Synthetic Datasets

Three group of experiments are implemented using the synthetic datasets. We first compare our fuzzy method, the baselines, and some famous methods. The second and third groups of experiments are executed to find out the impact of the data's structure to the performance of the proposed method.

In each group of the experiments, three datasets with different numbers of clusters are generated, and each time, two of them are selected as a source domain, where all data are labeled, and a target domain, where only 1% data are labeled.

In the first group of experiments, the proposed method is compared with one baseline, the source model, and two famous methods in transfer learning, TCA and SA. There are three datasets: "3r", "4r" and "5r", and "3r" means this dataset is generated using three clusters. Since we have three datasets, six experiments are implemented. In Table 1, the datasets applied in each experiment are indicated in column one. For example, "5r to 4r" means the source and target datasets assigned in this experiment are "5r" and "4r", separately. The second to fifth columns show the RMSE of the four methods.

From Table 1, the performance of our method is better than the baseline, TCA, and SA, based on the smaller values of RMSE in the six experiments.

To validate the effectiveness of applying the target data without labels, the performances of the models constructed using or not using unlabeled target data $\boldsymbol{H}_U$ are compared. In addition, we consider two cases, where the structures of the data are different, to find out the impact of data's partition to the target model.

The boundaries of the clusters are very clear and ambiguous in the second and third groups of experiments. The results are shown in Tables 2 and 3, separately. Table 2

Table 1. Transferring results in different methods

Source to target	RMSE of models			
	Baseline	TCA	SA	Our method
5r to 4r	5.23 ± 0.00	7.88 ± 0.00	7.58 ± 0.00	0.68 ± 0.01
5r to 3r	3.67 ± 0.00	4.66 ± 0.00	4.65 ± 0.00	1.14 ± 0.05
4r to 3r	0.97 ± 0.00	2.19 ± 0.00	2.37 ± 0.00	0.65 ± 0.00
3r to 4r	0.94 ± 0.00	6.10 ± 0.01	6.36 ± 0.00	0.72 ± 0.01
3r to 5r	4.16 ± 0.00	3.25 ± 0.00	3.21 ± 0.00	1.57 ± 0.00
4r to 5r	4.67 ± 0.00	5.91 ± 0.02	5.45 ± 0.01	1.39 ± 0.01

compares the performance the target models, which are built using or not using unlabeled target data. Similarly, Table 3 compares the RMSE of the constructed target model with and without H_U. Lower values are shown in bold.

Table 2. Target model built using/not using H_U – second group

Source to target datasets	RMSE of the models	
	M^t (without H_U)	M^t (with H_U)
5r to 4r	1.0781 ± 0.0004	**1.0756 ± 0.0004**
5r to 3r	0.9352 ± 0.0057	**0.8962 ± 0.0083**
4r to 3r	2.1269 ± 0.2059	**2.0996 ± 0.1718**
3r to 4r	0.8876 ± 0.0009	**0.8457 ± 0.0005**
3r to 5r	2.5273 ± 0.0007	**2.5397 ± 0.0014**
4r to 5r	3.0755 ± 0.0110	**3.0614 ± 0.0037**

From the results in Tables 2 and 3, we can see that if the partition of data in the input space has obvious clusters, the use of unlabeled data could enhance the model's accuracy notably. But if the division of data's clusters is not clear, the application of H_U is not always an advanced result. This is because when the boundaries of input data are ambiguous, the labeled target data may fall into the junctions of the clusters, and the utilizing of H_U, finding the h-nearest unlabeled target data for each labeled target data, will lead to a poor performance of the target model.

4.2   Experiments on Real-World Datasets

Since most studies on transfer learning focus on classification problems, there are no publicly used datasets for the regression tasks. In order to validate our method, and compare with the existing methods, we select some datasets from UCI Machine Learning Repository, and modify them for the purpose of simulating transfer learning scenarios.

Two datasets "Protein tertiary structure" and "Housing" are considered. The "Protein tertiary structure" contains nines input variables to predict the RMSD-size of the residue, and the dataset is divided into two sub datasets as source and target for the

Table 3. Target model built using/not using H_U – third group

Source to target datasets	RMSE of the models	
	M^t (without H_U)	M^t (without H_U)
5r to 4r	**2.05 ± 0.30**	2.11 ± 0.31
5r to 3r	**2.45 ± 0.71**	2.69 ± 1.01
4r to 3r	3.00 ± 2.24	**2.39 ± 1.32**
3r to 4r	**1.01 ± 0.00**	1.05 ± 0.00
3r to 5r	5.50 ± 0.51	**5.45 ± 0.35**
4r to 5r	4.79 ± 0.25	**4.51 ± 0.62**

purpose of transfer learning. In the dataset "Housing", there are six features and the output is the "MEDV". Because it is difficult to determine the numbers of clusters for the high-dimensional datasets, we adopt a brute-force way to try several different numbers of clusters and close the one with the best performance. Table 4 gives the results of the above two datasets.

Table 4. Results for real-world datasets

Protein tertiary structure			Housing		
c	M^s on H_U	M^t on H_U	c	M^s on H_U	M^t on H_U
8	50.88 ±27.15	6.00 ± 0.01	5	1.40 ± 0.71	0.19 ± 0.00
9	48.90 ± 37.85	5.93 ± 0.01	6	3.11 ± 0.41	0.22 ± 0.01
10	43.32 ± 87.07	6.10 ± 0.01	7	2.41 ± 0.21	0.15 ± 0.00
11	36.84 ± 23.49	5.90 ± 0.01	8	2.51 ± 0.25	0.15 ± 0.01
12	54.41 ± 15.00	5.98 ± 0.00	9	1.60 ± 1.14	0.15 ± 0.00

The large values in "M^s on H_U" in Table 4 indicate the poor performance of source model in solving the target tasks. And the results in "M^t on H_U" validate the effectiveness of our method. We find that no obvious trend is shown with a change in the number of fuzzy rules. So, in the practical situations, we adopt the strategy of going through all numbers in the given range, and select the number of rules with best performance when determining the number of fuzzy rules is difficult.

5   Conclusions and Future Work

This work explores the knowledge contained in the unlabeled target data to improve the performance of the constructed model in solving the domain adaptation problems in regression tasks. The results validate our fuzzy method. Also, the low RMSE in real-world datasets shows the ability of our method in dealing with practical problems.

This method, however, exists a limitation that it works in the situation that the partition of data is obvious. The utilization of unlabeled target data does not show a significant advantage when the boundaries of the clusters in data are ambiguous.

How to expand the application scope of our method and explore more information from unlabeled target data will be considered in the future work.

Acknowledgment. This work was supported by the Australian Research Council under DP 170101623.

References

1. Nasrabadi, N.M.: Pattern recognition and machine learning. J. Electron. Imaging **16**(4), 049901 (2007)
2. Lu, J., Xuan, J., Zhang, G., Luo, X.: Structural property-aware multilayer network embedding for latent factor analysis. Pattern Recogn. **76**, 228–241 (2018)
3. Pan, S.J., Yang, Q.: A survey on transfer learning. IEEE Trans. Knowl. Data Eng. **22**(10), 1345–1359 (2010)
4. Lim, C.-H., Wan, Y., Ng, B.-P., See, C.-M.S.: A real-time indoor WiFi localization system utilizing smart antennas. IEEE Trans. Consum. Electron. **53**(2) (2007)
5. Xu, J., Ramos, S., Vázquez, D., López, A.M.: Domain adaptation of deformable part-based models. IEEE Trans. Pattern Anal. Mach. Intell. **36**(12), 2367–2380 (2014)
6. Long, M., Wang, J., Cao, Y., Sun, J., Philip, S.Y.: Deep learning of transferable representation for scalable domain adaptation. IEEE Trans. Knowl. Data Eng. **28**(8), 2027–2040 (2016)
7. Gönen, M., Margolin, A.A.: Kernelized Bayesian transfer learning. In: AAAI, pp. 1831–1839 (2014)
8. Klenk, M., Forbus, K.: Analogical model formulation for transfer learning in AP physics. Artif. Intell. **173**(18), 1615–1638 (2009)
9. Bengio, Y.: Deep learning of representations for unsupervised and transfer learning. In Proceedings of ICML Workshop on Unsupervised and Transfer Learning, pp. 17–36 (2012)
10. Lu, J., Behbood, V., Hao, P., Zuo, H., Xue, S., Zhang, G.: Transfer learning using computational intelligence: a survey. Knowl.-Based Syst. **80**, 14–23 (2015)
11. Shao, L., Zhu, F., Li, X.: Transfer learning for visual categorization: a survey. IEEE Trans. Neural Netw. Learn. Syst. **26**(5), 1019–1034 (2015)
12. Zuo, H., Zhang, G., Pedrycz, W., Behbood, V., Lu, J.: Fuzzy regression transfer learning in Takagi-Sugeno fuzzy models. IEEE Trans. Fuzzy Syst. **25**(6), 1795–1807 (2017)
13. Zuo, H., Zhang, G., Pedrycz, W., Behbood, V., Lu, J.: Granular fuzzy regression domain adaptation in Takagi-Sugeno Fuzzy models. IEEE Trans. Fuzzy Syst. **26**(2), 847–858 (2017)
14. Rasmussen, C.E.: The infinite Gaussian mixture model. In: Advances in Neural Information Processing Systems, pp. 554–560 (2000)

Reinforcement Learning

Reinforcement Learning Policy with Proportional-Integral Control

Ye Huang, Chaochen Gu$^{(\boxtimes)}$, Kaijie Wu, and Xinping Guan

Key Laboratory of System Control and Information Processing, MOE of China,
Shanghai Jiao Tong University, Shanghai 200240, China
{lutein,jacygu,kaijiewu,xpguan}@sjtu.edu.cn

Abstract. Deep Reinforcement Learning has made impressive advances
in sequential decision making problems recently. Constructive reinforcement learning (RL) algorithms have been proposed to focus on the policy
optimization process, while further research on different network architectures of the policy has not been fully explored. MLPs, LSTMs and
linear layer are complementary in their controlling capabilities, as MLPs
are appropriate for global control, LSTMs are able to exploit history
information and linear layer is good at stabilizing system dynamics. In
this paper, we propose a "Proportional-Integral" (PI) neural network
architecture that could be easily combined with popular optimization
algorithms. This PI-patterned policy network obtains the advantages of
integral control and linear control that are widely applied in classic control systems, improving the sample efficiency and training performance
on most RL tasks. Experimental results on public RL simulation platforms demonstrate the proposed architecture could achieve better performance than generally used MLP and other existing applied models.

Keywords: Reinforcement learning · Deep learning · Neural network
Control theory

1 Introduction

Recently, Deep Reinforcement Learning (DRL) has made notable advances
in solving representative benchmark problems, especially in simulated control
[11,22], continuous robot control [5,9,13], Go game [24], Atari games [15] and
other sequential decision making domains. Directing an agent to interact with
the environment, the policy network of DRL is of critical importance to achieve
maximum cumulative long time reward. Generally, Convolutional Neural Network (CNN) is applied in visual tasks such as high-dimensional control of robots
that utilizes raw visual images or videos as input. As for non-visual tasks, the
widely-used Multi-Layer Perceptron (MLP) is considered as a basic policy network structure for many DRL algorithms. However, inductive research on the
effectiveness of policy network architecture remains to be further explored. It's
necessary to draw importance on the policy architecture to improve agent's performance better.

© Springer Nature Switzerland AG 2018
L. Cheng et al. (Eds.): ICONIP 2018, LNCS 11303, pp. 253–264, 2018.
https://doi.org/10.1007/978-3-030-04182-3_23

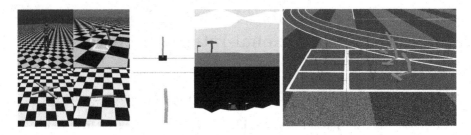

Fig. 1. Several popular reinforcement learning tasks implemented in public simulation platforms MuJoCo, OpenAI Gym and Roboschool. Including continuous control of simulated robots, classical control problems and games, etc.

In this work, we present an effective policy network architecture that is generic in handling benchmark RL problems from board games to simulated control tasks. Inspired by the Proportional-Integral Controller widely used in practical control systems, we introduce the memory mechanism of Long Short-Term Memory (LSTM) into policy network, in which characteristic could be found a clue to lead the agent exploits history information implicitly. With LSTM functioning as the integral controller, the "proportional" part is modelled as the linear projection of inputs. To better stabilize system dynamics, we use non-linear controller additionally. It's convenient to combine the proposed network with many existing DRL algorithms. The consolidation of linear, nonlinear and "integral" controllers could enhance the robustness and generalization of policy network compared with the typically applied MLP structure. Given current state, these three branches would evaluate respectively and then their results are combined to compute the final action.

Compared with generally applied MLPs policy networks, our history-concerned PI architecture could improve the performance of model on various RL tasks, especially on continuous control tasks. The key insight of our work is that the combination of control, memory mechanism and deep learning has distinct influence on the training efficiency and generalization ability. To validate the effectiveness of this policy network, extensive experiments are conducted on both classic control tasks as well as complex sequential decision making problems, such as pendulum control and humanoid walking, which are wrapped as standard RL environments in public simulation platforms such as MuJoCo [16], OpenAI Gym [3], Roboschool. We further perform different ablation experiments utilizing different policy optimization algorithms like Deep Deterministic Policy Gradient (DDPG) [11], Proximal Policy Optimization (PPO) [22], Actor Critic using Kronecker-Factored Trust Region (ACKTR) [31], etc. Our experimental results demonstrate that the proposed architecture is capable of enhancing model generalization as well as training efficiency compared with existing works.

In our paper, Sect. 2 introduces relevant researches about classic RL optimization methods and generally used network architectures, as well as the embedding of LSTM. In Sect. 3, we explain the proposed architecture in details and analyze its theoretical applicability. Experiment results are shown in Sect. 4, with several ablation experiments involved.

2 Related Works

Reinforcement learning problems are basically formulated as Markov Decision Process (MDP), in which an agent interacts with dynamic environment through trial-and-error [7]. Focusing on goal-directed learning, DRL algorithms are proved to be applicable for many sequential decision making problems in robotics [5,9,13], video games [15], simulations [11,21] and even self-driving systems [23]. Traditional approaches such as dynamic programming [10] and control methods fail to solve these challenges since the delayed feedback, unknown environment dynamics and the curse of high dimensions [7].

Constructive works on RL training process have been proposed in recent years. Generally, there are three main branches in RL, value-based, policy-based and actor-critic (combine both) methods. While classic value-based approaches such as SARSA [18] and Q-learning [28] have been shown unable to converge to policy for simple MDPs [26], recent model-free DRL algorithms have made dramatic advances in solving continuous control tasks. DQN [15] proposed by Google DeepMind has attracted great interest in the machine learning community, and for stochastic policy optimization, other policy-based algorithms such as Trust Region Policy Optimization (TRPO) [20], PPO [22], Asynchronous Advantage Actor-critic (A3C) [14], are effective in training the agents for accumulating more rewards through time. Policy gradient optimization methods only utilize states input for end-to-end training without any prior information.

However, further exploiting the applicability of different network architectures has not been fully studied. Most of the methods we discuss above adopt standard neural networks like MLPs, single LSTMs or autoencoders as policy network for the non-vision part, and pay their attention to optimization algorithms. Few works focus on using the internal structure in the policy parameterization to speed up learning process [30] and adding inductive bias to policy networks [27]. [27] proposed a novel network architecture named dueling architecture that represents separate estimators for state value function and state-dependent action advantage function respectively. Though splitting the Q-network into two streams, the Dueling Network can't deal with many continuous control tasks.

Similar work [17] proposes two applicable policy architectures: linear policy that maps from observations to actions, RBF policy that uses random Fourier features of the observations. These two architectures can achieve relatively promising performance on some continuous control tasks while still lacks generalization for most RL problems. Work [25] demonstrates that linear policy could make a complement to standard MLP network and this combination policy improves the sample efficiency, episodic reward and robustness. These relevant researches prove that the integration of linear and other specific architectures has potential for generating more effective models.

As a class of Recurrent Neural Networks (RNNs) architecture, LSTM is designed to learn temporal sequences and the long-term dependencies [12]. [2] presents a model-free RL-LSTM framework to solve non-Markovian tasks, and [29] uses LSTM to train an end-to-end dialog systems that is optimized with supervised learning and reinforcement learning. The inspirational application of LSTM in RL problems demonstrate that LSTM is capable of processing internal state information and exploring the long-term dependency between relevant events in other benchmark RL problems.

The idea of integrating LSTM with linear network and standard MLP network could be much similar to the traditional feedback control approach PI [1] which has been successfully applied to a variety of continuous control like robotics, unmanned air vehicles [19] and other automatic systems. Inspired by the widely used PI controller, we propose the "Proportional-Integral" policy network to represent the physical interpretations of control approach. Our architecture is easily to be combined with existing RL optimization algorithms and sufficient experiments show this policy network could achieve remarkable results on many benchmark tasks outperforming the results achieved by similar works [25].

3 Approach

3.1 Background

In the process of optimizing episodic reward while interacting with dynamic environment, the agent updates the policy π according to Bellman (Optimality) Equation. We formulate the standard RL environment of a sequential decision making problem as Markov Decision Process (MDP) defined by the tuple: $\mathcal{M} = \{\mathcal{S}, \mathcal{O}, \mathcal{A}, \mathcal{R}, \mathcal{P}, \gamma\}$, in which $\mathcal{S} \subseteq \mathbb{R}^n$ is an n-dimensional state space, $\mathcal{O}$ the observation space, $\mathcal{A} \subseteq \mathbb{R}^m$ an m-dimensional action space, $\mathcal{R}$ a bounded reward function, $\mathcal{P}$ a transition probability function, and $\gamma \in (0, 1]$ a discount factor.

At every time step t, the agent is given current state $s_t \in \mathcal{S}$ or observation $o_t \in \mathcal{O}$ and chooses one action a_t from finite action set $\mathcal{A}$ according to the policy $\pi_\theta(a_t|s_t)$ parameterized by θ. In problems with visual inputs, observation o_t is directly obtained from the environment and then processed by convolutional neural network to be fed into policy network. The performed action would affect the subsequent state iteratively because after action taken, the environment would return a reward value r and then transit to the next state s_{t+1} according to state transition probability matrix $\mathcal{P} = P(s_{t+1}|s_t, a_t)$. For example, in Atari domain, the player agent perceives current video as observation information, then chooses an action to perform and receives reward signal returned by game emulator.

The goal of RL is to learn an optimal policy that maximizes the total discounted reward through trading-off the exploration and exploitation. G_t is defined as the sum of discounted reward from time-step t:

$$G_t = R_{t+1} + \gamma R_{t+2} + \gamma^2 R_{t+3} + \cdots$$
$$= \sum_{k=0}^{\infty} \gamma^k R_{t+k+1} \tag{1}$$

where discount factor γ determines the present value of future rewards and values immediate reward above delayed reward, reward R at each time-step is a numerical number given by the environment.

We use state value function $V(s)$ to evaluate the long-term value of state s and action-state value function $Q(s,a)$ to figure out the value of state-action pair (s,a).

$$V^\pi(s) = \mathbb{E}[G_t|S_t = s]$$
$$= \mathbb{E}[R_{t+1} + \gamma V^\pi(S_{t+1})|S_t = s] \tag{2}$$

$$Q^\pi(s,a) = \mathbb{E}[G_t|S_t = s, A_t = a, \pi]$$
$$= \mathbb{E}[R_{t+1} + \gamma Q^\pi(s_{t+1}, a_{t+1})|S_t = s, A_t = a] \tag{3}$$

where policy π is parameterized by θ and experimentally implemented by neural networks.

3.2 Architecture

To apply accurate and optimal control, in this section we present a novel policy network architecture consisting of three independent branches: LSTM for exploiting hidden history information, nonlinear network for global control and linear network for stabilizing the system dynamics. The architecture of the proposed policy network is shown in Fig. 2.

Inspired by the Proportional-Integral Controller (PI) widely used in practical control systems, LSTM is adopted to utilize long-term encoded state information to control action at current time-step, which is similar to the control of the historic cumulative value of error used in PI. The aim of introducing control prior to policy is to eliminate the residual error in training process. Linear control policy has been proved effective for particular RL problems. In addition, we use basic MLP as nonlinear control network for its capability of global control and the promising performance on generic policy networks.

Given current state, three branches of policy $\pi_\theta(a_t|s_t)$ would evaluate respectively and then the results are combined to compute the resulting action at time t:

$$a_t = a_t^l + a_t^n + a_t^r \tag{4}$$

where a_t^l is the output of linear control network, a_t^n is the result of nonlinear policy module and a_t^r the time-dependent LSTM branch.

The key insight of our architecture is that the classic control prior knowledge combined with reinforcement learning has functioned practically well on continuous control tasks. To analyze the theoretical feasibility, we illustrate the generic task as traditional control problem. Let the desired current state denoted as s_t^d and the actual state at time step t as s_t, so the temporary error would be

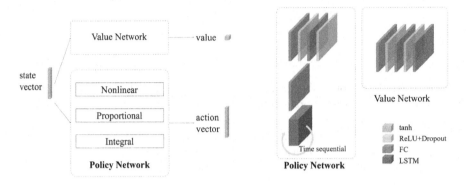

Fig. 2. The pipeline of reinforcement learning and the architecture of the proposed PI policy network.

$e_t = s_t - s_t^d$. According to control theory, the goal of control is to eliminate the error as much as possible.

In this formulation, given current state, the action should be:

$$
\begin{aligned}
a_t &= f_t^r + f_t^s(s_t, s_t^d) + f_t^e \\
&= f_t^r + f_t^s(s_t, s_t^d) + K_p \cdot (s_t - s_t^d)
\end{aligned}
\tag{5}
$$

where f_t^r is a history-concerned control term similar to the integral module in PI controller, and f_t^s is the nonlinear control branch formulated as the function of current state and desired state, f_t^e is the function of current error. As we stated before, this error function serves as a linear control module with K_p being the proportional terms for error e_t. In classic control theory, the proportional module is used for removing the gross error by applying the difference between the desired state and the measured state proportionally to the controlled variables. Furthermore, the nonlinear branch f_t^s works as global feedback control based on the predicted environment state s_t^d. "Integral" module f_t^r is utilized to eliminate the residual offset error by taking history error into account.

We further decompose the equation into:

$$
\begin{aligned}
a_t &= f_t^r + f_t^s(s_t, s_t^d) + K_p \cdot (s_t - s_t^d) \\
&= f_t^r + f_t^s + K_p \cdot s_t - K_p \cdot s_t^d \\
&= f_t^r + f_t^n + f_t^l
\end{aligned}
\tag{6}
$$

where we apply the transformation $f_t^n = f_t^s - K_p \cdot s_t^d$. The final control equation is totally the same as Eq. (4) we propose previously where f_t^l is denoted as linear control branch $K_p \cdot s_t$.

Experiment results demonstrate that both linear and nonlinear policy could achieve promising performance, as shown in Fig. 3. On some specific RL tasks like Humanoid, linear policy could obtain comparable effect with baseline MLP network while on more tasks, it fails to perform sound results. The linear control

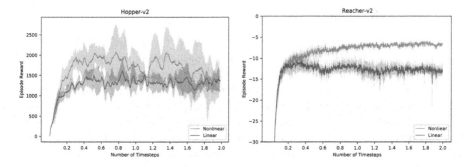

Fig. 3. Averaged learning curves of linear and nonlinear policy network of 5 sets of random seeds.

module f_t^l is implemented as 1 linear layer $K_p \cdot s_t + b$ where the gain matrix K_p and bias b are hyper-parameters need to be learned.

As part of our policy network, LSTM stores states information from previous time steps. This concept of holding long-term encoded information to control current action is similar to control the historic cumulative value of error used in PI, while the specific implementation and practical implication are quite distinct in some degree. In MDP, immediate states would play greater roles than delayed ones, which is in accordance with the internal states of LSTM. That's why we adopt LSTM as the most effective component of our policy network. In the proposed policy network, we use 1 LSTM layer with 64 hidden states to operate sequential data. The results of PI policy with different number of LSTM layer and hidden states are shown in experiment section.

The nonlinear network is implemented as generic MLP, and we also give the experimentally result that single nonlinear policy could acquire. Generally the individual nonlinear network works effectively for most of the RL tasks, which confirms the necessity of combing this nonlinear policy with the other two streams to further exploit.

4   Experiment

We conduct sufficient experiments on various benchmark RL tasks and widely used simulation environments to validate the applicability and effectiveness of the proposed policy network architecture. We mainly compare the training results with generic MLP policy and similar SCN policy network proposed by [25] under the same conditions. In addition, ablation experiments about the three policy architecture individually and the complement network are performed to confirm the capability. All these experiments are conducted under the guidance of RL reproducibility study [6]. In this section, experiment details and results are clearly explained.

4.1 RL Environments

Generally used RL environments including OpenAI Gym, MuJoCo, Roboschool that contain diverse RL tasks such as Atari games, continuous robot control and classic control problems are shown in Fig. 1. These simulation platforms are built with different physics engines and parameters, thus we could perform adequate validation experiments on available tests as many as possible.

Some standard test environments such as Humanoid-v1 and Swimmer-v1 are implemented in both MuJoCo and Gym. For example, Humanoid-v1 makes a three-dimensional bipedal robot walk forward without falling over. The state of this task is a 47 dimensional vector containing the position and velocity information. The action consists of a discrete 17 dimensional torque control vector over every joint of the humanoid robot.

4.2 Experimental Setup

As indicated before, we apply the proposed policy architecture to baseline RL algorithms PPO, ACKTR, A3C on popular benchmark tasks that have been widely used in the study of DRL. The test tasks consist of complicated continuous control problems, simplified classical control problems as well as Atari games. We mainly use the HalfCheetah-v2 and Hopper-v2 implemented in MuJoCo for their stable and contrasting dynamics.

The comparison experiments are a series of policy networks trained from scratch, including our control network, generic multilayer perceptron (MLP) and Structured Control Net (SCN). To avoid bias, these three policy networks are trained using the same algorithms with fixed hyper-parameters during the training. Applicable training algorithms PPO, ACKTR are implemented from OpenAI Baselines [4]. PPO is optimized by Adam optimizer [8] with initial learning rate as 3e−4, ϵ term as 1e−8. Particularly, in PPO generalized advantage estimation is used with $\tau = 0.95$ and the clip parameters is 0.2. In addition, ACKTR uses KFAC optimizer proposed by [31], and the learning rate is 3e−4, momentum parameter is 0.9.

In order to confirm the fairness, all the experiments we conduct use the same set of random seeds, and the depicted learning curves are obtained by averaging the evaluation results over five different random seeds from 1 to 5 respectively. We train these networks for 2M timesteps over every tasks, and the mini-batch size is fixed to 32, with fine-tuned reinforcement learning parameter discount factor $\lambda = 0.99$. We implement these experiments in PyTorch using 12 cores with Nvidia GeForce GTX 1060.

4.3 Results

In this section, we test three models: the proposed PI policy network, a baseline MLP and Structured Control Network (SCN) to compare their performances. The evaluation metrics widely applied in the reinforcement learning studies

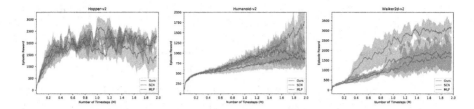

Fig. 4. Episode reward learning curves of the comparative methods: PI network, generic MLP and SCN, averaged on five sets of random seeds.

include the learning curves of cumulative reward along timesteps, the maximum reward and average reward over a fixed number of timesteps (Fig. 4).

The architecture of MLP used in these comparative experiments is a fully connected layer with two hidden layers, each of which consists 64 units and is activated by tanh function. This standard MLP-64 architecture is generally used in many algorithms [22,31]. According to the experiment details described in [25], the SCN is implemented as the combination of a generic three layer MLP (remove the bias of the last linear layer) and a linear layer. We adopt tanh as the activation function for the MLP used in SCN, and the two hidden layers of MLP have 64 units respectively. As shown in Fig. 2, the proposed architecture contains a LSTM with hidden size 64 and two nonlinear layers attached to the end of LSTM, a generic MLP-16 nonlinear network, and a simple linear network. The number of parameters of these three policy networks are quite approaching. For fairness comparison, the value network for all these optimization algorithms and models is fixed to be a three layer nonlinear network with one output dimension.

Comparison of Performance: The learning curves of the episode rewards are shown in Fig. 3. More accurately, the average reward and final episode reward are chosen to represent the model performance, as presented in Table 1. We only show the PPO training results since the ACKTR results are quite similar.

We compute the improvement of average episode reward to depict the training efficiency. In this evaluation, our model achieved 122% averaged reward improvement compared to generic MLP, and 126% to SCN.

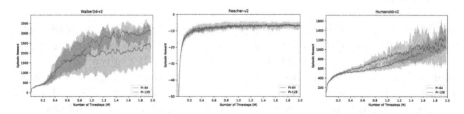

Fig. 5. Episode reward learning curves of the comparative methods: 1 LSTM layer with 64 hidden size PI and 1 LSTM layer with 128 hidden size PI, averaged on five sets of random seeds.

Table 1. Average and final episode rewards on various RL tasks of our model, SCN and MLP model.

Task	Average reward			Final reward		
	Ours	SCN	MLP	Ours	SCN	MLP
Ant	256	370	**489**	984	1205	**1583**
HalfCheetah	**1395**	1129	1390	**2356**	1181	2277
Hopper	**1961**	1854	1646	**2274**	2272	1449
Humanoid	**892**	789	662	**1762**	1007	866
Reacher	**−9.0**	−10.9	−9.3	−6.7	−7.3	**−6.6**
Swimmer	28	29	**34**	37	33	**49**
Walker2d	**2075**	1052	1191	**3133**	1678	1895

4.4   Ablation Experiments

To test the comparative effectiveness of the three modules of PI network respectively, we conduct ablation experiments that test each module separated from a fully trained PI model. We compare these branches' performances with an independently trained linear policy and a nonlinear policy (MLP) with the same size of linear and nonlinear branches in PI architecture. The training curves of single linear and nonlinear policy have been shown in Fig. 3. Since branches of PI policy network are jointly trained, we only compare the test rewards. The result of linear network contained in the learned PI model outperforms effectively in simple control tasks such as Pendulum and Walker2d when compared with single trained linear network, and trained MLP branch achieves slightly better rewards compared to separately trained MLP.

We also use LSTM with different number of hidden states and layers to evaluate the effect of proportional branch. As show in Fig. 5, LSTM simply with more hidden sizes couldn't achieve better performance on many tasks. Though more layers LSTM presents slight improvement compared with single layer LSTM, it contains much more parameters to compute.

5   Conclusion

In this paper, a simple but effective reinforcement learning policy network architecture is proposed to introduce control theory into reinforcement learning control tasks. In general RL problems (formulated as MDP), given current state information, the three branches of our network predict next action respectively, which would be combined to compute the final action. This Proportional-Integral architecture exploits the advantage of LSTM, linear control and nonlinear control, with LSTM taking advantage of history information, linear control stabilizing system dynamics, nonlinear branch serving as global controller. Sufficient comparative and ablation experiments demonstrate the proposed model outperform existing models on various RL tasks especially continuous control tasks.

References

1. Ang, K.H., Chong, G., Li, Y.: PID control system analysis, design, and technology. IEEE Trans. Control. Syst. Technol. **13**(4), 559–576 (2005)
2. Bakker, B.: Reinforcement learning with long short-term memory. In: Advances in Neural Information Processing Systems, pp. 1475–1482 (2002)
3. Brockman, G., et al.: OpenAI Gym (2016)
4. Dhariwal, P., et al.: OpenAI Baselines (2017). https://github.com/openai/baselines
5. Haarnoja, T., Pong, V., Zhou, A., Dalal, M., Abbeel, P., Levine, S.: Composable deep reinforcement learning for robotic manipulation. arXiv preprint arXiv:1803.06773 (2018)
6. Henderson, P., Islam, R., Bachman, P., Pineau, J., Precup, D., Meger, D.: Deep reinforcement learning that matters. arXiv preprint arXiv:1709.06560 (2017)
7. Kaelbling, L.P., Littman, M.L., Moore, A.W.: Reinforcement learning: a survey. J. Artif. Intell. Res. **4**, 237–285 (1996)
8. Kingma, D.P., Ba, J.: Adam: a method for stochastic optimization. CoRR abs/1412.6980 (2014). http://arxiv.org/abs/1412.6980
9. Levine, S., Finn, C., Darrell, T., Abbeel, P.: End-to-end training of deep visuomotor policies. J. Mach. Learn. Res. **17**(1), 1334–1373 (2016)
10. Lewis, F.L., Vrabie, D.: Reinforcement learning and adaptive dynamic programming for feedback control. IEEE Circuits Syst. Mag. **9**(3), 32–50 (2009)
11. Lillicrap, T.P., et al.: Continuous control with deep reinforcement learning. US Patent App. 15/217,758, 26 January 2017
12. Lipton, Z.C., Berkowitz, J., Elkan, C.: A critical review of recurrent neural networks for sequence learning. arXiv preprint arXiv:1506.00019 (2015)
13. Mahmood, A.R., Korenkevych, D., Komer, B.J., Bergstra, J.: Setting up a reinforcement learning task with a real-world robot. arXiv preprint arXiv:1803.07067 (2018)
14. Mnih, V., et al.: Asynchronous methods for deep reinforcement learning. In: International Conference on Machine Learning, pp. 1928–1937 (2016)
15. Mnih, V., et al.: Human-level control through deep reinforcement learning. Nature **518**(7540), 529 (2015)
16. Plappert, M., et al.: Multi-goal reinforcement learning: challenging robotics environments and request for research (2018)
17. Rajeswaran, A., Lowrey, K., Todorov, E.V., Kakade, S.M.: Towards generalization and simplicity in continuous control. In: Advances in Neural Information Processing Systems, pp. 6553–6564 (2017)
18. Rummery, G.A., Niranjan, M.: On-line Q-learning using connectionist systems, vol. 37. University of Cambridge, Department of Engineering (1994)
19. Salih, A.L., Moghavvemi, M., Mohamed, H.A., Gaeid, K.S.: Modelling and PID controller design for a quadrotor unmanned air vehicle. In: 2010 IEEE International Conference on Automation Quality and Testing Robotics (AQTR), vol. 1, pp. 1–5. IEEE (2010)
20. Schulman, J., Levine, S., Abbeel, P., Jordan, M., Moritz, P.: Trust region policy optimization. In: International Conference on Machine Learning, pp. 1889–1897 (2015)
21. Schulman, J., Moritz, P., Levine, S., Jordan, M., Abbeel, P.: High-dimensional continuous control using generalized advantage estimation. arXiv preprint arXiv:1506.02438 (2015)

22. Schulman, J., Wolski, F., Dhariwal, P., Radford, A., Klimov, O.: Proximal policy optimization algorithms. arXiv preprint arXiv:1707.06347 (2017)
23. Shalev-Shwartz, S., Ben-Zrihem, N., Cohen, A., Shashua, A.: Long-term planning by short-term prediction. arXiv preprint arXiv:1602.01580 (2016)
24. Silver, D., et al.: Mastering the game of go with deep neural networks and tree search. Nature **529**(7587), 484–489 (2016)
25. Srouji, M., Zhang, J., Salakhutdinov, R.: Structured control nets for deep reinforcement learning. arXiv preprint arXiv:1802.08311 (2018)
26. Sutton, R.S., McAllester, D.A., Singh, S.P., Mansour, Y.: Policy gradient methods for reinforcement learning with function approximation. In: Advances in Neural Information Processing Systems, pp. 1057–1063 (2000)
27. Wang, Z., Schaul, T., Hessel, M., Van Hasselt, H., Lanctot, M., De Freitas, N.: Dueling network architectures for deep reinforcement learning. arXiv preprint arXiv:1511.06581 (2015)
28. Watkins, C.J., Dayan, P.: Q-learning. Mach. Learn. **8**(3–4), 279–292 (1992)
29. Williams, J.D., Zweig, G.: End-to-end LSTM-based dialog control optimized with supervised and reinforcement learning. arXiv preprint arXiv:1606.01269 (2016)
30. Wu, C., et al.: Variance reduction for policy gradient with action-dependent factorized baselines. arXiv preprint arXiv:1803.07246 (2018)
31. Wu, Y., Mansimov, E., Grosse, R.B., Liao, S., Ba, J.: Scalable trust-region method for deep reinforcement learning using Kronecker-factored approximation. In: Advances in Neural Information Processing Systems, pp. 5285–5294 (2017)

Data-Efficient Reinforcement Learning Using Active Exploration Method

Dongfang Zhao, Jiafeng Liu, Rui Wu$^{(\boxtimes)}$, Dansong Cheng, and Xianglong Tang

School of Computer Science and Technology,
Harbin Institute of Technology, Harbin, China
`simple@hit.edu.cn`

Abstract. Reinforcement learning (RL) is an effective method to control dynamic system without prior knowledge. One of the most important and difficult problem in RL is how to improve data efficiency. PILCO is a state-of-art data-efficient framework which uses Gaussian Process (GP) to model dynamic. However, it only focuses on optimizing cumulative rewards, and does not consider the accuracy of dynamic model which is an important factor for controller learning. To further improve the data-efficiency of PILCO, we propose an active exploration version of PILCO (AEPILCO) which utilizes information entropy to describe samples. In policy evaluation stage, we incorporate information entropy criterion into long term sample prediction. With the informative policy evaluation function, our algorithm obtains informative policy parameters in policy improvement stage. Using the policy parameters in real execution will produce informative sample set which is helpful to learn accurate dynamic model. Thus our AEPILCO algorithm improves data efficiency through learning an accurate dynamic model by actively selecting informative samples with information-entropy criterion. We demonstrate the validity and efficiency of the proposed algorithm for several challenging controller problems involving cart-pole, pendubot, double-pendulum and cart-double-pendulum. The proposed AEPILCO algorithm can learn controller using less trials which is verified by both theoretical analysis and experimental results.

Keywords: Reinforcement learning · Information entropy · PILCO
Data efficiency

1 Introduction

Reinforcement Learning (RL) is a developing field in machine learning, which is also an efficient method for autonomous learning in robotics and control without prior knowledge. Differing from traditional supervised learning and unsupervised learning, which is typically learning from static training samples, reinforcement

Supported by National Science Foundation of China (Grant NO. 61672190, No. 61370162).

L. Cheng et al. (Eds.): ICONIP 2018, LNCS 11303, pp. 265–276, 2018.
https://doi.org/10.1007/978-3-030-04182-3_24

learning learns through interacting with environment autonomously. Generally, several interactions are required to collect knowledge about environment before learning to control. For realistic dynamic system which is sensitive to time and computation incremental, Too much interactions may also bring security risk. Thus, the required number of interactions should be taken into consideration when reinforcement learning method is applied into real robot system.

Reinforcement learning can be formalized as Markov Decision Process (MDP). Agent continually executes action to interact with environment and finally achieve explicit. These interaction behaviors make agent translates from present state to next state according to transition probability model and policy. Environment observation and feedback rewards obtained from interaction are used to learning transition probability and proper policy until a predefined system target is achieved. Data-efficient reinforcement learning is to find a proper policy which can maximize the cumulative rewards with a minimal number of interactions [2]. In data-efficient reinforcement learning, no prior knowledge about environment is the most fundamental challenge to achieve data efficiency. It is difficult to select an optimal policy for agent to control without an accurate dynamic model. In addition, exploration and exploitation tradeoff remains challenging for control systems.

Various effective algorithms to solve data-efficient problem are available in literatures [7,8]. The previous proposed data-efficiency algorithms are mostly based on model-based structure considering that model-based method has natural advantage than model-free method in dealing with data-efficiency [5]. Dyna [10] and Dyna-2 [9] are classical papers in model based reinforcement learning domain. There are many algorithms from the viewpoint of model structure and stochastic optimal control [4,8]. Alternatively, there are a few recent papers combining model-based learning with deep nets. [6] Moreover, Probabilistic Inference and Learning COntrol(PILCO) algorithm is an excellent framework to achieve data efficiency [2,3], PILCO use probability dynamic model instead of a single determinate model. And this probability description learns the uncertainty of dynamic model which is a important challenge of model-based reinforcement learning method. However, PILCO focuses on maximizing cumulative rewards to learn optimal policy parameters, and does not consider the accuracy of dynamic model which is an important factor when learning controller.

Motivated by the aforementioned limitations, in this paper we propose active exploration PILCO (AEPILCO) algorithm. We improve the typical PILCO with considering the influence of dynamic model. The key of our proposed algorithm is selecting sample set which is helpful to train dynamic model better. Due to the accurate dynamic model, optimal policy parameter will be learned and target will be achieved faster. To achieve this idea, information entropy is introduced to describe sample uncertainty. Samples with high uncertainty are more helpful to train accurate dynamic model. Thus, data efficiency is achieved in terms of carefully learn dynamic model. Simulation experiments on several challenging control problems verify the better data-efficient performance of AEPILCO.

2   The PILCO Framework

PILCO considers a dynamic system with continuous state $\mathbf{x}$ and action $\mathbf{u}$.

$$\mathbf{x}_{t+1} = f(\mathbf{x}_t, \mathbf{u}_t) \tag{1}$$

The states transition is considered as markov process. Given state $\mathbf{x}_t$, a dynamic system will transfer to state $\mathbf{x}_{t+1}$ with action $\mathbf{u}_t$ according to dynamic model f which describes the transition probability $p(\mathbf{x}_{t+1}|\mathbf{x}_t, \mathbf{u}_t)$. Define state-action vector $\tilde{\mathbf{x}}_t = [\mathbf{x}_t, \mathbf{u}_t]$. The set of tuples $< \tilde{\mathbf{x}}_t, \mathbf{x}_{t+1} >$ is defined as a sample. The state transitions obeys dynamic model. With the determined dynamic model, once $\tilde{\mathbf{x}}_{t-1}$ is known, $\mathbf{x}_t$ is easy to compute. Such that the current sample can be described using the probability of next state $p(\mathbf{x}_t)$ instead of $\tilde{\mathbf{x}}_{t-1}$.

In PILCO, the dynamic model is implemented as a gaussian process which is completely specified by its mean function and covariance function. According the definition of gaussian process transition probability. When system is in a determinate state $\mathbf{x}$, the predicted state $\mathbf{x}_*$ with action $\mathbf{u}$ obeys normal distribution.

$$p(\mathbf{x}_*|\mathbf{x}, \mathbf{u}) = N(m(\mathbf{x}, \mathbf{u}), \Sigma(\mathbf{x}, \mathbf{u})) \tag{2}$$

Squared Exponential kernel function is selected as covariance function:

$$k(\mathbf{p}, \mathbf{q}) = \sigma^2 \exp(-\frac{\|\mathbf{p} - \mathbf{q}\|^2}{2l^2}) + \delta\sigma_\varepsilon^2 \tag{3}$$

where noise variance σ_ε, latent function variance σ and lenght-scale l are Gaussian Process hyper-parameters which is needed to be learned. The initial training input is $D = \{\tilde{\mathbf{x}}_1, ..., \tilde{\mathbf{x}}_n\}$. The corresponding training target is its next state set $\{\mathbf{x}_2, ..., \mathbf{x}_{n+1}\}$ that the system will transition to. δ is Kronecker delta function which is one if the two input $\mathbf{p}, \mathbf{q}$ is equal and zero otherwise. With constant interaction with environment, there will more and more samples join the training data set. Thus agent can update hyper-parameters through retrain dynamic model with the new training data to predict next state. Moreover, PILCO assumes the action selection obey normal distribution when system is in a determinant state.

$$p(\mathbf{u}|\mathbf{x}) = N(\mu(\mathbf{x}), \sigma(\mathbf{x})) \tag{4}$$

In initialization stage, assuming policy selection obeys initial mean $\mu_0 = 0$ and covariance $\Sigma_0 = 1$, with this initialization, the system execute on real environment for initial training dataset D. Subsequently, agent use this training dataset to learn hyper-parameters of gaussian process. In policy evaluation stage, looking forward t steps to evaluate policy parameter θ with the learned dynamic model and parameterized policy function $\pi(\theta)$. Policy evaluation accumulates cost function value of t predicted steps and get $J^\pi(\theta)$. Consider that the simulating and gradient computation is too complex, PILCO makes some assumptions and simplification to realize it including first-order markov process and looking forward one-step. In experiment of PILCO, authors verified the simplification of the calculation would not affect too much. According to one-step prediction with determinate input derived in [11], the mean and variance is given by Eq. (5)

$$
\begin{aligned}
m(\mathbf{x}_*) &= k(\mathbf{x}_*, \mathbf{x}_i)(\mathbf{K} + \sigma_\varepsilon^2 \mathbf{I})^{-1}\mathbf{y} \\
\Sigma(\mathbf{x}_*) &= k(\mathbf{x}_*, \mathbf{x}_*) - k(\mathbf{x}_*, \mathbf{x}_i)(\mathbf{K} + \sigma_\varepsilon^2 \mathbf{I})^{-1}k(\mathbf{x}_i, \mathbf{x}_*)
\end{aligned}
\tag{5}
$$

where $k(\cdot, \cdot)$ is covariance function. $\mathbf{K}$ is covariance matrix of training inputs. $\mathbf{x}_i$ is element of training inputs $\mathbf{X} = \{\tilde{\mathbf{x}}_1, ..., \tilde{\mathbf{x}}_n\}$. $\mathbf{y} = [\mathbf{x}_2...\mathbf{x}_{n+1}]$ is training target. σ_ε is noise variance which is learned in dynamic model learning process and $\mathbf{I}$ is Kronecker delta matrix. In policy improvement stage, PILCO Minimizes $J^\pi(\theta)$ and (Broyde Fletcher Goldfarb Shanno)BFGS policy gradient method is utilized to get new policy parameters θ^* through computing $\mathrm{d}J^\pi(\theta)/\mathrm{d}\theta$. Then execute the current optimal policy to generate new samples and return to train new dynamic model. And repeat this processes until predefined task achieved.

3   Active Exploration PILCO

In this section, we detailly describe our active exploration PILCO algorithm. The typical bayesian reinforcement learning framework PILCO uses Gaussian Process to model the dynamic system. And update policy parameters through minimizing the mean of accurate reward which is estimated by the distance between current state and target state, then the agent focuses on exploiting current existing policy parameter to interact with environment. However, PILCO does not consider the accuracy of dynamic model. To deal with this problem, we propose an active exploration PILCO algorithm to achieve data-efficiency through actively selecting samples which is helpful to learn a more accurate dynamic model. Specifically, information entropy is utilized to describe long-term predicted samples. In the following subsections, We will firstly propose entropy-based sample description method and analysis. Subsequently, we analysis how to use information entropy method to achieve data efficiency in policy evaluation and improvement stages. At last, we describe the entire processes of the proposed algorithm and analysis the difference between PILCO and our AEPILCO.

3.1   Entropy-Based Sample Description

Consider that dynamic model is learned using interaction samples which depend on parameterized policy function. The policy parameter is updated depends on accumulated reward of t step simulations. Thus the key of learning accurate dynamic model is generating informative simulated samples in policy evaluation stage. In the following, we will introduce information entropy to describe simulated samples. According to gaussian process assumption, the transition probability $p(\mathbf{x}_{t+1}|\mathbf{x}_t, \mathbf{u}_t)$ which describes system transition from state $\mathbf{x}_t$ to state $\mathbf{x}_{t+1}$ with action $\mathbf{u}_t$ obeys normal distribution. The information entropy of predicted state distribution $p(\mathbf{x}_{t+1}|\mathbf{x}_t, \mathbf{u}_t)$ can be described as $- \int p(\mathbf{x}_{t+1}|\mathbf{x}_t, \mathbf{u}_t) \log p(\mathbf{x}_{t+1}|\mathbf{x}_t, \mathbf{u}_t) d\mathbf{x}_{t+1}$. This entropy describes uncertainty of variable $\mathbf{x}_{t+1}$. High information entropy means high uncertainty. Information entropy is used to describe the predicted sample. When predicting samples for efficient reinforcement learning algorithms, agent should select the samples with largest uncertainty. Because the sample set with higher information entropy is helpful to learn a more accurate dynamic model, the learned model based on these high uncertainty samples has stronger generalization ability. Therefore, we select the sample $\tilde{\mathbf{x}}_t$ which can obtain optimal predict state $\mathbf{x}_{t+1}$ with largest entropy. Thus the information entropy criterion:

$$\mathbf{x}_H^* = \arg\max_{\mathbf{x}_{t+1}} - \int p(\mathbf{x}_{t+1}|\mathbf{x}_t, \mathbf{u}_t) \log p(\mathbf{x}_{t+1}|\mathbf{x}_t, \mathbf{u}_t) d\mathbf{x}_{t+1} \tag{6}$$

where $\mathbf{x}_H^*$ is the optimal sample with highest information entropy we needed to sampling. $\mathbf{x}_{t+1}$ ranges over all possible states. $p(\mathbf{x}_{t+1}|\mathbf{x}_t, \mathbf{u}_t)$ describes the probability of state-action vector $< \mathbf{x}_t, \mathbf{u}_t >$ translates to state $\mathbf{x}_{t+1}$. From the assumption of Gaussian Process normal distribution and the first-order markov process, the posterior probability $p(\mathbf{x}_{t+1}|\mathbf{x}_t, \mathbf{u}_t)$ is specified by its mean and covariance function described in Eq. (5).

3.2   Policy Evaluation

In AEPILCO framework, the previous real generated samples are used to learn a basic dynamic model. Subsequently, this dynamic model is utilized to predict a sequence of simulation samples $\mathbf{x}_1, \mathbf{x}_2, ..., \mathbf{x}_t$. Our objective is to maximize the information entropy of entire predicted state distribution. Since every predicted state obeys normal distribution, the entropy of multivariate normal distribution having probability $p(\mathbf{x}_{t+1}|\mathbf{x}_t, \mathbf{u_t})$ can described as continuous integration which is given by Eq. (7).

$$H(p) = - \underbrace{\int_{-\infty}^{\infty} \int_{-\infty}^{\infty} \cdots \int_{-\infty}^{\infty}}_{t} p(\mathbf{x}_{t+1}|\mathbf{x}_t, \mathbf{u}_t) \log p(\mathbf{x}_{t+1}|\mathbf{x}_t, \mathbf{u}_t) d\mathbf{x}_{t+1} \tag{7}$$

where $\mathbf{x}_{t+1}$ is element of state set $\mathbf{X} = (\mathbf{x}_1, \mathbf{x}_2, \cdots, \mathbf{x}_t)'$ which is waiting to be simulated. According to entropy expression for multivariate normal distribution derived in Ref. [1]. The entropy of sample set in Eq. (7) can be rewritten as

$$H(p) = \frac{N}{2} + \frac{N}{2} \ln(2\pi) + \frac{1}{2} \ln\left(|\mathbf{\Sigma}|\right) \tag{8}$$

where $\mathbf{\Sigma}$ is covariance matrix of predicted sample set. $|\cdot|$ denotes determinant and N is the number of samples. To maximum $H(\mathbf{X})$ in Eq. (8), we just need to maximize $\ln\left(|\Sigma(\mathbf{X})|\right)$. Thus the active exploration optimistic term describes as:

$$J^E(\mathbf{X}) = \ln\left(|\mathbf{\Sigma}(\mathbf{X})|\right) \tag{9}$$

The active exploration optimization term focuses on generating informative samples. And adding this term into policy evaluation stage will helpful for learning accurate dynamic model. Consider that typical PILCO policy evaluation objective function only focuses on the distance between current state and target state without consider the accuracy of dynamic model which can contribute to improve data efficiency either. Such that we add Eq. (9) into policy evaluation of typical PILCO. Then the AEPILCO policy evaluation objective function is

$$J^{\pi}(\boldsymbol{\theta}) = \sum_{t=0}^{T} \mathrm{E}_{\mathbf{x}_t} [c(\mathbf{x}_t)] + \alpha \sum_{t=0}^{T} \ln \left[\boldsymbol{\Sigma}(\mathbf{x}_t) \right] \tag{10}$$

we call the added optimization function item as active exploration term because it helps for exploration. The covariance in Eq. (9) contains uncertainty information of dynamic model. Typical PILCO is a exploitation-greedy algorithm. Thus adding Eq. (9) is helpful for balancing exploration and exploitation.

3.3 Policy Improvement

In the following, we derive how to achieve policy improvement with our added active exploration item in Eq. (9). In AEPILCO, the objective function in policy evaluation is a sum of our active exploration item and typical cost function which has been derived in PILCO. Thus, we only need compute the derivation of Eq. (9) on policy parameters

$$\frac{dJ^E(\theta)}{d\theta} = \frac{d\ln|\boldsymbol{\Sigma}(\mathbf{x})|}{d|\boldsymbol{\Sigma}(\mathbf{x})|} \frac{d|\boldsymbol{\Sigma}(\mathbf{x})|}{d\theta} \tag{11}$$

where

$$\frac{d\ln|\boldsymbol{\Sigma}(\mathbf{x})|}{d|\boldsymbol{\Sigma}(\mathbf{x})|} = \frac{1}{|\boldsymbol{\Sigma}(\mathbf{x})|} \tag{12}$$

$$\frac{d|\boldsymbol{\Sigma}(\mathbf{x})|}{d\theta} = \frac{d|\boldsymbol{\Sigma}(\mathbf{x})|}{d\boldsymbol{\Sigma}(\mathbf{x})} \frac{d\boldsymbol{\Sigma}(\mathbf{x})}{d\theta} \tag{13}$$

according the differential of matrix derivation to matrix

$$\frac{d|\boldsymbol{\Sigma}(\mathbf{x})|}{d\boldsymbol{\Sigma}(\mathbf{x})} = |\boldsymbol{\Sigma}| \, \boldsymbol{\Sigma}^{-1} \tag{14}$$

Thus, the derivation of added active exploration term on policy parameter $\boldsymbol{\theta}$ depends on $|\boldsymbol{\Sigma}(\mathbf{x})|$, $|\boldsymbol{\Sigma}| \, \boldsymbol{\Sigma}^{-1}$ and $\partial \boldsymbol{\Sigma}(\mathbf{x})/\partial \boldsymbol{\theta}$ derived in Eqs. (12)–(14).

The covariance of predicted samples is derived in PILCO [3] by moment matching approximation method which computes the first two moments of the predictive distribution exactly.

The derivation of covariance to policy parameter is similar to $\partial \boldsymbol{\mu}(\mathbf{x})/\partial \boldsymbol{\theta}$ which is derived in PILCO [3]. For $\partial \boldsymbol{\Sigma}(\mathbf{x})/\partial \boldsymbol{\theta}$, we compute the derivative

$$\frac{d\boldsymbol{\Sigma}(\mathbf{x}_t)}{d\theta} = \frac{d\boldsymbol{\Sigma}}{dp(\mathbf{u}_{t-1})} \frac{dp(\mathbf{u}_{t-1})}{d\theta} = \frac{d\boldsymbol{\Sigma}}{d\mu_{\mathbf{u}}} \frac{d\mu_{\mathbf{u}}}{d\theta} + \frac{d\boldsymbol{\Sigma}}{d\boldsymbol{\Sigma}_{\mathbf{u}}} \frac{d\boldsymbol{\Sigma}_{\mathbf{u}}}{d\theta} \tag{15}$$

where $\partial\Sigma/\partial\boldsymbol{\mu}_u$ and $\partial\Sigma/\partial\Sigma_u$ are propagated by long-term prediction and computed by moment matching approximation method which is derived in [3]. The $\partial\boldsymbol{\mu}_u/\partial\boldsymbol{\theta}$ and $\partial\Sigma_u/\partial\boldsymbol{\theta}$ in Eq. (15) are derivations of mean and covariance of action at $t-1$ to policy parameter. The computation depends on the presentation of policy which is introduced in PILCO [3].

Algorithm 1 summarizes the entire process of our active exploration PILCO. Compared to typical PILCO, AEPILCO add active exploration term into objective function $J^{\pi}(\theta)$ in policy evaluation stage. The PILCO framework update policy depends on the evaluation of predicted simulations which is calculated from dynamic model. While in our AEPILCO, For learning a more accurate dynamic model, we extend the policy evaluation objective function by adding a optimization term which is used to describe the predicted samples with information entropy. Moreover, adding the optimization term is also helpful for exploration as it can describes the sample variance. This feature is analyzed in Subsect. 3.2. Thus, our AEPILCO can use minimal interactions to achieve a predefined target with the accurate dynamic model and exploration and exploitation balance.

Algorithm 1. Active Exploration PILCO

1: **Init:** Initialize random policy π with parameter $\theta \sim N(\mu_0, \Sigma_0)$. Execute policy on real system to gather training data D.

2: **Repeat:**

3:     learn GP dynamic model f using all data.

4:     **Repeat:**

5:         prediction: simulate system for $p(\mathbf{x}_1)...p(\mathbf{x}_t)$

6:         policy   evaluation: approximate   inference,   get $J^{\pi}(\theta)$ = $\sum_{t=0}^{T} E_{\mathbf{x}_t}[c(\mathbf{x}_t)] + \alpha \sum_{t=0}^{T} \ln[\Sigma(\mathbf{x}_t)].$

7:         policy improvement: BFGS based policy improvement, get $\mathrm{d}J^{\pi}(\theta)/\mathrm{d}\theta$.

8:     **until** convergence, return new policy parameter θ^*

9: $\pi(\theta) \leftarrow \pi(\theta^*)$

10: execute policy on real system to gather training data.

11: **until** task achieved.

Experiments and Analysis. In this section, we evaluate our AEPILCO algorithm described in Sect. 3 on several challenging control tasks including benchmark problems and high dimensional state space problems. We utilize simulated scenarioes provided by PILCO software package (http://mloss.org/software/view/508/) to verify our algorithm. The package provides six simulators of implemented scenarios as demonstration. All of them focus on the fundamental of solving the differential equations of nonlinear dynamic systems. Detail derivation of their physical model can be seen in PILCO software package. We use four of these predesigned scenarios with different dimensions of state space to verify our algorithm. Firstly, a comparison experiment to evaluate the dynamic model learned by AEPILCO and PILCO is presented in Sect. 3.4. Subsequently,

we design a comparison Finally, we provide a parameter selection experiment
in Sect. 3.5. Lastly, time consuming experiment and analysis are presented in
Sect. 3.5. All the experiments were done on a PC with the same hardware i5
CPU(2.57 GHz), 8 GB RAM and operating system WIN-10.

3.4   Dynamic Model Efficiency Experiment

This experiment is designed to evaluate the dynamic model learned by
AEPILCO. Since the simulated samples are generated by the learned dynamic
model. And real executed samples are based on the real dynamic model. Thus
the distance of cost function values between simulated and real execute samples
can reflects the similarity of dynamic models. Small distance means that the
learned dynamic model is closer to real model.

Figure 1 shows cost function value comparison of four scenarios including
cart pole, pendubot, double pendulum and cart double pendulum. The horizontal axises are steps of AEPILCO convergence trial. The vertical axises are
corresponding cost function value in each step. In order to see clearly, we show
cost function values of each step in convergent trial. The blue and red lines are
PILCO predicted mean simulation and real execute cost function value, respectively. The yellow and purple lines are our AEPILCO predicted mean simulation
and real execute cost function value, respectively. Specially, simulated samples
depend on mean and covariance, the predicted mean simulations shown in Fig. 1
are mean cost value of simulated samples. In general, Fig. 1 intuitively show that
the yellow and purple lines are more similar than blue and red lines which means
the dynamic model creating yellow and purple lines is more accurate.

We use Euclidean distance to quantitatively evaluate the distance of predicted simulation and real execute cost function values. Comparison results of
pendulum and other high dimensional problems are shown in Table 1. The first
and second rows provide the dimensions of state and action space. The third and
last bold rows are the distance between real execution cost and simulated cost
using PILCO and AEPILCO respectively. For all control problems, AEPILCO
show better performance on learning dynamic model. This is mainly because
our algorithm consider the accuracy of dynamic model and add active exploration item in policy evaluation stage. Consider that dynamic model is learned
using real interaction samples which depend on policy. In policy improvement
stage, policy parameters is calculated according to objective function which is
determined by rewards of simulation sample. In AEPILCO, the simulation samples are most informative due to information entropy descriptor. Thus we can
obtain informative simulation samples and policy parameters which is helpful
for learning accurate dynamic model.

3.5   Parameter Selection Experiment

This experiment is designed to analysis how parameter α in Eq. 10 affects convergence speed, meanwhile, select proper parameters for each scenarios and evaluate
the data efficiency of AEPILCO.

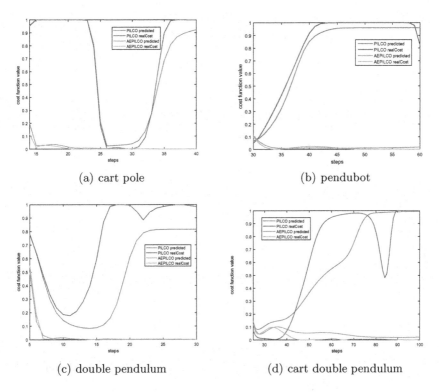

(a) cart pole

(b) pendubot

(c) double pendulum

(d) cart double pendulum

Fig. 1. Comparison in terms of cost function value at convergence trial with AEPILCO for (a) cart pole(trial #7), (b) pendubot(trial #10), (c) double pendulum(trial #6) and (d) cart double pendulum(trial #26). Predicted cost function values are produced by simulation samples and realCost cost function value are produced by real execute samples. The yellow and purple lines are more similar than blue and red lines which means the dynamic model creating yellow and purple lines is more accurate. (Color figure online)

Table 1. Comparison of dynamic model distances on four high dimension problems

	Cart pole	Pendubot	Double pendulum	Cart double pendulum
State space	$\mathbb{R}^4$	$\mathbb{R}^4$	$\mathbb{R}^4$	$\mathbb{R}^6$
Action space	$\mathbb{R}$	$\mathbb{R}$	$\mathbb{R}^2$	$\mathbb{R}$
PILCO [3]	0.3874	0.3433	1.7250	1.9974
AEPILCO	**0.1013**	**0.0635**	**0.0787**	**0.2946**

We evaluate the effection of parameter α for four scenarios. For each task, we choose fixed α. Figure 2 shows convergence speed comparison with different parameter α on cart pole, pendubot, double pendulum and cart double pendulum. The horizontal axises are number of trials. The vertical axises are mean cost value of last several steps at each trial. Blue lines show PILCO with no

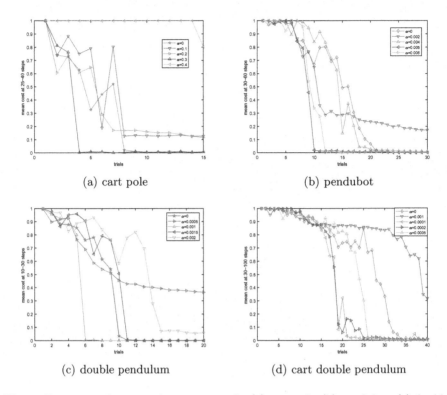

(a) cart pole   (b) pendubot

(c) double pendulum   (d) cart double pendulum

Fig. 2. Comparison in terms of parameter α for (a) cart pole, (b) pendubot, (c) double pendulum and (d) cart double pendulum. Subfigures show mean cost function values at each trial.

exploration. Others are results using AEPILCO with fixed α. The mean cost decrease over with trials runs. Once mean cost at last few steps at each trial approaches to zero, the agent achieve target state.

According derivation of AEPILCO in Sect. 3, the larger of α the more exploration. However, too much exploration may lead to divergent. Therefore, we use a adaptive parameter selection function:

$$\alpha = \alpha_0 e^{-wt} \tag{16}$$

where α_0 is the initial value of α, t is the number of trials, and w is setted according to different scenarios. This adaptive parameter selection function enable the agent exploration more at the first few trials and exploitation more when it approaches the target.

Table 2 shows the needed trials to achieve targets of each scenarios. The bold numbers are trails needed by AEPILCO. The first row is the results of PILCO. Results illustrate that our AEPILCO algorithm performs better than PILCO which is a no exploration algorithm. And AEPILCO with variable α achieves better performance than others on all scenarios. Moreover, the data-efficiency is also reflected in Fig. 1. In Fig. 1(a), (c) and (d), when PILCO's cost value jumps from 0.01 to 1, the AEPILCO has already achieves problem target state, and the cost function value tends to 0. In Fig. 1(b), AEPILCO achieves convergence state on pendubot while the cost function value using PILCO is still no sign of convergence. The last row in Table 2 summarizes the best improvement scale compared to typical PILCO. In general, for all scenarios, the AEPILCO with variable α has best performance. This is mainly because our algorithm select informative simulations and policy parameter at each trial which results in learning accurate dynamic model. Moreover, the adaptive parameter selection method makes the agent exploration more at beginning of interaction. When it get near to target the agent will exploitation more.

Table 2. Results of convergence trials using typical PILCO, AEPILCO with fixed α and AEPILCO with variable α for four high dimension problems

	Cart pole	Pendubot	Double pendulum	Cart double pendulum
PILCO [3]	8	15	10	34
AEPILCO with fixed α	7	**10**	6	**26**
AEPILCO with variable α	4	9	5	24
Best improvement	50%	40%	50%	29%

In terms of computational efficiency, the computation complexity of AEPILCO does not grow too much compared to typical PILCO. Because adding the active exploration term into policy evaluation stage corresponds to compute one more inverse of covariance matrix in Eq. (14). Consider the small amount of matrix, computation complexity will not significantly increase. Time comparison results are shown in Table 3. The first two rows show the total time needed to achieve target in each scenarios. The average time for each trial is shown in the last two rows. Although the average time using AEPILCO is lager than PILCO, the total time with AEPILCO is smaller than PILCO due to high data-efficiency of AEPILCO.

Table 3. Total time comparison with PILCO and AEPILCO algorithm (first two rows). Average time comparison for each trial with PILCO and AEPILCO algorithm (last two rows).

	Cart pole	Pendubot	Double pendulum	Cart double pendulum
PILCO total	≈27s	≈56s	≈ 38s	≈ 134s
AEPILCO total	≈18s	≈41s	≈3s	≈112s
PILCO average	≈3.4s	≈3.7s	≈ 3.8s	≈ 3.9s
AEPILCO average	≈4.5s	≈4.6s	≈4.6s	≈4.7s

4 Conclusion

In this paper, AEPILCO, an active exploration version of data-efficient framework PILCO is proposed to further improve data efficiency. Our algorithm utilizes information entropy criterion to select the most informative sample set to learn a more accurate dynamic model. Compared to previous data-efficient framework PILCO, our algorithm take the accuracy of dynamic model into consideration which is helpful to improve data-efficiency in controller learning task. Moreover, our AEPILCO algorithm can balance exploration and exploitation, because the active exploration item in policy evaluation objective function consists of covariance of predicted samples which describes the amount of exploration. In summary, more accurate dynamic model and more balance between exploration and exploitation can effectively improve data efficiency. Simulation experimental results on several challenging control problems verify the effectiveness and data-efficiency of our AEPILCO algorithm.

References

1. Ahmed, N.A., Gokhale, D.: Entropy expressions and their estimators for multivariate distributions. IEEE Trans. Inf. Theory **35**(3), 688–692 (1989)
2. Deisenroth, M., Rasmussen, C.E.: PILCO: a model-based and data-efficient approach to policy search. In: Proceedings of the 28th International Conference on Machine Learning, ICML2011, pp. 465–472. ACM, Bellevue (2011)
3. Deisenroth, M.P., Fox, D., Rasmussen, C.E.: Gaussian processes for data-efficient learning in robotics and control. IEEE Trans. Pattern Anal. Mach. Intell. **37**(2), 408–423 (2015)
4. Fabisch, A., Metzen, J.H.: Active contextual policy search. J. Mach. Learn. Res. **15**(1), 3371–3399 (2014)
5. Lai, T.L., Robbins, H.: Asymptotically efficient adaptive allocation rules. Adv. Appl. Math. **6**(1), 4–22 (1985)
6. Levine, S., Finn, C., Darrell, T., Abbeel, P.: End-to-end training of deep visuomotor policies. J. Mach. Learn. Res. **17**(1), 1334–1373 (2016)
7. Ng, A.Y., et al.: Autonomous inverted helicopter flight via reinforcement learning. In: Ang, M.H., Khatib, O. (eds.) Experimental Robotics IX. STAR, vol. 21, pp. 363–372. Springer, Heidelberg (2006). https://doi.org/10.1007/11552246_35
8. Pan, Y., Theodorou, E., Kontitsis, M.: Sample efficient path integral control under uncertainty. In: Advances in Neural Information Processing Systems, pp. 2314–2322 (2016)
9. Silver, D., Sutton, R.S., Mller, M.: Sample-based learning and search with permanent and transient memories. In: International Conference on Machine Learning, ICML2008, pp. 968–975. ACM, Helsinki (2008)
10. Sutton, R.S.: Dyna, an integrated architecture for learning, planning, and reacting. ACM Sigart Bull. **2**(4), 160–163 (1991)
11. Williams, C.K.: Gaussian Processes for Machine Learning. The MIT Press, pp. 7–30. Massachusetts Institute of Technology (2006)

Averaged-A3C for Asynchronous Deep Reinforcement Learning

Song Chen, Xiao-Fang Zhang$^{(\boxtimes)}$, Jin-Jin Wu, and Di Liu

School of Computer Science and Technology, Soochow University, Suzhou, China
xfzhang@suda.edu.cn

Abstract. In recent years, Deep Reinforcement Learning (DRL) has achieved unprecedented success in high-dimensional and large-scale space tasks. However, instability and variability of DRL algorithms have an important effect on their performance. To alleviate this problem, the Asynchronous Advantage Actor-Critic (A3C) algorithm uses the advantage function to update the policy and value network, but there still remains a certain variance in the advantage function. Aiming to reduce the variance of the advantage function, we propose a new A3C algorithm called Averaged Asynchronous Advantage Actor-Critic (Averaged-A3C). Averaged-A3C is an extension of the A3C algorithm, by averaging previously learned state value estimates to calculate the advantage function, which contributes to a more stable training procedure and improved performance. We evaluate the performance of the new algorithm through some games on the Atari 2600 and MuJoCo environment. Experimental results show that the Averaged-A3C algorithm effectively improves the performance of Agent and the stability of training process compared to the original A3C algorithm.

Keywords: Deep reinforcement learning
Asynchronous Advantage Actor-Critic · Advantage function · Average

1 Introduction

Deep Reinforcement Learning (DRL) combining Deep Learning (DL) [1] and Reinforcement Learning (RL) [2] is a new research hot spot in the artificial intelligence field. At present, RL has achieved remarkable results in the fields of simulation and game theory et al. [3]. However, in the complicated task of high-dimensional state space, the traditional RL algorithm does not perform well and the data needs to go through complex manual pretreatment. In order to solve this problem, using deep neural networks for effective recognition of high-dimensional state space, it is possible to make the RL algorithms more effective in the complex state tasks. In detail, Mnih et al. combined the Convolutional Neural Network (CNN) in deep neural networks with the Q-learning algorithm [4] in reinforcement learning for the first time, and proposed a Deep Q-Network (DQN) [5].

© Springer Nature Switzerland AG 2018
L. Cheng et al. (Eds.): ICONIP 2018, LNCS 11303, pp. 277–288, 2018.
https://doi.org/10.1007/978-3-030-04182-3_25

In DRL, DQN and its variants adopt the experience replay mechanism [6] to improve the efficiency of samples. Thus, many researchers focus on the optimization of experience replay mechanism. For example, to accelerate the speed of training and reduce computation, Minh et al. proposed Asynchronous Deep Reinforcement Learning (ADRL) [7], which improves the traditional experience replay by adopting the asynchronous updating methods. In ADRL, the Asynchronous Advantage Actor-Critic (A3C) algorithm uses the estimate of the advantage function [8] to update the policy and value network. However, the advantage function has a comparable variance and introduces bias [9], resulting in necessitation of much more samples. Furthermore, bias may cause the algorithm to fail to converge, or to converge to a poor solution that is not a local optimum. To solve this problem, John Schulman et al. proposed the Generalized Advantage Estimator (GAE) algorithm [9]. GAE uses the advantage function parameterized by $\gamma_{GAE} \in [0,1]$ and $\lambda \in [0,1]$ to significantly reduce variance while maintaining a tolerable level of bias.

In this paper, different from the idea of GAE and inspired by the idea of Averaged-DQN [10] and asynchronous updating, a new A3C algorithm called Averaged Asynchronous Advantage Actor-Critic (Averaged-A3C) is proposed. To reduce the variance of the advantage function used in A3C, Averaged-A3C is based on the idea of averaging previously learned state value estimates. The averaging methods can reduce variance which leads to stability, then improves Agent's performance. In addition, we provide experimental results on selected games of Atari 2600 and some continuous tasks of MuJoCo. Experimental results show that Averaged-A3C algorithm can effectively improve the stability of training process and the performance of Agent because of the averaging of the advantage function.

The rest of the paper is organized as follows. In Sect. 2 we discuss relevant related work. Section 3 elaborates on the RL and A3C algorithms. In Sect. 4 we present the Averaged-A3C algorithm. Section 5 provides an empirical evaluation of the Averaged-A3C algorithm both in several of the Atari 2600 and MuJoCo games. Section 6 closes the paper with our conclusions and possible future work.

2   Related Work

DRL methods can be divided into model-based reinforcement learning methods and model-free reinforcement learning methods. Moreover, model-free DRL methods include action value fitting methods and policy gradient methods. The action value fitting methods use the estimate of action value to update the policy [2]. Such as Deep Q-learning algorithm, it is one of the action value fitting methods. The policy gradient methods directly improve the policy by updating the parameters in the direction of the policy gradient [11]. Such as Actor-Critic (AC) algorithm [12], it is an important policy gradient method.

Most of DRL models introduce an experience replay mechanism. Since the importance of different training samples varies and experience replay uses the sampling method of equal probability, causing useful samples cannot be effectively used for the model training. To handle this problem, Schaul et al. proposed a deep reinforcement learning with prioritized experience replay algorithm [13] based on priority replay sampling. The idea is to use the information such as the immediate reward and time difference (TD error) as the priority of the samples in the experience buffer pool.

DRL algorithms based on the experience replay mechanism including DQN [5] and DDQN [14] have achieved great success in some high-dimensional state space tasks, such as Atari 2600 games. However, the experience replay mechanism has some inherent limitations. The premise of using the experience replay mechanism is that a large amount of storage space is needed to store the training samples, so that the demand for storage space of such DRL algorithms is significantly increased. And the experience replay mechanism leads to a large amount of computation and needs to use specialized hardware acceleration, such as graphics processor.

In response to this problem, Mnih et al. combined asynchronous methods with deep reinforcement learning and proposed Asynchronous Deep Reinforcement Learning (ADRL). ADRL replaces the traditional experience replay mechanism by using the asynchronous method, so that the DRL algorithm no longer needs to store a large number of training samples. Instead of updating the parameters of the network model after each interaction with the environment, Agent calculates the cumulative loss after interacting with the environment several times and updates the parameters by using the gradient descent method, which reduces the computational cost of the DRL algorithm. The policy gradient method of A3C in ADRL uses the advantage function to calculate policy gradient estimates and A3C achieves the best performance in ADRL.

Different from this, John Schulman et al. proposed a family policy gradient methods called the generalized advantage estimator (GAE) [9] which can significantly reduce variance by using parameterized advantage function. To optimize control policies with guaranteed monotonic improvement, John Schulman et al. also developed a practical algorithm, called Trust Region Policy Optimization (TRPO) [15] by making several approximations to the theoretically-justified scheme. GAE and TRPO can be combined in an effective way. To further simplify TRPO, a new method called proximal policy optimization (PPO) [16] was proposed. PPO enables multiple epochs of minibatch updates and it has some of the benefits of TRPO, but it is much simpler to implement, more general, and has better sample complexity.

In this paper, motivated by the idea of Averaged-DQN proposed by Oron Anschel et al. [10], we present a new Averaged-A3C algorithm to reduce the variance of the advantage function used in A3C.

3   Background

In this section, we elaborate on relevant RL background and specifically on the A3C algorithm.

3.1   Reinforcement Learning

In a standard reinforcement learning setting, Agent interacts with the environment in multiple discrete time steps. At each time step t, Agent receives a state s_t and selects an action a_t from the set of possible actions A according to the policy π. The policy π is a mapping from states s_t to actions a_t. As a feedback, Agent receives the next state s_{t+1} which the environment enters into after performing this action a_t and gets a scalar reward r_t. In the state s_t, Agent gets the expected return $R_t = \sum_{i=0}^{\infty} \gamma^i r_{t+i}$ which represents the total accumulated return from the time step t with a discount factor $\gamma \in (0, 1]$. The final goal of Agent is to maximize the expected return R_t obtained in the state s_t, and to obtain an optimal policy.

The state-action value functions $Q^\pi(s, a)$ is defined as the expected return that Agent gets in the state s_t and executes the action a_t following the policy π. The corresponding formula is shown in Eq. (1):

$$Q^\pi(s, a) = E^\pi[R_t | s_t = s, a_t = a] \tag{1}$$

The state value function $V^\pi(s)$ represents the expected return that Agent receives in the state s_t according to the given policy π, its formula is shown in Eq. (2):

$$V^\pi(s) = E^\pi[R_t | s_t = s] \tag{2}$$

In the classical reinforcement learning, the action value function will eventually converge and get the optimal policy by iterating Bellman equation constantly. For large-scale state space, Agent is trained by learning a parameterized state-action value function $Q(s, a|\theta)$ and θ is the parameter of the state-action value function.

3.2   Asynchronous Advantage Actor-Critic

The A3C algorithm combines deep neural networks with AC algorithms and uses an asynchronous method to update both the policy and the value function. A3C does not require complex preprocessing of the raw data and enables end-to-end learning.

There are two parts in A3C: one is value network $V(s|\theta_v)$, where θ_v is the parameter of the value network. The other part is policy network $\pi(a_t|s_t; \theta)$, where θ is the parameter of the policy network. The policy network is used to calculate the action a_t taken in the given state s_t, and this action a_t is evaluated by the value network.

The A3C algorithm uses the same weight for each state-action pair when updating policy network parameters. That is, each state action pair is treated equally. However, the importance of the state-action pairs is different. Some state-action pairs can obtain high return values, while some state-action pairs have relatively lower return values. Treating them equally will ignore the fact that they receive different returns. In order to make full use of this effective information, the A3C algorithm introduces an advantage function which is used to evaluate the advantage of the current state-action pair. It is expressed as $A\left(s_t, a_t|\theta, \theta_v\right)$, as shown in Eq. (3):

$$A\left(s_t, a_t|\theta, \theta_v\right) = \sum_{i=0}^{n-1} \gamma^i r_{t+i} + \gamma^n V\left(s_{t+n}|\theta_v\right) - V\left(s_t|\theta_v\right) \tag{3}$$

In the A3C algorithm, the policy and the value function are updated after every t_{max} actions or when a terminal state is reached. And in Eq. (3), n varies from state to state and is upper-bounded by t_{max}. The gradient calculation of the policy function and the value function of A3C is shown in Eq. (4) and (5), respectively.

$$d\theta = \nabla_\theta log\pi\left(a_t|s_t; \theta\right) A\left(s_t, a_t|\theta, \theta_v\right) \tag{4}$$

$$d\theta_v = \frac{\partial\left(R - V\left(s_t|\theta_v\right)\right)^2}{\partial\theta_v} \tag{5}$$

where R is defined as the return that received by taking action a_t according to the policy in the state s_t.

4   Averaged-A3C

Our averaged variant of the A3C algorithm is presented in Algorithm 1. The algorithm, which we call Averaged-A3C, is an extension of the A3C algorithm. Averaged-A3C uses the K previously learned state values estimates to produce the advantage function. Thus the averaged advantage function is shown as follow:

$$A\left(s_t, a_t|\theta, \theta_v\right) = \sum_{i=0}^{n-1} \gamma^i r_{t+i} + \gamma^n \frac{1}{K}\sum_{k=1}^{K} V\left(s_{t+n}|\theta_{v_k}\right) - \frac{1}{K}\sum_{k=1}^{K} V\left(s_t|\theta_{v_k}\right) \tag{6}$$

During the training process, we preserve the K previously learned parameters θ_v of value network. The difference between Eqs. (3) and (6) is that we use K averaged state values to estimate the state at time step $t+n$ and t. Then they are used to calculate the advantage function. The reason why we use K learned previously learned state values estimates to produce the advantage function is that it can provide more accurate estimates of state values, which has been proved in [10] and thus leads to reduce the variance of the advantage function.

The policy and the value function are updated after every t_{max} actions or when a terminal state is reached. The update performed by Averaged-A3C can be calculated as $\nabla_{\theta'} log\pi\left(a_t|s_t; \theta'\right) A\left(s_t, a_t|\theta, \theta_v\right)$, where $A\left(s_t, a_t|\theta, \theta_v\right)$ is calculated by Eq. (6). So the biggest difference between A3C and Averaged-A3C is that Averaged-A3C uses the averaged advantage function to update the policy and value network. As with the value-based methods, Averaged-A3C relies on parallel actor-learners and accumulated updates for improving training stability like the A3C algorithm. And the parameters θ of policy and θ_v of the value function are separate, Averaged-A3C uses a CNN network that has one softmax output for

Algorithm 1. Averaged Asynchronous Advantage Actor-Critic - pseudocode for each actor-learned thread

Initialize: initialize global shared parameter vectors θ and θ_v, global shared counter $T = 0$, thread-specific parameter vectors θ' and θ_v', thread step counter t, asynchronous max step size t_{max}, average size K

1: Initialize thread step counter $t \leftarrow 1$
2: **repeat**
3:     Reset gradients:$d\theta \leftarrow 0$ and $d\theta_v \leftarrow 0$
4:     Synchronize thread-specific parameters $\theta' = \theta$ and $\theta_v' = \theta_v$
5:     $t_{start} = t$
6:     Receive state s_t
7:     **repeat**
8:         Perform action a_t according to policy $\pi(a_t|s_t;\theta')$
9:         Get reward r_t and new next state s_{t+1}
10:        $t \leftarrow t + 1$
11:        $T \leftarrow T + 1$
12:    **until** terminal state s_t **or** $t - t_{start} == t_{max}$
13:    $R = 0$ for terminal state s_t
14:    $R = \frac{1}{K}\sum_{k=1}^{K} V(s_t|\theta_{v_k}')$ for non-terminal state s_t
15:    **for** $i \in \{t-1, ..., t_{start}\}$ **do**
16:        $R \leftarrow r_i + \gamma R$
17:        Accumulate gradients wrt θ':
18:        $d\theta \leftarrow d\theta + \nabla_{\theta'} log\pi(a_i|s_i;\theta')\left(R - \frac{1}{K}\sum_{k=1}^{K} V(s_i|\theta_{v_k}')\right)$
19:        Accumulate gradients wrt θ_v':
20:        $d\theta_v \leftarrow d\theta_v + \frac{\partial\left(R - \frac{1}{K}\sum_{i=1}^{K} V(s_i|\theta_{v_k}')\right)^2}{\partial\theta_v'}$
21:    **end for:**
22:    Perform an asynchronous update of θ using $d\theta$ and of θ_v using $d\theta_v$
23: **until** $T > T_{max}$

the policy $\pi(a_t|s_t;\theta)$ and one linear output for the value function $V(s_t|\theta_v)$, with all non-output layers shared.

Compared to A3C, the computational effort of Averaged-A3C is K-fold more forward passes through the value network while calculating the advantage function of actions. Using K previously learned value networks to produce the advantage function of current action leads to reduce the variance of the advantage function. So Averaged-A3C can stabilize the training process and improve the performance of Agent.

5   Experiment

In this section, we first introduce the platform used in the experiment and the parameter settings of the experiment. Then we evaluate the performance of the A3C, A3C-GAE and Averaged-A3C algorithms on some selected Atari 2600 and MuJoCo games. A3C-GAE uses the idea of GAE to calculate the advantage function in A3C, while Averaged-A3C uses the K previously learned state values estimates to produce the advantage function in A3C. Finally, we have carried out a detailed analysis of stability and performance of the proposed Averaged-A3C algorithms in discrete action space and continuous action space.

The experiments were designed to solve the following questions:

(1) Can Averaged-A3C improve the learned polices quality compared to A3C and A3C-GAE?

(2) How does the number K of averaged advantage function affect the performance of Averaged-A3C?

5.1 Experimental Environment and Setup

This article uses the Atari 2600 and MuJoCo game environment in the OpenAI Gym as the experimental environment. OpenAI Gym is an open source toolkit that provides a wide variety of Atari 2600 and MuJoCo game interfaces. The study of Mnih et al. [7] has indicated that in most Atari 2600 and MuJoCo games, A3C is significantly superior to DRL algorithms, such as DQN and DDQN. In our experiment, we select four Atari 2600 games and four MuJoCo games to test the performance of A3C, A3C-GAE and Averaged-A3C.

In order to compare the performance of different algorithms, all algorithms in this paper use the same set of parameters and the Adam gradient descent method. The parameters of the Adam gradient descent method are set as follows: $\eta = 0.001$, $\beta_1 = 0.9$, $\beta_2 = 0.99$, $\varepsilon = 0.001$. All of the above algorithms have a discount factor of $\gamma = 0.99$. The asynchronous update of network parameters is performed in the following way: the parameters of the shared network model are updated every 20 steps (frames) or at the end of the episode.

The network architecture used in A3C, A3C-GAE and Averaged-A3C is the same as that in [7], including a convolutional layer with 16 filters of size 8 × 8 with stride 4 followed by a convolutional layer with 32 filters of size 4 × 4 with stride 2, then followed by a fully connected layer with 256 hidden units. A3C and Averaged-A3C have a single linear output unit representing the action-value for each action, and two other set of outputs - a softmax output with one entry per action representing the probability of selecting the action, and a single linear output representing the value function. Moreover, both of the Atari and MuJoCo experiments use the same setup in [7].

The parameter of γ_{GAE} used in A3C-GAE is set as $\gamma_{GAE} = 0.99$ and the parameter of λ is set as $\lambda = 0.5$. There are three types of the number of averaging size K in Averaged-A3C. They are $K = 5$, $K = 10$, or $K = 15$. For Atari experiments, 1,000 training periods (Epoch) are used as the training period, including 80,000 steps. Thus, a total of 80,000,000 steps are trained. As for MuJoCo experiments, 200 training periods (Epoch) are conducted, which consists of 100 episodes per training period. All experiments in this article use 8 threads to accelerate model training.

5.2   Results of Atari

We first compared the rewards of A3C, A3C-GAE and Averaged-A3C during each epoch of training the Agent to play Atari 2600 games, including Seaquest, Q*bert, BeamRider, and Alien. The results are shown in Fig. 1. It is indicated that Averaged-A3C outperforms A3C and A3C-GAE on all of four Atari 2600 games. In addition, Averaged-A3C is more stable than A3C and A3C-GAE during the training process.

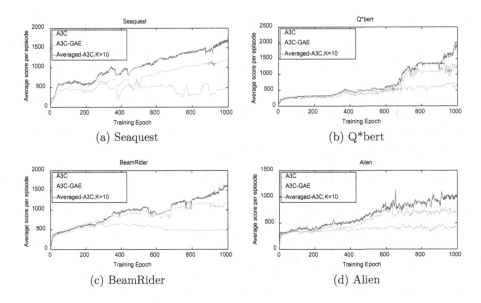

(a) Seaquest   (b) Q*bert

(c) BeamRider   (d) Alien

Fig. 1. Comparisons of A3C, A3C-GAE and averaged-A3C for Atari games

Then we compared the performance of Averaged-A3C with different values of K, that is $K = 5, 10, 15$, respectively. The results are shown in Fig. 2. It can be seen that, when the number of averaged state values K is 15, the corresponding Averaged-A3C algorithm outperforms other algorithms with smaller K. As a result, We can turn to the conclusion that increasing the number of averaged state values K leads to better performance and stability in Averaged-A3C.

Furthermore, we also compared the training time of A3C, A3C-GAE and Averaged-A3C with various K. The results are summarized in Table 1. It is illustrated that the training time of Averaged-A3C increases as the number K increases. Since Averaged-A3C needs the calculation of average, it requires more training time, but considering its performance improvement, the cost is acceptable.

To confirm that Averaged-A3C can perform well after training, we compared the performance of A3C, A3C-GAE and Averaged-A3C on four games after training 80M frames. For each game, the training completed model will be tested

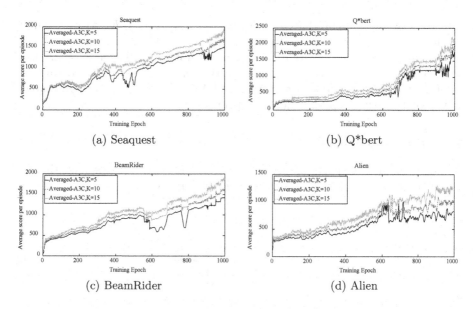

Fig. 2. Comparisons of different values of K for Atari games

Table 1. Average Epoch training time of A3C, A3C-GAE and averaged-A3C

Algorithm	A3C	A3C-GAE	Averaged-A3C (K = 5)	Averaged-A3C (K = 10)	Averaged-A3C (K = 15)
Training time (s)	90	122.4	129.6	172.8	223.2

Table 2. Average score of A3C, A3C-GAE and averaged-A3C after training

GAME	A3C	A3C-GAE	Averaged-A3C (K = 5)	Averaged-A3C (K = 10)	Averaged-A3C (K = 15)
SEAQUEST	965.2 (±251.4)	2504.3 (±350.5)	2412.5 (±280.2)	2857.3 (±326.4)	3251.4 (±285.3)
Q*BERT	2356.5 (±380.8)	4801.6 (±656.5)	4709.3 (±691.2)	4910.6 (±725.2)	5134.8 (±816.5)
BEAMRIDER	942 (±245.6)	2403.1 (±537.9)	2386.5 (±611.5)	2539.4 (±545.7)	2863.7 (±436.8)
ALIEN	1125.4 (±328.3)	1630.3 (±286.4)	1548.5 (±342.9)	1743.2 (±395.2)	2043.3 (±372.4)

200 times, and the initial state of each game is set to a different state, which fully guarantees the diversity of test results. Each test will receive a score that represents the average reward for each episode of the game. The results of the average scores of each model in four games after 200 tests are shown in Table 2. It's indicated that the performance of Averaged-A3C becomes better with the

increasing of the number of K. In addition, with increasing K, Averaged-A3C will have a better performance than A3C and A3C-GAE after training.

To sum up, in Atari environment, Averaged-A3C is more effective than A3C and A3C-GAE both in the training and testing process. Besides, increasing the number of averaged state values in Averaged-A3C results in better performance and stability with acceptable increased training time cost.

5.3   Results of MuJoCo

We also examined the performance of Averaged-A3C in some tasks of continuous action space. These tasks were simulated in the MuJoCo environment.

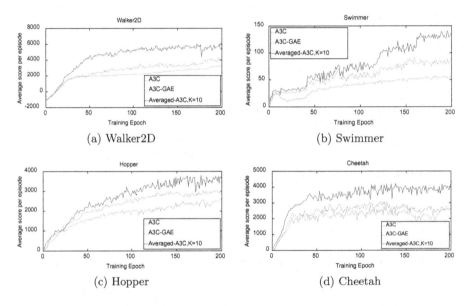

(a) Walker2D   (b) Swimmer

(c) Hopper   (d) Cheetah

Fig. 3. Comparisons of A3C, A3C-GAE and averaged-A3C for MuJoCo games

We selected four MuJoCo games including Walker 2D, Swimmer, Hopper, and Cheetah to compare the performance of A3C, A3C-GAE and Averaged-A3C during the training process. Figure 3 shows the average score per episode of these three algorithms during the training epoch. Figure 4 shows the performance of Averaged-A3C with different K. The observations in Figs. 3 and 4 are consistent to those of Atari. As a sequence, we can also turn to the conclusions that Averaged-A3C performs better than A3C and A3C-GAE, besides, the performance of Averaged-A3C depends on the number of averaged state values.

After conducting two sets of comparative experiments, we confirmed that Averaged-A3C can indeed achieve better performance than A3C and A3C-GAE in Atari 2600 and MuJoCo games during the training and testing process. Moreover, the performance of Averaged-A3C is affected by the number of averaged

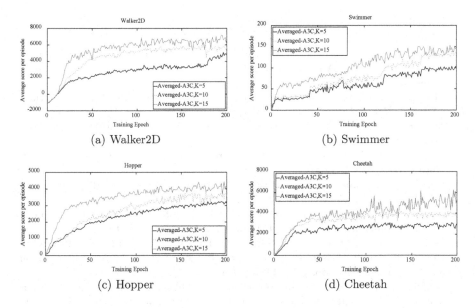

Fig. 4. Comparisons of different values of K for MuJoCo games

state values K. The larger the number of K is, the better performance and stability Agent can obtain. However, in Averaged-A3C, the suitable value of K should be chosen according to the computational resources since the larger number of K will require for much more training time cost.

6 Conclusion

In this paper, the Averaged-A3C algorithm is proposed, which uses K previously learned state values to calculate the advantage function with the purpose of reducing the variance of the advantage function efficiently. We have demonstrated that Averaged-A3C can indeed stabilize training and improve performance in several games of Atari and MuJoCo. In addition, we also conduct some experiments to study the effect of K on the Averaged-A3C algorithm. Empirical experiments results show that increasing the number of averaging previously learned state values leads to better performance. Averaged-A3C is a simple extension that can be easily integrated with other A3C variants such as A3C-LSTM.

In future work, investigations on the selection of K will be conducted. It means that we may dynamically learn how many previously learned state values should be used to average for best performance. One simple suggestion may be to correlate the number of K with the state TD-error. The other way is to make K become a dynamic learnable parameter with the use of neural networks. Finally, incorporating averaging techniques within policy-gradient based methods such as GAE and TRPO methods may further improve the performance of these algorithms.

References

1. LeCun, Y., Bengio, Y., Hinton, G.: Deep learning. J. Nature. **521**, 436–444 (2015)
2. Sutton, R.S., Barto, A.G.: Reinforcement Learning: An Introduction. MIT press, Cambridge (1998)
3. Silver, D., et al.: Mastering the game of go without human knowledge. J. Nature. **550**, 354–359 (2017)
4. Watkins, H., Cornish, C.J.: Learning from Delayed Rewards. King's College, Cambridge (1989)
5. Mnih, V., et al.: Human-level control through deep reinforcement learning. J. Nature. **518**, 529–533 (2015)
6. Lin, L.J.: Programming robots using reinforcement learning and teaching. In: AAAI Conference on Artificial Intelligence, pp. 781–786 (1991)
7. Mnih, V., et al.: Asynchronous methods for deep reinforcement learning. In: International Conference on Machine Learning, pp. 1928–1937 (2016)
8. Bellemare, M.G., Ostrovski, G., Guez, A., Thomas, P.S., Munos, R.: Increasing the action gap: new operators for reinforcement learning. In: AAAI Conference on Artificial Intelligence, pp. 1476–1483 (2016)
9. Schulman, J., Moritz, P., Levine, S., Jordan, M., Abbeel, P.: High-dimensional continuous control using generalized advantage estimation. arXiv:1506.02438 (2015)
10. Anschel, O., Baram, N., Shimkin, N.: Averaged-DQN: variance reduction and stabilization for deep reinforcement learning. In: International Conference on Machine Learning, pp. 176–185 (2017)
11. Silver, D., Lever, G., Heess, N., Degris, T., Wierstra, D., Riedmiller, M.: Deterministic policy gradient algorithms. In: International Conference on Machine Learning, pp. 387–395 (2014)
12. Konda, V.R., Tsitsiklis, J.N.: Actor-critic algorithms. In: Advances in Neural Information Processing Systems, pp. 1008–1014 (2000)
13. Schaul, T., Quan, J., Antonoglou, I., Silver, D.: Prioritized Experience Replay. arXiv:1511.05952 (2015)
14. Van Hasselt, H., Guez, A., Silver, D.: Deep reinforcement learning with double Q-learning. In: Association for the Advance of Artificial Intelligence, pp. 2094–2100 (2016)
15. Schulman, J.., Levine, S., Moritz, P., Jordan, M.I., Abbeel, P.: Trust region policy optimization. In: International Conference on Machine Learning, pp. 1889–1897 (2015)
16. Schulman, J., Wolski, F., Dhariwal, P., Radford, A., Klimov, O.: Proximal policy optimization algorithms. arXiv:1707.06347 (2017)

Deep Reinforcement Learning
for Multi-resource Cloud Job Scheduling

Jianpeng Lin[1], Zhiping Peng[2($\boxtimes$)], and Delong Cui[2]

[1] Guangdong University of Technology, Guangzhou, China
linjianpeng0615@163.com
[2] College of Computer and Electronic Information,
Guangdong University of Petrochemical Technology, Maoming, China
pengzp@foxmail.com, delongcui@163.com

Abstract. The resource scheduling problem in the cloud environment has always been a difficult and hot research field of cloud computing. The difficult problem of online decision-making tasks for resource management in a complex cloud environment can be solved by combining the excellent decision-making ability of reinforcement learning and the strong environmental awareness ability of deep learning. This paper proposes a multi-resource cloud job scheduling strategy in cloud environment based on Deep Q-network algorithm to minimize the average job completion time and average job slowdown. The experimental results show that the scheduling strategy is better than the scheduling strategy based on the standard policy gradient algorithm, and accelerate the convergence speed.

Keywords: Cloud computing · Deep reinforcement learning · Job scheduling

1 Introduction

Resource scheduling is a difficult and hot topic in the field of cloud computing. A good resource allocation and scheduling strategy can effectively use resources to increase the economic benefits of suppliers while ensuring the QoS (Quality of Services, QoS). Cloud computing resource scheduling is actually a multi-constraint, multi-objective optimization NP-hard problem. The traditional method to solve the decision problem is to design an efficient heuristic algorithm [1–3] with guaranteed performance under certain conditions, which is not very versatile and practical. Therefore, reinforcement learning [4] (RL), as a model-free learning method that can realize online adaptive decision-making, has been applied to solve some resource allocation problems in cloud computing systems [5–7]. Experimental results show that the key attributes of the RL method are applicable to complex cloud computing systems. RL-based agents do not need to perform a priori modeling of related attributes (such as workload, state transition, performance, etc.) of the underlying system in the process of learning optimal resource allocation decisions and implementing online system control.

Researchers [8–10] abstract the resource scheduling in the cloud environment as a sequential decision problem, turn the decision problem into an objective function optimization problem. The paper [8, 9] proposed a novel resource allocation scheme

© Springer Nature Switzerland AG 2018
L. Cheng et al. (Eds.): ICONIP 2018, LNCS 11303, pp. 289–302, 2018.
https://doi.org/10.1007/978-3-030-04182-3_26

based on reinforcement learning and queuing theory, which proved that it is superior to the common resource allocation method in terms of SLA conflict avoidance and user cost. In [10], a new job scheduling scheme based on SLA constraint-based reinforcement learning is proposed to minimize the completion time and average waiting time (AWT) under VM resources and deadlines, and adopt parallel multi-agent parallel technology to balance the exploration process. This method accelerates the convergence speed of the Q learning algorithm and achieves obvious optimization results. However, the reinforcement learning algorithm is difficult to deal with the resource scheduling problem in the complex high-dimensional state space, and the convergence speed is slow.

The current problem of resource management in the cloud computing field is often manifested as the difficulty of online decision-making tasks in a complex environment, and the appropriate solution depends on the understanding of the system's resource environment and job state. Inspired by [11] deep reinforcement learning has achieved breakthroughs in end-to-end learning in games and control. We believe that the combination of the interactive trial-and-error mechanism of reinforcement learning and the powerful nonlinear generalization ability of deep neural networks will be a good idea to solve the problem of cognitive decision-making in complex states. The paper [12] transforms multi-resource job scheduling into multi-resource task packing, abstracts resources and job state as images to represent the system's state space, as a neural network input, outputs the probability distribution of actions. The author trained the model using a standard policy gradient algorithm DeepRM to obtain a multi-resources job scheduling model under the cloud environment. The experimental results show that the strategy can adapt to different environments and performs better than most classic heuristic algorithms.

Based on the research [12] mention above, we propose a strategy which combines the characteristics of cloud job scheduling and deep Q-network algorithm, to solve the multi-resource cloud job scheduling in a complex cloud environment. In this paper, we improve the model by optimizing the calculation method of action's reward, using the incremental ε-greedy exploration method, and using a convolutional pooling layers input layer. The experimental results demonstrate significant improvement in the learning curve after adopting these approaches.

The cloud platform system model adopted in this paper is shown in Fig. 1. The system model consists of three parts (job pool, resource cluster, and intelligent resource controller). The job pool is used to cache different job requests from different types of users. The system abstracts various resources in the cloud environment into the form of cluster resources, which is convenient for users to select corresponding services. The intelligent resource controller is a key component of the system. It is responsible for configuring the job into the resource cluster according to the scheduling strategy while ensuring the user's service quality and satisfying the job resource request. The controller includes job monitor and resource monitor, scheduling strategy, and user quality of service (QoS) constraints. The monitor is responsible for collecting state information for the job pool and resource cluster. The scheduling strategy is generated by deep reinforcement learning model. QoS constraints ensure the reliability and efficiency of cloud computing services. In summary, controller and monitors coordinate with each other to implement intelligent configuration resources while obeying two constraints: one must meet the user's QoS requirements, and the other is that the resource usage must be less than the total amount of resources available in the system.

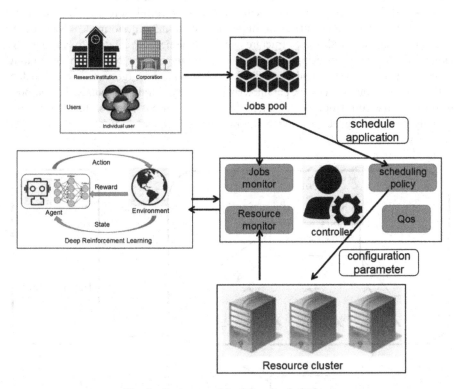

Fig. 1. System model of the cloud platform

2 Background

In recent years, the DeepMind team dedicated to deep reinforcement learning research has successfully combined the decision-making capabilities of reinforcement learning with the deep neural network comprehension capabilities. They proposes a Deep Q-network algorithm [13] which use a deep convolutional neural network to fit the optimal action-value function and successfully implements an end-to-end reinforcement learning algorithm that learns strategies directly from high-dimensional sensor inputs.

The algorithm has excellent performance in game control experiments, and the model can learn a variety of different tasks universally. Inspired by this, we try to apply the DQN algorithm to the cloud computing resource scheduling field. Therefore, we abstract the system's resources and job state into images, as the input of the network, and then combine the action reward function to train the network model to generates an adaptive scheduling strategy. More details as follow.

DQN mainly uses two key technologies: First, it uses experience replay. By storing the training data in the experience memory and then using the mini-batch training method [14]. The random sampling method reduces the correlation of the data samples, making the samples independent and improving the training performance. Second, the fixed target network with same structure as the online network, is used to calculate the target Q value instead of using the pre-updated online network directly. In the training process, we adopt a delay update method which update the target network parameters with the current online network parameter values every C training rounds. The purpose is to reduce the correlation between target value and current value, so that the training stability and convergence are better. Its framework is shown in Fig. 2.

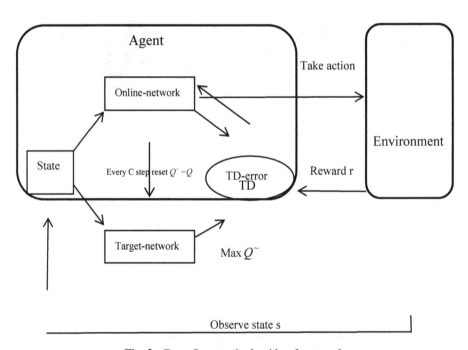

Fig. 2. Deep Q-network algorithm framework

Throughout the entire task, agent continuously explores with the environment, according to the Q value of each action generated by the online network, selects actions based on the incremental ε-greedy policy, and generates a series of states, actions as well as rewards. The goal is to maximize the expected cumulative discount reward. We use a deep convolutional network to fit the optimal action-value function.

$$Q^*(s,a) = \max_\pi E\left[r_t + \gamma r_{t+1} + \gamma^2 r_{t+2} + \ldots | s_t = s, a_t = a, \pi\right] \tag{1}$$

Which is the maximum sum of reward r_t discounted by γ at each time, and the behavioral strategy $\pi = P(a|s)$ indicates that action a is selected in the state s. In the training process, the mini-batch training method is used. Each training round randomly selects M experiences from the experience memory $D_i = \{e_1, e_2, \ldots, e_i\}$. The state s is used as the input of the online network, and obtained the actual Q value of action a, and the next state s' as input to the target network, obtain the maximum Q value of all actions in the target network, and use the following loss function to update the Q network.

$$L_i(\theta_i) = E_{(s,a,r,s') \sim D(M)}\left[\left(r + \gamma \max_{a'} Q(s', a'; \theta_i^\sim) - Q(s, a; \theta_i)\right)^2\right] \tag{2}$$

in which γ is a discount factor determining the agent's horizon, θ_i is the parameter of the Q network for the i-th iteration, and $\theta_i^\sim$ is the network parameter used to calculate the target value for the i-th iteration. This network parameter uses a delay update method. Update the target network parameters once every C training iterations, copy the current Q network parameter values to the target $Q^\sim$ network, and update the parameters of the Q network using the gradient descent method:

$$\theta \leftarrow \theta + \alpha\left(r + \gamma \max_{a'} Q(s', a'; \theta_i^\sim) - Q(s, a; \theta_i)\right) \tag{3}$$

where α is the learning rate, $r + \gamma \max_{a'} Q(s', a'; \theta_i^\sim) - Q(s, a; \theta_i)$ determines the update direction. In our experimental design, according to the characteristics of the job scheduling process, we use a slight variant [15] that reduces the variance of the gradient estimates by subtracting a baseline value from each return v_t. More details as follow.

3 Model

We consider a server cluster with multiple physical servers that offer d types of resources with relate to the cloud resource allocation in this paper. A two-tuple set $\{R_{cpu}, R_{mem}\}$ is associated with cluster indicating the total amount of CPU resources and memory resources in the cluster. T-the duration of the resource. The job will arrive online one after another according to the Poisson process. The scheduler chooses to schedule one or more waiting jobs at each timestep. We assume that the resource demand of each job is known upon arrival. What's more, the resource profile of each job i is represented by the vector $r_i = (r_{i,cpu}, r_{i,mem}, ..., r_{i,d})$ of resources requirements, $r_{i,cpu}$ represents the number of CPU resources that job i needs to occupy, $r_{i,mem}$ represents the number of memory resources that job i needs to occupy and T_i-the duration of the job i. The experiment does not set the job preemption mechanism, which means that the job will always occupy resources from the start of execution until completion. During resource configuration, one prerequisite should be met $\sum_i r_{i,cpu} \cdot T_i \leq R_{cpu} \cdot T$, $\sum_i r_{i,mem} \cdot T_i \leq R_{mem} \cdot T$. Moreover, we represent the server cluster as a single collection of resources without considering effects of machine fragmentation. While these aspects play an important role in actual job scheduling, this simpler model contains the basic elements of multi-resource scheduling and can verify the effectiveness of deep reinforcement learning methods in cloud resource scheduling.

3.1   State Space

We represent the state of the system - the current allocation of cluster resources and the resource requirements of jobs in the waiting queue - as distinct images (As shown in Fig. 3). The cluster images show the allocation of each resource to jobs which have been scheduled for service in the next T timesteps. Different colors in the cluster state image represent different jobs. For instance, the red job in cluster state image needs to

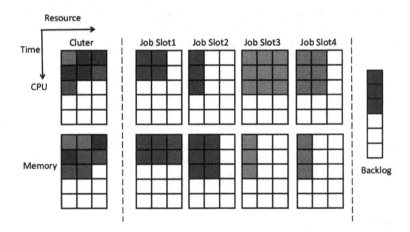

Fig. 3. State space (Color figure online)

occupy two units CPU, one unit memory, and the duration of three timesteps. The job slot images represent the resource requirements of waiting for scheduled jobs. For instance, job slot 1 requires two units CPU, three unit memory, and the duration of two timesteps. The state space will be represented as a binary matrix (Colored squares represent 1. White squares represent 0) as input to the neural network. Therefore, the state space can only fix the attributes of the M jobs that are waiting to be scheduled, and the jobs that are not selected into the scheduling queue will be backlogged in the backlog, waiting to be transferred to the scheduling queue. At the same time, this method limits the scale of the action space, improving the efficiency of the model learning process.

3.2   Action Space

At each timestep, the scheduler will admit more than one jobs into the cluster for service in the scheduling queue containing M jobs. The action space represented as $\{0, 1, 2, ..., M\}$. When $a = i$ indicates the scheduling job i, $a = 0$ indicates that the agent does not wish to schedule further jobs in the current timestep. When the selection of this moment is completed, the cluster images move upward one timestep. The scheduler will allocate the appropriate resources for the scheduled jobs and update the system's cluster resource state. However, if the remaining cluster resources cannot satisfy the requirement of the scheduled job, the job will be rejected. Subsequently, the same number of jobs from the backlog, will be added to the scheduling queue to keep the action space linear in M.

3.3   Objectives and Reward Functions

The reward function guides the Agent to continuously explore with the environment and towards the optimization objective. Different objective need to design different reward functions. In this paper, we adapt the average job slowdown and average job completion time as the system objective. Therefore, we set the slowdown for each job is $S_j = C_j / T_j \ (S_j \geq 1)$ and the corresponding reward function is $R = \sum_{j \in J} \frac{-1}{T_j}$, where C_j is the actual completion time (between arrival and completion of execution), T_j represents the ideal completion time, J is the current set of scheduled jobs and waiting jobs in the system. Moreover, the reward function for the average job completion time designed as: $R = -|J|$ where J is the number of unfinished jobs in the system in the current timestep.

3.4 Network Structure

We add convolutional layers and pooled layers to the network structure to enhance the feature extraction capabilities of the network for the state space and optimize the training. The specific network structure and parameters are shown in the Table 1.

Table 1. Network structure

Layer	Convolutional (input)	Max-pool	Fully-connected (output)
Input size	120*20 = 2400	118*18 = 2124	59*9 = 531
Filter size	2×2	2×2	——
Stride	(1,1)	(2,2)	——
#Filters	8	——	——
Activation	Relu	——	——
Output size	118*18 = 2124	59*9 = 531	11 (queue length + 1)

4 Training Process

In the training process, we use 100 different arrival sequence of job, called jobsets, each containing 60 jobs. In each training iteration, we simulate $N = 20$ Monte Carlo for each jobset to explore the probabilistic space of possible actions using the current policy. The episode terminates when all jobs finish executing. Furthermore, we record the current state, action, reward and next state information for all timesteps of each episode, and use the resulting data to compute the discounted cumulative reward v_t, at each timestep of each episode. In particular, in order to increase Agent's initial exploration of the state space, we use incremental ε-greedy policy to select actions (the initial value of ε is 0.7, the maximum is 0.9, and the increase of each training round is 0.001). When all job episodes of jobset are completed, we calculate the average of the cumulative discount rewards at the same timestep for the different job episodes of the jobset, as the baseline value b_t, and then subtract baseline value from the discounted cumulative reward v_t as the evaluation of the action $\Delta r_t = v_t - b_t$. Finally, we store the state information s_t, action a_t, action value Δr_t, and next state information s_{t+1} at each timestep t of the 20 job episodes of the jobset as an experience information $(s_t, a_t, \Delta r_t, s_{t+1})$, and store them in the experience memory D. Until experiences in the experience memory reaches the set number, the mini-batch training method is adopted, $M = 32$ pieces of experiences are randomly selected from D, and the Q network parameters are updated using formula (2) with the learning rate of 0.001. Each C training iterations copies the parameter value of the current Q network to the target $Q^\sim$ network and updates the target network parameters once. Detailed training process pseudo-code is shown as follow:

Initialize replay memory D to capacity M

Initialize action-value function Q with random weights θ

Initialize target action-value function $Q^\sim$ with random weights $\theta^\sim = \theta$

for each iteration:

    for each jobset:

        π_θ = incremental ε-greedy

        run episode i=1,2,...,N:

$$\left\{ s_1^i, a_1^i, r_1^i, s_2^i, ..., s_{L_i}^i, a_{L_i}^i, r_{L_i}^i, s_{L_i+1}^i \right\} \sim \pi_\theta$$

        Compute returns: $v_t^i = \sum_{s=t}^{L_i} \gamma^{s-t} r_s^i$

        for t = 1 to L:

            compute baseline: $b_t = \dfrac{1}{N} \sum_{i=1}^{N} v_t^i$

            for i = 1 to N:

                $\Delta r_t^i = v_t^i - b_t^i$

                Store transition $\left(s_t^i, a_t^i, \Delta r_t^i, s_{t+1}^i \right)$ in D

            end

        end

    end

    sample random mini-batch of transition $\left(s_t^i, a_t^i, \Delta r_t^i, s_{t+1}^i \right)$ from D

    update the network parameters θ

$$L_i(\theta_i) = E_{(s,a,r,s') \sim D(M)} \left[\left(r + \gamma \max_{a'} Q(s',a';\theta_i^\sim) - Q(s,a;\theta_i) \right)^2 \right]$$

    every C steps reset $Q^\sim = Q$

end

5   Experiments and Analysis

5.1   Experimental Parameters

In the experiment, we use the workload setting similar to that in [12]. There are two types resources with capacity $\{1r, 1r\}$ in the cluster. Moreover, we used 100 different arrival sequence jobsets as the training set and 20 jobsets as the test set (not appearing in the training set). Each jobset contains 60 jobs (80% of the jobs with duration uniformly chosen between $1t$ and $3t$; the remaining are chosen uniformly from $10t$ to $15t$). Each job has a primary resource which is selected randomly, and the rest is the secondary resource. The demand for primary resources is between $0.25r$ and $0.5r$, and the demand for secondary resources is between $0.05r$ and $0.1r$. Specifically, the job will arrive online one after another according to the Poisson process. So we can change the load of the cluster from 10% to 190% by controlling the arrival rate of the job. In each training iteration, We simulate $N = 20$ episodes for each jobset. The episode terminates when all jobs finish executing. For every 20 training iterations, we validate the model with the test set and record the experimental data.

5.2   Result Analysis

In this section, we will use experiments to analyze and compare the effects of the DQN algorithm used in this paper and the classical heuristic Shortest Job First algorithm, Teris* algorithm [16], the policy gradient DeepRM in [12] to optimize the objective. The experimental results are as follows:

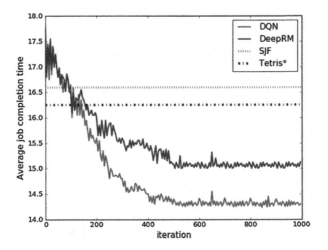

Fig. 4. The average job completion time

As shown in Fig. 4, we can observe that in the first 100 training iterations, the DQN and DeepRM curves are highly volatile and unstable, and the average job completion time is higher than the heuristic SJF algorithm and the Teris* algorithm. After 200 training iterations, the curve gradually stabilized and converged., and the average job completion time was significantly lower than the SJF algorithm and the Teris* algorithm. Furthermore, DQN algorithm outperforms DeepRM by reducing the average job completion time by up to 5.2%.

The curve in Fig. 5 shows that as the training progresses, the average total reward and the maximum total reward obtained by the agent to complete the scheduling task continuously increase until it converges. In addition, the average reward curve gradually moves closer to the maximum reward curve. The curve of Fig. 4 and the curve of Fig. 5 have synchronicity in convergence, indicating that the value of the reward to the agent is continuously increasing as the scheduling strategy continuously learns and optimizes towards the objective.

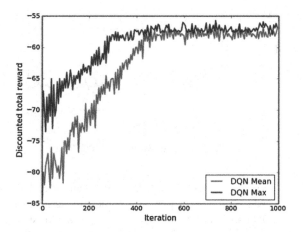

Fig. 5. Discounted total reward

Figure 6 shows the different trend of average job slowdown between 4 algorithms under different loads. As can be seen from Fig. 6, under low load conditions, there are little different between various algorithms with low job slowdown. However, When the load reaches more than 90%, it can be clearly seen that the average slowdown of the DeepRM and DQN with slow growth rate, is significantly lower than the heuristic

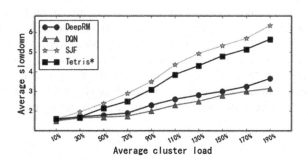

Fig. 6. Job slowdown under different loads

algorithms SJF and Tetris*. What's more, Fig. 7 shows that DeepRM and DQN algorithms tend to converge after 200 rounds of training at the load of 130%, and the average slowdown is less than SJF and Tetris*. Meanwhile, DQN converges faster than DeepRM, and gets smaller average slowdown.

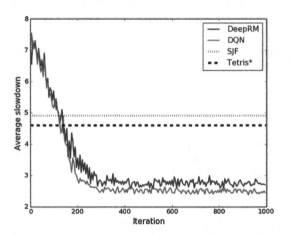

Fig. 7. Job slowdown under 130% load

The box plot of Fig. 8 shows the average job slowdown for each job length. We see that the average slowdown for short jobs is significantly large than long jobs with Tetris*, but as for DQN and DeepRM, the average slowdown for short jobs is less than long jobs. As can be seen from [16], the scheduling strategy adopted by Tetris* is to make full use of the available resources to allocate to more jobs. Therefore, when the cluster load is relatively large, the short job must wait for the release of cluster resources and cannot be scheduled in a short time. According to the slowdown formula, we can see that waiting for the same time, the short jobs has a greater increase in slowdown than the long jobs. Moreover, the short jobs has a relatively large proportion, so the average job slowdown is large. DeepRM and DQN learned from experience that reducing the slowdown of short jobs will help reduce overall slowdown. Therefore, when scheduling the job, the agent will be more willing to allocate resources to short jobs, reducing the waiting time for short jobs, and reducing the overall job slowdown.

In summary, the convolutional neural networks we adapted can capture the features of the job and system resource states more effectively. Moreover, we improve the DQN by optimizing the calculation method of action's reward and using the incremental ε-greedy exploration method. The experimental results also prove that the improved DQN algorithm can learn scheduling strategies directly from experiences faster and batter, and thus achieve faster convergence speed and better optimization effect than DeepRM.

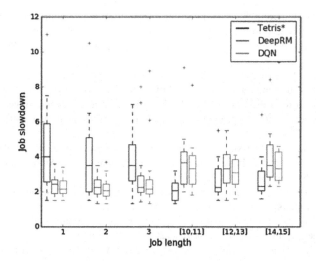

Fig. 8. Job slowdown for each job length

6   Conclusions and Prospects

This paper adopts the currently popular deep reinforcement learning Deep-Q network algorithm to solve the problem of multi-resource cluster scheduling in cloud computing. We adopt the convolutional neural network's ability to perceive the system resources, job state features, and reinforcement learning decision-making capabilities to solve online awareness decision problems in complex cloud environments. The experimental results show that the algorithm can learn resource scheduling strategies directly from experience, and it is more versatile. It outperforms most classical heuristic algorithms in performance, and it is more effective and convergent than policy gradient algorithm.

In the future, in order to make the research more practical, we will consider more factors that may occur in actual resource scheduling, such as multi-objective optimization, dependencies between cloud jobs, fragmentation management of cluster resources, and resource scheduling of multi-queues and multi-clusters and so on.

Acknowledgements. The work presented in this paper was supported by National Natural Science Foundation of China (61772145, 61672174).

References

1. Wang, T., Liu, Z., Chen, Y., Xu, Y., Dai, X.: Load balancing task scheduling based on genetic algorithm in cloud computing. In: Proceedings of the 12th International Conference on Dependable, Autonomic and Secure Computing, pp. 146–152 (2014)
2. Singh, S., Chana, I.: Resource provisioning and scheduling in clouds. QoS perspective. J. Supercomput. **72**, 926–960 (2016)

3. Zuo, L., Shu, L., Dong, S.: A multi-objective optimization scheduling method based on the ant colony algorithm in cloud computing. IEEE Access **3**, 2687–2699 (2017)
4. Sutton, R.S., Barto, A.G.: Reinforcement Learning: An Introduction. MIT Press, Cambridge (1998)
5. Dutreilh, X., Kirgizov, S., Melekhova, O.: Using reinforcement learning for autonomic resource allocation in clouds: towards a fully automated workflow, pp. 67–74 (2011)
6. Barrett, E., Howley, E., Duggan, J.: Applying reinforcement learning towards automating resource allocation and application scalability in the cloud. Concurr. Comput. Pract. Exp. **25**, 1656–1674 (2013)
7. Galstyan, A., Czajkowski, K., Lerman, K.: Resource allocation in the grid using reinforcement learning. In: International Joint Conference on Autonomous Agents and Multiagent Systems, pp. 1314–1315 (2004)
8. Peng, Z., Cui, D., Zuo, J.: Random task scheduling scheme based on reinforcement learning in cloud computing. Cluster Comput. **18**, 1595–1607 (2015)
9. Peng, Z., Cui, D., Zuo, J.: Research on cloud computing resources provisioning based on reinforcement learning. Math. Prob. Eng. **2015**, 1–12 (2015)
10. Peng, Z., Cui, D., Ma, Y., Xiong, J., Xu, B., Lin, W.: A reinforcement learning-based mixed job scheduler scheme for cloud computing under SLA constraint. In: International Conference on Cyber Security and Cloud Computing, pp. 142–147 (2016)
11. Mnih, V., Kavukcuoglu, K., Silver, D.: Human-level control through deep reinforcement learning. Nature **518**, 529 (2015)
12. Mao, H., Alizadeh, M., Menache, I.: Resource management with deep reinforcement learning. In: ACM Workshop on Hot Topics in Networks, pp. 50–56 (2016)
13. Mnih, V., Kavukcuoglu, K., Silver, D.: Playing Atari with deep reinforcement learning. Computer Science (2013)
14. Hinton, G.: Overview of mini-batch gradient descent. Neural Networks for Machine Learning. https://www.coursera.org/learn/neural-networks. Accessed 13 June 2018
15. Schulman, J., Levine, S., Moritz, P.: Trust region policy optimization. In: Computer Science, pp. 1889–1897 (2015)
16. Grandl, R., Ananthanarayanan, G., Kandula, S.: Multi-resource packing for cluster schedulers. ACM Sigcomm Comput. Commun. Rev. **44**(4), 455–466 (2014)
17. Liu, Q., Zhai, J.W., Zhang, Z.Z.: A survey on deep reinforcement learning. Chin. J. Comput. **40**, 1–28 (2018)

Accelerating Spatio-Temporal Deep Reinforcement Learning Model for Game Strategy

Yifan Li and Yuchun Fang[✉]

School of Computer Engineering and Science, Shanghai University,
Shanghai, China
ycfang@shu.edu.cn

Abstract. In recent years, deep reinforcement learning has developed rapidly. Many deep reinforcement learning models are applied in various simple game environments. There are many applications with environments far more complex than simple games. Hence, the performance of the deep reinforcement learning model should be improved in many aspects. In this paper, we explore the effect of fast training and enhancing spatio-temporal representation in deep reinforcement learning model. For the former aspect, we propose to utilize the depthwise separable Convolutional Neural Network (CNN) to accelerate deep reinforcement learning model. For the latter aspect, we introduce the convolutional long short-term memory network (ConvLSTM) to improve the expression ability of spatio-temporal feature. We verify the models in the experiments of StarCraft II [1], a game strategy with a complex environment for reinforcement learning. All of the agents learn a certain level game strategy, such as 'siege' and 'searching'. The experimental results show that depth-wise separable CNN has a good effect in shortening training time and the ConvLSTM has better spatial and temporal feature representation ability to improve the performance of the agents.

Keywords: Deep learning · Reinforcement learning · StarCraft II
Game strategy

1 Introduction

Deep learning [2] originates from the artificial neural network. It is inspired by the hierarchical network structure of inferential data in the human brain. Multilayer perceptrons and back propagation algorithms were proposed. With the development of computing hardware resources, deep learning has made great progress in the fields of image analysis [3, 4], video analysis [5], natural language processing [6, 7], and speech recognition [8, 9]. The basic idea of deep learning is to process data through multi-layered network structures and nonlinear transformations to form abstract high-level features and expose distributed data representations [10]. Therefore, deep learning focuses on the perception and expression of data.

Reinforcement learning is inspired by the organism's ability to respond to environmental stress and effective adaptation. It uses trial and error mechanisms to interact with the surrounding environment and uses learning strategies that maximize cumulative

© Springer Nature Switzerland AG 2018
L. Cheng et al. (Eds.): ICONIP 2018, LNCS 11303, pp. 303–312, 2018.
https://doi.org/10.1007/978-3-030-04182-3_27

rewards to learn optimal strategies [11]. It has been applied in robot control [12], games [13, 14], simulation [15], industrial manufacturing [16], optimization and scheduling [17, 18] and other fields. The goal of reinforcement learning is to obtain maximum cumulative rewards. In order to achieve this goal, on the one hand, it needs to "explore". It explores the environment during the learning process, fully grasps the environmental information, then finds a status with higher reward, and on the other hand, it needs to "use". It uses the historical experience that has been learned to select the highest-reward action and shifts the entire system to a better state. Therefore, reinforcement learning focuses on strategies for learning to solve problems.

Deep learning has a strong ability to perceptual expression but lacks a certain decision-making ability. Reinforcement learning has strong decision-making ability, but its ability in perception is lacking. Therefore, the combination of the two has complementary advantages and provides ideas for solving the cognitive and decision-making problems in complex environments. The DeepMind team combines deep learning with perceptual capabilities and reinforcement learning with decision-making capabilities, and proposes deep reinforcement learning, forming a new research direction in the field of artificial intelligence. In many challenging areas, the DeepMind team constructed and implemented a deepening learning model at the human expert level. These models' construction of their own knowledge system and learning of the environment all come directly from the original input signal, without any related domain knowledge. Deep reinforcement learning is a very versatile end-to-end sensing and control system.

At present, deep reinforcement learning has been applied in games [19–21], machine vision [22, 23], robot control [24–26], parameter optimization [27, 28] and other fields.

The efficiency of sample learning in deep reinforcement learning is extremely low, which leads to the long training time of agents. The training process of an agent is similar to that of video processing, and it also has the possibility of improving the feature extraction in time and space.

In this paper, we study the strategy of accelerating the speed of model learning and improving the model's performance from the perspective of deep neural networks.

This paper uses depthwise separable convolutional networks which effectively shortens the training time of some agents in the case where the performance of the agent is not greatly reduced, and convolutional LSTM networks which improve the performance of the agent in the game from the perspective of the neural network for deep reinforcement training. Section 2 reviews related work in building models which shorten the agent's learning time and improve the performance of the agent. Section 4 describes two methods depthwise separable convolutional networks and convolutional LSTM network. Section 5 describes the game performance by using methods of Sect. 4. Section 6 closes with a summary and conclusion.

2 Related Work

There has been a rising interest in deep reinforcement learning in recent years. Most of the deep reinforcement learning algorithms can be included under the actor-critic framework (AC) which is a traditional reinforcement learning algorithm. The deep reinforcement learning algorithm of agents in AC is divided into two parts: Actor module and Critic module. The Actor module inputs the state of the environment and then outputs the action according to the learning strategy. The Critic module is adjusted according to historical information and feedback (reward), and then the Critic module affects the strategy of the Actor module.

DeepMind extends the AC to deep reinforcement learning and introduces TD-error computation advantage. According to the idea of Asynchronous Reinforcement Learning, The Asynchronous Advantage Actor-Critic is proposed (A3C) [29]. A3C has greatly shortened the learning time of deep reinforcement learning by making full use of hardware resources.

Unsupervised reinforcement and auxiliary learning (UNREAL) [34] based on the A3C is another way to train the deep reinforcement learning tasks. It trains multiple auxiliary tasks while training the A3C model, and multiple tasks complement each other to accomplish the set goals. The UNREAL accelerates learning speed and improves performance by setting multiple auxiliary tasks. Auxiliary and target tasks are trained simultaneously by sharing weights.

This paper uses two kinds of deep reinforcement learning algorithms. The first one is Advantage Actor-Critic (A2C) which is a synchronous version of A3C. It collects data synchronously with all threads while collecting the state of the environment. The second one is Proximal Policy Optimization (PPO) [30]. It is based on the Trust Region Policy Optimization (TRPO) [31]. TRPO has disadvantages that are hard to ignore. It takes a lot of effort to debug. And PPO is an approximate solution to TRPO. It achieves a balance between ease of implementation, sampling complexity, and effort required for debugging.

3 Game Environment

StarCraft II is the real-time strategy (RTS) game launched by Blizzard Entertainment. It is the sequel of StarCraft I, and it is one of the most successful RTS games. In 2017, Blizzard Entertainment co-operated with DeepMind to launch the SC2LE [1], StarCraft II learning environment.

The environment has a huge search space. The StarCraft II game's unit limit is 400. If the game uses maps which size is 128 pixels * 128 pixels, without considering other details, there is about 101685 search space. StarCraft II's search space is 10 orders of magnitude higher than the 10^{170} search space of Go. It's too difficult.

In this paper, we use StarCraft II's mini-game as a training environment with three maps.

The first map is 'Defeat Roaches'. There are 9 Marines on one map, and 4 Roaches on the other side of the map. The rewards are obtained by defeating the Roaches by the Marines. The best game strategy required the Marines to defeat the Roaches one by one.

The second map is 'Find and Defeat Zerglings'. There are 3 Marines on the map and some Zerglings that are fixed locations. The rewards are awarded by defeating the Zerglings by the Marines. The best game strategy requires efficient exploration.

The third map is 'Defeat Zerglings And Banelings'. On the map, there are 9 Marines, a group of 6 Zerglings and 4 Banelings on the other side of the map. The rewards are obtained by defeating the Zerglings and Banelings by using the Marines. The best game strategy is to allocate reasonably the targets that Marines need to attack.

4 Method

4.1 Depthwise Separable Convolutional Neural Network

As we all know, deep reinforcement learning tasks take a long time to train. In particular, the policy gradient method has very low sampling efficiency and requires several millions or even billions of time steps to learn a simple task. Depthwise separable convolutions in MobileNets [32] can greatly reduce the computational cost of convolution operations. The part of the deep reinforcement learning model that observes the environment consists of a convolutional neural network. We introduce depthwise separable convolutions into deep reinforcement learning and test its influence on the speed of agent's training.

The sizes of convolution kernel used in this paper are 5 * 5 and 3 * 3, so the theoretical computational cost of the network with depthwise separable convolutions is 0.1 times than that of the standard CNN.

In the deep reinforcement learning model training, the depthwise separable convolutional neural network shortens the training time of the agent without substantially affecting the performance of the agent in the game. The time model token by training 100 M timesteps reduced from 96 h to 84 h. As shown in Table 1. The results of the agent that is compared with the baseline given by DeepMind on StarCraft II minigame [1] which was trained 800 M timesteps are shown in Fig. 1.

Table 1. Training time of standard CNN and depthwise separable CNN

Network	Training time
Standard CNN	96 h
Depthwise separable CNN	84 h

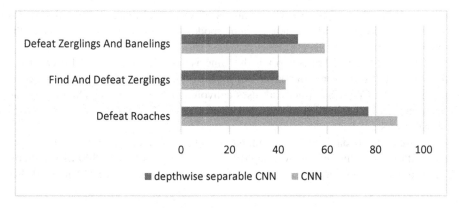

Fig. 1. Game performance of standard CNN and depthwise separable CNN. It only shows the mean score of the networks which is trained by the A2C algorithm. The specific values are in Table 2. The map1 is 'Defeat Roaches', the map2 is 'Find and Defeat Zerglings', and the map3 is 'Defeat Zerglings And Banelings'.

Table 2. Game performance of standard CNN and depthwise separable CNN

Network	Algorithm	Metric	Score of map1	Score of map2	Score of map3
CNN [1]	A3C [1]	MEAM	**100**	**45**	62
		MAX	355	**56**	**251**
CNN [1]	A2C	MEAM	89	43	59
		MAX	355	44	247
CNN [1]	PPO	MEAM	93	44	**63**
		MAX	**361**	55	243
Depthwise separable CNN	A2C	MEAM	77	40	48
		MAX	327	54	236
Depthwise separable CNN	PPO	MEAM	88	42	54
		MAX	347	52	243

4.2 Convolutional Long Short-Term Memory Network

The feedforward architecture without memory is sufficient for simple tasks. In the complex tasks, the LSTM module is usually introduced to enable the network to make the agent have memory capabilities and have a better perception of temporal and spatial features.

But the standard LSTM network is fully connected LSTM network, and its internal gates rely on a computational approach that is similar to a feed-forward neural network, which does not consider the spatial correlation and contains a large amount of spatial data redundancy. To solve this problem, Shi X et al. proposed a convolutional LSTM network (ConvLSTM) [33] that not only has the temporal modeling capabilities of standard LSTM network but also extracts spatial features like a convolutional neural network. It has the ability to extract features of space and time. We introduced this module into the deep reinforcement learning model to improve the performance of the model in the game.

The core of ConvLSTM network is the same as that of standard LSTM networks. It takes the output of the previous layer as the input of the next layer. The difference of them is that the ConvLSTM network uses convolution operations, which not only can obtain temporal relationships of data, but also can extract spatial features like convolutional layers.

We train agents on StarCraft II minigame and compare the game performances of them with the baseline given by DeepMind [1] which was trained 800 M timesteps. The results are shown in Fig. 2. The results show that the ConvLSTM network has

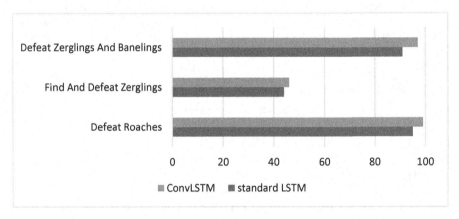

Fig. 2. Game performance of standard LSTM network and ConvLSTM network. It only shows the mean score of the networks which is trained by the A2C algorithm. The specific values are in Table 3, the map1 is 'Defeat Roaches', the map2 is 'Find and Defeat Zerglings', and the map3 is 'Defeat Zerglings And Banelings'.

Table 3. Game performance of standard LSTM network and ConvLSTM network

Network	Algorithm	Metric	Score of map1	Score of map2	Score of map3
Standard LSTM	A3C [1]	MEAN	98	44	96
		MAX	373	57	444
Standard LSTM	A2C	MEAN	95	44	91
		MAX	367	57	444
Standard LSTM	PPO	MEAN	98	45	93
		MAX	373	57	423
ConvLSTM	A2C	MEAN	**99**	**46**	97
		MAX	375	59	462
ConvLSTM	PPO	MEAN	96	45	**98**
		MAX	**377**	**60**	**473**

better temporal and spatial expression compared with the standard LSTM network, and the former has better performance.

5 Game Performance

In the minigame under the Starcraft II Reinforcement learning environment, we select three maps to perform deep reinforcement learning model training and explores the application of deep reinforcement learning in the game strategy. Each agent uses 32 threads for training and the training time is 100 M time steps.

In the map 'Defeat Roaches', the Marines learned the strategy of 'siege': the marines first kill an enemy unit and kill one by one. As shown in Fig. 3(1). Agents first attack the top enemy collectively and then attack other enemies after killing.

In the map 'Find and Defeat Zerglings', the Marines learned the strategy of 'searching': the marines explore unknown positions in clockwise order. As shown in Fig. 3(2). In the minimap, we can see clearly that the agent first searches the lower left corner, then searches the upper left corner, then searches the upper right corner.

In the map 'Defeat Zerglings And Banelings', the Marines learned the strategy of 'attacking by order': the marines first kill enemy units with higher damage ability, and then kill other enemy units. As shown in Fig. 3(3). The agent first attacks the Banelings with greater attack power but ignores the Zerglings that has already been close to itself.

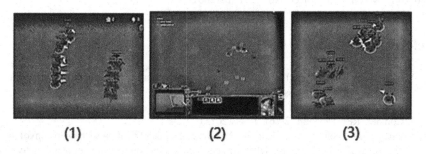

Fig. 3. Game strategy

In three maps, all agents learn the game strategies to achieve better performance. These game strategies are common game strategies in the StarCraft II game.

The average score and maximum score of the depthwise separable CNN are low, but it does not affect the learning of the game strategy in the case of effectively reducing the training time of the agent.

6   Conclusion

This paper focuses on the accelerating spatio-temporal deep reinforcement learning model for game strategy. We train the agents on the mini-game of StarCraft II's learning environment which is a complex learning environment. In this paper, the depthwise separable CNN is applied to deep reinforcement learning model. Compared with the standard CNN used in deep reinforcement learning model before, the training time is shortened by more than 1/9. But, in terms of game strategy and performance, depthwise separable CNN has little impact on game strategy, except slightly reducing the score of the game. We also use ConvLSTM network in deep reinforcement learning model to enhance the perception of spatial-temporal comprehensive features of agents. Then we obtain better performance of the agents that used ConvLSTM than agents' which used standard LSTM network. This indicates that introducing ConvLSTM to replace LSTM in deep reinforcement learning model is helpful to the extraction of environmental features. Finally, we carried out some experiments to verify the above two points. The results of experiments show that these two conclusions are correct. From the perspective of the deep neural network, we have realized the shortening of the training time for the deep reinforcement learning model and the improvement of the performance of the deep reinforcement learning model.

Acknowledgements. The work is funded by the Shanghai Undergraduate Student Innovation Project, the National Natural Science Foundation of China (No. 61170155) and the Shanghai Innovation Action Plan Project (No. 16511101200).

References

1. Vinyals, O., et al.: StarCraft II: A New Challenge for Reinforcement Learning. https://arxiv.org/abs/1708.04782. Accessed 16 Aug 2017
2. Yu, K., Jia, L., Chen, Y., Xu, W.: Deep learning: yesterday, today, and tomorrow. J. Comput. Res. Develop. **20**(6), 1349 (2013)
3. Krizhevsky, A., Sutskever, I., Hinton, G.E.: ImageNet classification with deep convolutional neural networks. In: International Conference on Neural Information Processing Systems, pp. 1097–1105. Curran Associates Inc. (2012)
4. Russakovsky, O., Deng, J., Su, H., Krause, J., Satheesh, S.: ImageNet large scale visual recognition challenge. Int. J. Comput. Vis. **115**(3), 211–252 (2015)
5. Karpathy, A., Toderici, G., Shetty, S., Leung, T., Sukthankar, R., Li, F.F.: Large-scale video classification with convolutional neural networks. In: IEEE Conference on Computer Vision and Pattern Recognition, pp. 1725–1732. IEEE Computer Society (2014)
6. Cho, K., Merrienboer, B.V., Gulcehre, C., Bahdanau, D., Bougares, F., Schwenk, H., Bengio, Y.: Learning phrase representations using RNN encoder-decoder for statistical machine translation. In: Proceedings of Conference on Empirical Methods in Natural Language Processing, Doha, Qatar, pp. 1724–1734 (2014)
7. Yang, Z., Tao, D.P., Zhang, S.Y., Jin, L.W.: Similar handwritten Chinese character recognition based on deep neural networks with big data. J. Commun. **35**(9), 184–189 (2014)

8. Graves, A., Mohamed, A.R., Hinton, G.: Speech recognition with deep recurrent neural networks. In: IEEE International Conference on Acoustics, Speech and Signal Processing, pp. 6645–6649. IEEE (2013)

9. Li, Y., Zhang, J., Pan, D., Hu, D.: A study of speech recognition based on RNN-RBM language model. J. Comput. Res. Develop. **51**(9), 1936–1944 (2014)

10. Sun, Z.J., Xue, L., Xu, Y.M., Wang, Z.: Overview of deep learning. Appl. Res. Comput. **29** (8), 2806–2810 (2012)

11. Sutton, R.S., Barto, A.G.: Reinforcement Learning: An Introduction, Bradford Book. MIT Press, Cambridge (2005). IEEE Transactions on Neural Networks 16(1), 285–286

12. Kober, J., Peters, J.: Reinforcement learning in robotics: a survey. Int. J. Robot. Res. **32**(11), 1238–1274 (2013)

13. Tesauro, G.: TD-Gammon, a self-teaching backgammon program, achieves master-level play. Neural Comput. **6**(2), 215–219 (1989)

14. Kocsis, L., Szepesvári, C.: Bandit based Monte-Carlo planning. In: Fürnkranz, J., Scheffer, T., Spiliopoulou, M. (eds.) ECML 2006. LNCS (LNAI), vol. 4212, pp. 282–293. Springer, Heidelberg (2006). https://doi.org/10.1007/11871842_29

15. Fu, Q.M., Liu, Q., Wang, H., Xiao, F., Yu, J., Li, J.: A novel off policy Q(λ) algorithm based on linear function approximation. Chin. J. Comput. **37**(3), 677–686 (2014)

16. Gao, Y., Zhou, R.Y., Wang, H., Cao, Z.X.: Study on an average reward reinforcement learning algorithm. Chin. J. Comput. **30**(8), 1372–1378 (2007)

17. Wei, Y.Z., Zhao, M.Y.: A reinforcement learning-based approach to dynamic job-shop scheduling. Acta Autom. Sin. **31**(5), 765–771 (2005)

18. Ipek, E., Mutlu, O., Carunana, R.: Self-optimizing memory controllers: a reinforcement learning approach. In: International Symposium on Computer Architecture, pp. 39–50. IEEE (2008)

19. Mnih, V., et al.: Playing atari with deep reinforcement learning. Computer Science, pp. 201–220 (2013)

20. Mnih, V., et al.: Human-level control through deep reinforcement learning. Nature **518** (7540), 529–533 (2015)

21. Silver, D.: Mastering the game of Go with deep neural networks and tree search. Nature **529** (7587), 484–489 (2016)

22. Oh, J., Guo, X., Lee, H., Lewis, R., Singh, S.: Action-conditional video prediction using deep networks in atari games. In: Proceedings of the Neural Information Processing Systems, Montreal, Canada, pp. 2863–2871 (2015)

23. Caicedo, J.C., Lazebnik, S.: Active object localization with deep reinforcement learning. In: IEEE International Conference on Computer Vision, pp. 2488–2496. IEEE Computer Society (2015)

24. Lillicrap, T.P.: Continuous control with deep reinforcement learning. Comput. Sci. **8**(6), A187 (2016)

25. Duan, Y., Chen, X., Houthooft, R., Schulman, J., Abbeel, P.: Benchmarking deep reinforcement learning for continuous control, pp. 1329–1338 (2016)

26. Gu, S., Lillicrap, T., Sutskever, I., Levine, S.: Continuous deep Q-Learning with model-based acceleration. In: Proceeding of ICML 2016 Proceedings of the 33rd International Conference on International Conference on Machine Learning, vol. 48, pp. 2829–2838 (2016)

27. Hansen, S.: Using deep Q-learning to control optimization hyperparameters. https://arxiv.org/abs/1602.04062v2. Accessed 19 Jun 2016

28. Andrychowicz, M.: Learning to learn by gradient descent by gradient descent. In: Proceedings of the Conference on Neural Information Processing Systems, Barcelona, Spain, pp. 3981–3989 (2016)

29. Mnih, V.: Asynchronous methods for deep reinforcement learning. In: Proceedings of the International Conference on Machine Learning, New York, USA, pp. 1928–1937 (2016)

30. Schulman, J., Wolski, F., Dhariwal, P., Radford, A., Klimov, O.: Proximal Policy Optimization Algorithms. https://arxiv.org/abs/1707.06347v2. Accessed 28 Aug 2017

31. Schulman, J., Levine, S., Moritz, P., Jordan, M.I., Abbeel, P.: Trust Region Policy Optimization. Computer Science, pp. 1889–1897 (2015)

32. Howard, A.G., et al.: MobileNets: efficient convolutional neural networks for mobile vision applications. https://arxiv.org/abs/1704.04861. Accessed 17 Apr 2017

33. Shi, X., Chen, Z., Wang, H., Yeung, D.Y., Wong, W., Woo, W.: Convolutional LSTM Network: a machine learning approach for precipitation nowcasting. In: International Conference on Neural Information Processing Systems, pp. 802–810. MIT Press (2015)

34. Jaderberg, M., et al.: Reinforcement Learning with Unsupervised Auxiliary Tasks. https://arxiv.org/abs/1611.05397. Accessed 16 Nov 2016

ASD: A Framework for Generation of Task Hierarchies for Transfer in Reinforcement Learning

Jatin Goyal[1]([⊠]), Abhijith Madan[1], Akshay Narayan[2], and Shrisha Rao[1]

[1] International Institute of Information Technology, Bangalore, India
{jatin.goyal056,abhijith.m}@iiitb.org, shrao@ieee.org
[2] National University of Singapore, Singapore, Singapore
anarayan@comp.nus.edu.sg

Abstract. We present ASD (Action, Sequence, and Divide), a new framework for Hierarchical Reinforcement Learning (HRL). Present HRL methods construct the task hierarchies but fail to avoid exploration when tasks are to be performed in a particular sequence, resulting in the agent needlessly exploring all permutations of the tasks. When the task hierarchies are used as an ASD framework, the RL agent encounters better constraints, preventing it from pursuing policies that are not valid, thus enabling the agent to achieve the optimal policy faster. The hierarchies created using the methods explained in this paper can be used to solve new episodes of the same environment, as well as similar instances of the problem. The hierarchies generated with an ASD framework can be used to establish an ordering of tasks. The objective is to not only to complete the tasks but also give the agent insights into the sequence of tasks that need to be performed in order to correctly solve a problem. We present an algorithm to generate the hierarchies as an ASD framework. The algorithm has been evaluated on some of the standard RL domains, namely, Taxi and Wargus, and is found to give correct results.

1 Introduction

A standard Reinforcement Learning (RL) [14] problem assumes that an agent starts afresh with zero knowledge about the environment. The agent [1] starts exploring and accumulating rewards for the various actions it takes and devises a policy to solve the problem by either maximizing or minimizing the rewards it gets, depending on the problem. In large environments, a substantial amount of time is spent by the agent to discover the optimal solution.

In HRL [4,15], the agent exploits the environment structure to generate the task hierarchies [10]. These hierarchies can be used to transfer the knowledge from one domain to another which speeds up the overall learning phase, but the agent has no knowledge as to what is the sequence of tasks that needs to be performed, resulting in the agent needlessly evaluating every permutation of the tasks.

© Springer Nature Switzerland AG 2018
L. Cheng et al. (Eds.): ICONIP 2018, LNCS 11303, pp. 313–325, 2018.
https://doi.org/10.1007/978-3-030-04182-3_28

We propose the ASD framework to address the problems. ASD is an approach to HRL which divides actions into three types of nodes: Action, Sequence, and Divide. Scaling down a version of the same problem by generating a task hierarchy, it helps solve problems in large domains. When the hierarchy generated in smaller problems is fed as input to similar larger problems, it brings about a significant improvement over the training cost of the new problem.

The generation of hierarchies as an ASD framework has two steps. First, we access the solutions of the smaller problem which are stored in the form of a Directed Acyclic Graph (DAG) [16] to hold the sequence of operations it needs to perform. By using this structure, we generate a hierarchical sequence of tasks (the Component Hierarchy) for that episode. The DAGs can be generated by simulating a much smaller and simpler instance of the environment and letting the agent perform some episodes by interacting with it (optimal policies of these episodes can be stored in the form of a DAG). Second, we amalgamate results from different episodes of the problem. This is done by merging component hierarchies from different episodes into a single hierarchy.

Algorithm 1 constructs a component hierarchy when a solution to the RL problem is given in the form of a CAT (causally annotated trajectory) [11]. We parse the CAT by identifying the task hierarchies based on which subtasks should be accomplished in what order and construct a hierarchical structure. The inner nodes of the hierarchy consist of high-level tasks. The leaf nodes of the hierarchy are the primitive actions that need to be taken in order to complete the high-level tasks. Some sample component hierarchies are shown in Figs. 3 and 4.

Algorithm 2 merges two component hierarchies into a single hierarchy. Here, the three types of nodes: Action, Sequence and Divide, are used to merge the hierarchies. We start from the root nodes of the two trees and check which type of node is present in it. Based on the type of node we merge the children such that the sequence of execution of the subtasks is maintained in the merged hierarchy.

We tested our algorithms with the Taxi problem [8] and the Wargus domain [11], and accurately generated the task hierarchies. For the case where the environment is stochastic in nature, the problem is solved as a Markov Decision Problem [2]. However, the algorithms mentioned in this paper do not depend on the nature of the environment, i.e., it can either be deterministic or probabilistic [6]. The algorithms can also handle conjunctive goals (both goals can be performed with or without any order) as well as disjunctive goals (the overall task being considered complete if either one of the subtask is completed), which is a significant improvement over the HI-MAT Algorithm [11] which worked on a single DAG.

The key difference between HI-MAT and ASD is that HI-MAT gives task hierarchies using only one CAT, which may not explore the complete dynamics of the environment, whereas ASD uses multiple CATs, makes individual hierarchies, and then merges all the results to get one complete task hierarchy.

2   Background and Related Work

HRL allows the agent to exploit the domain structure by splitting a task into subtasks and solving them, which speeds the overall learning phase [1,13].

To generate a component hierarchy, we require optimal policies for numerous instances. The policies are defined in the form of a CAT [11]. A CAT is a directed acyclic graph that gives information about how the variables of an environment are changed with the actions that are being taken. Consider the Taxi problem in Fig. 1, whose objective is to pickup a passenger from a specific location and then drop off the passenger at a particular destination. The pickup and drop off locations are given as inputs to the agent, the variable *pass.loc* changes at two locations, once during pickup, i.e., when moving from start to pickup and then, at drop off, i.e. when moving from pickup to drop off. Similarly, *pass.dest* is changed only after the agent has picked the passenger and then dropped off at the required destination. These CATs reflect how some of the tasks contains smaller subtasks, as seen in Fig. 1.

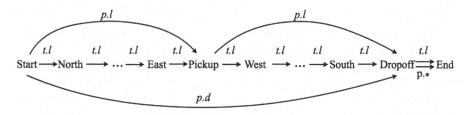

Fig. 1. CAT for taxi problem, where p.l = pass.loc, t.l = taxi.loc and p.d = pass.dest [11]

HEXQ [7], VISA [9], and HI-MAT [11] are some existing approaches to get the required task hierarchies. HEXQ generates the hierarchies based on the frequency of change in values of variables. VISA uses Dynamic Bayesian Networks [3,12] (DBNs) to get the required hierarchies. HI-MAT uses CATs to get the required Hierarchies for MAXQ [4] decomposition. In HI-MAT, the CAT and DBN models are applied to previously solved RL tasks to induce the MAXQ task hierarchies.

3   Component Hierarchies

ASD is an abbreviation of Action, Sequence and Divide, these being the three types of nodes. These nodes are used in the component hierarchy by identifying the sequence of subtasks which need to be completed in order to accomplish a task. Each node has information about the variable that is changing when an action is performed by the agent with the help of a label which is either *Sequence, Divide* or *Action*. It also contains a task equation [11] which signifies the task that the node needs to accomplish, the task equation also behaves as an identifier to differentiate it from other nodes.

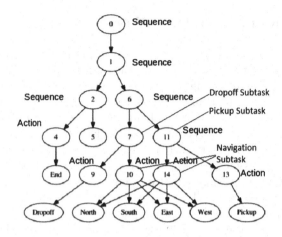

Fig. 2. Component hierarchy for taxi problem [11]

Action

Action nodes signify that there are no more subtasks left, i.e., it is a primitive action. To complete the task signified by this Action node, the agent interacts with the environment using the primitive actions available to it (in the hierarchies). The end state and the start state are provided in the parent node of this subtask. The agent gives the optimal policy and total optimal reward for reaching the end state. For example, in the component hierarchy for the Taxi problem shown in Fig. 2, the navigation task has only four primitive actions (North, South, East, West) which can be used to complete the task.

By marking the Action nodes in the ASD framework, the agent uses only those actions which are available as children of that Action node. This prevents performing some unnecessary computations (like computing costs for pickup and dropoff for a navigation task), which speeds up the overall process for getting the optimal policy. For example, in Taxi problem, the agent tries not to pickup or dropoff while on route to the destination.

Sequence

Sequence nodes signify that the subtasks need to be performed sequentially. For example in Fig. 2, the entire problem is marked as a Sequence task, with two subtasks, pickup (task 11 in Fig. 2) and dropoff (task 7 in Fig. 2), which needs to be performed in that order alone. Like Actions, Sequence nodes also prevent the performing of some unnecessary computations to improve the performance of the system. An agent on coming across a *Sequence* task knows what needs to be performed first, and does not complete the second task before the first task, as in the case of traditional Q-learning agents.

Divide

Divide nodes signify that although the entire task of this node has been split into subtasks, there is no required order for them to be executed, hence the optimal policy can be any permutation of the subtasks in the hierarchy, with the present node as root. For example, in the Wargus domain problem, the agent can mine gold first or chop wood, both being valid ways to do it, but the agent needs to identify the order of the subtasks, so as to get the optimal rewards. In the Taxi problem, there are no Divide tasks since tasks are always in a Sequence and no permutations are allowed.

4   Generating Component Hierarchies

We first automate the discovery of the task hierarchies and generate multiple component hierarchies (smaller task hierarchies) for different instances of the task. Next, we amalgamate these component hierarchies into a single merged hierarchy. The hierarchies once generated need not be computed again; they can be stored and reused repeatedly unless the dynamics of the environment is changed; for instance, by adding a new variable. One such change would be, in the Taxi problem, if we add a new action like Refuel which should be performed before the fuel of the taxi runs out. This would affect the sequence of operations that should be performed and would require retraining, but no retraining will be required if the change is to rearrange the blocked grids in the environment or similar, which does not affect the dynamics of the environment.

Automatic Discovery of Task Hierarchies

The algorithm needs the optimal policies of the source task, stored in the form of a causally annotated trajectory [11] (CAT) to create task hierarchies. Since a single CAT may not use all the actions available to it, we need multiple such CAT which cover the entire dynamics of the environment by using all possible actions in a valid sequence.

An example of a CAT is shown in Fig. 1 for the Taxi problem. These trajectories contain information about the variables that are being changed on every action, and how these change in values of variables tends to completion of the given task.

A CAT is a directed acyclic graph. The CATs are made in such a way that they all have a *Start* and an *End* node. First, the shortest path is calculated from the *Start* node to the *End* node using Depth First Search (DFS) [5]. Also, the in-degree of the *End* node is calculated and stored in the variable *indegree* (line 4). If *indegree* is equal to 2 (line 4), then we mark the entire task as a *Sequence* task since there can be only one way to complete this subtask. If it is more than 2 (line 17), then we mark the entire subtask as a *Divide* subtask since there can be more than one permutation of subtasks that are valid, and if it is equal to 1 (line 11), we mark the subtask as an *Action* subtask.

Algorithm 1. generateASDHierarchy(Ω, ω, S, V) is used to generate a component hierarchy from a CAT. Where Ω is the CAT, ω is the root node of the component hierarchy, S is the start node of the CAT and V is the end node of the CAT.

Input: CAT Ω, ASDNode ω, Start S and End V nodes of the CAT
Output: ASD Hierarchy

1 minCostPath $= \delta(S, V) = min(W(p)$ S$\rightarrow$ V)
2   where $W(p)$ is the Cost of path $= \sum_{i=1}^{k} w(v_{i-1}, v_i)$
3 indegree $=$ inDegree(V) //finds the number of incoming edges of a node
4 **if** *indegree.equalsTo(2)* **then**
5 $\quad$ remEdge $=$ minCostPath.remove(0) //remove the first edge from the list of edges
6 $\quad$ $S =$ remEdge[0]
7 $\quad$ $V =$ remEdge[1]
8 $\quad$ Create a new node of type "Sequence" and add it to the root node
9 $\quad$ generateASDHierarchy(Ω, ω, S, V)
10 **else if** *indegree.equalsTo(1)* **then**
11 $\quad$ Create a new node of type "Action" and add it to the root node
12 $\quad$ **foreach** *Action $a_i \in \Omega$* **do**
13 $\quad\quad$ add a_i to *node* as a child
14 $\quad$ ω.add(node)
15 **else**
16 $\quad$ Create a new node of type "Divide" and add it to the root node
17 $\quad$ $V = V$.parent
18 $\quad$ generateASDHierarchy(Ω, ω, S, V)

Algorithm 2. MergeHierarchies(ω_1, ω_2) is used to merge two component hierarchies generated by Algorithm 1. Here ω_1 and ω_2 are the two component hierarchies.

Input: ASDNode ω_1, ASDNode ω_2
Output: Merged ASD hierarchy

1 **if** *(ω_1.type.equalsTo("Actions") AND ω_2.type.equalsTo("Actions"))* **then**
2 $\quad$ ω_2.children $= \omega_1$.children $\cup \omega_2$.children
3 **if** *(ω_1.type.equalsTo("Divide") OR ω_2.type.equalsTo("Divide"))* **then**
4 $\quad$ ω_1.type $=$ "Divide"
5 $\quad$ ω_2.type $=$ "Divide"
6 $\quad$ **foreach** *child $c_i \in \omega_1$* **do**
7 $\quad\quad$ **if** *($\neg\omega_2$.contains(c_i))* **then**
8 $\quad\quad\quad$ ω_2.add(c_i)
9 **if** *(ω_1.type.equalsTo("Sequence") AND ω_2.type.equalsTo("Sequence"))* **then**
10 $\quad$ **foreach** *child $c_i \in \omega_1$* **do**
11 $\quad\quad$ **if** *($\neg\omega_2$.contains(c_i))* **then**
12 $\quad\quad\quad$ ω_2.add(c_i)
13 $\quad\quad$ **else**
14 $\quad\quad\quad$ $MergeHierarchies(\omega_1.child[i], \omega_2.child[i])$
15 **return** ω_2

Case 1: If the type of task is *Sequence*, we first remove the minimum path present in the CAT and update the *Start* and *End* nodes with the beginning and ending edges of the removed edge. We create a new node of type *Sequence* and add it to the hierarchy. The function *generateASDHierarchy(Start, End)* is recursively called for the updated *Start* and *End* nodes. Note that these subtasks are solved in the order of the edges appearing in the shortest path, and when the agent sees the node as a *Sequence* task, it executes these subtasks in that particular order alone.

Case 2: If the type of task is *Action*, then we simply read all the nodes appearing in the path from *Start* to *End*, and add each one of them as a child of this task. The children are a set of primitive actions that need to be performed to accomplish the overall task of this node.

Case 3: If the type of task is *Divide*, then the function *generate ASDHierarchy(Start, End)* is recursively called *indegree* − 1 times. For each time the function is called, the *End* node is updated to the starting node of the edge connecting to the *End* node. This is also explained in Sect. 5. When the agent sees the node as a *Divide* node, it tries all permutations of the subtasks and chooses the optimal policy.

This algorithm ends when it has reached the action path, as there are no more recursive calls from there. The Action path is the sequence of primitive actions that are taken by the agent to generate the optimal policy, and since it is the longest path in the trajectory, it is always the last one selected for building the hierarchies.

Incomplete Hierarchies

Algorithm 1 relies on optimal policies stored for making the hierarchies, but an optimal policy stored may not contain all the primitive actions in the environment. It may not even contain an entire subtask because that was not needed for that particular optimal policy though it might be needed for other episodes of the same environment. Therefore the hierarchies should cover the entire dynamics of the system. For example, in Fig. 3, the Hierarchy is of a Taxi problem which does not contain the complete dynamics of the Taxi environment, as it does not contain "West" as a primitive action, also for subtask `pickup`, only "North" is available for navigation, which shows incomplete information about the environment. To overcome this problem, we give Algorithm 2 for merging multiple hierarchies of the same environment.

The function *MergeHierarchies()* in Algorithm 2 takes two hierarchies from different instances of the same environment as inputs. The function recursively compares the nodes and their children. First, the function compares the type (Action, Sequence or Divide) of both the root nodes.

Case 1: If both nodes are of type *Action*, the function *MergeHierarchies* merges the primitive actions list of both the root nodes and store in the primitive actions list of the root node in *Hierarchy2*.

Case 2: If either of the nodes is a *Divide* node, mark the node in *Hierarchy2* as *Divide*. If a subtask of the root node in *Hierarchy1* is found in *Hierarchy2*, call *MergeHierarchies(Hierarchy1, Hierarchy2)* using the similar nodes as root. For every child in the list of subtasks of the root node in *Hierarchy1*, not present in children list in *Hierarchy2*, append those tasks to the children list of the root node of *Hierarchy2*.

Case 3: If both the nodes are of type *Sequence*, then children of both the nodes are compared. Since these tasks have to be executed in a sequence, the comparison done in the reverse order of execution. If the list from *Hierarchy1* contains some subtask which is not present in *Hierarchy2*, these subtasks are appended to the start of the list in *Hierarchy2* (because the function gives *Hierarchy2* as output). Also for the subtask found identical in both the lists of subtasks, call *MergeHierarchies(Hierarchy1, Hierarchy2)* using the identical node as the root nodes.

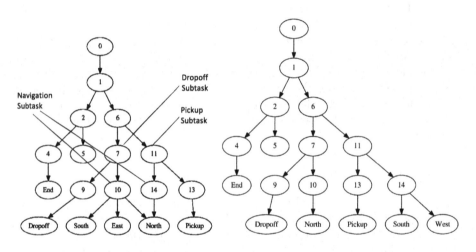

Fig. 3. Hierarchy 1   **Fig. 4.** Hierarchy 2

5   Evaluation of ASD in Standard RL Domains

Taxi Problem

Consider the 5×5 single agent Taxi Cab domain [11]; the CAT for a solved episode is given in Fig. 1. As given in Algorithm 1, for the *Start* and *End* node, the shortest path is calculated and it is stored; in this case, it is *Start* → *dropoff* → *End*. Consecutively, the number of edges incoming to the end node is also calculated, i.e., the in-degree of the end node which is 2 in this case. According to Algorithm 1, we mark this task as *Sequence* node and divide the entire problem into two subtasks, 1. *Start* → *dropoff*, 2. *dropoff* → *End*

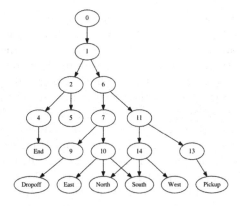

Fig. 5. Merged result for Hierarchy 1 in Fig. 3 and Hierarchy 2 in Fig. 4

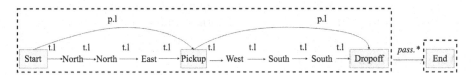

Fig. 6. Dividing tasks in a CAT according to Algorithm 1

as shown in Fig. 6. According to Algorithm 1, the edge *Start* → *dropoff* is removed and *generateASDHierarchy*(*Start*, *dropoff*) is called. Algorithm 1 is followed, finding shortest path (with *dropoff* as the end node), which is *Start* → *pickup* → *dropoff*, and the in-degree of the end node is calculated which is 2 in this case. We then mark this subtask as *Sequence* and further divide it into subtasks as shown in Fig. 7.

The edge *Start* → *pickup* is removed and *generateASDHierarchy* (*Start*, *pickup*) is called recursively. Now the in-degree of the end node is 1 and as a result, this subtask is marked as *Action* node, and all the actions occurring on the path from the *Start* node to *End* node are added to this node as its children. After *Actions* there are no more recursive calls, hence *generateASDHierarchy*(pickup, dropoff) is called and the same steps are followed, and consecutively, component hierarchies are formed. Similar steps are followed and the entire task of picking up and dropping off the passenger is divided into subtasks, which are executed according to the node type, as is mentioned in Sect. 3.

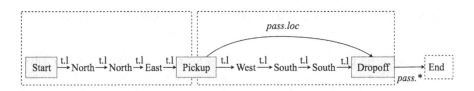

Fig. 7. Dividing tasks of Fig. 6

322   J. Goyal et al.

Similarly, two task hierarchies are generated, as shown in Figs. 3 and 4, from two different CATs. It can be observed that neither of them contains all the primitive actions available to the agent. Algorithm 2 gives the protocol to merge the knowledge of both the hierarchies. Starting from the root node (0) in both the trees, nodes at each level are compared in both hierarchies. If both the nodes are *Sequence* then we move to the next level, and if one of them is *Divide* then the resulting node is also *Divide*. In the case of Figs. 3 and 4, both are *Sequence*, so we move to node 1 in both the hierarchies; both are *Sequence*, so we move to the next level. This level contains multiple nodes in both the hierarchies; first the union of both the nodes is done and stored as the result. If two nodes contains the same task equation, then *MergeHierarchies(node1, node2)* is implemented, which follows the Algorithm 2 again. When the iteration reaches the Action nodes, the union of the primitive actions in both the nodes is done resulting in a node which contains actions from both the nodes and is stored in the resultant hierarchy. As it is seen in Fig. 5, all the primitive actions in *Hierarchy1* and *Hierarchy2* in Figs. 3 and 4 are merged to form action nodes in Fig. 5.

Wargus Domain Problem

Consider a 5×5 grid for Wargus resource gathering-problem [11] where the agent is the peasant and its objective is to chop wood, mine gold and deposit it at the City center. The optimal policy is the steps taken to complete the task with maximum reward.

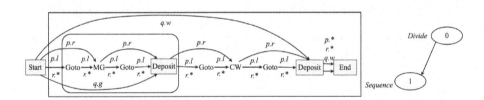

Fig. 8. Solved RL CAT for Wargus domain

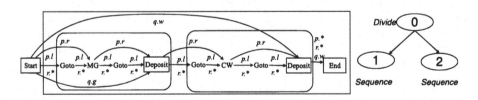

Fig. 9. Dividing tasks in Fig. 8 and making hierarchies according to Algorithm 1

For generating the component hierarchies for this problem, access to solved instances of Wargus domain are required, an example is shown in Fig. 8, here *p.l* stands for the peasant's location, *p.r* is the peasant's resource, *r.w* and *r.g* stands for region to chop the wood and region to mine for gold. *q.w* and *q.g* stands for quota for wood and gold respectively. Similar to the previous example, first the shortest path between *Start* and *End* is calculated, as also the number of edges connecting to the *End*, which is 3 in this case. As a result this task is marked as *divide*, the edge *Start → Deposit* is removed and the procedure *generateASDHierarchy(Start, Deposit)* is recursively called. The task hierarchies are shown in Figs. 8, 9 and 10. Note when calling *generateASDHierarchy(Start, Deposit)* for the second *Deposit* node, the *Start* node is changed to *Goto* in *Deposit → Goto*. This is because while dividing the task into subtasks, first task was completed at the first *Deposit* node, and the second task started from there, but unlike the Taxi problem, there is no edge from first *Deposit* node to the second one, because both the goals can be achieved in either order. So when the agent uses the merged hierarchy for solving problems, and reaches the *Divide* node, it evaluates rewards for both the actions and then decides which one to choose based on the policy that gives the higher reward. Before *generateASDHierarchy(Start, Deposit)* is called, the edge *Start → Deposit* is removed, after which the shortest path is calculated, and number of edges connecting to the *End* node, which is 2 in this case, as a result of which, this task is marked as *Sequence*, as shown in Figs. 8 and 9, similar to the Taxi Cab domain problem, and the final task hierarchies are generated as shown in Fig. 10. Also when a task is marked as *Divide*, the subtask associated with the last edge of the shortest path, is not used for any input in task hierarchies, in this case, the tasks associated with both the *Deposit → End* edges are never used to make the hierarchies, because these edges signify the completion of that task.

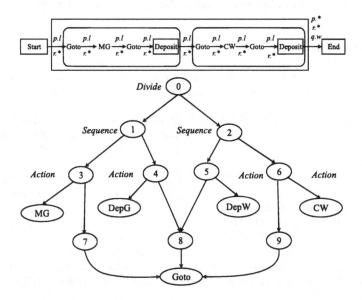

Fig. 10. Dividing tasks and generating task hierarchies

In certain instances, it is possible that the CAT may contain only one of the subtasks (chop wood or mine gold). In such cases, the component hierarchies would be incomplete. Also Algorithm 1 would mark this subtask as a *Sequence* node. But when this component hierarchy is merged with a component hierarchy generated for a CAT shown in Fig. 8, then according to Algorithm 2, it is clear that when either of the two tasks at the same level are *Divide* while merging, then the resulting task will also be a *Divide* task. The main function of Algorithm 2 is to learn about the tasks and the actions available from multiple task hierarchies.

6   Observation and Conclusion

Both Algorithms 1 and 2 were tried on the Taxi domain problem and the Wargus problem, and the result is that the hierarchies obtained contained the subtasks in the right order. The CATs required to generate the Hierarchies may come from a smaller version of the environment. For example, in the Taxi domain problem, the required simulations were to be done on a 50×50 grid, and we know the optimal policies need to come from the same environment. However, we noticed, that if we input CAT from a grid of a 5×5, the hierarchies generated are in accordance to the 50×50 grid. This is because irrespective of the size of the domain, the sequence of actions performed to achieve a task would remain the same.

We presented an approach to automatically induce subtasks, which cover the entire dynamics of the system, from solved RL problems. We presented the algorithm to generate a component hierarchy from the CATs of all the solved instances of the problem. This will speed up the process of finding solutions in other instances of same domain. Also, these hierarchies are reusable.

References

1. Andre, D., Russell, S.: State abstraction for programmable Reinforcement Learning agents. In: Association for the Advancement of Artificial Intelligence, vol. 112, pp. 119–125 (2002)
2. Bai, A., Srivastava, S., Russell, S.: Markovian state and action abstractions for MDPS via hierarchical MCTS. In: Proceedings of the Twenty-Fifth International Joint Conference on Artificial Intelligence, IJCAI 2016, pp. 3029–3037. AAAI Press (2016)
3. Cao, F., Ray, S.: Bayesian Hierarchical Reinforcement Learning. In: Pereira, F., Burges, C.J.C., Bottou, L., Weinberger, K.Q. (eds.) Advances in Neural Information Processing Systems 25, pp. 73–81. Curran Associates Inc, New York (2012)
4. Dietterich, T.G.: Hierarchical Reinforcement Learning with the MAXQ value function decomposition. J. Artif. Intell. Res. **13**, 227–303 (2000)
5. Franciosa, P.G., Gambosi, G., Nanni, U.: The incremental maintenance of a depth-first-search tree in directed acyclic graphs. Inf. Process. Lett. **61**(2), 113–120 (1997)
6. Getoor, L., et al.: Introduction to Statistical Relational Learning. MIT press, Cambridge (2007)

7. Hengst, B.: Discovering hierarchy in Reinforcement Learning with HEXQ. In: Proceedings of the Nineteenth International Conference on Machine Learning, ICML 2002, pp. 243–250. Morgan Kaufmann Publishers Inc., San Francisco, CA, USA (2002)
8. Jardim, D., Nunes, L., Oliveira, S.: Hierarchical Reinforcement Learning: Learning sub-goals and state-abstraction. IEEE, July 2011
9. Jonsson, A., Barto, A.: Causal graph based decomposition of factored MDPS. J. Mach. Learn. Res. **7**, 2259–2301 (2006)
10. Knoblock, C.A.: Hierarchical problem solving. In: Knoblock, C.A. (ed.) Generating Abstraction Hierarchies. The Springer International Series in Engineering and Computer Science (Knowledge Representation, Learning and Expert Systems), vol. 214. Springer, Boston (1993). https://doi.org/10.1007/978-1-4615-3152-4_3
11. Mehta, N., Ray, S., Tadepalli, P., Dietterich, T.: Automatic discovery and transfer of MAXQ hierarchies. In: Proceedings of the 25th International Conference on Machine Learning, pp. 648–655. ACM (2008)
12. Ngo, V.A., Ngo, H., Wolfgang, E.: Monte carlo bayesian Hierarchical Reinforcement Learning. In: Proceedings of the 2014 International Conference on Autonomous Agents and Multi-agent Systems, AAMAS 2014, pp. 1551–1552. International Foundation for Autonomous Agents and Multiagent Systems, Richland, SC (2014)
13. Sutton, R., Precup, D., Singh, S.: Between mdps and semi-MDPS: a framework for temporal abstraction in reinforcement learning. Artif. Intell. **112**, 181–211 (1999)
14. Sutton, R.S., Barto, A.G.: Reinforcement Learning: An introduction, vol. 1. MIT press, Cambridge (1998)
15. Tadepalli, P., Dietterich, T.G.: Hierarchical explanation-based Reinforcement Learning. In: In Proceedings of the Fourteenth International Conference on Machine Learning, pp. 358–366. Morgan Kaufmann (1997)
16. VanderWeele, T.J., Robins, J.M.: Directed acyclic graphs, sufficient causes, and the properties of conditioning on a common effect. Am. J. Epidemiol. **166**(9), 1096–1104 (2007)

Driving Control with Deep and Reinforcement Learning in The Open Racing Car Simulator

Yuanheng Zhu[1,2(✉)] and Dongbin Zhao[1,2(✉)]

[1] Institute of Automation, Chinese Academy of Sciences, Beijing 100190, China
{yuanheng.zhu,dongbin.zhao}@ia.ac.cn
[2] School of Artificial Intelligence, University of Chinese Academy of Sciences, Beijing 100049, China

Abstract. Vision-based control is a hot topic in the field of computational intelligence. Especially the development of deep learning (DL) and reinforcement learning (RL) provides effective tools to this field. DL is capable of extracting useful information from images, and RL can learn an optimal controller through interactions with environment. With the aid of these techniques, we consider to design a vision-based robot to play The Open Racing Car Simulator. The system uses DL to train a convolutional neural network to perceive driving data from images of first-person view. These perceived data, together with the car's speed, are input into a RL-learned controller to get driving commands. In the end, the system shows promising performance.

Keywords: TORCS · Vision-based control · Reinforcement learning
Deep learning

1 Introduction

The Open Racing Car Simulator (TORCS)[1] is an open source 3D car racing simulator. It provides realistic experience with powerful physics engines and sophisticated 3D graphics. Players can not only drive cars in TORCS, but also design their own robots with intelligent techniques [8,17]. Based on computational intelligence (CI) techniques, numerous robots have been successfully developed by researchers to play TORCS [2,12].

Most robots use real measurements from the TORCS engine as input state, such as distance, angle, track shape, to name a few. These data are reliable and low-dimensional, but must be provided by the TORCS engine. In contrast, when humans play TORCS or drive real cars, they can perform well based on only drivers' view. In recent years, deep learning (DL) makes it feasible and easy to process high-dimensional images [4,6,7]. Essential features can be extracted with deep neural networks (DNNs). Inspired by that, in [3], authors try to predict

---

[1] http://torcs.sourceforge.net/.

© Springer Nature Switzerland AG 2018
L. Cheng et al. (Eds.): ICONIP 2018, LNCS 11303, pp. 326–334, 2018.
https://doi.org/10.1007/978-3-030-04182-3_29

driving data from the first-person view in TORCS. They collect images and data and put them into a convolutional neural network (CNN) to train network weights.

However, DL lacks the ability of interacting with external environment. To achieve vision-based control in complex systems like TORCS, researchers have been working on combining DL with reinforcement learning (RL). RL considers how to choose a series of actions to maximize the accumulated rewards from environment [5,9,14–16]. Researchers in [10,11,18] combine these two methods and propose deep reinforcement learning (DRL) to play Atari games. To achieve satisfying results, these DRL algorithms have to run plenty of trials through interactions with environment, and most early trials end up with failures. For vision-based autonomous driving, it is more reliable to separate action-decision and image-perception processes apart. A driving controller should be learned with only a small number of trials, and a perception module can be trained by data that are collected from skilled drivers in a safe condition.

In this paper, we aim to integrate the latest RL and DL methodologies together to design a vision-based self-driving robot in TORCS. First we use low-dimensional, ground-truth driving data provided by the TORCS engine to learn a driving controller with only a small number of trials. Then we train a CNN to perceive driving data from images of first-person view. After integrating the strategy and perception parts together, our robot only takes the first-person view and its own speed as input, and can drive successfully. Since the framework only involves collecting data from human drivers in a safe condition and learning controllers with a small number of trials, it is easy and reliable to extend the work to practical applications.

2 Problem Description in TORCS

The control variables in TORCS (Fig. 1) include $\mathbf{u}_t = [\delta_t, \tau_t]^T$. δ_t is the steering angle percentage, ranged by $[-1, 1]$. τ_t is the throttle or brake percentage, ranged by $[-1, 1]$. t specifies the time index. As for the gear control, we use an automatic transmission algorithm to shift the gear automatically.

For ease of analysis, the car dynamics is treated as a discrete-time system with a fixed step dT. The evolving variables, which we term as *inherent variables*, include $\mathbf{x}_t = [d_t, a_t, v_t]^T$, where d_t is the deviation distance (m), a_t is the deviation angle (rad), and v_t is the current speed (km/h). Its evolution is determined by command $\mathbf{u}_t$ and *dynamical variables* $\mathbf{y}_t = [d_t, a_t, v_t, \kappa_t]^T$ where κ_t is the road curvature (m^{-1}). The transition function is defined as $\mathbf{x}_{t+1} = f(\mathbf{y}_t, \mathbf{u}_t)$ and it is unknown to robot designers. Due to disturbance and sensor noise, observations of $\mathbf{x}_{t+1}$ are perturbed by noise, which here we assume as Gaussian noise $\varepsilon \sim \mathcal{N}(0, \Sigma_\varepsilon)$, where $\Sigma_\varepsilon = \mathrm{diag}(\sigma_{\varepsilon_d}, \sigma_{\varepsilon_a}, \sigma_{\varepsilon_v})$.

The following *cost* function evaluates the driving performance at each step

$$c_t = c(\mathbf{z}_t) = 1 - \exp\left(-\frac{1}{2b^2}\left[\omega_d d_t^2 + \omega_a a_t^2 + \omega_v(v_t - v_t')^2\right]\right) \qquad (1)$$

Fig. 1. Screenshot of TORCS.

where v_t' is an auxiliary variable that represents the desired speed (km/h) at the current position. $b, \omega_d, \omega_a, \omega_v$ are the cost coefficients. The *cost variables* $\mathbf{z}_t$ are composed of $\mathbf{z}_t = [d_t, a_t, v_t, \kappa_t, v_t']^T$. The long-term goal is to minimize the *return*, which is the sum of costs during a period of time $\min J = \min \mathbb{E}\left[\sum_{t=1}^{T} c_t\right]$.

The desired speed v_t' is widely used in the design of TORCS robots [1,13]. It is calculated based on track curvature and road friction coefficient. Due to page limit, we omit the calculation details and suggest readers to refer to [1,13].

The controller is constructed in the form of $\mathbf{u}_t = \pi(\mathbf{z}_t|p)$, where p represent controller parameters, and the target is to minimize J. Note that the controller needs all the variable information, including not only physical variables like d_t, a_t, κ_t, but also auxiliary variable v_t'. In the open-source TORCS, these data are available from the TORCS engine. In the next, we plan to learn the controller parameters by RL based on real variable information, and then perceive these variables from images by DL so that the robot can drive based on first-person view.

3   Learn Driving Controller by Modified PILCO

3.1   Gaussian Process Model

PILCO [5], short for Probabilistic Inference for Learning COntrol, is a model-based RL algorithm. In PILCO, the system dynamics is considered as a Gaussian Process (GP). Suppose we have collected a group of driving data $\{\mathbf{y}_t, \mathbf{u}_t\}$. We use $\tilde{\mathbf{y}}_t = [\mathbf{y}_t^T, \mathbf{u}_t^T]^T$ as training inputs, and the difference $\Delta\mathbf{x}_{t+1} = \mathbf{x}_{t+1} - \mathbf{x}_t + \varepsilon$ as training targets, where ε is Gaussian noise. The mean and variance of $\mathbf{x}_{t+1}$ now become $\mu_{t+1} = \mathbf{x}_t + \mathbb{E}_f[\Delta\mathbf{x}_{t+1}]$, $\Sigma_{t+1} = \mathrm{var}_f[\Delta\mathbf{x}_{t+1}]$.

Suppose there exist n training inputs, $\tilde{\mathbf{Y}} = [\tilde{\mathbf{y}}_1, \ldots, \tilde{\mathbf{y}}_n]$, and n training targets, $\Delta\mathbf{X} = [\Delta\mathbf{x}_1, \ldots, \Delta\mathbf{x}_n]$. Consider the scalar target $\Delta\mathbf{x}_i \in \mathbb{R}$ and deterministic test input $\tilde{\mathbf{y}}_*$. The predictive probability of test target $\Delta\mathbf{x}_*$ is Gaussian with mean and variance as

$$\mu_* = \mathbb{E}_f[\Delta \mathbf{x}_*] = \mathbf{k}_*^T (\mathbf{K} + \sigma_\epsilon^2 \mathbf{I})^{-1} \Delta \mathbf{X} = \mathbf{k}_*^T \beta \tag{2}$$

$$\sigma_*^2 = \mathrm{var}_f[\Delta \mathbf{x}_*] = k_{**} - \mathbf{k}_*^T (\mathbf{K} + \sigma_\epsilon^2 \mathbf{I})^{-1} \mathbf{k}_* \tag{3}$$

where $\mathbf{k}_* = k(\tilde{\mathbf{Y}}, \tilde{\mathbf{y}}_*)$, $k_{**} = k(\tilde{\mathbf{y}}_*, \tilde{\mathbf{y}}_*)$, $\beta = (\mathbf{K} + \sigma_\epsilon^2 \mathbf{I})^{-1} \Delta \mathbf{X}$, and $\mathbf{K}$ is the Gram matrix with entries $K_{ij} = k(\tilde{\mathbf{y}}_i, \tilde{\mathbf{y}}_j)$. Here the kernel function k selects the squared exponential (SE) kernel

$$k(\tilde{\mathbf{y}}_1, \tilde{\mathbf{y}}_2) = \alpha^2 \exp\left(-\frac{1}{2}(\tilde{\mathbf{y}}_1 - \tilde{\mathbf{y}}_2)^T \Lambda^{-1} (\tilde{\mathbf{y}}_1 - \tilde{\mathbf{y}}_2)\right) \tag{4}$$

where α and Λ are function parameters. These parameters can be learned by evidence maximization.

When test input is distributed, the target distribution is complicated but we can still approximate it as a GP. Still consider the scalar target, i.e. $\Delta x_i \in \mathbb{R}$, and suppose the test input satisfies $\tilde{\mathbf{y}}_* \sim \mathcal{N}(\mu, \Sigma)$. The target distribution is approximated by Gaussian $\Delta \mathbf{x}_* \sim \mathcal{N}(\mu_*, \sigma_*^2)$ where

$$\mu_* = \beta^T \mathbf{q}, \sigma_*^2 = \alpha^2 - \mathrm{tr}\left((\mathbf{K} + \sigma_\epsilon^2 \mathbf{I})^{-1} \tilde{\mathbf{Q}}\right) + \beta^T \tilde{\mathbf{Q}} \beta - \mu_*^2$$

$$\mathbf{q} = [q_1, \dots, q_n]^T, q_i = \alpha^2 |\Sigma \Lambda^{-1} + \mathbf{I}|^{-\frac{1}{2}} \exp\left(-\frac{1}{2}(\tilde{\mathbf{y}}_i - \mu)^T (\Sigma + \Lambda)^{-1} (\tilde{\mathbf{y}}_i - \mu)\right)$$

and $\tilde{\mathbf{Q}}$ is a $n \times n$ matrix with entries

$$\tilde{Q}_{ij} = \frac{k(\tilde{\mathbf{y}}_i, \mu) k(\tilde{\mathbf{y}}_j, \mu)}{|2\Sigma \Lambda^{-1} + \mathbf{I}|^{\frac{1}{2}}} \exp\left((\tilde{\rho}_{ij} - \mu)^T (\Sigma + \frac{1}{2}\Lambda)^{-1} \Sigma \Lambda^{-1} (\tilde{\rho}_{ij} - \mu)\right)$$

and $\tilde{\rho}_{ij} = \frac{1}{2}(\tilde{\mathbf{y}}_i + \tilde{\mathbf{y}}_j)$. The above results of scalar input can be easily extended to multivariate case, so we omit it here.

One drawback of GP is its computational complexity. If the data set is large, the training and predicting processes will be slow and unsuitable for real applications. We discretize the input space into non-overlapping equal-sized cells, and each cell can store at most one data [19]. In this way, the stored data are naturally separated and a sparse training set is obtained.

3.2 Return Evaluation

The driving controller π is specified to a linear controller with saturation

$$\mathbf{u}_t = \pi(\mathbf{z}_t) = \mathbf{u}_{\max} \mathrm{sat}(\mathbf{w}\mathbf{z}_t + \mathbf{b}) \tag{5}$$

where the saturation function is defined by $\mathrm{sat}(a) = \frac{1}{8}(9\sin(a) + \sin(3a))$. $\mathbf{u}_{\max}$ indicates the maximum command values. For simplicity, we denote $\mathbf{p} = \{\mathbf{w}, \mathbf{b}\}$.

With the controller structure in (5), it is feasible to compute mean and variance of control variables $\mathbf{u}_t$ with a Gaussian distributed input $\mathbf{z}_t$. Similarly, the probability of control variables is approximated by Gaussian with the calculated mean and variance. If we split the distribution of dynamical variables $\mathbf{y}_t$ from $\mathbf{z}_t$ and combine with action $\mathbf{u}_t$, the Gaussian distribution of input $\tilde{\mathbf{y}}_t = [\mathbf{y}_t^T, \mathbf{u}_t^T]^T$ is known. With the trained GP model, the probability of next-step $\mathbf{x}_{t+1}$ is predicted. Combined with the cost function given in (1), the expected cost of $\mathbf{z}_{t+1}$ is analyzed by $\mathbb{E}[c_{t+1}] = \int c(\mathbf{z}_{t+1})p(\mathbf{z}_{t+1})d\mathbf{z}_{t+1}$ if we further specify the desired velocity v'_{t+1}.

The above process can be repeated for the next $(T-1)$ steps. Given road curvatures $\kappa_t, \dots, \kappa_{t+T}$ and desired velocities $v'_t, \dots, v'_{t+T}$, the distributions of $\mathbf{x}_{t+1}, \dots, \mathbf{x}_{t+T}$ and the corresponding costs are calculated. The estimated return of a starting state $\mathbf{x}_t$ under the current controller is analyzed and is related to the controller parameters $\mathbf{p}$ in the form $J = \sum_{k=t+1}^{t+T} \mathbb{E}[c(\mathbf{z}_k)] \propto \mathbf{p}$.

3.3   Policy Gradient Search

With the analytic solution of J, we calculate the gradient of J towards the controller parameters $\mathbf{p}$. Then policy gradient search is followed to adjust $\mathbf{p}$ to minimize J. However, computing J needs not only a starting state $\mathbf{x}_t$, but also external variables $\kappa_{t+1}, \dots, \kappa_{t+T}$ and $v'_{t+1}, \dots, v'_{t+T}$. We define multiple scenarios with different starting states $\mathbf{x}$ and different κ, v' for policy gradient search in order to gain comprehensive performance. Once the gradient is calculated, $\mathbf{p}$ can be trained by many optimization methods to minimize J.

4   RL Experiment Results

Now we apply the modified PILCO algorithm in TORCS. The track we used for learning is *CG Track 3* and it is marked by lane lines to mimic real-world roads with one lane as illustrated in Fig. 1. The slowdown deceleration a_b selects $2\,\mathrm{m/s}^2$. The discrete-time step selects 0.1 s. The Gaussian noise ε of the observed $\mathbf{x}_t$ satisfies $\varepsilon \sim \mathcal{N}(0, \Sigma_\varepsilon)$, where $\Sigma_\varepsilon = \mathrm{diag}([0.01, 0.01, 1.5]^2)$. The bounds of control actions are $\mathbf{u}_{\max} = [1,1]^T$. The width b in cost function selects 0.4. The importance weights are set to $\omega_d = 1, \omega_a = 1, \omega_v = 400$. When computing the gradient $\mathrm{d}J/\mathrm{d}\mathbf{p}$ with GP model, the future steps T choose 30. To store GP training data, each dimension is divided by 20 between its lower and upper bounds according to experimental experience. 7 scenarios are defined for calculating the gradient, including straight cases and turning cases with different velocities, deviation distances, and curvatures.

After 6 trials, the controller is able to complete the track and the learning stops. Trajectories of states and actions using the final learned controller are plotted in Fig. 2. For comparison, the desired velocity is also plotted along with the real velocity. Small deviations only occur at the moment when the road changes from one segment to another. And the deviations are regulated in a short time.

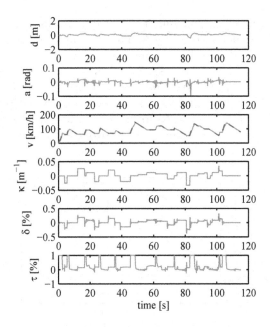

Fig. 2. Trajectories of states and actions using the final learned controller by modified PILCO. The blue solid line in figure 'v' represents the car's velocity, while the red dash-dot line represents the desired velocity. (Color figure online)

5   Perceive Driving Data from Images by DL

In the above section, we learn a driving controller with the full access to state variables provided by TORCS engine. As mentioned above, these data can be perceived from the driver's view, except the car velocity that is known to the car. Inspired by the work of [3], in this section we use a CNN to predict the driving data from images.

First we let a human player drive the car, and store images and driving data every 0.1 s. The images are directly captured from the first-person view with the size of $3 \times 210 \times 280$ (RGB). The driving data include deviation distance, deviation angle, road curvature, and desired velocity. To increase the diversity of data set and improve the generalization of network, the car is driven on different tracks with different backgrounds and lanes. At last we collect a total of 53139 images and driving data as the train set and 10699 images and driving data as the test set.

The network uses the same architecture given in [6], except the output layer is adjusted to suit our needs. To speed up the learning process, we use the results of [6] to initialize the network weights. The network is trained using stochastic gradient descent with a batch size of 64, a momentum of 0.9, and a weight decay of 0.0005. The learning rate is initialized to 0.01 and is dropped by a factor of 0.9 every 8000 steps.

Network is trained for a maximum of 100000 iterations. The curves of train loss and test loss are depicted in Fig. 3. The loss curves drop dramatically once the network starts training. That is because [6] has trained the network on a large data set of real-world images, and we use their results to initialize our network. The shallow layers have already had a high level of feature extraction. With more iterations, the train loss vibrates occasionally but the test loss keeps dropping slightly. The prediction performance of the trained CNN is illustrated in the next section where we combine the network with the driving controller.

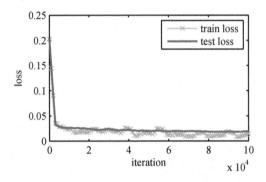

Fig. 3. Loss curves of CNN along iterations.

6   Combination of RL and DL in TORCS

Now we combine the visual perception trained by DL with the controller learned by RL, and apply them in TORCS to drive on the track of *CG track 3*[2]. CNN outputs are plotted in Fig. 4. For comparison, the ground-truth values are presented in the same figures. The predictions generally match the true values. But it is noted that the curves are not as smooth as those produced by the controller with the full access to driving data in Fig. 2. Some noticeable vibrations occur in d and a. This phenomenon is caused by CNN errors. There are small differences between the predicted values and the true values. Prediction errors disturb the controller to output right commands, and sometimes even make the car move to the opposite directions. Fortunately the prediction errors are small, so the car will not leave the track in spite of occasionally inaccurate commands.

---

[2] Video results are available in https://www.youtube.com/watch?v=hUpuE7qL5NQ.

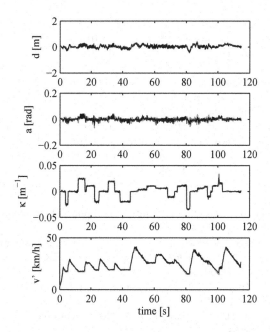

Fig. 4. Trajectories of predicted and true values in TORCS by vision-based robot. The black thin lines indicate the predicted values by CNN, while the green thick lines indicate the true values. (Color figure online)

7   Conclusion

In this paper, we first use a modified PILCO algorithm to learn a driving controller with full access to the TORCS engine. The algorithm learns a satisfactory controller with just several trials. Then we train a CNN to perceive driving data from images of first-person view in a supervised learning manner. After combining the two parts together, we get a vision-based robot for TORCS. It takes images and car's velocity as input, and drives the car well on the road.

Acknowledgments. This work is partly supported by the Beijing Science and Technology Plan under Grants Z181100008818075, National Natural Science Foundation of China (NSFC) under Grants No. 61603382, No. 61573353, No. 61533017, and the National Key Research and Development Program of China under Grant No. 2016YFB0101003.

References

1. Cardamone, L., Loiacono, D., Lanzi, P.L.: Learning drivers for TORCS through imitation using supervised methods. In: 2009 IEEE Symposium on Computational Intelligence and Games, pp. 148–155 (2009)
2. Cardamone, L., Loiacono, D., Lanzi, P.L.: On-line neuroevolution applied to the open racing car simulator. In: 2009 IEEE Congress on Evolutionary Computation, pp. 2622–2629 (2009)

3. Chen, C., Seff, A., Kornhauser, A., Xiao, J.: DeepDriving: learning affordance for direct perception in autonomous driving. In: 2015 IEEE International Conference on Computer Vision (ICCV), pp. 2722–2730 (2015)
4. Chen, Y., Zhao, D., Lv, L., Zhang, Q.: Multi-task learning for dangerous object detection in autonomous driving. Inf. Sci. **432**, 559–571 (2018)
5. Deisenroth, M.P.: Efficient Reinforcement Learning Using Gaussian Processes. KIT Scientific Publishing, Karlsruhe (2010)
6. Krizhevsky, A., Sutskever, I., Hinton, G.E.: ImageNet classification with deep convolutional neural networks. In: Pereira, F., Burges, C., Bottou, L., Weinberger, K. (eds.) Advances in Neural Information Processing Systems 25, pp. 1097–1105. Curran Associates, Inc., Red Hook (2012)
7. Lecun, Y., Bengio, Y., Hinton, G.: Deep learning. Nature **521**(7553), 436–444 (2015)
8. Loiacono, D., et al.: The 2009 simulated car racing championship. IEEE Trans. Comput. Intell. AI Games **2**(2), 131–147 (2010)
9. Loiacono, D., Prete, A., Lanzi, P.L., Cardamone, L.: Learning to overtake in TORCS using simple reinforcement learning. In: IEEE Congress on Evolutionary Computation, pp. 1–8 (2010)
10. Mnih, V., et al.: Human-level control through deep reinforcement learning. Nature **518**(7540), 529–533 (2015)
11. Mnih, V., et al.: Playing Atari with deep reinforcement learning. CoRR abs/1312.5602 (2013)
12. Muñoz, J., Gutierrez, G., Sanchis, A.: Controller for TORCS created by imitation. In: 2009 IEEE Symposium on Computational Intelligence and Games, pp. 271–278 (2009)
13. Muñoz, J., Gutierrez, G., Sanchis, A.: A human-like TORCS controller for the simulated car racing championship. In: Proceedings of the 2010 IEEE Conference on Computational Intelligence and Games, pp. 473–480 (2010)
14. Shao, K., Zhu, Y., Zhao, D.: Cooperative reinforcement learning for multiple units combat in StarCraft. In: 2017 IEEE Symposium Series on Computational Intelligence (SSCI), pp. 1–6 (2017)
15. Shao, K., Zhu, Y., Zhao, D.: StarCraft micromanagement with reinforcement learning and curriculum transfer learning. IEEE Trans. Emerg. Top. Comput. Intell. 1–12 (2018)
16. Sutton, R.S., Barto, A.G.: Reinforcement Learning: An Introduction. MIT Press, Cambridge (1998)
17. Wymann, B., Dimitrakakisy, C., Sumnery, A., Guionneauz, C.: TORCS: the open racing car simulator (2015)
18. Zhao, D., Wang, H., Shao, K., Zhu, Y.: Deep reinforcement learning with experience replay based on SARSA. In: 2016 IEEE Symposium Series on Computational Intelligence (SSCI), pp. 1–6 (2016)
19. Zhao, D., Zhu, Y.: MEC - a near-optimal online reinforcement learning algorithm for continuous deterministic systems. IEEE Trans. Neural Netw. Learn. Syst. **26**(2), 346–356 (2015)

An Adaptive Box-Normalization Stock Index Trading Strategy Based on Reinforcement Learning

Yingying Zhu, Hui Yang, Jianmin Jiang$^{(\boxtimes)}$, and Qiang Huang$^{(\boxtimes)}$

College of Computer Science and Software Engineering, Shenzhen University,
Shenzhen 518060, China
{jianmin.jiang,jameshq}@szu.edu.cn

Abstract. Financial time series prediction and stock trading strategy have always been the focus of research due to the generous returns. Stock box theory is a classic investment strategy, which has been studied by investors and scholars for many years. In this paper, we propose an adaptive box-normalization (ABN) stock trading strategy based on reinforcement learning (RL), which improves the original box theory. In our ABN strategy, the stock market data is independently normalized inside each oscillation box. Given the data of each box, support vector regression (SVR) is applied to predict the maximum rise range and maximum fall range within a certain period in the future. Meanwhile, the genetic algorithm (GA) is employed to optimize the input features of SVR via the mean square error (MSE) of prediction. We construct the trading strategies by Q-learning for the trading of single-stock and two-stock portfolio. Finally, the trigger threshold of oscillation box is dynamically adjusted according to the volatility of the stock price. Extensive experiments support that our proposed strategy performs well on different stock indices and achieves promising results.

Keywords: Stock prediction · Box theory · Reinforcement learning
Genetic algorithm · Asset allocation · Quantitative trading

1 Introduction

Stock market prediction has been an important issue in the fields of finance, engineering and mathematics due to its potential financial gain. There has been so much work done on ways to predict stock price [1,6]. With the great advancement of computer science, many recent works have utilized machine learning methods, such as neural networks (NN) [7], Bayesian approach [9] and support vector machine (SVM) [14], to analyze financial time series. Most work can generally be divided into two purposes. One is to forecast the future trend or exact price of stocks [5]. Another is to construct a special quantitative trading strategy with certain signals produced from other models [3].

© Springer Nature Switzerland AG 2018
L. Cheng et al. (Eds.): ICONIP 2018, LNCS 11303, pp. 335–346, 2018.
https://doi.org/10.1007/978-3-030-04182-3_30

Stock box theory is a classical quantitative trading strategy. It was first proposed in [11] and has been verified in the real stock market for many years. Afterwards, scholars carried out a lot of research about the box theory. Wen proposed an intelligent trading system by combining stock box theory and SVR [15]. Two SVR estimators are first utilized to make forecasts of the upper bound and lower bound of the oscillation box. Then a trading strategy based on the two bound forecasts is constructed to make trading decisions. With the development of deep learning, Zhu employed deep belief networks (DBN) to predict the upper and lower bounds of the oscillation box [17]. Lately, Zhang proposed a new status box method to predict stock trends [16]. The status box packages some stock points into three categories of boxes: up box, down box and flat box, which indicates different stock status. Then the specific trading strategy is formulated according to the status of these boxes.

It is worth mentioning that in the study of stock box theory, stock data is usually normalized globally in the time range of testing. But in terms of a specific box, the data range outside the box has slight effect on the prediction results. What's more, because of inflation or economic development, the price of stock index is rising year by year, which leads to a larger range of the stock data, and the price may break through the highest critical value in the real market. As a result, the data beyond the critical value can not be predicted well by prediction models. It brings difficulties to the prediction and practical application of box theory. In addition, the traditional box theory usually adopts fixed trigger threshold to produce trading signals. But when the box moving forward along time, the trigger threshold of the box is sensitive to different market conditions, which will lead to obvious differences in the results of transaction. Finally, different types of box trading strategies are usually made artificially. Many attempts and parameter debugging are needed, which takes up a lot of time.

To solve the problem above, We propose an adaptive box-normalization (ABN) stock index trading strategy, which generates trading strategy intelligently through reinforcement learning (RL). Different from the traditional box theory, we present a box-normalization method in each oscillation box, so that the prediction model, support vector regression (SVR), can pay more attention to the pattern and trend inside the box body. At the same time, we adjust the forecast targets to the maximum rise range (MRR) and maximum fall range (MFR) in the future time range. In the selection of indicators, we apply the genetic algorithm (GA) to optimize the input features of SVR. After determining the basic parameters of prediction model, we utilize Q-learning to generate the transaction strategy automatically based on the definition of the state of oscillation box, including the single-stock transaction and the asset allocation of two stock indices. Finally, we dynamically adjust the trigger threshold of the box according to the current volatility of price, so that the box can adapt to the market condition. Our ABN framework is proved to be effective on various stock indices and achieves spectacular returns. Some experiments are illustrated in this paper.

The remainder of this paper is structured as follows. In Sect. 2, we briefly review related work in stock box theory and RL. In Sect. 3, we describe the architecture and detailed design of the framework. Then the experiments and the corresponding analysis are shown in Sect. 4. Finally, some concluding remarks are drawn in Sect. 5.

2 Related Work

2.1 Stock Box Theory

The stock box theory is a powerful tool in quantitative investment. The basic idea of the stock box theory is that the stock price is supposed generally oscillates within a certain period of time, which is called oscillation box. The price will fall when it is close to the upper bound of the box and will rise when it close to the lower bound of the box. The essence of box theory is when the stock price effectively breaks the upper bound or the lower bound of the price box, the price will enter another oscillation box. It means the price will start an upward or downward trend in another box, so it is the high time to buy or sell the stock. However, the application of box theory is usually based on experience of investors. The difficulties are the way to identify the price box and how to make corresponding strategies according to the state of oscillation box.

Wen proposed an automatic decision support system combining box theory and SVR [15]. The SVR was used to make forecasts of the top and bottom of the oscillation box. Then the trading strategy based on box theory was constructed to make trading decisions. They investigated the performance of supposed system on individual stocks with different movement patterns. The experimental results showed a promising performance. In [17], researchers attempted to apply DBN to forecast the upper and lower bounds of the oscillating box and achieved promising results. In [16], stock data points were packaged in some successive status boxes based on the duration and oscillation of the initial turning points. These status boxes were classified into three categories: up box, down box and flat box, which represent the stock quotation being in the rising trend, falling trend and steady state in different time interval respectively.

2.2 Stock Trading Strategy Based on RL

RL is often employed to train the quantitative trading models. Common practices of RL include determining trading strategies and asset allocation strategies. However, stock investment is not a suitable application scenario for RL because investors' behavior will not affect the stock market and will not directly lead to the transfer of states. Researchers usually take two options to address this situation. One solution is to combine the states of asset with market conditions as an integral state of RL. In [10], the asset allocation strategy optimized with Q-learning is shown to be equivalent to a policy computed by dynamic programming. The approach is then tested on the task to invest liquid capital in the German stock market. In [8], a new stock trading method that incorporates dynamic asset allocation in a RL framework was presented. They utilized the temporal information from both stock recommendations and the ratio of the stock fund over the asset. Experimental results using the Korean stock market showed that the proposed method outperformed other fixed asset-allocation strategies. An alternative is to maximize only the immediate reward for each individual state. We only need a method to assign the best action to a specific situation. In [4], a decision support algorithm which used RL in order to improve the economic benefits of the basic seasonality strategy was proposed. Their states of RL included only the stock market situations and they only maximized the profit of each order to build their final strategy. In [13], researchers considered a two-asset personal retirement portfolio and proposed several RL agents for trading portfolio assets. They didn't take assets into the consideration of states as well.

3   Architecture of ABN Framework

The architecture of the ABN framework is first briefly outlined in this section. The block diagram of the framework is shown in Fig. 1. First, stock index data and various technical indicator data is normalized in each certain box, then the normalized box data is used as input features for SVR model to predict the MRR and MFR in the future period of time. At the same time, we apply GA to optimize the input features of SVR. Then, we design the states, actions and Reward of Q-learning according to the value and trend of MRR, and train the final trading strategy. Accordingly, the values of MFR are considered as stop loss signals for transactions. Finally, according to the volatility of stock price, we dynamically adjust the trigger threshold of the box to adapt to different market conditions.

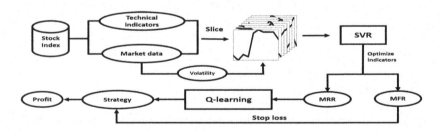

Fig. 1. The block diagram of ABN framework

3.1   Predict MRR and MFR with SVR

In the process of moving forward the oscillation box along time, the continuous price data and technical indicator data are normalized in each oscillation box, as shown in Fig. 2. The width of the box is identified as N, and the prediction time range is defined as M. The prediction values are MRR and MFR in the future M days. In this way of data preprocessing, we reorganize the input data of the prediction model in each box units.

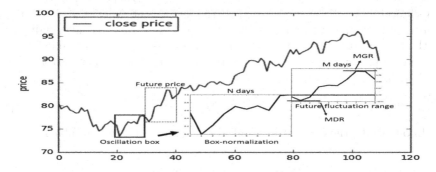

Fig. 2. Schematic diagram of box-normalization method

SVR is employed as the prediction model in our box-normalization method. SVR performs linear regression in the high dimension feature space using ϵ - insensitivity loss. At the same time, SVR strives to reduce model complexity by minimizing $||\omega||^2$. This can be described by introducing slack variables ξ_i and ξ_i^*, where $i = 1, ...n$, to measure the deviation of training sample outside ϵ - sensitive zone.

$$\frac{1}{2}||\omega||^2 + C\sum_{i=1}^{n}(\xi_i + \xi_i^*) \tag{1}$$

$$min \begin{cases} y_i - f(x_i, \omega) \leq \epsilon + \xi_i^* \\ f(x_i, \omega) - y_i \leq \epsilon + \xi_i \\ \xi_i, \xi_i^* \geq 0, i = 1...n \end{cases} \tag{2}$$

This optimization problem can transform into the dual problem and solution is given by

$$f(x) = \sum_{i=1}^{n_{sv}}(\alpha_i - \alpha_i^*)K(x_i, x) \tag{3}$$

$$Subject \quad to \quad 0 \leq \alpha_i^* \leq C, 0 \leq \alpha_i \leq C$$

Where n_{sv} is the number of support vectors and K is the kernel function. In order to prevent overfitting and enhance the generalization ability of the prediction model, we apply k-fold cross validation approach and grid search to optimize the meta-parameters C and g of SVR.

3.2   Optimize Input Feature with GA

The data in each oscillation box contains stock market data and technical indicator data. Research demonstrated that the technical indicators can improve the accuracy of the stock forecasts compared to that made with the original series of closing prices [12]. In many studies of stock price prediction, the choice of technical indicators is often based on the investment experience. There is no well-perform method that can measure the impact of various technical indicators on the forecast results. For this reason, we utilize GA to select suitable features in common-use indicators to minimize the prediction mean square error (MSE) of SVR. In our GA, the fitness function is defined as:

$$J = -log(P_{MSE}) \tag{4}$$

In our study, the candidate technical indicators are listed in Table 1:

Table 1. Candidate indicators

Indicator	Description	Indicator	Description
Open price	Daily open price	Close price	Daily close price
High price	Daily highest price	Low price	Daily lowest price
Volume	Daily turnover of stock	MACD	Moving average convergence and divergence
MA_6	6-day moving average	MA_{12}	12-day moving average
RSI_6	6-day relative strength index	RSI_{12}	12-day relative strength index
K	Stochastic index K	D	Stochastic index D

The chromosomes of GA are represented in binary code, and the structure of chromosome is shown in Fig. 3. Each bit represents one indicator. When the value of the gene is 1, the corresponding indicator is selected into the feature subset. Otherwise, the feature subset doesn't contain that corresponding indicator. We adopt roulette wheel selection method to choose the better father chromosomes. In the first step of our GA, we randomly generate a pre-defined number of chromosomes in the population of 20. The crossover rate and mutation rate are set at 0.9 and 0.1, respectively. The iteration of generations is limited at 20.

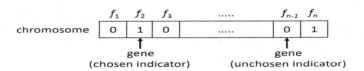

Fig. 3. Representation of chromosomes

3.3 Construct Trading Strategy Based on RL

After determining the definition of oscillation box and the parameters of SVR, a completed trading strategy should be constructed. Instead of designing trading algorithm artificially based on experience, we apply Q-learning to automatically construct trading strategy according to the states of oscillation box. RL emphasizes learning the best investment decision through trial and error. The essence of RL is the design of states, actions, and the selection of reward. In our research, the states of Q-learning are divided according to the characteristics of MRR predicted through each oscillation box. The tree structure in Fig. 4 shows our design of states.

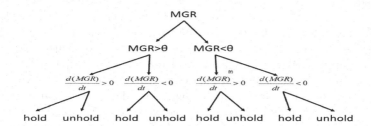

Fig. 4. State binary tree of Q-learning

In the figure above, θ is regarded as the threshold of MRR to provide important signals for the buying or selling points of the trading strategy. It is an important judgment for future trends from the pattern of box data. $\frac{d(MRR)}{dt}$ is the slope of MRR, which indicates consistent information about the future trend forecasting through the continuous box. 'Hold' indicates whether investors hold the stock index. The states of the oscillation box will transform only when the above three signals change. Therefore, the time that each state maintains during the movement of the oscillation box is uncertain.

The Q-learning of our design includes only three kinds of actions, buy, sell and sit. And only one share of stock index is traded at a time. The profit of the transaction is taken as the reward of our Q-learning. We always maximize only the immediate reward for each state in this trading example. As a result, the reward of each state will be maximized at the end of training.

As for the portfolio of two different stock indices, we extend a branch on the original state tree for each MRR of two indices. The actions of Q-learning change to the ratio of asset allocated to the two indices, exactly 0:0, 1:0, 0:1 and 1:1, while reward remains the same.

4 Experiment

In the hope of testing the feasibility of our ABN trading strategy, several typical experiments are implemented in this Section. Further more, the implementation platform is carried out via LibSVM expected to construct SVR models [2]. The basic experiments are developed on the S&P 500 Index ETF (SPY) data from 1993 to 2013, 80% as the training set and 20% as the test set.

4.1   Experiments with Constant Trigger Threshold of Box

In this section, we set the width of box to a fixed value and discuss the case of our ABN strategy when N is 10. First of all, in order to highlight the advantages of our box-normalization method, we compare it with the global normalization in the prediction stage. Figure 5 illustrates the regression effects of different normalization modes intuitively. In Fig. 5(a), when the training set and the test set are normalized as a whole, the data in the test set may exceed the critical value of the training set, which is very common in the stock market. This will cause the predictions fail to fit the data in the test set. As shown in Fig. 5(b), normalizing the training set and test set separately can improve this situation. But in practical applications, it is difficult to determine the normalization range of data in any time frame. Different time ranges will lead to different levels of data stretching, causing large deviations in continuous data. The box-normalization method can effectively avoid the occurrence of these two problem. As shown in Fig. 5(c), the box-normalization method only focuses on the data in the box to predict the future trend, and does not require other data assistance.

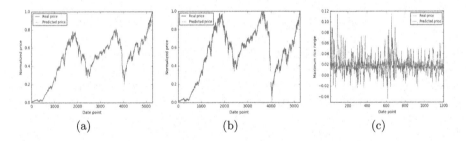

Fig. 5. Different methods of data normalization

The input features of SVR are optimized with different values of M. We count the number of times that different indicators are selected by GA. The results show that high price, close price, RSI_6 and $MACD$ are more important to the prediction of MSE than other indicators. So we use them as fixed input features of SVR in other experiments.

At the beginning of Q-learning, we set the random parameter γ to 100%, so that the selection of action in each state is completely random. During the training process, the γ gradually decrease to 10% with the number of iterations. We test the training situation when $N = 10, M = 10, \theta = 0.02$. As shown in Fig. 6(a), the y-axis shows the total returns at the end of the transaction and the x-axis shows the number of learning iterations. After 100 iterations, the number of total returns during the transaction tends to be stable and converges to a relatively large value. It indicates that the state-action pairs in the Q table are basically determined. Since there is still 10% randomness in trading simulation, it will have an indeterminate effect on the subsequent states and transaction results.

We use the trading strategy trained by Q-learning to conduct trading simulations on the training set and the test set. The results of the transaction are shown in Fig. 6(b) and (c). Our strategy earns significantly higher returns than a simple buy-and-hold strategy. In addition, the effect of parameter θ in the training set and test set on the final returns is shown in Fig. 6(d). It can be observed that the optimal values of θ in the training set and the test set are approximately the same, so we can determine the optimal value of θ in the training set without worrying about overfitting problem.

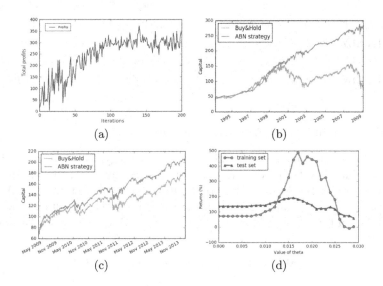

Fig. 6. Construct one-stock strategy by Q-learning

In the asset allocation problem of two stock indices, we collected market data from different industries in the United States from the year 1999 to 2018 to conduct our experiments. Figure 7(a) shows the price curves of the two indexes. Figure 7(b) and (c) show the returns of the strategy conducted on the training set and test set, where the strategy is learned using Q-learning. The y-axis represents the returns of the stock portfolio. Our strategy achieves promising returns on training sets and test sets.

The returns of different stock portfolios on the test set are listed in Table 2. In the table, we illustrate the description of the two stock indices, returns of each index, returns of equal asset for two indices, and the returns of our strategy. It can be noted that our strategy can achieve higher returns than holding two stock indices equally.

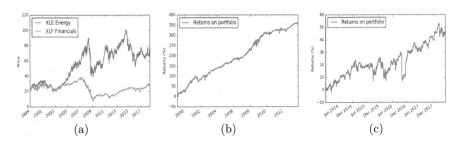

Fig. 7. Results of two-stock strategy

Table 2. Results on two-stock trading simulation

Portfolio	Description	Returns of XLE	Returns of index2	Average returns	Returns of ABN
XLE&XLF	Energy&Financials	−14.8%	30.9%	8.05%	45.1%
XLE&XLY	Energy&Cyclicals	−14.8%	59.5%	22.3%	31.9%
XLE&XLK	Energy&Technology	−14.8%	92.4%	38.8%	81.1%
XLE&XLV	Energy&Health	−14.8%	53.1%	19.1%	39.4%

4.2   Experiments with Adaptive Trigger

Prior to the improvement of the adaptive box strategy, we experimented with the best value of N for predicting different time ranges M on the training set, so as to minimize the MSE of the MRR within the predicted M days. And for different M values, trading simulations are carried out. The results are shown in Table 3. When the value of M is 10, the returns on SPY index reach the largest.

Table 3. Results on different values of M

Predict time range M	2	4	6	8	10	12	14	16	18	20
Optimal N	14	16	14	12	12	12	12	12	12	12
MSE (10^{-4})	1.77	2.41	3.06	3.63	4.12	4.59	5.13	5.72	6.35	6.95
Returns (%)	218.4	226.6	217.8	238.9	**244.2**	240.6	233.0	214.7	209.1	204.1

The volatility of the stock price is calculated in a quarterly cycle, and the volatility is expressed by the mean square deviation of price data. The change in the volatility of SPY is shown in Fig. 8(a). Usually, when the volatility of stock price is large, a more frequent trading strategy should be adopted. Conversely, in the case of a relatively stable stock price, take a low-frequency trading strategy. So when the volatility of stock price is lower than 0.02, we make the trigger threshold θ of oscillating box 0.17; when the volatility is higher than 0.02, θ

takes 0.14. Figure 8(b) shows the capital curve for our ABN strategy tested on SPY. The purple line in the figure is the return of the classic dual moving average (DMA) strategy. Our strategy gains significantly higher returns than the DMA strategy.

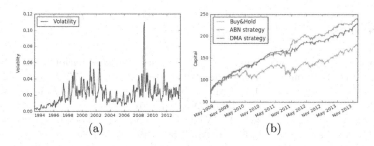

(a)   (b)

Fig. 8. Stock volatility and ABN trading simulation

Finally, we use the forecasted MFR value to determine stop loss and reduce the drawdown of capital during the transaction. When MFR is less than a specified threshold, if the index is held, it will be sold immediately. We utilize the final ABN strategy to conduct trading simulation on multiple stock indices, and the results are shown in Table 4. It demonstrates whether it is a single-stock or two-stock index transaction, our strategy can obtain relatively high returns.

Table 4. Trading simulations on other indices use ABN strategy

Index	Return of Buy&Hold	Return of ABN strategy	Portfolio	Average return	Return of ABN strategy
QQQ	265.4%	335.4%	XLE&XLF	8.05%	56.7%
SMH	196.2%	368.0%	XLE&XLY	22.3%	47.6%
EFA	119.6%	298.9%	XLE&XLK	38.8%	90.7%
IWM	231.4%	366.5%	XLE&XLF	19.1%	66.0%

5   Conclusion

In this paper, we improve the classic stock box theory and propose an ABN stock trading strategy. Compared with the global normalization method, our box normalization method illustrates better results for MRR and MFR predictions. After optimizing the input features of SVR by GA, we utilize Q-learning to construct trading strategies, including the strategy of single-stock and two-stock portfolio. After establishing the basic architecture of our strategy, we dynamically adjust the width of oscillation box based on the volatility of stock price. Finally, our ABN strategy can obtain generous returns in the trading of single-stock and two-stock portfolio.

Acknowledgements. This work was supported by: (i) National Natural Science Foundation of China (Grant No. 61602314); (ii) The Natural Science Foundation of Guangdong Province of China (Grant No. 2016A030313043); (iii) Fundamental Research Project in the Science and Technology Plan of Shenzhen (Grant No. JCYJ20160331114551175).

References

1. Balvers, R., Wu, Y., Gilliland, E.: Mean reversion across national stock markets and parametric contrarian investment strategies. J. Financ. **55**(2), 745–772 (2000)
2. Chang, C.C., Lin, C.J.: LibSVM: a library for support vector machines. ACM Trans. Intell. Syst. Technol. (TIST) **2**(3), 27 (2011)
3. Chiang, W.C., Enke, D., Wu, T., Wang, R.: An adaptive stock index trading decision support system. Expert Syst. Appl. **59**, 195–207 (2016)
4. Eilers, D., Dunis, C.L., Mettenheim, H.J.V., Breitner, M.H.: Intelligent trading of seasonal effects: a decision support algorithm based on reinforcement learning. Decis. Support. Syst. **64**(3), 100–108 (2014)
5. Enke, D., Thawornwong, S.: The use of data mining and neural networks for forecasting stock market returns. Expert Syst. Appl. **29**(4), 927–940 (2005)
6. Fama, E.F.: The behavior of stock-market prices. J. Bus. **38**(1), 34–105 (1965)
7. Guresen, E., Kayakutlu, G., Daim, T.U.: Using artificial neural network models in stock market index prediction. Expert Syst. Appl. **38**(8), 10389–10397 (2011)
8. Jangmin, O., Lee, J., Lee, J.W., Zhang, B.T.: Adaptive stock trading with dynamic asset allocation using reinforcement learning. Inf. Sci. **176**(15), 2121–2147 (2006)
9. Malagrino, L.S., Roman, N.T., Monteiro, A.M.: Forecasting stock market index daily direction: a Bayesian network approach. Expert Syst. Appl. **105**, 11–22 (2018)
10. Neuneier, R.: Optimal asset allocation using adaptive dynamic programming. In: Advances in Neural Information Processing Systems, pp. 952–958 (1995)
11. Nicolas, D.: How I Made Two Million Dollars in the Stock Market. BN Publishing (2007)
12. Oriani, F.B., Coelho, G.P.: Evaluating the impact of technical indicators on stock forecasting. In: 2016 IEEE Symposium Series on Computational Intelligence (SSCI), pp. 1–8. IEEE (2016)
13. Pendharkar, P.C., Cusatis, P.: Trading financial indices with reinforcement learning agents. Expert Syst. Appl. **103**, 1–13 (2018)
14. Wang, G.J., School, B.: Time series forecast of stock price based on the PSO-LSSVM predict model. Sci. Technol. Ind. **10**, 135–140 (2017)
15. Wen, Q., Yang, Z., Song, Y., Jia, P.: Automatic stock decision support system based on box theory and SVM algorithm. Expert Syst. Appl. **37**(2), 1015–1022 (2010)
16. Zhang, X.D., Li, A., Pan, R.: Stock trend prediction based on a new status box method and AdaBoost probabilistic support vector machine. Appl. Soft Comput. **49**, 385–398 (2016)
17. Zhu, C., Yin, J., Li, Q.: A stock decision support system based on DBNs. J. Comput. Inf. Syst. **10**(2), 67–79 (2014)

Heterogeneous Multi-task Learning of Evaluation Functions for Chess and Shogi

Shanchuan Wan[1(✉)] and Tomoyuki Kaneko[1,2]

[1] The University of Tokyo, Tokyo, Japan
{swan,kaneko}@graco.c.u-tokyo.ac.jp
[2] PRESTO, Japan Science and Technology Agency, Kawaguchi, Japan

Abstract. Using advanced deep learning methods, artificial intelligence is able to achieve unprecedented high performance in playing complex board games. However, in conventional practice, models for different games require separate training with domain-specific datasets, which is not conducive to enable the full use of the correlation between tasks and may cause unnecessary consumption of computing resources. This paper presents a novel multi-task learning framework for the training of deep-convolutional-neural-network-based evaluation functions for two heterogeneous but related games – chess and shogi. Experimental results show that the application of the proposed framework improved the prediction accuracy for both networks with limited training steps.

Keywords: Neural networks · Multi-task learning · Computer chess

1 Introduction

For many years, researchers have been committed to developing and improving artificial intelligence techniques for popular board games including chess, shogi (a Japanese chess game) and Go. Modern game-playing programs are commonly driven by an elaborated game tree search algorithm with a high-accuracy evaluation function, which is the main learning objective of this paper.

In early studies, evaluation functions were simply linear combinations of hand-crafted features. With the development of computing devices and machine learning techniques, accurate evaluation functions composed of deep neural networks have been introduced into game programming and achieved great success. Through the training with millions of self-play games, deep-convolutional-neural-network-based (DCNN-based) computer programs have already been proven to be capable of beating the best human players in the world [1].

However, the application of deep neural networks also introduces huge training time and computational costs. It is a common practice to conduct training processes for different networks individually, but this is not always the optimal approach. When high-performance networks are required for solving related

© Springer Nature Switzerland AG 2018
L. Cheng et al. (Eds.): ICONIP 2018, LNCS 11303, pp. 347–358, 2018.
https://doi.org/10.1007/978-3-030-04182-3_31

tasks in multiple domains, the training costs increase at least linearly along with the increasing number of tasks. A feasible solution to control training costs while improving evaluation accuracy is to apply joint learning to evaluation functions to fully exploit the inner relationship among multiple games. Through this type of training, knowledge and expertise can be directly exchanged via common features, while the over-fitting problem of deep networks can be further alleviated.

In this context, we present a novel multi-task learning framework, where two DCNN-based evaluation functions for chess and shogi share a part of convolutional layers and are trained simultaneously. Experimental results demonstrate that the proposed framework improves the prediction accuracy for both networks compared with those trained individually.

2 Related Work

2.1 Neural Networks and Board Games

In recent studies, the terms "value network" and "policy network" have been used to denote neural-network-based (NN-based) evaluation and move prediction functions, respectively. The value network is trained to predict the winner of a specific game position, and the policy network learns to predict an expert move in supervised learning or a self-play move in reinforcement learning. The two networks together constitute the core of modern game-playing programs.

Silver et al. demonstrated through the success of AlphaGo [2] that the introduction of DCNN-based policy networks is beneficial for implementing a high-quality game simulation, and that of value networks is beneficial for acquiring a more consistent and comprehensive understanding towards every single position in Go. In their later studies of AlphaGo Zero [1] and AlphaZero [3], parameters of the two networks were incorporated into a jointly trained multi-task model and succeeded in achieving a higher prediction accuracy in less training time.

There have also been many other valuable studies focused on applying neural networks to board games. Lai developed the chess program Giraffe [4] using a modified version of the TD-LEAF(λ) algorithm [5] and NN-based evaluation functions, and David et al. trained a multi-layer perceptron network in their DeepChess [6] program to select the better of two candidate positions. However, to the best of our knowledge, the current studies on board games have progressed independently, and no successful multi-game learning is available for reference.

Among the widely known board games, chess shares many key elements in definitions and playing rules with its variants, which suggests that their evaluation/prediction functions share common knowledge, and conducting joint training could potentially lead to positive effects in both functions. Moreover, considering that evaluation functions directly determine the quality of search results and have a uniform output format to enable more common features to be shared, we present a cross-domain multi-task learning framework for training the evaluation functions (value networks) for chess and shogi, and conducted preliminary experiments to examine the effectiveness of the proposed framework.

2.2   Uniformity Regularization

The uniformity regularization (UR) network was originally proposed in our previous work [7] for improving the prediction accuracy of evaluation functions,
while in this paper, an extended version of the method is applied to a novel
joint learning framework, and is essential for extracting common features from
multiple games.

On the basis of a plain value network composed of a DCNN-based feature
extractor and a fully connected output layer, UR essentially adds another output
layer that shares the same feature extractor, as shown in Fig. 1. The original output layer is trained to give an evaluation score towards a certain position, while
the newly added layer is trained to discriminate if two successive positions form a
transition caused by an expert move. Then, through a uniformity regularization
loss applied to the output values of the discriminator, both the extracted features
and the evaluation scores of the value network can be regularized simultaneously.

In this paper, we also adopted this basic network structure and extended
the uniformity regularization loss to enable it to facilitate extracting common
features from multiple task domains.

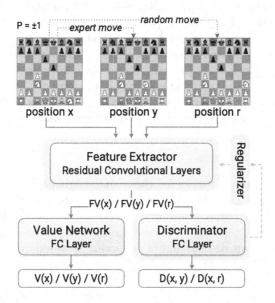

Fig. 1. Value network with uniformity regularization

2.3   Multi-task Learning

According to the format of input data, multi-task learning (MTL) tasks can be
classified into homogeneous and heterogeneous tasks.

The training and test data in **homogeneous MTL** have a uniform feature
domain, and may be processed by multiple feature extractors or output layers

to solve multiple tasks. Past experience shows that applying homogeneous MTL can be advantageous in terms of predictive performance relative to learning these tasks independently [8]. All the methods adopted in AlphaGo Zero [1], Alpha Zero [3] and the original UR network [7] are classified as homogeneous MTL.

The data in **heterogeneous MTL** may originate from various task domains, with significantly varied feature representations. Usually, without an appropriate feature transfer, the predictive performance improved through joint learning is negligible or negative, even though there are strong correlations between the tasks themselves. This presents a challenge for researchers conducting MTL among different games.

3 Learning Framework

3.1 Data Representation

We utilized the design of DCNNs in Alpha Zero [3]. Training positions in chess and shogi are randomly selected from preset game records and encoded as 13-channel and 43-channel images, respectively. Each of the first 12 and 28 channels corresponds to a type of piece. All pixels are set as 0 by default. For each grid that contains a piece, the corresponding pixel of that channel is set as 1, otherwise. The 29th to 42nd channels in shogi data represent prisoner pieces. The 13th and the 43rd channels in the two types of data denote the next player to move.

Since the two games have different board sizes, to unify the dimensions of intermediate-level feature maps, 1 extra row and 1 extra column with blank data are inserted into the left and the bottom borders of the chess board, respectively. Thus, the image size in every channel of input data is always 9×9.

3.2 Network Architecture

As shown in Fig. 2, the networks for chess and shogi share an identical design. Both networks accept a multi-channel image as input, automatically extract features through 32 standard convolutional residual layers [9] and a global average pooling layer, and output move discrimination or position evaluation results through local and joint discriminators and value networks, which are made up of a fully-connected layer. All hidden layers are activated by rectified linear unit functions. The output of the local and joint discriminators is constrained into the range $[-1, 1]$ by a $\tanh(x)$, while no constraint is applied to the output of the value networks.

All the 32 convolutional layers are equally divided into 3 main stages – low-level, intermediate-level and high-level feature extraction. We let the parameters of the intermediate layers be shared by the two networks, ensuring common features can be effectively exchanged.

In this paper, we defined 3 types of experimental models, the details about their network architectures are listed in Table 1.

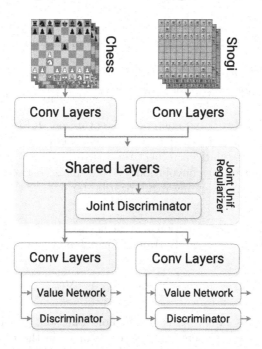

Fig. 2. Multi-task learning framework for chess and shogi

3.3 Learning Objectives

Training data are randomly sampled from game records and pre-processed, so that every instance in the training dataset contains a quadruple $\langle x, y, r, p \rangle$. x is a game position randomly selected from the game records. y and r are the subsequent positions after a recorded move (e.g., an expert move in supervised training) and a random but legal move respectively. p represents the player to move at the turn of position x, where $p = 1$ and -1 for the first and the second player, respectively.

Network models for chess and shogi share identical architecture and objective functions, and are trained simultaneously by using the same momentum optimizers. We also attempted to train all parameters alternatively by using multiple optimizers in previous experiments, in which high-speed training could be temporarily realized – top-1 accuracy increased over 0.27% and 0.41% with 100 million training instances in chess and shogi, respectively – but the model eventually converged to a relatively poor accuracy due to unstable gradients and the abuse of duplicate data.

The learning objectives for each game are defined as follows.

Overall Objective. The network parameters for all games are simultaneously optimized according to an integrated objective function $J_{\mathrm{MTL}}(\theta)$.

$$
\begin{aligned}
J_{\mathrm{MTL}}(\theta) =& J_{\mathrm{V}}^{\mathrm{chess}}(\theta) + J_{\mathrm{D}}^{\mathrm{chess}}(\theta) + J_{\mathrm{UR}}^{\mathrm{chess}}(\theta) + J_{\mathrm{JD}}^{\mathrm{chess}}(\theta) + J_{\mathrm{JUR}}^{\mathrm{chess}}(\theta) \\
&+ J_{\mathrm{V}}^{\mathrm{shogi}}(\theta) + J_{\mathrm{D}}^{\mathrm{shogi}}(\theta) + J_{\mathrm{UR}}^{\mathrm{shogi}}(\theta) + J_{\mathrm{JD}}^{\mathrm{shogi}}(\theta) + J_{\mathrm{JUR}}^{\mathrm{shogi}}(\theta),
\end{aligned}
\tag{1}
$$

where θ denote the network parameters to be optimized, $J_{\mathrm{V}}(\theta)$, $J_{\mathrm{D}}(\theta)$ and $J_{\mathrm{UR}}(\theta)$ are the loss functions for the value network, the local discriminator network, and the local uniformity regularizer, respectively. $J_{\mathrm{JD}}(\theta)$ is for the joint discriminator added to the output of the convolutional layers shared between the two games, and $J_{\mathrm{JUR}}(\theta)$ is its corresponding regularization term.

Value Network. The loss function of value networks is defined as the following. All value networks are separately trained for different games with independent parameters.

$$
J_{\mathrm{V}}(\theta) = \frac{1}{n} \sum_{i=1}^{n} \sigma(p_i(\mathrm{V}(r_i) - \mathrm{V}(y_i))),
\tag{2}
$$

where $\sigma(x) = 1/(1 + \exp(-x))$ is the sigmoid function, n is the batch size of training data, and p_i represents the player to move in the ith instance. Equation (2) is based on the objective function in the training of the evaluation function of the shogi program *Bonanza* [10], which won first place at the World Computer Shogi Championship in 2006 and 2013. This function encourages that the value of the position after an expert move $\mathrm{V}(y_i)$ should be better than that after a random move $\mathrm{V}(r_i)$. This function is also used in comparison training in chess [11].

Discriminator (Local). The loss function of the discriminator network is defined as the difference between the certainty of a move played in a real game record y and that of a random move r at position x. A local discriminator has its independent parameters, and is trained exclusively for only one game.

$$
J_{\mathrm{D}}(\theta) = \frac{1}{n} \sum_{i=1}^{n} [\mathrm{D}(\langle x_i, y_i \rangle) - \mathrm{D}(\langle x_i, r_i \rangle)) - 2]^2,
\tag{3}
$$

where $\langle x_i, y_i, r_i \rangle$ represents the ith instance in that batch. The minimization of Eq. (3) encourages $\mathrm{D}(\langle x_i, y_i \rangle)$ to be 1 and $\mathrm{D}(\langle x_i, r_i \rangle))$ to be -1.

Discriminator (Joint). Besides the local discriminator, there is a newly added joint discriminator (JD) for regularizing the output of the shared convolutional layers, which also adopts Eq. (3) as its objective function. Different with local discriminators, the output of the joint discriminator is denoted as $\mathrm{JD}(x)$. JD processes different data in the two games with the same network parameters.

Uniformity Regularizer (Local). The loss function of the local uniformity regularizer for each game is defined based on reducing the difference in the output results of the local discriminator network. Local uniformity regularizes are trained separately for each game, without shared parameters.

$$J_{\text{UR}}(\theta) = \frac{1}{n} \sum_{i=1}^{n} [(\text{D}(\langle x_i, y_i \rangle) - \overline{\text{D}}(\langle \mathbf{X}, \mathbf{Y} \rangle))^2 \\ + (\text{D}(\langle x_i, r_i \rangle) - \overline{\text{D}}(\langle \mathbf{X}, \mathbf{R} \rangle))^2], \tag{4}$$

where $\langle \mathbf{X}, \mathbf{Y} \rangle$ and $\langle \mathbf{X}, \mathbf{R} \rangle$ denote 2 sets of moves that contain all expert and random moves trained in past training steps. $\overline{\text{D}}(\langle \mathbf{X}, \mathbf{Y} \rangle)$ and $\overline{\text{D}}(\langle \mathbf{X}, \mathbf{R} \rangle)$ are the exponential moving average (with an initial value 0 and a decrease coefficient $\alpha = 0.999$ in this paper) of the discriminator's output with respect to expert and random moves, respectively. Equation (4) is a stable version of the original UR [7]. Since all instances in training batches have been randomly shuffled in advance, Eq. (4) is equivalent to measuring and minimizing the difference between the discrimination results of two randomly sampled positive (negative) samples, aiming to reduce the variance of discrimination results and maintain the consistency of extracted features.

Uniformity Regularizer (Joint). The loss function the joint uniformity regularizer (JUR) is basically identical to its local version, but its parameters are shared with networks for both games. The learning objective of the JUR is to indirectly unify all extracted intermediate-level features in different games by regularizing the discrimination results of the JD.

$$J_{\text{JUR}}(\theta) = \frac{1}{n} \sum_{i=1}^{n} [(\text{JD}(\langle x_i, y_i \rangle) - \overline{\text{JD}}(\langle \mathbf{X}, \mathbf{Y} \rangle))^2 \\ + (\text{JD}(\langle x_i, r_i \rangle) - \overline{\text{JD}}(\langle \mathbf{X}, \mathbf{R} \rangle))^2], \tag{5}$$

where $\overline{\text{JD}}(\langle \mathbf{X}, \mathbf{Y} \rangle)$ and $\overline{\text{JD}}(\langle \mathbf{X}, \mathbf{R} \rangle)$ are 2 exponential moving average values of the joint discriminator's output with respect to expert and random moves recently trained, as defined in Eq. (4).

In heterogeneous MTL, feature representations can hardly be efficiently unified by simply sharing the network parameters. To solve this problem, we apply Eq. (5) to enable the features extracted in previous convolutional layers to be uniformly discriminated no matter what game they come from, which aims to shorten the distance between different feature spaces.

4 Experiments

4.1 Datasets

We collected 783,129 games from the computer chess database CCRL 40/40[1] on January 14, 2018 and 868,161 games from the computer shogi server Floodgate[2]

---

[1] http://www.computerchess.org.uk/ccrl/4040.
[2] http://wdoor.c.u-tokyo.ac.jp/shogi/index-e.html.

on December 31, 2017. In total, 750,000 game records were randomly selected from each database and pre-processed for training, and the remaining were used for testing. Only moves made by the winner were adopted in the training and testing datasets. All positions in the datasets were randomly shuffled in advance.

4.2   Training Configurations

Experimental models were trained in 4 processes on a single machine with two NVIDIA 1080Ti GPUs. A stochastic gradient descent optimizer with a momentum rate of 0.9 was used. The learning rate started at 0.001 and decays every 10,000 steps with a base of 0.99. The batch size n was set as 128. The weights in neural networks were randomly initialized by a normal distribution with mean 0 and variance 1.

4.3   Evaluation Metrics

The **top-k accuracy** of predicting expert moves was adopted to evaluate the performance of a value network in our experiments. Specifically, for a given position x and its subsequent position y after an expert move, the output of a value network was accurate if $V(y)$ was in the k highest scores among all subsequent positions of x when $p = 1$, or in the k lowest scores when $p = -1$.

4.4   Models

We defined 3 types of experimental models based on the architecture shown in Fig. 2, with the model size of A $\approx$ B < C. The definitions of the feature extraction parts in 3 types of models are listed in Table 1. The number of neurons in the value, discriminator networks of each model was the output size shown in the table. The input size of the joint discriminator equaled to the number of channels in the last shared layer.

The parameters from the 9th to the 24th layer in type A, and the parameters from the 13th to the 24th layer in type B and C were **shared** for chess and shogi, while the first and the last several layers remained **unshared** to extract game-specific information. There were an additional global average pooling layer and an output head for the joint discriminator and joint uniformity regularizer after the 24th layer in type A, B, and C.

The first several unshared layers in every type were for extracting low-level features from the input game board, and unifying the feature representation for different games. The input size of the first parameter-sharing layer was 5×5 in A, and 3×3 in B and C. Receiving a smaller feature map as the input of shared layers in B and C was considered as beneficial for the unification of feature representation. We set the number of output channels for type A, B, and C to be 128, 256, and 384, respectively. The number of channels in every layer of type C was twice of the number in B. This suggests that type B and C should have a better ability to extract high-level features from the output of shared layers, and the feature extraction ability of C should be further better than B.

Table 1. Details of feature extraction in experimental models

	Layer	Type A	Type B	Type C
	input	[H=9, W=9, C=13] (chess), [9, 9, 43] (shogi)		
convolutional layers	$1^{st} - 5^{th}$	[K=3 × 3, S=1, C=128]		[3 × 3, 1, 192]
	6th	[3 × 3, 1, 128]	[3 × 3, 2, 128]	[3 × 3, 2, 192]
	$7^{th} - 9^{th}$		[3 × 3, 1, 128]	[3 × 3, 1, 192]
	10th	[3 × 3, 2, 128]	[3 × 3, 2, 128]	[3 × 3, 2, 192]
	$11^{th} - 17^{th}$	[3 × 3, 1, 128]	[3 × 3, 1, 128]	[3 × 3, 1, 192]
	18th	[3 × 3, 2, 128]		
	$19^{th} - 24^{th}$	[3 × 3, 1, 128]		
	bypass to	global_avg_pool → flattening → joint discriminator		
	25th	[3 × 3, 1, 128]	[3 × 3, 1, 128]	[3 × 3, 1, 192]
	$26^{th} - 32^{nd}$		[3 × 3, 1, 256]	[3 × 3, 1, 384]
	pooling	global average pooling		
	flattening	[1, 1, 128]	[1, 1, 256]	[1, 1, 384]
	output to	value network, local discriminator		

For every type, 2 models were trained for comparison.

- **Baseline** models were trained with a local UR network for playing a single game (in Fig. 1) without any shared parameters between different games.
- **Proposed** models were trained with the proposed MTL framework, in which a part of parameters was shared for two games and regularized by the proposed joint uniformity regularizer in Eq. (5).

4.5   Results

There were 6 experimental models trained with 200 million instances (600 million game positions) for every game. Their prediction accuracies and training curve are shown in Table 2 and Fig. 3. The prediction accuracies and loss values were recorded every 3 min, and smoothed in the figures with an exponential moving average (decrease coefficient $\alpha = 0.99$).

It can be found from the table that the proposed heterogeneous MTL framework performed well with type-B and type-C models, but was not compatible with type-A models. The results suggest that the effects of the MTL of multiple games were highly related to the network architecture. To take full advantage of the proposed framework, it is important to apply it to a larger model and insert sufficient unshared layers before and after the shared parameters.

After applying the proposed framework, the training of networks was accelerated, and the top-1 accuracies of predicting expert moves in both chess and shogi were improved in type-B and type-C models. To a better understanding of

356   S. Wan and T. Kaneko

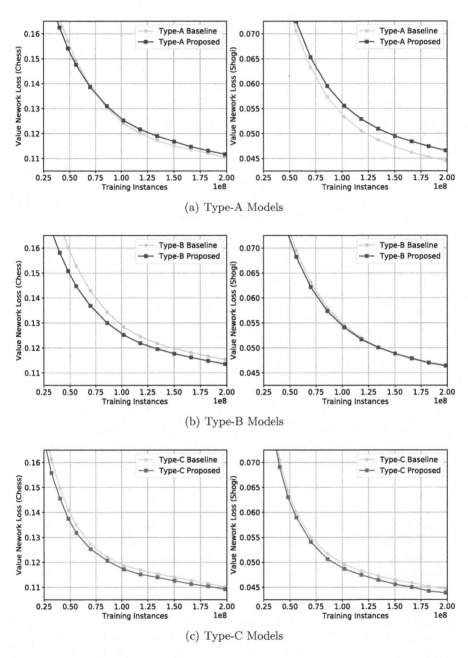

(a) Type-A Models

(b) Type-B Models

(c) Type-C Models

Fig. 3. Validation loss of value networks in experimental models (left and right models are trained for chess and shogi, respectively)

Table 2. Prediction accuracies of experimental models

Accuracy	Model	Chess			Shogi		
		Type-A	Type-B	Type-C	Type-A	Type-B	Type-C
Top-1	Baseline	**37.17%**	36.09%	36.88%	**43.40%**	42.42%	43.10%
	Proposed	36.85%	**36.50%**	**37.10%**	42.40%	**42.46%**	**43.63%**
	Δ	−0.32%	+0.41%	+0.22%	−1.01%	+0.03%	+0.52%
Top-5	Baseline	**81.28%**	80.18%	81.02%	**84.60%**	**83.80%**	84.30%
	Proposed	80.81%	**80.49%**	**81.30%**	83.73%	83.69%	**84.67%**
	Δ	−0.47%	+0.30%	+0.28%	−0.87%	−0.11%	+0.37%

Table 3. Top-1 prediction accuracies of stockfish

Nodes	Top-1 Acc	Δ	Nodes	Top-1 Acc	Δ
≤1,000	32.96%	−	≤6,000	39.94%	+0.78%
≤2,000	35.22%	+2.26%	≤7,000	40.72%	+0.78%
≤3,000	36.93%	+1.71%	≤8,000	41.19%	+0.47%
≤4,000	38.15%	+1.21%	≤9,000	41.62%	+0.43%
≤5,000	39.16%	+1.01%	≤10,000	42.12%	+0.50%

the experimental results, we tested the performance of *Stockfish* [12] – one of the most advanced open-source chess programs – with about 333,000 positions randomly sampled from the same test dataset as the other experimental networks. The top-1 accuracies of stockfish were measured by comparing its "bestmove" with each expert move in the test dataset, through the universal chess interface (UCI) protocol implemented by python-chess[3]. The results are summarized in Table 3. Since the accuracy of our network for chess listed in Table 2 is compatible with the test dataset sampled for *Stockfish*, our chess evaluation function reached a decent performance even without search. The top-1 accuracies in type-B and type-C models increased from 36.09% and 36.88% to 36.50% and 37.10% were approximately equivalent to searching extra **9.5%** (2,507 → 2,746) and **5.7%** (2,970 → 3,140) nodes in the game tree of *Stockfish*, respectively.

We also investigated the effects of adding an additional discriminator to a part of convolutional layers for local UR networks as the joint UR, but no significant accuracy increase could be observed. This test result suggested that simply enlarging the scale of loss functions and adding joint learning tasks were not always effective in heterogeneous MTL, and the accuracies increased in the above experiments were actually benefited from the partially shared feature extraction and the joint uniformity regularization in the proposed framework.

---

[3] https://python-chess.readthedocs.io/en/v0.22.1/index.html.

5   Conclusion

Training deep neural networks for games requires a lot of computing resources. At this time, there are still a few studies on how to improve training efficiency by making full use of common knowledge among related games. We proposed a novel heterogeneous multi-task learning (MTL) framework in this paper, and conducted experiments of jointly trained evaluation functions for playing chess and shogi. Preliminary experimental results demonstrated that the proposed framework is beneficial for accelerating convergence rates and improving the performances of both games with limited training costs. However, the networks we tested in this paper were still not the optimal architecture for the MTL. In our future work, we are planning to conduct comparison experiments systematically to figure out the application scope of the proposed framework and the optimal configuration for it.

Acknowledgement. A part of this work was supported by JSPS KAKENHI Grant Number 16H02927 and by JST, PRESTO.

References

1. Silver, D., et al.: Mastering the game of go without human knowledge. Nature **550**(7676), 354 (2017)
2. Silver, D., et al.: Mastering the game of go with deep neural networks and tree search. Nature **529**(7587), 484–489 (2016)
3. Silver, D., et al.: Mastering chess and shogi by self-play with a general reinforcement learning algorithm. arXiv preprint arXiv:1712.01815 (2017)
4. Lai, M.: Giraffe: using deep reinforcement learning to play chess. arXiv preprint arXiv:1509.01549 (2015)
5. Baxter, J., Tridgell, A., Weaver, L.: Learning to play chess using temporal-differences. Mach. Learn. **40**(3), 242–263 (2000)
6. David, O.E., Netanyahu, N.S., Wolf, L.: DeepChess: end-to-end deep neural network for automatic learning in chess. In: Villa, A.E.P., Masulli, P., Pons Rivero, A.J. (eds.) ICANN 2016. LNCS, vol. 9887, pp. 88–96. Springer, Cham (2016). https://doi.org/10.1007/978-3-319-44781-0_11
7. Wan, S., Kaneko, T.: Building evaluation functions for chess and shogi with uniformity regularization networks. In: IEEE Conference on Computational Intelligence and Games (2018)
8. Evgeniou, T., Pontil, M.: Regularized multi-task learning. In: Proceedings of the Tenth ACM SIGKDD International Conference on Knowledge Discovery and Data Mining, pp. 109–117. ACM (2004)
9. He, K., Zhang, X., Ren, S., Sun, J.: Deep residual learning for image recognition. In: Proceedings of the IEEE Conference on Computer Vision and Pattern Recognition, pp. 770–778 (2016)
10. Hoki, K., Kaneko, T.: Large-scale optimization for evaluation functions with minimax search. Mach. Learn. **49**, 527–568 (2014)
11. Tesauro, G.: Comparison training of chess evaluation functions. In: Machines that Learn to Play Games, pp. 117–130. Nova Science Publishers, Inc. (2001)
12. Romstad, T., Costalba, M., Kiiski, J.: Stockfish: A Strong Open Source Chess Engine (2018). https://stockfishchess.org

Reinforcement Learning Based Dialogue Management Strategy

Tulika Saha$^{(\boxtimes)}$, Dhawal Gupta, Sriparna Saha, and Pushpak Bhattacharyya

Department of Computer Science and Engineering,
Indian Institute of Technology Patna, Patna, India
sahatulika15@gmail.com, dhawal.gupta.iitp@gmail.com,
sriparna.saha@gmail.com, pushpakbh@gmail.com

Abstract. This paper proposes a novel Markov Decision Process (MDP) to solve the problem of learning an optimal strategy by a Dialogue Manager for a flight enquiry system. A unique representation of state is presented followed by a relevant action set and a reward model which is specific to different time-steps. Different Reinforcement Learning (RL) algorithms based on classical methods and Deep Learning techniques have been implemented for the execution of the Dialogue Management component. To establish the robustness of the system, existing Slot-Filling (SF) module has been integrated with the system. The system can still generate valid responses to act sensibly even if the SF module falters. The experimental results indicate that the proposed MDP and the system hold promise to be scalable across satisfying the intent of the user.

Keywords: Dialogue management strategy · Reinforcement learning
Markov decision process · Slot-filling

1 Introduction

Dialogue systems are characterized as chat bots with which humans interact on a turn-by-turn basis wherein natural language plays an essential role in the communication [6]. With the massive development in the field of Natural Language Understanding (NLU), it has now become feasible to develop dialogue systems for many task oriented applications. The role of the dialogue management component in such application oriented systems is to interact with the user in a way that helps the user to complete the task which the system is meant to deliver. The prime objective of the dialogue management module is to manage the progress of a conversation which trivially involves the following tasks [11]: (i) to elicit necessary information from the user and to determine whether the information obtained is adequate enough to facilitate communication with an external application, (ii) communicating with an external application such as a database to retrieve information that is to be communicated to the user, (iii) presenting the information to the user which is retrieved based on the data elicited.

© Springer Nature Switzerland AG 2018
L. Cheng et al. (Eds.): ICONIP 2018, LNCS 11303, pp. 359–372, 2018.
https://doi.org/10.1007/978-3-030-04182-3_32

There are distinct and diverse roles that different Dialogue Managers (DMs) fulfill which can be grouped under strategic flow-control and tactic flow-control DM. In strategic flow-control, the DM needs to learn an action-selection strategy, i.e., what action should be taken at a given time-step of the dialogue. These DMs find their applications in Topic Tracking, Form Filling [19], General Planning [1] scenarios etc. Whereas in tactic flow-control, the DM along with the normal functioning of the dialogue need to make some tactical conversational decisions that have some impacts in the quality of the dialogue for example, error handling mechanisms [3], control initiative and learned tactics [15] etc. Initially, dialogue management strategy was supposedly rule-based [10] where high degree of human intervention was required to design the component by hand. The inclusion of Reinforcement Learning (RL) [7] in its entirety minimizes the use of hand-crafting and reduces human effort to the point where the complete system becomes automated from end to end. Motivated by the recent advances in RL [2], especially the growth and availability of various Deep Reinforcement Learning (DRL) algorithms [13] that allow for learning policies and features parallelly, this paper presents a traditional and DRL methodology for automatically learning a policy by a DM in a task oriented framework.

A number of existing works suggest considering dialogue design as a MDP [8] which explicitly means defining a stochastic environment having its own finite representation of states, set of plausible actions, an acceptable reward model. Hence, the goal of such an abstraction is to learn a policy that maximizes the measure of the reward model. In this paper, a new MDP has been proposed and developed for a flight enquiry system. Since, the proposed MDP (discussed in Sect. 4.1) has an extensive usage of confidence scores (or say probabilities) from the NLU module more specifically a slot-filling (SF) module, an existing SF module has been used for the identification of different slots. Q-learning [18], which is a model-free reinforcement learning algorithm has been employed in its traditional and DRL variants to learn a policy. The traditional version is basically the table implementation of the said algorithm. In the DRL approach, the Deep Q-network (DQN), Double Deep Q-network (DDQN), DQN with Prioritized Experience Replay (DQN-PER) and DDQN with Prioritized Experience Replay (DDQN-PER) algorithms have been employed. A detailed analysis of the policy learnt by the Virtual Agent (VA) on all the approaches is presented.

The key contributions of this paper are the following:

- A novel MDP is proposed in terms of states, actions and reward model which reduces the complexity of the problem by separating the policy and the SF aspect of the dialogue unlike earlier works as explained in Sect. 2.
- Integration of an already existing SF module to identify the slots with the RL framework to establish the robustness of the learnt dialogue strategy.
- The paper aims to establish the fact that even if the SF module which is used to extract relevant information from the user's utterance is not very robust and has shortcomings, the designed system can still meet the input requirements of the user from the execution of the policy learnt.

2   Related Work

This section provides a brief description on the works done so far on RL based Dialogue Management Strategy followed by the motivation behind solving this problem.

2.1   Background

This section presents a brief survey on Reinforcement Learning based Dialogue Management Strategy.

In [21], authors proposed a method in which a Spoken Dialogue System, named ELVIS, is developed to choose a dialogue strategy to interact with the users. The reinforcement learning module implements the Q-learning algorithm in its traditional front and the performance modeling module uses the PARADISE evaluation framework to learn the reward function used in the reinforcement learning. In [9], authors presented a reinforcement learning approach for learning an optimal dialogue strategy for a spoken dialogue system named NJFun, which helps users in finding fun places in New Jersey. They first implemented NJFun system using the EIC (Exploratory for Initiative and Confirmation) dialogue strategy. The dialogues obtained after executing this strategy were used to build an empirical MDP and then an optimal strategy was learnt in this MDP. In [17], authors developed a common software tool named Reinforcement Learning for Dialogue Systems (RLDS) for a MDP structure and had implemented it on dialogue corpora from two different real-time dialogue systems, TOOT and ELVIS. In [4], authors developed an easy and open-sourced dialogue system using DRL for the restaurant domain without the use of the NLU module. It employs the DQN algorithm for its implementation. In [5], authors proposed a fast DRL approach that uses a network of DQN agents that skips weight updates during exploitation of actions.

2.2   Motivation

From the literature, it is evident that several works done earlier in the context of dialogues had shortcomings. The applications developed earlier based on the traditional RL approach had tremendous amount of human labor and interference involved right from manual hand-crafting of the rules to carry out experiments to train the agent. Performing large scale experiments to establish the robustness of the learnt strategy was a cumbersome process. State tracking was difficult because the representation of the states in the MDP was complex as more number of variables with varied range were used to capture the information in a particular time-step. Recent works, which employed Deep RL technique for the problem, incorporate vocabulary of the system as state representation without the use of the SF module. So, even if the VA learns an optimal policy, its usability is restricted because of its dependence on the vocabulary and hence is not scalable.

Motivated by the inadequacy of the existing system and approaches, this paper presents a concrete and a concise representation of the MDP. We also integrate the system with an existing SF module which liberates the system from any constraint of the vocabulary and passes on this responsibility to the competency of the SF model.

3   Reinforcement Learning for Dialogue Strategy

This section presents an overview of Reinforcement Learning followed by a short description of a dialogue strategy.

3.1   An Introduction to Reinforcement Learning

Any RL setup can be typically modeled as a MDP [18] which is defined as:

(i) a set of finite states $S = \{s_i\}$,
(ii) a set of finite actions $A = \{a_i\}$,
(iii) a probability transition model $P(s, a, s')$,
(iv) a reward model $R(s, a, s')$ that corresponds to the immediate reward given to the agent for selecting an action a in state s and henceforth making a transition to the next state s'.

So, the MDP is solved to obtain an optimal policy π^*. A policy π is defined as $\pi(s) \rightarrow a$ which is a mapping from states to actions that typically depicts the behavior of the agent. π^* represents an optimal policy which maximizes the cumulative reward at the end of an episode.

In this paper, we have employed DQN and it's various variants as mentioned below:

– DQN [13]
– DDQN [20]
– DQN - PER
– DDQN - PER [16]

3.2   Dialogue Strategy

A dialogue strategy is simply defined by the following:

(i) how to control the flow of a dialogue,
(ii) how a system will respond with queries to keep the dialogue meaningful and rational.

The agent should be able to ask relevant questions based on the present context of the conversation. Here, Reinforcement Learning framework [18] offers a good way of solving the problem where the current situation of the conversation can be represented as a state in the RL framework, the type of question asked can be treated as the action taken by the agent. User satisfaction can be treated as a reward function. As it is difficult to quantify user satisfaction and it varies from person to person, it is tough to formulate the reward function in the current context. To circumvent the problem we have based our reward functions on certain pretexts that are common to satisfaction from a certain aspect of the dialogue.

4   Proposed Methodology

This section presents the proposed MDP followed by the experiments conducted in two different frameworks, a short description of the SF module integrated, followed by the working of the proposed system.

4.1   Proposed MDP

The VA or the Dialogue Manager (presented here) is based on a flight enquiry system where the user has a single intent of booking a flight and wants to know about available flight options as per his/her preference. The task of the VA is to elicit necessary information from the user. It needs to fill the slots to make a valid database query so as to provide necessary and apt information based on the data elicited. Slots are basically defined as the important information that are present in the user utterances. An example of an user utterance with its valid slots is shown in Table 1.

Table 1. An example utterance with its valid slots

Utterance	I	want	to	fly	from	Pittsburgh	to	Denver
Slot	O	O	O	O	O	deptCity	O	arrCity

Table 2. Slots to be elicited

SLOTS	deptCity	arrCity	deptTime	depDay	class
DESCRIPTION	Departure City	Arrival City	Depart Time	Depart Date	Class of the flight

The necessary slots to be filled for this particular task are described in Table 2. The state space is represented as a tuple of five variables:

$$[\textbf{ deptCity } \textbf{ arrCity } \textbf{ depDay } \textbf{ deptTime } \textbf{ class }]$$

These five variables correspond to confidence scores of different slots which are basically the probability values outputted from the SF module representing the confidence of the module in predicting a particular slot label. Its permissible set of values ranges from 0 to 1.

Table 3. Action set

(a)

TYPE	ASK						
ACTION	askdeptCity	askarrCity	askdeptTime	askdepDay	askclass	askDeptandArr	askDateTime
SLOTS FILLED	deptCity	arrCity	deptTime	depDay	class	deptCity, arrCity	depDay, deptTime

(b)

TYPE	REASK/CONFIRM					SALUTATION
ACTION	reaskdeptCity	reaskarrCity	reaskdeptTime	reaskdepDay	reaskclass	closing_conversation
SLOTS FILLED	deptCity	arrCity	deptTime	depDay	class	-

The action space constitutes of 13 actions categorized in three different classes shown in Table 3a and b. As seen in the table, apart from having actions to fill individual slots, there are hybrid actions to fill two slots at the same time. Reask/Confirm actions act as a tool to fill up any capability lacked by the SF module in terms of the confidence in understanding the information given by the user as slots and tend to present a more natural conversational experience.

Simple Reward Model. Initially for this particular work, the reward model formulated was in lines with a typical computer game scenario where credits are assigned only at the end of an episode depending on the win or loss situation. So, originally a simple and a straight-forward reward model was assumed which is as follows:

- *Case 1:* The reward function at any other time-step except at the terminating or closing step was -0.01, i.e., a negative reward is given to control the number of steps to be taken for a particular dialogue conversation.
- *Case 2:* The reward function at the terminating time-step is subject to a checking condition. The condition was to check if the confidence scores of all the slots are greater than a particular threshold set to be 0.7. If the checking condition is satisfied, the agent gets a reward of $+1$.
- *Case 3:* If the checking condition isn't satisfied, the agent was given a reward of -1.

After conducting a set of experiments with this particular reward model (results of which are presented in Sect. 5), it was analyzed that with such an elementary reward function, the agent or the learning algorithm didn't converge well enough or didn't converge at all to learn an optimal dialogue strategy. Thus, it was inferred that a simple reward model like this wasn't sufficient or adequate enough for a complicated scenario as that of dialogues where user behavior or demand is unpredictable. So, there was a need to devise a new, more meaningful and relevant reward model.

Proposed Reward Model. Therefore, the new reward model is designed in a way such that the immediate credits assigned to the agent in different instances of the dialogue are attributed differently so as to make the agent understand what it needs to do distinctively at different time-steps of the conversation. It is described as:

- *Case 1:* The reward function at any other time-step except at the terminating or closing step is as follows:

$$R(s, a, s') = (w_1 * (\| \overrightarrow{NS} \|_1 - \| \overrightarrow{CS} \|_1)) - (w_2) \tag{1}$$

where $\overrightarrow{NS}$ corresponds to the state vector for the new state s'. $\| \overrightarrow{NS} \|_1$ is the summation of the confidence scores of all the state variables in the state vector which is obtained after taking an action a in state s. $\overrightarrow{CS}$ corresponds to the state vector for the current state s. $\| \overrightarrow{CS} \|_1$ is the summation of the confidence scores of all the state variables in the state vector for state s. w_1 is the weight over the difference of the summation of the two state vectors in state s and s'. w_1 is used to encourage the agent to act in a way so as to increase it's confidence on the acquired slots. w_2 is basically used to encourage useful communication and discourage unnecessary iteration. Here, $w_1 = 8$ and $w_2 = 1$.
- *Case 2:* The reward function at the terminating time-step is subject to a checking condition (mentioned below). If the checking condition is satisfied, the agent gets the reward as follows:

$$R(s, a, s') = V * w_1 * \| \overrightarrow{CS} \|_1 \tag{2}$$

where V is the value obtained from the checking condition (greater than zero). $\overrightarrow{CS}$ and w_1 are same as described earlier.
- *Case 3:* If the checking condition isn't satisfied, the reward function is:

$$R(s, a, s) = -w_1 * (\| \overrightarrow{EV} \|_1 - \| \overrightarrow{CS} \|_1) \tag{3}$$

where $\overrightarrow{EV}$ is the state vector for the expected value. $\| \overrightarrow{EV} \|_1$ is the summation of the maximum expected confidence scores of different slots that adds up to be equal to 5.

The checking criteria is as follows: if the confidence scores of three individual slots are greater than a threshold set to 0.7, partial credit of $1/3$ is assigned to the reward function, i.e., the value of V is 0.33 and thus the value of V increases based on the number of slots satisfying the condition which is 0.67 and 1 for four and five slots, respectively. If less than three slots satisfy the criteria at the end of the dialogue, the value of V is zero and hence the checking condition is not fulfilled. The system is trained on a pseudo-environment mimicking the confidence values of the SF module and hence a threshold of 0.7 is fixed. Later, the learned policy which is trained on the pseudo-environment is tested with a real SF module (mentioned in Sect. 4.2).

All these proposed reward functions for this particular MDP are motivated by the fact that it is necessary for the agent to learn different and permissible actions at different time-steps given the state of the dialogue and hence it needs to be credited and penalized differently for the right and the wrong actions picked up. The agent should be capable of picking up the conversation from any point to act so as to acquire only the necessary information required without wasting any number of iterations. Also, the reward model is designed to be generic so that it finds its usability across varied domains with no changes at all. It simply depends upon the domain it is used for and the slots on which the domain operates.

4.2   Implementation

Two variants of RL technique have been used to learn an optimal policy. One is based on traditional table implementation of the Q-learning algorithm. The second is the DRL based Q-learning technique, i.e., DQN, DDQN, DQN-PER and DDQN-PER algorithms.

Q-Learning: Q-Table Approach. In the table version, initially all the state-action values or the $Q(s, a)$ values are initialized to zero and as the dialogue progresses and rewards are observed for the picked up actions, the Q-table values are updated using the Bellman Eq. [18]. The value of γ or the discount factor in the equation is set to be 0.9 and the training is done for 1,00,000 dialogues. The value of epsilon is set to be 0.1 which allows a very small degree of randomness in the action-selection process. During testing, the agent picks up an action having the highest $Q(s, a)$ value given a state s based on the Q-table converged during training. The results are presented in Sect. 5.

Deep Reinforcement Learning. In the DRL implementation, the architecture of the neural network is as follows: 5 nodes are used in the input layer (corresponding to the size of the state vector), followed by one hidden layer with 75 nodes and 13 nodes in the output layer corresponding to the action set. The activation function used in the hidden layer is Rectified Linear Units to normalize their weights. The DQN, DDQN, DQN-PER, DDQN-PER algorithms are employed on the developed MDP to learn the optimal dialogue strategy. In some of the experiments, for a better convergence of the learning algorithm, constrained set of actions were used rather than the entire set of actions [4]. These constrained set of actions for a given state are obtained by using the SVM classifier which presents actions with the top five probability values trained from the dummy data prepared manually based on the state vector values. Hence, the Q-learning updates are applied only on this valid set of actions. The other parameters of the model are: discount factor $(\gamma) = 0.7$, minimum epsilon $= 0.15$, experience replay size $= 100000$, batch size $= 32$. The training is done for 200000 dialogues. Selection of hyper-parameters has been done after careful experimentation. Setting higher value of discount factor was not resulting in the proper convergence of the algorithm, also lower epsilon values were not sufficiently exploring the state spaces, causing the agent to get stuck in local optima. Experience replay size and batch sizes is set according to the computational resources that are available at hand. The results are presented in Sect. 5.

Slot-Filling Module. To extract relevant information from the user's utterance, i.e., to identify the necessary slots so as to have a relevant flight inquiry query, an existing SF module [12] has been used. It is a deep learning model which uses a simple Recurrent Neural Network at its core. This particular SF module is trained on the ATIS dataset [14] which contains user utterances relating to the Air Travel Information System, wherein only a single intent of *flight* as per the dataset is used with a fixed number of slots. So, the slots identified with the help of this module are in direct sync with the slots the proposed system aims to fill. The necessary slots identified, along with the probability scores of the predicted labels are used by the VA for further processing.

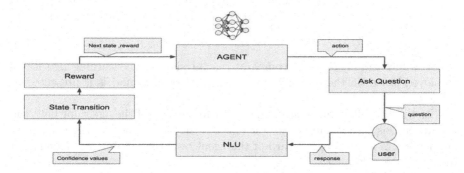

Fig. 1. Flow diagram for the proposed system

Working of the VA. The functioning of the flight enquiry system is as follows: The VA picks up an action based on the state using the Q-table (for the traditional implementation) or the neural network (for the DRL implementation). The action selected by the VA is presented in a text format to the user. The user's reply based on the VA's response is fed to the existing SF module to identify necessary slots mentioned above in the Sect. 4.1 and hence produce confidence scores to be taken as a state input by the VA. Based on the actions, rewards are generated and the model is trained to pick up right actions given a state. The flow diagram for the proposed system is shown in Fig. 1.

5   Results and Discussion

Results of different algorithms employed in various experimental set-up are presented here.

The learning curves of the VA for the best performing set-up during training followed by the verbalization of the best policy learnt during the testing phase are presented below. Also, the following metrics were used to measure the performance of the system for various algorithms employed: average episodic reward which is the cumulative reward through all the time-steps at the end of a dialogue, average dialogue length which is basically the average system actions per

Table 4. Quantitative Comparison of different algorithms and reward functions

	Algorithm	Average episodic reward	Average dialogue length	Training time (in hrs)
Simple reward model	DQN with SVM	−6.89 ± 5.62	673.45 ± 564.02	71.97
	DDQN with SVM	−8.51 ± 5.40	791.65 ± 529.12	93.65
	DDQN-PER with SVM	−13.51 ± 9.15	1342.3 ± 915.20	52.56
	DDQN-PER	−11.26 ± 9.17	1039.07 ± 933.62	112.12
Proposed reward model	DQN with SVM	−313.25 ± 308.63	367.52 ± 315.03	40.71
	DDQN with SVM	−273.52 ± 271.97	330.52 ± 278.67	54.2
	DQN-PER with SVM	−131.80 ± 181.13	183.2 ± 182.07	18.47
	DQN-PER	-569.85 ± 469.48	589.03 ± 479.09	16.83
	DDQN-PER with SVM	**57.20 ± 7.99**	**7.67 ± 0.53**	**20.82**
	DDQN-PER	**50.07 ± 8.11**	**8.09 ± 1.06**	**17.74**

dialogue and training time[1]. A comparative analysis of all the learning algorithms employed based on these metrics on two sets of reward models is presented in Table 4.

For the Q-table implementation, the learning curve of the agent for both the reward models is shown in Fig. 2. The figure clearly indicates that there has been an increase in the average reward of the agent over time and hence it managed to learn a decent policy.

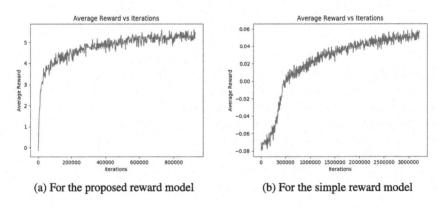

(a) For the proposed reward model   (b) For the simple reward model

Fig. 2. Learning Curve of the Agent based on the Q-Table implementation

It is evident from Table 4 that because of the larger length of the dialogues and lower episodic reward, the simple reward model failed to converge on any of the DRL variant algorithms employed. Figure 3 shows the learning curve of the agent for the two best performing DRL implementations on the proposed reward model, i.e., DDQN-PER and DDQN-PER with SVM. The highest average reward attained in both these models is that of **7** which is better compared

---

[1] Ran on Intel(R) Xeon(R) CPU E5-2650 v4 @ 2.20 GHz, 251 GB RAM.

to its equivalent Q-table implementation where the maximum average reward attained is of **5**. It is also visible from Table 4 that the average episodic reward for both these models are **50.07 ± 8.11** and **57.20 ± 7.99** respectively. The average dialogue lengths for both these models are **8.09 ± 1.06** and **7.67 ± 0.53** respectively. Therefore, it is observed that the average episodic reward is maximum and the average dialogue length is small for these two learning algorithms, which are favorable for any dialogue conversation, as the task of the agent is to maximize its cumulative reward at the end of an episode and it should learn to do so in minimal number of time steps. Also, the training time for these algorithms is comparatively less, implying that the learning algorithm indeed manages to converge on this reward model. It is also evident from the table that other DRL algorithms such as DQN with SVM, DDQN with SVM, DQN-PER, DQN-PER with SVM failed to converge on this model as well because of the limitations of these learning algorithms [13,20]. A sample conversation to demonstrate the best

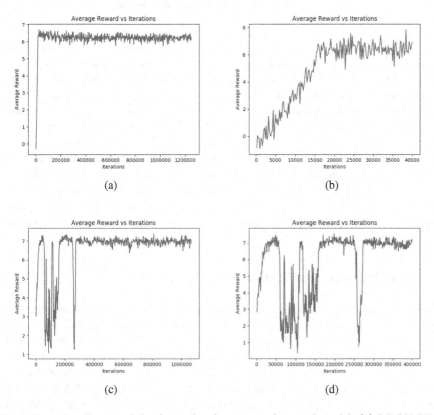

Fig. 3. Learning Curves of the Agent for the proposed reward model: (a) DDQN-PER curve for 1200000 iterations, (b) DDQN-PER curve for 40000 iterations, (c) DDQN-PER curve with SVM for 1000000 iterations, (d) DDQN-PER curve with SVM for 400000 iterations

dialogue strategy learnt is shown in Fig. 4. From the conversation, it is evident that the agent is interactive. It focuses on filling its slot with higher confidence. So, if the SF module has lower confidence in understanding a particular slot, the agent with the help of reask actions acts so as to confirm from the user, the authenticity of the identified slot.

```
State : [0. 0. 0. 0. 0.]
Agent > Hello How may I help you?
User  > flights from pittsburgh to denver on fifteenth june at afternoon
State : [1. 1. 0.558 1. 0.]
Agent[askclass] > Please specify the class of flight?
User  > business class
State : [1. 1. 0.558 1. 1.]
Agent[reaskdepDay] > Are you travelling on fifteenth june?
User  > yes
State : [1. 1. 1. 1. 1.]
Agent[closing_conversation] > The flights from pittsburgh to denver
on fifteenth june at afternoon via business class are JW345, US298
Thanks for using the flight attendant.
```

Fig. 4. Verbalization of the Policy learnt

Instead of using a dictionary-based state representation, the system utilizes a slot-based state representation scheme. So, the representation of the state is unique. Merit of the system lies in the integration of SF module with dialogue management strategy. Current set of experiments clearly illustrate that the proposed MDP indeed enables the agent to learn a dialogue strategy. The snapshot of the conversation at the testing time evidently proves that the learning process has the merit to rectify the shortcomings of the SF segment. We are unable to present any comparison of our approach with the state-of-the-art because as per the best of our knowledge we have not come across an airline enquiry chat-bot system employing RL algorithms. No literature in our view addresses the slot-filling approach of the state vector based on probability values of an external system and tries to tackle the uncertainties of that system which our proposed model can successfully handle.

6   Conclusions and Future Work

This paper presents a MDP for a flight enquiry system in terms of defining state space, action set and reward model. Traditional Q-learning as well as its DRL variants are implemented to learn an optimal policy. The learning curves of the VA during training followed by the quantitative analysis based on several metrics are presented for both the approaches. The verbalization of the best dialogue strategy learnt is presented and hence it can be inferred that the VA is managed to learn a sensible policy to control the flow of the dialogue which justifies the correctness of the proposed MDP.

An extensive study needs to be carried out to measure the robustness of the system. Also, the current system focuses on processing only a single intent of the user, i.e., finding flight availabilities. Multiple intents based on the flight domain can be incorporated to increase the scope of the system. An effort can be made to benchmark the system against different SF modules to prove its strength, i.e., given a poor SF module, the VA will be able to control the dialogue in an intelligent manner.

References

1. Allen, J.F., Byron, D.K., Dzikovska, M., Ferguson, G., Galescu, L., Stent, A.: Toward conversational human-computer interaction. AI Magaz. **22**(4), 27 (2001)
2. Arulkumaran, K., Deisenroth, M.P., Brundage, M., Bharath, A.A.: A brief survey of deep reinforcement learning. arXiv preprint arXiv:1708.05866 (2017)
3. Bohus, D., Rudnicky, A.I.: Error handling in the ravenclaw dialog management framework. In: Proceedings of the Conference on Human Language Technology and Empirical Methods in Natural Language Processing, pp. 225–232. Association for Computational Linguistics (2005)
4. Cuayáhuitl, H.: SimpleDS: a simple deep reinforcement learning dialogue system. In: Jokinen, K., Wilcock, G. (eds.) Dialogues with Social Robots. LNEE, vol. 999, pp. 109–118. Springer, Singapore (2017). https://doi.org/10.1007/978-981-10-2585-3_8
5. Cuayáhuitl, H., Yu, S., et al.: Deep reinforcement learning of dialogue policies with less weight updates (2017)
6. Fraser, N.: Assessment of interactive systems. In: Handbook of Standards and Resources for Spoken Language Systems, pp. 564–615. Mouton de Gruyter (1998)
7. Kaelbling, L.P., Littman, M.L., Moore, A.W.: Reinforcement learning: a survey. J. Artif. Intell. Res. **4**, 237–285 (1996)
8. Levin, E., Pieraccini, R., Eckert, W.: Using markov decision process for learning dialogue strategies. In: Proceedings of the 1998 IEEE International Conference on Acoustics, Speech and Signal Processing, vol. 1, pp. 201–204. IEEE (1998)
9. Litman, D.J., Kearns, M.S., Singh, S., Walker, M.A.: Automatic optimization of dialogue management. In: Proceedings of the 18th Conference on Computational Linguistics, vol. 1, pp. 502–508. Association for Computational Linguistics (2000)
10. McTear, M.F.: Modelling spoken dialogues with state transition diagrams: experiences with the CSLU toolkit. In: Fifth International Conference on Spoken Language Processing (1998)
11. McTear, M.F.: Spoken dialogue technology: enabling the conversational user interface. ACM Comput. Surv. (CSUR) **34**(1), 90–169 (2002)
12. Mesnil, G., He, X., Deng, L., Bengio, Y.: Investigation of recurrent-neural-network architectures and learning methods for spoken language understanding. In: Interspeech, pp. 3771–3775 (2013)
13. Mnih, V., et al.: Playing atari with deep reinforcement learning. arXiv preprint arXiv:1312.5602 (2013)
14. Price, P.J.: Evaluation of spoken language systems: the atis domain. In: Speech and Natural Language: Proceedings of a Workshop Held at Hidden Valley, Pennsylvania, 24–27 June 1990
15. Rieser, V., Lemon, O.: Reinforcement learning. In: Reinforcement Learning for Adaptive Dialogue Systems, pp. 29–52. Springer, Heidelberg (2011). https://doi.org/10.1007/978-3-642-24942-6
16. Schaul, T., Quan, J., Antonoglou, I., Silver, D.: Prioritized experience replay. arXiv preprint arXiv:1511.05952 (2015)

17. Singh, S.P., Kearns, M.J., Litman, D.J., Walker, M.A.: Reinforcement learning for spoken dialogue systems. In: Advances in Neural Information Processing Systems, pp. 956–962 (2000)
18. Sutton, R.S., Barto, A.G.: Reinforcement Learning: An Introduction, vol. 1. MIT press, Cambridge (1998)
19. Traum, D.: Approaches to dialogue systems and dialogue management. Lecture Notes, University of Southern California (2008). http://people.ict.usc.edu/~traum/ESSLLI08
20. Van Hasselt, H., Guez, A., Silver, D.: Deep reinforcement learning with double q-learning. In: AAAI, vol. 16, pp. 2094–2100 (2016)
21. Walker, M.A.: An application of reinforcement learning to dialogue strategy selection in a spoken dialogue system for email. J. Artif. Intell. Res. **12**, 387–416 (2000)

Other Learning Approaches

A Family of Maximum Margin Criterion for Adaptive Learning

Miao Cheng[1(✉)], Zunren Liu[1], Hongwei Zou[2], and Ah Chung Tsoi[3]

[1] Department of Computer Science, Qingdao University, Qingdao, China
`mewcheng@gmail.com, liuzunren@126.com`
[2] Division of Information Technology, Chongqing Branch of China Merchants Bank,
Chongqing, China
`hongwei_z@cmbchina.com`
[3] Department of Computer Science and Software Engineering,
University of Wollongong, Wollongong, Australia
`act@uow.edu.au`

Abstract. In recent years, pattern analysis plays an important role in data mining and recognition, and many variants have been proposed to handle complicated scenarios. In the literature, it has been quite familiar with high dimensionality of data samples, but either such characteristics or large data sets have become usual sense in real-world applications. In this work, an improved maximum margin criterion (MMC) method is introduced firstly. With the new definition of MMC, several variants of MMC, including random MMC, layered MMC, 2D^2 MMC, are designed to make adaptive learning applicable. Particularly, the MMC network is developed to learn deep features of images in light of simple deep networks. Experimental results on a diversity of data sets demonstrate the discriminant ability of proposed MMC methods are component to be adopted in complicated application scenarios.

Keywords: Maximum margin criterion (MMC) · Adaptive learning
Variants of MMC · MMC network

1 Introduction

As a promising step, feature extraction has become an important approach to data mining and pattern recognition. And traditional methods usually suffer from intrinsic limitations from characteristics of original data. The first one refers to high-dimensionality of samples that hinders efficient calculation, and the outstanding solutions come down to direct approach to scatter matrices decomposition. Furthermore, there arise broad interests in large-scale data mining in many real-world applications, and this pushes new challenge for feature analysis and reduction. In terms of such demands, it has become a vivid research topic to devise improved learning methods to conduct high-dimensional data with large amount meanwhile.

© Springer Nature Switzerland AG 2018
L. Cheng et al. (Eds.): ICONIP 2018, LNCS 11303, pp. 375–387, 2018.
https://doi.org/10.1007/978-3-030-04182-3_33

In the literature, principal component analysis (PCA) [1] and linear discriminant analysis (LDA) [2,3] have become popular methods for pattern analysis in statistical learning theory. To address high-dimensional problem of original data, there has been a common sense that reaches the eigen-decomposition of scatter matrices with calculational tricks of sub-matrices' multiplications [4]. Besides traditional ratio LDA, there is another approach to do discriminant analysis with subtraction formalism, i.g., maximum margin criterion (MMC), while null denominator problem can be avoided [5,6]. Nevertheless, just like an old Chinese saying goes, "There would be something else in loss if something was obtained". The calculational trick of sub-matrices is unavailable for MMC anymore, and extra calculations are to be involved in general. In the previous work of ours, a direct solution is proposed to handle the calculational limitation of large discriminant scatter of high-dimensional data for MMC [7]. And discriminant analysis can be proceeded straightforward while both S_b and S_w scatters are considered together with preserved efficiency.

In this work, the direct MMC framework is further developed to conduct adaptive learning, and insensitive to high-dimensionality problem of data in any scenarios. Furthermore, several extensions of MMC are proposed to conduct adaptive classification of different categories of data. The rest of this paper is organized as follows. The background knowledge of direct MMC is reviewed in Sect. 2, while the calculation efficiency is discussed in theory, followed by the details of an improved MMC in Sect. 3. And then, several extensions of MMC is proposed for applications of different scenarios. A set of comparison experiments on discriminant learning are given in Sect. 4. Finally, the conclusion is draw in Sect. 5.

2   Direct Maximum Margin Criterion

The original MMC considers the substraction of discriminant scatters of original data. Given data set $X = [x_1, x_2, \cdots, x_n] \in R^{d \times n}$ in c classes, the S_b and S_w scatters are defined as

$$S_b = \sum_{i=1}^{c} n_i (m_i - m)(m_i - m)^T = XL_bX^T$$
$$S_w = \sum_{i=1}^{c} \sum_{j=1}^{n_i} (x_{ij} - m_i)(x_{ij} - m_i)^T = XL_wX^T \tag{1}$$

where m_i and m respectively denote the mean data of i-th class and whole data, and n_i denotes the sample amount of i-th class. Besides, the discriminant scatters can also be described in graph formula with definition of Laplacian matrices L_b and L_w [8,9]. As a result, MMC solves the following quadratic optimization objective to find the ideal w,

$$J(w) = w^T (S_b - \gamma S_w) w = w^T XLX^T w. \tag{2}$$

Here, γ indicates the trade-off parameter to balance between-class and within-class scatters, and L refers to the graph Laplacian of MMC [7]. Obviously, the solution to such objective can be reached via eign-decomposition of J, such as

$$(S_b - \gamma S_w)\, w = XLX^T w = \lambda w. \tag{3}$$

Compared with traditional LDA framework, it is able to avoid rank calamity of scatters and exceed the restricted bounding of the category of samples. Nevertheless, it is hardly to be adopted to high-dimensional data, as eigen analysis of large matrix usually leads to overflow of memory. In light of kernel-view idea, we proposed a direct approach to efficient discriminant analysis [7], and the whole procedure is given in Algorithm 1.

Algorithm 1. The proposed MMC algorithm

Input: Given data points $X \in R^{d \times n}$ in c classes, the desired reduced
      dimensionality r.
Output: Projective directions, $w_i, i = 1, 2, \cdots, r$.
1. Calculate between-class scatter S_b and within-class scatter S_w with given
  data;
2. **if** *The dimensionality of samples $< \theta$* **then**
    2.1 Perform spectral decomposition on $S_b - \gamma S_w$, and obtain projective
    directions $w_i, i = 1, 2, \cdots, r$.
  else
      2.2 Construct the sample kernel matrix $K = X^T X$, perform spectral
      decomposition on KLK, and obtain $E^T KLKE = \Lambda$;
      2.3 Calculate the SVD of $XE\Lambda^{-\frac{1}{2}}$, and obtain orthogonal matrix U and
      singular matrix $s_i, i = 1, 2, \cdots, n$;
      2.4 Set columns of U to $w_i, i = 1, 2, \cdots, r$ in reverse order.
  end
end

Though MMC involves the similar discriminant scatters with LDA, the proposed direct approach is quite distinctive compared with traditional ideas. To avoid extra branches of execution, an dimensional threshold is added in MMC. If the sample dimension is less than the given threshold, standard procedure of MMC would be proceeded. On the contrary, an efficient calculational idea would be adopted, and sample kernel $K = X^T X$ is constructed to reach the kernel scatter. As a consequence, the original MMC problem is transformed to

$$J(e) = e^T X^T (S_b - \gamma S_w) X e = e^T KLKe, \tag{4}$$

where K denotes the sample kernel, and e is the resulting orthogonal directions of MMC. It is noticeable that, the size of decomposed matrix $\mathbb{R}^{d \times d}$ is reduced to $\mathbb{R}^{n \times n}$. Then, the final results can be obtained via calculational tricks of matrix decomposition.

The most distinguishing points mainly come from step 3 and 4. In step 3, the SVD is proceeded on $XE\Lambda^{-\frac{1}{2}}$, namely,

$$XE\Lambda^{-\frac{1}{2}} = USV^T, \tag{5}$$

Algorithm 2. The RMMC algorithm

Input: Given data points $X \in R^{d \times n}$ in c classes, the desired reduced
dimensionality r, the dimensional threshold θ for efficient calculation,
and the number of selected samples t.

Output: Projective directions, $w_i, i = 1, 2, \cdots, r$.

1. Calculate between-class scatter S_b and within-class scatter S_w with given
data;

2. **if** *The dimensionality of samples* $< \theta$ **then**

2.1 Perform spectral decomposition on $S_b - \gamma S_w$, and obtain projective
directions $w_i, i = 1, 2, \cdots, r$.

else

2.2 Randomly select t samples $A = [x_{a1}, x_{a2}, \cdots, x_{at}]$ from whole data,
construct the sample kernel matrix $M = A^T X$, and discriminant
scatter MLM^T.

2.3 Follow the similar steps of MMC, and obtain $w_i, i = 1, 2, \cdots, r$.

end

end

---

which is different from traditional LDA that generally do similar operation on within-class scatter S_w. Furthermore, the obtained orthogonal vectors $u_i, i = 1, 2, \cdots, n$ is sorted in descending order corresponding to discriminant power, though they are actually adhere to the largest singular values. Thereafter, the final projective directions w_i needs to be reverse vectors of u_i in step 4. In addition, a dimensional threshold is absorbed into original MMC for adaptive dimensionality reduction. That is, the efficient calculation approach would be referred if dimensionality of original data is larger than given threshold.

It is demonstrated that, the proposed MMC method is competent to deal with linear supervised learning in general, while calculational efficiency is preserved. The computational cost mainly depends on $O(n^3)$ for spectral decomposition, compared with $O(d^3)$ of original MMC. For convenience, such approach is called MMC directly in this context. Nevertheless, there also exists some exceptions that n is still large for direct calculation of big data, and efficiency is unable to be reached. In terms of this limitation, an improved MMC is designed in this work to make supplement of the previous work of ours. The main improvement refers to construction of sample kernel in MMC, and a subset of whole samples are picked up to form the kernel matrix for following step [10]. Suppose that there are t samples are selected, the whole procedure is summarized as random MMC (RMMC) in Algorithm 2. Obviously, the computational complexity reduces to $O(t^3)$ in this improved procedure, and the theoretical basis can be derived.

3   Adaptive Learning of MMC

As progress of information technology, there are diversity of handled data categories and application scenarios. The goal of adaptive learning is to exploit the

---

Algorithm 3. The single LMMC algorithm

Input: Given data points $X \in R^{d \times n}$ in c classes, the desired reduced
      dimensionality r, the dimensional threshold θ for efficient calculation,
      the median dimension m, and the number of selected samples t.

Output: Projective directions, $w_i, i = 1, 2, \cdots, r$.

1. Calculate between-class scatter S_b and within-class scatter S_w with given
  data;

2. **if** *The dimensionality of samples* $< \theta$ **then**
    2.1 Perform spectral decomposition on $S_b - \gamma S_w$, and obtain projective
    directions $w_i, i = 1, 2, \cdots, r$.

   else
      2.2 Construct the random projection matrix $P \in \mathbb{R}^{d \times g}$, randomly select
      t samples $A = [x_{a1}, x_{a2}, \cdots, x_{at}]$ from whole data, and calculate the
      projection matrix $B = PP^T$.
      2.3 Construct the sample kernel matrix $M = A^T BX$, and discriminant
      scatter MLM^T.
      2.3 Follow the similar steps of MMC, and obtain $w_i, i = 1, 2, \cdots, r$.
   end

end

---

intrinsic patterns of data with different analysis demands in a unified framework
as possible.

3.1 Layered MMC and 2D MMC

With a multi-layer structure, it is believed that the hidden features can be
exploited by enlarging original ones from data [11]. More specifically, a median
layer is added to transform each data into a much high-dimensional space, and
reduced in following steps with general feature learning. Supposed that there is
a given data x, the transformation can be formalized as

$$x \mapsto h(x), \mathbb{R}^d \mapsto \mathbb{R}^g \tag{6}$$

where $h(\cdot)$ denotes the data transformation from original space with dimensionality d to a much higher dimensionality g in general. Obviously, such approach is
quite identical with kernel learning framework, and can be conduced as a median
learning step, e.g.,

$$h(x) = Bx, B = PP^T. \tag{7}$$

Here, $P \in \mathbb{R}^{d \times g}$ denotes the linear transformation directions. For the more
specific scenarios, it can be defined as a collaborative learning combined with a
nonlinear mapping $f(x)$ and a linear projection and P, which has been employed
in extreme learning machine (ELM) [11]. Surprisingly, it is learned that there
is a little disparity among different layered approaches for MMC. The whole
procedure of layered MMC is given as LMMC algorithm.

On the other hand, there are lots of real-world applications intuitively refer to
multi-dimensional media information, e.g., images, videos, which rely on surface

---

Algorithm 4. The 2D^{2}MMC algorithm

Input: Given each 2D data $x_i \in \mathbb{R}^{\alpha \times \beta}, i = 1, 2, \cdots, n$, in c classes, the desired reduced dimensionality l and r, the dimensional threshold θ for efficient calculation.

Output: Bi-directional projective directions,
$$p_i, q_j, i = 1, 2, \cdots, l, j = 1, 2, \cdots, r.$$

1. Calculate 2D scatters of column and row (left and right) directions:
$S_{bl} \in \mathbb{R}^{d_1 \times d_1}, S_{br} \in \mathbb{R}^{d_2 \times d_2}, S_{wl} \in \mathbb{R}^{d_1 \times d_1}, S_{wr} \in \mathbb{R}^{d_2 \times d_2}$ with given 2D data;

2. Calculate row projective directions:

if *The length of height $d_1 < \theta$* **then**

    2.1 Perform spectral decomposition on $S_{bl} - \gamma S_{wl}$, and obtain projective directions $p_i, i = 1, 2, \cdots, l$.

else

    2.2 Construct the randomly reduced scatters with lower height as done in MMC.

    2.3 Follow the similar steps of MMC, and obtain $p_i, i = 1, 2, \cdots, l$.

end

end

3. Calculate column projective directions:

if *The length of width $d_2 < \theta$* **then**

    3.1 Perform spectral decomposition on $S_{br} - \gamma S_{wr}$, and obtain projective directions $q_i, i = 1, 2, \cdots, r$.

else

    3.2 Construct the randomly reduced scatters with shorter width as done in MMC.

    3.3 Follow the similar steps of MMC, and obtain $q_i, i = 1, 2, \cdots, r$.

end

end

---

of 2D-dimensional space. In order to handle those kinds of data directly, some 2D based methods are to make learning succinct, e.g., two-dimensional PCA [12,13], two-dimensional LDA [14–16]. In general, 2D raw data $x \in \mathbb{R}^{m \times n}$ is involved to find reflecting information between rows of images, e.g., $y = xV$. As the original 2D methods that calculate the single direction for feature extraction [12,14], two-directional & two-dimensional methods, e.g., 2D^{2}PCA [13] and 2D^{2}LDA [15,16], are proposed to address the limitation of single-directional learning.

In terms of this consideration, 2D^{2}MMC is devised as a natural extension of original MMC. The main difference between 2D^{2}MMC and original ones is on the fact that 2D data are referred in construction of scatters, while certain steps need to be modified correspondingly. For a given 2D data $x \in \mathbb{R}^{d_1 \times d_2}$, it aims to find bi-directional projections $P \in \mathbb{R}^{d_1 \times l}$ and $Q \in \mathbb{R}^{d_2 \times r}$, and yields a smaller 2D data $y \in \mathbb{R}^{l \times r}$, e.g.,

$$y = P^T x Q. \tag{8}$$

Algorithm 5. The single L2D^{2}MMC algorithm

Input: Given each 2D data $x_i \in \mathbb{R}^{\alpha \times \beta}, i = 1, 2, \cdots, n$, in c classes, the desired reduced dimensionality l and r, the dimensional threshold θ for efficient calculation.

Output: Bi-directional projective directions,
$p_i, q_j, i = 1, 2, \cdots, l, j = 1, 2, \cdots, r.$

1. Calculate 2D scatters of column and row (left and right) directions:
$S_{bl} \in \mathbb{R}^{d_1 \times d_1}, S_{br} \in \mathbb{R}^{d_2 \times d_2}, S_{wl} \in \mathbb{R}^{d_1 \times d_1}, S_{wr} \in \mathbb{R}^{d_2 \times d_2}$ with given 2D data;

2. Construct the random matrix $P \in \mathbb{R}^{h1 \times d1}$ and $Q \in \mathbb{R}^{h2 \times d2}$, and transform original scatters into high-dimensional spaces.

3. Calculate row projective directions:
if *The length of height $d_1 < \theta$* **then**
> 3.1 Perform spectral decomposition on $S_{bl} - \gamma S_{wl}$, and obtain projective directions $p_i, i = 1, 2, \cdots, l.$

else
>> 3.2 Follow the similar steps of RMMC, and perform the economical calculations with randomly select sample data. Obtain
>> $p_i, i = 1, 2, \cdots, l.$

end

end

4. Calculate column projective directions:
if *The length of width $d_2 < \theta$* **then**
> 4.1 Perform spectral decomposition on $S_{br} - \gamma S_{wr}$, and obtain projective directions $q_i, i = 1, 2, \cdots, r.$

else
>> 4.2 Follow the similar steps of RMMC, and perform the economical calculations with randomly select sample data. Obtain
>> $q_i, i = 1, 2, \cdots, r.$

end

end

The calculation of P and Q is mainly based on construction of 2D scatters with respect to MMC, e.g.,

$$
\begin{aligned}
S_{bl} &= \sum_{i=1}^{c} n_i \left(m_i^2 - m^2\right) \left(m_i^2 - m^2\right)^T \\
S_{wl} &= \sum_{i=1}^{c} \sum_{j=1}^{n_i} \left(x_{ij}^2 - m_i^2\right) \left(x_{ij}^2 - m_i^2\right)^T \\
S_{br} &= \sum_{i=1}^{c} n_i \left(m_i^2 - m^2\right)^T \left(m_i^2 - m^2\right) \\
S_{wr} &= \sum_{i=1}^{c} \sum_{j=1}^{n_i} \left(x_{ij}^2 - m_i^2\right)^T \left(x_{ij}^2 - m_i^2\right)
\end{aligned}
\tag{9}
$$

Here, S_{bl}, S_{wl}, S_{br} and S_{wr} indicate the 2D scatters with respect to left and right directions, m_i^2 and m^2 denote the 2D data of the i-th intra-class mean and the total mean respectively, and x_{ij}^2 denotes the j-th 2D data belonging to i-th class. Then, the desired P and Q can be obtained with standard process of

MMC, e.g.,

$$J(p) = p^T (S_{bl} - \gamma S_{wl}) p$$
$$J(q) = q^T (S_{br} - \gamma S_{wr}) q. \tag{10}$$

and the whole procedure is summarized as $2D^2$MMC algorithm.

Similarly, it is also feasible to employ a dimension threshold to ensure the calculational efficiency, especially if large 2D data are referred, e.g., high-resolution images. The related dimension of directional side is reduced with data kernel if the original length (α or β) is larger than threshold θ. By an example, this can be done with randomly selected rows (or columns) of 2D sample data. Furthermore, it is straightforward to extend original 2D methods to layered ones. Due to limited space, only the layered $2D^2$MMC is discussed here. The main branches are quite similar to LMMC algorithm, that is, row and column projections are transformed into high-dimensional space firstly, and followed by MMC approach. The differences come from the handling of high dimensionality of mapped 2D data, which can also be conducted with solution of RMMC similarly. Instead, only one sample data is generally selected to apply economical calculations, and partial columns (or rows) are employed. The single layered-$2D^2$MMC (L$2D^2$MMC) is summarized in Algorithm 5. As a consequence, multi-layered $2D^2$MMC can be deduced easily, and is able to be proceeded in hierarchical structures of sequential networks.

3.2 MMC Network

Inspired by convolution neural network (CNN), PCA Network (PCA-Net) is able to learn classification features of images with a very simple deep learning network [17]. Nevertheless, the composition of PCA-Net is only the very basic data processing components: cascaded principal component analysis (PCA), binary hashing, and block-wise histograms. In the PCA-Net architecture, PCA is adopted to learn multistage filter banks, followed by simple binary hashing and block histograms for indexing and pooling. With easy and efficient implementation, PCA-Net has been widely employed to learn deep features of objects. Obviously, it is straightforward to extend PCA-Net to MMC, e.g., MMC Network (MMC-Net).

For each image data $x_i \in \mathbb{R}^{m \times n}$, an image patch is taken around each pixel with size of $k_1 \times k_2$ as the manners of local binary patterns (LBP) [18]. As a consequence, there are $m \times n$ vectorized patches picked up from x_i, i.e., $x_{i,1}, x_{i,2}, \cdots, x_{i,mn} \in R^{k_1 k_2}$. Assume that the mean-removed patches of each image is indicated by $\widetilde{X}$. Then, the class mean and intra-class scatter of k-th category with n_k images can be defined as

$$m_k = \frac{1}{n_k} \sum_{i \in c_k} \widetilde{X}_i$$
$$S_\phi = \frac{1}{n_k} \sum_{i \in c_k} \left(\widetilde{X}_i - m_k\right)\left(\widetilde{X}_i - m_k\right)^T. \tag{11}$$

Similarly, the inter-class scatter of image patches belonging to different categories can be defined as

$$S_{\psi} = \frac{1}{n_c} \sum_{c=1}^{c} (m_k - m)(m_k - m)^T,$$ (12)

where m indicates the mean of class means. With the repeated PCA-Net stage, the output is composed of hashing and histogram of input images.

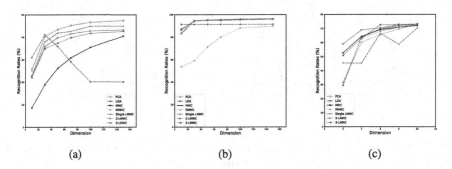

(a)   (b)   (c)

Fig. 1. The results of different methods on three data sets. (a) Experimental results on SUN database. (b) Experimental results on MNIST database. (c) Experimental results on STL-10 data set.

4 Experiments

In this section, several experiments are performed to evaluate the performance of proposed MMC methods[1]. First of all, the ability of linear feature extraction is tested, and three data sets, namely SUN scene categorization database[2] [19], MNIST digit database[3] [20], and STL-10 data set [22], are involved.

In the SUN database [19], the deep features of each image are extracted by keras toolkit[4] with pre-learned VGG-16 model of imagenet, and a 512 dimensional feature is obtained to describe each image. Among all categories, random 100 classes are selected to be employed in experiments, and random half images of each class are used for training and testing, respectively. Among image data of each digit in the MNIST data set [20], 2,000 images in training data are randomly selected to form training set, while 500 images in testing data are used for testing stage. As a consequence, the total training set are organized by 20,000 images, while testing set contains 5,000 images. Furthermore, the simple sparse coding features of MNIST data are adopted to make an improvement for classification [21]. For STL-10 data set [22], the deep representation of each image

---

[1] The implementations are available at: http://mch.one/resources.
[2] http://vision.princeton.edu/projects/2010/SUN.
[3] http://yann.lecun.com/exdb/mnist.
[4] https://keras.io.

with target coding is employed in experiments [23], and a 255 dimensional feature is adopted for each data. Similarly, separate 2,000 and 500 data from each training and testing categories are randomly selected to be training and testing set correspondingly.

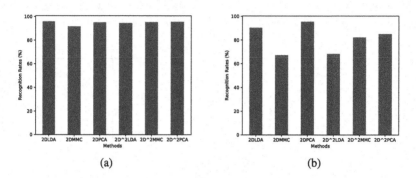

(a)   (b)

Fig. 2. The results of different 2D methods on three data sets. (a) Experimental results on ALOI database. (b) Experimental results on MNIST database.

The results on different data sets are shown in Fig. 1. For simplicity, the amount of randomly selected samples are set to be double quantity of reduced dimension r for RMMC algorithm. In terms of results, PCA can present stable results for SUN and MNIST data sets, but cannot learn discriminative information with unsupervised features. Similarly, LDA is only up to discriminant analysis for two data sets. The results of RMMC gets close to the best ones, while MMC is incompetent to pattern analysis compared with other linear methods for SUN database. Especially, MMC is hardly to be proceeded for MNIST in our experiments with both high dimensionality and large sample amount, which can be accomplished by RMMC instead. Furthermore, RMMC is able to reach results approximate to MMC in most cases, but much more efficiency can be preserved. For the layered MMC algorithms, stable performance are still available, and deeper layers lead to better recognition performance except for SUN data set.

In the second experiment, the discriminant ability of 2D features are evaluated, while the ALOI[5] and MNIST databases are involved. In the ALOI data set, [24], the whole data are combined while the original order of data is disordered, and then a subset of 50 categories are randomly selected to be involved into experiments. For each category of object, separate 18 and 54 images are randomly picked up to form a small training set compared with testing set of remaining images. For each digit of MNIST, 2,000 and 500 images from training and testing sets are randomly selected to be 2D data, respectively. Since most methods give the close results in different dimensions, the bar chart is adopted to illustrated the results in Fig. 2.

---

[5] http://aloi.science.uva.nl.

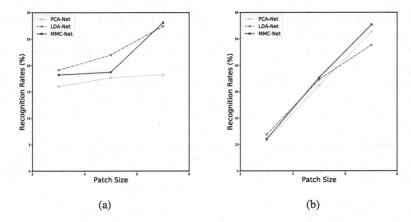

(a)   (b)

Fig. 3. The results of different 2D methods on three data sets. (a) Experimental results on MNIST data set. (b) Experimental results on ALOI data set.

As experimental results shown, the discriminant ability of different 2D methods are quite close to each other for ALOI data set. And the best result is contributed by 2DLDA with 95.89%, followed by 95.63% of 2D^{2}PCA. Nevertheless, it is noticeable that, 2DMMC can obtain close result 91.7%, and 2D^{2}MMC is able to reach 95.37%. In other words, MMC methods can attain similar performance to other methods. For MNIST database, 2DPCA presents the best result of 95.44%, and other methods are hardly to reach above 90%. On the contrary, the results from 2D^{2}LDA and 2DMMC are pessimistic among all algorithms, but hopefully, 2D^{2}MMC can still get close recognition results to 2D^{2}PCA method.

In the third experiment, different dimensionality reduction methods are adopted to learn deep neural network structures, i.e., PCA, LDA and MMC Networks, are involved. Two data sets, MNIST and ALOI, are involved in this experiment. For each digit in MNIST, 100 images are randomly selected from training and testing sets, while random 30 categories are selected from ALOI data set with reshaped size of 30 × 30. To reduce calculational complexity, three stages only are employed to learn the filter banks, and number of filters are set to be eight. With different size of patch sizes, it is able to disclose the intrinsic affection on subspace neural networks. The experimental results of patch sizes of 3, 4, 5 on two data sets are shown in Fig. 3. In terms of the results, there are few differences among three methods, and both LDA-Net and MMC-Net can reach better results in the stage of dimensionality reduction compared with PCA-Net. Furthermore, it seems that quite similar performance can be obtained with small patch sizes.

5   Conclusion

As a classical learning method, MMC is quite popular in various fields of data mining and pattern analysis, as well as its ubiquitous applications in intelligent computing. In this work, a direct MMC approach is given firstly, and then several variants of MMC are designed for adaptive learning, i.g., random MMC, layered MMC, 2D^2 based MMC. Inspired by PCA Network, a MMC network method is proposed to make simple deep learning applicable. Experiments on several data sets demonstrate comparable performance of proposed methods for applications of different categories of data types, and it is component to learn the associated patterns for adaptive recognition.

Acknowledgements. The authors would like to thank Universität zu Lübeck for sparse coding data set of MNIST, and the Chinese University of Hong Kong for target coding data set of STL-10. The corresponding author of this work is Dr. Miao Cheng.

References

1. Turk, M.A., Pentland, A.P.: Eigenfaces for recognition. J. Cogn. Neurosci. **3**(1), 195–197 (1981)
2. Belhumeur, P.N., Hespanha, J.P., Kriegman, D.J.: Eigenfaces vs. Fisherfaces: recognition using class specific linear projection. IEEE Trans. Patt. Anal. Machine Intell. **19**(7), 711–720 (1997)
3. Bishop, C.M.: Pattern Recognition and Machine Learning. Springer, New York (2011)
4. Yu, H., Yang, J.: A direct LDA algorithm for high-dimensional data - with application to face recognition. Patt. Recog. **19**(7), 2067–2070 (2001)
5. Li, H., Jiang, T., Zhang, K.: Efficient and robust feature extraction by maximum margin criterion. IEEE Trans. Neural Netw. **17**(1), 157–165 (2006)
6. Liu, J., Chen, S., Tan, X., et al.: Comments on 'Efficient and robust feature extraction by maximum margin criterion'. IEEE Trans. Neural Netw. **18**(6), 1862–1864 (2007)
7. Cheng, M., Tang, Y.Y., Pun, C.-M.: Nonparametric feature extraction via direct maximum margin alignment. In: Proceedings of IEEE International Conference on Machine Learning and Applications, Hawaii, USA, pp. 585–591 (2010)
8. Cheng, M., Fang, B., Tang, Y.Y., et al.: Incremental embedding and learning in the local discriminant subspace with application to face recognition. IEEE Trans. Syst. Man. Cybern. Part C Appl. Rev. **40**(5), 580–591 (2010)
9. Yan, S., Xu, D., Zhang, B., et al.: Graph embedding and extensions: a general framework for dimensionality reduction. IEEE Trans. Patt. Anal. Mach. Intell. **29**(1), 40–51 (2007)
10. Cheng, M., Tsoi, A.C.: CRH: a simple benchmark approach to continuous hashing. In: Proceedings of IEEE Global Conference on Signal and Information Processing, Orlando, USA, pp. 1076–1080 (2015)
11. Huang, G.-B., Zhou, H., Ding, X., et al.: Extreme learning machine for regression and multiclass classification. IEEE Trans. Syst. Man Cybern. Part B Cybern. **42**(2), 513–529 (2012)
12. Yang, J., Zhang, D., Frangi, A.F., et al.: Two-dimensional PCA: a new approach to appearance-based face representation and recognition. IEEE Trans. Patt. Anal. Mach. Intell. **26**(1), 131–137 (2004)
13. Kong, H., Li, X., Wang, L., et al.: Generalized 2D principal component analysis. In: Proceedings of IEEE International Joint Conference on Neural Networks, Montreal, Canada (2005)

14. Li, M., Yuan, B.: 2D-LDA: a statistical linear discriminant analysis for image matrix. Patt. Recog. Lett. **26**(5), 527–532 (2005)
15. Xiong, H., Swamy, M.N.S., Ahmad, M.O.: Two-dimensional FLD for face recognition. Patt. Recog. **38**(7), 1121–1124 (2005)
16. Ye, J., Janardan, R., Li, Q.: Two-dimensional linear discriminant analysis. In: NIPS (2004)
17. Chan, T.-H., Jia, K., Gao, S., et al.: PCANet: a simple deep learning baseline for image classification? IEEE Trans. Patt. Anal. Mach. Intell. **24**(12), 5017–5032 (2015)
18. Ojala, T., Pietikainen, M., Maenpaa, T.: Multiresolution gray-scale and rotation invariant texture classification with local binary patterns. IEEE Trans. Patt. Anal. Mach. Intell. **24**(7), 971–987 (2002)
19. Xiao, J., Ehinger, K.A., Hays, J., et al.: SUN database: exploring a large collection of scene categories. IJCV **119**(1), 3–22 (2016)
20. Lecun, Y., Bottou, L., Bengio, Y., et al.: Gradient-based learning applied to document recognition. Proc. IEEE **86**(11), 569–571 (1998)
21. Labusch, K., Barth, E., Martinetz, T.: Simple method for high-performance digit recognition based on sparse coding. IEEE Trans. Neural Netw. **19**(11), 1985–1989 (2008)
22. Coates, A., Lee, H., Ng, A.Y.: An analysis of single-layer networks in unsupervised feature learning. In: Proceedings of International Conference on Artificial Intelligence and Statistics, Ft. Lauderdale, USA, pp. 215–223 (2011)
23. Yang, S., Luo, P., Loly, C.C., et al.: Deep representation learning with target coding. In: AAAI, Austin Texas, USA, pp. 3848–3854 (2015)
24. Geusebroek, J.M., Burghouts, G.J., Smeulders, W.M.: The amsterdam library of object images. IJCV **61**(1), 103–112 (2005)

Multi-task Manifold Learning Using Hierarchical Modeling for Insufficient Samples

Hideaki Ishibashi[1], Kazushi Higa[2], and Tetsuo Furukawa[3(✉)]

[1] The Institute of Statistical Mathematics, Tachikawa 190-8562, Japan
ishibashi.hideaki@ism.ac.jp
[2] HORIBA Ltd., Kyoto 601-8510, Japan
kazushi.higa@horiba.com
[3] Kyushu Institute of Technology, Kitakyushu 808-0196, Japan
furukawa@brain.kyutech.ac.jp

Abstract. In this paper, we propose a method for multi-task manifold learning. For a set of tasks of dimensionality reduction, the aim of the method is to model each given dataset as a manifold, and map it to a low-dimensional space. For this purpose, we use a hierarchical manifold modeling approach. Thus, while each data distribution is represented by a manifold model, the obtained models are further modeled by a higher-order manifold in a function space. The higher-order model mediates the information transfer between tasks, and as a result, the performance of each task is improved. The results of simulations show that the proposed method can estimate manifolds approximately, even in cases in which a tiny number of samples are provided for each task.

Keywords: Multi-task learning · Multi-task unsupervised learning
Manifold learning · Hierarchical modeling · Multi-level modeling

1 Introduction

Multi-task learning is a paradigm of machine learning that aims to improve performance by simultaneously learning similar tasks [2, 28]. Many studies have been conducted on multi-task learning, particularly supervised learning. By contrast, there have been few studies on multi-task unsupervised learning, and only a few studies have been conducted on multi-task clustering [28]. To date, few works have been reported on dimensionality reduction, particularly in the context of non-linear manifold learning. The purpose of this study is to develop a method for multi-task manifold learning. We focus in particular on scenarios in which the number of data samples is too small to estimate manifolds and the assistance of other tasks is indispensable.

This work was supported by JSPS KAKENHI Grant Number 18K11472 and ZOZO Research.

© Springer Nature Switzerland AG 2018
L. Cheng et al. (Eds.): ICONIP 2018, LNCS 11303, pp. 388–398, 2018.
https://doi.org/10.1007/978-3-030-04182-3_34

A typical example is face image modeling. It is well known that face images are modeled by a manifold [3,23]. To estimate a face manifold, we typically need a sufficient number of photographs taken from various viewpoints with various expressions that cover the manifold entirely. However, in practice, it is typically difficult to obtain such an exhaustive image set of a single person. Instead, we typically have a huge number of photographs of other people. Thus, we have many image sets of various people, each of which consists of a small number (i.e., insufficient number) of photographs. In such a scenario, our aim is to improve modeling performance by transferring the information between tasks.

To achieve the above, we use a hierarchical modeling approach in this study. Thus, while each given dataset is modeled by a manifold, the manifold models are further modeled by a higher-order manifold in a function space. This higher-order model mediates information transfer among the given tasks, thereby improving the performance of manifold modeling. The proposed method consists of hierarchically coupled manifold models based on the kernel smoother (kernel-smoother-based manifold modeling: KSMM), referred to as the hierarchical KSMM (H-KSMM).

The remainder of the paper is structured as follows: The problem is formulated in Sect. 2, and related work is introduced in Sect. 3. The proposed method is presented in Sect. 4 and experiment results to verify it are described in Sect. 5. A discussion of the results and the conclusions of this paper are provided in the final section.

2   Problem Formulation

Suppose we have I tasks. Thus, we have I datasets $\{\mathcal{S}_1, \ldots, \mathcal{S}_I\}$ in high-dimensional space $\mathcal{X} = \mathbb{R}^{D_{\mathcal{X}}}$, each of which consists of N_i samples. The entire dataset is denoted as $\mathcal{S} = \bigcup_i \mathcal{S}_i = \{\mathbf{x}_n\}_{n=1}^N$, where $N = \sum_i N_i$. We also describe the entire dataset using matrix $\mathbf{X} = (\mathbf{x}_n^{\mathsf{T}}) \in \mathbb{R}^{N \times D_{\mathcal{X}}}$. Furthermore, let i_n be the task index of sample n and $\mathcal{N}_i$ the index set of samples that belong to task i.

When such datasets are provided, our first aim is to map the data to low-dimensional space $\mathcal{Z} = \mathbb{R}^{D_{\mathcal{Z}}}$. Thus, the first aim is to estimate $\{\mathbf{z}_n\}$ that corresponds to $\{\mathbf{x}_n\}$. Our second aim is to model each data distribution using a nonlinear manifold. Thus, for the ith dataset, the method models $\mathbf{x} \,|\, \mathbf{z} \sim \mathcal{N}\big(f_i(\mathbf{z}), \beta^{-1}\mathbf{I}\big)$, where $f_i : \mathcal{Z} \to \mathcal{X}$ is a smooth embedding from $\mathcal{Z}$ to $\mathcal{X}$. Then, the image of f_i becomes a nonlinear manifold $\mathcal{M}_i = f_i(\mathcal{Z})$ in $\mathcal{X}$. In this work, f_i is referred to as the 'task model.' Note that $\{f_i\}$ belongs to the same function space $\mathcal{F}$, because $\mathcal{X}$ and $\mathcal{Z}$ are common to all tasks in this paper.

To achieve the above aims, the following hierarchical model is assumed in this work. Suppose that $\mathcal{Y}$ is another low-dimensional space for task sets, and all task models $\{f_i\}$ are assigned to $\{\mathbf{y}_i\}$ as low-dimensional representations. Suppose further that $g : \mathcal{Y} \to \mathcal{F}$ is a smooth embedding that satisfies $f_i = g[\mathbf{y}_i]$. Thus, the task models are further modeled by manifold $\mathcal{L} = g(\mathcal{Y})$ in function space $\mathcal{F}$. Then, all datasets are modeled as $\mathbf{x} \,|\, \mathbf{z}, \mathbf{y} \sim \mathcal{N}\big(F(\mathbf{z}, \mathbf{y}), \beta^{-1}\mathbf{I}\big)$, where $F : \mathcal{Z} \times \mathcal{Y} \to \mathcal{X} : (\mathbf{z}, \mathbf{y}) \mapsto \big(g[\mathbf{y}]\big)(\mathbf{z})$. In this paper, F is referred to as a '*general model.*' Under these assumptions, the aim of multi-task manifold learning is then to estimate $\{\mathbf{z}_n\}$, $\{\mathbf{y}_i\}$, and F simultaneously.

3   Related Work

To date, few studies have reported multi-task learning in the context of dimensionality reduction tasks, subspace methods, and manifold learning. To the best of our knowledge obtained from a survey, multi-task principal component analysis is the only development in the literature that is expressly aimed at the multi-task learning of subspace methods [27]. However, by extending the scope of our survey, we can locate related methods in the field of hierarchical modeling (or multi-level modeling) that aim to obtain higher-order models of tasks [5]. Although hierarchical modeling does not aim to improve the performance of tasks, the areas of hierarchical modeling and multi-task learning overlap, where the former is sometimes used as an approach to the latter [9,10,29].

Among methods for unsupervised hierarchical modeling, the higher rank of self-organizing maps (SOM^2) is the most relevant work to this study [6,7]. SOM^2 has been applied to several problems in multi-task learning, such as face images of various people [14], nonlinear dynamical systems with latent state variables [21,22], the shapes of various objects [25,26], and members of various groups [12,13]. In this sense, SOM^2 is one of the earliest examinations of multi-task unsupervised learning for nonlinear subspaces.

Although SOM^2 works like a multi-task learning method, it remains challenging to estimate manifolds when the number of samples per task is small. Moreover, SOM^2 has several limitations that originate from SOM itself, such as poor manifold representation using discretized nodes and the brute force optimization of latent variables. In this paper, we attempt to eliminate such limitations from SOM^2 by replacing it with KSMM and extending it for the multi-task learning paradigm.

4   Proposed Method

KSMM is used as the building block of hierarchical manifold modeling in the proposed method. In this section, we first describe KSMM and introduce the proposed method, called H-KSMM.

4.1   Kernel-Smoother-Based Manifold Modeling (KSMM)

Generally, nonlinear methods for dimensionality reduction are categorized into two groups [17]. The first consists of methods that project data points *from* a high-dimensional space (data space) *to* a low-dimensional space (visualization space). Most dimensionality reduction methods are in this group. By contrast, the second group consists of methods that estimate the mapping *from* a low-dimensional space (latent space) *to* a high-dimensional space (visible space).

As the latter group of methods aim to model the data distribution using a manifold, we refer to the group as *manifold modeling*. Representative methods of manifold modeling are generative topographic mapping (GTM) [1] and the Gaussian process latent variable model (GPLVM) [17,18], which originate from self-organizing maps (SOMs) [16]. To estimate a smooth manifold, GTM and GPLVM use the Gaussian process, whereas SOM uses a kernel smoother.

KSMM uses a kernel smoother, such as the original SOM, instead of a Gaussian process because this makes it easier to extend SOM^2 to H-KSMM. Moreover, to the best of our knowledge, the kernel smoother stabilizes manifold modeling to a greater extent than the Gaussian process, particularly in challenging conditions, such as the case that we consider.

Although not by this particular name, KSMM has been proposed in many studies as a theoretical generalization of SOM [4,8,11,20,24]. According to these studies, the cost function of KSMM is given by

$$E = \frac{\beta}{2} \sum_n \int h(\mathbf{z}, \mathbf{z}_n) \left\| \mathbf{x}_n - f(\mathbf{z}) \right\|^2 p(\mathbf{z}) \, d\mathbf{z}. \tag{1}$$

In (1), $h(\mathbf{z}, \mathbf{z}')$ is a non-negative smoothing kernel defined on $\mathcal{Z}$, which is typically $h(\mathbf{z}, \mathbf{z}') = \mathcal{N}\left(\mathbf{z} \mid \mathbf{z}', \lambda_{\mathcal{Z}}^2 \mathbf{I}\right)$. The prior of $\mathbf{z}$ is a uniform distribution on a unit square space, that is, $p(\mathbf{z}) = 1$ for $\mathbf{z} \in [-1/2, +1/2]^{D_z}$; otherwise, $p(\mathbf{z}) = 0$. In this study, nonlinear mapping f is represented parametrically using orthonormal basis functions (e.g., normalized Legendre polynomials). Thus, $f(\mathbf{z} \mid \mathbf{V}) = \mathbf{V}^{\mathrm{T}} \boldsymbol{\varphi}^{\mathrm{T}}(\mathbf{z})$, where $\boldsymbol{\varphi} = (\varphi_1, \ldots, \varphi_L)^{\mathrm{T}}$ is the basis set and $\mathbf{V} \in \mathbb{R}^{L \times D_x}$ is the coefficient matrix.

Nonlinear mapping f and latent variables $\{\mathbf{z}_n\}$ are alternately updated, as in a generalized expectation maximization algorithm. To update f, coefficient matrix $\mathbf{V}$ is calculated as $\mathbf{V} = \mathbf{A}^{-1}\mathbf{B}\mathbf{X}$, where

$$\mathbf{A} = \int \boldsymbol{\varphi}(\mathbf{z}) \, \boldsymbol{\varphi}^{\mathrm{T}}(\mathbf{z}) \, \overline{h}(\mathbf{z}) \, p(\mathbf{z}) \, d\mathbf{z} \tag{2}$$

$$\mathbf{B} = \int \boldsymbol{\varphi}(\mathbf{z}) \, \mathbf{h}(\mathbf{z})^{\mathrm{T}} \, p(\mathbf{z}) \, d\mathbf{z}, \tag{3}$$

where $\mathbf{h}(\mathbf{z}) = \left(h(\mathbf{z}, \mathbf{z}_1), \ldots, h(\mathbf{z}, \mathbf{z}_N)\right)^{\mathrm{T}}$ and $\overline{h}(\mathbf{z}) = \sum_n h(\mathbf{z}, \mathbf{z}_n)$. By contrast, $\{\mathbf{z}_n\}$ are updated using a gradient method so that the value of the objective function (1) is reduced.

4.2 Hierarchical KSMM (H-KSMM)

H-KSMM consists of two hierarchically coupled KSMMs: a lower-KSMM and higher-KSMM. The lower KSMM estimates each task model, whereas the higher-KSMM estimates the general model.

In H-KSMM, task information is transferred in two ways. The first involves forming a *weighted mixture of the sample datasets*. If task i' is a neighbor of task i in latent space $\mathcal{Y}$, then sample set $\mathcal{S}_{i'}$ is merged into target set $\mathcal{S}_i$ as an auxiliary sample set with a larger weight. By contrast, if task i'' is far from task i in $\mathcal{Y}$, then $\mathcal{S}_{i''}$ is merged into $\mathcal{S}_i$ with a small (or zero) weight. Let us denote the weight of sample n of task i_n with respect to target task i as ρ_{in} $(0 \leq \rho_{in} \leq 1)$. Typically, $\rho_{in} \equiv \rho(\mathbf{y}_i, \mathbf{y}_{i_n}) = \exp\left[-\frac{1}{2\lambda_\rho^2}\|\mathbf{y}_i - \mathbf{y}_{i_n}\|^2\right]$, where λ_ρ determines the size of the neighborhood for data mixing. By contrast, the second way of transferring task information involves forming a *weighted mixture of the task models* among neighboring tasks, that is, the kernel smoothing of the task models.

The H-KSMM algorithm is as follows:

Step 1: Suppose $\{\mathbf{z}_n\}$ and $\{\mathbf{y}_i\}$ have been estimated in a preceding calculation loop (or initialized randomly in the first loop). In Step 1, ρ_{in} is calculated as described above.

Step 2: To obtain task models $\{f_i\}$, corresponding coefficient matrices $\{\mathbf{V}_i\}$ are calculated by $\mathbf{V}_i = \mathbf{A}_i^{-1}\mathbf{B}_i\mathbf{X}$, where

$$\mathbf{A}_i = \int \boldsymbol{\varphi}(\mathbf{z})\,\boldsymbol{\varphi}^{\mathrm{T}}(\mathbf{z})\,\overline{h}_i(\mathbf{z})\,p(\mathbf{z})\,d\mathbf{z} \tag{4}$$

$$\mathbf{B}_i = \int \boldsymbol{\varphi}(\mathbf{z})\,\mathbf{h}_i(\mathbf{z})^{\mathrm{T}}\,p(\mathbf{z})\,d\mathbf{z}. \tag{5}$$

In (5) (4), $\mathbf{h}_i(\mathbf{z}) = \left(\rho_{i1}h(\mathbf{z}, \mathbf{z}_1), \ldots, \rho_{iN}h(\mathbf{z}, \mathbf{z}_N)\right)^{\mathrm{T}}$, and $\overline{h}_i(\mathbf{z}) = \sum_n \rho_{in}h(\mathbf{z}, \mathbf{z}_n)$. The coefficient matrices are collectively expressed as third-order tensor $\underline{\mathbf{V}} = (\mathbf{V}_i) \in \mathbb{R}^{I \times L \times D_x}$.

Step 3: To obtain general model F, coefficient tensor $\underline{\mathbf{W}} \in \mathbb{R}^{I \times L \times D_x}$ is calculated by

$$\underline{\mathbf{W}} = \underline{\mathbf{V}} \times_1 \left(\mathbf{C}^{-1}\mathbf{D}\right) \tag{6}$$

$$\mathbf{C} = \int \boldsymbol{\psi}(\mathbf{y})\,\boldsymbol{\psi}^{\mathrm{T}}(\mathbf{y})\,\overline{k}(\mathbf{y})\,p(\mathbf{y})\,d\mathbf{y} \tag{7}$$

$$\mathbf{D} = \int \boldsymbol{\psi}(\mathbf{y})\,\mathbf{k}^{\mathrm{T}}(\mathbf{y})\,p(\mathbf{y})\,d\mathbf{y}, \tag{8}$$

where $\mathbf{k}(\mathbf{y}) = \left(k(\mathbf{y}, \mathbf{z}_1), \ldots, k(\mathbf{y}, \mathbf{y}_I)\right)^{\mathrm{T}}$, $\overline{k}(\mathbf{y}) = \sum_i k(\mathbf{y}, \mathbf{y}_i)$, and $k(\mathbf{y}, \mathbf{y}')$ and $\boldsymbol{\psi}(\mathbf{y})$ are the smoothing kernel and basis functions for the higher-KSMM, respectively. Symbol $\times_m$ denotes the tensor–matrix product of the mth mode. Then, the general model can be represented as

$$F(\mathbf{z}, \mathbf{y}) = \underline{\mathbf{W}} \times_1 \boldsymbol{\psi}(\mathbf{y}) \times_2 \boldsymbol{\varphi}(\mathbf{z}). \tag{9}$$

Step 4: Using a gradient method, latent variables $\{\mathbf{y}_i\}$ are updated so that the approximated cost function of the higher-KSMM decreases in value. The approximated cost function is given by

$$E(\mathbf{y}_i) = \frac{\beta}{2} \sum_{n \in \mathcal{N}_i} \| \mathbf{x}_n - F(\mathbf{z}_n, \mathbf{y}_i \mid \underline{\mathbf{W}}) \|^2 . \tag{10}$$

The integral with respect to $\mathbf{y}$ is omitted to simplify the calculation. Such an approximation is commonly used in SOM and KSMM literatures.

Step 5: Finally, latent variables $\{\mathbf{z}_n\}$ are updated using the gradient method so that the approximated cost function of the lower-KSMM decreases. The cost function is given by

$$E(\mathbf{z}_n) = \frac{\beta}{2} \| \mathbf{x}_n - F(\mathbf{z}_n, \mathbf{y}_{i_n}) \|^2 . \tag{11}$$

These five steps are repeated until the calculation converges. During the iterations, the length constant of the smoothing kernels is gradually reduced to avoid local minima.

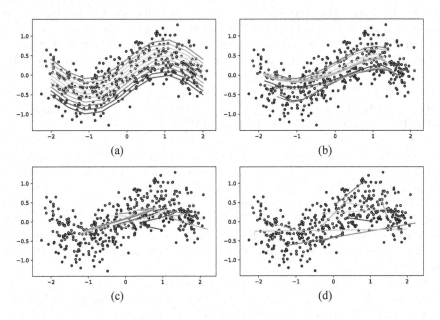

Fig. 1. Results of the artificial dataset. A total of 200 tasks and 2 samples/task were used for training, and 10 of 200 manifolds are shown in the figures. (a) Ground truth. (b) H-KSMM (multi-task learning). (c) SOM^2 (multi-task learning). (d) KSMM (single-task learning).

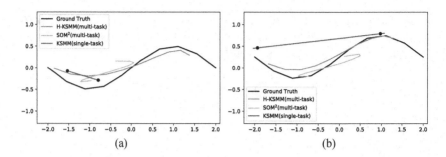

Fig. 2. Two representative tasks extracted from Fig. 1. The two black markers represent the data for the task.

5   Experimental Results

5.1   Artificial Datasets

The performance of the proposed method was examined using an artificial dataset. We used one-dimensional (1D) sinusoidal shape manifolds with different biases, which are embedded into ten-dimensional (10D) space. The solid curves in Fig. 1(a) are the examples of the manifolds. (To draw the figure, the manifolds in 10D space are projected to the principal 2D space. The rest of the eight dimensions correspond to noise.) To generate the sample sets, we added 10D Gaussian noise $\varepsilon \sim \mathcal{N}(\mathbf{0}_{10}, \sigma^2 \mathbf{I}_{10})$, where $\sigma = 0.2$. For the training dataset, we prepared 200 tasks, each of which consisted of N_i samples (N_i was common to all tasks) generated randomly. We compared the results of H-KSMM (the proposed method), SOM^2, and a single task on KSMM[1].

A representative result is shown in Fig. 1. In this case, each task has only two samples. Thus, it is impossible to estimate the manifold shape using single-task learning (Fig. 1(d)). Surprisingly, the proposed algorithm was able to capture the outlines of the manifold shapes (Fig. 1(b)). To show details, two of the 200 tasks are shown in Fig. 2. Because only two samples were provided to the task, single-task KSMM estimated the manifold as a straight-line segment that connected two data points. By contrast, H-KSMM was able to reproduce the sinusoidal manifold shape, although its marginal area was truncated because there were insufficient samples.

We assessed learning performance quantitatively using two methods: the root mean square error (RMSE) between the test data and manifold, and mutual information (MI) of the true and estimated latent variables. RMSE evaluates the error in visible space $\mathcal{X}$, whereas MI evaluates accuracy in latent space $\mathcal{Z}$. Figure 3(a) and (b) show the RMSE and MI measured using the test data on

---

[1] For a fair comparison, we modified SOM^2 so that it could represent a continuous mapping using basis functions in the same manner as KSMM. Thus, it should be rather referred to as KSMM^2. By this modification, the result shown for SOM^2 is better than that of the original.

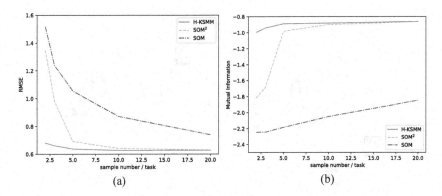

Fig. 3. Generalization performance of existing tasks on the test data. The horizontal axis denotes samples/task for training. (a) Root mean square error between the data and models. (b) Mutual information between the true and estimated latent variables.

the given tasks, respectively. The results show that H-KSMM exhibited excellent performance, particularly when the number of samples/task was small.

Using the general model, it is not only possible to estimate the manifolds of the given tasks, but also possible to predict manifolds of unseen tasks. Figure 4 shows the RMSE and MI for 100 new tasks. The results show that H-KSMM has a high generalization capability, even for new tasks.

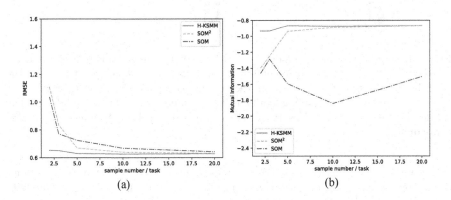

Fig. 4. Generalization performance on new tasks. The horizontal axis denotes samples/task for training. (a) Root mean square error between the data and models. (b) Mutual information between the true and estimated latent variables.

5.2 Face Image Datasets

We applied the proposed method to face image modeling. The dataset used was a subset of the extended Cohn–Kanade (CK+) face image database [15,19]. The data used in the experiment consisted of image sequences of 78 people, where

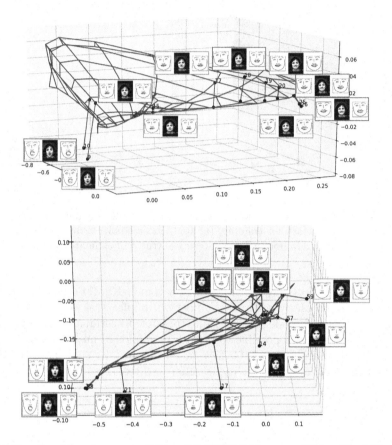

Fig. 5. Results of face image modeling of H-KSMM. Face manifolds of 2 of 78 people are shown. The red boxes represent the training data (5 samples per task) and the green boxes represent the test data. In each box, the original face image is displayed at the center and the corresponding landmark face is indicated on the left-hand side. The landmark face reconstructed by H-KSMM is indicated on the right-hand side. (Color figure online)

each sequence began with a neutral expression and proceeded to a distinct emotional expression. The dataset thus contained a large number of intermediate expressions. In this study, we used four types of sequences: anger, fear, happiness, and surprise. We also used landmark data as features. Thus, each face datum was represented by a 136-dimension vector that corresponded to the 2D coordinates of 68 landmarks. To construct the training data, we sampled five images randomly from each person. Thus, the entire dataset consisted of 78 tasks, each of which consisted of five samples. Note that two expressions were often missing in each task, and it was nearly impossible to estimate the face manifolds using single-task learning.

The results are shown in Fig. 5, which represents the face manifolds estimated by H-KSMM depicted in 3D space spanning the first three principal components.

H-KSMM represented the training data well (indicated by red boxes) and reproduced the test data successfully (indicated by green boxes). Thus, these face manifolds successfully represented various facial expressions, even though the data provided were insufficient.

6   Discussion and Conclusion

In this paper, we proposed a method for multi-task manifold modeling based on the hierarchical modeling approach. Characteristics of the method are two means of information transfer: *the weighted mixture of sample datasets* and *the weighted mixture of task models*. The latter method of information transfer is mediated by a higher-order model in hierarchical modeling; that is, the former method of information transfer was executed *before* the manifold modeling of each task, whereas the latter was executed *after* manifold modeling. Providing a theoretical basis for these means of information transfer will form the focus of our future work in this area.

References

1. Bishop, C.M., Svensen, M., Williams, C.K.I.: GTM: the generative topographic mapping. Neural Comput. **10**, 215–234 (1998)
2. Caruana, R.: Multitask learning. Mach. Learn. **28**, 41–75 (1997)
3. Chang, Y., Hu, C., Feris, R., Turk, M.: Manifold based analysis of facial expression. Image Vis. Comput. **24**(6), 605–614 (2006)
4. Cheng, Y.: Convergence and ordering of Kohonen's batch map. Neural Comput. **9**(8), 1667–1676 (1997)
5. Dedrick, R.F., et al.: Multilevel modeling: a review of methodological issues and applications. Rev. Educ. Res. **79**(1), 69–102 (2009)
6. Furukawa, T.: SOM of SOMs: self-organizing map which maps a group of self-organizing maps. In: Duch, W., Kacprzyk, J., Oja, E., Zadrożny, S. (eds.) ICANN 2005. LNCS, vol. 3696, pp. 391–396. Springer, Heidelberg (2005). https://doi.org/10.1007/11550822_61
7. Furukawa, T.: SOM of SOMs. Neural Netw. **22**(4), 463–478 (2009)
8. Graepel, T., Burger, M., Obermayer, K.: Self-organizing maps: generalizations and new optimization techniques. Neurocomputing **21**, 173–190 (1998)
9. Han, L., Zhang, Y.: Learning multi-level task groups in multi-task learning. In: Proceedings of the National Conference on Artificial Intelligence, vol. 4, pp. 2638–2644 (2015)
10. Han, L., Zhang, Y.: Learning tree structure in multi-task learning. In: Proceedings of the ACM SIGKDD International Conference on Knowledge Discovery and Data Mining, vol. 2015, pp. 397–406, August 2015
11. Heskes, T., Spanjers, J.J., Wiegerinck, W.: EM algorithms for self-organizing maps. In: IJCNN (6), pp. 9–14 (2000)
12. Ishibashi, H., Furukawa, T.: Hierarchical tensor SOM network for multilevel-multigroup analysis. Neural Process. Lett. **47**, 1011–1025 (2017)

13. Ishibashi, H., Shinriki, R., Isogai, H., Furukawa, T.: Multilevel–multigroup analysis using a hierarchical tensor SOM network. In: Hirose, A., Ozawa, S., Doya, K., Ikeda, K., Lee, M., Liu, D. (eds.) ICONIP 2016. LNCS, vol. 9949, pp. 459–466. Springer, Cham (2016). https://doi.org/10.1007/978-3-319-46675-0_50

14. Jiang, J., Zhang, L., Furukawa, T.: Improving the generalization of fisherface by training class selection using SOM^2. In: King, I., Wang, J., Chan, L.-W., Wang, D.L. (eds.) ICONIP 2006, Part II. LNCS, vol. 4233, pp. 278–285. Springer, Heidelberg (2006). https://doi.org/10.1007/11893257_31

15. Kanade, T., Cohn, J., Tian, Y.: Comprehensive database for facial expression analysis. In: Proceedings of 4th IEEE International Conference on Automatic Face and Gesture Recognition, FG 2000, pp. 46–53 (2000)

16. Kohonen, T.: Self-organized formation of topologically correct feature maps. Biol. Cybern. **43**(1), 59–69 (1982)

17. Lawrence, N.D.: Probabilistic non-linear principal component analysis with gaussian process. J. Mach. Learn. Res. **6**, 1783–1816 (2005)

18. Lawrence, N.D.: Gaussian process latent variable models for visualisation of high dimensional data. In: Neural Information Processing Systems (NIPS), vol. 16, pp. 329–336 (2003)

19. Lucey, P., Cohn, J.F., Kanade, T., Saragih, J., Ambadar, Z., Matthews, I.: The extended Cohn-Kanade Dataset (CK+): a complete dataset for action unit and emotion-specified expression. In: 2010 IEEE Computer Society Conference on Computer Vision and Pattern Recognition - Workshops, pp. 94–101, June 2010

20. Luttrell, S.P.: Self-organization: A derivation from first principle of a class of learning algorithms. In: IEEE Conference on Neural Networks, pp. 495–498 (1989)

21. Ohkubo, T., Tokunaga, K., Furukawa, T.: RBF×SOM: an efficient algorithm for large-scale multi-system learning. IEICE Trans. Inf. Syst. **E92–D**(7), 1388–1396 (2009)

22. Ohkubo, T., Furukawa, T., Tokunaga, K.: Requirements for the learning of multiple dynamics. In: Laaksonen, J., Honkela, T. (eds.) WSOM 2011. LNCS, vol. 6731, pp. 101–110. Springer, Heidelberg (2011). https://doi.org/10.1007/978-3-642-21566-7_10

23. Shan, C., Gong, S., McOwan, P.W.: Appearance manifold of facial expression. In: Sebe, N., Lew, M., Huang, T.S. (eds.) HCI 2005. LNCS, vol. 3766, pp. 221–230. Springer, Heidelberg (2005). https://doi.org/10.1007/11573425_22

24. Verbeek, J., Vlassis, N., Krose, B.: Self-organizing mixture models. Neurocomputing **63**, 99–123 (2005)

25. Yakushiji, S., Furukawa, T.: Shape space estimation by SOM^2. In: Lu, B.-L., Zhang, L., Kwok, J. (eds.) ICONIP 2011, Part II. LNCS, vol. 7063, pp. 618–627. Springer, Heidelberg (2011). https://doi.org/10.1007/978-3-642-24958-7_72

26. Yakushiji, S., Furukawa, T.: Shape space estimation by higher-rank of SOM. Neural Comput. Appl. **22**(7–8), 1267–1277 (2013)

27. Yamane, I., Yger, F., Berar, M., Sugiyama, M.: Multitask principal component analysis. In: Durrant, R.J., Kim, K.E. (eds.) Proceedings of The 8th Asian Conference on Machine Learning, The University of Waikato, Hamilton, New Zealand, 16–18 November 2016. Proceedings of Machine Learning Research, vol. 63, pp. 302–317. PMLR (2016)

28. Zhang, Y., Yang, Q.: An overview of multi-task learning. Natl. Sci. Rev. **5**, 30–43 (2018)

29. Zweig, A., Weinshall, D.: Hierarchical regularization cascade for joint learning. In: 30th International Conference on Machine Learning, ICML 2013, no. Part 2, pp. 1074–1082 (2013)

Enhanced Metric Learning via Dempster-Shafer Evidence Theory

Ying Li[1,2], Yabo Zhang[1], and Yaxin Peng[3]([✉])

[1] School of Computer Engineering and Science,
Shanghai University, Shanghai 200444, China
[2] Shanghai Institute for Advanced Communication and Data Science,
Shanghai University, Shanghai 200444, China
[3] Department of Mathematics, School of Science, Shanghai University,
Shanghai 200444, China
yaxin.peng@shu.edu.cn

Abstract. Metric learning is a hot topic in machine learning. A proper learned metric can measure the similarity between samples better and hence significantly improves the performance of machine learning algorithm. In this paper, we propose a novel enhanced distance metric learning method via Dempster-Shafer (D-S) evidence theory. We consider each instance as an independent source of evidence and combine these pieces of evidence by using Dempster's rule. Firstly, with reference to the D-S theory, we construct the balanced weight function corresponding to each instance in the metric. Secondly, the novel competitive-cost function is given, which can improve classifier accuracy by narrowing the inner-class distance and increasing the inter-class distance. Finally, we implement a series of experiments on classification by using UCI and face recognition data sets. Experimental results validate that the proposed method can significantly improve the performance of the classifier and the robustness of the algorithm.

Keywords: Metric learning · Dempster-Shafer (D-S) evidence theory
Classification · Competitive-cost function

1 Introduction

Distance metric learning in feature space has always been a hot issue in machine learning and pattern recognition in recent years. Many pattern recognition and machine learning applications involve feature space classification and clustering. The similarity between samples determines the performance of a variety of machine learning algorithms such as K-Nearest Neighbor (K-NN) classification [1], Radial Basis Function Network [2] and Support Vector Machine [3] (e.g., In the K-NN classifier, the key is to identify the set of labeled instance that is closest to a given test instance in the feature space - involving the estimation of a distance metric). In order to describe the similarity between the samples,

© Springer Nature Switzerland AG 2018
L. Cheng et al. (Eds.): ICONIP 2018, LNCS 11303, pp. 399–410, 2018.
https://doi.org/10.1007/978-3-030-04182-3_35

a mapping function based on the sample input feature space is very important. By learning from samples, this appropriate mapping function may be the key to the successful application of these algorithms.

Based on the above analysis, Mahalanobis distance metric learning method [4–8] can improve the classification accuracy by changing the relative position of the samples in the feature space. The goal of distance metric learning algorithm is to find a proper transformation of the feature space. Using the prior information in label form, the relevant dimensions are emphasized, while irrelevant ones are discarded. That is to say, samples belonging to the same class should be close to each other, while those from different classes should be farther apart. In the literature, a large number of metric learning methods have been proposed and performed well across various learning tasks, such as relevant component analysis (RCA) [9], principal component analysis (PCA) [10], distance metric learning of large margin nearest neighbor (LMNN) [11], information-theoretic metric learning (ITML) [12], logistic discriminant-based metric learning (LDML) [13], sparse distance metric learning (SDML) [14], keep it simple and straightforward metric learning (KISS) [15], least squared-residual metric learning (LSML) [16] and an intrinsic approach [17].

As the above analysis indicates, the main task of metric learning is to learn a distance function that can reflect the characteristics of the sample space through the training data. Under this distance function, the samples of the similar are close together, and vice versa. Since the target distance function of metric learning is indispensable for various learners, traditional metric learning theories combined with other fields in machine learning have formed many research hotspots. This combination will have a significant impact on machine learning research. In this paper, we propose a new distance metric learning method via D-S theory, and apply the gradient descent method to solve the corresponding optimization problem.

The paper is organized as follows. In Sect. 2, we review the basics of D-S theory and describe the relation of our work. Section 3 describes our algorithm model and gives the optimization objective function. The optimization algorithm is introduced in Sect. 4. Section 5 shows the performance comparison between our proposed method and other metric learning methods. Finally, we conclude our main contributions in Sect. 6.

2   Preliminaries

In this section, we introduce some notations and preliminaries. D-S evidence theory [18], originated in the 1960s, uses the discernment frame Θ to represent the proposition set of interest, and defines a mass function $m : 2^{\Theta} \to [0, 1]$ as the basic probability assignment function on the discernment frame. The function satisfies the following two conditions:

$$m(\emptyset) = 0, \quad \sum_{A \subseteq \Theta} m(A) = 1 \tag{1}$$

For all proposition sets, the belief function $Bel()$ and the plausibility function $Pl()$ are defined as:

$$Bel(A) = \sum_{B \subseteq A} m(B), \quad \forall A \subseteq \Theta$$

$$Pl(A) = 1 - Bel(\bar{A}), \quad \forall A \subseteq \Theta \tag{2}$$

That is, the belief function of A is the sum of the belief degrees of each subset in A, indicating the minimum degree of uncertainty support to the establishment of the proposition. The plausibility function indicates the extent that the proposition A is not denied. $[Bel(A), Pl(A)]$ represents the uncertainty interval of proposition set A.

Dempster also defined the well-known Dempster's rule of combination. Let m_1 and m_2 be two mass functions derived from independent items of evidence. They can be fused via Dempster's rule to induce a new mass function $m_1 \oplus m_2$ defined as:

$$(m_1 \oplus m_2)(A) = K^{-1} \sum_{A_i \cap B_i = A} m_1(A_i) m_2(B_i) \tag{3}$$

where

$$K = 1 - \sum_{A_i \cap B_i = \emptyset} m_1(A_i) m_2(B_i) \tag{4}$$

The evidential K-NN [19] classification rule addresses the problem of classifying an unseen pattern on the basis of its nearest neighbors in a recorded data set is addressed from the point of view of Dempster-Shafer theory. Each neighbor x_j of sample x_i to be classified is considered as an item of evidence that supports certain hypotheses regarding the class membership of that pattern. The degree of support m_i^j is defined as a function of the distance d_i^j between these two vectors:

$$m_i^j(y_q) = \alpha_0 f_q(d_i^j)$$

$$m_i^j(\Theta) = 1 - \alpha_0 f_q(d_i^j) \tag{5}$$

where the index q indicates that the influence of d_i^j may depend on the class of x_i, and y_q is the label of x_j and Θ is a collection of all class labels.

The evidence of the k nearest neighbors is then pooled by means of Dempster's rule of combination:

$$m_i^{\Gamma_q}(\{y_q\}) = 1 - \prod_{j \in \Gamma_q} (1 - \alpha_0 f_q(d_i^j))$$

$$m_i^{\Gamma_q}(\Theta) = \prod_{j \in \Gamma_q} (1 - \alpha_0 f_q(d_i^j)) \tag{6}$$

This approach provides a global treatment of such issues as ambiguity and distance rejection, imperfect knowledge regarding the class membership of training patterns. The effectiveness of this scheme as compared to the voting and distance-weighted K-NN procedures is demonstrated using several sets of simulated and real-world data.

Based on the above work, Evidential Dissimilarity Metric Learning (EDML) algorithm [20] was proposed that uses a low-dimensional transformation of the input space to learn appropriate metrics M. As the result, the learned metric M can lead $\boldsymbol{x}_i$ only close to samples from the same class in the transformed space, thus protecting the classification performance of the evidential K-NN method. Based on EDML, we construct the competitive-cost function, which can improve traditional K nearest classifier accuracy by narrowing the inner-class distance and increasing the inter-class distance.

3   Metric Learning with Competitive-Cost Function

Let $\{(\boldsymbol{x}_i, y_i)\}_{i=1}^n$ be the labeled samples, where $\boldsymbol{x}_i = (x_i^1, \ldots, x_i^d)^T \in \mathbb{R}^d$ is a d dimensional column vector and $y_i \in \{c_1, \ldots, c_m\}$ is the corresponding class label. Our goal is to learn a positive semi-definite matrix $M \in \mathbb{R}^{d \times d}$, which will be used to compute the following squared distance:

$$D(\boldsymbol{x}_i, \boldsymbol{x}_j) = (\boldsymbol{x}_i - \boldsymbol{x}_j)^T M (\boldsymbol{x}_i - \boldsymbol{x}_j) \tag{7}$$

For training example $\boldsymbol{x}_j$, w_i^j denotes the weight of $\boldsymbol{x}_i$, which can be simply defined as:

$$w_i^j = 1 - \exp\{-\lambda D(\boldsymbol{x}_i, \boldsymbol{x}_j)\} \tag{8}$$

where λ is a given parameter that controls the distance between sample pairs.

Define a training samples set $\Gamma_q (q = 1, \ldots, m)$, where the samples in this set belong to the same class label c_q. We can refer to Dempster's rule to deduce a global importance value for all training samples in Γ_q:

$$w_i^{\Gamma_q} = \prod_{j \in \Gamma_q} [1 - \exp\{-\lambda D(\boldsymbol{x}_i, \boldsymbol{x}_j)\}] \tag{9}$$

For $q = 1, \ldots, m$, the global weight $w_i^{\Gamma_q}$ quantifies the evidence refined from the training examples that supports the assertion $y_i = c_q$. In other words, it can be viewed as a calculation of the unreliability of the hypothesis $y_i = c_q$. The corresponding weight should be close to zero if the true value of y_i is c_q; in contrast, for all $r \neq q$, $w_i^{\Gamma_r}$ should be close to one. With above analysis, we construct the competitive-cost function of the model. The competitive-cost function consists of two terms, one for attracting samples with the same label and the other for rejecting alien labels and making them more separated. The two terms have a clear competitive relationship because the former narrows the sample point directly, and the other is the opposite. We discuss each term in order.

The first term of competitive-cost function, penalizes the larger distance between each input and its similar sample, can be given by:

$$\varepsilon_{pull} = \frac{1}{2}(\omega_i^{\Gamma_q})^2 \tag{10}$$

The gradient of this term produces a pull in the input space of the linear transformation to attract similar samples, which will penalize the distance of all same labels and eventually bring the samples together as much as possible in order to improve the performance of the classifier.

The second term of the competitive-cost function penalizes the small distance between the different classes of samples. By defining the form of multiplication, all samples that are dissimilar from the target sample category are away from it. The second term of the competitive-cost function ε_{push} is given by:

$$\varepsilon_{push} = \frac{1}{2}(1 - \prod_{r \neq q}^{m} \omega_i^{\Gamma_r})^2 \tag{11}$$

The gradient of this term produces a pushing force that will push the counterfeiter.

We combine the two terms ε_{pull} and ε_{push} into a predictive cost function for the training sample point (x_i, y_i) based on the assumptions presented above. Since these two terms have a competitive effect - not only attract isomorphism, but also exclusion isomerism. We introduce the weighted parameter $\mu \in [0, 1]$ to balance these goals:

$$\varepsilon_i = \sum_{q=1}^{m} \delta_{i \cdot q} \cdot \{(1 - \mu)\varepsilon_{pull} + \mu\varepsilon_{push}\} \tag{12}$$

where

$$\begin{cases} \delta_{i \cdot q} = 1 & \text{if } y_i = c_q \\ \delta_{i \cdot q} = 0 & \text{if } y_i \neq c_q \end{cases} \tag{13}$$

When $y_i = c_q$, minimizing ε_i can force both ε_{pull} and ε_{push} to approach zero as much as possible. Although in our experience, the result of minimizing the cost function is not sensitive to the value of the parameter μ, we usually choose it by cross validation. In the following experiments, the value $\mu = 0.5$ works best.

For all training sample points, our goal is to find an optimal metric matrix M. For this purpose, we use the cumulative sum of the smallest form, which can be defined as:

$$\arg \min_{M} \varepsilon = \frac{1}{n} \sum_{i=1}^{n} \varepsilon_i \tag{14}$$
$$\text{s.t.} \quad M \succeq 0$$

4   Optimization

There is a constraint on the semi-definite matrix domain for the objective function Eq. 14. It can be solved by standard solvers, such as CPLEX and MOSEK. These standard semi-definite problem solvers require an expensive computational cost. Fortunately, the domain of M is the cone of a positive semi-definite matrix, which can be factorized as $M = L^T L$. Hence, we can reformulate Eq. 7 as:

$$D(\boldsymbol{x}_i, \boldsymbol{x}_j) = ||L(\boldsymbol{x}_i - \boldsymbol{x}_j)||_2^2 \tag{15}$$

So the objective function Eq. 14 can be rewritten as follows

$$\arg\min_L \; \varepsilon = \frac{1}{n}\sum_{i=1}^{n}\varepsilon_i \tag{16}$$

This is an "unconstrained" metric learning problem, so we use the gradient descent method to find the optimal metric for this problem.

$$\frac{\partial \varepsilon_i}{\partial L} = \sum_{q=1}^{m}\delta_{i\cdot q}\left\{(1-\mu)\frac{\partial\varepsilon_{pull}}{\partial L} + \mu\frac{\partial\varepsilon_{push}}{\partial L}\right\} \tag{17}$$

In which, for $q = 1,\ldots,m$, the partial derivative $\frac{\partial\varepsilon_{pull}}{\partial L}$ of ε_{pull} to L can be expressed as follows:

$$\begin{aligned}
\frac{\partial\varepsilon_{pull}}{\partial L} &= \omega_i^{\Gamma_q}\frac{\partial\omega_i^{\Gamma_q}}{\partial L}\\
&= \omega_i^{\Gamma_q}\sum_{j\in\Gamma_q}(\frac{\partial\omega_i^j}{\partial L}\prod_{l\in\Gamma_q\backslash j}\omega_i^l)\\
&= \omega_i^{\Gamma_q}\sum_{j\in\Gamma_q}\left(2\lambda(1-\omega_i^j)L(\boldsymbol{x}_i - \boldsymbol{x}_j)(\boldsymbol{x}_i - \boldsymbol{x}_j)^T\prod_{l\in\Gamma_q\backslash j}\omega_i^l\right)
\end{aligned} \tag{18}$$

As above, $\frac{\partial\varepsilon_{push}}{\partial L}$ can be calculated as:

$$\begin{aligned}
\frac{\partial\varepsilon_{push}}{\partial L} &= (1-\prod_{r\neq q}^{m}\omega_i^{\Gamma_r})(-\frac{\partial(\prod_{r\neq q}^{m}\omega_i^{\Gamma_r})}{\partial L})\\
&= (1-\prod_{r\neq q}^{m}\omega_i^{\Gamma_r})(-\sum_{r\neq q}^{m}(\frac{\partial\omega_i^{\Gamma_r}}{\partial L}\prod_{s\neq r,q}^{m}\omega_i^{\Gamma_s}))\\
&= (1-\prod_{r\neq q}^{m}\omega_i^{\Gamma_r})\left(-\sum_{r\neq q}^{m}((\sum_{j\in\Gamma_r}\frac{\partial\omega_i^j}{\partial L}\prod_{l\in\Gamma_r\backslash j}\omega_i^l)\prod_{s\neq r,q}^{m}\omega_i^{\Gamma_s})\right)\\
&= (1-\prod_{r\neq q}^{m}\omega_i^{\Gamma_r})\left(-\sum_{r\neq q}^{m}((\sum_{j\in\Gamma_r}2\lambda(1-\omega_i^j)L\right.\\
&\quad\left.(\boldsymbol{x}_i - \boldsymbol{x}_j)(\boldsymbol{x}_i - \boldsymbol{x}_j)^T)\prod_{l\in\Gamma_r\backslash j}\omega_i^l)\prod_{s\neq r,q}^{m}\omega_i^{\Gamma_s})\right)
\end{aligned} \tag{19}$$

In order to speed up the convergence process, the metric matrix obtained by the classical metric learning method can be used instead of the identity matrix as the initialization. We refer to the proposed method as Metric Learning via D-S theory (MLDST). Its pseudo-code implementation is given in Algorithm 1. The inputs to the algorithm are the starting matrix L, the data matrix, and the corresponding parameters.

Algorithm 1. MLDST

Input: the data matrix $X \in \mathbb{R}^{d \times n}$; the label matrix $Y \in \mathbb{R}^{1 \times n}$; the parameter λ; the budget iteration number T; the weighted parameter $\mu \in [0, 1]$; the learning rate η.
Output: the metric matrix $M = L^T L$.
1: Initialize $L \in \mathbb{R}^{l \times d}$; $\lambda = 1$; $\mu = 0.50$; $\eta = 1e - 6$.
2: **for** t = 1,...,T **do**
3:     gradient dG = 0
4:     **for** i = 1,...,n **do**
5:         comput $\frac{\partial \varepsilon_{push}}{\partial L}$ by solving (18).
6:         comput $\frac{\partial \varepsilon_{pull}}{\partial L}$ by solving (19).
7:         comput $\frac{\partial \varepsilon_i}{\partial L}$ by solving (17).
8:         dG = dG + $\frac{\partial \varepsilon_i}{\partial L}$.
9:     **end for**
10:    $L = L - \eta \times dG$
11: **end for**

---

5   Experimental Results

In order to verify the performance of the proposed method, we present a series of experiments in this section. The experiments contain two parts. In the first part, several data sets from the UCI Machine Learning Repository were used. The other one is the face recognition dataset. We compare the performance of the proposed method with RCA, PCA, LMNN, ITML, LDML, SDML and LSML. In addition, a baseline experiment, i.e., the Euclidean method, is conducted by using k-NN classifier with the Euclidean distance.

5.1   UCI Data Sets

We test our proposed method on four real-world datasets from UCI repository: Iris, Wine, Seeds and Digits. In our experiments, each data set is randomly divided into two subsets: we selected 70% of the sample as the training set, while the rest as the test set. In order to explain the difference of various metric learning algorithms on data sets, we run ten times on each data set and then take the average correct rate.

Comparison and Parameter Setting. The performances of our method are compared with the following methods.

(1) Euclidean: Without metric learning.
(2) RCA: The number of chunks is chosen from 50 to 100.
(3) PCA: We reserve 85 to 95 percent of the principal components.
(4) LMNN: The parameter μ is chosen increasingly from 0.1 to 1 with interval 0.1.
(5) ITML: The identity matrix is set as the initial metric matrix. The γ is chosen from 10^{-4} to 10^4.
(6) SDML: The balance parameter η is chosen increasingly from 0.1 to 1 with interval 0.1 and the sparsity parameter $\mu = 0.01$.
(7) LSML: The prior matrix M_0 is set as an identity matrix.
(8) MLDST: For our proposed method, we set $\lambda = 1$ and $\mu = 0.5$, as many experiments demonstrate that the performance of these two parameters is better.

The results of best classification accuracy for each method are listed in Table 1, in which results obtained by Euclidean distance are also presented as one baseline for comparison. Table 1 shows that our proposed method has a good performance on most data sets.

Table 1. Comparison with other metric learning algorithms on the UCI repository.

Method	UCI repository accuracy (ave +/− std)			
	Iris	Wine	Seeds	Digits
	[150/4/3]	[178/13/3]	[210/7/3]	[1797/64/10]
Euclidean	93.24(+/−0.62)	72.02(+/−7.72)	87.61(+/−7.12)	97.51(+/−1.75)
RCA	95.33(+/−5.25)	93.23(+/−2.71)	94.36(+/−3.95)	94.49(+/−0.62)
PCA	94.16(+/−4.19)	75.23(+/−3.36)	90.84(+/−4.41)	96.15(+/−2.74)
LMNN	96.67(+/−4.47)	91.45(+/−6.02)	94.28(+/−2.85)	98.27(+/−1.51)
ITML	95.14(+/−3.71)	96.86(+/−3.11)	92.51(+/−2.15)	98.24(+/−0.67)
LDML	94.68(+/−5.11)	96.65(+/−4.79)	93.21(+/−3.38)	98.15(+/−2.13)
SDML	97.18(+/−2.17)	92.72(+/−4.61)	91.23(+/−6.52)	98.01(+/−3.30)
LSML	95.87(+/−3.56)	94.11(+/−6.36)	93.87(+/−3.73)	97.21(+/−1.94)
MLDST	**97.33**(+/−4.42)	**97.01**(+/−5.03)	**94.76**(+/−3.34)	**98.35**(+/−2.39)
Average	95.86	92.64	92.54	97.37

Detailed Analysis of the Parameter λ. Figure 1 shows that the parameter λ controls the effect of the distance between the sample points based on the cost function. The value should be adjusted for different data. Under normal circumstances, a sufficiently small parameter λ may cause the distance between the pair too large, which will affect the subsequent classification processing. However if the λ is too large, it may also have a negative effect. Experiments show that when λ is between 0.01 and 1, better results are obtained.

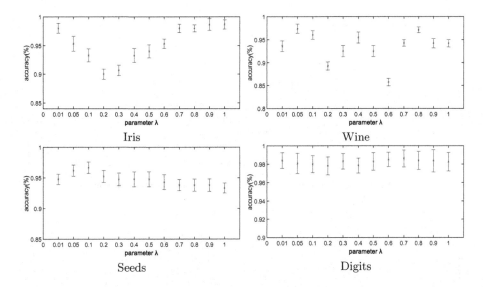

Fig. 1. Average accuracy of the K-NN classification on the UCI repository with regard to the parameter λ.

Detailed Analysis of the Nearest Neighbors K. We also analyze the influence of parameter K of the K-NN classification on the accuracy of the data set. In this subsection, the procedure of experiment is almost the same as that in Subsection 4.2. The only difference is that in this experiment, we set $\lambda = 1$ and $\mu = 0.5$, the number of neighbors K is selected from 1 to 15 in order to classify the test samples in the transformed space. What is more, we compare the results of the Euclidean distance when the values of K and the data distributions are same. The results with different data sets are shown in Fig. 2. Figure 2 shows that the accuracy by applying our proposed method is higher than applying the distance of the Euclidean input space when the values of K are same. In addition, the change of K has a little effect on the result, which implies that our proposed method is robust to K.

Data Set 2-D Visualization Results. Figure 3 shows the distributions of wine data sets under the learned metrics where different clusters are represented by different colors and shapes. The first line from left to right is the Euclidean metric, PCA, LMNN, and ITML, respectively. The second line from left to right, the methods used are LDML, SDML, LSML, and our method, respectively. To show the distributions under the corresponding metrics, we project them to 2 dimensional spaces by PCA method. Visually, the linear separabilities of data are improved in different degrees under our learned metric. It is also validated in Table 1 that the classification accuracies on this data set are obviously improved.

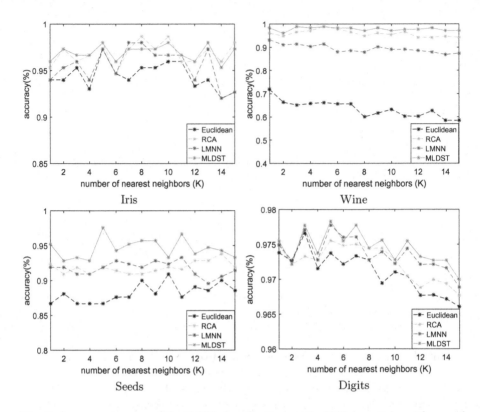

Fig. 2. Average accuracy of the K-NN classification on the UCI repository with regard to the number of nearest neighbors K.

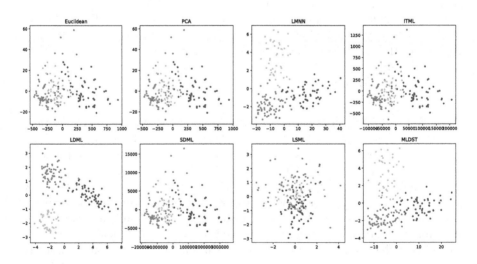

Fig. 3. Data set 2-dimension visualization results.

5.2 Face Recognition Data Sets

We also present the proposed model with two real face recognition data sets, ORL data set and Yale data set. ORL face recognition data set includes 40 subjects and 10 images per subject. For some subjects, the images were taken from different times, lighting, facial expressions (open/closed eyes, smiling/not smiling) and facial details (glasses/no glasses). All the images were taken against a dark homogeneous background with the subjects in an upright, frontal position (with tolerance for some side movement). Yale face recognition data set contains 165 grayscale images in 15 individuals. There are 11 images per subject, one per different facial expression or configuration: center-light, with glasses, happy, left-light, no glasses, normal, right-light, sad, sleepy, surprised, and wink. For these data, we subsampled the images to 32×32 pixels and the dimensionality is further reduced to 64 by PCA. Average error rates with each face recognition data sets are presented in Table 2.

Table 2. Comparison with other metric learning algorithms on the face recognition dataset.

	Data	Euclid	RCA	LMNN	ITML	LDML	MLDST
Average error	ORL	10.25(+/-3.43)	7.51(+/-4.89)	9.38(+/-3.01)	11.32(+/-2.62)	13.27(+/-5.34)	**5.12(+/-4.65)**
rate (%)	YaleB	38.57(+/-10.66)	13.02(+/-4.32)	11.81(+/-1.91)	12.55(+/-3.84)	10.93(+/-2.99)	**8.57(+/-3.13)**

6 Conclusion

In this paper, we have proposed an enhanced metric learning method via D-S evidence theory, and applied the gradient descent method to solve the corresponding optimization problem. We test our method in two aspects. Firstly, the proposed method is applied with four data sets from UCI repository. The experiment results indicate the effectiveness of this method. Then the method is applied with two face recognition data sets. The results show that our proposed method can improve the accuracy of face recognition effectively.

Acknowledgments. This work was supported in part by Shanghai Sailing Program (16YF1404000), the National Natural Science Foundation of China under Grants Nos. 11771276, 11471208 and 11601315.

References

1. Cover, T.M., Hart, P.E.: Nearest neighbor pattern classification. IEEE Trans. Inf. Theory. **13**, 21–27 (1967)
2. Broomhead, D.S., Lowe, D.: Multivariable functional interpolation and adaptive networks. J. Comp. Syst. **2**(3), 321–355 (1988)
3. Cortes, C., Vapnik, V.: Support vector networks. J. Mach. Learn. **20**(3), 273–297 (1995)

4. Xing, E.P., Jordan, M.I., Russell, S.J., Ng, A.Y.: Distance metric learning with application to clustering with side-information. In: Advances in Neural Information Processing Systems, pp. 521–528 (2003)
5. Liao, S.C., Hu, Y., Zhu, X.Y., Li, S.Z.: Person re-identification by local maximal occurrence representation and metric learning. In: IEEE Conference on Computer Vision and Pattern Recognition, pp. 2197–2206. IEEE Computer Society, Boston (2015)
6. Li, Z., Chang, S.Y., Liang, F., Huang, T.S., Cao, L.L., Smith, J.R.: Learning locally-adaptive decision functions for person verification. In: IEEE Conference on Computer Vision and Pattern Recognition. vol. 9, pp. 3610–3617. IEEE Computer Society, Portland (2013)
7. Zheng, W.S., Gong, S.G., Xiang, T.: Reidentification by relative distance comparison. IEEE Trans. Pattern Anal. Mach. Intell. **35**(3), 653–668 (2013)
8. Hu, L.F., Hu, J., Ye, Z., Shen, C.M., Peng, Y.X.: Performance analysis for SVM combining with metric learning. Neural Process. Lett. **3**, 1–12 (2018)
9. Bar-Hillel, A., Hertz, T., Shental, N., Weinshall, D.: Learning distance functions using equivalence relations. In: International Conference on Machine Learning, pp. 11–18. ACM, Atlanta (2003)
10. Jolliffe, I.T.: Principal Component Analysis. Springer, New York, vol. 87, no. 100, pp. 41–64 (2005)
11. Weinberger, K.Q., Saul, L.K.: Distance metric learning for large margin nearest neighbor classification. J. Mach. Learn. Res. **10**(1), 207–244 (2009)
12. Davis, J.V., Kulis, B., Jain, P., Sra, S., Dhillon, I.S.: Information-theoretic metric learning. In: International Conference on Machine Learning, vol. 227, pp. 209–216. ACM, Corvalis (2007)
13. Guillaumin, M., Verbeek, J., Schmid, C.: Is that you? metric learning approaches for face identification. In: International Conference on Computer Vision, vol. 30, pp. 498–505. IEEE, Sydney (2011)
14. Qi, G.J., Tang, J., Zha, Z.J., Chua, T.S., Zhang, H.J.: An efficient sparse metric learning in high-dimensional space via l1-penalized log-determinant regularization. In: International Conference on Machine Learning. vol. 382, pp. 841–848. ACM, Montreal (2009)
15. Kostinger, M., Hirzer, M., Wohlhart, P., Roth, P.M., Bischof, H.: Large scale metric learning from equivalence constraints. In: IEEE Conference on Computer Vision and Pattern Recognition, pp. 2288–2295. IEEE Computer Society, Providence (2012)
16. Liu, E.Y., Guo, Z., Zhang, X., Jojic, V., Wang, W.: Metric learning from relative comparisons by minimizing squared residual. In: IEEE International Conference on Data Mining, vol. 5, pp. 978–983. IEEE Computer Society, Brussels (2012)
17. Ying, S.H., Wen, Z.J., Shi, J., Peng, Y.X., Peng, J.G., Qiao, H.: Manifold preserving: an intrinsic approach for semisupervised distance metric learning. IEEE Trans. Neural Netw. Learn. Syst. **29**(7), 2731–2742 (2017)
18. Shafer, G.: A Mathematical Theory of Evidence. Princeton University Press, Princeton (1976)
19. Denœux, T.: A k-nearest neighbor classification rule based on Dempster-Shafer theory. IEEE Trans. Syst. Man Cybern. **25**(5), 804–813 (1995)
20. Lian, C.F., Su, R., Denœux, T.: Dissimilarity metric learning in the belief function framework. IEEE Trans. Fuzzy. Syst. **24**(6), 1555–1564 (2016)

InsightGAN: Semi-Supervised Feature Learning with Generative Adversarial Network for Drug Abuse Detection

Guangzhen Liu, Jun Hu, An Zhao, Mingyu Ding, Yuqi Huo, and Zhiwu Lu[✉]

Beijing Key Laboratory of Big Data Management and Analysis Methods,
School of Information, Renmin University of China, Beijing 100872, China
luzhiwu@ruc.edu.cn

Abstract. We present a novel generative adversarial network (GAN) model, called InsightGAN, for drug abuse detection. Our model is inspired by two closely related works on machine learning for healthcare applications: (1) drug abuse detection has been solved by machine learning with plentiful data from social media (where face pictures can be easily obtained); (2) facial characteristics have been explored in mental disorder diagnosis (drug addiction is also a mental disorder). In this paper, we adopt deep learning to extract discriminative facial features for drug abuse detection. However, in this application, the face pictures with ground-truth labels are far from sufficient for training a deep learning model. To alleviate the scarcity of labelled data, we thus propose a semi-supervised facial feature learning model based on GAN. Moreover, we also develop a robust algorithm for training our InsightGAN. Experimental results show the promising performance of our InsightGAN.

Keywords: Drug abuse detection · Deep learning · Social media

1 Introduction

Machine learning has been employed in many healthcare problems [8,19], due to the recent advances in this field [17]. Since illegal drug abuse is one of the fastest spreading public health problems in the world [15,16], drug abuse detection has attracted worldwide researchers from multiple fields [4,6,14,30], including those from machine learning [5,11,23,26].

Our present work on drug abuse detection is inspired by two closely related works [1,23]. In [23], a drug abuse detection system was developed by machine learning with plentiful data from social media (i.e. Twitter). Due to Twitter's wide reach and effective openness, this system acts as a bridge between people who need help in dealing with drug addictions and the services they need. This is crucial for reversing the growing epidemic of drug abuse [15,16], given that people tend to hide their addictions from others for the criminalization and stigma of illicit drug use [23]. In this paper, instead of analyzing the texts from

© Springer Nature Switzerland AG 2018
L. Cheng et al. (Eds.): ICONIP 2018, LNCS 11303, pp. 411–422, 2018.
https://doi.org/10.1007/978-3-030-04182-3_36

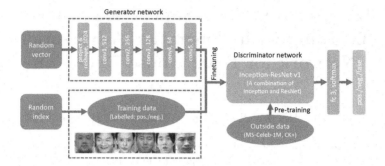

Fig. 1. Flowchart of our InsightGAN model for drug abuse detection.

Twitter, we choose to explore face pictures (easily obtained from social media) in drug abuse detection. This choice is supported by the interesting findings from [1]. In this previous work, face pictures were used for the diagnosis of one typical mental disorder, i.e., autism spectrum disorder (ASD), given that facial characteristics of children with ASD were shown to have distinct differences from those of typically developing children. Similar findings have also been reported in [3]. Since drug addiction is also a mental disorder [30] and deep learning is one of the most powerful tools in machine learning [13], we employ deep learning to extract discriminative facial features for drug abuse detection in this paper.

Note that deep learning has a distinct limitation in healthcare applications. Specifically, the ground truth labels of medical data are often very expensive to access, and thus we are generally provided with a small labelled set for model training. For example, in [23], only 300 tweets were manually classified into drug abuse tweets and non-abuse tweets by two professors and three students who have expertise in health informatics. Since the scarcity of training data tends to cause the overfitting of a deep learning model [18], we need to overcome this issue when leveraging deep learning in drug abuse detection.

Considering the significant advantage of generative adversarial network (GAN) in deep feature learning [9,24], we propose a semi-supervised facial feature learning model based on GAN to alleviate the scarcity of labelled data for drug abuse detection, which is called InsightGAN in this paper. As compared to the conventional GAN that learns deep features in an unsupervised manner, our InsightGAN is trained with *both labelled and generated examples* for semi-supervised feature learning. Because the generated examples are used for model training, our InsightGAN is shown to be able to alleviate the scarcity of labelled data. Although machine learning has been widely used in drug abuse detection [5,11,23,26], deep learning (including GAN) has been rarely used for this task.

Since Inception-ResNet v1 [31] is one of the most advancing convolutional neural network (CNN) models [12,29,32], we use it as the discriminator network in our InsightGAN model, as shown in Fig. 1. Moreover, we design the generator network using a series of five fractionally-strided convolutions (see Fig. 1), which is similar to [24]. Given that our focus is drug abuse detection (i.e. classification) rather than face generation, the discriminator network in our InsightGAN model is much more complicated than the generator network. In addition, by pre-training the extremely large discriminator network with outside data such as MS-Celeb-1M [10] and CK+ [20], we develop a robust algorithm for training our InsightGAN model. Many semi-supervised GAN training strategies [7,25] exist; in this paper, only the strategy of [21] is employed by our algorithm. Note that our model is trained in an end-to-end manner and a non-transductive classifier can be obtained for drug abuse detection when our model is well-trained.

To evaluate our InsightGAN model, we conduct extensive experiments of drug abuse detection (DAD) on a face dataset (denoted as DAD-Face), which is collected from a local hospital and also from the web (e.g., the we-media on the drug use topic, and drug abuse news of celebrities). The experimental results show that our InsightGAN model generally yields better result than the Inception-ResNet v1 model when the same number of labelled data are provided for drug abuse detection. This validates that the generated examples obtained by GAN indeed help to improve the performance of Inception-ResNet v1.

Our main contributions are summarized as follows: (1) We have proposed a semi-supervised GAN model for drug abuse detection, which can be used to boost the drug safety surveillance systems through Twitter. (2) We have made the first contribution to exploring deep learning in extracting discriminative features for drug abuse detection, to the best of our knowledge. (3) We have developed a robust algorithm for training our InsightGAN model by pre-training the discriminator network (Inception-ResNet v1) with outside data.

2   Related Work

2.1   Drug Abuse Detection

Illegal drug abuse is one of the fastest spreading public health problems in the world [15,16]. To reverse the growing epidemic of drug abuse, great efforts have been made on drug safety surveillance and drug abuse prevention. In particular, drug abuse detection has attracted worldwide researchers from multiple fields including medical science, chemistry, physics, and computer science [2,4,6,14, 22,30]. Specifically, a series of drug abuse testing methods have been proposed by analyzing different types of medical data, which are collected from oral fluid [6], urine [2], blood [22], and hair [4]. However, these drug abuse testing methods have a limitation in real-world applications: it is of high cost to collect the biological samples, since people tend to hide their addictions from others. This is mainly due to the criminalization and stigma of illicit drug use [23]. As a result, it becomes very difficult to identify vulnerable individuals, who may take an overdose of illicit drugs, when only these testing methods are employed.

To provide a bridge between people who need help in dealing with drug addictions and the services they need, many researchers have paid their attention to developing drug safety surveillance systems through Twitter [5,11,23,26]. Due to the wide reach and effective openness of social media (e.g. Twitter), the spread of information about the use of illegal drugs can be detected in social media. In particular, in [23], a drug abuse detection system was developed by machine learning with the texts from Twitter (labelled as drug abuse tweets and non-abuse tweets). In this paper, instead of analyzing the texts from Twitter, we choose to explore face pictures in drug abuse detection and propose an Insight-GAN model for drug abuse detection. Since the face pictures are easily obtained from social media, our InsightGAN model can be used to boost the drug safety surveillance systems through Twitter.

2.2 Semi-Supervised Learning with GAN

GAN has been shown to yield exciting results in many applications [9,24]. Due to the adversarial training of the generator and discriminator networks, GAN can provide the ability of both data generation and classification. In this paper, since our focus is drug abuse detection (i.e. classification) rather than face generation, the discriminator network (Inception-ResNet v1) in our InsightGAN model is much more complicated than the generator network. Considering the scarcity of labelled data provided for drug abuse detection, we adopt two model training strategies: (1) Semi-supervised learning is employed. That is, our InsightGAN model is trained with both labelled and generated examples. In the literature, there exist many semi-supervised learning strategies [7,25] for training a GAN model. In this paper, only the strategy of [21] is adopted in our training algorithm. (2) Outside data are explored in model training. Specifically, MS-Celeb-1M [10] and CK+ [20] are used to pre-train the discriminator network (i.e. Inception-ResNet v1). In summary, we have developed a robust algorithm for training our InsightGAN model in an end-to-end manner.

3   The Proposed Model

3.1   Face Preprocessing

Similar to previous work on face recognition, we first preprocess the original large face pictures to obtain standard faces as follows:

- **Face Detection.** We first detect a single face or multiple faces from each original large picture using FaceNet [27].
- **Facial Keypoint Detection.** For each detected face, we detect 68 facial keypoints for face alignment and cropping, using the Caffe open library available at https://github.com/qiexing/face-landmark-localization.
- **Face Alignment & Cropping.** With the 68 keypoints, we align each face using a 2D affine transformation and then crop it to 160*160 pixels.

3.2   Model Architecture

For the detection of illegal drug abuse, we design a semi-supervised GAN model based on Inception-ResNet v1 [31], which is a typical CNN model. The network architecture of our InsightGAN model is illustrated in Fig. 1. It can be seen that our InsightGAN model consists of two main networks: (1) the discriminator network, which uses Inception-ResNet v1 as a basic CNN model; (2) the generator network, which is composed of five fractionally-strided convolutions, similar to [24]. The details of the two networks in our model are given as follows:

- **Discriminator Network.** The discriminator network is generally inherited from the original Inception-ResNet v1 model. In this paper, we make modifications in two aspects: (1) The fully-connected layer of Inception-ResNet v1 is reduced to three softmax output units, with one unit for each of the three classes [pos., neg., fake]; (2) In the input layer, half of each mini-batch includes the generated examples outputted by the generator network, and the other half contains examples from the DAD-Face dataset. This setting keeps unchanged when training the discriminator network, while only the first half of mini-batch is used for training the generator network.
- **Generator Network.** The design of the generator network is inspired by [24]. As shown in Fig. 2, the generator network consists of one project and reshape layer and five fractionally-strided convolutional layers. Note that no pooling layers are used here. Specifically, the input of the generator network is a 100-dimensional random vector z drawn from a uniform distribution, and the other six layers of the generator network have the following output sizes: [5 × 5 pixels, 1024 channels], [10 × 10 pixels, 512 channels], [20 × 20 pixels, 256 channels], [40 × 40 pixels, 128 channels], [80 × 80 pixels, 64 channels], [160 × 160 pixels, 3 channels]. The generated example is denoted as $G(z)$.

Note that the discriminator network is much more complicated than the generator network. This is consistent with the fact that our focus is drug abuse detection (but not face generation). Since one fully-connected layer has been added at the end of the discriminator network for the classification of [pos., neg., fake], we can train our model in an end-to-end manner. Moreover, since we are provided only with labelled and generated examples (without unlabelled examples), we can obtain a non-transductive classifier for drug abuse detection, when our model is well-trained.

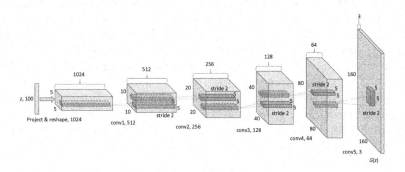

Fig. 2. The architecture of the generator network in our InsightGAN model for drug abuse detection. A 100-dimensional random vector z is drawn from a uniform distribution. A series of five fractionally-strided convolutions then convert this random vector into a color image $G(z)$ of 160 × 160 pixels.

Algorithm 1. InsightGAN Training Algorithm

Input: The labelled examples
            Parameters I, k, m
Output: D, G
1. Pre-train Inception-ResNet v1 with MS-Celeb-1M;
2. Finetune Inception-ResNet v1 with CK+;
3. Initialize the discriminator using the finetuned Inception-ResNet v1 model;
for $i = 1$ **to** I **do**
> **for** $j = 1$ **to** k **do**
>> 4. Draw m noise samples $\{z^{(1)}, ..., z^{(m)}\}$ from $p_z(z)$;
>> 5. Draw m examples $\{x^{(1)}, ..., x^{(m)}\}$ from data generating distribution $p_{\text{data}}(x)$, with their labels being $\{y^{(1)}, ..., y^{(m)}\}$;
>> 6. Update the discriminator by ascending its stochastic gradient on the combined mini-batch of size $2m$:

$$\nabla_{\theta_d} \frac{1}{m} \sum_{s=1}^{m} [\log D(x^{(s)}|y^{(s)}) + \log(1 - \sum_{y=1}^{2} D(G(z^{(s)})|y))];$$

end
7. Draw m noise samples $\{z^{(1)}, ..., z^{(m)}\}$ from $p_z(z)$;
8. Update the generator by descending its stochastic gradient on the mini-batch of size m:

$$\nabla_{\theta_g} \frac{1}{m} \sum_{s=1}^{m} \log(1 - \sum_{y=1}^{2} D(G(z^{(s)})|y)).$$

end
return D, G.

3.3 Adversarial Training

We formulate the problem of adversarial training of our InsightGAN model as follows. To learn the generator's distribution $p_g(x)$ over data x, we define a noise prior $p_z(z)$ on the noise vector z, and then represent a mapping to data space as $G(z; \theta_g)$, where G is a differentiable function represented by the generator network with parameters θ_g. Moreover, we define the class label $y^{(x)}$ of data x as: $y^{(x)} = 1$ if x comes from the positive class, and $y^{(x)} = 2$ if x comes from the negative class. We thus represent the output of the discriminator network with parameters θ_d as: $D(x|y^{(x)}; \theta_d)$, where $D(x|y^{(x)})$ is the probability that x comes from class $y^{(x)}$ rather than p_g. The adversarial training of D and G can be formulated as the following minimax problem:

$$\min_G \max_D V(G, D) = \mathbb{E}_{x \sim p_{\text{data}}(x)} [\log D(x|y^{(x)})]$$

$$+ \mathbb{E}_{z \sim p_z(z)} [\log(1 - \sum_{y=1}^{2} D(G(z)|y))] \tag{1}$$

where $V(G, D)$ is the objective function for adversarial training, and $p_{\text{data}}(x)$ is the data generating distribution (other than the generator's distribution $p_g(x)$).

Note that the above minimax optimization problem is different from those minimax problems defined in [9,21,24], since three types of examples (positive/negative/fake) are included. However, following the adversarial training algorithms developed in these previous works, we can similarly solve Eq. (1) using an iterative numerical approach. Specifically, we alternate between k steps of optimizing D and one step of optimizing G, where the stochastic gradient descent method is employed. This ensures that D is maintained near its optimal solution, so long as G changes slowly enough.

Although the adversarial training approach can alleviate the scarcity of training data, the DAD-Face dataset is still "small" for training our InsightGAN model, especially when the extremely large CNN model (i.e. Inception-ResNet v1) is used as the discriminator network. As a remedy, we explore the MS-Celeb-1M [10] and CK+ [20] datasets as outside data for model training. In particular, MS-Celeb-1M is a large-scale face dataset of 100 K subjects and 10M face pictures, and CK+ is a facial expression dataset of 2,977 face pictures from eight emotion categories. In this paper, the MS-Celeb-1M and CK+ datasets are used for pre-training and finetuning of the discriminator network, respectively.

By considering the adversarial training and pre-training steps together, the complete algorithm for training our InsightGAN model is shown in Algorithm 1.

3.4 Test Process

Once our model has been well trained, we can evaluate its performance on the test set of DAD-Face. Specifically, each test face is first preprocessed to standard face, and then inputted into the discriminator to be classified as positive or negative (but not corresponding to the fake class). Moreover, our later experiments also show that our InsightGAN model is very efficient during the test process.

4 Experimental Evaluation

4.1 Data Collection

For performance evaluation, we construct a face dataset called DAD-Face[1], which consists of 1,581 face pictures (pos./neg. = 784/797). To make this dataset as large as possible, we have collected the face pictures not only from a local hospital but also from the web (e.g., the we-media on the drug use topic, and drug abuse news of celebrities). Note that the dataset collected in this way would inevitably have noise. As a remedy, we have made great effort on quality assurance during data collection, i.e., each case has been checked by two experts and three students who have expertise in health informatics. In addition, our project has been confirmed by the Ethics Committee of the local hospital.

---

[1] https://github.com/JoinGitHubing/drugIdentification.

4.2 Experimental Setup

To evaluate the performance of our InsightGAN model in drug abuse detection, we make use of the following five measures: accuracy (ACC), sensitivity (SEN), specificity (SPE), positive predictive value (PPV), and negative predictive value (NPV). Given the number of true positives (TP), false negatives (FN), false positives (FP) and true negatives (TN), the five metrics are defined as follows:

$$ACC = (TP + TN)/(TP + FN + TN + FP) \tag{2}$$

$$SEN = TP/(TP + FN), \quad SPE = TN/(TN + FP) \tag{3}$$

$$PPV = TP/(TP + FP), \quad NPV = TN/(TN + FN) \tag{4}$$

For our InsightGAN model, we randomly initialize the fully-connected layer (at the end of the discriminator) by drawing weights from a zero-mean Gaussian distribution with standard deviation 0.01, and initialize the bias to 0. All the other layers of the discriminator are initialized by Inception-ResNet v1 trained with MS-Celeb-1M and CK+. Moreover, for all the convolutional layers of the generator, we adopt the Xavier initialization. The learning rate is set to 0.0001. The two parameters about iterations are set as $I = 10,000$ and $k = 1$. A weight decay of 0.0 is used. In addition, we train InsightGAN with 2 GPUs (each with batch size $m = 32$) in the TensorFlow framework.

4.3 Comparison to Alternative Detection Methods

To show the effectiveness of our InsightGAN model for drug abuse detection, we make comparison among four closely related methods: (1) InsightGAN – the proposed model shown in Fig. 1; (2) Inception-ResNet v1 – the original model proposed in [31]; (3) DCGAN+SVM – the method that utilizes DCGAN [24] (with the same generator and discriminator as InsightGAN) for unsupervised feature learning and SVM for drug abuse detection; (4) Hand-Craft Features+SVM – the method that uses SVM for classification with hand-craft features. Note that the first two methods adopt supervised/semi-supervised feature learning, while the last two methods adopt unsupervised feature learning. In this paper, we extract hand-craft features by combining two typical methods [28,33]: (1) All the pairwise distances among the 68 keypoints (detected from each face picture) are computed to form a 2,278-dimensional feature vector; (2) The Gabor filtering (with 6 scales and 4 orientations) is performed on each face picture and then the mean Gabor values within the 3*3 neighborhood of each keypoint are computed to form a 1,632-dimensional feature vector; (3) The two groups of hand-craft features are concentrated into a 3,910-dimensional feature vector for drug abuse detection with SVM. Note that these two groups of hand-craft features have been widely used for healthcare applications based on face recognition.

Table 1 shows the results obtained by different detection methods with different traning/test splits for drug abuse detection. We can make the following observations: (1) By making an overall evaluation, our InsightGAN model not only performs the best in all cases, but also yields gradually larger gains over the other detection methods when the size of training data decreases. In fact, the performance of our InsightGAN model degrades the most slowly as less labelled data is provided for model training. This means that our InsightGAN model indeed helps to alleviate the scarcity of labelled data. In other words, we have clearly demonstrated the effectiveness of semi-supervised GAN model in drug abuse detection. (2) The gains achieved by our InsightGAN model over Inception-ResNet v1 (see InsightGAN vs. Inception-ResNet v1) provide further evidence that semi-supervised GAN is effective for drug abuse detection, since the two deep learning models utilize the same number of training data. (3) The supervised/semi-supervised feature learning methods (i.e. InsightGAN and Inception-ResNet v1) generally outperform the unsupervised feature learning methods (i.e. DCGAN+SVM and Hand-Craft Features+SVM), due to more labelled data used for feature learning. (4) The unsupervised feature learning method (i.e. DCGAN+SVM) yields comparable results with respect to Hand-Craft Features+SVM, which shows the advantage of GAN in unsupervised feature learning for drug abuse detection.

Table 1. Comparison among four detection methods with different training/test splits for drug abuse detection. All five metrics (%) are used.

Training set (pos./neg.)	Test set (pos./neg.)	Methods	ACC	SEN	SPE	PPV	NPV
25/25	759/772	InsightGAN (Ours)	**86.0**	83.7	**88.3**	**87.5**	84.7
		Inception-ResNet v1	74.3	**87.7**	52.5	66.6	**88.3**
		DCGAN+SVM	78.5	69.5	87.4	84.3	74.5
		Hand-Craft Features+SVM	76.7	63.4	80.8	74.0	71.4
50/50	734/747	InsightGAN (Ours)	**87.7**	**89.1**	**86.4**	**86.5**	**89.0**
		Inception-ResNet v1	84.2	83.2	73.1	77.7	79.1
		DCGAN+SVM	80.4	80.7	80.2	80.1	80.8
		Hand-Craft Features+SVM	82.0	78.9	85.0	83.8	80.4
100/100	684/697	InsightGAN (Ours)	**89.9**	**87.1**	**92.7**	**92.1**	**88.0**
		Inception-ResNet v1	84.4	84.8	76.8	79.5	82.5
		DCGAN+SVM	82.5	80.6	84.3	83.4	81.6
		Hand-Craft Features+SVM	83.8	84.9	82.6	82.8	84.8
500/500	284/297	InsightGAN (Ours)	**95.6**	**94.9**	**96.4**	**96.1**	**95.2**
		Inception-ResNet v1	94.3	94.0	94.6	94.3	94.3
		DCGAN+SVM	86.1	87.5	84.9	84.4	87.8
		Hand-Craft Features+SVM	90.0	88.7	91.2	90.6	89.4

Figure 3 presents the results obtained by our InsightGAN model with different initializations: (1) Random – the random initialization is used; (2) MS-Celeb-1M - our model is initialized with the outside data MS-Celeb-1M; (3) MS-Celeb-1M & CK+ – our model is initialized with both MS-Celeb-1M and CK+. The training set (pos./neg. = 500/500) is used for adversarial training after model initialization. It can be seen that: (1) Our model with more powerful initialization leads to better results for drug abuse detection. (2) Our model with random initialization still yield promising results, validating the effectiveness of our adversarial training.

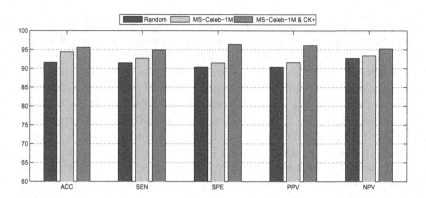

Fig. 3. The results obtained by our InsightGAN model with different initializations. The training set (pos./neg. = 500/500) is used.

We finally provide the training and test time of our InsightGAN model for drug abuse detection. The following computer is employed: 2 Intel Xeon E5-2609 v3 CPUs (each with 1.9 GHz and 6 cores), 2 Titan X GPUs (each with 12G memory), and 96G RAM. For the training set (pos./neg. = 500/500), the time of training InsightGAN is 112 min. Moreover, the time of processing a test face is 0.03 second, i.e., our model can provide real-time detection of drug abuse.

4.4   Comparison to Alternative CNN Models

In this paper, we employ Inception-ResNet v1 as the discriminator of our Insight-GAN model for drug abuse detection. Besides this complicated CNN model, any other CNN model can also be used to design our network architecture. In the following, we compare Inception-ResNet v1 to ResNet-101 [12] and VGG-16 [29] by directly applying them to drug abuse detection (without combination to any other models). The same setting is adopted for all CNN models, i.e., each model is initialized using the same outside data (MS-Celeb-1M and CK+).

The comparative results are shown in Table 2, where only a single training/test split is considered. It can be seen that: (1) The three CNN models generally yield comparable performance in the task of drug abuse detection. (2) Inception-ResNet v1 achieves slight improvements over the well-known VGG-16

Table 2. Comparison among the three CNN models directly used for drug abuse detection with the training set (pos./neg. = 500/500).

CNN models	ACC	SEN	SPE	PPV	NPV
Inception-ResNet v1	**94.3**	**94.0**	**94.6**	**94.3**	**94.3**
ResNet-101	93.6	93.7	93.6	93.4	94.0
VGG-16 (VGGFace)	93.2	90.8	93.6	93.2	91.6

that has been widely used for face recognition. Therefore, although the architecture of Inception-ResNet v1 is more complicated than that of VGG-16, we prefer to Inception-ResNet v1 as the discriminator in this paper.

5 Conclusion

In this paper, we have proposed a novel InsightGAN model for drug abuse detection. To alleviate the scarcity of labeled data, we have designed a semi-supervised GAN model and also developed a robust training algorithm. Experimental results show the superior performance of our model. In the future work, we will combine the tweets and also the faces from Twitter for drug abuse detection.

Acknowledgements. This work was partially supported by National Natural Science Foundation of China (61573363), and the Fundamental Research Funds for the Central Universities and the Research Funds of Renmin University of China (15XNLQ01).

References

1. Aldridge, K., George, I.D., Cole, K.K., et al.: Facial phenotypes in subgroups of prepubertal boys with autism spectrum disorders are correlated with clinical phenotypes. Mol. Autism **2**(1), 15 (2011)
2. Alnajjar, A., Idris, A.M., Multzenberg, M., Mccord, B.: Development of a capillary electrophoresis method for the screening of human urine for multiple drugs of abuse. J. Chromatogr. B **856**(1–2), 62–67 (2007)
3. Austin, J.R., Takahashi, T.N., Duan, Y.: Distinct facial phenotypes in children with autism spectrum disorders and their unaffected siblings. In: International Meeting for Autism Research (2012)
4. Baciu, T., Borrull, F., Aguilar, C., Calull, M.: Recent trends in analytical methods and separation techniques for drugs of abuse in hair. Analytica Chimica Acta **856**, 1–26 (2015)
5. Coloma, P.M., Becker, B., Sturkenboom, M.C., van Mulligen, E.M., Kors, J.A.: Evaluating social media networks in medicines safety surveillance: two case studies. Drug Saf. **38**(10), 921–30 (2015)
6. Cone, E.J., Huestis, M.A.: Interpretation of oral fluid tests for drugs of abuse. Ann. New York Acad. Sci. **1098**(1), 51–103 (2010)
7. Dai, Z., Yang, Z., Yang, F., Cohen, W., Salakhutdinov, R.: Good semi-supervised learning that requires a bad GAN. arXiv Preprint arXiv:1705.0978 (2017)
8. Esteva, A., et al.: Dermatologist-level classification of skin cancer with deep neural networks. Nature **542**, 115–118 (2017)
9. Goodfellow, I., Pouget-Abadie, J., Mirza, M., Xu, B., et al.: Generative adversarial nets. In: NIPS, pp. 2672–2680 (2014)
10. Guo, Y., Zhang, L., Hu, Y., He, X., Gao, J.: MS-Celeb-1M: a dataset and benchmark for large-scale face recognition. In: Leibe, B., Matas, J., Sebe, N., Welling, M. (eds.) ECCV 2016. LNCS, vol. 9907, pp. 87–102. Springer, Cham (2016). https://doi.org/10.1007/978-3-319-46487-9_6
11. Hanson, C.L., Cannon, B., Burton, S., Giraudcarrier, C.: An exploration of social circles and prescription drug abuse through Twitter. J. Med. Int. Res. **15**(9), e189 (2013)

12. He, K., Zhang, X., Ren, S., Sun, J.: Deep residual learning for image recognition. In: CVPR, pp. 770–778 (2016)

13. Hinton, G.E., Salakhutdinov, R.: Reducing the dimensionality of data with neural networks. Science **313**, 504–507 (2006)

14. Huestis, M.A., Smith, M.L.: Modern analytical technologies for the detection of drug abuse and doping. Drug Discovery Today Technol. **3**(1), 49–57 (2007)

15. Ingraham, C.: Heroin deaths surpass gun homicides for the first time, CDC data shows. The Washington Post (2016). Accessed 8 Dec 2016

16. Jia, Z., et al.: Tracking the evolution of drug abuse in China, 2003-10: a retrospective, self-controlled study. Addiction **110**(S1), 4–10 (2015)

17. LeCun, Y., Bengio, Y., Hinton, G.E.: Deep learning. Nature **521**, 436–444 (2015)

18. Lee, J.G., Jun, S., Cho, Y.W., et al.: Deep learning in medical imaging: general overview. Korean J. Radiol. **18**(4), 570–584 (2017)

19. Long, E., Lin, H., et al.: An artificial intelligence platform for the multihospital collaborative management of congenital cataracts. Nat. Biomed. Eng. **1**, 0024 (2017)

20. Lucey, P., Cohn, J.F., Kanade, T., Saragih, J., Ambadar, Z., Matthews, I.: The extended Cohn-Kanad dataset (CK+): a complete dataset for action unit and emotion-specified expression. In: CVPR Workshops, pp. 94–101 (2010)

21. Odena, A.: Semi-supervised learning with generative adversarial networks. In: ICML 2016 Workshop on Data-Efficient Machine Learning (2016)

22. Peters, F.T., Kraemer, T., Maurer, H.H.: Drug testing in blood: validated negative-ion chemical ionization gas chromatographicc-mass spectrometric assay for determination of amphetamine and methamphetamine enantiomers and its application to toxicology cases. Clin. Chem. **48**(9), 1472–1485 (2002)

23. Phan, N., Chun, S.A., Bhole, M., Geller, J.: Enabling real-time drug abuse detection in Tweets. In: ICDE Workshop (2017)

24. Radford, A., Metz, L., Chintala, S.: Unsupervised representation learning with deep convolutional generative adversarial networks. arXiv Preprint arXiv:1511.06434 (2015)

25. Salimans, T., Goodfellow, I., Zaremba, W., Cheung, V., Radford, A., Chen, X.: Improved techniques for training GANs. In: NIPS, pp. 2234–2242 (2016)

26. Sarker, A., et al.: Social media mining for toxicovigilance: automatic monitoring of prescription medication abuse from Twitter. Drug Saf. **39**(3), 231–240 (2016)

27. Schroff, F., Kalenichenko, D., Philbin, J.: FaceNet: a unified embedding for face recognition and clustering. In: CVPR, pp. 815–823 (2015)

28. Shen, L., Bai, L.: A review on Gabor wavelets for face recognition. Patt. Anal. Appl. **9**(2–3), 273–292 (2006)

29. Simonyan, K., Zisserman, A.: Very deep convolutional networks for large-scale image recognition. arXiv preprint arXiv:1409.1556 (2014)

30. Stolle, M., Sack, P.M., Thomasius, R.: Substance abuse in children and adolescents - early detection and intervention. Dtsch Arztebl **104**(28–29), A2061–A2070 (2007)

31. Szegedy, C., Ioffe, S., Vanhoucke, V., Alemi, A.: Inception-v4, Inception-ResNet and the impact of residual connections on learning. In: AAAI, pp. 4278–4284 (2017)

32. Szegedy, C., Vanhoucke, V., Ioffe, S., Shlens, J., Wojna, Z.: Rethinking the inception architecture for computer vision. In: CVPR, pp. 2818–2826 (2016)

33. Zhai, G., Ren, F., Zhang, G., Evison, M.: Facial shape analysis based on Euclidean distance matrix analysis. In: International Conference on Biomedical Engineering and Informatics, pp. 1896–1900 (2011)

Sparse Feature Learning Using Ensemble Model for Highly-Correlated High-Dimensional Data

Ali Braytee[1]([✉]), Ali Anaissi[3], and Paul J. Kennedy[2]

[1] School of Biomedical Engineering, University of Technology Sydney,
Ultimo, Australia
{Ali.Braytee,Paul.Kennedy}@uts.edu.au
[2] School of Software, University of Technology Sydney, Ultimo, Australia
[3] The University of Sydney, Sydney, NSW 2006, Australia
Ali.Anaissi@sydney.edu.au

Abstract. High-dimensional highly correlated data exist in several domains such as genomics. Many feature selection techniques consider correlated features as redundant and therefore need to be removed. Several studies investigate the interpretation of the correlated features in domains such as genomics, but investigating the classification capabilities of the correlated feature groups is a point of interest in several domains. In this paper, a novel method is proposed by integrating the ensemble feature ranking and co-expression networks to identify the optimal features for classification. The main advantage of the proposed method lies in the fact, that it does not consider the correlated features as redundant. But, it shows the importance of the selected correlated features to improve the performance of classification. A series of experiments on five high dimensional highly correlated datasets with different levels of imbalance ratios show that the proposed method outperformed the state-of-the-art methods.

Keywords: Feature selection · High-dimensional data
Feature correlation

1 Introduction

In the era of high-throughput technologies, the term "big data" is coined to reflect the amount of the data increasingly being generated in many fields. The available data exceeds the ability of the existing machine learning algorithms to analyse it. The complexities and challenges of data in some fields are reflected in the generated datasets. One of these types of complex structures is high-dimensional data which have a relatively low number samples, known as "the curse of dimensionality" problem or $p >> n$. The problem of the curse of dimensionality has become increasingly common in several domains, especially in biomedicine and genomics applications. Furthermore, the dilemma is

L. Cheng et al. (Eds.): ICONIP 2018, LNCS 11303, pp. 423–434, 2018.
https://doi.org/10.1007/978-3-030-04182-3_37

exacerbated by the presence of highly correlated features and the imbalanced data problem. In machine learning, many feature selection algorithms have been proposed to select the important features and eliminate the unimportant ones. However, most of these existing algorithms follow an individual feature ranking approach which discards the existence of the correlated features including SVM-RFE [9], LASSO [7,17]. Surprisingly, a few feature ranking algorithms based on correlated features are proposed in the literature [5]. Feature selection for predictive models in the presence of high-dimensional and imbalanced data with many highly correlated covariates is a challenging problem that affects many disciplines. Initially, researchers were unaware of the importance of correlated covariates in interpreting predictive models. However, recent studies have been conducted to interpret groups of highly correlated features to identify significant functional modules, to improve classification accuracy, and to reflect on the semantic components of these features.

Recently, we proposed an ensemble SVM (ESVM-RFE) algorithm [2] for individual feature ranking in high-dimensional data. The ESVM-RFE uses the ensemble strategy with the SVM [4] classifier as the base learning model. It uses the binary SVM as a decision boundary to separate two classes, defined by solving a quadratic optimization problem. The decision boundary is specified by a subset of critical training samples named support vectors that lie on the edge. Ensemble techniques have the advantage of handling the problem of the curse of dimensionality and reducing the potential of over-fitting the training data. The ESVM-RFE follows the ensemble and bagging concepts of random forest and adopts a backwards elimination strategy. Also, it handles the problem of imbalanced datasets by constructing roughly balanced bootstrap samples or bootstrap samples biased to the minority class.

In this paper, a novel sparse feature learning algorithm (SFL-ESVM) is proposed to handle the correlated features in high-dimensional data. The SFL-ESVM algorithm consists of three components: first, it generates isolated feature modules based on the network structure of the data. Each module contains the correlated features, and the correlation between the modules is low. Second, our previous study of the ESVM-RFE algorithm [2] is used to select the most important features within each module. Finally, the selected features are aggregated and again, ESVM-RFE is applied to select the optimal features of the modules. Specifically, **the contributions** of this paper are:

1. Propose an effective ensemble feature ranking method using co-expression networks to select optimal features for classification.
2. Provide comprehensive evaluations of our method on real-world high dimensional imbalanced datasets which show the advantages of our method.

2   Related Work

Feature selection process is considered as a prerequisite step for many data mining high dimensional datasets including genomic data. It reduces the number of dimensions by selecting a certain number of features or genes which able to explain the differences between patients in regards to the type of the disease [14]. In fact, many benefits are observed from achieving a feature selection include the ability of better understand the data with less informative features, reducing the complexity and computation time of the learning model, removing the noisy features and others. There are three types of feature selection approaches as defined by Sayes et al [14]: filter, wrapper and embedded approaches. The main difference among them that filter approach is independent of any classification algorithm. However, wrapper and embedded use the classification algorithm in the feature selection process. Wrapper evaluates the goodness of features using the classification algorithm and embedded performs the feature selection during the learning process.

Many feature selection algorithms have been proposed to select the important features and eliminate the unimportant ones. However, most of these existing algorithms follow an individual feature ranking approach which discards the existence of the correlated features. Surprisingly, a few feature ranking algorithms based on correlated features are proposed in the literature. The main two approaches which proposed in literature that consider the correlated features are: the sparse models and feature clustering methods. Sparse models including group LASSO [11] and fused SVM [13] suffer from a correlation bias during the feature weighting process, because they assign the weights based on the group size [18]. Therefore, features which belong to a big group may receive small weights. Furthermore, they are considered as parametric methods which need to set some parameters beforehand, which is not guaranteed to hold in practical applications [6]. Feature clustering methods determine the group features using clustering methods, then select a limited number of features to train the models. It is reported that this approach may remove the correlation bias [12]. But, several issues were found in this approach: firstly, the features are clustered using the standard parametric clustering methods which needs to optimize the numbers of clusters parameter. Secondly, the feature importance scores are unstable due to a single ranking of the features in the proposed model. Finally, the problem of class imbalance problem is not handled by the existing feature selection at the presence of highly correlated features, which may assigns a larger weights for the features which predict the majority class. Recent work, called the fuzzy forests method, has been proposed by [6] which uses recursive feature elimination random forests to select the features from the correlated feature blocks. Fuzzy forests depends on random forest feature selection which has a high computational complexity in terms of running time compared to the feature selection method using the support vector machine [2]. Furthermore, the fuzzy forests method does not take into account the imbalanced data problem which may generate features which are biased towards the majority class. A drawback of supervised clustering methods is that they do not identify the correlated features to improve the classification performance along with the interpretation.

3 The Proposed Method SFL-ESVM

In this section, SFL-ESVM is proposed as a feature module learning framework. Subsection 3.1 shows how to cluster the co-expression networks to generate the feature modules. Subsection 3.2 reviews the ESVM-RFE algorithm, and finally, presents the proposed SFL-ESVM algorithm.

3.1 Clustering Co-Expression Networks

A very widely used method to cluster the co-expression biological networks is hierarchical clustering and in particular, the weighted gene co-expression network analysis (WGCNA) algorithm. The WGCNA is initially developed to find the relevant biological modules by detecting a network of highly correlated genes [3]. The gene co-expression network generated by WGCNA can be clustered into groups of highly interconnected nodes.

The WGCNA uses a similarity function such as Pearson correlation to construct a correlated similarity network between the genes. Then, the similarity network is transformed into an adjacency network by taking the absolute value of the similarity network entries and raising it to the power β. This step indicates the strong correlation among genes and rejects the weak ones. Scale-free topology criterion is used to choose the best value of parameter β. Next, the modules are identified by searching for strongly connected genes which is known as high topological overlap. After constructing the topological overlap network for all pairs of genes, the hierarchical clustering algorithm uses this information to identify the modules of correlated genes. The WGCNA has the advantage that it does need to set the number of clusters in advance.

3.2 Review of ESVM-RFE

The ESVM-RFE [2] ranks the features by constructing an ensemble of SVM models in each iteration of SVM-RFE using a random bootstrap subset from the training set. Then, it aggregates all the feature rankings as an ensemble vote. The least important features are eliminated based on multiple votes in each iteration. This process is repeated until a specified number of features is reached.

3.3 The Proposed SFL-ESVM Based on the Co-Expression Feature Network

The proposed SFL-ESVM does not consider the correlated features as redundant which must be removed. For example, in microarray gene expression data, genes that have either similar genomic locations or molecular functions are assumed to co-function and are highly correlated [18]. The correlation issue negatively impacts the classical feature selection algorithms which follow an individual feature ranking process.

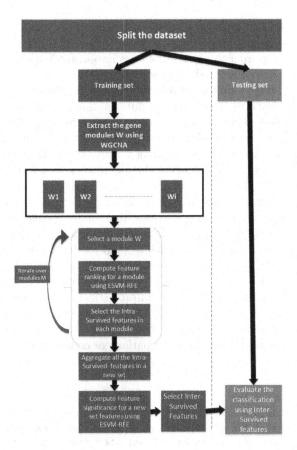

Fig. 1. The flowchart of SFL-ESVM

The SFL-ESVM algorithm aims to achieve a feature selection in the presence of the correlated features. It follows a backwards feature elimination method. The flowchart of the SFL-ESVM algorithm is shown in Fig. 1. It is divided into two phases: an intra-screening phase and an inter-screening phase. The intra-screening phase is composed of multiple steps: firstly, it constructs a feature co-expression network which captures the correlation between the features. This step can be achieved using WGCNA or any other graph clustering method. Then, the modules M_i of the correlated features are extracted from the feature co-expression network using hierarchical clustering. The screening phase is applied on each module M_i to filter out the unimportant features. It is known as intra-screening because it operates on each module independently. For each correlated feature module M_i, ESVM-RFE is used to generate weights for the features. This is described in Algorithm 1. It starts with the entire set of features in the module, and in each iteration, an ensemble SVM is trained by taking bootstrap samples from the training dataset. The feature weights are estimated using the absolute value of coefficients of the support vectors for each SVM model. The estimated feature weights are aggregated from the ensemble of SVM models and ranked in decreasing order to remove the least important features with the small weights. Features are eliminated over multiple iterations on each module until a specific threshold of the number of selected features is reached.

The selected features from each module in the intra-screening phase are aggregated and passed as an input to the inter-screening phase. The inter-screening phase in the SFL-ESVM algorithm is to capture the interaction among the modules. It aggregates all the surviving features from the previous phase and applies one more ESVM-RFE to select the global surviving features.

The proposed SFL-ESVM algorithm is an appropriate solution to reduce the correlation bias in the presence of imbalanced data. It differs from other feature selection algorithms in the following ways: first, it makes use of the ensemble SVM to reduce the influence of correlation bias, because in each iteration, the ranking decision is generated from multiple SVM models on different bootstraps and it is not related to the module size. Second, feature ranking in each iteration is achieved on equal bootstrap samples to mitigate the effect of the imbalance class problem. Third, bagging ensembles improve the performance in the presence of small sample size. Fourth, it uses WGCNA to estimate the network structure of the data, and consequently, estimate the correlated features. Finally, stability is targeted by achieving multiple perturbations of constructing SVM models in each iteration.

Input: training data X
Class labels y
parameter : $inter\text{-}d$;   // Number of selected features between modules
$intra\text{-}d$;   // Number of selected features from each module
b ;   // Size of ensemble SVM in each iteration
E ;   // The % of features to eliminate at each iteration
$bagSize$;   // Balanced bootstrap from training dataset

$modules \leftarrow \text{WGCNA}(X)$;   // the interconnected features using WGCNA
$l \leftarrow \text{length}(modules)$;
for $i \leftarrow 1$ **to** l **do**
$\quad M \leftarrow modules_i$;
$\quad data \leftarrow X(,M)$;
$\quad$;   // M is the correlated genes in each module
$\quad intra\text{-}features \leftarrow \text{ESVMRFE}(data, y, b, E, intra\text{-}d, bagSize)$;
$\quad intra\text{-}Set \leftarrow intra\text{-}Set \cup intra\text{-}features$;
end
$selectedData = trainingdata[, intra\text{-}Set]$;
$inter\text{-}features \leftarrow \text{ESVMRFE}(selectedData, y, b, E, inter\text{-}d, bagSize)$;
// inter-features: the surviving features between the modules using Algorithm ESVM-RFE
Output: $inter\text{-}features$

Algorithm 1. Sparse Feature Learning algorithm (SFL-ESVM)

Function *ESVMRFE (data, class, b, E, d, bagSize)*

    Input: $surviveIndexes = seq(1 : ncol(data))$
           $n = nrow(data)$

    for $d \leftarrow 1$ **to** $length(surviveIndexes)$ **do**

        $m = length(surviveIndexes)$;
        $survive = m - m \times E$;   // survive: number of features in the current iteration
        $ensRes = matrix(n, b)$;   // ensRes: feature's weight of each SVM model
        for $i \leftarrow 1$ **to** b **do**

            $bag \leftarrow bootstrap(data, bagSize)$;
            $bagClass \leftarrow bootstrap(class, bagSize)$;
            $model \leftarrow svm(bag[, survivingIndexes], bagClass)$;
            $weightVector \leftarrow transpose(model\$coefs)\% * \%model\$SV$;
            // Compute the weight vector
            $featureWeight \leftarrow weightVector * weightVector$;   // Compute ranking criteria
            $ensRes \leftarrow merge(ensRes, featureWeight)$;   // Accumulate feature's weight

        end

        $totalWeight = rowSum(ensRes)$;   // Aggregate feature's weight
        $sortedWeight \leftarrow sort(totalWeight)$;   // Sort the total feature's weight by decreasing order
        $sortedIndexes \leftarrow index(sortedWeight)$;
        $surviveIndexes \leftarrow surviveIndexes[sortedIndexes[1 : survive]]$;
        // Eliminate features with smallest weight

    end

    Output: $selectedData = data[, surviveIndexes]$

Algorithm 2. ESVM-RFE for feature learning

4   Experiments

In this section, the experimental evaluations on high-dimensional, highly correlated datasets are reported. This section analyses and compares the classification performance of the proposed SFL-ESVM against the state-of-the-art algorithms namely SVM-RFE [9], Fuzzy Forests [6], and Hybrid L1/2 L2 regularization (HLR) [10]. SVM-RFE is evaluated as a baseline method, Fuzzy Forests as an ensemble feature ranking algorithm for correlated features, and HLR from the point of view of a sparse model for dimensionality reduction in the presence of correlated features. It is important to note that the main purpose of these experiments is to evaluate the potential of the proposed SFL-ESVM algorithm to improve the classification performance in the presence of a large number of correlated features.

Without loss of generality, linear SVM is used as a classifier to evaluate the performance of the selected features from the compared algorithms. The performance is measured by the widely used metric AUC under the receiver operating characteristic (ROC) analysis. The optimal tuning parameters of the

SFL-ESVM, Fuzzy Forests, HLR and SVM-RFE approaches were identified by five-fold cross-validation on the training set. The datasets are divided at random such that approximately 75% is used as a training set and 25% as a test set. The datasets are z-score normalized.

4.1   Datasets

The experiments are conducted on one dataset collected from The Children's Hospital at Westmead, and four public datasets. The details of these datasets are summarised in Table 1. The common characteristics of these datasets are highly dimensional, highly correlated, have a small number of samples and some of them are imbalanced. A stratified random sampling function (stratified) in R is applied on the evaluated datasets to split the data into a training and testing set, with a quarter of the dataset considered as a testing set and the reminder as a training set.

Table 1. Datasets

Dataset	#Attributes	#Instances	Source
Childhood Leukaemia	22277	60	TB-CHW
DLBCL-FSCC	7129	77	[15]
Prostate cancer	6033	102	[16]
ALL/AML	7129	73	[8]
Breast cancer	8141	295	[19]

4.2   Results and Discussion

The goal of this section is to evaluate the performance of the selected features from the compared algorithms on the real-world datasets. In the following experiments, for a fair comparison of all algorithms, the AUC accuracy is estimated using the .632+ bootstrap method [1] with 100 bootstrap samples. For each bootstrap sample, AUC accuracy is obtained on the test dataset.

Figure 2 shows the AUC evaluated on the test dataset across a different number of features. The figures present the results of up to 100 features because the evaluated datasets contain a small number of samples which needs a small number of features to avoid over-fitting. As shown in Fig. 2, the proposed SFL-ESVM algorithm outperforms the state-of-the-art feature selection methods in most feature sets in all datasets. The AUC classification performance is further investigated based on the best number of selected features. As shown in Table 2, several statistical measures are included, namely minimum, maximum, first quartile, third quartile, median and mean on 100 bootstrap samples on the test data. For example, as shown in Table 2 for the DLBCL-FSCC dataset, the best AUC is achieved for the compared algorithms: SFL-ESVM, Fuzzy Forest, HLR and

Table 2. The quartile and mean values of AUC accuracies of the compared algorithms on the evaluated datasets at the best number of features.

Dataset	Method	Min	1st Qu	Median	Mean	3rd Qu	Max	Best features
Childhood leukaemia	SVM-RFE	0.250	0.400	0.500	0.509	0.600	0.800	64
	Fuzzy forest	0.250	0.387	0.450	0.448	0.500	0.700	48
	HLR	0.081	0.331	0.381	0.370	0.431	0.681	30
	SFL-ESVM	**0.500**	**0.600**	**0.700**	**0.692**	**0.750**	**0.850**	24
DLBCL -FSCC	SVM-RFE	0.585	0.741	0.811	0.797	0.848	0.904	78
	Fuzzy forest	0.647	0.743	0.810	0.800	0.854	0.897	74
	HLR	0.679	0.834	0.904	0.891	0.942	0.997	74
	SFL-ESVM	**0.833**	**0.900**	**0.966**	**0.939**	**1.00**	**1.00**	50
Prostate	SVM-RFE	0.562	0.601	0.639	0.639	0.678	0.716	64
	Fuzzy forest	0.744	0.783	0.821	0.829	0.860	0.898	68
	HLR	0.690	0.730	0.730	0.745	0.769	0.807	74
	SFL-ESVM	**0.807**	**0.884**	**0.884**	**0.886**	**0.923**	**0.923**	34
Breast	SVM-RFE	0.505	0.659	0.710	0.701	0.738	0.818	24
	Fuzzy forest	0.471	0.544	0.624	0.618	0.669	0.760	30
	HLR	0.462	0.570	0.633	0.618	0.678	0.741	70
	SFL-ESVM	**0.556**	**0.677**	**0.727**	**0.716**	**0.772**	0.818	30
ALL/AML	SVM-RFE	0.683	0.754	0.754	0.784	0.826	0.826	32
	Fuzzy forest	0.804	0.834	0.876	0.859	0.876	0.905	50
	HLR	0.800	0.800	0.841	0.827	0.841	0.871	68
	SFL-ESVM	**0.815**	**0.928**	**0.958**	**0.953**	**1.00**	**1.00**	46

SVM-RFE is 50, 74, 74, and 78 features respectively. This clearly shows that the proposed algorithm achieves better results than the compared algorithms using different statistical measures. Furthermore, the proposed algorithm SFL-ESVM obtained the best accuracy results compared to the others with a small number of features in most datasets, which leads to less computational complexity during the training process. It can also be observed that the classification results of the SFL-ESVM algorithm tend to be stable after increasing the selected features above approximately 50 features in the evaluated datasets. This indicates the stability and capability of SFL-ESVM to select a lower percentage of features and realise good accuracy results.

On the other hand, the experimental results indicate that the feature selection methods that handle correlated features such as the proposed SFL-ESVM, Fuzzy forest, and HLR, perform better than SVM-RFE which does not consider the correlated features and achieves individual feature ranking. Therefore, it demonstrates the importance of handling correlated features in high-dimensional datasets to improve the performance of the classifiers. Finally, a statistical t-test is also conducted between the vector results of the proposed algorithm against state-of-the-art methods under the null hypothesis that AUC on vectors of the used method is not significantly different to SFL-ESVM. The p-value is lower than 0.05 which rejects the null hypothesis.

432   A. Braytee et al.

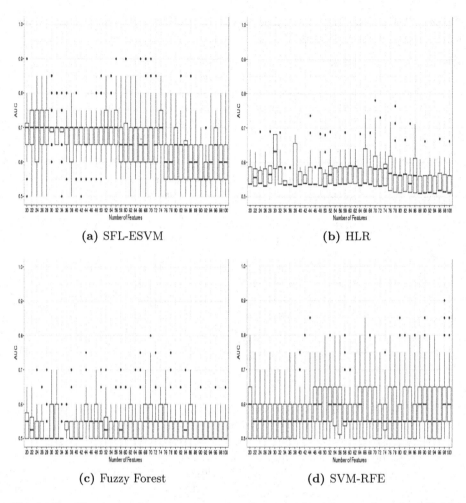

(a) SFL-ESVM

(b) HLR

(c) Fuzzy Forest

(d) SVM-RFE

Fig. 2. Classification performance comparison between algorithms evaluated on Childhood Leukaemia dataset using the 0.632+ bootstrap method with 100 bootstrap samples across a different number of features

A further investigation is made of the selected features using the proposed SFL-ESVM from the ALL/AML dataset to see if the proposed algorithm can define separated clusters based on ALL and AML class outcomes. To do this, Singular Value Decomposition (SVD) is applied to the original ALL/AML training set using all the features and project the testing samples using the first three principal components. Then, the testing samples are visualised with different shapes for the ALL and AML samples. A similar process is applied on the training set with the top 46 features selected by the proposed SFL-ESVM feature selection algorithm. Without loss of generality, the top 46 features are used in these figures. As shown in Fig. 3, it is clear that the clusters of the ALL and AML

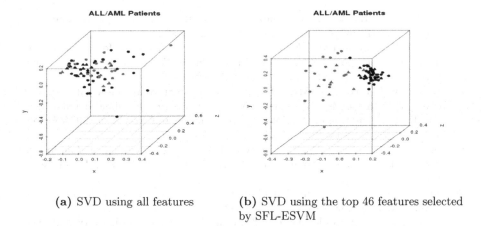

(a) SVD using all features

(b) SVD using the top 46 features selected by SFL-ESVM

Fig. 3. SVD on ALL/AML dataset to show the clusters of ALL and AML patients. Black = ALL, red = AML, circle = training samples, and triangle = testing samples (Color figure online)

classes in Fig. 3b are well separated compared to the clusters in Fig. 3a, which overlap. This example confirms that efficacy of the proposed algorithm to select the optimal features in the presence of complex datasets which importantly, are able to explain the differences between different classes.

5   Conclusion

This paper proposed a novel algorithm to select the best features in the presence of highly correlated features that improves the classification performance. The proposed SFL-ESVM does not consider the correlated features as redundant, rather it selects the top correlated features from each feature module using the ESVM-RFE algorithm. Then, it aggregates all features from different modules and again applies ESVM-RFE to rank the combined features. The proposed algorithm can improve the classification accuracy in the presence of very complex datasets. These datasets contain high-dimensional, highly correlated features and a low number of samples. Extensive experiments are conducted on different datasets. Our results show the high-performing quality of the proposed method on benchmark datasets which outperformed the-state-of-the-art methods.

References

1. Ambroise, C., McLachlan, G.J.: Selection bias in gene extraction on the basis of microarray gene-expression data. PNAS **99**(10), 6562–6566 (2002)
2. Anaissi, A., Goyal, M., Catchpoole, D.R., Braytee, A., Kennedy, P.J.: Ensemble feature learning of genomic data using support vector machine. PLOS ONE **11**(6), 1–17 (2016)

3. Bin, Z., Steve, H.: A general framework for weighted gene co-expression network analysis. Stat. Appl. Gen. Mol. Biol. **4**(1), 11–28 (2005)

4. Boser, B.E., Guyon, I.M., Vapnik, V.N.: A training algorithm for optimal margin classifiers. In: Proceedings of the Fifth Annual Workshop on Computational Learning Theory, pp. 144–152. ACM, Pittsburgh (1992)

5. Braytee, A., Liu, W., Kennedy, P.J.: Supervised context-aware non-negative matrix factorization to handle high-dimensional high-correlated imbalanced biomedical data. In: 2017 International Joint Conference on Neural Networks (IJCNN), pp. 4512–4519. IEEE, Anchorage (2017)

6. Conn, D., Ngun, T., Li, G., Ramirez, C.: Fuzzy forests: extending random forests for correlated, high-dimensional data. UCLA Biostatistics Working Paper Series (2015)

7. Cui, C., Wang, D.: High dimensional data regression using lasso model and neural networks with random weights. Inf. Sci. **372**, 505–517 (2016)

8. Golub, T.R., et al.: Molecular classification of cancer: class discovery and class prediction by gene expression monitoring. Science **286**(5439), 531–537 (1999)

9. Guyon, I., Weston, J., Barnhill, S., Vapnik, V.: Gene selection for cancer classification using support vector machines. Mach. Learn. **46**(1), 389–422 (2002)

10. Huang, H.H., Liu, X.Y., Liang, Y.: Feature selection and cancer classification via sparse logistic regression with the hybrid L1/2 + 2 regularization. PLOS ONE **11**(5), 1–15 (2016)

11. Meier, L., Van De Geer, S., Bühlmann, P.: The group LASSO for logistic regression. J. R. Stat. Soc. Seri. B (Stat. Methodol.) **70**(1), 53–71 (2008)

12. Park, M.Y., Hastie, T., Tibshirani, R.: Averaged gene expressions for regression. Biostatistics **8**(2), 212–227 (2007)

13. Rapaport, F., Barillot, E., Vert, J.P.: Classification of arrayCGH data using fused SVM. Bioinformatics **24**(13), i375–i382 (2008)

14. Saeys, Y., Inza, I., Larrañaga, P.: A review of feature selection techniques in bioinformatics. Bioinformatics **23**(19), 2507–2517 (2007)

15. Shipp, M.A., et al.: Diffuse large B-cell lymphoma outcome prediction by gene-expression profiling and supervised machine learning. Nat. Med. **8**(1), 68–74 (2002)

16. Singh, D., et al.: Gene expression correlates of clinical prostate cancer behavior. Cancer Cell **1**(2), 203–209 (2002)

17. Tibshirani, R.: Regression shrinkage and selection via the LASSO. J. R. Stat. Soc. Ser. B (Methodological) **58**(1), 267–288 (1996)

18. Tolosi, L., Lengauer, T.: Classification with correlated features: unreliability of feature ranking and solutions. Bioinformatics **27**(14), 1986–1994 (2011)

19. Van De Vijver, M.J., et al.: A gene-expression signature as a predictor of survival in breast cancer. New Engl. J. Med. **347**(25), 1999–2009 (2002)

Task and Instance Quadratic Ordering for Active Online Multitask Learning

Jing Zhao[1][(✉)], Shaoning Pang[1], Iman Tabatabaei Ardekani[1], Yuji Sekiya[2],
and Daisuke Miyamoto[3]

[1] Unitec Institute of Technology, Auckland, New Zealand
{jzhao,ppang,iardekani}@unitec.ac.nz
[2] The University of Tokyo, Tokyo, Japan
sekiya@nc.u-tokyo.ac.jp
[3] Nara Institute of Science and Technology, Ikoma, Nara Prefecture, Japan
daisu-mi@is.naist.jp

Abstract. For online multitask learning (oMTL), when a chunk of tasks consisting of multiple related instances is received in one batch, the learner normally has the chance to actively order these tasks to improve the learning efficiency. This paper proposes a quadratic ordering method for active oMTL, where instance ordering is integrated into task ordering by taking each instance in one task. The proposed task and instance quadratic ordering is able to facilitate oMTL better than single task ordering. The orderings derived in this paper can be incorporated into any individual oMTL algorithms for active oMTL. The performance evaluations on four real-word datasets demonstrate the benefits of the proposed algorithms.

Keywords: Online Multitask Learning (oMTL) · Active oMTL
Quadratic ordering · Task ordering · Instance ordering

1 Introduction

Multitask learning (MTL) explores a set of offline machine learning systems which are able to learn a number of related tasks in one batch [1]. Considering the successive tasks in real world applications often arrive one after another over an extended period of time. Thus online multitask learning (oMTL) is more desirable in practice, since it aims to model systems which can learn multiple related tasks in real time by sharing common information among them. As compared to the traditional online single task learning (oSTL), oMTL often achieves better generalization performance across all tasks than separately learning each task. Effectively using task relatedness rather than simply ignoring it makes oMTL dramatically outperform oSTL in many real-world applications [2].

Typical oMTL algorithms do not control the order in which they learn the tasks and instances. However, the ordering effects exist for oMTL when a chunk of tasks are presented for learning. The different ordered sequences of these tasks

© Springer Nature Switzerland AG 2018
L. Cheng et al. (Eds.): ICONIP 2018, LNCS 11303, pp. 435–447, 2018.
https://doi.org/10.1007/978-3-030-04182-3_38

and instances can lead to the different learning consequences. For the same training data in two different sequences, such ordering of tasks and instances allows one to favor the learning more than the learning with random selection. In the literature, oMTL with ordering or selection is called active oMTL [3]. The key of active oMTL is to find from all permutations of the training data the one that yields the best performance.

Neither task ordering nor instance ordering by itself is an optimal solution to active oMTL. Instead, it is desirable to perform instance ordering to facilitate the learning of each individual task. Motivated by this, we propose task and instance quadratic ordering for active oMTL. This work focuses on how to simultaneously order tasks and instances efficiently in the online learning scenario. The philosophy of our methods is to conduct quadratic ordering to enhance learning performance. First, the strategies should be used to perform the instance ordering on the training data. Then, a task ordering is operated on tasks available so far. To assess the proposed four quadratic ordering methods, we apply them to the existing oMTL approaches, the experimental results demonstrate that they can learn training data more efficiently than only task ordering or only instance ordering. In particular, using these active learning algorithms, the system can acquire a clear further reduction of necessary training data for achieving a particular level of performance, as compared to the passive learning algorithms.

In general, the contributions of this paper are:

1. This is the first work, to our best knowledge, where instances and tasks ordering are jointly investigated independent to the learning process of MTL, rather than incorporated into an oMTL learner.
2. An efficient task and instance quadratic ordering algorithm is developed to solve the proposed objective function, and the convergence of the algorithm can be guaranteed. Experimental results on four real-world datasets demonstrate the effectiveness of the proposed approach.

2   Related Work

2.1   Instance Ordering

Instance ordering has been investigated mainly for single task learning (STL). Tong et al. [4] proposed instance ordering for the learning of support vector machine (SVM) [5]. They used the duality between parameter space and feature space to reduce version space as much as possible at each query. Considerable gains are reported on learning effectiveness in both inductive and transductive settings. Bengio et al. [6] explored curriculum learning by implementing the strategy of learning the easiest instances first and then incrementally processing harder instances. Their experiments reveal that such strategies lead to faster training and higher prediction quality.

For active MTL, Kumar et al. [7] addressed the active learning of a useful MTL tool named latent variable models and proposed an iterative self-paced learning algorithm for instance ordering, where each iteration simultaneously chooses easy instances and learns a new parameter vector. The issue of active instance selection in oMTL was addressed in [8]. Saha et al. proposed an adaptive framework of oMTL, where an adaptive relationship matrix is built to evaluate the relatedness among multiple tasks. By taking into account the task relatedness, the informativeness of an incoming instance can be quantified using the adaptive relationship matrix. In this way, the learner can employ this matrix to actively choose the most informative instances. In essence, the approach transforms MTL to STL problem by merging the instances of each task into a single large task, and then performs only instance ordering on the merged task.

2.2 Task Ordering

Task ordering is an alternative approach that orders training data to maximize learning performance across all tasks. It typically involves the estimation of tasks similarity and tasks relatedness.

Ruvolo et al. [3] described a scenario, where the next task can be selected from a pool of candidate tasks. Two general methods for active task selection were proposed: one is based on information maximization, where the criteria for choosing the next task focus on maximizing expected information gain about the shared basis; the other is based on model performance, in which the next task is selected in terms of minimizing the worst-case fit of shared basis to each candidate task. These approaches are used in the algorithm ELLA proposed in [9].

In [10], we proposed two task ordering algorithms: QR-decomposition Ordering and Minimal-loss Ordering. The QR-decomposition Ordering measures the within-task distance of the training data, and chooses the next task with the shortest within-task distance. The Minimal-loss Ordering computes the predictive loss of the learned model, and selects the next task with the minimal loss. Based on [10], we propose four new quadratic ordering algorithms which combine task ordering with instance ordering in this paper.

3   Task and Instance Ordering

In the oMTL scenario, the training task or instance usually arrives one-at-a-time, and the learner has no control on the sequence in which learning tasks are presented for training. However, when a chunk of training data which consists of multiple related tasks is received in one batch, the learner obviously has the chance to actively order these tasks to improve the learning efficiency.

We assume the problem of ordering data in the following setting: at the tth iteration, the learner receives the training data for k tasks, which are indexed as $\{T_1, T_2, \cdots, T_k\}$; and each task consists of u_t instances, which are indexed as $\{1, 2, \cdots, u_t\}$. To achieve better performance, we attempt to make the training data be learned in a particular order, which means a new permutation of $\{T_1, T_2, \cdots, T_n\}$ and a corresponding permutation of $\{1, 2, \cdots, u_t\}$ for each task.

For the rest of the paper, superscripts denote the variables related to a particular task, for instance, $X^{(T_1)}$ and $y^{(T_1)}$ are related to task T_1; and subscripts represent the variables related to a certain instance, for example, x_i and y_i are related to instance i.

The global loss $\mathcal{L}$ consists of the cost caused by instance ordering $\mathcal{L}^I$ on each task available so far, and the loss on task ordering $\mathcal{L}^T$. It can be formulated as

$$\mathcal{L} = \sum_{j=1}^{k} \mathcal{L}_j^I + \mathcal{L}^T. \tag{1}$$

The objective of the proposed task and instance quadratic ordering is to seek a permutation of newly received training instances by minimizing the above cumulative loss.

In Eq. (1), $\mathcal{L}^T$ is determined by the error rate of task labelling/classification for all instances received at t iteration. This implies that, an instance may still be given a correct task label even if it is mis-classified within a task. Thus we have $\mathcal{L}^T \leq \sum_{j=1}^{k} \mathcal{L}_j^I$. Further, task ordering is very much dependent on the learner's knowledge over each individual task, as it relies on the estimation of task relatedness. In other words, reducing the first term gives normally a smaller second term, especially when task ordering and instance ordering share the same ordering approach, and task learning and instance learning within task use the same learner.

4   Proposed Quadratic Ordering

For task ordering, the effectiveness of QR-decomposition (QR) and Minimal-loss (ML) has been demonstrated in [10]. In this work, we further investigate the use of QR and ML for instance ordering, and develop four quadratic ordering algorithms, which include QR-QR, QR-ML, ML-QR, and ML-QR ordering.

4.1   QR-Decomposition Task Ordering

The QR decomposition for task ordering [10] is to compare the within-task distance of the training data, in which QR decomposition of centroid matrix A of training data X is employed: $A = QR$, where Q refers to an orthogonal matrix, and R denotes an upper triangular matrix.

Suppose that at the m-th iteration we receive training data $(X_{new}^{(t)}, y_{new}^{(t)})$, t denotes task $t \in \{T_1, T_2, \cdots, T_n\}$. Given a data matrix $X_{old}^{(t)} = [A_1^{(t)}, \cdots, A_k^{(t)}] \in \mathbb{R}^{d \times n}$ with $A_i^{(t)} \in R^{d \times u_i}$ (write $X_{old}^{(t)} = 0$ when t is new), where $A_i^{(t)}$ represents the previously received training data for task t, u_i denotes the number of instances contained in $A_i^{(t)}$, and d is feature dimension. Suppose $C^{(t)} = Q^{(t)} R^{(t)}$ is the QR decomposition of the centroid matrix $C^{(t)} = [m_1^{(t)}, \cdots, m_k^{(t)}]$ and $H_w^{(t)} = [H_1^{(t)}, \cdots, H_k^{(t)}]$, where $H_i^{(t)} = [A_i^{(t)} - m_i^{(t)} e_i^T]$ with $e_i \in (1, \cdots, 1)^T \in \mathbb{R}^{u_i}$.

Let the column vector $X_{new}^{(t)}$ to be given by

$$X_{new}^{(t)} = [x_1^{(t)}, x_2^{(t)}, \cdots, x_{u_t}^{(t)}],$$

and

$$w^{(t)}(x_i) = \frac{\|(Q^{(t)})^T(x_i - m_j)\|^2}{n_j + 1} \quad \text{or} \tag{2}$$

$$w^{(t)}(x_i) = \frac{\|(H_w^{(t)})^T(I - Q^{(t)}(Q^{(t)})^T)x_i)\|^2}{\|(I - Q^{(t)}(Q^{(t)})^T)x_i\|^2}, \tag{3}$$

accordingly, as x_i lies in the jth class of $X_{old}^{(t)}$ or a new class.

The criterion for QR-decomposition ordering to choose the next task is summarized as,

$$t_{next} = \underset{t \in \{T_1, T_2, \cdots, T_n\}}{argmin} w^{(t)}(X_{new}^{(t)}), \tag{4}$$

where

$$w^{(t)}(X_{new}^{(t)}) = \frac{1}{u_t} \sum_{i=1}^{u_t} w^{(t)}(x_i). \tag{5}$$

Note that the above calculation does not require any knowledge of learning system, thus QR-decomposition task ordering is a leaner independent task ordering approach.

4.2 Minimal-Loss Task Ordering

The minimal-loss for task ordering is to calculate the predictive loss of learned model, and select the next task that causes the minimal loss [10]. Here, a shared basis L are used for modeling task relatedness and sharing useful information among multiple tasks [9]. The model of task t is represented as a parameter vector $\theta^{(t)}$ that is a linear combination of the columns of shared basis L according to the weight vector $s^{(t)}$: $\theta^{(t)} = Ls^{(t)}$.

Suppose that at the m-th iteration we receive training data $(X_{new}^{(t)}, y_{new}^{(t)})$. Now, let us define

$$X^{(t)} = [X_{old}^{(t)} X_{new}^{(t)}] \quad \text{or} \quad X^{(t)} = X_{new}^{(t)},$$

$y^{(t)} = (y_{old}^{(t)}; y_{new}^{(t)})$ or $y^{(t)} = y_{new}^{(t)}$ according as t is an old or new task. Also, let us consider

$$X^{(t)} = [x_1^{(t)}, x_2^{(t)}, \cdots, x_{u_t}^{(t)}],$$

as a column vectors and define

$$F^{(t)}(\theta) = \frac{1}{n_t} \sum_{i=1}^{n_t} \mathcal{L}(f(x_i^{(t)}; \theta), y_i^{(t)}), \tag{6}$$

$$D^{(t)} = \frac{1}{2} \nabla_{\theta,\theta}^2 (F^{(t)}(\theta))|_{\theta = \theta^{(t)}}, \tag{7}$$

where $\mathcal{L}$ is a known loss function and f is the prediction function. Let

$$\ell(L, s, \theta, D) = \mu \|s\|_1 + \|\theta - Ls\|_D^2. \tag{8}$$

The strategy for Minimal-loss Ordering method to choose the next task is as follows:

$$t_{next} = \mathop{argmin}_{t \in \{T_1, T_2, \cdots, T_n\}} G(L_{m+1}^{(t)}), \tag{9}$$

where

$$G(L) = \hat{g}_m(L) = \lambda \|L\|_F^2 + \frac{1}{T} \sum_{i=1}^{T} \ell(L, s^{(t)}, \theta^{(t)}, D^{(t)}). \tag{10}$$

4.3   QR-QR Ordering

In the proposed QR-QR ordering, QR-decomposition is utilized to conduct task ordering on training data, then the same algorithm is applied for instance ordering within each chosen task (called QR instance ordering). QR instance ordering measures the within-instance distance of the training data, and selects the next instance with the shortest within-class distance. We consider one instance as a special case of a task that contains only one instance. By Eq. 4, we derive the strategy for QR instance ordering to choose the next instance i_{next} as,

$$i_{next} = \mathop{argmin}_{i \in \{1, 2, \cdots, u_t\}} w^{(t)}(x_i), \tag{11}$$

where

$$w^{(t)}(x_i) = \frac{\|(Q^{(t)})^T (x_i - m_j)\|^2}{n_j + 1} \quad \text{or}$$

$$w^{(t)}(x_i) = \frac{\|(H_w^{(t)})^T (I - Q^{(t)}(Q^{(t)})^T) x_i)\|^2}{\|(I - Q^{(t)}(Q^{(t)})^T) x_i\|^2},$$

accordingly, as x_i lies in the jth class of $X_{old}^{(t)}$ or a new class. The procedure of QR-QR ordering at the mth iteration is summarized in Algorithm 1.

Algorithm 1. QR-QR Ordering Algorithm

Require: training data $(X_{new}^{(t)}, y_{new}^{(t)})$
Ensure: next task t_{next} and next instance i_{next}
1: /* Choose next task using QR-decomposition */
2:
$$t_{next} = *arg\,min_{t\in\{T_1,T_2,\cdots,T_n\}}\, w^{(t)}(X_{new}^{(t)})\quad \#\,Equation\,4$$
3: /* For each task t, choose next instance using QR-decomposition */
4:
$$i_{next} = *arg\,min_{i\in\{1,2,\cdots,u_t\}}\, w^{(t)}(x_i)\quad \#\,Equation\,11$$

4.4 QR-ML Ordering

In the proposed QR-ML ordering, QR-decomposition is firstly utilized for task ordering, the Minimal-loss criterion is applied to the instance ordering within each selected task (called ML instance ordering).

Specifically, QR-decomposition task ordering determines the next task t_{next} to be learned by Eq. 4. Within the selected task, the QR-ML ordering conducts further instance ordering using Minimal-loss criterion, which measures the predictive loss of the learning model and selects the next instance with the minimal loss. Again, we consider one instance as one task. We straightforwardly extend minimal-loss task ordering, Eq. 9, for instance ordering as,

$$i_{next} = \mathop{argmin}_{i \in \{1,2,\cdots,u_t\}} G_i(L_{m+1}^{(t)}), \tag{12}$$

where

$$G_i(L) = \lambda\|L\|_F^2 + \ell_i(L, s^{(t)}, \theta^{(t)}, D^{(t)}), \tag{13}$$

$$\ell_i(L, s^{(t)}, \theta^{(t)}, D^{(t)}) = \mu\|s\|_1 + \|\theta - Ls\|_{D_i^{(t)}}^2, \tag{14}$$

$$D_i^{(t)} = \frac{1}{2}\nabla_{\theta,\theta}^2((F_i^{(t)}(\theta))|_{\theta=\theta^{(t)}}, \tag{15}$$

$$F_i^{(t)}(\theta) = \mathcal{L}(f(x_i^{(t)};\theta), y_i^{(t)}). \tag{16}$$

The procedure of QR-ML ordering at the m-th iteration is summarized in Algorithm 2.

4.5 ML-QR Ordering

In the proposed ML-QR ordering, Minimal-loss is firstly employed for the task ordering of training data, then QR instance ordering is applied to ranking instances within each selected task.

Algorithm 2. QR-ML Ordering Algorithm

Require: training data $(X_{new}^{(t)}, y_{new}^{(t)})$
Ensure: next task t_{next} and next instance i_{next}
1: /* Choose next task using QR-decomposition */
2:
$$t_{next} = * \arg min_{t \in \{T_1, T_2, \cdots, T_n\}} w^{(t)}(X_{new}^{(t)}) \quad \# \text{ Equation } 4$$
3: /* For each task t, choose next instance using Minimal-loss */
4:
$$i_{next} = * \arg min_{i \in \{1, 2, \cdots, u_t\}} G_i(L_{m+1}^{(t)}) \quad \# \text{ Equation } 12$$

Algorithm 3. ML-QR Ordering Algorithm

Require: training data $(X_{new}^{(t)}, y_{new}^{(t)})$
Ensure: next task t_{next} and next instance i_{next}
1: /* Choose next task using Minimal-loss */
2:
$$t_{next} = * \arg min_{t \in \{T_1, T_2, \cdots, T_n\}} G(L_{m+1}^{(t)}) \quad \# \text{ Equation } 9$$
3: /* For each task t, choose next instance using QR-decomposition */
4:
$$i_{next} = * \arg min_{i \in \{1, 2, \cdots, u_t\}} w^{(t)}(x_i) \quad \# \text{ Equation } 11$$

From Eq. 9, minimal-loss task determines the next task t_{next} to be learned. Within the selected task, the ML-QR ordering conducts further instance ordering using QR-decomposition criterion. The same as above, one instance is viewed as one task, QR task ordering is extended for instance ordering as Eq. 11, by which we choose the next instance i_{next} with the shortest within-class distance. The procedure of ML-QR ordering at the m-th iteration is given in Algorithm 3.

4.6 ML-ML Ordering

In the proposed ML-ML ordering, Minimal-loss is utilized for task ordering, then ML instance ordering is performed on the instances of each selected task.

From Eq. 9, minimal-loss task ordering determines the next task t_{next} to be learned. Within the selected task, ML-ML ordering conducts further instance ordering by using ML criterion. Similarly, one instance is viewed as one task, ML task ordering is extended for instance ordering as Eq. 12, by which we choose the next instance i_{next} with the shortest within-class distance. The procedure of ML-ML ordering at the m-th iteration is given in Algorithm 4.

5 Experimental Results

We evaluated the proposed four quadratic ordering algorithms by comparing them against four other task ordering approaches: InfoMax [3], Diversity [3], QR-decomposition (QR) [10] and Minimal-loss (ML) [10]. We applied these eight

---

Algorithm 4. ML-ML Ordering Algorithm

Require: training data $(X_{new}^{(t)}, y_{new}^{(t)})$
Ensure: next task t_{next} and next instance i_{next}
1: /* Choose next task using Minimal-loss */
2:
$$t_{next} = *arg\,min_{t \in \{T_1, T_2, \cdots, T_n\}} G(L_{m+1}^{(t)}) \quad \# Equation\ 9$$
3: /* For each task t, choose next instance using Minimal-loss */
4:
$$i_{next} = *arg\,min_{i \in \{1,2,\cdots,u_t\}} G_i(L_{m+1}^{(t)}) \quad \# Equation\ 12$$

---

methods to an existing lifelong learning algorithm (ELLA) [9] to assess their performance.

5.1 Datasets

We conducted oMTL experiments for each algorithm on four real-world datasets: Computer Survey, London Schools, Land Mine Detection and Facial Expression Recognition. Computer Survey is a regression dataset which has already been widely used for the assessment of MTL algorithms [11]. London Schools is another regression dataset which is from Inner London Education Authority, and which has been studied in many previous MTL and oMTL works [9]. Land Mine Detection is a classification dataset [9]. It aims to discriminate whether or not a land mine is present in an area based on radar images. Facial Expression Recognition is also a classification dataset which comes from a recent facial expression recognition challenge [12].

5.2 Experimental Procedure

As in [10], for each dataset, we randomly split data for 20 times in its predefined proportion of training to testing. We repeated active oMTL experiments on the produced data split for 1,000 times to smooth out variability. The average results are reported for each task ordering method. In parameters setting, we followed [3] to maximize performance on the evaluation tasks averaged over all the task ordering methods, and used a grid-search method to choose the value of the parameter k in $\{1, 2, \cdots, 10\}$, and the ridge term Γ from the set $\{e^{-5}, e^{-3}, e^{-1}, e^1\}$. The value of λ and μ are specified as e^{-5} and 1 respectively through cross validation experiments.

For performance evaluation, we set observed oMTLs to achieve and maintain a certain level of performance, then estimate, as in [3], how many less tasks (in terms of the percentage to the total number of tasks) are demanded as compared to the number of tasks required for oMTL on random task order. The oMTL performance for classification tasks is measured by the area under the ROC curve (AUC), and regression tasks by the negative root mean squared error (-rMSE). A positive score on % Less Tasks Required indicates the ordering method has

higher learning efficiency than random task ordering, a negative score displays
that it is less efficient, and a score of 0 reveals that it has no improvement in
learning efficiency.

5.3   Results

We compared the proposed data ordering methods: QR-QR, QR-ML, ML-ML
and QR-QR with four existing methods: InfoMax [3], QR-decomposition (QR)
[10], Diversity [3] and Minimal-loss (ML) [10]. We measured the less tasks
required during the learning process, the average less tasks required, and the
final less tasks required, as compared to random data ordering.

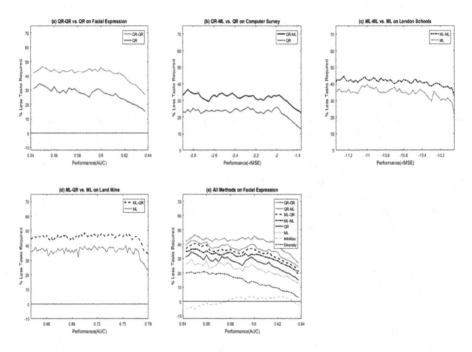

Fig. 1. The results of task ordering on oMTLs. Each plot shows the accuracy achieved
by each method versus the oMTL efficiency (in terms of the number of tasks, and in
comparison to random task ordering).

Figure 1 shows the oMTL experimental results of less tasks required during
the learning process for different data ordering algorithms on four real-world
datasets. As we can see, all eight data ordering methods achieve more or less
learning efficiency gain over the random data ordering, which demonstrates the
oMTL with data ordering is always more efficient than that without ordering. In
particular, the plots from QR-QR, QR-ML, ML-QR and ML-ML are displayed
on the top of the plots from QR, ML, InfoMax and Diversity for all datasets,

which reveals that the proposed quadratic ordering algorithms have completely outperformed the existing task ordering methods in all datasets. However, those top four algorithms are very competitive to each other. The ML-ML method wins on the classification oMTL of the Land Mine dataset and the regression oMTL of the Computer Survey dataset; whereas the QL-QL algorithm is the most efficient for the classification oMTL of the Facial Expression dataset and the regression oMTL of the London School dataset. The best performance on each dataset is achieved by either ML-ML or QL-QL, which reveals that using the same method for task and instance ordering can minimize the learning loss and increase the ability of generalization.

Table 1. Average Less Tasks Required over all performance levels.

Method		Average %Less Tasks Required (Standard Deviation)			
		Facial Expression	Land Mine	London School	Computer Survey
Quadratic	QR-QR	**41.4(±2.9)**	40.8(±3.4)	**49.3(±3.0)**	29.0(±3.3)
	QR-ML	36.4(±3.1)	42.9(±2.9)	47.1(±3.2)	31.5(±3.5)
	ML-QR	34.0(±2.8)	45.5(±3.5)	43.9(±3.3)	33.5(±3.2)
	ML-ML	31.4(±3.0)	**48.2(±3.4)**	41.0(±3.3)	**36.7(±3.4)**
Single	QR	27.1(±3.8)	33.5(±4.4)	37.0(±3.6)	22.6(±3.1)
	ML	22.7(±3.4)	36.1(±3.2)	34.5(±4.3)	24.5(±2.9)
	InfoMax	0.5(±2.6)	5.1(±3.7)	29.8(±6.8)	12.9(±3.8)
	Diversity	14.6(±5.1)	29.4(±4.1)	21.0(±3.1)	18.5(±3.4)
Diff.		+52.7%	+33.5%	+33.2%	+49.7%
Average		+42.2%			

Table 1 presents the results, which are measured in terms of the percent less tasks required for oMTL with data ordering, and averaged across all performance levels. Table 2 displays the less tasks required by data ordering at the highest performance level. In these tables, the mean and standard deviations are reported, numbers in bold represent the best performance on the column dataset. As we can see, the results of these two tables reveal the same behavior as that of Fig. 1 that, InfoMax, Diversity, QR and ML were dominated in all four datasets by the proposed QR-QR, QR-ML, ML-QR and ML-ML ordering. Furthermore, Tables 1 and 2 give the performance difference between the best proposed method and the best existing methods, and its average over all four datasets. As seen, the proposed data ordering methods are in general over 40% more efficient than the existing methods for all performance level experiemtns. But, the superiority is increased to over 60% when the highest performance level experiment is counted. The maximum superiority of QR-QR to QR reaches 79.4%.

Table 2. Less Tasks Required by task ordering at the highest performance level.

Method		Final %Less Tasks Required (Standard Deviation)			
		Facial Expression	Land Mine	London School	Computer Survey
Quadratic	QR-QR	**27.1(±3.0)**	28.2(±3.1)	**39.1(±3.2)**	20.5(±3.4)
	QR-ML	24.1(±3.2)	31.3(±3.3)	37.0(±3.5)	22.4(±3.4)
	ML-QR	21.2(±3.0)	34.0(±3.2)	35.1(±3.4)	24.3(±3.5)
	ML-ML	19.1(±3.3)	**37.0(±3.2)**	32.2(±3.1)	**26.3(±3.2)**
Single	QR	15.1(±3.5)	21.2(±3.9)	25.0(±3.8)	12.8(±3.3)
	ML	13.0(±3.2)	23.1(±3.5)	21.2(±3.3)	15.5(±3.9)
	InfoMax	-2.2(±2.9)	-6.9(±3.5)	0.5(±3.8)	2.0(±3.6)
	Diversity	3.1(±4.1)	8.2(±3.1)	9.0(±3.3)	8.3(±3.2)
Diff.		+79.4%	+60.1%	+56.4%	+69.6%
Average		+66.3%			

6   Conclusion

The novel task and instance quadratic ordering algorithms proposed by this paper can be efficiently used for active online multitask learning. The focus of the algorithms is on how to jointly order tasks and instances effectively in the online setting. This paper explores the usage of QR-decomposition (QR) and Minimal-loss (ML) for instance ordering, and designs four quadratic algorithms: QR-QR, QR-ML, ML-QR, and ML-QR ordering. Experimental comparative tests and quantitative performance evaluations on four real-word datasets demonstrate that the proposed algorithms outperform all existing task ordering methods for active oMTL. The best performance on each dataset is obtained by either ML-ML or QL-QL, which indicates that the identical strategy for task and instance ordering can minimize the learning loss and improve the generalization. The orderings derived in this paper deliver a generic data ordering approach independent to any particular oMTL algorithms, classification or regression. In practice, they can be incorporated into any specific oMTL process for active oMTL.

References

1. Ando, K., Zhang, T.: A framework for learning predictive structures from multiple tasks and unlabeled data. J. Mach. Learn. Res. **6**(2), 1817–1853 (2005)
2. Cavallanti, G., Cesabianchi, N., Gentile, C.: Linear algorithms for online multitask classification. J. Mach. Learn. Res. **11**, 2901–2934 (2010)
3. Ruvolo, P., Eaton, E.: Active task selection for lifelong machine learning. In: 27th AAAI Conference on Artificial Intelligence, pp. 862–868. Springer, Washington (2013)
4. Tong, S., Koller, D.: Support vector machine active learning with applications to text classification. J. Mach. Learn. Res. **2**(8), 45–66 (2001)

5. Steinwart, I., Christmann, A.: Support Vector Machines. Springer Science & Business Media, USA (2008)
6. Bengio, Y., Louradour, J., Collobert, R., Weston, J.: Curriculum learning. In: 26th International Conference on Machine Learning, pp. 41–48. ACM, USA(2009)
7. Kumar, M., Packer, B., Koller, D.: Self-paced learning for latent variable models. Neural Inf. Process. Syst. **23**(5), 1189–1197 (2010)
8. Saha, A., Rai, P., Daume, H., Venkatasubramanian, S.: Active online multitask learning. In: 27th International Conference on Machine Learning, pp. 1123–1131. Citeseer, Israel (2010)
9. Ruvolo, P., Eaton, E.: Ella: An efficient lifelong learning algorithm. In: 30th International Conference on Machine Learning, pp. 507–515. ACM, Atlanta (2013)
10. Pang, S., An, J., Zhao, J., Li, X., Ban, T., Inoue, D., Sarrafzadeh, A.: Smart task orderings for active online multitask learning. In: Proceedings of 2014 SIAM International Conference on Data Mining. SIAM, Pennysylvania (2014)
11. Argyriou, A., Evgeniou, T., Pontil, M.: Convex multi-task feature learning. Mach. Learn. **73**(3), 243–272 (2008)
12. Valstar, M., Jiang, B., Pantic, M., Scherer, K.: The first facial expression recognition and analysis challenge. In: 9th IEEE International Conference on Automatic Face Gesture Recognition, pp. 921–926. IEEE, California (2011)

Learning from Titles to Recommend Keywords for Academic Papers

Huifang Ma[1,2(✉)], Fang Liu[1], Qin Xia[1], and Li Yu[1]

[1] College of Computer Science and Engineering,
Northwest Normal University, Lanzhou 730070, China
mahuifang@yeah.net
[2] Guangxi Key Laboratory of Trusted Software,
Guilin University of Electronic Technology, Guilin 541004, China

Abstract. With the increasing number of scientific papers, it is difficult for researchers to locate the most relevant and important keywords from the vast majority of papers and establish the research focus and preliminaries. Based on the commonly accepted assumption that the title of a document is always elaborated to reflect the content of a document and consequently keywords tend to be closely related to the title, a keyword ranking from paper titles involving both real-time and authoritativeness is presented in this paper. We suggest exploring paper titles as a weighted hypergraph and random walk is performed, which considers weights of both hyper-edges and hyper-vertices to model short documents social features as well as discriminative weights respectively, while measuring the centrality of words in the hyper-graph to obtain the recommended keywords. Experimental results demonstrate that the proposed approach is robust for extracting keywords from short texts.

Keywords: Extraction · Weighted Hyper-graph · Weighting strategy
Word correlation · Random walk

1 Introduction

With the great advancement of technology and the continuous enrichment of human knowledge, the number of scientific literature has increased rapidly. We need a specific keyword extraction algorithm to introduce the research hotspots and basic knowledge. The extracted keywords are not only able to reveal the knowledge composition and structure of the current field but also combine the real-time character and the authoritativeness. This is a well-studied problem given the complete text, however, in many cases, due to copyright privileges, research papers databases do not have the complete text, only metadata, such as the title and abstract [1]. It is well agreed that the title itself is not only a high degree of summary of the research content, but also the main form of knowledge concept expression and communication. Therefore, titles have a similar role with keyword and they are both elaborated to reflect the content of a document. Therefore, terms in the titles are often appropriate to be keywords. In this paper we study the problem of predicting keywords appropriate for scientific papers, using only the paper titles.

© Springer Nature Switzerland AG 2018
L. Cheng et al. (Eds.): ICONIP 2018, LNCS 11303, pp. 448–459, 2018.
https://doi.org/10.1007/978-3-030-04182-3_39

The related work mainly involves the following two research areas: keyword extraction and application based on hyper-graph. Our work mainly focuses on unsupervised keyword extraction methods which can be roughly divided into three categories: keyword extraction based on statistical features, keyword extraction based on topic model and keyword extraction based on graph. The statistical features based approaches take statistical information into account, such as n-gram statistics, word frequency, which can be domain-independent and do not require training data [2]. However, some important low-frequency words and semantic characteristics of topic distribution are often neglected. Some researchers [3] considered the co-occurrence degree and correlation between words to overcome the above limitations. Hua et al. [4] proposed the co-occurrence distance to punish those word pairs that appear together but far apart. Topic modeling has been proven to be useful for automatic topic discovery from a huge volume of texts. Latent Dirichlet Allocation (LDA) is a widely used topic model to discover subject topic, hot topic and development trend in scientific and technical intelligence analysis, which have demonstrated great success on long texts [5, 6]. Compared with long texts, short texts such as paper titles, have their own characteristics and LDA cannot work very well on short texts [7, 8].

At present, text documents are always represented as an undirected graph, where the vertices of the graph contains words of the document and the edges are assigned values based on a statistical measure of similarity between the two vertices. Willyan et al. [9] proposed a keyword extraction method for tweet collections and applied centrality measures for finding the relevant vertices. The limitation of the conventional pairwise graph based modeling is its inability to completely capture n-ary association among multiple words. It is obvious that richer information should be considered into the conventional graph. Wang et al. [10] proposed to take advantage of hyper-graph for summarization, where sentences and their associations are modeled as a hyper-graph, i.e. a generalization of the conventional graph, which is able to formulate more types of relationships. Zhou et al. have proposed an extension for defining random walks on hyper-graphs, which combines the weights of destination vertices and hyper-edges in a probabilistic manner to accurately capture transition probabilities [11].

In this paper, we propose a keyword extraction approach based on weighted hyper-graph random walk. Two basic issues addressed in this paper are: (1) how to assign weights for hyper-edges and (2) how to assign weights for hyper-vertices for a particular hyper-edge. Hence, the main contributions of our work are: (1) an appropriate hyper-edge weighting strategy is investigated to characterize both authority and real-time factor; (2) the correlation degree, co-occurrence distance of each pair hyper-vertices in a specific hyper-edge, and co-occurrence degree are established for hyper-vertex weighting. Finally, the random walk process is performed on the hyper-graph where the surfer could differentiate between destination vertices within a hyper-edge depending on their features. Once the term significance scores are ready, we rank terms in descending order of the calculated scores and the sorted terms from which the highest ranked terms are chosen into a recommending list.

2 Weighting Strategy for Hyper-edge and Hyper-vertex

In our work, the construction of hyper-graph is straightforward, paper titles are regarded as hyper-edges, and the terms in the title are regarded as hyper-vertices in the particular hyper-edge. In essence, this model regards title d_i as a bag of word model which is composed of different terms, such as $d_i = \{v_1, v_2,...,v_s\}$, and the collection set of these titles $D = \{d_1, d_2, ... d_m\}$ is the lexical hyper-graph.

2.1 The Weighting Hyper-graph Model

Let $HG(V, E)$ be a hyper-graph with the hyper-vertex set V and the set of hyper-edges E, which can be considered as a generalization of the conventional graph. A hyper-edge e is a subset of V where $\cup_{e \in E} e = V$. A hyper-edge e is to be incident with v when $v \in e$. Given a hyper-graph, its matrix representation $H \in R^{|V| \times |E|}$, called the incidence matrix, with its entries of either 1 or 0, i.e. if $v \in e$, $h(v, e) = 1$, $h(v, e) = 0$ otherwise.

Let $WHG(V, E, w(e), w(v, e))$ be a weighted hyper-graph where $w(e):e \rightarrow R^+$ is the hyper-edge weight, and $w(v, e): v_e \rightarrow R^+$ is the weight of a hyper-vertex v on a particular hyper-edge e. The incidence matrix $H_{w|V| \times |E|}$ of the weighted hyper-graph is denoted as:

$$h_w(v, e) = \begin{cases} w(v, e), & v \in e \\ 0, & v \notin e \end{cases} \tag{1}$$

The hyper-vertex degree and hyper-edge degree are defined as:

$$d(v) = \sum_{e \in E} w(e) h(v, e) \tag{2}$$

$$d(e) = \sum_{v \in V} w(v_e) h(v, e) \tag{3}$$

Let D_v and D_e denote the diagonal matrices containing the node and the hyper-edge degrees respectively, and W denote the diagonal matrix containing the hyper-edge weights. Compared with the conventional hyper-graph, the weighted hyper-graph possesses weight for both hyper-edge and hyper-vertex. This section will detail the specific weighting strategy of hyper-edge and hyper-vertex. It is worth noting that the d_i is synonymous with e, which represents a particular hyper-edge.

2.2 Weighting Strategy for Hyper-edge

The authority, real-time nature and contribution rate of the selected research objects in the current field of study are extremely important factors for scientific research. On the one hand, according to the source of the literature, we can determine its authority, namely $R_{\text{paper-rank}}(d_i)$. Based on the classification of the literature's importance provided by China Computer Federation (CCF) [12], different ranks of literature C are defined, and its corresponding $R_{\text{paper-rank}}(d_i)$ is defined as:

$$R_{\text{paper-rank}(d_i)} = \begin{cases} 1, & d_l \in A \\ 2/3, & d_l \in B \\ 1/3, & d_l \in C \end{cases} \qquad (4)$$

On the other hand, temporal attributes of a research paper is an important dimension to understand evolving topics and keywords. Besides, the number of times cited by others is actually a key indicator of the relative importance of a work in science. We therefore measure the temporal effect and the number of citations as a ranking function $R_{time\text{-}quote}$ as follows:

$$R_{\text{time-quote}}(d_i) = e^{-\frac{(c-y_i)+1}{k+1}} \qquad (5)$$

here c, and y_i represent the current time and publication time of the literature respectively, k stands for the number of citations. The weight of hyper-edge is calculated by integrating $R_{\text{paper-rank}}$ and $R_{\text{time-quote}}$ as follows:

$$w(d_i) = \lambda R_{\text{paper-rank}}(d_i) + (1 - \lambda)R_{\text{time-quote}}(d_i) \qquad (6)$$

here, the smoothing factor λ is used to adjust the relative importance of $R_{\text{paper-rank}}(d_i)$ and $R_{\text{time-quote}}(d_i)$. We experimented with different values for λ which will be discussed in the experiment section. The final title rank $w(d_i)$ will be embedded in the hyper-graph as a hyper-edge weight to reflect title's importance over keywords.

2.3   Weighting Strategy for Hyper-vertex

In this paper, we attempt to seek the weighting strategy for hyper-vertex on a particular hyper-edge, which considers the correlation degree, co-occurrence distance of each pair of hyper-vertices in a specific hyper-edge, and co-occurrence degree. Table 1 summarizes the notations that will be used for hyper-vertex weighting and Fig. 1 shows and example of our weighting scheme.

Table 1. Definition of Each Notation for the Hyper-Vertex weighting

Notation	Definition	Notation	Definition
v_i	hyper-vertex v_i, that is a word	$iw(v_i, d_l)$	initial weight of v_i in d_l
$co\_d_l(v_i, v_j)$	co-occurrence degree of v_i and v_j in d_l	$w(v_i, d_l)$	hyper-vertex v_i's weight in d_l
$co(v_i, v_j)$	co-occurrence degree of v_i and v_j	$N_{nei}(v_i)$	the number of words shared with v_j
$ucor_i(v_i, v_j)$	unilateral correlation degree of v_i	$n(d_l)$	co-occurrence number of v_i and v_j in d_l
$cor(v_i, v_j)$	correlation degree between v_i and v_j	$tf(d_l, v_i)$	the number of v_i appears in d_l
$cow(v_i, d_l)$	correlation weight of hyper-vertex v_i in d_l	$df(v_i)$	the number of documents that contains v_i

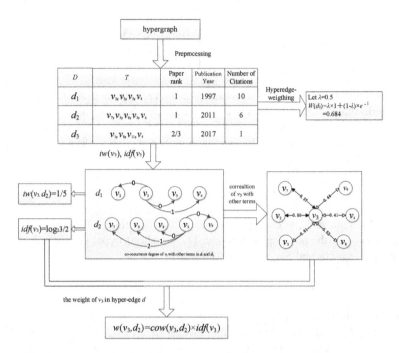

Fig. 1. An example of weighting strategy for hyper-edge and hyper-vertex

Given a specific hyper-edge d_l, the hyper-vertex v_i and v_j and their co-occurrence $co\_d_l(v_i, v_j)$ in d_l is defined as:

$$co\_d_l(v_i, v_j) = n_{d_l}(v_i, v_j) \times e^{-dist_{d_l}(v_i,v_j)} \qquad (7)$$

where $n_{d_l}(v_i, v_j)$ denotes the number of co-occurrence of hyper-vertex v_i and v_j in d_l $dist_{d_l}(v_i, v_j)$ is the number of interval words between v_i and v_j in the d_l, which is used to penalize long distance co-occurrence.

The co-occurrence of v_i and v_j on the entire corpus can be defined as:

$$co(v_i, v_j) = \sum_{l=1}^{m} co\_d_l(v_i, v_j) \qquad (8)$$

The correlation degree between v_i and v_j can be further investigated by $co(v_i, v_j)$, since this correlation degree is asymmetric, the unilateral correlation degree between v_i and v_j is defined by a modified tf-idf formula as follows:

$$ucor_i(v_i, v_j) = \frac{co(v_i, v_j)}{\sum_{q=1}^{n} co(v_i, v_q)} \times \log_2 \frac{n}{N_{nei}(v_j)} \qquad (9)$$

Here, $N_{nei}(v_j)$ is the number of co-occurrence neighbors of v_j. The former part reflects the probability that humans think of v_i when v_j is observed. The latter part punishes those v_j that co-occur with almost every other terms, this is similar to *idf*. It's obvious that v_j should rarely appear with other words except v_i if v_j is important to v_i.

The unilateral correlation degree between v_i and v_j is asymmetric, we then define the correlation degree between v_i and v_j in a symmetric way, which is essentially the average value of unilateral correlation:

$$cor(v_i, v_j) = \frac{ucor_i(v_i, v_j) + ucor_j(v_j, v_i)}{2} \tag{10}$$

Term frequency is always adopted as local features for document, however, its probability equals to 1 in most cases for short text, which indicates that the local feature of short text will be neglected. In this paper, a new local weighting scheme of hyper-vertex in a hyper-edge, namely correlation weight $cow(vi, d_l)$ is defined. Hyper-vertex with more and higher correlations to the others in a particular hyper-edge are more important to the hyper-edge and should be assigned a bigger weight. Therefore, we introduce the new hyper-vertex weighting scheme, which is based on the degree of the hyper-vertex correlation with other tokens in the same hyper-edge.

For each hyper-vertex, we define its initial weight $iw(v_i, d_l)$ as the term frequency of v_i in the current d_l as:

$$iw(v_i, d_l) = \frac{tf(d_l, v_i)}{|d_l|} \tag{11}$$

The correlation weight can be obtained by combining the correlation degree with initial weight. The correlation weight of hyper-vertex v_i in a specific hyper-edge d_l can be defined as:

$$cow(v_i, d_l) = iw(v_i) + \frac{\sum_{j=1}^{|d_l|} iw(v_i, d_l) \times cor(v_i, v_j)}{|d_l|} \tag{12}$$

The correlation weight denotes the reliability and importance of the hyper-vertex in the hyper-edge. In order to make the features as discriminative as possible, we further combine the correlation weight with global statistic weights in the weighting scheme and finally the weight of hyper-vertex v_i is in a specific hyper-edge d_l can be calculate as:

$$w(v_i, d_l) = cow(v_i, d_l) \times idf(v_i) = cow(v_i, d_l) \times \log_2 \frac{m}{df(v_i)} \tag{13}$$

3 Random Walk Process on Hyper-graph

A random walk is a mathematical object, known as a stochastic or random process, which describes a path that consists of a succession of random steps on some mathematical space. In a simple graph, it is essentially the transition between vertices by starting at a given vertex and moving to another neighboring vertex after each discrete time step and then the randomly selected point sequence forms a random walk on the graph. The process can be modeled as a finite Markov chain M over a set of states $\{s_1, s_2, ..., s_n\}$. The transition matrix $P_{|V| \times |V|}$ is always defined with its entries $P(u, v) = Prob(s_{t+1} = v|s_t = u)$ indicating that the chain M will be at v at time $t + 1$ given that it was observed at u at time t, and for any vertex $\Sigma v\, p\,(u, v) = 1$.

The hyper-graph random walk process can also be defined as a Markov chain where the hyper-vertex set is the state set of the chain similar to a simple graph. At each time step the surfer moves in the incident hyper-edge to another hyper-vertex. Therefore, a more general random walk is required for hyper-graph. Bellaachia et al. [13] have extended the random walk on weighted hyper-graph. For a hyper-graph with both weighted hyper-edges and weighted hyper-vertex, the random walk process is as follows: starting from hyper-vertex u, a hyper-edge e incident with the current hyper-vertex u is chosen proportional to the hyper-edge weight $w(e)$. Then, a hyper-vertex v is selected proportional to its weight $w(e, v)$ within the hyper-edge e. Next, we can calculate the transition matrix P as follows:

$$P(u, v) = \sum_{e \in E} w(e) \frac{h(u, e)}{\sum_{\hat{e} \in E} w(\hat{e})} \frac{h_w(v, e)}{\sum_{\hat{v} \in e} h_w(\hat{v}, e)} \tag{14}$$

Or in matrix notation:

$$P = D_v^{-1} H W_e D_e^{-1} H_w^T$$

Where $h_w(v, e)$ is the weight of the destination hyper-vertex v in hyper-edge e. D_v is the diagonal matrix of the weighted degree of hyper-vertices as in formula (2). H is the incidence matrix of a non-weighted hyper-graph. W_e is the diagonal matrix of the hyper-edge weights. D_e is the diagonal matrix for weighted degree of hyper-edges as in formula (3). H_w is the incidence matrix of the weighted hyper-graph.

After calculating the transition matrix P, we now explain the random walk process as follows. First, we assign the equal positive significance score to all the hyper-vertices. All the nodes then spread their significance scores out to their nearby neighbors via the hyper-graph. The weights of the transitions between any two nodes are defined by P in the following way:

$$\overrightarrow{v}_{(i+1)} = \alpha P^T \overrightarrow{v}_{(i)} + (1 - \alpha) \overrightarrow{e}/n \tag{15}$$

Where a is a parameter that specifies the proportion of how much a hyper-vertex should learn from its neighbors, and how much it should learn as of equal importance as the initial scores. According to the experience, the value of a is set as 0.15.

The propagation process is repeated until a global stable state is achieved. Then all the hyper-vertices have obtained their final significance scores. Once the term significance scores are ready, we rank terms in descending order of the calculated scores and the top-K sorted terms from which the highest ranked terms are picked as the selected keywords set.

4 Experimental Performance and Analysis

In this section, we report experimental results. We first describe the experimental data, and then explain the evaluating metrics. Finally, we evaluate the performance of our approach to keyword extraction for short texts and compare our algorithms with other baselines.

4.1 Datasets

To the best of our knowledge, there has been no standard dataset to evaluate keyword extraction for paper titles. Hence, we built a ground truth dataset from DBLP [14], which provides open bibliographic information on major computer science journals and proceedings. According to the CCF specification, the recommended conferences and journals are selected from which paper titles are obtained, coming from TPAMI, JMLR, AAAI, NIPS, COLT, ECCV, PAKDD, ACML et al. And then 10 categories are organized and each class contains approximately 1000 short texts as the experimental data. The initial data format collected is as follows: the title of each paper, the year of publication, the rank of the journal/conference, the number of citations. In the pre-processing step, stop-words are removed from the titles and the Porter's stemmer is adopted [15] to remove common morphological and inflectional endings from English words.

4.2 Evaluation Metrics

To quantitatively measure the performance of the ranking method, we take advantage of some common evaluation metrics that are used in the information retrieval literature such as Precision (abbreviated as *Pr*), Recall (abbreviated as *Re*) and F1-measure precision. However, there is no 'correct' set of keywords for a given document set, not even humans may agree on the keywords they extract. Therefore, to assess the performance of the proposed algorithm, the following methodology was adopted: three experts are invited to suggest an unspecified number of keywords from the documents. And then the intersection of these sets for each topic is determined, these new datasets contain the relevant documents denoted as {Relevant}. Finally, the evaluation metrics can be defined as:

$$\mathrm{Pr} = \frac{|\{Relevant\} \cap \{Retrieved\}|}{|\{Retrieved\}|} \tag{16}$$

$$Re = \frac{|\{Relevant\} \cap \{Retrived\}|}{|\{Relevant\}|} \qquad (17)$$

$$F - measure = 2 \times \frac{Pr \times Re}{(Pr + Re)} \qquad (18)$$

here Pr is equal to the number of keywords that appear in at least one of the human lists and the number of relevant documents retrieved, Re is ratio of keywords in the intersection set of the three human lists and the number of relevant documents. Therefore, the term in {Relevant} for Pr appears at least once in the list of keywords given by three experts, while the term in {Relevant} for Re appear at the intersection set of keywords given by the three experts. {Retrieved} represents the keyword set for each retrieval, and here we set the value of |{Retrieved}| to 10. Table 2 shows an example of the sets of keywords suggested by each expert for the topic 'Artificial Intelligence and Pattern Recognition'. The common keywords (intersection) among these sets are highlighted in bold.

Table 2. Example of Keyword sets suggested by each human expert for the topic 'Artificial Intelligence and Pattern Recognition'.

Expert1	**pattern recognition** | **clustering** | **classification** | **deep learning** | **natural language** | keyword | feature | neural network | optimization | fuzzy set | forecast | solve | Big data | weighting | **recommendation** | information network
Expert2	**deep learning** | heterogeneous | information retrieval | key word | **recommendation** | **classification** | **clustering** | fuzzy set | optimization | **pattern recognition** | learning strategy | neural network | decision tree | **natural language**
Expert3	**pattern recognition** | neural network | **recommendation** |**clustering** | **classification** | key word | feature | optimization | forecast | **natural language** | data mining | **deep learning**

4.3   Experimental Results and Discussion

4.3.1   Parameter Tuning

Parameters λ is the learning factor tuning the influence of $R_{(paper\text{-}rank)}$ (d_i) and $R_{time\text{-}quote}$ (d_i). When $\lambda = 0$, it means that the weight of paper rank is zero and only time and the number of citations are considered. As we can see from Fig. 2, the performance is improved in most cases in terms of all evaluation metrics and reaches a peak at $\lambda = 0.75$. This suggests these two aspects are complementary and the contribution from the time and quote factor is more salient. It is reasonable because time information is more sensitive than rank aspects.

4.3.2   Comparison with Other Approaches

In order to verify the effectiveness of the proposed keyword extraction algorithm based on weighted hyper-graph random walk, we select the TF × IDF method, key word extraction method based on the LDA, keyword extraction method based on graph TKG2|W1/F|CE [9], keyword extraction method based on weighted hyper-graph

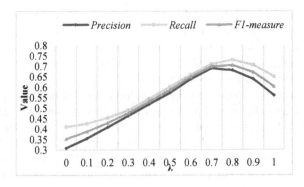

Fig. 2. Experimental result with variation of λ

random walk without considering co-occurrence distance (COW-dist) × IDF as the baseline to compare the experimental results.

Table 3 summarizes the keywords extracted of different algorithms. Keywords over a gray background match the set of relevant keywords, and those printed in gray appear in at least of the sets proposed by humans. Precision, Recall and F-measure are calculated, respectively. The number of topics for LDA is set as 10 while the TKG configurations have used closeness (CC) as centrality measure. The results presented in the table show that our approach exhibits the best performance among all the five approaches.

Table 3. Summary of the keywords extraction results for all methods

	TF×IDF	LDA	TKG₂|W^(1/F)|C_E	(COW-dist)×IDF	COW×IDF
1	encrypt	security	detect	detect	detect
2	decrypt	encrypt	restore	encrypt	encrypt
3	algorithm	decrypt	anonymous	decrypt	cloud
4	security	network	encrypt	cloud	quantum
5	defense	model	decrypt	protect	invasion
6	improve	network vulnerability	network vulnerability	trust relationship	dynamic
7	key	agreement	dynamic	dynamic	defense
8	question	algorithm	public key	defense	sensor
9	optimize	identity	cloud	attack	distributed
10	analysis	forge	trouble	forecast	optimize
Pr	40%	30%	60%	60%	70%
Re	44.44%	22.22%	44.44%	55.56%	77.78%
F1- measure	41.90%	25.29%	51.06%	57.93%	73.78%

Figure 3 shows the average performance of various algorithms respectively. We can see that our method outperforms all the baselines in all three measures. The possible reasons can be summarized as follows. The *TF × IDF* weighting method and LDA based method are more suitable for long text, the occurrence number of each term in specific title tends to be the same, and the efficiency of *TF × IDF* is greatly reduced. Besides, topic models experience a large performance degradation over short texts because of data sparsity, impeding the generation of discriminative document-topic distributions, and the resultant topics are less semantically coherent. The relationships among the objects are more complicated than simple pair-wise ones. The limitation that one can observe is its inability to completely capture correlations among multiple terms, ignoring the more types of relationships including both group ones and social attributes. The weighted hyper-graph random walk based approach is not only able to reveal the high-order relations between document-term, term-term, but also considers the correlation degree, co-occurrence distance of each pair of hyper-vertices in a specific hyper-edge, and co-occurrence degree for hyper-vertex weighting.

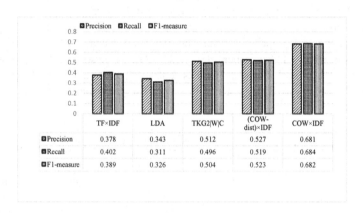

Fig. 3. Comparison of algorithm performance on English data sets

5   Conclusions and Future Work

In order to facilitate scientific researchers quickly and effectively select the most relevant and important keywords from vast majority of papers, and therefore establish the research focus and preliminaries, a keyword ranking method involving both real-time and authoritativeness is presented. In this paper, a weighted hyper-graph is constructed where hyper-vertices represent weighted terms and weighted hyper-edges measure the semantic relatedness of both binary relations and nary relations among terms. The model not only takes into account the social information to measure the importance of hyper-edge but also establish the weighting scheme for a hyper-vertex in a particular hyper-edge. Finally, the significance scores of hyper-vertices are calculated via random walk on the hyper-graph. We believe that our approach can be expected to reach a better performance for keyword extraction task if more relations among terms can be discovered and incorporated in the hyper-graph. In our future work, we will further investigate the discovery and computation of various relationship existing among terms.

Acknowledgement. This work is supported by the National Natural Science Foundation of China (No. 61762078, 61363058, 61862058), Gansu province college students' innovation and entrepreneurship training program (201610736041), and Guangxi Key Laboratory of Trusted Software (No. kx201705).

References

1. Blank, I., Rokach, L., Shani, G.: Leveraging the citation graph to recommend keywords. In: 7th ACM Conference on Recommender Systems, pp. 359-362. ACM, Hong Kong (2013)
2. Erra, U., Senatore, S., Minnella, F., Caggianese, G.: Approximate TF–IDF based on topic extraction from massive message stream using the GPU. Inf. Sci. Int. J. **292**(C), 143–161 (2015)
3. Ma, H., Xing, Y., Wang, S., Li, M.: Leveraging term co-occurrence distance and strong classification features for short text feature selection. In: Li, G., Ge, Y., Zhang, Z., Jin, Z., Blumenstein, M. (eds.) KSEM 2017. LNCS (LNAI), vol. 10412, pp. 67–75. Springer, Cham (2017). https://doi.org/10.1007/978-3-319-63558-3_6
4. Hua, W., Wang, Z., Wang, H., Zhou, X.F.: Short text understanding through lexical-semantic analysis. In: 31st International Conference on Data Engineering, pp. 495–506. IEEE Computer Society, Seoul (2015)
5. Blei, D.M., Ng, A.Y., Jordan, M,I.: Latent dirichlet allocation. J. Mach. Learn. Res. **3**(1), 993–1022 (2012)
6. Saeidi, R., Astudillo, R., Kolossa, D.: Uncertain LDA: including observation uncertainties in discriminative transforms. IEEE Trans. Pattern Anal. Mach. Intell. **38**(7), 1479–1488 (2015)
7. Qiang, J., Chen, P., Wang, T., Wu, X.: Topic modeling over short texts by incorporating word embeddings. In: Kim, J., Shim, K., Cao, L., Lee, J.-G., Lin, X., Moon, Y.-S. (eds.) PAKDD 2017. LNCS (LNAI), vol. 10235, pp. 363–374. Springer, Cham (2017). https://doi.org/10.1007/978-3-319-57529-2_29
8. Li, C.L., Duan, Y., Wang, H.R., Zhang Z.Q.: Enhancing topic modeling for short texts with auxiliary word embeddings. ACM Trans. Inf. Syst. **36**(2), 11:1–11:30 (2017)
9. Abilhoa, W.D., Castro, L.N.: A keyword extraction method from twitter messages represented as graphs. Appl. Math. Comput. **240**(4), 308–325 (2014)
10. Wang, W., Li, S.J., Li, W.J., Wei, F.R.: Exploring hypergraph-based semi-supervised ranking for query-oriented summarization. Inf. Sci. Int. J. **237**(13), 271–286 (2013)
11. Zhou, D., Huang, J.: Learning with hypergraphs: clustering, classification, and embedding. In: 20th International Conference on Neural Information Processing Systems, pp. 1601–1608. British Columbia (2006)
12. CCF. http://www.ccf.org.cn/sites/ccf/paiming.jsp
13. Bellaachia, A., Mohammed, A.: HG-RANK: A Hypergraph-based keyphrase extraction for short documents in dynamic genre. In: 4th Workshop on Making Sense of Microposts Co-located with the 23rd International World Wide Web Conference, pp. 42–49. Microposts2014, Seoul (2014)
14. DBLP Dataset. http://dblp.uni-trier.de/xml/. Accessed 20 April 2016
15. Porter, M.F.: An Algorithm for Suffix Stripping. Readings in Information Retrieval. 1st edn. Morgan Kaufmann Publishers (1997)

Zero-Shot Learning with Superclasses

Yuqi Huo, Mingyu Ding, An Zhao, Jun Hu, Ji-Rong Wen, and Zhiwu Lu$^{(\boxtimes)}$

Beijing Key Laboratory of Big Data Management and Analysis Methods
School of Information, Renmin University of China, Beijing 100872, China
luzhiwu@ruc.edu.cn

Abstract. Zero-shot learning (ZSL) can be regarded as transfer learning from seen classes to unseen ones so that the later can be recognized without any training samples. Its main difficulty lies in that there often exists a large domain gap between the seen and unseen class domains. Inspired by the fact that an unseen class is not strictly 'zero-shot' (thus easier to recognize) if it falls into a superclass that consists of one or more seen classes, we propose a new ZSL model, termed ZSL with superclasses (ZSLS), that leverages the superclasses as the bridge between seen and unseen classes to narrow the domain gap. By generating the superclasses with k-means clustering over all seen and unseen class prototypes, we formulate ZSLS as a min-min optimization problem. An efficient iterative algorithm is also developed for model optimization. Extensive experiments show that our model achieves the state-of-the-art results.

Keywords: Zero-shot learning · Domain gap · Min-min optimization

1 Introduction

Although there have been emerging advances in large-scale object recognition in the past five years, most existing object recognition models (particularly deep learning based ones) require hundreds of image samples to be collected from each object class. This condition is often hard to satisfy, e.g., some object classes are rare themselves. More worse, these object recognition models are shown to be easy to attack [13] even if sufficient labeled samples are provided. Therefore, the small sample size problem has recently been revisited in computer vision.

One solution to the small sample size problem is zero-shot learning (ZSL) [3,25,26]. ZSL aims to recognize a set of new/unseen classes without any training samples by exploiting the knowledge distilled from seen classes. All existing ZSL models assume that each class name is embedded in a semantic space, such as attribute space [14,17] or word vector space [7,28]. In this space, the names of both seen and unseen classes are embedded as high dimensional vectors called class prototypes, and their semantic relationship can be measured by the distance between these vectors. With this semantic attribute/word space and a visual feature space representing the appearance of an image, a ZSL model typically learns a projection function so that both feature and semantic vectors

© Springer Nature Switzerland AG 2018
L. Cheng et al. (Eds.): ICONIP 2018, LNCS 11303, pp. 460–472, 2018.
https://doi.org/10.1007/978-3-030-04182-3_40

are embedded in the same space. This projection is learned with the seen class training samples only. But once learned, it is used to project the unseen class samples, and a test sample is assigned to the nearest unseen class prototype.

The main difficulty of ZSL lies in that there often exists a large domain gap between the seen and unseen classes. Specifically, considering the seen and unseen classes as two domains, the projection function is learned from the seen class domain but applied to the unseen class domain. Containing completely different classes, a big domain gap exists; as a result, the projection is often biased towards the seen class prototypes, i.e., the well-known projection domain shift problem [8]. This is particularly acute when ZSL is carried out in a more realistic setting, e.g., the recently proposed generalized ZSL setting [4], under which both seen and unseen class samples need to be recognized during the test time. The bias towards seen class domain makes most test samples being classified as seen classes even if they belong to unseen classes.

To tackle the projection domain shift [8] caused by the domain gap, a number of ZSL models resort to transductive learning with not only the training set of labelled seen class data but also the test set of unlabelled unseen class data. According to whether the predicted labels of the test set are iteratively used for model learning, existing transductive ZSL models can be divided into two categories: (1) The first category [8,10,23,32] first constructs a graph in the semantic space and then transfers to the test set by label propagation. A variant is the structured prediction model [36] which employs a Gaussian parametrization of the unseen class domain label predictions. (2) The second category [11,15,18,27,30,33] involves using the predicted labels of the unseen class data in an iterative model update/adaptation process as in self-training [31]. However, rare attention has been paid to the combination of these two categories of transductive ZSL models in a unified framework. The reason may be that they seem not directly related and thus are difficult to integrate.

In this paper, a unified framework is proposed to combine the above two categories of transductive ZSL models, termed ZSL with superclasses (ZSLS). Specifically, we align the seen class and unseen class domains by exploring shared superclasses. The idea is simple: an unseen class is not strictly 'zero-shot' (thus easier to recognize) if it falls into a superclass that contains one or more seen classes. In this work, we take a data driven approach without the need for manually defined taxonomy. That is, the superclasses are generated by k-means clustering in the semantic space. Additionally, by focusing on learning a projection function and keeping the same linear regression as in the ridge regression model [26], we formulate transductive ZSL as a min-min optimization problem. An efficient iterative algorithm is also developed for model optimization.

Our contributions are: (1) A novel transductive ZSL model is proposed which aligns the seen and unsee class domains using superclasses shared across domains. (2) We formulate transductive ZSL as a mini-min optimization problem with a simple linear formulation that can be solved by an efficient iterative algorithm. Extensive experiments show that our model yields state-of-the-art results. The gain over alternative models is even bigger under the generalised ZSL setting.

2   Related Work

Projection Learning. Relying on how the projection function is established, existing ZSL models can be organised into three categories: (1) The first category learns a projection function from a visual feature space to a semantic space (i.e. in a forward projection direction) by employing conventional regression/ranking models [1,17] or deep neural network regression/ranking models [7,28]. (2) The second category chooses the reverse projection direction [15,26,27,34], i.e. from the semantic space to the feature space, to alleviate the hubness problem suffered by nearest neighbour search in a high dimensional space. (3) The third category learns an intermediate space as the embedding space, where both the feature space and the semantic space are projected to [3,35]. An exception is the semantic autoencoder proposed in [16] which can be regarded as a combination of the first and second categories. Our ZSLS model falls into the second category, but it is additionally formulated for transductive learning and ZSL with superclasses to address the domain gap problem.

Generalized ZSL. In the area of ZSL, the standard setting takes only unseen classes for test process. However, the generalized ZSL setting [2,4,7,22,25,28] makes a different assumption that the test samples come from both seen and unseen classes. This is clearly more suitable for real-world application scenarios, but also induces larger challenge into ZSL, precisely because of the projection domain shift that existing transductive ZSL models attempt to tackle. In particular, as shown in [3], since the projection is learned using the seen classes only, during test, most of the test images from unseen classes would be projected to be close to the seen class prototypes, and thus misclassified. To address this projection bias problem, novelty detection [28] has been used as a preprocessing step to predict whether a test sample is from seen/unseen classes. Alternatively, calibrated stacking [3] has also been proposed to postprocess the results of ZSL. Our ZSLS model is naturally suitable for generalized ZSL: First, its transductive learning formulation enables us to adapt the projection towards the unseen class domain. Second, the formulation of ZSL with superclasses can leverage the superclasses as the bridge to recognize both seen and unseen classes.

ZSL with Superclasses. There has been little attention on ZSL with superclasses. Two exceptions are: (1) [12] learns the relation between attributes and superclasses for sematic embedding; (2) [19] uses the taxonomy to define the semantic representation of each object class. Note that these two methods have a limitation that the manually-defined taxonomy must be provided at advance. In this paper, our approach is more flexible by generating the superclasses automatically with k-means clustering over all seen/unseen class prototypes.

3   The Proposed Framework

3.1   Problem Definition for ZSL

Let $\mathcal{S} = \{s_1, ..., s_p\}$ denote a set of seen classes and $\mathcal{U} = \{u_1, ..., u_q\}$ denote a set of unseen classes, where p and q are the total numbers of seen and unseen classes, respectively. These two sets of classes are disjoint, i.e. $\mathcal{S} \cap \mathcal{U} = \phi$. Similarly, $\mathbf{Y}_s = [\mathbf{y}_1^{(s)}, ..., \mathbf{y}_p^{(s)}] \in \mathbb{R}^{k \times p}$ and $\mathbf{Y}_u = [\mathbf{y}_1^{(u)}, ..., \mathbf{y}_q^{(u)}] \in \mathbb{R}^{k \times q}$ denote the corresponding seen and unseen class semantic representations (e.g. k-dimensional attribute vector). We are given a set of labelled training images $\mathcal{D}_s = \{(\mathbf{x}_i^{(s)}, l_i^{(s)}, \mathbf{y}_{l_i^{(s)}}^{(s)}) : i = 1, ..., N_s\}$, where $\mathbf{x}_i^{(s)} \in \mathbb{R}^{d \times 1}$ is the d-dimensional visual feature vector of the i-th image in the training set, $l_i^{(s)} \in \{1, ..., p\}$ is the label of $\mathbf{x}_i^{(s)}$ according to $\mathcal{S}$, $\mathbf{y}_{l_i^{(s)}}^{(s)}$ is the semantic representation of $\mathbf{x}_i^{(s)}$, and N_s denotes the total number of labeled images. Let $\mathcal{D}_u = \{(\mathbf{x}_i^{(u)}, l_i^{(u)}, \mathbf{y}_{l_i^{(u)}}^{(u)}) : i = 1, ..., N_u\}$ denote a set of unlabelled test images, where $\mathbf{x}_i^{(u)} \in \mathbb{R}^{d \times 1}$ is the d-dimensional visual feature vector of the i-th image in the test set, $l_i^{(u)} \in \{1, ..., q\}$ is the unknown label of $\mathbf{x}_i^{(u)}$ according to $\mathcal{U}$, $\mathbf{y}_{l_i^{(u)}}^{(u)}$ is the unknown semantic representation of $\mathbf{x}_i^{(u)}$, and N_u denotes the total number of unlabeled images. The goal of zero-shot learning is to predict the labels of test images by learning a classifier $f : \mathcal{X}_u \to \mathcal{U}$, where $\mathcal{X}_u = \{\mathbf{x}_i^{(u)} : i = 1, ..., N_u\}$.

Existing ZSL models typically learn a projection function from a visual feature space to a semantic space. To alleviate the hubness problem commonly suffered by the nearest neighbour search in a high dimensional space, a reverse projection learning (RPL) model was proposed in [26] to project the semantic prototypes (stored in $\mathbf{Y}_s$) into the feature space as follows:

$$\min_{\mathbf{W}} \sum_{i=1}^{N_s} \|\mathbf{x}_i^{(s)} - \mathbf{W}\mathbf{y}_{l_i^{(s)}}^{(s)}\|_2^2 + \lambda\|\mathbf{W}\|_F^2, \tag{1}$$

where $\mathbf{W} \in \mathbb{R}^{d \times k}$ is a projection matrix, and λ is a regularization parameter. When the best projection matrix $\mathbf{W}^*$ is learnt, we can project an unseen semantic prototype $\mathbf{y}_j^{(u)}$ from the test set into the feature space as $\hat{\mathbf{x}}_j^{(u)} = \mathbf{W}^*\mathbf{y}_j^{(u)}$. The nearest neighbor search is then performed in the feature space to predict the label of a test image $\mathbf{x}_i^{(u)}$: $l_i^{(u)} = \arg\min_j \|\mathbf{x}_i^{(u)} - \hat{\mathbf{x}}_j^{(u)}\|_2^2$.

3.2 ZSL with Superclasses

Although the above RPL model is shown to obtain impressive results [15,26, 32,34], it cannot tackle the projection domain shift. We thus choose to improve the original RPL model in two aspects: (1) A transductive learning strategy is followed by exploiting the unlabeled unseen class data to adapt the projection toward the unseen class domain for ZSL; (2) The superclass prototypes are leveraged in model formulation, instead of the seen/unseen class prototypes, to align the seen and unsee class domains using superclasses shared across domains.

Let $\mathbf{Z} = [\mathbf{z}_1, ..., \mathbf{z}_r] \in \mathbb{R}^{k \times r}$ denote the r superclass prototypes which are represented with the cluster centers generated by k-means clustering over all unseen/seen class prototypes $[\mathbf{Y}_s, \mathbf{Y}_u]$. The superclass label of a training seen class sample $\mathbf{x}_i^{(s)}$ is denoted as $\pi(l_i^{(s)})$, where $\pi(\cdot)$ is the clustering mapping function from the seen/unseen class prototypes to the superclass prototypes. With this superclass representation, our model is formulated as:

$$\min_{\widehat{\mathbf{W}}} \sum_{i=1}^{N_s} \|\mathbf{x}_i^{(s)} - \widehat{\mathbf{W}}\mathbf{z}_{\pi(l_i^{(s)})}\|_2^2 + \lambda\|\widehat{\mathbf{W}}\|_F^2 + \gamma \sum_{i=1}^{N_u} \min_j \|\mathbf{x}_i^{(u)} - \widehat{\mathbf{W}}\mathbf{z}_j\|_2^2, \quad (2)$$

where $\widehat{\mathbf{W}} \in \mathbb{R}^{d \times k}$ is the projection matrix from the superclass semantic representation to the visual feature presentation, and γ is a weighting coefficient that controls the importance of the first and third terms (which correspond to the losses on the seen and unseen class samples respectively).

Different from existing transductive ZSL models that belong to either of the two categories relying on how the test unseen class data is exploited for ZSL, we unify these two categories within a single framework as proposed in Eq. (2), i.e., our model integrates transductive projection learning and superclasses shared across domain aligning to tackle the projection domain shift problem. Moreover, we focus on learning a projection function for ZSL, i.e., a single task is considered in our model. In contrast, most existing transductive ZSL models have to solve two or more subtasks at once, including projection learning, label prediction, and semantic embedding. This requires one or more intermediate variables to be introduced into the ZSL models, complicating the optimization problem. We believe that focusing on projection learning only with a simple formulation enables our model to better overcome the projection domain shift problem.

3.3   Model Optimization

Note that the third term of the objective function in Eq. (2) is a sum of minimums and it is thus difficult to solve the optimization problem in this equation. In the following, we develop an iterative solver to model optimization.

Given the projection matrix $\widehat{\mathbf{W}}^{(t)}$ at iteration t during model optimization, we define $\mathbf{g}_i^{(t)} = [g_{i1}^{(t)}, ..., g_{ir}^{(t)}]^T$ for the test unseen class sample $\mathbf{x}_i^{(u)}$ ($i = 1, ..., N_u$), where $g_{ij}^{(t)} = \|\mathbf{x}_i^{(u)} - \widehat{\mathbf{W}}^{(t)}\mathbf{z}_j\|_2^2$ ($j = 1, ..., r$). For the minimum function $\min \mathbf{g}_i^{(t)}$, we define its gradient $\zeta_i^{(t)} = [\zeta_{i1}^{(t)}, ..., \zeta_{ir}^{(t)}]^T$ with respect to $\mathbf{g}_i^{(t)}$ as:

$$\zeta_{ij}^{(t)} = \begin{cases} 1/n_i^{(t)}, & \text{if } g_{ij}^{(t)} = \min \mathbf{g}_i^{(t)} \\ 0, & \text{otherwise} \end{cases}, \quad (3)$$

where $n_i^{(t)}$ is the number of $g_{ij}^{(t)}$ ($j = 1, ..., r$) that satisfy $g_{ij}^{(t)} = \min \mathbf{g}_i^{(t)}$. With the Taylor expansion, we obtain:

$$\min_j \|\mathbf{x}_i^{(u)} - \widehat{\mathbf{W}}^{(t+1)}\mathbf{z}_j\|_2^2 = \min \mathbf{g}_i^{(t+1)} \approx \min \mathbf{g}_i^{(t)} + \zeta_i^{(t)^T}(\mathbf{g}_i^{(t+1)} - \mathbf{g}_i^{(t)})$$

$$= \zeta_i^{(t)^T}\mathbf{g}_i^{(t+1)} = \sum_{j=1}^{r} \zeta_{ij}^{(t)}\|\mathbf{x}_i^{(u)} - \widehat{\mathbf{W}}^{(t+1)}\mathbf{z}_j\|_2^2. \quad (4)$$

The objective function in Eq. (2) at iteration $t+1$ can thus be estimated as:

$$\mathcal{G}(\widehat{\mathbf{W}}^{(t+1)}) = \sum_{i=1}^{N_s} \|\mathbf{x}_i^{(s)} - \widehat{\mathbf{W}}^{(t+1)} \mathbf{z}_{\pi(l_i^{(s)})}\|_2^2 + \lambda \|\widehat{\mathbf{W}}^{(t+1)}\|_F^2$$

$$+ \gamma \sum_{i=1}^{N_u} \sum_{j=1}^{r} \zeta_{ij}^{(t)} \|\mathbf{x}_i^{(u)} - \widehat{\mathbf{W}}^{(t+1)} \mathbf{z}_j\|_2^2. \tag{5}$$

Let $\frac{\partial \mathcal{G}(\widehat{\mathbf{W}}^{(t+1)})}{\partial \widehat{\mathbf{W}}^{(t+1)}} = 0$, we obtain a linear equation:

$$\widehat{\mathbf{W}}^{(t+1)} \mathbf{A}^{(t)} = \mathbf{B}^{(t)}, \tag{6}$$

$$\mathbf{A}^{(t)} = \sum_{i=1}^{N_s} \mathbf{z}_{\pi(l_i^{(s)})} \mathbf{z}_{\pi(l_i^{(s)})}^T + \lambda I + \gamma \sum_{j=1}^{r} \sum_{i=1}^{N_u} \zeta_{ij}^{(t)} \mathbf{z}_j \mathbf{z}_j^T, \tag{7}$$

$$\mathbf{B}^{(t)} = \sum_{i=1}^{N_s} \mathbf{x}_i^{(s)} \mathbf{z}_{\pi(l_i^{(s)})}^T + \gamma \sum_{j=1}^{r} \sum_{i=1}^{N_u} \zeta_{ij}^{(t)} \mathbf{x}_i^{(u)} \mathbf{z}_j^T. \tag{8}$$

Let $\alpha = \gamma/(1+\gamma) \in (0,1)$ and $\beta = \lambda/(1+\gamma)$, we have:

$$\widehat{\mathbf{A}}^{(t)} = (1-\alpha) \sum_{i=1}^{N_s} \mathbf{z}_{\pi(l_i^{(s)})} \mathbf{z}_{\pi(l_i^{(s)})}^T + \beta I + \alpha \sum_{j=1}^{r} \sum_{i=1}^{N_u} \zeta_{ij}^{(t)} \mathbf{z}_j \mathbf{z}_j^T, \tag{9}$$

$$\widehat{\mathbf{B}}^{(t)} = (1-\alpha) \sum_{i=1}^{N_s} \mathbf{x}_i^{(s)} \mathbf{z}_{\pi(l_i^{(s)})}^T + \alpha \sum_{j=1}^{r} \sum_{i=1}^{N_u} \zeta_{ij}^{(t)} \mathbf{x}_i^{(u)} \mathbf{z}_j^T. \tag{10}$$

In this paper, we empirically set $\beta = 0.01$. The linear equation in Eq. (6) is then reformulated as follows:

$$\widehat{\mathbf{W}}^{(t+1)} \widehat{\mathbf{A}}^{(t)} = \widehat{\mathbf{B}}^{(t)}. \tag{11}$$

The proposed algorithm for ZSL with superclasses is outlined in Algorithm 1. Note that any ZSL model can be used to obtain the initial projection matrix $\widehat{\mathbf{W}}^{(0)}$. In this paper, we choose the RPL model [26] for this initialization. Once the optimal projection matrix $\widehat{\mathbf{W}}^*$ is found by the proposed algorithm, we first project the semantic prototypes of superclasses into the feature space, and then predict the superclass label of a test sample $\mathbf{x}_i^{(u)}$ as: $\arg\min_j \|\mathbf{x}_i^{(u)} - \widehat{\mathbf{W}}^* \mathbf{z}_j\|_2^2$.

We provide the time complexity analysis of the proposed algorithm as follows. Concretely, the computation of $[\zeta_{ij}^{(t)}]_{N_u \times r}$, $\widehat{\mathbf{A}}^{(t)}$, and $\widehat{\mathbf{B}}^{(t)}$ has a time complexity of $O(rN_u)$, $O(k^2 N_s + k^2 N_u)$, and $O(dkN_s + dkN_u)$, respectively. Here, the sparsity of $[\zeta_{ij}^{(t)}]$ is used to reduce the cost of computing $\widehat{\mathbf{A}}^{(t)}$ and $\widehat{\mathbf{B}}^{(t)}$. Moreover, since $\widehat{\mathbf{A}}^{(t)} \in \mathbb{R}^{k \times k}$ and $\widehat{\mathbf{B}}^{(t)} \in \mathbb{R}^{d \times k}$, solving Eq. (11) has a time complexity of $O(dk^2)$. To sum up, the time complexity of one iteration is $O(rN_u + (d+k)k(N_s + N_u) + dk^2)$ $(d, k, r \ll (N_s + N_u))$. Given that the proposed algorithm is shown to converge very quickly $(t < 10)$ and the superclass generation with k-means clustering has a time complexity of $O(r(p+q))$, it is efficient even for large-scale ZSL problems.

Algorithm 1. ZSL with Superclasses

Input: Training and test sets $\mathcal{D}_s, \mathcal{X}_u$
            Seen and unseen class prototypes $\mathbf{Y}_s, \mathbf{Y}_u$
            Parameters α, r
Output: $\widehat{\mathbf{W}}^*$
1. Initialize $t = 0$;
2. Generate the r superclass prototypes $\mathbf{Z}$ by k-means clustering over $[\mathbf{Y}_s, \mathbf{Y}_u]$;
3. Initialize $\widehat{\mathbf{W}}^{(0)}$ with the RPL model;
repeat
    4. Compute $\zeta_{ij}^{(t)}$ with Eq. (3);
    5. Compute $\widehat{\mathbf{A}}^{(t)}$ and $\widehat{\mathbf{B}}^{(t)}$ with Eqs. (9)–(10);
    6. Update $\widehat{\mathbf{W}}^{(t+1)}$ by solving Eq. (11);
    7. Set $t = t + 1$;
until a stopping criterion is met
8. $\widehat{\mathbf{W}}^* = \widehat{\mathbf{W}}^{(t)}$.

Algorithm 2. Full ZSL Algorithm

Input: Training and test sets $\mathcal{D}_s, \mathcal{X}_u$
            Seen and unseen class prototypes $\mathbf{Y}_s, \mathbf{Y}_u$
            Parameters α, r
Output: labels of test samples
1. Solve Eq. (2) for ZSL with superclasses using Algorithm 1;
2. Generate $\mathcal{N}(\mathbf{x}_i^{(u)})$ for each test sample $\mathbf{x}_i^{(u)}$;
3. Solve Eq. (12) with a similar iterative algorithm;
4. Predict the unseen class label of each test sample $\mathbf{x}_i^{(u)}$.

3.4 Full ZSL Algorithm

The results of ZSL with superclasses can be used for the original ZSL task as follows. First, we predict the top 5 superclass labels of each test unlabelled unseen sample $\mathbf{x}_i^{(u)}$ with the optimal projection matrix $\widehat{\mathbf{W}}^*$ learned by Algorithm 1. Second, derived from the top 5 superclass labels of $\mathbf{x}_i^{(u)}$, we obtain the set of the most possible unseen class labels $\mathcal{N}(\mathbf{x}_i^{(u)})$ according to the k-means clustering results of superclass generation. Third, we learn the projection function for the original ZSL task by solving the following optimization problem:

$$\min_{\mathbf{W}} \sum_{i=1}^{N_s} \|\mathbf{x}_i^{(s)} - \mathbf{W}\mathbf{y}_{l_i^{(s)}}^{(s)}\|_2^2 + \lambda\|\mathbf{W}\|_F^2 + \gamma \sum_{i=1}^{N_u} \min_{j \in \mathcal{N}(\mathbf{x}_i^{(u)})} \|\mathbf{x}_i^{(u)} - \mathbf{W}\mathbf{y}_j^{(u)}\|_2^2. \quad (12)$$

We can develop an iterative solver similar to Algorithm 1 (with the same α), and the only difference is that the gradient is computed with the constraint $j \in \mathcal{N}(\mathbf{x}_i^{(u)})$. The full ZSL algorithm is summarized in Algorithm 2.

Table 1. Five benchmark datasets used for performance evaluation. Notations: 'SS' – semantic space, 'SS-D' – the dimension of semantic space, 'A' – attribute, and 'W' – word vector. The two splits of SUN are separated by '|'.

Dataset	# images	SS	SS-D	# seen/unseen
AwA	30,475	A	85	40/10
CUB	11,788	A	312	150/50
aPY	15,339	A	64	20/12
SUN	14,340	A	102	707/10|645/72
ImNet	218,000	W	1,000	1,000/360

4   Experiments

4.1   Datasets and Settings

Datasets. We select five widely-used benchmark datasets for performance evaluation. Four of them are of medium-size: Animals with Attributes (AwA) [17], CUB-200-2011 Birds (CUB) [29], aPascal&Yahoo (aPY) [6], and SUN Attribute (SUN) [21]. One large-scale dataset is ILSVRC2012/2010 (ImNet), where the 1,000 classes of ILSVRC2012 are used as seen classes and 360 classes of ILSVRC2010 (not included in ILSVRC2012) are used as unseen classes, as in [9]. The details of these benchmark datasets are given in Table 1.

Semantic Spaces. We form the semantic space with attributes for the four medium-scale datasets, all of which provide the attribute annotations. The semantic representation based on word vectors is used for the large-scale ImNet. We train a skip-gram text model on a corpus of 4.6 M Wikipedia documents to obtain the word2vec word vectors.

Visual Features. All recent ZSL models use visual features extracted by deep CNN models. In this paper, we extract the GoogLeNet features which are the 1,024-dimensional activations of the final pooling layer as in [16].

Evaluation Metrics. (1) **Standard ZSL**: For the four medium-scale datasets, we compute the multi-way classification accuracy as in previous works. For the large-scale ImNet dataset, the flat hit@5 classification accuracy is computed as in [9], where hit@5 means that a test sample is classified to a 'correct label' if it is among the top 5 labels. (2) **Generalized ZSL**: three metrics are defined: (1) acc_s – the accuracy of classifying the samples from seen classes to all seen/unseen classes; (2) acc_u – the accuracy of classifying the samples from unseen classes to all seen/unseen classes; (3) HM – the harmonic mean of acc_s and acc_u.

Parameter Settings. Our full ZSL model has only two free parameters to tune: $\alpha \in (0,1)$ (Eqs. (9)–(10)) and r (see Step 2 in Algorithm 1). As in [16], the two parameters are selected by class-wise cross-validation using the training data. Moreover, out of the five datasets, only the CUB and SUN datasets have multiple seen/unseen splits. We take the same 4 splits used in [1] for CUB and the same 10 splits used in [3] for SUN, and report the average accuracies.

Compared Methods. A wide range of existing ZSL models are selected for comparison. For either of the two ZSL settings (i.e. standard and generalized), we pay more attention to the latest and representative ZSL models that have achieved the state-of-the-art results in the area of ZSL.

Table 2. Comparative ZSL classification accuracies (%) on the four medium-scale datasets under the standard ZSL setting. For the SUN dataset, the results are obtained for the 707/10 and 645/72 splits respectively, separated by '|'.

Model	SS	Trans.?	AwA	CUB	aPY	SUN
DeViSE [7]	A	N	56.7	33.5	–	– | –
USE [12]	A	N	46.4	–	–	– | –
DAP [17]	A	N	60.1	–	38.2	72.0|44.5
ESZSL [24]	A	N	75.3	48.7	24.3	82.1|18.7
RPL [26]	A	N	80.4	52.4	48.8	84.5| –
SJE [1]	A+W	N	73.9	51.7	–	– |56.1
JLSE [35]	A	N	80.5	42.1	50.4	83.8| –
SynC [3]	A	N	72.9	54.7	–	– |62.7
SAE [16]	A	N	84.7	61.4	55.4	91.5|65.2
LESD [5]	A	N	82.8	56.2	58.8	88.3| –
SCoRe [19]	A	N	82.8	59.5	–	– | –
AMP [10]	A+W	Y	66.0	–	–	– | –
UDA [15]	A+W	Y	75.6	40.6	–	– | –
Li et al. [18]	A	Y	40.1	–	24.7	– | –
SS-Voc [9]	A	Y	78.3	–	–	– | –
SMS [11]	A	Y	78.5	–	39.0	82.0| –
SP-ZSR [36]	A	Y	92.1	55.3	69.7	89.5| –
SSZSL [27]	A	Y	88.6	58.8	49.9	86.2| –
DSRL [32]	A	Y	87.2	57.1	56.3	85.4| –
TSTD [33]	A	Y	90.3	58.2	–	– | –
BiDiLEL [30]	A	Y	95.0	62.8	–	– | –
Our full model	A	Y	**96.2**	**64.0**	**83.9**	**93.5|67.6**

4.2 Results Under Standard ZSL Setting

Comparative Evaluation. The comparative results under the standard ZSL setting are shown in Tables 2 and 3(a). For comprehensive comparison, both

Table 3. (a) Comparative accuracies (%) on the large-scale ImNet under the standard ZSL setting. (b) Comparative results (%) of generalized ZSL on AwA and CUB.

Model	SS	Trans.?	hit@5
DeViSE [7]	W	N	12.8
ConSE [20]	W	N	15.5
AMP [10]	W	Y	13.1
SS-Voc [9]	W	Y	16.8
SAE [16]	W	N	27.2
Ours	W	Y	**30.8**

(a)

Model	AwA			CUB		
	acc_s	acc_u	HM	acc_s	acc_u	HM
ConSE [20]	75.9	9.5	16.9	69.9	1.8	3.5
APD [22]	43.2	**61.7**	50.8	23.4	**39.9**	29.5
GAN [2]	**81.3**	32.3	46.2	**72.0**	26.9	39.2
SAE [16]	67.6	43.3	52.8	36.1	28.0	31.5
Ours	67.8	58.7	**62.9**	41.6	38.0	**39.7**

(b)

transductive and non-transductive ZSL models are included. We have the following observations: (1) Our full ZSL model achieves the best results on all five datasets, showing that the combination of transductive projection learning and superclasses shared across domain aligning is indeed effective for tackling the projection domain shift. (2) For the four medium-scale datasets (see Table 2), the improvements obtained by our model over the second-best model vary from 1.2% to 14.2%. This actually creates new baselines in the area of ZSL, given that most of these competitors adopt far more complicated nonlinear (even deep) models. (3) For the large-scale ImNet (see Table 3(a)), our model achieves a 3.6% improvement over the strongest competitor [16]. This demonstrates that our model scales up to large-scale ZSL problems.

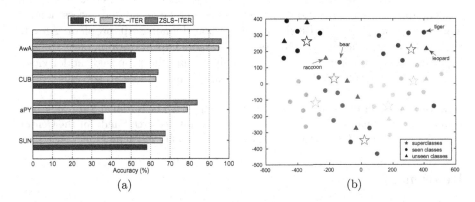

Fig. 1. (a) The ablation study results on the four medium-scale datasets. (b) The t-SNE visualization of the superclasses ($r = 7$) on the AwA dataset.

Ablation Study. Our full ZSL model (denoted as ZSLS-ITER) has two simplified versions: (1) Without exploring the superclasses in the iterative algorithm (Algorithm 1)), our full ZSL model degrades to the contentional ZSL model (denoted as ZSL-ITER); (2) For $\alpha = 0$, our ZSL-ITER model further degrades to the RPL model [26]. We compare the three models under the same standard

ZSL setting. The results in Fig. 1(a) show that: (1) Our ZSL-ITER model yields significant gains over RPL, ranging from 8% to 43%. This provides evidence that the transductive projection learning induced in our model indeed enables us to narrow the projection domain shift. (2) Our ZSLS-ITER model achieves about 1–5% improvements over ZSL-ITER, validating the effectiveness of superclasses shared across domain aligning.

4.3   Results Under Generalized ZSL Setting

Generalized ZSL has drawn much attention recently, which assumes that the test set contains samples from both seen and unseen classes. We follow the same setting of [4], i.e., 20% of the samples from seen classes are held out and then mixed with the samples from unseen classes. The comparative results on the AwA and CUB datasets are presented in Table 3(b). It can be seen that: (1) Different ZSL models have a distinct trade-off between acc_s and acc_u, and the overall performance is thus measured by the HM metric. (2) Our model clearly obtains the best overall performance, mainly because of ZSL with superclasses. This is also supported by the t-SNE visualization of the superclasses in Fig. 1(b), where an unseen class tends to be semantically related to a seen class within the same superclass.

5   Conclusion

In this paper, we have proposed a novel model for ZSL with superclasses. In our model, transductive projection learning and superclasses shared across domain aligning are integrated to tackle the projection domain shift problem. An efficient iterative algorithm is also developed for model optimization. The extensive experiments on five benchmark datasets show that the proposed model yields state-of-the-art results under the standard and generalized ZSL settings.

Acknowledgements. This work was partially supported by National Natural Science Foundation of China (61573363), and the Fundamental Research Funds for the Central Universities and the Research Funds of Renmin University of China (15XNLQ01).

References

1. Akata, Z., Reed, S., Walter, D., Lee, H., Schiele, B.: Evaluation of output embeddings for fine-grained image classification. In: CVPR, pp. 2927–2936 (2015)
2. Bucher, M., Herbin, S., Jurie, F.: Generating visual representations for zero-shot classification. In: ICCV Workshops, pp. 2666–2673 (2017)
3. Changpinyo, S., Chao, W.L., Gong, B., Sha, F.: Synthesized classifiers for zero-shot learning. In: CVPR, pp. 5327–5336 (2016)
4. Chao, W.L., Changpinyo, S., Gong, B., Sha, F.: An empirical study and analysis of generalized zero-shot learning for object recognition in the wild. In: ECCV, pp. 52–68 (2016)

5. Ding, Z., Shao, M., Fu, Y.: Low-rank embedded ensemble semantic dictionary for zero-shot learning. In: CVPR, pp. 2050–2058 (2017)

6. Farhadi, A., Endres, I., Hoiem, D., Forsyth, D.: Describing objects by their attributes. In: CVPR, pp. 1778–1785 (2009)

7. Frome, A., et al.: DeViSE: a deep visual-semantic embedding model. In: NIPS, pp. 2121–2129 (2013)

8. Fu, Y., Hospedales, T.M., Xiang, T., Gong, S.: Transductive multi-view zero-shot learning. TPAMI 37(11), 2332–2345 (2015)

9. Fu, Y., Sigal, L.: Semi-supervised vocabulary-informed learning. In: CVPR, pp. 5337–5346 (2016)

10. Fu, Z., Xiang, T., Kodirov, E., Gong, S.: Zero-shot object recognition by semantic manifold distance. In: CVPR, pp. 2635–2644 (2015)

11. Guo, Y., Ding, G., Jin, X., Wang, J.: Transductive zero-shot recognition via shared model space learning. In: AAAI, pp. 3494–3500 (2016)

12. Hwang, S.J., Sigal, L.: A unified semantic embedding: Relating taxonomies and attributes. In: NIPS, pp. 271–279 (2014)

13. Ilyas, A., Engstrom, L., Athalye, A., Lin, J.: Query-efficient black-box adversarial examples. arXiv preprint arXiv:1712.07113 (2017)

14. Kankuekul, P., Kawewong, A., Tangruamsub, S., Hasegawa, O.: Online incremental attribute-based zero-shot learning. In: CVPR, pp. 3657–3664 (2012)

15. Kodirov, E., Xiang, T., Fu, Z., Gong, S.: Unsupervised domain adaptation for zero-shot learning. In: ICCV, pp. 2452–2460 (2015)

16. Kodirov, E., Xiang, T., Gong, S.: Semantic autoencoder for zero-shot learning. In: CVPR, pp. 3174–3183 (2017)

17. Lampert, C.H., Nickisch, H., Harmeling, S.: Attribute-based classification for zero-shot visual object categorization. TPAMI 36(3), 453–465 (2014)

18. Li, X., Guo, Y., Schuurmans, D.: Semi-supervised zero-shot classification with label representation learning. In: ICCV, pp. 4211–4219 (2015)

19. Morgado, P., Vasconcelos, N.: Semantically consistent regularization for zero-shot recognition. In: CVPR, pp. 6060–6069 (2017)

20. Norouzi, M.: Zero-shot learning by convex combination of semantic embeddings. In: ICLR (2014)

21. Patterson, G., Xu, C., Su, H., Hays, J.: The sun attribute database: Beyond categories for deeper scene understanding. IJCV 108(1), 59–81 (2014)

22. Rahman, S., Khan, S.H., Porikli, F.: A unified approach for conventional zero-shot, generalized zero-shot and few-shot learning. arXiv preprint arXiv:1706.08653 (2017)

23. Rohrbach, M., Ebert, S., Schiele, B.: Transfer learning in a transductive setting. In: NIPS, pp. 46–54 (2013)

24. Romera-Paredes, B., Torr, P.H.S.: An embarrassingly simple approach to zero-shot learning. In: ICML, pp. 2152–2161 (2015)

25. Scheirer, W.J., de Rezende Rocha, A., Sapkota, A., Boult, T.E.: Toward open set recognition. TPAMI 35(7), 1757–1772 (2013)

26. Shigeto, Y., Suzuki, I., Hara, K., Shimbo, M., Matsumoto, Y.: Ridge regression, hubness, and zero-shot learning. In: ECML-PKDD, pp. 135–151 (2015)

27. Shojaee, S.M., Baghshah, M.S.: Semi-supervised zero-shot learning by a clustering-based approach. arXiv preprint arXiv:1605.09016 (2016)

28. Socher, R., Ganjoo, M., Manning, C.D., Ng, A.: Zero-shot learning through cross-modal transfer. In: NIPS, pp. 935–943 (2013)

29. Wah, C., Branson, S., Welinder, P., Perona, P., Belongie, S.: The caltech-ucsd birds-200-2011 dataset. Technical report CNS-TR-2011-001, California Institute of Technology (2011)
30. Wang, Q., Chen, K.: Zero-shot visual recognition via bidirectional latent embedding. IJCV **124**(3), 356–383 (2017)
31. Xu, X., Hospedales, T., Gong, S.: Transductive zero-shot action recognition by word-vector embedding. IJCV **123**(3), 309–333 (2017)
32. Ye, M., Guo, Y.: Zero-shot classification with discriminative semantic representation learning. In: CVPR, pp. 7140–7148 (2017)
33. Yu, Y., Ji, Z., Li, X., Guo, J., Zhang, Z., Ling, H., Wu, F.: Transductive zero-shot learning with a self-training dictionary approach. arXiv preprint arXiv:1703.08893 (2017)
34. Zhang, L., Xiang, T., Gong, S.: Learning a deep embedding model for zero-shot learning. In: CVPR, pp. 2021–2030 (2017)
35. Zhang, Z., Saligrama, V.: Zero-shot learning via joint latent similarity embedding. In: CVPR, pp. 6034–6042 (2016)
36. Zhang, Z., Saligrama, V.: Zero-shot recognition via structured prediction. In: ECCV, pp. 533–548 (2016)

Active Learning Methods with Deep Gaussian Processes

Jingjing Fei, Jing Zhao, Shiliang Sun$^{(\boxtimes)}$, and Yan Liu

Department of Computer Science and Technology, East China Normal University,
3663 Zhongshan Road, Shanghai 200241, People's Republic of China
{jzhao,slsun}@cs.ecnu.edu.cn

Abstract. Active learning is an effective method to reduce the learning time, space and economic costs in the whole training procedure. It aims to select more informative points from the unlabeled data pool, label them and add them into the training set, which helps to improve the performance of learning models. Learning models and active learning strategies are two essential elements in the framework of active learning. Probabilistic models such as Gaussian processes are often used as learning models for active learning, which have achieved promising results attributed to their predictive uncertainty. In order to well model complex data and characterize uncertainty, we employ deep Gaussian processes (DGPs) as learning models, based on which active learning strategies are made. Specifically, we design appropriate active learning strategies based on DGPs for solving binary and multi-class classification tasks, respectively. The experiments on educational and non-educational text classification and handwritten digit recognition demonstrate the effectiveness of the proposed active learning methods.

Keywords: Active learning · Probabilistic model
Deep Gaussian processes · Predictive uncertainty

1 Introduction

Enough labeled data as training set are often required in many supervised machine learning applications such as text classification and image recognition, which will cost lots of labor. Active learning provides an effective framework, under which learning models can learn from small amounts of data, select informative points, label them and add them into the training set. Active learning is an iterative process, and it hopes that the performance of the learning model will be better and better through this process. Therefore, the ability of learning models and the rationality of active learning selection strategies are the keys to determine the effect of the whole active learning.

Some existing active learning methods have proved to be useful for machine learning classification problems. The selection strategies are usually developed by considering the characters of learning models. The support vector machine

© Springer Nature Switzerland AG 2018
L. Cheng et al. (Eds.): ICONIP 2018, LNCS 11303, pp. 473–483, 2018.
https://doi.org/10.1007/978-3-030-04182-3_41

(SVM) and Gaussian processes (GPs) are the representatives of the deterministic and probabilistic models, respectively, on which some active learning methods have been developed. Specifically, for SVMs, the classification results and margin information are employed for active learning where the points nearest to the interface are selected [1–3]. For GPs, the prediction results and predictive uncertainty from GPs are employed for active learning where the points nearest to the interface and the points with the lowest predictive confidence are selected [2,4–7]. Besides predictions of learners, the manifold information [2] and spatial properties [8] of the unlabeled data have also been exploited for active learning.

In the literatures, active learning based on probability models obtained more encouraging results than that based on deterministic models since probabilistic models provide the uncertainty of prediction as a part of the inference process [9–11]. The GP is one of the most famous probabilistic models, which provides an effective Bayesian method for solving nonlinear regression and classification tasks [12–15]. In order to improve the ability of modeling the data with complicated structures, deep Gaussian processes (DGPs) as a kind of deep probabilistic models have been proposed, and are empirically demonstrated to be more powerful than GPs. For learning DGPs more accurately and efficiently, various inference methods have been developed [16–19]. Therefore, active learning methods with DGPs are expected to be feasible and effective. On one hand, unsupervised DGPs have the ability of extracting features, which will help active learning to obtain extra information from the unlabeled data. On the other hand, the supervised DGPs may have better classification performance for complex data, which will help active learning to obtain more accurate discriminative information.

In this paper, we propose appropriate active learning methods by employing the advantages of DGPs. Specifically, we design active learning strategies fitting well in DGPs for solving binary and multi-class classification problems, respectively. In the proposed methods, information from the DGPs such as the distances from points to the interface, the predictive confidence and the entropy are reasonably used. In addition, we employ the characteristic of the unlabeled data by using an unsupervised DGP to extract features. The experiments for educational text classification and handwritten digit recognition show promising results, and provide inspiration for further research of active learning.

2   DGPs as Learning Models

2.1   Gaussian Processes

The GP supposes that any finite collection of random variables still subjects to a Gaussian distribution [12]. It models the mapping from inputs to outputs as a GP with the mean function and covariance function. Suppose that the data points $\{\mathbf{x}_n, y_n\}_{n=1}^{N}$ constitute a training set $\mathcal{D}$. The output y_n is assumed to be generated from the corresponding latent function $f(\mathbf{x}_n)$ with an independent Gaussian noise, i.e., $y_n = f(\mathbf{x}_n) + \epsilon_n, \epsilon_n \sim \mathcal{N}(\mathbf{0}, \delta_n^2 \mathbf{I})$, and the latent function is a GP, $f(\mathbf{x}) \sim \mathcal{GP}(\mathbf{0}, k(\mathbf{x}, \mathbf{x}'))$, which captures the dependency and characteristics of data. The covariance function $k(\mathbf{x}, \mathbf{x}')$ is the key to characterize the mapping from inputs to outputs. For example, the automatic relevance determination (ARD) kernel function $k(\mathbf{x}_i, \mathbf{x}_j) = \sigma_f^2 e^{-\frac{1}{2}\sum_{q=1}^{Q} w_q(x_{i,q}-x_{j,q})^2}$ with kernel parameters $\theta = \{\sigma_f^2, w_1, ... w_Q\}$ can characterize the effect of input data from different dimensions on outputs, which is often employed to construct the covariance matrix. Figure 1(a) shows the graphical model of a GP.

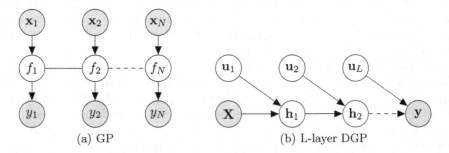

(a) GP   (b) L-layer DGP

Fig. 1. (a) shows the graphical model of a GP with N training points. (b) shows the graphical model of an L-layer DGP, where $\{\mathbf{u}_i\}_{i=1}^{L}$ are the inducing points.

Given the GP assumption for the training data, the model can be learned by maximizing the marginal likelihood $p(\mathbf{y}|\mathbf{X})$. Then the prediction distribution of a new point $\mathbf{x}^*$ can further be derived as $p(f^*|\mathbf{X}, \mathbf{y}, \mathbf{x}^*) = \mathcal{N}(\mu^*, \Sigma_f^*)$ with

$$\mu^* = \mathbf{k}(\mathbf{x}^*, X)^T (K(X, X) + \delta_n^2 \mathbf{I})^{-1} \mathbf{y},$$
$$\Sigma_f^* = \mathbf{k}(\mathbf{x}^*, \mathbf{x}^*) - \mathbf{k}(\mathbf{x}^*, X)(K(X, X) + \delta_n^2 \mathbf{I})^{-1} \mathbf{k}(X, \mathbf{x}^*). \tag{1}$$

2.2 Deep Gaussian Processes

The DGP makes the GP deeper by adding more latent layers $\{\mathbf{h}_l\}_{l=1}^{L-1}$, each of which acts as the output of the above layer and the input of the next layer,

$$p(\mathbf{y}|\mathbf{h}_{L-1}) = \mathcal{N}(\mathbf{h}_{L-1}, \delta_n^2 \mathbf{I}),$$
$$p(\mathbf{h}_l|\mathbf{h}_{l-1}) = \mathcal{GP}(\mathbf{h}_l; 0, k(\mathbf{h}_{l-1}, \mathbf{h}'_{l-1})), l = 2, ..., L-1,$$
$$p(\mathbf{h}_1|X) = \mathcal{GP}(\mathbf{h}_1; 0, k(\mathbf{x}, \mathbf{x}')). \tag{2}$$

DGPs often introduce inducing variables $\mathbf{u}_l$ with inducing locations $\mathbf{z}_l{}^1$ to release the burden of computation. Figure 1(b) shows the graphical model of an L-layer DGP with inducing points.

By introducing inducing points in DGPs, several approximate inference methods for DGPs were developed, such as variational inference and expectation propagation (EP). The EP employs $q(\mathbf{u}) \propto p(\mathbf{u}) \prod_{n=1}^{N} \tilde{t}_n(\mathbf{u})$ to approximate true

---

[1] $\mathbf{z}_l$ will be omitted in our paper to simplify the notation.

posterior and $\{\tilde{t}_n(\mathbf{u})\}_{n=1}^N$ are the approximate data factors. In EP, approximate marginal likelihood can be expressed as

$$\log p(\mathbf{y}|\boldsymbol{\Theta}) \approx \mathcal{F}(\boldsymbol{\Theta}) = \phi(\theta) - \phi(\theta_{prior}) + \sum_{n=1}^N \log \widetilde{\mathcal{Z}}_n, \tag{3}$$

where $\log \widetilde{\mathcal{Z}}_n = \log \mathcal{Z}_n + \phi(\theta^{\backslash n}) - \phi(\theta)$, and $\boldsymbol{\Theta}$ denotes all the model parameters. ϕ is the log normalizer of a Gaussian distribution. $\theta, \theta^{\backslash n}$ and θ_{prior} are the natural parameters of the distributions $q(\mathbf{u}), q^{\backslash n}(\mathbf{u})^2$ and $p(\mathbf{u})$, respectively, and $\mathcal{Z}_n = \int p(y_n|\mathbf{u}, \mathbf{X}_n) q^{\backslash n}(\mathbf{u}) d\mathbf{u}$. Furthermore, EP stores all the approximate data factors, which costs $\mathcal{O}(NLM^2)$ memory.

To reduce the expensive memory of EP, the stochastic approximate EP (SEP) [20] assumes that all the data factors are tied and uses $q(\mathbf{u}) \propto p(\mathbf{u}) g(\mathbf{u})^N$ to approximate $p(\mathbf{u}|X, \mathbf{y})$ where $g(\mathbf{u})$ can be seen as an average data factor. In practice, SEP was found to perform almost as well as full EP while largely reducing EP's memory from $\mathcal{O}(NLM^2)$ to $\mathcal{O}(LM^2)$. The SEP approximate method uses the following energy function as the objective likelihood,

$$\mathcal{F}(\boldsymbol{\Theta}) = (1 - N)\phi(\theta) + N\phi(\theta^{\backslash 1}) - \phi(\theta_{prior}) + \sum_{n=1}^N \log \mathcal{Z}_n, \tag{4}$$

where $\theta, \theta^{\backslash 1}$ and θ_{prior} are the natural parameters of the distributions $q(\mathbf{u}), q^{\backslash 1}(\mathbf{u})^3$ and $p(\mathbf{u})$, respectively, and $\mathcal{Z}_n = \int p(y_n|\mathbf{u}, \mathbf{X}_n) q^{\backslash 1}(\mathbf{u}) d\mathbf{u}$. The SEP can make DGP scalable, in which the propagation and moment-matching process consumes $\mathcal{O}(NLM^2)$ time complexity for all data points. The last term of the objective function (4) is the sum of $\{\log \mathcal{Z}_n\}_{n=1}^N$, which allows using stochastic optimization and decreases the computational complexity substantially to $\mathcal{O}(N_b LM^2)$ where N_b denotes the mini-batch size. With the approximate posterior distribution optimized, the prediction distribution for a new point $\mathbf{x}^*$ can be expressed as

$$p(y^*|\mathbf{x}^*, \mathbf{X}, \mathbf{y}) \simeq \int p(y^*|\mathbf{x}^*, \mathbf{u}) q(\mathbf{u}|\mathbf{X}, \mathbf{y}) d\mathbf{u}. \tag{5}$$

Remarks. On one hand, variables X in the DGP can be either observed which makes the DGP a supervised model, or unobserved which makes the DGP a unsupervised model. Particularly for the unsupervised DGP, the observed variable $\mathbf{y}$ represents observations with original features, and the latent variable X is to be learned. Unsupervised DGPs can be used to extract abstract features from unlabeled data. On the other hand, the likelihood of output $\mathbf{y}$ conditional on $\mathbf{f}$, i.e., $p(\mathbf{y}|\mathbf{f})$ can be various, which will be suitable for different tasks. We will introduce the specific likelihood for dealing with the binary classification and multi-class classification problems.

---

[2] The $q^{\backslash n}(\mathbf{u})$ is the variational cavity distribution of $\mathbf{u}$ and $q^{\backslash n}(\mathbf{u}) = q(\mathbf{u})/\tilde{t}_n(\mathbf{u})$.

[3] The $q^{\backslash 1}(\mathbf{u})$ is the variational cavity distribution of $\mathbf{u}$ and $q^{\backslash 1}(\mathbf{u}) = q(\mathbf{u})/g(\mathbf{u})$..

2.3   Deep Gaussian Process Classification

Binary Classification. As GP regression model has the ability of handling binary classification tasks, we directly use the DGP regression model for binary classification. Particularly, in the DGP regression model, the prediction distribution $p(y^*|\mathbf{x}^*, \mathbf{X}, \mathbf{y})$ is non-Gaussian, and we use the sign of $\mathcal{E}(y^*)$ as the prediction label, where $\mathcal{E}(y^*) = \mathbf{argmax}\ p(y^*|\mathbf{x}^*, \mathbf{X}, \mathbf{y})$.

Multi-class Classification. For multi-class classification, the vector of latent function values at the ith point for all C classes is introduced, $\mathbf{f}_i = (f_i^1, ..., f_i^C)$. Then a softmax can be used to express $p(y_i|\mathbf{f}_i)$, that is,

$$p(y_i = c|\mathbf{f}_i) = \exp(f_i^c)/\sum\nolimits_{c'} \exp(f_i^{c'}). \tag{6}$$

After deriving the prediction distribution for f^{*c} which is defined as $p(f^{*c}|\mathbf{x}^*, \mathbf{X}, \mathbf{y})$, we use a softmax mapping on the expectation of $p(f^{*c}|\mathbf{x}^*, \mathbf{X}, \mathbf{y})$ referred to as $\mathcal{E}(f^{*c})$ to calculate the probability of prediction labels.

$$p(y^* = c|\mathbf{f}^*) = \exp(\mathcal{E}(f^{*c}))/\sum\nolimits_{c'} \exp(\mathcal{E}(f^{*c'})). \tag{7}$$

3   Active Learning Methods with DGPs

3.1   Active Learning Strategies for Binary Classification

For GP binary classification, minimizing the ratio of absolute prediction mean and prediction variance is an effective strategy. This is equivalent to selecting the most uncertain point for the current classifier, where the cumulative distribution value of the point belonging to one class should be nearest to 0.5, i.e., $p(y^* \geq 0) \rightarrow 0.5$. For DGPs, inspired by the same idea, we design the following specific selection criterion,

$$\hat{\mathbf{x}} = \operatorname{argmin}_{\mathbf{x}^*} |m(\mathbf{x}^*)|/\sqrt{\sigma(\mathbf{x}^*)}, \tag{8}$$

where $m(\mathbf{x}^*)$ represents the prediction mean and $\sigma(\mathbf{x}^*)$ represents the prediction variance. Note that the difference between DGPs and GPs is that the prediction distribution of DGPs is non-Gaussian. However in DGPs, the prediction mean and variance can still be calculated. This criterion is rational as it chooses the points that are nearest to the interface and have lowest predictive confidence. This criterion combines the information from prediction mean and prediction variance, which is referred to as DGP-MeanVar.

3.2   Active Learning Strategies for Multi-class Classification

Max entropy and max variation-ratio are two commonly used strategies for multi-class active learning selection. For our DGP learning models, they are expressed as two detailed selection criteria.

1. Choose the points that maximize the predictive entropy (max entropy), that is, $\hat{\mathbf{x}} = \text{argmax}_{\mathbf{x}^*} \mathbb{H}(y^* | \mathbf{x}^*, \mathcal{D})$ with

$$\mathbb{H}(y^* | \mathbf{x}^*, \mathcal{D}) = -\sum_c p(y^* = c | \mathbf{x}^*, \mathcal{D}) \log p(y^* = c | \mathbf{x}^*, \mathcal{D}). \qquad (9)$$

This criterion is defined as DGP-Entropy.
2. Choose the points that maximize the variation ratios (max variation-ratio), that is, $\hat{\mathbf{x}} = \text{argmax}_{\mathbf{x}^*} \ variation - ratio[\mathbf{x}^*]$, with

$$variation - ratio[\mathbf{x}^*] = 1 - \max_{y^*} p(y^* | \mathbf{x}^*, \mathcal{D}). \qquad (10)$$

This criterion is defined as DGP-VarRatio.

3.3 Constructing Additional Features Using Unlabeled Data by DGP

As introduced in Sect. 2, the DGP can be used as unsupervised methods. In the framework of active learning, there are large amounts of unlabeled data. We use the unsupervised DGP to extract features from unlabeled data. In this case, the unobserved variable X in the DGP is to be integrated out, and the posterior distribution of X is to be inferred. The DGP can be trained through maximizing the marginal distribution of observations $p(Y)$, in which the approximate posterior distribution of hidden layer variables $p(\mathbf{h}_l | X, Y)$ can be optimized. As latent layers in DGPs express abstract features of complex data, we use the expectation of approximate posterior distribution for top hidden layer variables X as additional features, and splice them with original features Y to form new features. For active learning, we use a DGP classifier as learning models on newly constructed features and employ the corresponding active learning strategies. We refer this learning model as DGPfea.

4   Experiments

We conduct experiments on two different classification tasks to demonstrate the effectiveness of active learning methods based on DGPs. One is educational and non-educational text classification which is a binary classification, and the other is handwritten digit recognition which is a multi-class classification.

4.1   Data Description

We evaluate our methods on two datasets, educational text and handwritten digit data. The educational text data are collected from the Internet and labeled manually. After some conventional text processing like word stemmer, stop word removal and phases segments on the original text, the word2vec[4] is employed to

---

[4] Word2vec is an efficient tool for Google to represent the words as real value vectors. The python program can be achieved using the gensim toolkit.

represent words as continuous vectors. The resulting text data are composed of 2663 instances with each one having 50 attributes. The handwritten digit data is a public dataset called USPS, which includes 10 (0–9) digits. We conduct experiments on 2007 instances with each one having 256 attributes.

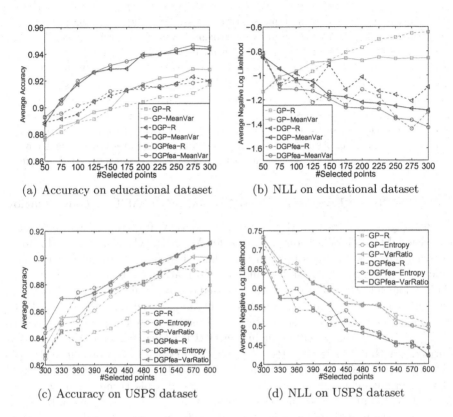

(a) Accuracy on educational dataset   (b) NLL on educational dataset

(c) Accuracy on USPS dataset   (d) NLL on USPS dataset

Fig. 2. (a) and (b) show the average accuracy and NLL on educational dataset, respectively. (c) and (d) show the average accuracy and NLL on USPS dataset, respectively. "NLL" is short for "Negative Log Likelihood".

4.2 Experimental Settings

We compare the proposed DGP based methods with GP based methods. The random selection counterpart for GP recorded as GP-R and DGP recorded as DGP-R are used as baselines. For text binary classification, we randomly select 50 points as initial labeled training set and leave 666 (a quarter of the entire data) as test data. We select 25 points to label them each time and go on for ten times. For handwritten digit recognition, we randomly select 300 points as initial labeled training set and leave 1007 as test data. We select 30 points to label them each time, and go on for ten times. All the experiments are repeated for five times, and all the reported results are the average values.

Table 1. Average results with the standard derivation on educational dataset. The above is the average accuracy. The below is the average negative log likelihood.

Points	GP-R	GP-MeanVar	DGP-R	DGP-MeanVar	DGPfea-R	DGPfea-MeanVar
50	87.76 ± 0.72	87.76 ± 0.72	88.89 ± 0.87	88.89 ± 0.87	$\mathbf{89.32 \pm 1.28}$	89.32 ± 1.28
100	88.93 ± 0.85	88.99 ± 0.65	89.50 ± 1.36	91.73 ± 1.52	90.17 ± 0.63	$\mathbf{92.00 \pm 1.17}$
150	89.87 ± 0.62	89.93 ± 1.28	90.88 ± 1.79	92.87 ± 0.77	91.22 ± 0.99	$\mathbf{93.14 \pm 1.27}$
200	90.42 ± 0.22	91.73 ± 1.59	91.63 ± 1.29	$\mathbf{93.99 \pm 1.00}$	91.40 ± 0.99	93.81 ± 0.61
250	90.85 ± 0.61	92.33 ± 1.09	91.78 ± 1.39	94.14 ± 0.43	91.72 ± 1.01	$\mathbf{94.35 \pm 0.66}$
300	91.68 ± 1.04	92.82 ± 1.07	91.95 ± 1.15	94.37 ± 0.47	91.96 ± 0.88	$\mathbf{94.51 \pm 0.65}$
50	$\mathbf{-1.13 \pm 0.16}$	$\mathbf{-1.13 \pm 0.16}$	-0.85 ± 0.06	-0.85 ± 0.06	-0.82 ± 0.09	-0.82 ± 0.09
100	-1.05 ± 0.05	-1.01 ± 0.09	-0.98 ± 0.11	-1.03 ± 0.05	-1.03 ± 0.27	$\mathbf{-1.11 \pm 0.47}$
150	-0.88 ± 0.08	-0.88 ± 0.08	-0.92 ± 0.23	-1.17 ± 0.07	-1.14 ± 0.39	$\mathbf{-1.20 \pm 0.57}$
200	-0.76 ± 0.03	-0.88 ± 0.05	-1.01 ± 0.53	-1.22 ± 0.07	-1.12 ± 0.45	$\mathbf{-1.27 \pm 0.62}$
250	-0.69 ± 0.02	-0.86 ± 0.05	-1.16 ± 0.58	-1.25 ± 0.05	-1.34 ± 0.66	$\mathbf{-1.36 \pm 0.56}$
300	-0.64 ± 0.05	-0.86 ± 0.03	-1.10 ± 0.58	-1.29 ± 0.06	-1.30 ± 0.59	$\mathbf{-1.43 \pm 0.40}$

4.3 Results for Educational and Non-educational Text Classification

Figure 2(a) shows the trend of average accuracy for educational and non-educational text classification. The DGPfea-MeanVar which uses the additional abstract features performs best on the whole, because additional features provide more comprehensive and abundant information for training. When using the random selection strategy, methods based on DGPs including DGPfea-R and DGP-R obtain better performance than methods based on GPs, (i.e., GP-R). Similarly, when employing the strategy combined with prediction mean and variance, methods based on DGPs including DGPfea-MeanVar and DGP-MeanVar outperform that based on GPs, (i.e., GP-MeanVar). In a word, using DGPs as learning models are better than using GPs. The reasons are listed. First, DGPs can automatically construct complex and flexible kernels that work well for real-world datasets. Second, the mapping from inputs to outputs of the DGP is non-Gaussian, which is a more general modeling choice. The learning strategy based on prediction mean and variance is better than the random selection strategy as the former can select uncertain points. In addition, as the number of selected points increases, the average accuracy of the classification becomes higher. When the number of selected points is relatively large, such as 300, the classification accuracy converges. Figure 2(b) shows the trend of average negative log likelihood (NLL) on educational dataset. The lower NLL of DGPs over GPs shows a better fitting result, which is own to the deep structure of DGPs and rational active learning strategies based on them. We also list the detailed average values with the standard deviation in Table 1. The best experimental results are in bold, all of which are obtained by DGPs.

4.4 Results for Handwritten Digit Recognition

Figure 2(c) shows the trend of average accuracy for handwritten digit recognition. We omit the results of the methods, which directly use DGPs as learning models

Table 2. Average results with the standard derivation on USPS dataset. The above is the average accuracy. The below is the average negative log likelihood.

Points	GP-R	GP-Entropy	GP-VarRatio	DGPfea-R	DGPfea-Entropy	DGPfea-VarRatio
300	82.67 ± 0.78	82.67 ± 0.78	82.67 ± 0.78	$\mathbf{84.77 \pm 0.64}$	84.77 ± 0.64	$\mathbf{84.77 \pm 0.64}$
360	83.57 ± 1.65	85.31 ± 1.81	85.63 ± 1.48	84.66 ± 0.58	$\mathbf{87.42 \pm 0.64}$	86.96 ± 0.91
450	85.48 ± 0.80	87.85 ± 1.29	88.11 ± 0.38	87.99 ± 0.80	89.13 ± 0.96	$\mathbf{89.19 \pm 1.52}$
510	86.47 ± 1.25	88.58 ± 0.87	88.92 ± 1.40	88.84 ± 1.13	89.54 ± 0.57	$\mathbf{89.73 \pm 0.67}$
600	87.93 ± 1.06	88.82 ± 1.17	90.02 ± 0.57	90.05 ± 0.64	91.07 ± 0.97	$\mathbf{91.10 \pm 1.09}$
300	0.72 ± 0.01	0.72 ± 0.01	0.72 ± 0.02	$\mathbf{0.67 \pm 0.10}$	$\mathbf{0.67 \pm 0.10}$	$\mathbf{0.67 \pm 0.10}$
360	0.65 ± 0.02	0.66 ± 0.02	0.65 ± 0.02	0.60 ± 0.04	$\mathbf{0.54 \pm 0.02}$	0.57 ± 0.06
450	0.58 ± 0.02	0.58 ± 0.02	0.56 ± 0.01	0.51 ± 0.04	0.54 ± 0.11	$\mathbf{0.49 \pm 0.02}$
510	0.56 ± 0.03	0.55 ± 0.02	0.55 ± 0.01	0.48 ± 0.04	0.48 ± 0.04	$\mathbf{0.47 \pm 0.02}$
600	0.50 ± 0.01	0.50 ± 0.02	0.49 ± 0.01	0.44 ± 0.03	0.42 ± 0.03	$\mathbf{0.42 \pm 0.02}$

since they got bad performance. We suspect that the current DGP multi-class classifiers with softmax are not suitable for small amount of complex data, and we will improve the classifiers in the future work. Figure 2(d) shows the trend of average NLL on USPS dataset. The NLL of DGPs is lower than that of GPs. This fits the fact that the model will be more flexible with a deeper structure and additional feature extraction. Seen from the Fig. 2(c) and (d), DGPfea-VarRatio performs best on the whole. The methods with DGPfea as learning models using additional features are better than those using original features. This might be because the extracted features obtained by using additional unlabeled data give better representations for the complex data. Additionally, the methods with strategies of maximizing VarRatio are better than those of maximizing entropy. Table 2 shows the detailed average value with the standard deviation for multi-class classification tasks.

5 Conclusion

In this paper, we proposed active learning methods based on DGPs to solve binary and multi-class classification. We evaluated our methods through experiments on educational text classification and handwritten digit recognition. The promising results have demonstrated the effectiveness of the proposed methods. The experimental results also point the potential of DGPs for active learning on complex data. Particularly, the ability of DGPs to construct abstract features will help active learning to employ information from unlabeled data. In the future work, we will study a good combination of unsupervised DGP and supervised DGP for active learning, which may lead to more performance improvements.

Acknowledgments. The first two authors Jingjing Fei and Jing Zhao are joint first authors. The corresponding author is Shiliang Sun. This work is sponsored by NSFC Project 61673179, Shanghai Sailing Program, and Shanghai Knowledge Service Platform Project (No. ZF1213).

References

1. Huang, S., Jin, R., Zhou, Z.: Active learning by querying informative and representative examples. IEEE Trans. Pattern Anal. Mach. Intell. **36**(10), 1936–1949 (2014)
2. Zhou, J., Sun, S.: Gaussian process versus margin sampling active learning. Neurocomputing **167**(1), 122–131 (2015)
3. Gavves, E., Mensink, T., Tommasi, T., Snoek, C.G.M., Tuytelaars, T.: Active transfer learning with zero-shot priors: reusing past datasets for future tasks. In: Proceedings of the IEEE International Conference on Computer Vision, pp. 2731–2739. IEEE, New York (2015)
4. Seeger, M.: Gaussian processes for machine learning. Int. J. Neural Syst. **14**(2), 69–106 (2004)
5. Ma, Y., Sutherland, D., Garnett, R., Schneider, J.: Active pointillistic pattern search. In: Proceedings of the 18th International Conference on Artificial Intelligence and Statistics, pp. 672–680. JMLR, Cambridge (2015)
6. Liu, Q., Sun, S.: Sparse multimodal Gaussian processes. In: Sun, Y., Lu, H., Zhang, L., Yang, J., Huang, H. (eds.) IScIDE 2017. LNCS, vol. 10559, pp. 28–40. Springer, Cham (2017). https://doi.org/10.1007/978-3-319-67777-4_3
7. Luo, C., Sun, S.: Variational mixtures of Gaussian processes for classification. In: Proceedings of the 26th International Joint Conference on Artificial Intelligence, pp. 4603–4609. Morgan Kaufmann, San Francisco (2017)
8. Mosinskadomanska, A., Sznitman, R., Glowack, P., Fua, P.: Active learning for delineation of curvilinear structures. In: Proceedings of the IEEE Conference on Computer Vision & Pattern Recognition, pp. 5231–5239. IEEE, New York (2015)
9. Lu, J., Zhao, P., Steven, S.C.H.: Online passive aggressive active learning and its applications. In: Proceedings of the Asian Conference on Machine Learning, pp. 266–282. JMLR, Cambridge (2014)
10. Yang, Y., Ma, Z., Nie, F., Chang, X., Hauptmann, A.G.: Multi-class active learning by uncertainty sampling with diversity maximization. Int. J. Comput. Vis. **113**(2), 113–127 (2015)
11. Lewenberg, Y., Bachrach, Y., Paquet, U., Rosenschein, J.: Knowing what to ask: A Bayesian active learning approach to the surveying problem. In: Proceedings of the 31st AAAI Conference on Artificial Intelligence, pp. 1396–1402. AAAI, San Francisco (2017)
12. Rasmussen, C.: Gaussian Processes for Machine Learning. MIT Press, Cambridge (2006)
13. Nickisch, H., Rasmussen, C.: Approximations for binary Gaussian process classification. J. Mach. Learn. Res. **9**(10), 2035–2078 (2008)
14. Kim, H., Ghahramani, Z.: Bayesian Gaussian process classification with the EM-EP algorithm. IEEE Trans. Pattern Anal. Mach. Intell. **28**(12), 1948–1959 (2006)
15. Zhao, J., Sun, S.: Variational dependent multi-output Gaussian process dynamical systems. J. Mach. Learn. Res. **17**(1), 1–36 (2016)
16. Lawrence, N., Moore, A.: Hierarchical Gaussian process latent variable models. In: Proceedings of the 24th International Conference on Machine Learning, pp. 481–488. ACM, New York (2007)
17. Damianou, A., Lawrence, N.: Deep Gaussian processes. In: Proceedings of the 16th International Conference on Artificial Intelligence and Statistics, pp. 207–215. JMLR, Cambridge (2013)

18. Dai, Z., Damianou, A., González, J., Lawrence, N.: Variational auto-encoded deep Gaussian processes. Comput. Sci. **14**(9), 3942–3951 (2015)
19. Bui, T., Hernández-Lobato, D., Hernandez-Lobato, J., Li, Y., Turner, R.: Deep Gaussian processes for regression using approximate expectation propagation. In: Proceedings of the 33rd International Conference on Machine Learning, pp. 1472–1481. ACM, New York (2016)
20. Li, Y., Hernández-Lobato, J., Turner, R.: Stochastic expectation propagation. Adv. Neural Inf. Process. Syst. **28**(1), 2323–2331 (2015)

Transductive Learning with String Kernels for Cross-Domain Text Classification

Radu Tudor Ionescu$^{(\boxtimes)}$ and Andrei Madalin Butnaru

University of Bucharest, 14 Academiei, Bucharest, Romania
raducu.ionescu@gmail.com, butnaruandreimadalin@gmail.com

Abstract. For many text classification tasks, there is a major problem posed by the lack of labeled data in a target domain. Although classifiers for a target domain can be trained on labeled text data from a related source domain, the accuracy of such classifiers is usually lower in the cross-domain setting. Recently, string kernels have obtained state-of-the-art results in various text classification tasks such as native language identification or automatic essay scoring. Moreover, classifiers based on string kernels have been found to be robust to the distribution gap between different domains. In this paper, we formally describe an algorithm composed of two simple yet effective transductive learning approaches to further improve the results of string kernels in cross-domain settings. By adapting string kernels to the test set without using the ground-truth test labels, we report significantly better accuracy rates in cross-domain English polarity classification.

Keywords: Transductive learning · Domain adaptation
Cross-domain classification · String kernels · Sentiment analysis
Polarity classification

1 Introduction

Domain shift is a fundamental problem in machine learning, that has attracted a lot of attention in the natural language processing and vision communities [2,6,11,13,29,30,32,37,39,40,42]. To understand and address this problem, generated by the lack of labeled data in a target domain, researchers have studied the behavior of machine learning methods in cross-domain settings [12,13,29] and came up with various domain adaptation techniques [6,11,28,39]. In cross-domain classification, a classifier is trained on data from a source domain and tested on data from a (different) target domain. The accuracy of machine learning methods is usually lower in the cross-domain setting, due to the distribution gap between different domains. However, researchers proposed several domain adaptation techniques by using the unlabeled test data to obtain better performance [5,14,16,25,37]. Interestingly, some recent works [13,18] indicate that string kernels can yield robust results in the cross-domain setting without any

© Springer Nature Switzerland AG 2018
L. Cheng et al. (Eds.): ICONIP 2018, LNCS 11303, pp. 484–496, 2018.
https://doi.org/10.1007/978-3-030-04182-3_42

domain adaptation. In fact, methods based on string kernels have demonstrated impressive results in various text classification tasks ranging from native language identification [22–24,36] and authorship identification [34] to dialect identification [4,18,21], sentiment analysis [13,35] and automatic essay scoring [7]. As long as a labeled training set is available, string kernels can reach state-of-the-art results in various languages including English [7,13,23], Arabic [4,17,18,24], Chinese [35] and Norwegian [24]. Different from all these recent approaches, we use unlabeled data from the test set in a transductive setting in order to significantly increase the performance of string kernels. In our recent work [19], we proposed two transductive learning approaches combined into a unified framework that improves the results of string kernels in two different tasks. In this paper, we provide a formal and detailed description of our transductive algorithm and present results in cross-domain English polarity classification.

The paper is organized as follows. Related work on cross-domain text classification and string kernels is presented in Sect. 2. Section 3 presents our approach to obtain domain adapted string kernels. The transductive transfer learning method is described in Sect. 4. The polarity classification experiments are presented in Sect. 5. Finally, we draw conclusions and discuss future work in Sect. 6.

2   Related Work

2.1   Cross-Domain Classification

Transfer learning (or domain adaptation) aims at building effective classifiers for a target domain when the only available labeled training data belongs to a different (source) domain. Domain adaptation techniques can be roughly divided into graph-based methods [6,31–33], probabilistic models [30,42], knowledge-based models [3,12,16] and joint optimization frameworks [28]. The transfer learning methods from the literature show promising results in a variety of real-world applications, such as image classification [28], text classification [14,25,42], polarity classification [11,30–33] and others [8].

General Transfer Learning Approaches. Long et al. [28] proposed a novel transfer learning framework to model distribution adaptation and label propagation in a unified way, based on the structural risk minimization principle and the regularization theory. Shu et al. [39] proposed a method that bridges the distribution gap between the source domain and the target domain through affinity learning, by exploiting the existence of a subset of data points in the target domain that are distributed similarly to the data points in the source domain. In [37], deep learning is employed to jointly optimize the representation, the cross-domain transformation and the target label inference in an end-to-end fashion. More recently, Sun et al. [40] proposed an unsupervised domain adaptation method that minimizes the domain shift by aligning the second-order statistics of source and target distributions, without requiring any target labels.

Chang et al. [6] proposed a framework based on using a parallel corpus to calibrate domain-specific kernels into a unified kernel for leveraging graph-based label propagation between domains.

Cross-Domain Text Classification. Joachims [25] introduced the Transductive Support Vector Machines (TSVM) framework for text classification, which takes into account a particular test set and tries to minimize the error rate for those particular test samples. Ifrim et al. [16] presented a transductive learning approach for text classification based on combining latent variable models for decomposing the topic-word space into topic-concept and concept-word spaces, and explicit knowledge models with named concepts for populating latent variables. Guo et al. [14] proposed a transductive subspace representation learning method to address domain adaptation for cross-lingual text classification. Zhuang et al. [42] presented a probabilistic model, by which both the shared and distinct concepts in different domains can be learned by the Expectation-Maximization process which optimizes the data likelihood. In [1], an algorithm to adapt a classification model by iteratively learning domain-specific features from the unlabeled test data is described.

Cross-Domain Polarity Classification. In recent years, cross-domain sentiment (polarity) classification has gained popularity due to the advances in domain adaptation on one side, and to the abundance of documents from various domains available on the Web, expressing positive or negative opinion, on the other side. Some of the general domain adaptation frameworks have been applied to polarity classification [1,6,42], but there are some approaches that have been specifically designed for the cross-domain sentiment classification task [2,11–13,26,30–33]. To the best of our knowledge, Blitzer et al. [2] were the first to report results on cross-domain classification proposing the structural correspondence learning (SCL) method, and its variant based on mutual information (SCL-MI). Pan et al. [32] proposed a spectral feature alignment (SFA) algorithm to align domain-specific words from different domains into unified clusters, using domain-independent words as a bridge. Bollegala et al. [3] used a cross-domain lexicon creation to generate a sentiment-sensitive thesaurus (SST) that groups different words expressing the same sentiment, using unigram and bigram features as [2,32]. Luo et al. [30] proposed a cross-domain sentiment classification framework based on a probabilistic model of the author's emotion state when writing. An Expectation-Maximization algorithm is then employed to solve the maximum likelihood problem and to obtain a latent emotion distribution of the author. Franco-Salvador et al. [12] combined various recent and knowledge-based approaches using a meta-learning scheme (KE-Meta). They performed cross-domain polarity classification without employing any domain adaptation technique. More recently, Fernández et al. [11] introduced the Distributional Correspondence Indexing (DCI) method for domain adaptation in sentiment classification. The approach builds term representations in a vector space common to both domains where each dimension reflects its distributional correspondence to a highly predictive term that behaves similarly across domains. A graph-based approach for sentiment classification that models the relatedness of different domains based on shared users and keywords is proposed in [31].

2.2 String Kernels

In recent years, methods based on string kernels have demonstrated remarkable performance in various text classification tasks [7,10,13,18,23,27,34]. String kernels represent a way of using information at the character level by measuring the similarity of strings through character n-grams. Lodhi et al. [27] used string kernels for document categorization, obtaining very good results. String kernels were also successfully used in authorship identification [34]. More recently, various combinations of string kernels reached state-of-the-art accuracy rates in native language identification [23] and Arabic dialect identification [18]. Interestingly, string kernels have been used in cross-domain settings without any domain adaptation, obtaining impressive results. For instance, Ionescu et al. [23] have employed string kernels in a cross-corpus (and implicitly cross-domain) native language identification experiment, improving the state-of-the-art accuracy by a remarkable 32.3%. Giménez-Pérez et al. [13] have used string kernels for single-source and multi-source polarity classification. Remarkably, they obtain state-of-the-art performance without using knowledge from the target domain, which indicates that string kernels provide robust results in the cross-domain setting without any domain adaptation. Ionescu et al. [18] obtained the best performance in the Arabic Dialect Identification Shared Task of the 2017 VarDial Evaluation Campaign [41], with an improvement of 4.6% over the second-best method. It is important to note that the training and the test speech samples prepared for the shared task were recorded in different setups [41], or in other words, the training and the test sets are drawn from different distributions. Different from all these recent approaches [13,18,23], we use unlabeled data from the target domain to significantly increase the performance of string kernels in cross-domain text classification, particularly in English polarity classification.

3 Transductive String Kernels

String Kernels. Kernel functions [38] capture the intuitive notion of similarity between objects in a specific domain. For example, in text mining, string kernels can be used to measure the pairwise similarity between text samples, simply based on character n-grams. Various string kernel functions have been proposed to date [23,27,38]. Perhaps one of the most recently introduced string kernels is the histogram intersection string kernel [23]. For two strings over an alphabet Σ, $x, y \in \Sigma^*$, the intersection string kernel is formally defined as follows:

$$k^\cap(x,y) = \sum_{v \in \Sigma^p} \min\{\text{num}_v(x), \text{num}_v(y)\}, \tag{1}$$

where $\text{num}_v(x)$ is the number of occurrences of n-gram v as a substring in x, and p is the length of v. The spectrum string kernel or the presence bits string kernel can be defined in a similar fashion [23].

Transductive String kernels. We present a simple and straightforward approach to produce a transductive similarity measure suitable for strings. We take the following steps to derive transductive string kernels. For a given kernel (similarity) function k, we first build the full kernel matrix K, by including the pairwise similarities of samples from both the train and the test sets. For a training set $X = \{x_1, x_2, ..., x_m\}$ of m samples and a test set $Y = \{y_1, y_2, ..., y_n\}$ of n samples, such that $X \cap Y = \emptyset$, each component in the full kernel matrix is defined as follows:

$$K_{ij} = k(z_i, z_j), \tag{2}$$

where z_i and z_j are samples from the set $Z = X \cup Y = \{x_1, x_2, ..., x_m, y_1, y_2, ..., y_n\}$, for all $1 \leq i, j \leq m + n$. We then normalize the kernel matrix by dividing each component by the square root of the product of the two corresponding diagonal components:

$$\hat{K}_{ij} = \frac{K_{ij}}{\sqrt{K_{ii} \cdot K_{jj}}}. \tag{3}$$

We transform the normalized kernel matrix into a radial basis function (RBF) kernel matrix as follows:

$$\tilde{K}_{ij} = exp\left(-1 + \hat{K}_{ij}\right). \tag{4}$$

Each row in the RBF kernel matrix $\tilde{K}$ is now interpreted as a feature vector. In other words, each sample z_i is represented by a feature vector that contains the similarity between the respective sample z_i and all the samples in Z. Since Z includes the test samples as well, the feature vector is inherently adapted to the test set. Indeed, it is easy to see that the features will be different if we choose to apply the string kernel approach on a set of test samples Y', such that $Y' \neq Y$. It is important to note that through the features, the subsequent classifier will have some information about the test samples at training time. More specifically, the feature vector conveys information about how similar is every test sample to every training sample. We next consider the linear kernel, which is given by the scalar product between the new feature vectors. To obtain the final linear kernel matrix, we simply need to compute the product between the RBF kernel matrix and its transpose:

$$\ddot{K} = \tilde{K} \cdot \tilde{K}'. \tag{5}$$

In this way, the samples from the test set, which are included in Z, are used to obtain new (transductive) string kernels that are adapted to the test set at hand.

4   Transductive Kernel Classifier

We next present a simple yet effective approach for adapting a one-versus-all kernel classifier trained on a source domain to a different target domain. Our

---

Algorithm 1: Transductive Kernel Algorithm

---

1 **Input:**
2 $\mathcal{X} = (X, T) = \{(x_i, t_i) \mid x_i \in \mathbb{R}^q, t_i \in \{1, 2, ..., c\}, i \in \{1, 2, ..., m\}\}$ – the training set of m training samples and associated class labels;
3 $Y = \{y_i \mid y_i \in \mathbb{R}^q, i \in \{1, 2, ..., n\}\}$ – the set of n test samples;
4 k – a kernel function;
5 r – the number of test samples to be added in the second round of training;
6 $\mathcal{C}$ – a binary kernel classifier.

7 **Domain-Adapted Kernel Matrix Computation Steps:**
8 $Z \leftarrow \{x_1, x_2, ..., x_m, y_1, y_2, ..., y_n\}$;
9 $K \leftarrow \mathbf{0}_{m+n}$; $\hat{K} \leftarrow \mathbf{0}_{m+n}$; $\tilde{K} \leftarrow \mathbf{0}_{m+n}$; $\ddot{K} \leftarrow \mathbf{0}_{m+n}$;
10 **for** $z_i \in Z$ **do**
11 $\quad$ **for** $z_j \in Z$ **do**
12 $\quad\quad$ $K_{ij} \leftarrow k(z_i, z_j)$;

13 **for** $i \in \{1, 2, ..., m + n\}$ **do**
14 $\quad$ **for** $j \in \{1, 2, ..., m + n\}$ **do**
15 $\quad\quad$ $\hat{K}_{ij} \leftarrow \dfrac{K_{ij}}{\sqrt{K_{ii} \cdot K_{jj}}}$;
16 $\quad\quad$ $\tilde{K}_{ij} \leftarrow exp\left(-1 + \hat{K}_{ij}\right)$;

17 $\ddot{K} = \tilde{K} \cdot \tilde{K}'$;

18 **Transductive Kernel Classifier Steps:**
19 $T_{OVA} \leftarrow 2 \cdot \mathbf{1}_c(T, :) - 1$;
20 $i_{train} \leftarrow 1 : m$;
21 $i_{test} \leftarrow m + 1 : m + n$;
22 **for** $s \in \{1, 2\}$ **do**
23 $\quad$ $\ddot{K}_{train} \leftarrow \ddot{K}(i_{train}, i_{train})$;
24 $\quad$ $\ddot{K}_{test} \leftarrow \ddot{K}(i_{test}, i_{train})$;
25 $\quad$ $S_{OVA} \leftarrow \mathbf{0}_{n,c}$;
26 $\quad$ **for** $i \in \{1, 2, ..., c\}$ **do**
27 $\quad\quad$ $(\alpha, b) \leftarrow$ the dual weights of $\mathcal{C}$ trained on $\ddot{K}_{train}$ with the labels $T_{OVA}(:, i)$;
28 $\quad\quad$ $S_{OVA}(:, i) \leftarrow \ddot{K}_{test} \cdot \alpha + b$;
29 $\quad$ $P \leftarrow \mathbf{0}_{n,1}$; $S \leftarrow \mathbf{0}_{n,1}$;
30 $\quad$ **for** $i \in \{1, 2, ..., n\}$ **do**
31 $\quad\quad$ $P_i \leftarrow argmax(S_{OVA}(i, :))$;
32 $\quad\quad$ $S_i \leftarrow max(S_{OVA}(i, :))$;
33 $\quad$ **if** $s = 1$ **then**
34 $\quad\quad$ $i_{sort} \leftarrow$ sort S in descending order and return the sorted indexes;
35 $\quad\quad$ $i_{keep} \leftarrow i_{sort}(1 : r)$;
36 $\quad\quad$ $P_{keep} \leftarrow P(i_{keep})$;
37 $\quad\quad$ $T \leftarrow T \cup P_{keep}$;
38 $\quad\quad$ $T_{OVA} \leftarrow 2 \cdot \mathbf{1}_c(T, :) - 1$;
39 $\quad\quad$ $i_{train} \leftarrow i_{train} \cup i_{test}(i_{keep})$;

40 **Output:**
41 $P = \{p_i \mid p_i \in \{1, 2, ..., c\}, i \in \{1, 2, ..., n\}\}$ – the set of predicted labels for the test samples in Y.

---

transductive kernel classifier (TKC) approach is composed of two learning iterations. Our entire framework is formally described in Algorithm 1.

Notations. We use the following notations in the algorithm. Sets, arrays and matrices are written in capital letters. All collection types are considered to be indexed starting from position 1. The elements of a set S are denoted by s_i, the elements of an array A are alternatively denoted by $A(i)$ or A_i, and the elements of a matrix M are denoted by $M(i,j)$ or M_{ij} when convenient. The sequence $1, 2, ..., n$ is denoted by $1 : n$. We use sequences to index arrays or matrices as well. For example, for an array A and two integers i and j, $A(i : j)$ denotes the sub-array $(A_i, A_{i+1}, ..., A_j)$. In a similar manner, $M(i : j, k : l)$ denotes a sub-matrix of the matrix M, while $M(i, :)$ returns the i-th row of M and $M(:, j)$ returns the j-th column of M. The zero matrix of $m \times n$ components is denoted by $\mathbf{0}_{m,n}$, and the square zero matrix is denoted by $\mathbf{0}_n$. The identity matrix is denoted by $\mathbf{1}_n$.

Algorithm Description. In steps 8–17, we compute the domain-adapted string kernel matrix, as described in the previous section. In the first learning iteration (when $s = 1$), we train several classifiers to distinguish each individual class from the rest, according to the one-versus-all (OVA) scheme. In step 27, the kernel classifier $\mathcal{C}$ is trained to distinguish a class from the others, assigning a dual weight to each training sample from the source domain. The returned column vector of dual weights is denoted by α and the bias value is denoted by b. The vector of weights α contains m values, such that the weight α_i corresponds to the training sample x_i. When the test kernel matrix $\ddot{K}_{test}$ of $n \times m$ components is multiplied with the vector α in step 28, the result is a column vector of n positive or negative scores. Afterwards (step 34), the test samples are sorted in order to maximize the probability of correctly predicted labels. For each test sample y_i, we consider the score S_i (step 32) produced by the classifier for the chosen class P_i (step 31), which is selected according to the OVA scheme. The sorting is based on the hypothesis that if the classifier associates a higher score to a test sample, it means that the classifier is more confident about the predicted label for the respective test sample. Before the second learning iteration, a number of r test samples from the top of the sorted list are added to the training set (steps 35-39) for another round of training. As the classifier is more confident about the predicted labels P_{keep} of the added test samples, the chance of including noisy examples (with wrong labels) is minimized. On the other hand, the classifier has the opportunity to learn some useful domain-specific patterns of the test domain. We believe that, at least in the cross-domain setting, the added test samples bring more useful information than noise. We would like to stress out that *the ground-truth test labels are never used in our transductive algorithm.* Although the test samples are required beforehand, their labels are not necessary. Hence, our approach is suitable in situations where unlabeled data from the target domain can be collected cheaply, and such situations appear very often in practice, considering the great amount of data available on the Web.

5   Polarity Classification

Data Set. For the cross-domain polarity classification experiments, we use the second version of Multi-Domain Sentiment Dataset [2]. The data set contains Amazon product reviews of four different domains: Books (B), DVDs (D), Electronics (E) and Kitchen appliances (K). Reviews contain star ratings (from 1 to 5) which are converted into binary labels as follows: reviews rated with more than 3 stars are labeled as positive, and those with less than 3 stars as negative. In each domain, there are 1000 positive and 1000 negative reviews.

Baselines. We compare our approach with several methods [3,12,13,15,32,40] in two cross-domain settings. Using string kernels, Giménez-Pérez et al. [13] reported better performance than SST [3] and KE-Meta [12] in the multi-source domain setting. In addition, we compare our approach with SFA [32], CORAL [40] and TR-TrAdaBoost [15] in the single-source setting.

Table 1. Multi-source cross-domain polarity classification accuracy rates (in %) of our transductive approaches versus a state-of-the-art baseline based on string kernels [13], as well as SST [3] and KE-Meta [12]. The best accuracy rates are highlighted in bold. The marker * indicates that the performance is significantly better than the best baseline string kernel according to a paired McNemar's test performed at a significance level of 0.01.

Method	DEK→B	BEK→D	BDK→E	BDE→K
SST [3]	76.3	78.3	83.9	85.2
KE-Meta [12]	77.9	80.4	78.9	82.5
$K_{0/1}$ [13]	82.0	81.9	83.6	85.1
$K_\cap$ [13]	80.7	80.7	83.0	85.2
$\ddot{K}_{0/1}$	82.9	83.2*	84.8*	86.0*
$\ddot{K}_\cap$	82.5	82.9*	84.5*	86.1*
$\ddot{K}_{0/1}$ + TKC	**84.1***	**84.0***	**85.4***	86.9*
$\ddot{K}_\cap$ + TKC	83.8*	83.5*	85.0*	**87.1***

Evaluation Procedure and Parameters. We follow the same evaluation methodology of Giménez-Pérez et al. [13], to ensure a fair comparison. Furthermore, we use the same kernels, namely the presence bits string kernel ($K_{0/1}$) and the intersection string kernel ($K_\cap$), and the same range of character n-grams (5–8). To compute the string kernels, we used the open-source code provided by Ionescu et al. [20,23]. For the transductive kernel classifier, we select $r = 1000$ unlabeled test samples to be included in the training set for the second round of training. We choose Kernel Ridge Regression [38] as classifier and set its regularization parameter to 10^{-5} in all our experiments. Although Giménez-Pérez et al. [13] used a different classifier, namely Kernel Discriminant Analysis, we observed that Kernel Ridge Regression produces similar results ($\pm 0.1\%$) when we employ the same string kernels. As Giménez-Pérez et al. [13], we evaluate our approach in two cross-domain settings. In the multi-source setting, we train the models on all domains, except the one used for testing. In the single-source setting, we train the models on one of the four domains and we independently test the models on the remaining three domains.

Table 2. Single-source cross-domain polarity classification accuracy rates (in %) of our transductive approaches versus a state-of-the-art baseline based on string kernels [13], as well as SFA [32], CORAL [40] and TR-TrAdaBoost [15]. The best accuracy rates are highlighted in bold. The marker * indicates that the performance is significantly better than the best baseline string kernel according to a paired McNemar's test performed at a significance level of 0.01.

Method	D→B	E→B	K→B	B→D	E→D	K→D
SFA [32]	79.8	78.3	75.2	81.4	77.2	**78.5**
CORAL [40]	78.3	-	-	-	-	73.9
TR-TrAdaBoost [15]	74.7	69.1	70.6	79.6	71.8	74.4
$K_{0/1}$ [13]	82.0	72.4	72.7	81.4	74.9	73.6
$K_\cap$ [13]	82.1	72.4	72.8	81.3	75.1	72.9
$\ddot{K}_{0/1}$	83.3*	74.5*	74.3*	83.0*	76.9*	74.9*
$\ddot{K}_\cap$	83.2*	74.2*	74.0*	82.8*	76.4*	75.1*
$\ddot{K}_{0/1}$ + TKC	**84.9***	**78.5***	**76.6***	84.0*	**79.6***	76.4*
$\ddot{K}_\cap$ + TKC	84.5*	**78.5***	75.8*	**84.2***	79.1*	76.5*
Method	B→E	D→E	K→E	B→K	D→K	E→K
SFA [32]	73.5	76.7	85.1	79.1	80.8	86.8
CORAL [40]	76.3	-	-	-	-	83.6
TR-TrAdaBoost [15]	74.9	75.9	83.1	77.8	75.7	83.7
$K_{0/1}$ [13]	71.3	74.4	83.9	74.6	75.4	84.9
$K_\cap$ [13]	71.8	74.5	84.4	74.9	75.1	84.9
$\ddot{K}_{0/1}$	74.0*	76.0*	85.4*	77.6*	77.3*	86.0*
$\ddot{K}_\cap$	74.2*	75.9*	85.2*	77.6*	77.3*	85.9*
$\ddot{K}_{0/1}$ + TKC	76.6*	**77.1***	**86.4***	**79.6***	**80.9***	**87.0***
$\ddot{K}_\cap$ + TKC	**76.7***	76.8*	**86.4***	79.4*	80.5*	**87.0***

Results in Multi-source Setting. The results for the multi-source cross-domain polarity classification setting are presented in Table 1. Both the transductive presence bits string kernel ($\ddot{K}_{0/1}$) and the transductive intersection kernel ($\ddot{K}_\cap$) obtain better results than their original counterparts. Moreover, according to the McNemar's test [9], the results on the DVDs, the Electronics and the Kitchen target domains are significantly better than the best baseline string kernel, with a confidence level of 0.01. When we employ the transductive kernel classifier (TKC), we obtain even better results. On all domains, the accuracy rates yielded by the transductive classifier are more than 1.5% better than the best baseline. For example, on the Books domain the accuracy of the transductive classifier based on the presence bits kernel (84.1%) is 2.1% above the best baseline (82.0%) represented by the intersection string kernel. Remarkably, the improvements brought by our transductive string kernel approach are statistically significant in all domains.

Results in Single-Source Setting. The results for the single-source cross-domain polarity classification setting are presented in Table 2. We considered all possible combinations of source and target domains in this experiment, and we improve the results in each and every case. Without exception, the accuracy rates reached by the transductive string kernels are significantly better than the best baseline string kernel [13], according to the McNemar's test performed at a confidence level of 0.01. The highest improvements (above 2.7%) are obtained when the source domain contains Books reviews and the target domain contains Kitchen reviews. As in the multi-source setting, we obtain much better results when the transductive classifier is employed for the learning task. In all cases, the accuracy rates of the transductive classifier are more than 2% better than the best baseline string kernel. Remarkably, in four cases (E→B, E→D, B→K and D→K) our improvements are greater than 4%. The improvements brought by our transductive classifier based on string kernels are statistically significant in each and every case. In comparison with SFA [32], we obtain better results in all but one case (K→D). Remarkably, we surpass the other state-of-the-art approaches [15, 40] in all cases.

6   Conclusion

In this paper, we presented two domain adaptation approaches that can be used together to improve the results of string kernels in cross-domain settings. We provided empirical evidence indicating that our framework can be successfully applied in cross-domain text classification, particularly in cross-domain English polarity classification. Indeed, the polarity classification experiments demonstrate that our framework achieves better accuracy rates than other state-of-the-art methods [3, 12, 13, 15, 32, 40]. By using the same parameters across all the experiments, we showed that our transductive transfer learning framework can bring significant improvements without having to fine-tune the parameters for each individual setting. Although the framework described in this paper can be generally applied to any kernel method, we focused our work only on string kernel approaches used in text classification. In future work, we aim to combine the proposed transductive transfer learning framework with different kinds of kernels and classifiers, and employ it for other cross-domain tasks.

References

1. Bhatt, S.H., Semwal, D., Roy, S.: An iterative similarity based adaptation technique for cross-domain text classification. In: Proceedings of CONLL, pp. 52–61 (2015)
2. Blitzer, J., Dredze, M., Pereira, F.: Biographies, bollywood, boomboxes and blenders: domain adaptation for sentiment classification. In: Proceedings of ACL, pp. 187–205 (2007)
3. Bollegala, D., Weir, D., Carroll, J.: Cross-domain sentiment classification using a sentiment sensitive thesaurus. IEEE Trans. Knowl. Data Eng. **25**(8), 1719–1731 (2013)
4. Butnaru, A.M., Ionescu, R.T.: UnibucKernel reloaded: first place in Arabic dialect identification for the second year in a row. In: Proceedings of VarDial Workshop of COLING, pp. 77–87 (2018)
5. Ceci, M.: Hierarchical text categorization in a transductive setting. In: Proceedings of ICDMW, pp. 184–191, December 2008
6. Chang, W.C., Wu, Y., Liu, H., Yang, Y.: Cross-domain kernel induction for transfer learning. In: Proceedings of AAAI, pp. 1763–1769, February 2017
7. Cozma, M., Butnaru, A., Ionescu, R.T.: Automated essay scoring with string kernels and word embeddings. In: Proceedings of ACL, pp. 503–509 (2018)
8. Daumé III, H.: Frustratingly easy domain adaptation. In: Proceedings of ACL, pp. 256–263 (2007)
9. Dietterich, T.G.: Approximate statistical tests for comparing supervised classification learning algorithms. Neural Comput. **10**(7), 1895–1923 (1998)
10. Escalante, H.J., Solorio, T., Montes-y-Gómez, M.: Local histograms of character n-grams for authorship attribution. In: Proceedings of ACL: HLT, vol. 1, pp. 288–298 (2011)
11. Fernández, A.M., Esuli, A., Sebastiani, F.: Distributional correspondence indexing for cross-lingual and cross-domain sentiment classification. J. Artif. Intell. Res. **55**(1), 131–163 (2016)
12. Franco-Salvador, M., Cruz, F.L., Troyano, J.A., Rosso, P.: Cross-domain polarity classification using a knowledge-enhanced meta-classifier. Knowl. Based Syst. **86**, 46–56 (2015)
13. Giménez-Pérez, R.M., Franco-Salvador, M., Rosso, P.: Single and cross-domain polarity classification using string kernels. In: Proceedings of EACL, pp. 558–563, April 2017
14. Guo, Y., Xiao, M.: Transductive representation learning for cross-lingual text classification. In: Proceedings of ICDM, pp. 888–893, December 2012
15. Huang, X., Rao, Y., Xie, H., Wong, T.L., Wang, F.L.: Cross-domain sentiment classification via topic-related TrAdaBoost. In: Proceedings of AAAI, pp. 4939–4940 (2017)
16. Ifrim, G., Weikum, G.: Transductive learning for text classification using explicit knowledge models. In: Fürnkranz, J., Scheffer, T., Spiliopoulou, M. (eds.) PKDD 2006. LNCS (LNAI), vol. 4213, pp. 223–234. Springer, Heidelberg (2006). https://doi.org/10.1007/11871637_24
17. Ionescu, R.T.: A fast algorithm for local rank distance: application to arabic native language identification. In: Arik, S., Huang, T., Lai, W.K., Liu, Q. (eds.) ICONIP 2015. LNCS, vol. 9490, pp. 390–400. Springer, Cham (2015). https://doi.org/10.1007/978-3-319-26535-3_45
18. Ionescu, R.T., Butnaru, A.: Learning to identify arabic and german dialects using multiple kernels. In: Proceedings of VarDial Workshop of EACL, pp. 200–209 (2017)

19. Ionescu, R.T., Butnaru, A.M.: Improving the results of string kernels in sentiment analysis and Arabic dialect identification by adapting them to your test set. In: Proceedings of EMNLP (2018)

20. Ionescu, R.T., Popescu, M.: Native language identification with string kernels. In: Ionescu, R.T., Popescu, M. (eds.) Knowledge Transfer between Computer Vision and Text Mining. ACVPR, pp. 193–227. Springer, Cham (2016). https://doi.org/10.1007/978-3-319-30367-3_8

21. Ionescu, R.T., Popescu, M.: UnibucKernel: an approach for Arabic dialect identification based on multiple string kernels. In: Proceedings of VarDial Workshop of COLING, pp. 135–144 (2016)

22. Ionescu, R.T., Popescu, M.: Can string kernels pass the test of time in native language identification? In: Proceedings of the 12th Workshop on Innovative Use of NLP for Building Educational Applications, pp. 224–234 (2017)

23. Ionescu, R.T., Popescu, M., Cahill, A.: Can characters reveal your native language? A language-independent approach to native language identification. In: Proceedings of EMNLP, pp. 1363–1373, October 2014

24. Ionescu, R.T., Popescu, M., Cahill, A.: String kernels for native language identification: insights from behind the curtains. Comput. Linguist. **42**(3), 491–525 (2016)

25. Joachims, T.: Transductive inference for text classification using support vector machines. In: Proceedings of ICML, pp. 200–209 (1999)

26. Li, T., Sindhwani, V., Ding, C., Zhang, Y.: Knowledge transformation for cross-domain sentiment classification. In: Proceedings of SIGIR, pp. 716–717 (2009)

27. Lodhi, H., Saunders, C., Shawe-Taylor, J., Cristianini, N., Watkins, C.J.C.H.: Text classification using string kernels. J. Mach. Learn. Res. **2**, 419–444 (2002)

28. Long, M., Wang, J., Ding, G., Pan, S.J., Yu, P.S.: Adaptation regularization: a general framework for transfer learning. IEEE Trans. Knowl. Data Eng. **26**(5), 1076–1089 (2014)

29. Lui, M., Baldwin, T.: Cross-domain feature selection for language identification. In: Proceedings of IJCNLP, pp. 553–561 (2011)

30. Luo, K.H., Deng, Z.H., Yu, H., Wei, L.C.: JEAM: a novel model for cross-domain sentiment classification based on emotion analysis. In: Proceedings of EMNLP, pp. 2503–2508 (2015)

31. Nelakurthi, A.R., Tong, H., Maciejewski, R., Bliss, N., He, J.: User-guided cross-domain sentiment classification. In: Proceedings of SDM (2017)

32. Pan, S.J., Ni, X., Sun, J.T., Yang, Q., Chen, Z.: Cross-domain sentiment classification via spectral feature alignment. In: Proceedings of WWW, pp. 751–760 (2010)

33. Ponomareva, N., Thelwall, M.: Semi-supervised vs. cross-domain graphs for sentiment analysis. In: Proceedings of RANLP, pp. 571–578, September 2013

34. Popescu, M., Grozea, C.: Kernel methods and string kernels for authorship analysis. In: Proceedings of CLEF (Online Working Notes/Labs/Workshop), September 2012

35. Popescu, M., Grozea, C., Ionescu, R.T.: HASKER: an efficient algorithm for string kernels. Application to polarity classification in various languages. In: Proceedings of KES, pp. 1755–1763 (2017)

36. Popescu, M., Ionescu, R.T.: The story of the characters, the DNA and the native language. In: Proceedings of the Eighth Workshop on Innovative Use of NLP for Building Educational Applications, pp. 270–278, June 2013

37. Sener, O., Song, H.O., Saxena, A., Savarese, S.: Learning transferrable representations for unsupervised domain adaptation. In: Proceedings of NIPS, pp. 2110–2118 (2016)
38. Shawe-Taylor, J., Cristianini, N.: Kernel Methods for Pattern Analysis. Cambridge University Press, Cambridge (2004)
39. Shu, L., Latecki, L.J.: Transductive domain adaptation with affinity learning. In: Proceedings of CIKM, pp. 1903–1906. ACM (2015)
40. Sun, B., Feng, J., Saenko, K.: Return of frustratingly easy domain adaptation. In: Proceedings of AAAI, pp. 2058–2065 (2016)
41. Zampieri, M., et al.: Findings of the VarDial evaluation campaign 2017. In: Proceedings of VarDial Workshop of EACL, pp. 1–15 (2017)
42. Zhuang, F., Luo, P., Yin, P., He, Q., Shi, Z.: Concept learning for cross-domain text classification: a general probabilistic framework. In: Proceedings of IJCAI, pp. 1960–1966 (2013)

Overcoming Catastrophic Forgetting with Self-adaptive Identifiers

Fangzhou Xiong[1,2], Zhiyong Liu[1,2,3,4(✉)], and Xu Yang[1,2]

[1] The State Key Lab of Management and Control for Complex Systems,
Institute of Automation, Chinese Academy of Science, Beijing 100190, China
zhiyong.liu@ia.ac.cn
[2] School of Artificial Intelligence, University of Chinese Academy of Sciences
(UCAS), Beijing 100049, China
[3] Centre for Excellence in Brain Science and Intelligence Technology,
Chinese Academy of Sciences, Shanghai 200031, China
[4] Cloud Computing Center, Chinese Academy of Sciences,
DongGuan 523808, GuangDong, China

Abstract. Catastrophic forgetting is a tough issue when the agent faces the sequential multi-task learning scenario without storing previous task information. It gradually becomes an obstacle to achieve artificial general intelligence which is generally believed to behave like a human with continuous learning capability. In this paper, we propose to utilize the variational Bayesian inference method to overcome catastrophic forgetting. By pruning the neural network according to the mean and variance of weights, parameters are vastly reduced, which mitigates the storage problem of double parameters required in variational Bayesian inference. Based on this lightweight version, autoencoders trained on different tasks are employed to self-adaptively match the corresponding task parameters to tackle sequential multi-task learning problem. We show experimentally on several fundamental datasets that the proposed method can perform substantial improvements without catastrophic forgetting over other classic methods especially in the setting where the probability distributions between tasks present more different.

Keywords: Variational Bayesian inference · Pruning · Autoencoder

1 Introduction

As a core component in artificial general intelligence (AGI), lifelong learning [1] has gradually become an essential skill to address a variety of tasks like a human being to learn. Traditional learning methods in machine learning community usually require all task data collected in advance to train the model, while it is difficult in real-world settings: tasks may not be provided simultaneously. After learning new tasks, the agent is prone to forget the old ones without accessing to previous data. This is called catastrophic forgetting in sequential multi-task

© Springer Nature Switzerland AG 2018
L. Cheng et al. (Eds.): ICONIP 2018, LNCS 11303, pp. 497–505, 2018.
https://doi.org/10.1007/978-3-030-04182-3_43

setting where the neural network tends to forget the weights learned in previous tasks after training on subsequent ones [2], which is a basic challenge to realize AGI.

The main dilemma that we face is to make the learned model adapt to new data without forgetting knowledge learned on the previously visited tasks. The majority of classic solutions for this problem suffer from some disadvantages. For example, fine-tune [3] behaves obliviousness property for old tasks since it only uses the optimal settings of old tasks to help initialize and study for the new tasks. As for feature extraction [4], it gives priority to reuse features obtained from the old tasks, which will present sub-optimal results for the new tasks. These methods can not achieve good performances for sequentially given tasks.

Whereas recent advances in machine learning have provided multiple ways to overcome catastrophic forgetting across a variety of domains [5,6]. Fernando et al. [7] propose an ensemble of neural networks to recombine different modules within a single network PathNet to complete different tasks. Serra et al. [8] employ a task-based hard attention mechanism to preserve previous tasks' information without affecting the current task's learning. Besides, [5] is the first to introduce Distillation Networks and fine-tune technique to enable learning without forgetting. According to this basic framework, [9] designs an extra undercomplete autoencoder to preserve the information on which the previous tasks are mainly relying. These methods, more or less, all need to specify which task to perform during the test phase, which is to say additional task identifiers have to be supplied to assign the corresponding parameters for corresponding tasks. For instance, the last fully-connected layers in [5] have to be indicated manually for corresponding tasks.

Fortunately, some other algorithms recently have received much attention, which can perform different tasks without identifiers. Elastic weight consolidation (EWC) [2], an algorithm analogous to synaptic consolidation for artificial neural networks, is proposed to reduce the plasticity of weights that are vital to previously learned tasks. It only studies a set of parameters via Fisher information to finish all tasks, which is an elegant approach in Bayesian framework to overcome catastrophic forgetting. Additionally, Lee et al. [10] present the incremental moment matching (IMM) to incrementally match the moment of Gaussian posterior distribution of different tasks in Bayesian neural networks. Nevertheless, these kind of approaches only achieve good performances on similar tasks. When the difference between task distributions becomes more larger, they are prone to forget more information about previously learned tasks.

In this paper, our motivation is to address catastrophic forgetting problem in a more realistic scenario where the gap between given tasks behaves more larger than traditional settings. For this end, a Bayesian framework is presented to remember the posterior distributions of different tasks, which actually is equivalent to an ensemble of different neural networks [11]. Meanwhile, variational inference method is introduced to approximate the posterior distributions so that each task could be trained to their optimal values. To be more specific, bottom layers near to the input are shared to catch common features between tasks, while top layers near to the output are trained individually for personalized solutions. Rather than integrating those solutions to one representation, we consider to keep their individual representations with additional task identifiers which are realized by autoencoders trained together with tasks so that they are able to help select corresponding top layers, i.e., corresponding tasks. We hence could sequentially conduct different tasks with automatically specifying parameters, which makes the algorithm behave more intelligent. More importantly, when the gap between different tasks presents more larger, the correct top layers for corresponding tasks will be selected more easily through these trained autoencoders. There is even no mismatching between autoencoders and tasks, as long as the reconstruction errors produced by each autoencoder present more different when the task distributions are great of difference. In addition, to make a lightweight network as well as select autoencoders more convenient, a pruning technique based on the mean and variance of network weights is adopted to vastly reduce the number of parameters in the network.

In summary, the main contribution of this paper is to present a self-adaptive identifier for conducting sequentially given tasks without catastrophic forgetting. Specifically, the proposed method achieves better performances when the new task is more different than previously learned tasks. The proposed algorithm can effectively utilize autoencoders with pruned weights to automatically match the corresponding parameters, enabling a more intelligent approach to perform sequential multiple tasks without indicating task identifiers so as to overcome catastrophic forgetting. The rest of the paper is organized as follows: Sect. 2 presents the proposed work. Comparative experimental results are presented and discussed in Sect. 3. Finally, concluding remarks and future work suggestions are outlined in Sect. 4.

2   The Proposed Method

The proposed method is built upon the framework of variational Bayesian inference [11], where a reparameterization trick with unbiased Monte Carlo gradients is utilized to optimize the parametric distribution. The weight w of neural network could be transformed as:

$$w = \mu + \sigma\epsilon, \tag{1}$$

where $\epsilon \sim N(0,1)$ and $\theta = (\mu, \sigma)$ which comprises the mean μ and standard deviation σ of the network. Obviously, it requires double memory and resources than other Bayesian inference methods with neural networks, e.g. EWC algorithm, which only considers to optimize the mean of each weight.

Therefore, it is natural to decrease the overhead of the system by means of pruning the neural network based on these two parameters. More precisely, only the top layers near to the output are pruned which helps reduce the system overhead. And the bottom layers should be complete and frozen once the first task has been learned since the common features between different tasks are constructed based on these layers. At the same time, autoencoders for different

---

Algorithm 1. Pruned network with self-adaptive identifiers

1: Training Phase
2: **for** index $i = 1, 2, \ldots, n$ **do**
3:    Train task i with stochastic gradient descent
4:    Train autoencoder AE_i, and record reconstruction error E_i
5:    **if** $i = 1$ **then**
6:       Frozen bottom layers forever, and record as $N1$
7:    **end if**
8:    Prune top layers, and record as $N2_i$
9: **end for**
10: ──
11: Testing Phase
12: Given a task,
13:    Select task identifier $j = \mathrm{argmin}_i E_i$ $(i = 1, 2, \ldots, n)$
14: Perform task with network $N1 + N2_j$

---

tasks are trained with the same training data. According to the pruned network, autoencoders are utilized to help select corresponding top layers via reconstruction errors. Together with the frozen bottom layers, different networks are generated for corresponding tasks. We hence could automatically conduct different tasks without specifying which parameters to load. The proposed algorithm is presented in Algorithm 1, and we give a detailed explanation from two parts in the upcoming subsections.

2.1 Network Pruning

In this subsection, we introduce the mean and standard deviation of weights to show how to prune the neural network. Specifically speaking, the signal-to-noise ratio (SNR) is employed to address this problem, which is calculated as:

$$\mathrm{SNR} = \frac{|\mu|}{\sigma}. \tag{2}$$

By employing a large SNR which implies a large mean and a small standard deviation, we are supposed to achieve a positive effect in measuring the significance of weights so that the network will be pruned reasonably.

Given the SNR of weights w in descending order and pruning ratio k that describes the number of weights to be removed, we can initialize the mask as:

$$\lambda = \mathrm{SNR}[length(w) \cdot (1 - \mathrm{k})]$$
$$\mathrm{mask} = \mathbb{1}(\mathrm{SNR} \geq \lambda), \tag{3}$$

where λ records the threshold of weights to be pruned. Then the weights can be updated with the mask:

$$w = w \cdot \mathrm{mask}. \tag{4}$$

After pruning the network, the system overhead has been decreased. If we select a high level of pruning ratio, the storage space could be vastly saved.

More importantly, the remaining weights that we care about can perform tasks with almost no performance degradation, which can be testified by subsequent experiments. Since there are multiple sets of parameters to learn for different tasks, it is more valuable to save the system overhead by pruning network.

2.2 Task Indicating

Aimed for sequential multi-task learning, bottom layers near to the input are designed to be shared between tasks. As for top layers pruned and updated by (4), we consider to match them to the corresponding tasks automatically.

Autoencoders here based on reconstruction errors are utilized for task identifiers. Furthermore, autoencoders are trained concurrently with normal task learning, and their optimal weights are usually produced by minimizing the distance between the inputs and their corresponding reconstructions. Concretely, the under-complete autoencoders (i.e., requiring the dimension of the code is smaller than the dimension of the input) are adopted to train for different tasks. In our setting, a two-layer network with a sigmoid activation function in the hidden layer is used for each task.

After training autoencoders, the minimal reconstruction error among all tasks is believed to describe the most potential autoencoder that is capable of representing the task identifier. When conducting different tasks, we could reasonably select the corresponding autoencoder as the task identifier for these pruned top layers. Actually, the automation of selecting task identifier is realized by comparing the reconstruction errors between tasks. If the gap between different task distributions presents more larger to some extent, the correct autoencoder for corresponding task will be picked out more easily, which means the proposed method could handle this situation better to guard against catastrophic forgetting.

3 Experiment

3.1 Experiment Setting

The proposed method is evaluated with a fully-connected neural network [784-800-800-10], whose first layer will be frozen after training the first task. Besides, each layer is activated by a ReLU function except the last one with softmax function used for classification, whose basic architecture is also adopted in [10]. However, we introduce the mean and standard deviation of each weight to implement the variational Bayesian learning. The datasets used in our experiments are summarized in Table 1. The MNIST dataset comprises 28×28 images of handwritten digits. The Shuffled MNIST dataset contains the same images to MNIST but whose input pixels of images are shuffled with a random permutation. The Split MNIST dataset is constructed by splitting MNIST into 0–4 and 5–9 as two tasks respectively.

Table 1. Datasets used in the experiment: name, number of classes and number of train and test samples.

Dataset	Classes	Train	Test
MNIST	10	60000	10000
Shuffled MNIST	10	60000	10000
Split MNIST (0–4)	5	30630	5105
Split MNIST (5–9)	5	29370	4895

3.2   Experimental Results and Discussion

There are two experiments conducted to show the algorithm's performance on sequential multi-task learning. One experiment employs MNIST as the first task and Shuffled MNIST as the second task, and the other invokes Split MNIST to produce two tasks respectively. The proposed method is compared with recently published algorithms EWC and IMM, which can perform tasks in a set of parameters without catastrophic forgetting.

Shuffled MNIST: As Shuffled MNIST is shuffled from MNIST, its distribution is more similar than Split MNIST with respect to MNIST. We first consider the Shuffled MNIST experiment to evaluate the proposed method. Table 2 illustrates the comparable results.

Table 2. Results of Shuffled MNIST experiment.

Method	Hyperparameter	Test accuracy
EWC	$\lambda = 20$	98.20[a]
IMM	$\alpha = 0.33$	98.30[a]
OUR	k = 0.5	98.25

[a] Optimal experimental results are cited from [10]

These approaches achieve similar results on tasks which share similar distributions. For the EWC and IMM, they address catastrophic forgetting problems in a set of parameters. For our approach, the network is pruned using SNR with the probability of 50% except the frozen layers, which results in actually only one set of parameters to learn. Instead of studying two sets of parameters for two tasks separately, this pruning technique contributes to save the storage space except two parameters (mean and standard deviation) which are inevitable overhead in variational Bayesian learning. Overall, we achieve comparable results compared to the EWC and IMM algorithms.

Split MNIST: Now we consider a more difficult situation where the distributions of two tasks present more different. First, the proposed method is evaluated

Table 3. Results of Split MNIST experiment with k = 0.5.

Method	Hyperparameter	Test accuracy
EWC	$\lambda = 20$	52.72
IMM	$\alpha = 0.33$	94.12
OUR	k = 0.5	98.58

under the same pruning ratio, and Table 3 witnesses the results with other two approaches.

In fact, the second task in Split MNIST experiment follows a more different distribution than in the Shuffled MNIST experiment. It could be easily verified in Table 3 by the test accuracy of the EWC which achieves only 52.72% compared to 98.20% in Shuffled MNIST experiment. The IMM obtains a better result since it employs mixed posteriors with transfer learning techniques. The proposed method obtains the best performance due to the individually trained autoencoders whose reconstruction errors indicate corresponding tasks to conduct. If the gap between task distributions presents more larger, the propose method will benefit more from these different distributions contrasted to other approaches. Actually, the autoencoders are employed to study reconstruction errors in different data distributions. When the task behaves more different than previous ones, its reconstruction error is more likely to be different than others. Therefore, the correct autoencoder could be picked out more easily to serve as the task identifier. Apparently, the proposed method achieves the best performance with self-adaptive autoencoders in this experiment.

Next, different pruning ratio values are designed to state the significance of SNR for pruning network. Table 4 presents the relevant results.

Table 4. Results of Split MNIST experiment with different pruning ratio k.

Pruning ratio k	0.25	0.5	0.75	0.95
Test accuracy	98.60	98.58	98.24	97.60

As the pruning ratio increases, more weights in the network are reset to zero values, which brings a more lightweight neural network. It is natural to suppose that the gradually decreased performances will present since the pruning ratio increases. However, according to Table 4, although the 95% weights of network in top layers are replaced with zero, there is still 97.60 average accuracy for the Split MNIST experiment. Obviously, the drawback in this pruned network is that more system overhead is incurred by the standard deviation of each weight in the networks. Nevertheless, it illustrates that the variational Bayesian method with pruning technique is capable of handing different tasks with almost no performance degradation. In addition, the concurrently trained autoencoders assure that the correct matching of different tasks' parameters, which is a significant step for performing sequential multiple tasks without catastrophic forgetting.

4 Conclusions

In this paper, a Bayesian framework is adopted together with variational method to approximate the posterior distributions of different tasks. In order to resolve the catastrophic forgetting problem, we utilize autoencoders as task identifiers to self-adaptively select the corresponding parameters in sequential multi-task scenario. Meanwhile, pruning technique based on SNR values contributes to a more lightweight network, which greatly benefits for the parameters reducing in traditional variational Bayesian learning. The proposed method is testified by two classical experiments, and more different tasks will be expanded to study the issue of overcoming catastrophic forgetting using the task similarity and difference in our future works.

Acknowledgments. This work was supported in part by the National Key Research and Development Plan of China under Grants 2017YFB1300202 and 2016YFC0300801, in part by the NSFC under Grants U1613213, 61627808, 61503383, 61210009, 91648205, 61702516, and 61473236, in part by the MOST under Grant 2015BAK35B00 and Grant 2015BAK35B01, in part by the Guangdong Science and Technology Department under Grant 2016B090910001.

References

1. Thrun, S.: Lifelong learning algorithms. In: Thrun, S., Pratt, L. (eds.) Learning to Learn, pp. 181–209. Springer, Boston (1998). https://doi.org/10.1007/978-1-4615-5529-2_8
2. Kirkpatrick, J., et al.: Overcoming catastrophic forgetting in neural networks. Proc. Natl. Acad. Sci. **114**(13), 3521–3526 (2017)
3. Girshick, R., Donahue, J., Darrell, T., Malik, J.: Rich feature hierarchies for accurate object detection and semantic segmentation. In: Proceedings of the IEEE Conference on Computer Vision and Pattern Recognition, pp. 580–587 (2014)
4. Donahue, J., et al.: DeCAF: a deep convolutional activation feature for generic visual recognition. In: Proceedings of the International Conference on Machine Learning, pp. 647–655 (2014)
5. Li, Z., Hoiem, D.: Learning without forgetting. IEEE Trans. Pattern Anal. Mach. Intell. (2017)
6. Xiong, F., et al.: Guided policy search for sequential multitask learning. IEEE Trans. Syst. Man Cybern. Syst. (2018)
7. Fernando, C., et al.: PathNet: Evolution channels gradient descent in super neural networks. arXiv preprint arXiv:1701.08734 (2017)
8. Serrà, J., Surís, D., Miron, M., Karatzoglou, A.: Overcoming catastrophic forgetting with hard attention to the task. In: Proceedings of the International Conference on Machine Learning (2018)
9. Ep Triki, A.R., Aljundi, R., Blaschko, M., Tuytelaars, T.: Encoder based lifelong learning. In: Proceedings of the IEEE International Conference on Computer Vision, pp. 1320–1328 (2017)

10. Lee, S.W., Kim, J.H., Jun, J., Ha, J.W., Zhang, B.T.: Overcoming catastrophic forgetting by incremental moment matching. In: Advances in Neural Information Processing Systems, pp. 4655–4665 (2017)
11. Blundell, C., Cornebise, J., Kavukcuoglu, K., Wierstra, D.: Weight uncertainty in neural networks. Proc. Int. Conf. Mach. Learn. **37**, 1613–1622 (2015)

Learning, Storing, and Disentangling Correlated Patterns in Neural Networks

Xiaolong Zou[1], Zilong Ji[2], Xiao Liu[2], Tiejun Huang[1], Yuanyuan Mi[3]
Dahui Wang[2,4], and Si Wu[1(✉)]

[1] School of Electronics Engineering and Computer Science,
IDG/McGovern Institute for Brain Research,
Peking University, Beijing 100871, China
siwu@pku.edu.cn
[2] State Key Laboratory of Cognitive Neuroscience & Learning, Beijing Normal
University, Beijing 100875, China
[3] Institute for Neurointelligence, School of Medicine,
Chongqing University, Chongqing, China
[4] School of Systems Science, Beijing Normal University, Beijing 100875, China

Abstract. The brain encodes object relationship using correlated neural
representations. Previous studies have revealed that it is a difficult task
for neural networks to process correlated memory patterns; thus, strategies based on modified unsupervised Hebb rules have been proposed.
Here, we explore a supervised strategy to learn correlated patterns in a
recurrent neural network. We consider that a neural network not only
learns to reconstruct a memory pattern, but also holds the pattern as an
attractor long after the input cue is removed. Adopting backpropagation
through time to train the network, we show that the network is able to
store correlated patterns, and furthermore, when continuously morphed
patterns are presented, the network acquires the structure of a continuous
attractor neural network. By inducing spike frequency adaptation in the
neural dynamics after training, we further demonstrate that the network
has the capacities of anticipative tracking and disentangling superposed
patterns. We hope that this study gives us insight into understanding
how neural systems process correlated representations for objects.

Keywords: Neural network · Correlated patterns
Continuous attractor neural network · Backpropagation through time
Spike frequency adaptation

1 Introduction

In reality, the brain needs to encode not only the identities of objects, e.g.,
whether an animal is cat or dog, but also the relationships between objects, e.g.,
cat and dog are both mammalian but belong to different categories. The experimental data has indicated that the categorical relationships between objects are

---

X. Zhu and Z. Ji—Equal contribution.

L. Cheng et al. (Eds.): ICONIP 2018, LNCS 11303, pp. 506–514, 2018.
https://doi.org/10.1007/978-3-030-04182-3_44

encoded in the correlated neural representations of the objects, in term of that for objects belonging to the same category, their neural representations have larger correlations than those of objects belonging to different categories [1,2]. Interestingly, in an artificial deep neural network (DNN) trained by ImageNet, the correlation between object representations (measured by the overlap between activities in the representation layer, i.e., the one before the read-out layer) also reflects the semantic similarity between the objects [3,4]. To understand how a neural system encodes the relationship between objects, it is important to understand how neural networks learn, store, and retrieve correlated memory patterns.

A large volume of theoretical studies has, however, pointed out that it is not a trivial task for a neural network to process correlated memory patterns [5–8]. These studies, which are based on the classical Hopfield model that constructs neuronal connections according to the unsupervised Hebb rule, have shown that the correlations between patterns deteriorate memory retrieval dramatically, leading to that the Hopfield network is unable to support a large memory capacity [5]. To overcome this flaw, several strategies have been proposed, which include: (1) a novelty-based method [6], which considers that neuronal connections are modified only when a novel pattern is presented (the novelty is defined according to that the pattern can be retrieved or not by the current network structure); (2) a popularity-based method [7], which modifies the Hebb rule by reducing the contributions of those popular neurons that are active in many memory patterns to avoid overwhelmed learning of the connections of those neurons; (3) an orthogonalization-based method [8], which orthogonalizes correlated patterns before applying the Hebb rule. All these methods are based on the unsupervised Hebb learning, and each of them works well in certain circumstances, but their biological plausibility has yet been properly justified.

In the present study, we explore the possibility of using a supervised strategy to train a recurrent neural network to learn correlated patterns. Specifically, we consider a computational task, in which the network not only learns to reconstruct the presented input pattern, but also holds the pattern as persistent activity long after the input is removed. Mathematically, this requires that the network holds the pattern as an attractor of its dynamics. We use backpropagation through time (BPTT) [9] to train the network and demonstrate that the network learns to store a number of highly correlated patterns. Moreover, we find that when a set of continuously morphed patterns are presented, the network acquires the structure of a continuous attractor neural network (CANN), a canonical model for neural information processing [10]. After training, we induce spike frequency adaptation (SFA), a popular negative feedback modulation [11], in the neural dynamics, and find that the network holds interesting computational properties, including anticipative tracking and the capacity of disentangling superposed patterns. We hope this study enriches our knowledge of how neural systems process correlated representations for objects.

2   The Model

As illustrated in Fig. 1, the network model we consider consists of three layers: input, recurrent, and output layers. Neurons in the recurrent layer are connected recurrently, whose dynamics are written as follows,

$$\tau\frac{du_i(t)}{dt} = -u_i(t) + \sum_{j=1}^{N} W_{ij}^{rec} x_j(t) + \sum_{k=1}^{N_{in}} W_{ik}^{in} I_k(t) + b_i + \sigma\xi_i(t), \qquad (1)$$

$$x_i(t) = \tanh\left[u_i(t)\right], \qquad (2)$$

where u_i, for $i = 1, \ldots, N$, is the synaptic input received by neuron i in the recurrent layer and x_i the activity of the neuron. τ is the time constant. N is the number of neurons in the recurrent layer. W_{ij}^{rec} denotes the recurrent connection strength from neuron j to i, W_{ik}^{in} the feedforward connection strength from input component k to neuron i, I_k the external input, and N_{in} the input dimension. b_i is a biased constant input received by neuron i. $\xi_i(t)$ represents Gaussian white noise of zero mean and unit variance, and σ the noise strength.

The neurons in the output layer read-out information by combining the neuronal activities in the recurrent layer linearly, which are written as

$$y_i(t) = \sum_{j=1}^{N} W_{ij}^{out} x_j(t), \qquad (3)$$

where y_i, for $i = 1, \ldots, N_{out}$, represents the activity of neuron i in the output layer, and W_{ij}^{out} the read-out connection strength from neuron j in the recurrent layer to neuron i in the output layer. $N_{out} = N_{in}$ holds in our model.

The Learning Procedure

Our goal is to train the network, such that the network holds the predefined memory patterns as attractors of its dynamics. To achieve this goal, we construct a learning task, which requires that the network output not only reconstructs the given input pattern, but also holds the pattern long after the input cue is removed. Mathematically, these two conditions enforce that the network learns to hold the input pattern as an attractor of its dynamics, which mimics the persistent activity observed in working memory in neural systems [12].

Let us consider that the network learns to memorize M patterns, referred to as $\mathbf{P}^i$, for $i = 1, \ldots, M$, hereafter. Denote T_{sti} the duration of presenting each memory pattern as an external input to the network, T_{seq} the duration of the network holding the memory pattern, and $T_{sti} << T_{seq}$ is imposed. For a memory pattern $\mathbf{P}^i$, the corresponding external input to the network is given by, $\mathbf{I}^i(t) = \mathbf{P}^i + \eta^i(t)$, for $0 < t < T_{sti}$ and otherwise $\mathbf{I}^i(t) = 0$. Here, η^i represent Gaussian white noises, which have the same dimensionality as the input and its elements are independently sampled from Gaussian distributions of zero mean and variances uniformly distributed in the range of $[0, 1]$. These noises are essential for robust learning. Denote the network output to be $\mathbf{Y}^i(t)$,

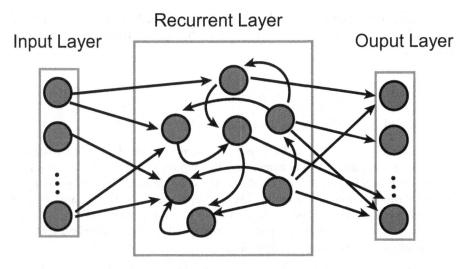

Fig. 1. Network structure. The network contains an input, a recurrent, and an output layers. Neurons in the recurrent layer are connected recurrently.

for $0 < t < T_{seq}$, when the pattern $\mathbf{P}^i$ is presented. The objective function of the learning task is written as,

$$L = \sum_{i=1}^{M} \int_{0}^{T_{seq}} \left[\mathbf{Y}^i(t) - \mathbf{P}^i\right]^2 dt, \quad \mathbf{I}^i(t) \neq 0, \quad \text{for } 0 < t < T_{sti}, \quad (4)$$

where $\mathbf{Y}^i(t) = f\left[\mathbf{I}^i(t)\right]$ represents the nonlinear function implemented by the network. We use the Euler method to discrete the network dynamics and the objective function, and adopt backpropagation through time (BPTT) to optimize the network parameters, including the connection weights $\mathbf{W}^{in}, \mathbf{W}^{rec}, \mathbf{W}^{out}$ and the bias terms $\mathbf{b}$. Before training, $\mathbf{W}^{out}$ are initialized to be zeros, $\mathbf{W}^{in}$ a Gaussian distribution of zero mean and unit variance, $\mathbf{W}^{rec}$ an orthogonal and normalized matrix, and $\mathbf{b}$ zeros.

Spike Frequency Adaptation

After training, we add spike frequency adaptation (SFA) in the neural dynamics to induce extra computational properties of the network. With SFA, Eq. (1) becomes

$$\tau \frac{du_i(t)}{dt} = -u_i(t) + \sum_{j=1}^{N} W_{ij}^{rec} x_j(t) + \sum_{k=1}^{N_{in}} W_{ik}^{in} I_k(t) + b_i + \sigma \xi_i(t) - v_i(t), \quad (5)$$

$$\tau_v \frac{dv_i(t)}{dt} = -v_i(t) + m u_i(t), \quad (6)$$

where $v_i(t)$ is the current induced by SFA, a negative feedback modulation widely observed in neural systems [11], whose effect is to suppress neuronal responses when they are too strong. τ_v is the time constant of SFA, and $\tau_v \gg \tau$ implies that SFA is a slow process compared to neural firing. The parameter m controls the amplitude of SFA.

3 Results

3.1 Learning to Memorize Correlated Patterns

To demonstrate that our model is able to learn correlated memory patterns, we chose handwriting digit numbers as the inputs (see Fig. 2). These image patterns are highly correlated (overlapped) and hence can not be memorized by the conventional Hopfield model. We test three unsupervised strategies, and found that the orthogonalization-based method accomplished the task, but the other two, novelty-based and popularity-based, failed. Our supervised strategy accomplished the task successfully (Fig. 2).

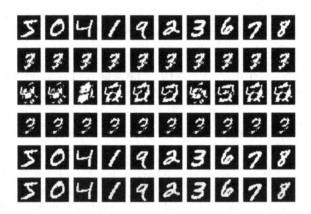

Fig. 2. Retrieval performances of different learning methods. From top to bottom: the original ten digit numbers from the dataset of Mnist, the retrieval of the conventional Hopfield model, the retrieval of the popularity-based method, the retrieval of the novelty-based method, the retrieval of the orthogonalization-based method, and the retrieval of our approach. Parameters: $\tau = 5, T_{sti} = 3, T_{seq} = 30, \sigma = 0.01, N = 200, N_{in} = N_{out} = 784$.

3.2 Disentangling Superposed Memory Patterns

After training the network to memorize ten digit numbers, we add SFA in the neural dynamics (see Eq. (5, 6)). In a real neural system, this corresponds to that during learning, SFA is either frozen or too slow and can be ignored compared to the fast synaptic plasticity. We check the network responses when an image of superposed two digit numbers is presented. As illustrated in Fig. 3, the network outputs the two digit numbers alternatively over time. The underlying mechanism is that: (1) through training, the network has learned to memorize the two digit numbers as its attractors; (2) when the ambiguous image is presented, the network receives two competing input cues and evolves into one of two attractors depending on biases; (3) because of the negative feedback from SFA, the network state becomes unstable gradually, and under the competition from the other cue, the network state moves into the other attractor; (4) this progress goes on, and the network state oscillates between two attractors. Our study suggests that a neural network can use negative feedback such as SFA to disentangle correlated patterns.

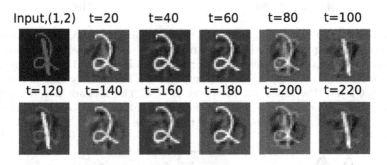

Fig. 3. Disentangling superposed correlated memory patterns. The network was first trained to memorize ten digit numbers as in Fig. 2. For convenience, we use colors to differentiate different digit numbers, but in practice gray images are used. The input is the superposed images of 1 and 2. The evolving of the network output over time is presented. Parameters are: $m = 3.4, \tau_v = 30$. Other parameters are the same as in Fig. 2.

3.3 Learning a CANN

We show that when continuously morphed patterns are memorized, the network acquires the structure of a CANN. Figure 4A displays the set of continuously morphed gaussian bumps to be memorized by the network. After training, the network leans to store each of them as attractors, in terms of that: (1) the network evolves to one-to-one mapped stationary state when each of gaussian bumps is presented (Fig. 4A); and (2) the network remains to be at the active state long after the input is removed (Fig. 4B).

Properties of the Network
We check that the learned network indeed has the good computational properties of CANNs. Figure 5 shows that the network has the properties: (1) mental rotation [13,14], the network exhibits the mental rotation behavior when the external inputs abrupt change (Fig. 5A); (2) travelling wave [15], the network holds a self-sustained travelling wave when SFA is strong enough (Fig. 5B); (3) anticipation tracking [15], the network is able to track a moving input anticipatively if SFA is strong enough (Fig. 5C).

512   X. Zou et al.

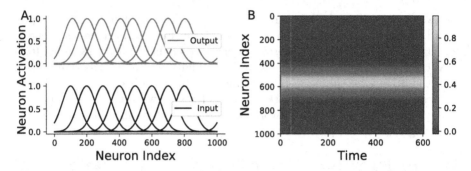

Fig. 4. Learning continuously morphed patterns to form a CANN. Totally 1000 morphed gaussian-bump-like patterns are constructed. (A) Lower panel: examples of input patterns; upper panel: examples of the learned network output. There are one-to-one correspondence between the inputs and outputs. (B) The activity map of neurons in the output layer. The input is removed at $T = 3$. The network state is sustained after the input is removed, indicating the existence of an attractor. Parameters are: $N_{in} = N_{out} = 1000$. Other parameters are the same as in Fig. 2.

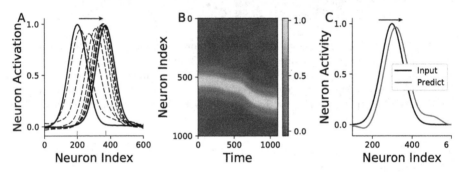

Fig. 5. Properties of the learned network. (A) Mental rotation. The network state is initially at pattern index 200. Under the drive of an external input at pattern index 360, the network state smoothly rotates from the initial to the target positions. (B) Travelling wave. The activity map of the output layer in the travelling state. $m = 0.4$. (C) Anticipate tracking. The black curve is the external moving input, and the red curve the network state which leads the moving input. $m = 0.4$. Parameters are: $\tau_v = 60, N_{in} = N_{out} = 1000$. Other parameters are the same as in Fig. 2. (Color figure online)

4   Conclusion

In the present study, we have investigated a supervised strategy to learn correlated patterns in neural networks, which are different from the unsupervised ones proposed in the literature. The key idea of our method is that we enforce the network to learn the memory patterns as its attractors. To achieve this goal, we require that the network not only learns to reconstruct the given input pattern, but also holds the pattern as persistent activity long after the input cue is removed. Using both synthetic and real data, we show that after training, the network is able to store highly correlated patterns and can also acquire the structure of a CANN if continuously morphed patterns are presented. Moreover, we induce SFA in the neural dynamics after training, and demonstrating that the network holds interesting computational properties, including anticipative tracking and the capacity of disentangling superposed patterns. We hope this study, as to a complement to other unsupervised approaches, enrich our knowledge of how neural systems process correlated representations for objects.

Acknowledgments. This work was supported by National Key Basic Research Program of China (2014CB846101), BMSTC (Beijing municipal science and technology commission) under grant No: Z161100000216143 (SW), Z171100000117007 (DHW & YYM); the National Natural Science Foundation of China (No: 31771146, 11734004, YYM), Beijing Nova Program (No: Z181100006218118, YYM).

References

1. Huth, A.G., Nishimoto, S., Vu, A.T., et al.: A continuous semantic space describes the representation of thousands of object and action categories across the human brain. Neuron **76**(6), 1210–1224 (2012)
2. Chang, L., Tsao, D.Y.: The code for facial identity in the primate brain. Cell **169**(6), 1013–1028 (2017)
3. Bengio, Y., Courville, A., Vincent, P.: Representation learning: a review and new perspectives. IEEE Trans. Pattern Anal. Mach. Intell. **35**(8), 1798–1828 (2013)
4. Yosinski, J., Clune, J., Nguyen, A., et al.: Understanding neural networks through deep visualization. Computer Science (2015)
5. Hertz, J., Krogh, A., Palmer, R.G.: Introduction to the Theory of Neural Computation. The Advanced Book Program (1991)
6. Blumenfeld, B., Preminger, S., Sagi, D., et al.: Dynamics of memory representations in networks with novelty-facilitated synaptic plasticity. Neuron **52**(2), 383–394 (2006)
7. Kropff, E., Treves, A.: Uninformative memories will prevail: the storage of correlated representations and its consequences. HFSP J. **1**(4), 249–262 (2007)
8. Zou, X., Ji, Z., Liu, X., Mi, Y., Wong, K.Y.M., Wu, S.: Learning a continuous attractor neural network from real images. In: Liu, D., Xie, S., Li, Y., Zhao, D., El-Alfy, E.S. (eds.) Neural Information Processing. ICONIP 2017. LNCS, vol. 10637. Springer, Cham (2017). https://doi.org/10.1007/978-3-319-70093-9_66
9. Pascanu, R., Mikolov, T., Bengio, Y.: On the difficulty of training recurrent neural networks. In: International Conference on Machine Learning. JMLR.org, III-1310 (2013)
10. Wu, S., Wong, K.Y.M., Fung, C.C.A., et al.: Continuous attractor neural networks: candidate of a canonical model for neural information representation. F1000Research, 5 (2016)
11. Gutkin, B., Zeldenrust, F.: Spike frequency adaptation. Scholarpedia **9**(2), 30643 (2014)
12. Curtis, C.E., D'Esposito, M., Curtis, C.E.: Persistent activity in the prefrontal cortex during working memory. Trends Cognit. Sci. **7**(9), 415–423 (2003)
13. Shepard, R.N., Metzler, J.: Mental rotation of three-dimensional objects. Science **171**(3972), 701–703(1971)

14. Fung, C.C.A., Wong, K.Y.M., Wu, S.: A moving bump in a continuous manifold: a comprehensive study of the tracking dynamics of continuous attractor neural networks. Neural Comput. **22**(3), 752 (2010)
15. Mi, Y., Fung, C.C.A., Wong, K.Y.M., et al.: Spike frequency adaptation implements anticipative tracking in continuous attractor neural networks. In: Advances in Neural Information Processing Systems, vol. 1, no. 3, pp. 505–513 (2014)

MultNet: An Efficient Network Representation Learning for Large-Scale Social Relation Extraction

Jun Yuan[1,2,3], Neng Gao[2,3], Lei Wang[2,3], and Zeyi Liu[3(✉)]

[1] School of Cyber Security, University of Chinese Academy of Sciences,
Beijing, China
[2] Data Assurance and Communications Security Center,
Chinese Academy of Sciences, Beijing, China
[3] Institute of Information Engineering, Chinese Academy of Sciences, Beijing, China
{yuanjun,gaoneng,wanglei,liuzeyi}@iie.ac.cn

Abstract. Network representation learning (NRL), which has become
an focus of current research, learns low-dimensional vertex representations to capture network information. However, conventional NRL models either largely neglect the rich semantic information on edges and fail
to extract good features of relations, or employ complex models that have
rather high space and time complexities. In this work, we present an efficient NRL model, MultNet, for Social Relation Extraction (SRE) task,
which evaluates the ability of NRL models on modeling the relationships
between vertices. We conduct extensive experiments on several public
data sets and experiments on SRE indicate that MultNet outperforms
other baseline models significantly.

Keywords: Network representation learning · Embedding
Social Relation Extraction

1 Introduction

Nowadays, networks are ubiquitous and the way to represent networks is crucial
for many downstream applications, such as vertex classification [6], clustering
[10] and information retrieval [19]. Network representation learning (NRL) is an
effective method to learn useful network representation, which embeds networks
into low-dimensional vector spaces. We denote embedding vector with the same
letters in boldface in this paper.

However, conventional network representation learning (NRL) models simply
regard each edge as a continuous or binary value when learning low-dimensional
vertex representations. In fact, there exists rich semantic information on edges.
For example, an edge between two authors in co-author network always indicates
common academic interests. Current models largely neglect these information
and fail to extract good features of relations.

L. Cheng et al. (Eds.): ICONIP 2018, LNCS 11303, pp. 515–524, 2018.
https://doi.org/10.1007/978-3-030-04182-3_45

Social Relation Extraction (SRE) task is proposed to evaluate the ability of NRL models on modeling relationships between vertices. Tu et al. [16] show that we can use key phrases extracted from the interactive text to represent social relations. Meanwhile, there are often multiple relational labels to demonstrate the complicated relation between two vertices. Formally, SRE is defined as follows:

Suppose there is a network, represented by $G = (V, E)$, where $E \subseteq (V \times V)$ is the edge set between vertices and V is the vertex set. Noted that edges in E are partially labeled, which are denoted as E_L. Specifically, $\forall e \in E_L$, the label set of e is denoted as $l = \{t^{(1)}, t^{(2)}, ...\}$, where every label $t^{(i)} \in l$ comes from T, a fixed label vocabulary. SRE aims to predict the labels on edges over unlabeled edges in E_U, where $E_U = E - E_L$ represents the unlabeled edge set.

SRE cannot be well tackled by conventional NRL models. Because, as mentioned in [16], less work consider the rich semantic information of edges and make elaborate predictions of relations on edges. TransNet [16] is a promising method proposed recently for SRE, achieving state-of-the-art predictive performance. However, relatively high time and space complexity prevent it from applying on large scale networks.

In this paper, we attempt to propose a novel method that can efficiently adopt rich information on edges and be applied on large scale networks. The basic idea is that we regard the social network G as a multi-relational graph and every label t in T that represents a type of relationship between vertices. Based on it, we propose a novel NRL model MultNet, which embeds both vertices and labels. MultNet has much lower time and space complexity than TransNet. Thus, our model is practical to be applied on large scale social networks. We make extensive experiments to evaluate MultNet, and results show that MultNet outperforms all baselines.

2   Related Works

Network representation learning or NRL can be traced back to the feature engineering for network analysis [9,13] and graph mining tasks [20]. Despite the success of these NRL models, they all employ shallow models. However, shallow models are difficult to effectively capture the highly non-linear structure in the networks [17]. Therefore, various deep NRL models have been proposed recently. Some works attempt to learn representations from local network structure, such as Deepwalk [8], LINE [11], node2vec [4] and SDNE [17]. Some works intend to learn the global structure and community patterns, such as CNRL [2] and MNMF [18]. Moreover, some works try to incorporate heterogeneous information into NRL. For example, TADW [22] and CANE [14] introduces text information into NRL, and MMDW [15] and DDRW [5] incorporate labelling information into NRL.

As far as we know, few researches consider to use the rich semantic information of edges and make elaborate predictions of relations on edges. TransNet [16] is a promising method for SRE, achieving state-of-the-art predictive performance. TransNet uses two vectors for every vertex, corresponding to its tail representation $\mathbf{u}_t$ and head representation $\mathbf{u}_h$. TransNet uses one-hot vector to represent label set between vertices and runs an auto-encoder to embed the one-hot vector into low-dimensional vector l. Then TransNet regards l as translation between the heads and tails and wants $\mathbf{u}_h + \mathbf{l} \approx \mathbf{v}_t$ when (u, v, l) holds. However, relatively high time and space complexity prevent it from applying on large scale networks. We will make detailed complexity analysis in Subsect. 3.3.

SRE is similar to relation extraction task in *Knowledge Graphs* (KGs). Knowledge representation learning (KRL) such as TransE [1] are the most widely used methods for relation extraction in KGs. In this paper, we also use the TransE as a baseline model. The difference between these two tasks is that there are always no well pre-defined relation types in SRE. Besides, the ratios of multi-labeled edges on SRE data sets are much larger than in KGs [16].

3 MultNet

In this paper, we focus on the problem of utilizing the rich semantic information on edges efficiently. We use a set of labels to represent the rich information on edges. Then we regard the social network G as a multi-relational graph and every label t in label vocabulary T that represents a type of relationship between vertices.

In this section, we describe the MultNet Architecture, along with the method for model training. Besides, we make complexity analysis on MultNet compared with some baselines.

3.1 Architecture

For each training instance (u, v, l) in data sets, we firstly split it into several triplets, *i.e.*, $(u, v, t^{(i)})$, and form new data sets, where $l = \{t^{(1)}, t^{(2)}, ...\}$, $t^{(i)} \in T$ and $u, v \in V$. Note that, unlike TransNet, the vector that represents a given vertex is the same when the vertex appears as the head or as the tail of a triplet.

The Architecture of MultNet is shown at Fig. 1. Given a training set S of triplets (u, v, t), our model learns embedding vectors of the vertices and the labels. The embeddings are denoted with the same letters in boldface characters. Since the edges in social network are always undirected, we assume every label is symmetric. That is to say, (u, v, t) is equal to (v, u, t). Therefore, we use weighted element-wise dot product (multiplicative operation) [21] to compose vertex and label embedding vectors and propose a novel model MultNet. Hence the score function are defined as follow:

$$f_t(u, v) = <\mathbf{u}, \mathbf{v}, \mathbf{t}> = \sum_{i=1}^{m} u_i v_i t_i, \tag{1}$$

where $\mathbf{u}, \mathbf{v}, \mathbf{t} \in \mathbb{R}^m$. The score is high if (u, v, t) holds, and low otherwise.

For general evaluation, we input the score of a triplet into a sigmoid function to represent the probability of (u, v, t) existing in G. Because the sigmoid function sets output between 0 and 1.

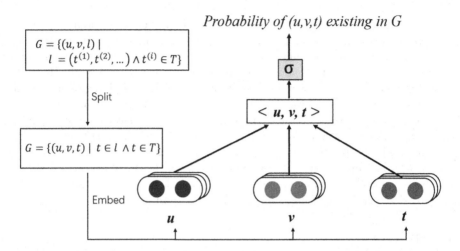

Fig. 1. The Architecture of MultNet. We split every labeled edges in social network G into several triplets and every label t represents a type of relationship. MultNet embeds both vertices and labels and uses multiplicative operation to get probability of triplets existing in G.

3.2 Training

To train MultNet, we minimize a common used margin-based ranking criterion over the training set

$$\mathcal{L} = \sum_{(u,v,t)\in S} \sum_{(u',v',t')\in S'} [f_t(u',v') - f_t(u,v) + \gamma]_+ + \lambda||\theta||_2^2, \qquad (2)$$

where $[x]_+ \triangleq \max(0, x)$, $\gamma > 0$ is a margin hyper-parameter, λ is the weight of regularization, Θ is the set of parameters, the corrupted triplet set S' is composed of training triplets with the head or tail or label replaced randomly. That is

$$S' = \{(u',v,t)|u' \in V\} \cup \{(u,v',t)|v' \in V\} \cup \{(u,v,t')|t' \in T\} \qquad (3)$$

This loss function is used to encourage discrimination between training triplets and corrupted triplets by favoring higher scores for training triplets than for corrupted triplets.

All vertex and label vectors are first initialized from a uniform distribution $U[-\frac{6}{\sqrt{m}}, \frac{6}{\sqrt{m}}]$ [3]. The process of minimizing the above loss function is carried out using stochastic gradient descent (SGD) with constant learning rate. The training process is stopped based on the performance of our method on the validation set.

3.3 Complexity Analysis

As shown in Table 1, we compare the number of parameters and computational complexity of various baselines with our model. In this table, we denote m as the embedding vector dimension, n as the node number of Huffman trees. From Table 1, we observe that MultNet has comparative space and time complexity as TransE, which is much lower than TransNet and Deepwalk. Moreover, considering the high number of edges, the time and space complexity of TransNet are both much higher than Deepwalk. Due to the low time and space complexity, MultNet can be applied on large-scale social networks easier than TransNet.

Table 1. Complexities (the number of parameters, time complexity) of models.

Model	#Parameters	Time complexity
Deepwalk	$O(m(n+1)|V|)$	$O(|V|\log|V|)$
TransE	$O((|V|+|E|)m)$	$O(|V|)$
TransNet	$O((|V|+|T||E|)m)$	$O(|V|+|T||E|)$
MultNet (this paper)	$O((|V|+|E|)m)$	$O(|V|)$

4 Experiment

To empirically evaluate the effectiveness of MultNet on modeling relationships between vertices, we compare our proposed model with several baselines on SRE task on data sets that provided by [16].

4.1 Data Sets

Tu et al. [16] automatically constructed three social network data sets from ArnetMiner. **ArnetMiner** [12] is an online academic website providing search and mining services for researcher social networks. In ArnetMiner, authors collaborate with different researchers on different topics, and the co-authored papers always reflect the elaborate relationships between them.

Table 2. Statistics of data sets. (ML indicates multi-label edges.)

Datasets	Vertices	Edges	Train	Test	Valid	Labels	ML proportion (%)
Arnet-S	187,939	1,619,278	1,579,278	20,000	20,000	100	42.46
Arnet-M	268,037	2,747,386	2,147,386	30,000	30,000	500	63.74
Arnet-L	945,589	5,056,050	3,856,050	60,000	60,000	500	61.68

Tu et al. constructed the co-authored network with labeled edges in the following steps. Firstly, they build the label vocabulary by collecting all the research interest phrases from the author profiles. Secondly, for each co-author relationship, they filtered out the labels in vocabulary in the abstracts of coauthored papers and view them as the ground truth labels of the edge. Thirdly, as all edges in co-author networks are undirected, they replaced every edge with two directed edges in opposite directions.

In addition, Tu et al. constructed three data sets in different scales, denoted as Arnet-S, Arnet-M (medium) and Arnet-L to better investigate the characteristics of different models. The details are shown in Table 2.

4.2 Baselines

We compare MultNet with TransNet [16] and the typical knowledge embedding model, TransE [1]. We use the codes provided by [16] to evaluate these two models on various data sets.

We also employ the following conventional models as baselines. For these NRL models, we follow the protocol of [16], treating SRE task as a multi-label classification task. That is, we concatenate the tail and head vertexes embeddings as the feature vector. Then we adopt one-vs-rest (ovr) logistic regression to train a multi-label classifier and implement it by Tensorflow.

Deepwalk [8] uses random walks over networks to yield random walk sequences. Based on the sampled sequences, Deepwalk employs Skip-Gram [7] model to generate network representation.

LINE [11] defines the first-order and second-order proximities of networks separately, attempting to optimizes the conditional and joint probabilities of edges in networks.

Node2Vec [4] extends Deepwalk with a biased random walk trick and explores the neighborhood structure more efficiently.

4.3 Results and Analysis

SRE aims to predict the missing label t for a triplet (u, v, t). In this task, the model is asked to rank a set of candidate labels from label vocabulary, instead of giving one best result. For each test triplet (u, v, t), we replace the label by all possible candidates and rank these label in descending order of probabilities calculated by $\sigma(f_t(u, v))$.

We report two measures as our evaluation metrics: the average rank of all correct labels (*Mean Rank*) and the proportion of correct labels ranked in top K (*Hits@K*). A good model should achieve lower *Mean Rank* and higher *Hits@K*. In fact, corrupted triplets, which is generated in the aforementioned process of removal and replacement, may also exist in G. These triplets should be considered as correct. Hence, we follow the evaluation protocol in [1] and remove the corrupted triplets included in train, valid and test sets before ranking. In this paper, we call the evaluation setting without removing operation as "Raw" and the other as "Filter".

In training, we select the margin γ among $\{0.5, 1, 2, 4\}$, the dimension of representation vectors m among $\{20, 50, 100, 200\}$, and the weight of regularization λ among $\{0, 0.0001, 0.0005, 0.001, 0.01\}$. Besides, we set the learning rate of SGD to 0.1 and the mini-batch size to 480. The best configurations obtained by the validation set are: on Arnet-S, $\gamma = 1$, $m = 100$ and $\lambda = 0.0005$; on Arnet-M, $\gamma = 0.5$, $m = 200$ and $\lambda = 0.0005$; on Arnet-L, $\gamma = 0.5$, $m = 200$ and $\lambda = 0.0005$. For all data sets, we traverse all training triplets for at most 100 iterations.

Table 3. Social Relation Extraction results on Arnet-S. ($\times$ 100 for *Hits@k*.)

Metric	Mean Rank		Hits@1		Hits@5		Hits@10	
	Raw	Filter	Raw	Filter	Raw	Filter	Raw	Filter
DeepWalk	20.11	18.87	12.78	18.60	35.80	39.42	50.10	51.99
LINE	24.76	23.14	11.20	14.87	31.56	32.86	43.96	45.75
Node2vec	19.55	18.56	12.84	18.22	35.60	39.07	49.56	51.66
TransE	6.21	5.29	38.15	54.89	77.66	81.77	86.34	88.46
TransNet	5.74	4.86	45.82	75.65	84.23	89.53	90.17	91.41
MultNet	**5.46**	**4.23**	**47.61**	**79.36**	**86.87**	**90.50**	**91.89**	**92.99**

Table 4. Social Relation Extraction results on Arnet-M. ($\times$ 100 for *Hits@k*.)

Metric	Mean Rank		Hits@1		Hits@5		Hits@10	
	Raw	Filter	Raw	Filter	Raw	Filter	Raw	Filter
DeepWalk	84.12	79.56	7.21	10.89	19.65	21.44	28.34	30.11
LINE	96.10	93.00	5.47	7.47	16.34	17.86	23.96	25.75
Node2vec	81.54	79.85	6.84	10.92	20.13	22.77	27.55	29.96
TransE	27.84	24.33	16.81	30.87	47.56	53.34	60.35	65.23
TransNet	25.94	23.56	27.32	57.69	65.43	73.53	76.23	78.11
MultNet	**24.40**	**21.20**	**27.50**	**59.12**	**66.43**	**74.22**	**78.12**	**80.19**

Tables 3, 4 and 5 shows the SRE evaluation results. From these tables we observe that:

1. MultNet achieves consistent improvement than all baselines on all data sets, while has much lower time and space complexity than TransNet. It indicates the effectiveness of MultNet on modeling and predicting relationships between vertices.
2. Conventional NRL models have poor performance on SRE task, because they all neglect rich semantic information over edges. It indicates the importance of considering the elaborate edge information.
3. MultNet has almost the same time and space complexity as TransE, while outperforms TransE significantly. It indicates the rationality of multiplicative operation on modeling relation between vertices.

Table 5. Social Relation Extraction results on Arnet-L. ($\times$ 100 for *Hits@k.*)

Metric	Mean Rank		Hits@1		Hits@5		Hits@10	
	Raw	Filter	Raw	Filter	Raw	Filter	Raw	Filter
DeepWalk	103.40	101.77	5.21	7.10	15.80	16.92	22.97	23.99
LINE	96.70	93.51	4.76	10.84	20.53	22.48	27.88	30.45
Node2vec	81.45	80.01	7.03	10.20	19.67	22.47	29.46	31.11
TransE	26.44	24.03	18.12	29.79	47.79	53.87	61.03	64.14
TransNet	25.88	22.96	27.64	58.45	66.13	74.35	75.71	79.48
MultNet	**25.07**	**21.86**	**28.71**	**59.86**	**68.12**	**75.50**	**76.09**	**81.00**

4. MultNet only has a small drop when the number of labels and percentage of multi-label edges turn larger, *e.g.* from 90% to 80% on *Hits@10*. This demonstrates stability of MultNet and the good ability to handle multi-label edges.

4.4 Parameter Sensitivity

For in-depth understanding our method, we investigate the parameter sensitivities on Arnet-S. The weight of regularization λ is crucial hyper-parameter in MultNet to prevent overfitting. When the optimal γ has been determined, we show the filtered *Hits@10* results in Fig. 2. From Fig. 2, we observe that the performance of MultNet rises quickly and then becomes stable. More specifically, MultNet can outperform TransE within 15 iterations and achieve stable performance around 0.9 at *Hits@10*. These results indicates the validity that we regard social network as multi-relational graph. λ should be around 0.001, when

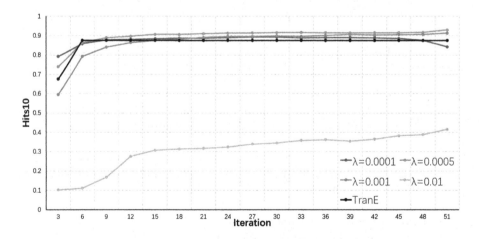

Fig. 2. Parameter sensitivity.

λ is relatively large, like 0.01, MultNet will be underfitting. Hence it should be careful about choosing λ when implementing MultNet.

5   Conclusion and Future Work

We propose a novel network representation learning model MultNet for social relation extraction task. MultNet reduces time complexity by embedding both vertices and labels of edges into low-dimensional continues vector space to avoid deep auto-encoder of TransNet. Our model uses multiplicative operation to model the symmetric relations between vertices. In addition, unlike TransNet, the vector that represents a given vertex is the same when the vertex appears as the head or as the tail of a triplet. As a result, MultNet has less time and space complexity and more flexibility than TransNet. Extensive experiments show that MultNet outperforms all baselines.

Following research directions will be explored in the future: (1) We will explore modeling heterogeneous networks, which always have various types of vertices. (2) We will explore different regularization methods to enhance Mult-Net. (3) We will explore some novel optimization methods to boost MulNet.

Acknowledgement. This work is supported by the National Key Research and Development Program of China (No. 2016YFB0800504), and National Natural Science Foundation of China (No. U163620068).

References

1. Bordes, A., Usunier, N., Weston, J., Yakhnenko, O.: Translating embeddings for modeling multi-relational data. In: NIPS, pp. 2787–2795 (2013)
2. Zeng, X., Liu, Z., Tu, C., Wang, H., Sun, M.: Community-enhanced network representation learning for network analysis. arXiv preprint arXiv:1611.06645 (2016)
3. Glorot, X., Bengio, Y.: Understanding the difficulty of training deep feedforward neural networks. In: AISTATS (2010)
4. Grover, A., Leskovec, J.: node2vec: scalable feature learning for networks. In: SIGKDD, pp. 855–864 (2016)
5. Li, J., Zhu, J., Zhang, B.: Discriminative deep random walk for network classification. In: ACL, pp. 1004–1013 (2016)
6. Lindamood, J., Heatherly, R., Kantarcioglu, M., Thuraisingham, B.: Inferring private information using social network data. In: WWW, pp. 1145–1146 (2013)
7. Mikolov, T., Sutskever, I., Chen, K., Corrado, G.S., Dean, J.: Distributed representations of words and phrases and their compositionality. In: NIPS, pp. 3111–3119 (2013)
8. Perozzi, B., Al-Rfou, R., Skiena, S.: DeepWalk: online learning of social representations. In: SIGKDD, pp. 701–710 (2014)
9. Roweis, S.T., Saul, L.K.: Nonlinear dimensionality reduction by locally linear embedding. Science **290**(5500), 2323–2326 (2000)
10. Shepitsen, A., Gemmell, J., Mobasher, B., Burke, R.: Personalized recommendation in social tagging systems using hierarchical clustering. In: RecSys, pp. 259–266 (2008)

11. Tang, J., Qu, M., Wang, M., Zhang, M., Yan, J., Mei, Q.: LINE: large-scale information network embedding. In: WWW, pp. 1067–1077 (2015)
12. Tang, J., Zhang, J., Yao, L., Li, J., Zhang, L., Su, Z.: ArnetMiner: extraction and mining of academic social networks. In: SIGKDD, pp. 990–998 (2008)
13. Tenenbaum, J.B., De Silva, V., Langford, J.C.: A global geometric framework for nonlinear dimensionality reduction. Science **290**(5500), 2319–2323 (2000)
14. Tu, C., Liu, H., Liu, Z., Sun, M.: CANE: context-aware network embedding for relation modeling. In: ACL, pp. 1722–1731 (2017)
15. Tu, C., Zhang, W., Liu, Z., Sun, M.: Max-Margin DeepWalk: discriminative learning of network representation. In: IJCAI, pp. 3889–3895 (2016)
16. Tu, C., Zhang, Z., Liu, Z., Sun, M.: TransNet: translation-based network representation learning for social relation extraction. In: International Joint Conference on Artificial Intelligence, pp. 2864–2870 (2017)
17. Wang, D., Cui, P., Zhu, W.: Structural deep network embedding. In: SIGKDD, pp. 1225–1234 (2016)
18. Wang, X., Cui, P., Wang, J., Pei, J., Zhu, W., Yang, S.: Community preserving network embedding. In: AAAI (2017)
19. Weiss, Y., Torralba, A., Fergus, R.: Spectral hashing. In: NIPS, pp. 1753–1760 (2008)
20. Yan, S., Xu, D., Zhang, B., Zhang, H.J., Yang, Q., Lin, S.: Graph embedding and extensions: a general framework for dimensionality reduction. IEEE Trans. Pattern Anal. Mach. Intell. **29**(1), 40 (2007)
21. Yang, B., Yih, W., He, X., Gao, J., Deng, L.: Embedding entities and relations for learning and inference in knowledge bases. In: ICLR (2015)
22. Yang, C., Liu, Z., Zhao, D., Sun, M., Chang, E.Y.: Network representation learning with rich text information. In: IJCAI, pp. 2111–2117 (2015)

Geometrical Formulation
of the Nonnegative Matrix Factorization

Shotaro Akaho[1]([✉]) [ID], Hideitsu Hino[2][ID], Neneka Nara[3], and Noboru Murata[3][ID]

[1] National Institute of Advanced Industrial Science and Technology,
Tsukuba, Ibaraki 305-8568, Japan
s.akaho@aist.go.jp
[2] The Institute of Statistical Mathematics, Tachikawa, Tokyo 190-8562, Japan
[3] Waseda University, Shinjuku, Tokyo 169-0072, Japan

Abstract. Nonnegative matrix factorization (NMF) has many applications as a tool for dimension reduction. In this paper, we reformulate the NMF from an information geometrical viewpoint. We show that a conventional optimization criterion is not geometrically natural, thus we propose to use more natural criterion. By this formulation, we can apply a geometrical algorithm based on the Pythagorean theorem. We also show the algorithm can improve the existing algorithm through numerical experiments.

Keywords: Information geometry · Dimension reduction
Topic model

1 Introduction

Nonnegative matrix factorization (NMF) [15] is a dimension reduction method in which data matrix X is approximated by a product of low rank matrices D and C, and all components of X, D, C are nonnegative. The NMF has been applied to many application areas such as computer vision, signal processing, and recommender systems [5,6,10,19]. The contribution of this paper is to reformulate NMF from an information geometrical viewpoint [2] instead of a conventional formulation. Information geometry has provided a unified interpretation for various kinds of machine learning algorithms [3]. In the case of NMF, we show that the problem is to find a projection onto a flat subspace of probability vectors. Based on the geometrical understandings, we propose a new geometrical projection algorithm. We also show the effectiveness of the algorithm through numerical experiments.

2 NMF and Topic Model

Suppose X is a given $d \times n$ matrix, the goal of NMF is to find a low rank approximating decomposition $X \simeq DC$ that minimizes some cost function, where D

Supported by JSPS KAKENHI Grant Number 16K16108, 17H01793.

and C are $d \times k$ and $k \times n$ matrices respectively and all components of D and C are nonnegative. In this paper, we assume $k(< \min(d, n))$ is fixed.

In order to deal with NMF within an information geometrical framework, first we consider the normalization of each column of X so that the sum of its components is 1, which makes it possible to regard the column as a probability vector. Let us introduce a column-wise normalization operator Π,

$$Q = \Pi[X], \quad Q_{ij} = \frac{X_{ij}}{\sum_{i'} X_{i'j}}. \tag{1}$$

In the NMF, there is a freedom of scale, we can assume D is normalized without loss of generality, i.e., $\Pi(D) = D$. In that case, if $X = DC$, it is easy to show that it holds [9]

$$\Pi[X] = \Pi[D]\Pi[C]. \tag{2}$$

Now we have a normalized version of the NMF problem, i.e., suppose we have a positive-valued matrix Q whose columns are normalized, the problem is to find a low rank approximation $Q \simeq PW$, where P and W are also positive-valued matrices whose columns are normalized. Hereafter we consider this normalized version of the problem. This special case of NMF is called "topic model", and it has its own applications such as datamining from big text data [4,14] and analysis of compositional data of rocks in geology [20].

The topic model is interpreted geometrically as follows. Let $\boldsymbol{q}_1, \boldsymbol{q}_2, \ldots, \boldsymbol{q}_n$ be the columns of Q, and $\boldsymbol{p}_1, \boldsymbol{p}_2, \ldots, \boldsymbol{p}_k$ be the columns of P. They all belong to the space of d-valued discrete distribution $p(X; \boldsymbol{p})$ representing $\Pr[X = i] = p_i, i = 1, \ldots, d$,

$$\mathcal{S} = \{p(X; \boldsymbol{p}) \mid \boldsymbol{p} = (p_1, p_2, \ldots, p_d)^{\mathrm{T}}, \sum_{i=1}^{d} p_i = 1, p_i > 0, i = 1, \ldots, d\}. \tag{3}$$

The set of vectors $\boldsymbol{p}_1, \boldsymbol{p}_2, \ldots, \boldsymbol{p}_k$ defines a subset $\mathcal{P} \subset \mathcal{S}$,

$$\mathcal{P} = \{p(X; \boldsymbol{p}) \mid \boldsymbol{p} = \sum_{j=1}^{k} w_j \boldsymbol{p}_j, \sum_{j=1}^{k} w_j = 1, w_j \geq 0, j = 1, \ldots, k\}, \tag{4}$$

whose parameter space forms a simplex. Each vector $\boldsymbol{q}_i \in \mathcal{S}$ is approximated by a point $\hat{\boldsymbol{q}}_i \in \mathcal{P}$ to minimize a certain loss function $l(\boldsymbol{q}_i, \hat{\boldsymbol{q}}_i)$. A simple choice of l is the squared loss

$$l_{sq}(\boldsymbol{q}, \hat{\boldsymbol{q}}) = \sum_{i=1}^{d} (q_i - \hat{q}_i)^2, \tag{5}$$

while we consider the Kullback-Leibler divergence in later discussion. In any case, when we fix P, we can define the optimal column vector $\boldsymbol{w}_j$ of W for each $\boldsymbol{q}_j$ so that the loss function is minimized. Then the problem of topic model is to optimize P as well, thus it can be written as minimizing the total amount of the loss function with respect to both P and W

$$L(P, W) = \sum_{j=1}^{n} l(\boldsymbol{q}_j, \hat{\boldsymbol{q}}_j). \tag{6}$$

The optimization is usually performed alternatively, which optimizes one of P and W while the other is fixed. In the following discussion, we first focus on the optimization of W with a fixed P.

3   Projection onto an Autoparallel Submanifold

Let us briefly review the notion of dually flat structure of the manifold S that is a special case of exponential family [2,3,16]. By differential geometrical discussion, S has dual affine connections, e-connection and m-connection, and there exist affine coordinate systems $\boldsymbol{\theta}$ and $\boldsymbol{\eta}$ with respect to each connection, which are called e-coordinate and m-coordinate respectively. In the case of the space of d-valued discrete distributions, e-coordinate and m-coordinate are given by

$$\theta_i = \log p_i - \log\left(1 - \sum_{i'=1}^{d-1} p_{i'}\right), \quad \eta_i = p_i, \quad i = 1, 2, \ldots, d-1. \tag{7}$$

An affine subspace of each coordinate system is called e-autoparallel submanifold and m-autoparallel submanifold. In particular, one dimensional case defines an e-geodesic and m-geodesic. Here let us consider the projection from a point $q \in S$ onto a subspace $M \in S$. The e-projection is defined as a point $p \in M$ such that an e-geodesic connecting q and p is orthogonal at p with respect to the Riemannian metric $G(\boldsymbol{\theta}(\boldsymbol{p}))$ whose i, j component is defined by

$$G_{ij}(\boldsymbol{\theta}(\boldsymbol{p})) = \mathrm{E}_X \left[\frac{\partial \log p(X; \boldsymbol{p})}{\partial \theta_i(\boldsymbol{p})} \frac{\partial \log p(X; \boldsymbol{p})}{\partial \theta_j(\boldsymbol{p})}\right], \tag{8}$$

where E_X denotes the expectation with respect to $p(X; \boldsymbol{p})$. The matrix $G(\boldsymbol{\theta}(\boldsymbol{p}))$ is known as the Fisher information matrix.

The following theorem gives a characterization of the projection.

Theorem 1 (Projection theorem [2]). *For a dually flat manifold S and a submanifold $M \in S$, the e-projection $\hat{q}$ from a point $q \in S$ onto M is given by a critical point of the Kullback-Leibler divergence,*

$$D[\hat{\boldsymbol{q}}, \boldsymbol{q}] = \sum_{i=1}^{d} \hat{q}_i (\log \hat{q}_i - \log q_i). \tag{9}$$

In particular, if M is an m-autoparallel submanifold, the e-projection is unique and is given by minimizing $D[\hat{\boldsymbol{q}}, \boldsymbol{q}]$. On the other hand, the m-projection is given by a critical point of the dual form of (9), $D[\boldsymbol{q}, \hat{\boldsymbol{q}}]$, and if M is an e-autoparallel submanifold, the m-projection is unique and is given by minimizing $D[\boldsymbol{q}, \hat{\boldsymbol{q}}]$.

This theorem suggests that if the subspace is m-autoparallel, the e-projection has a good property such as uniqueness.

In order to consider applying this framework to the topic model (normalized NMF), first let us define a linear subspace of the d-valued discrete distribution space $\mathcal{S}$,

$$\mathcal{M} = \{p(X;\boldsymbol{p}) \mid \boldsymbol{p} = \sum_{j=1}^{k} w_j \boldsymbol{p}_j, \sum_{j=1}^{k} w_j = 1, w_i > 0, i = 1,\ldots,d\}. \qquad (10)$$

Since $\mathcal{M}$ is a linear subspace of m-coordinate, $\mathcal{M}$ is an m-autoparallel submanifold of $\mathcal{S}$.

From the projection theorem, it is natural to take the e-projection onto $\mathcal{M}$, which has been proposed as the extension of PCA to the statistical manifold [1,7,18]. Note that $\mathcal{P}$ appeared in the topic model is a subset of $\mathcal{M}$, therefore it seems natural to take the e-projection also in the topic model. If the e-projection from $q \in \mathcal{S}$ onto $\mathcal{M}$ does not belong to $\mathcal{P}$, the point $\hat{q} \in \mathcal{P}$ minimizing the divergence $D[\hat{q}, q]$ is not the e-projection onto $\mathcal{P}$. However, even in such a case, we can show that $\hat{q}$ is a projection onto a boundary of $\mathcal{P}$, which is also a subset of m-autoparallel submanifold of $\mathcal{S}$ as summarized in the following proposition.

Proposition 1. *In the dually flat manifold $\mathcal{S}$, suppose k-simplex $\mathcal{P} \subset \mathcal{S}$ defined by a convex hull of $\boldsymbol{p}_i$ ($i = 1,\ldots,k$), then the point $\hat{q}$ that minimizes $D(\hat{q}, q)$ for $q \in \mathcal{S}$ is the e-projection from q to a k'-face of $\mathcal{P}$, where $k' \leq k$.*

Proof. Let the m-autoparallel submanifold $\mathcal{M} \subset \mathcal{S}$ that is defined by expanding $\mathcal{P}$. The e-projection $\hat{q}$ from q to $\mathcal{M}$ uniquely exists and it minimizes $D(\hat{q}, q)$. If $\hat{q} \in \mathcal{P}$, that is a desired point. Otherwise, the point $\hat{q}$ that minimizes $D(\hat{q}, q)$ lies on the boundary of $\mathcal{P}$. The boundary consists of faces that has lower dimension than $\mathcal{P}$. Each face $\mathcal{P}'$ is also a simplex, thus we can continue the above discussion recursively until the e-projection is included in a simplex. This proves the proposition.

In usual formulation of NMF or topic models, the divergence $D[q, \hat{q}]$ corresponding to the m-projection has been usually used in the topic model, since the m-projection is equivalent to the maximum likelihood estimation. However, from the discussion above, we see that the dual $D[\hat{q}, q]$ is a more natural loss function from an information geometrical viewpoint.

So far we have formulated the optimization of W with a fixed P, and now we describe that the optimization problem of P with a fixed W is also an e-projection onto an m-autoparallel submanifold. A given matrix Q can be considered as a point of the product space $\mathcal{S}^n$. On the other hand, a matrix PW with a fixed W is a linear subspace of $\mathcal{S}^n$ with respect to the m-coordinate (independent parameters among P), where P is constrained to be probability distribution, i.e., it is positive and column-wise normalized, Therefore, it is a subset $\mathcal{Q}$ of the m-autoparallel submanifold, and it is natural again to take the e-projection in $\mathcal{S}^n$, and the corresponding divergence is the sum of divergences $D(PW, Q) = \sum_{i=1}^{n} D(\hat{q}_i, q_i)$, where $\hat{q}_i$ is the i-th row of PW, i.e., the objective function is the same as the case of optimizing W with a fixed P. As similarly in the optimization of W with a fixed P, the e-projection does not necessarily lies in $\mathcal{Q}$ that is a polytope in general, but it can be characterized as an e-projection onto a face of $\mathcal{Q}$. The above discussion can be summarized as the following proposition, which is easily proved in the same way as Proposition 1.

Proposition 2. *In the product space $\mathcal{S}^n$, consider a convex polytope $\mathcal{Q} \subset \mathcal{S}$ defined by PW with a fixed W, then the point $\hat{P}$ that minimizes $D(\hat{P}W, Q)$ is the e-projection from Q to a face of $\mathcal{Q}$ whose dimension is equal or less than $\mathcal{Q}$.*

4   Geometrical Projection Algorithm

The merit of taking the e-projection is not only because of geometrical naturality. We can also apply a geometrical projection algorithm based on the generalized Pythagorean theorem that is a key theorem in the information geometry.

Theorem 2 (Generalized Pythagorean theorem [2]). *Suppose $\mathcal{S}$ be a dually flat manifold, and there are three points $\boldsymbol{p}, \boldsymbol{q}, \boldsymbol{r} \in \mathcal{S}$. If the e-geodesic connecting $\boldsymbol{p}$ and $\boldsymbol{q}$ and the m-geodesic connecting $\boldsymbol{q}$ and $\boldsymbol{r}$ are orthogonal at $\boldsymbol{q}$, then the following relation holds*

$$D[\boldsymbol{q}, \boldsymbol{p}] + D[\boldsymbol{r}, \boldsymbol{q}] = D[\boldsymbol{r}, \boldsymbol{p}]. \tag{11}$$

In the theorem, suppose $\boldsymbol{q} \in \mathcal{M}$ is the e-projection of $\boldsymbol{p}$, and $\boldsymbol{r}$ is another point of $\mathcal{M}$, then the Pythagorean relation holds for those three points.

Suppose $\boldsymbol{p}_1, \boldsymbol{p}_2, \ldots, \boldsymbol{p}_k$ are the columns of P and $\boldsymbol{q}$ is one of the columns of Q, and a current estimation of the projection point on the simplex is represented as $\hat{\boldsymbol{q}} = \sum_i \hat{w}_i \boldsymbol{p}_i$, where $\sum_i \hat{w}_i = 1, \hat{w}_i \geq 0$. Let us consider the value defined by

$$\gamma_i = D[\hat{\boldsymbol{q}}, \boldsymbol{q}] + D[\boldsymbol{p}_i, \hat{\boldsymbol{q}}] - D[\boldsymbol{p}_i, \boldsymbol{q}]. \tag{12}$$

From the generalized Pythagorean theorem, if $\gamma_i = 0$ for all i, it implies that $\hat{\boldsymbol{q}}$ is the m-projection of $\boldsymbol{q}$ onto $\mathcal{M}$. On the other hand, if $\gamma_i > 0$ then the m-projection is closer to $\boldsymbol{p}_i$ than $\hat{\boldsymbol{q}}$, and if $\gamma_i < 0$ then the m-projection is more distant from $\boldsymbol{p}_i$. The proposed algorithm is based on this idea and w_i is increased or decreased in accordance with the sign of γ_i. More specifically, the algorithm consists of the following steps for each $\boldsymbol{q} = \boldsymbol{q}_j, j = 1, \ldots, n$.

1. Initialize $\hat{w}_i, i = 1, \ldots, k$.
2. Calculate γ_i by (12).
3. Update $\hat{w}_i$ by

$$\hat{w}_i \leftarrow \hat{w}_i f(\gamma_i), \tag{13}$$

   where $f(u)$ is a monotonically increasing function that takes positive value and $f(0) = 1$. A typical choice of f is $f(u) = 1/(1 + \exp(-\alpha u))$, where α is a fixed constant controlling the learning step.
4. Normalize the weight by

$$\hat{w}_i \leftarrow \frac{\hat{w}_i}{\sum_{i'=1}^{k} \hat{w}_{i'}}. \tag{14}$$

The algorithm is originally proposed for the m-projection onto an e-autoparallel submanifold [13,17]. The algorithm only depends on the divergence values, and it is derivative-free.

Although the algorithm can be started from random values of $\hat{w}_i$, we apply the algorithm to the results of the existing algorithm proposed by Dhillon and Sra [8] in the numerical experiments of the next section. There are two reasons: one is that the optimization of P is not practical for geometrical algorithm since it is an optimization problem in much larger dimensional space and the feasible region $\mathcal{Q}$ is not as simple as $\mathcal{P}$ for the optimization of W, and the other is that the existing algorithm does not assume the normalized NMF, thus the normalization step is necessary, and we observed that it does not converge to the optimum and the geometrical algorithm can improve the performance.

The existing algorithm consists of the following update steps: to update W for each $\boldsymbol{w}_i, i = 1, \ldots, n$ and $j = 1, \ldots, k$,

$$w_{ji} \leftarrow w_{ji} \exp \left(\frac{[P^{\mathrm{T}} \log(\boldsymbol{q}_i/P\boldsymbol{w}_i)]_j}{[P^{\mathrm{T}}\mathbf{1}]_j} \right), \tag{15}$$

then the W is normalized, and to update P for each $\boldsymbol{p}_i, i = 1, \ldots, d$ and $j = 1, \ldots, k, l = 1, \ldots, n$,

$$p_{ij} \leftarrow p_{ij} \exp \left(\frac{[\log(\boldsymbol{q}_l/(\boldsymbol{p}_i^{\mathrm{T}} W))^{\mathrm{T}} W^{\mathrm{T}}]_j}{[\mathbf{1}^{\mathrm{T}} W^{\mathrm{T}}]_j} \right), \tag{16}$$

and then the P is normalized.

5   Experiments

We performed some numerical examples to examine the performance of the geometrical algorithm. Throughout the experiments, if there are small values less than $\epsilon = 10^{-10}$ in a given matrix Q, we replace them by ϵ (and then renormalize Q) in order to avoid numerical instability to calculate log function.

In order to evaluate the improvement of the proposed method, we define the measure of improvement by

$$I = \frac{L_d - L_g}{L_d}, \tag{17}$$

where L_d is a loss value Eq. (6) obtained by the existing method (Dhillon and Sra [8]) and L_g is a loss value of the proposed method (geometrical algorithm).

5.1   Synthetic Data

We randomly generate n probability vectors of dimension d by a Dirichlet distribution, and perform the existing algorithm and then apply the proposed algorithm.

Dependency of Number of Samples. In the first experiment, we fix $d = 50$ and the number of basis vectors $k = 10$, and change the number of samples n.

The experiments are performed for 10 times. Figure 1 shows the average value of the measure of improvement I for different number of samples. It is almost flat and decreases gradually.

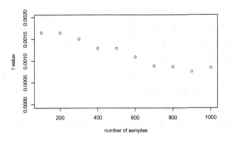

Fig. 1. Dependency of the number of samples

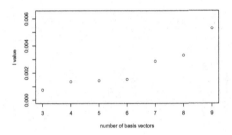

Fig. 2. Dependency of the number of basis vectors

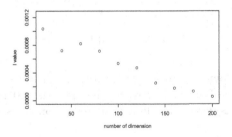

Fig. 3. Dependency of the dimensionality

Dependency of Number of Basis Vectors. In the second experiment, we fix $n = 1000$ and $d = 10$, and change the number of basis vectors k. Figure 2 shows the average of 10 times experiments. It seems the degree of improvements increases as k increases.

Dependency of Dimensionality. In the third experiment, we fix $n = 1000$ and $k = 10$, and change the dimensionality d. Figure 3 shows the average of 10 times experiments. It seems the degree of improvements decreases as k increases.

5.2 Real Data

Associated Press Data. Associated Press data [11] is based on 2246 articles published in Associated Press in US. For each article, the frequency of words (total 10473) are recorded, i.e., $n = 2246$ and $d = 10473$. The topic model has been applied to such a set of documents in order to categorize articles into topics. More specifically, each article q_j is decomposed into weighted sum of p_i, each of which is considered to represent a topic. Nonzero components of the data matrix are about 1.1% of all components. We fixed $k = 20$ and performed the algorithm. Figure 4 shows the values of the objective function, where the first 10 steps are alternating optimization of P and W by the existing algorithm and the last 1 step represents the optimizing W by the proposed geometrical algorithm until convergence. For this dataset, the proposed algorithm seems to improve the performance significantly.

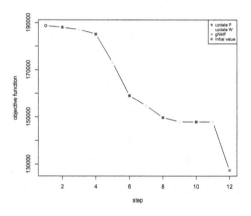

Fig. 4. Associated Press data

MovieLens Data. MovieLens data [12] is a data matrix of the 1–5 ratings of movies by 943 users. Each user gave ratings at least 20 movies and unrated movies are scored as zero. The dimension reduction for such data is used in

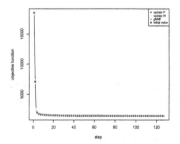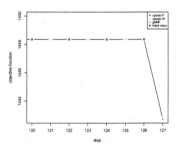

Fig. 5. MovieLens data

recommender systems. Based on the estimated ratings for the moves that have not rated yet by a user, the system can recommend the movie to the user. Here we regard $n = 943$ and $d = 1862$ and ratings are normalized so that the sum of ratings becomes one. We fixed $k = 100$ and performed the algorithm. Figure 5 shows the values of the objective function, where the first 126 steps are the existing algorithm and the last step is the optimization of W by the proposed algorithm. The improvement is slight but as shown in the right figure (magnified at the last part), it actually decreases the objective function.

6   Conclusion

We have presented geometrical formulation of the topic model as a special form of NMF. By natural formulation based on a dualistic structure, we apply the geometrical algorithm and showed it can improve the performance of the existing algorithm through numerical experiments.

There are two major problems that are not fully solved in this paper. One is the application of the geometrical algorithm for optimizing P with a fixed W. We have shown that the problem can be dealt with an e-projection problem, but it is a very high dimensional optimization and it is not practical as it is. We need to develop an efficient algorithm. The second is the theoretical guarantee of convergence of the geometrical algorithm. Empirically it converges in most cases, but it is necessary to clarify the condition of convergence.

References

1. Akaho, S.: The e-PCA and m-PCA: dimension reduction of parameters by information geometry. In: Proceedings of the 2004 IEEE International Joint Conference on Neural Networks, vol. 1, pp. 129–134. IEEE (2004)
2. Amari, S.: Differential-Geometrical Methods in Statistics. Springer, Heidelberg (1985). https://doi.org/10.1007/978-1-4612-5056-2D
3. Amari, S.: Information Geometry and Its Applications. AMS, vol. 194. Springer, Tokyo (2016). https://doi.org/10.1007/978-4-431-55978-8
4. Blei, D.M.: Probabilistic topic models. Commun. ACM **55**(4), 77–84 (2012)
5. Cho, Y.C., Choi, S.: Nonnegative features of spectro-temporal sounds for classification. Pattern Recognit. Lett. **26**(9), 1327–1336 (2005)
6. Cichocki, A., Zdunek, R., Phan, A.H., Amari, S.: Nonnegative Matrix and Tensor Factorizations: Applications to Exploratory Multi-way Data Analysis and Blind Source Separation. Wiley, Chichester (2009)
7. Collins, M., Dasgupta, S., Schapire, R.E.: A generalization of principal component analysis to the exponential family. In: NIPS, vol. 13, p. 23 (2001)
8. Dhillon, I.S., Sra, S.: Generalized nonnegative matrix approximations with Bregman divergences. In: NIPS, vol. 18 (2005)
9. Dong, B., Lin, M.M., Chu, M.T.: Nonnegative rank factorization—a heuristic approach via rank reduction. Numer. Algorithms **65**(2), 251–274 (2014)
10. Févotte, C., Bertin, N., Durrieu, J.L.: Nonnegative matrix factorization with the Itakura-Saito divergence: with application to music analysis. Neural Comput. **21**(3), 793–830 (2009)

11. Harman, D.: Overview of the first text retrieval conference (TREC-1). In: The First Text REtrieval Conference (TREC-1), pp. 1–20, no. 1 (1992)
12. Harper, F.M., Konstan, J.A.: The MovieLens datasets: history and context. ACM Trans. Interact. Intell. Syst. (TIIS) 5(4), 19 (2016)
13. Hino, H., Takano, K., Akaho, S., Murata, N.: Non-parametric e-mixture of density functions. In: Hirose, A., Ozawa, S., Doya, K., Ikeda, K., Lee, M., Liu, D. (eds.) ICONIP 2016. LNCS, vol. 9948, pp. 3–10. Springer, Cham (2016). https://doi.org/10.1007/978-3-319-46672-9_1
14. Hofmann, T.: Probabilistic latent semantic indexing. In: Proceedings of the 22nd Annual International ACM SIGIR Conference on Research and Development in Information Retrieval, pp. 50–57. ACM (1999)
15. Lee, D.D., Seung, H.S.: Algorithms for non-negative matrix factorization. In: Advances in Neural Information Processing Systems, pp. 556–562 (2001)
16. Nagaoka, H., Amari, S.: Differential geometry of smooth families of probability distributions. Technical report METR 82-7, University of Tokyo (1982)
17. Takano, K., Hino, H., Akaho, S., Murata, N.: Nonparametric e-mixture estimation. Neural Comput. 28(12), 2687–2725 (2016)
18. Watanabe, K., Akaho, S., Omachi, S., Okada, M.: Variational Bayesian mixture model on a subspace of exponential family distributions. IEEE Trans. Neural Netw. 20(11), 1783–1796 (2009)
19. Wohlmayr, M., Pernkopf, F.: Model-based multiple pitch tracking using factorial HMMs: model adaptation and inference. IEEE Trans. Audio Speech Lang. Process. 21(8), 1742–1754 (2013)
20. Yoshida, K., Kuwatani, T., Hirajima, T., Iwamori, H., Akaho, S.: Progressive evolution of whole-rock composition during metamorphism revealed by multivariate statistical analyses. J. Metamorph. Geol. 36(1), 41–54 (2018)

Information Geometric Perspective of Modal Linear Regression

Keishi Sando², Shotaro Akaho³ , Noboru Murata⁴ , and Hideitsu Hino¹⁽✉⁾

¹ The Institute of Statistical Mathematics, Tachikawa, Tokyo 190-8562, Japan
hino@ism.ac.jp
² University of Tsukuba, Tsukuba, Ibaraki 305-8573, Japan
³ National Institute of Advanced Industrial Science and Technology,
Tsukuba, Ibaraki 305-8568, Japan
⁴ Waseda University, Shinjuku, Tokyo 169-0072, Japan

Abstract. Modal linear regression (MLR) is a standard method for modeling the conditional mode of a response variable using a linear combination of explanatory variables. It is effective when dealing with response variables with an asymmetric, multi-modal distribution. Because of the nonparametric nature of MLR, it is difficult to construct a statistical model manifold in the sense of information geometry. In this work, a model manifold is constructed using observations instead of explicit parametric models. We also propose a method for constructing a data manifold based on an empirical distribution. The *em* algorithm, which is a geometric formulation of the EM algorithm, of MLR is shown to be equivalent to the conventional EM algorithm of MLR.

Keywords: Modal linear regression · Information geometry
EM algorithm

1 Introduction

In linear regression analysis, the conditional mean of a response variable y given predictor variable x is modeled using a linear predictor function of x. Unfortunately, a well-known least squares estimator for linear regression coefficients is highly sensitive to outliers. To alleviate this problem, numerous estimators such as robust M-estimators [5,6] have been developed. However, the consistency of the robust M-estimators requires the homoscedasticity and symmetricity of a conditional error distribution given a predictor. In reality, much data exist that do not follow these assumptions. One example is a conditional distribution of public spending given variables that reflect the social conditions, e.g., voter turnout, radio penetration, and bank deposits per capita. In [3], it was pointed out that the estimation cannot be consistent unless the data follow appropriate assumptions.

Supported by JST KAKENHI 16K16108, 17H01793 and JST CREST JPMJCR1761.

© Springer Nature Switzerland AG 2018
L. Cheng et al. (Eds.): ICONIP 2018, LNCS 11303, pp. 535–545, 2018.
https://doi.org/10.1007/978-3-030-04182-3_47

Modal linear regression (MLR) models a conditional mode of y given x by a linear predictor function of x. MLR relaxes the distribution assumptions for the conventional linear regression, and is robust to outliers compared to least squares estimates of linear regression coefficients. It was proven in [7] and [13] that their estimators for the MLR model were consistent even when the error distribution was asymmetric. For the above reasons, it is important to investigate the mode estimation methods and this has been done for many years.

In general, if a probability density function of a random variable X has a unique mode and is symmetric with respect to the mode, $\Pr(p - w \leq X \leq p + w)$ with fixed w is maximized when p is the mode. Based on this property, [8] proposed an estimator for the coefficients of MLR. An MLR model is formulated as follows:

$$y = x^\top \beta + \varepsilon, \quad \text{where} \quad \text{Mode}\,[\varepsilon; x] = 0. \tag{1}$$

The estimator proposed by [8] is consistent when there exists $w > 0$, and a probability density function of ε is symmetric in the range of $0 \pm w$. In [7], it is proved that the mode estimator for the coefficients of MLR is consistent even if the symmetry is not satisfied. In [13], an EM algorithm was proposed to estimate the coefficients of MLR.

Geometric formulations of statistical and machine learning algorithms can offer a deep understanding and improvement of these algorithms [1,10]. Motivated by the importance of such a geometric formulation and analysis of the modal linear regression, in this paper, we provide an information geometric perspective of MLR. In information geometry, we often construct a model manifold using a parametric distribution and regard the projection of an empirical distribution onto the model manifold as an estimation. In the case of linear regression, we construct a model manifold on the basis of the assumption that an error variable has a normal distribution. Because of the lack of a parametric distribution, it is difficult to construct a model manifold that corresponds to the MLR model using conventional approaches. There have been studies related to nonparametric models in information geometry. In [11], it was shown that a well-defined Banach manifold for probability measures can be constructed. In [12], a framework for a nonparametric e-mixture estimation was proposed. In this paper, the difficulty of constructing a model manifold is overcome by a nonparametric model, which is identified by a finite number of parameters. We propose the construction of a model manifold using observations, as is done when constructing an empirical distribution in conventional approaches. Our proposal gives a geometric view of the MLR model.

2   Modal Linear Regression

Let $x \in \mathbb{R}^p$, $y \in \mathbb{R}$ be a set of predictor variables and a response variable, respectively. Although least squares linear regression estimates a conditional mean of y given x, MLR estimates a conditional mode of y given x.

2.1 Formulation

Suppose that $\{x_i, y_i\}_{i=1}^N$ are i.i.d. observations. MLR is used to model a conditional mode of y given x by $\text{Mode}\,[y; x] = x^\top \beta$. Namely, y and x are related by Eq. (1). To estimate β, a loss function of the form

$$l(\beta; y, x) = -\phi_h\left(y - x^\top \beta\right),\tag{2}$$

is introduced [8], where $\phi_h(x) = \frac{1}{h}\phi\left(\frac{x}{h}\right)$, $\phi(\cdot)$ is a kernel function, and h is a bandwidth parameter. Minimizing the empirical loss allows us to estimate $\hat\beta$ of the linear coefficient:

$$\hat\beta = \max_\beta \frac{1}{N}\sum_{i=1}^N \phi_h(y_i - x_i^\top \beta).\tag{3}$$

In this paper, we assume that $\phi(\cdot)$ denotes a standard normal density function.

2.2 EM Algorithm for MLR

The modal expectation-maximization algorithm was proposed in [13] and consists of the following two steps, starting from an initial estimate $\beta^{(1)}$.
E-step: In this step, the purpose is to derive a surrogate function $g(\beta; \beta^{(k)})$:

$$\log\left[\frac{1}{N}\sum_{i=1}^N \phi_h\left(y_i - x_i^\top \beta\right)\right] = \log\left[\sum_{i=1}^N \pi_i^{(k)} \frac{\frac{1}{N}\phi_h\left(y_i - x_i^\top \beta\right)}{\pi_i^{(k)}}\right], \quad \text{by Jensen's inequality}$$

$$\geq \sum_{i=1}^N \pi_i^{(k)} \log\left[\frac{\frac{1}{N}\phi_h\left(y_i - x_i^\top \beta\right)}{\pi_i^{(k)}}\right] = g(\beta; \beta^{(k)}),\tag{4}$$

where $\pi_i^{(k)} = \frac{\phi_h(y_i - x_i^\top \beta^{(k)})}{\sum_{j=1}^N \phi_h(y_j - x_j^\top \beta^{(k)})}$, $\quad i = 1 \ldots N$.
M-step: In the M-step, the parameter β is updated to increase the value of $\frac{1}{N}\sum_{i=1}^N \phi_h\left(y_i - x_i^\top \beta\right)$:

$$\beta^{(k+1)} = \underset{\beta}{\operatorname{argmax}} \sum_{i=1}^N \pi_i^{(k)} \log \phi_h(y_i - x_i^\top \beta),\tag{5}$$

If $\phi(\cdot)$ is a standard normal density function, $\beta^{(k+1)}$ is

$$= \left(X^\top W_k X\right)^{-1} X^\top W_k y,\tag{6}$$

where $W_k = \operatorname{diag}\left(\pi_1^{(k)} \cdots \pi_N^{(k)}\right)$. The detailed derivation and property of the estimate $\hat\beta$ are found in [13].

3 Information Geometry

Information geometry [2] is a framework to describe spaces that consist of probability density functions by means of differential geometry. We consider the space

$\mathscr{S}$ of probability density functions (pdfs) in the class of the exponential family. In this case, one of the natural measures of dispersion of two pdfs is the Kullback-Leibler divergence $D^{(m)}(q||p)$, which is expressed as follows:

$$D^{(m)}(q||p) = D^{(e)}(p||q) = \int q(x) \log \frac{q(x)}{p(x)} dx.$$

Here, $D^{(e)}$ is called the e-divergence, and $D^{(m)}$ is called the m-divergence.

In information geometry, a statistical inference is often regarded as a projection of an empirical distribution onto a model manifold.

The expectation-maximization (EM) algorithm [4] is one of the methods used to find maximum likelihood estimates of parameters in a latent variable model. In information geometry, the exponential-mixture (em) algorithm [1] corresponds to the EM algorithm.

A manifold that consists of statistical models is called a model manifold and denoted by $\mathscr{M}$, and a manifold that consists of the empirical joint probability distributions of observable variables and latent variables is called a data manifold $\mathscr{D}$. The purpose of the em algorithm is to find the points $p^* \in \mathscr{M}$ and $q^* \in \mathscr{D}$ that minimize the KL-divergence from q^* to p^*. In order to achieve this goal, the em algorithm iterates the following two steps, starting from an initial guess $p^{(1)}$.

e-**step**: e-projection of $p^{(k)} \in \mathscr{M}$ onto the data manifold $\mathscr{D}$.

$$q^{(k)} = \underset{q \in \mathscr{D}}{\operatorname{argmin}} D^{(e)}(p^{(k)}||q).$$

m-**step**: m-projection of $q^{(k)} \in D$ onto the model manifold $\mathscr{M}$.

$$p^{(k+1)} = \underset{p \in \mathscr{M}}{\operatorname{argmin}} D^{(m)}(q^{(k)}||p).$$

4   Information Geometry of MEM Algorithm

This section introduces the modal EM (MEM) algorithm [9] as a basis for the information geometric formulation of MLR. The information geometric perspective for the MEM algorithm is useful in overcoming the difficulty of constructing manifolds for the MLR model. Let us consider Gaussian mixture models with known parameters. In general, even though all of the model parameters are known, it is difficult to express the mode of the Gaussian mixture model in a closed form. In order to obtain the mode, we need to resort to numerical optimization such as with the gradient ascent method.

The MEM algorithm is an iterative method to find a local mode of a mixture distribution in the following form:

$$f(x) = \sum_{i=1}^{K} \pi_i f_i(x), \quad x \in \mathbb{R}^p, \quad \begin{cases} \pi_i \geq 0, \quad \sum_{i=1}^{K} \pi_i = 1, \\ \\ f_i : \mathbb{R}^p \to \mathbb{R}, \ i = 1 \ldots K \text{ are pdfs,} \end{cases}$$

where all of the parameters in this model are known. The purpose of the MEM algorithm is to find the mode of $f(x)$, that is, $x^* = \operatorname{argmax}_x f(x)$. In [9], the proposal was made to iterate the following two steps starting with an initial estimate $x^{(1)}$.

E-step

$$p_i^{(k)} = \frac{\pi_i f_i(x^{(k)})}{f(x^{(k)})}, \quad i = 1 \ldots K. \tag{7}$$

M-step

$$x^{(k+1)} = \operatorname{argmax}_x \sum_{i=1}^{K} p_i^{(k)} \log f_i(x). \tag{8}$$

4.1 Information Geometric Formulation

To provide an information geometric perspective of the MEM algorithm, we add a latent variable $Z \in \{1 \ldots K\}$ to the mixture model $f(x) = \sum_{i=1}^{K} \pi_i f_i(x)$. The latent variable specifies a mixture component that yields an observation x. A joint pdf $g(x, z)$ is expressed as follows:

$$g(x, z) = \prod_{i=1}^{K} [\pi_i f_i(x)]^{\delta_i(z)}, \quad \text{where} \quad \begin{cases} \pi_i \geq 0, \quad \sum_{i=1}^{K} \pi_i = 1, \\ f_i, \quad i = 1 \ldots K \text{ are pdfs,} \\ \delta_i(z) = \begin{cases} 1 & i = z, \\ 0 & i \neq z. \end{cases} \end{cases} \tag{9}$$

The extension to a joint pdf expressed as Eq. (9) was introduced by [1] to treat mixture modeling in the framework of EM. In general, an empirical density function is constructed based on observations. For example, when observations $\{x_i\}_{i=1}^{N}$ are i.i.d., an empirical density function is defined as $\frac{1}{N} \sum_{i=1}^{N} \delta(x - x_i)$, where $\delta(\cdot)$ denotes the Dirac delta function.

In the formulation of the MEM algorithm, the construction of an empirical density function is nontrivial because there is no observation. In this paper, we interpret the formulation of the MEM algorithm as the problem of estimating the likelihood of the given model for a *pseudo-observation*, namely, we assume that "one observation is obtained" and propose to define an empirical density function $p(x) = \delta(x - m)$, where m denotes the pseudo-observation. We treat it as an unknown parameter. Introducing a latent variable $Z \in \{1 \ldots K\}$, we extend $p(x)$ to an empirical joint density function of X and Z as $h(x, z) = p(x)q(z \mid x)$. Introducing parameters $\{q_i\}_{i=1}^{K}$, the conditional density function $q(z \mid x)$ is modeled as follows:

$$q(z \mid x) = \sum_{i=1}^{K} q_i \delta_i(z), \quad \text{where} \quad q_i \geq 0, \quad \sum_{i=1}^{K} q_i = 1.$$

Then, the empirical joint density function $h(x, z \; ; m, q_1 \ldots q_K)$ is expressed as follows:

$$h(x, z \; ; m, q_1 \ldots q_K) = \sum_{i=1}^{K} q_i \delta(x - m) \delta_i(z), \tag{10}$$

where $q_i \geq 0$ and $\sum_{i=1}^{K} q_i = 1$. A data manifold $\mathscr{D}$ is defined as follows:

$$\mathscr{D} = \left\{ h(x, z \; ; m, q_1 \ldots q_K) \mid m \in \mathbb{R}^p, \quad q_i \geq 0, \quad \sum_{i=1}^{K} q_i = 1 \right\}. \tag{11}$$

Let $\mathscr{D}(m')$ be a subset of $\mathscr{D}$ restricted with $m = m'$ and $\mathscr{D}\left(\{q_i'\}_{i=1}^{K}\right)$ be a subset of $\mathscr{D}$ restricted with $q_i = q_i'$, $i = 1 \ldots K$.

$$\mathscr{D}(m') = \left\{ h(x, z \; ; m = m', q_1 \ldots q_K) \mid q_i \geq 0, \quad \sum_{i=1}^{K} q_i = 1 \right\},$$

$$\mathscr{D}\left(\{q_i'\}_{i=1}^{K}\right) = \{ h(x, z \; ; m, q_1 = q_1' \ldots q_K = q_K') \mid m \in \mathbb{R}^p \}.$$

We consider the formulation of the MEM algorithm to be the problem of estimating a likelihood for the observation given a model. Thus, let us consider the e-projection of a model $g(x, z)$ onto a data manifold $\mathscr{D}$, namely, $\min_{h \in \mathscr{D}} D^{(e)}(g\|h)$. We minimize $D^{(e)}(g\|h)$ by alternately optimizing m and $\{q_i\}_{i=1}^{K}$. The optimization problem with respect to $\{p_i\}_{i=1}^{K}$ is formulated as follows:

$$\min_{h \in \mathscr{D}(m^{(k)})} D^{(e)}(g\|h) \rightarrow \left| \begin{array}{l} \min_{q_1 \ldots q_K} \quad D^{(e)}\left(g\|h(\cdot, \cdot \; ; m = m^{(k)}, q_1 \ldots q_K)\right), \\[2mm] \text{s.t.} \quad \begin{cases} q_i \geq 0, \\ \sum_{i=1}^{K} q_i = 1. \end{cases} \end{array} \right. \tag{12}$$

Using the Lagrange multiplier method, the optimal solution for Eq. (12) is given as follows:

$$q_i^{(k)} = \frac{\pi_i f_i(m^{(k)})}{f(m^{(k)})}, \quad i = 1 \ldots K. \tag{13}$$

The optimization problem with respect to m is formulated as follows:

$$\min_{h \in \mathscr{D}\left(\left\{q_i^{(k)}\right\}_{i=1}^{K}\right)} D^{(e)}(g\|h), \tag{14}$$

which is equivalent to

$$\max_{m \in \mathbb{R}^p} \sum_{i=1}^{K} q_i^{(k)} \log f_i(m), \tag{15}$$

which is equivalent to Eq. (8).

We consider the MEM algorithm to be the problem of estimating the likelihood of a given model for a pseudo-observation and optimize two parameters: the pseudo-observation and latent variable of the mixture model. The e-projection of a model distribution $g(x,z)$ onto $\mathscr{D}(m^{(k)})$ gives the optimal $q_i^{(k)}$, $i = 1 \ldots K$, which is equal to Eq. (7) in the original MEM algorithm. The e-projection of model distribution $g(x,z)$ onto $\mathscr{D}\left(\left\{q_i^{(k)}\right\}_{i=1}^{K}\right)$ makes it possible to derive the optimal $m^{(k+1)}$, which is consistent with Eq. (8) in the original MEM algorithm.

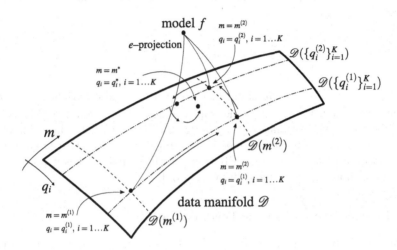

Fig. 1. Diagram of em algorithm corresponding to MEM algorithm

Figure 1 shows the iteration process of the em algorithm corresponding to the MEM algorithm. The proposed em algorithm corresponding to the MEM algorithm is different from a conventional em algorithm for latent parameter estimation. The difference is whether a model is unique or not. For example, in the case of the em algorithm for estimating the parameter of a Gaussian mixture model, the model includes unknown parameters. Thus, the em algorithm for a Gaussian mixture model consists of the e-projection and m-projection, where the former denotes the projection of a model onto a data manifold and the latter denotes the projection of an empirical distribution onto a model manifold. On the other hand, the em algorithm corresponding to MEM consists of the e-projection only. Therefore, the proposed em algorithm for MEM is not strictly the em algorithm, but we purposely call the proposed algorithm the em algorithm because the MEM algorithm is derived from the EM algorithm.

5   Information Geometry of MLR Algorithm

In this section, we analyze MLR from the information geometric perspective. We elucidate the source of the difficulty in constructing a model manifold and data manifold for the MLR model, and propose a framework to geometrically formulate the MLR model.

5.1 Constructing Manifolds

In order to elucidate the source of the difficulty in constructing manifolds for the MLR model, we consider the parameter estimation of a Gaussian mixture model as a specific example of statistical inferences in information geometry. Suppose that observations $\{x_i\}_{i=1}^{N}$ are i.i.d. and x_i has a Gaussian mixture distribution expressed as follows:

$$f(x;\mu,\Sigma) = \sum_{i=1}^{K} \pi_i g(x;\mu_i,\Sigma_i), \quad \pi_i \geq 0, \quad \sum_{i=1}^{K} \pi_i = 1,$$

where $g(x;\mu_i,\Sigma_i)$ is the Gaussian pdf with mean μ_i and covariance Σ_i. A data manifold is constructed based on the empirical density function $\frac{1}{N}\sum_{i=1}^{N}\delta(x-x_i)$.

In the parameter estimation of a Gaussian mixture model, a model manifold is constructed based on the parametric distribution. On the other hand, there is no assumption of parametric distributions in MLR. This makes it nontrivial to construct a model manifold and data manifold.

5.2 Information Geometric Formulation

To construct a model manifold for the MLR model, we consider (i) the assumption that $\mathrm{Mode}\,[\varepsilon;x] = 0$ and (ii) the form of the objective function of β for the MLR model, $\frac{1}{N}\sum_{i=1}^{N}\phi_h\left(y_i - x_i^\top\beta\right)$. With this assumption and fact, the optimization problem of Eq. (3) can be regarded as a maximization problem for a kernel density estimate at $\varepsilon = 0$ of a probability density function of ε. We propose to construct a model for the MLR as follows:

$$f(\varepsilon;\beta) = \frac{1}{N}\sum_{i=1}^{N}\phi_h\left(\varepsilon - \varepsilon_i(\beta)\right), \tag{16}$$

where $\varepsilon_i(\beta) = y_i - x_i^\top\beta$, $i = 1\ldots N$, and a variable ε denotes an error variable. We introduce a latent variable $Z \in \{1\ldots N\}$, which specifies a mixture component from which an observation is obtained. The joint density function of ε and Z is expressed as $g(\varepsilon,z;\beta) = \prod_{i=1}^{N}\left[\frac{1}{N}\phi_h\left(\varepsilon - \varepsilon_i(\beta)\right)\right]^{\delta_i(z)}$. A model manifold $\mathscr{M}$ is denoted by

$$\mathscr{M} = \{g(\varepsilon,z;\beta) \mid \beta \in \mathbb{R}^p\}. \tag{17}$$

In general, an empirical density function is often constructed based on observations. In this paper, observations are used for constructing a model manifold. Thus, from (i) the construction proposed in Sect. 4.1 and (ii) the assumption that Mode $[\varepsilon; x] = 0$, we propose to construct an empirical density function as follows:

$$p(\varepsilon) = \delta(\varepsilon - 0) = \delta(\varepsilon). \tag{18}$$

Introducing a latent variable $Z \in \{1 \ldots N\}$ to Eq. (18), we extend $p(\varepsilon)$ to an empirical joint density function of ε and Z as $h(\varepsilon, z) = p(\varepsilon)q(z \mid \varepsilon)$. Introducing parameters $\{q_i\}_{i=1}^{N}$, the conditional density function $q(z \mid \varepsilon)$ is modeled as follows:

$$q(z \mid \varepsilon) = \sum_{i=1}^{N} q_i \delta_i(z), \quad \text{where} \quad \begin{cases} q_i \geq 0, \\ \displaystyle\sum_{i=1}^{N} q_i = 1. \end{cases}$$

Then, the empirical joint density function $h(\varepsilon, z; q_1 \ldots q_N)$ is expressed as follows:

$$h(\varepsilon, z \,; q_1 \ldots q_N) = \sum_{i=1}^{N} q_i \delta(\varepsilon)\delta_i(z), \quad q_i \geq 0, \quad \sum_{i=1}^{N} q_i = 1. \tag{19}$$

A data manifold $\mathscr{D}$ is defined as follows:

$$\mathscr{D} = \left\{ h(\varepsilon, z \,; q_1 \ldots q_N) \mid q_i \geq 0, \quad \sum_{i=1}^{N} q_i = 1 \right\}. \tag{20}$$

Let us consider the e-projection of a model, whose parameters are $\beta^{(k)}$, onto the data manifold:

$$\min_{h \in \mathscr{D}} D^{(e)}(g(\cdot, \cdot; \beta^{(k)}) \| h) \rightarrow \quad \begin{aligned} &\min_{q_1 \ldots q_N} \quad D^{(e)}\left(g(\cdot, \cdot; \beta^{(k)}) \| h(\cdot, \cdot \,; q_1 \ldots q_N)\right), \\ &\text{s.t.} \quad q_i \geq 0, \quad \sum_{i=1}^{N} q_i = 1. \end{aligned} \tag{21}$$

An optimal solution for Eq. (21) is

$$q_i^{(k)} = \frac{\phi_h\left(y_i - x_i^\top \beta^{(k)}\right)}{\sum_{j=1}^{N} \phi_h\left(y_j - x_j^\top \beta^{(k)}\right)}, \quad i = 1 \ldots N, \tag{22}$$

which is equivalent to Eq. (4).

Then, let us consider the m-projection of an empirical joint density function, whose parameters are $q_i = q_i^{(k)}$, $i = 1 \ldots N$, onto the model manifold:

$$\min_{g \in \mathscr{M}} D^{(m)}(h(\cdot, \cdot \,; q_1 = q_1^{(k)} \ldots q_N = q_N^{(k)}) \| g). \tag{23}$$

The optimization problem expressed as Eq. (23) is equal to

$$\max_{\beta} \sum_{i=1}^{N} q_i^{(k)} \log \phi_h \left(y_i - x_i^\top \beta \right),\tag{24}$$

which is equivalent to Eq. (5). Figure 2 shows the process for updating the *em* algorithm corresponding to the MLR model, which iterates the e-projection and *m*-projection.

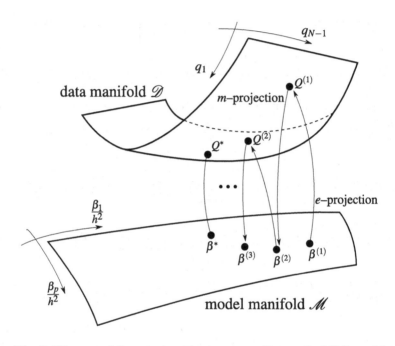

Fig. 2. Diagram of the *em* algorithm corresponding to the MLR model

6   Conclusion

In this paper, we proposed a method for constructing a model manifold and data manifold for the MLR model. Although a model manifold is often constructed based on a parametric distribution assumption, we proposed a method based on observations. In the manifolds constructed by the proposed approach, we formulated the *em* algorithm to estimate the coefficients of MLR models. The result showed that the e-projection from a model with a fixed $\beta^{(k)}$ onto the data manifold, and the m-projection from an empirical distribution with fixed parameters $q_i = q_i^{(k)}$, $i = 1 \ldots N$ onto the model manifold, led to the original E- and M-steps of the EM algorithm for the MLR parameter estimation [13].

This work is a purely theoretical one to shed light on the well-known MLR from the viewpoint of information geometry. The obtained results do not provide any computational improvement. Hence, we did not perform any computational experiments. Our future work will include further analyses of the MLR model based on the proposed geometric perspective, and the elucidation of its statistical characteristics, which include robustness, consistency, and effectiveness. Because MLR is based on the kernel density estimator, we believe that it would also be interesting to optimize the kernel bandwidth using ideas from information geometry. The theoretical evaluation of the robustness, consistency, and effectiveness will be followed by experimental evaluations of those properties, which might lead to novel algorithms with better properties.

References

1. Amari, S.: Information geometry of the EM and *em* algorithms for neural networks. Neural Netw. **8**(9), 1379–1408 (1995)
2. Amari, S., Nagaoka, H.: Methods of Information Geometry. American Mathematical Society (2000)
3. Baldauf, M., Silva, J.S.: On the use of robust regression in econometrics. Econ. Lett. **114**(1), 124–127 (2012)
4. Dempster, A.P., Laird, N.M., Rubin, D.B.: Maximum likelihood from incomplete data via the EM algorithm. J. Royal Stat. Soc. Ser. B, 1–38 (1977)
5. Hampel, F.R., Ronchetti, E.M., Rousseeuw, P.J., Stahel, W.A.: Robust Statistics - The Approach Based on Influence Functions. Wiley (1986)
6. Huber, P.J., Ronchetti, E.M.: Robust Statistics. Wiley (2011)
7. Kemp, G.C., Silva, J.S.: Regression towards the mode. J. Econometrics **170**(1), 92–101 (2012)
8. Lee, M.J.: Mode regression. J. Econometrics **42**(3), 337–349 (1989)
9. Li, J., Ray, S., Lindsay, B.G.: A nonparametric statistical approach to clustering via mode identification. J. Mach. Learn. Res. **8**, 1687–1723 (2007)
10. Murata, N., Takenouchi, T., Kanamori, T., Eguchi, S.: Information geometry of u-boost and Bregman divergence. Neural Comput. **16**(7), 1437–1481 (2004)
11. Pistone, G., Sempi, C.: An infinite-dimensional geometric structure on the space of all the probability measures equivalent to a given one. Ann. Statist. **23**(5), 1543–1561 (1995)
12. Takano, K., Hino, H., Akaho, S., Murata, N.: Nonparametric e-mixture estimation. Neural Comput. **28**(12), 2687–2725 (2016)
13. Yao, W., Li, L.: A new regression model: modal linear regression. Scand. J. Stat. **41**(3), 656–671 (2014)

Learning from Audience Intelligence: Dynamic Labeled LDA Model for Time-Sync Commented Video Tagging

Zehua Zeng[1,2,3], Cong Xue[3(✉)], Neng Gao[2,3], Lei Wang[3], and Zeyi Liu[3]

[1] School of Cyber Security, University of Chinese Academy of Sciences,
Beijing, China
[2] State Key Laboratory of Information Security,
Chinese Academy of Sciences, Beijing, China
[3] Institute of Information Engineering, Chinese Academy of Sciences, Beijing, China
zengzehua@is.ac.cn, {xuecong,gaoneng,wanglei,liuzeyi}@iie.ac.cn

Abstract. With the boom of online video uploading, video tagging becomes an important way for video indexing. However, text-based video tagging methods ignore either genre labels or temporal differences of videos, which makes results defective. Fortunately, a new type of videos called time-sync commented videos which contains large amounts of information commented by the users helps videos tagging. In this paper, we propose a supervised dynamic Latent Dirichlet Allocation model utilizing the variational topics of time-sync comments to extract both genre labels and keywords as tags. We also implement experiments on large scale real-world datasets and the effectiveness of our model are proved both in genre label classification and keyword extraction compared with baseline models.

Keywords: Video tagging · Time-sync commented videos
Bullet-screen comments · Keyword extraction
Multi-label classification

1 Introduction

In these years, millions of videos are uploaded everyday. To summarize these videos, tags of videos provide users a fast primary impression of the videos. However, the lack of tags and subjectivity of uploaders decrease both the quality of video tags and the experience of users. Fortunately, a new form of video called time-sync commented video become more and more popular in these years, and their data provided a new opportunity to solve video tagging problems better.

Time-sync commented videos are videos with real-time comments associated with playback time and content generated by users. When watching a video, users can comment the content at any time of the video and comments will overlay over the video directly, synchronized to the playback time. Other users can also see these comments at the same playback time. There is an example for

© Springer Nature Switzerland AG 2018
L. Cheng et al. (Eds.): ICONIP 2018, LNCS 11303, pp. 546–559, 2018.
https://doi.org/10.1007/978-3-030-04182-3_48

Time-sync commented videos in Fig. 1. This kind of comments are also called **bullet-screen comments** [5], which make it possible to learn the temporal semantic of videos, however, there are still some existing problems.

Most previous video tagging works which use bullet-screen comments [13,15, 16] only regard extracted keywords as tags. However, there are also some tags that can not be obtained from keywords such as genre labels tagged by uploaders. Thus, simply regarding extracted keywords as tags is insufficient to reflect the content of the videos. Genre labels such as "Comedy" or "Action" tagged by uploaders can reflect the main topic of the videos and sometimes maybe more interested by users. It is also important to tag videos by these genre labels. For videos that lack tags, using supervised method to learning genre labels as tags from other tagged videos is helpful.

There is a problem that traditional supervised multi-label classification models [6,18] ignore the minor labels. For example, the genre labels of movie *your name*[1] in *Internet Movie Database(IMDb)*[2] are "Drama","Fantasy" and "Romance", but other minor labels such as "Amuse" are omitted although there are many scenes of the movie are about "Amuse". In General, the temporal differences in videos make the content of bullet-screen comments various and uploaders always omit minor genre labels for brief, which also ignore the different focuses of some fragments beyond the storyline. But comments related to these omitted labels would disturb the learning of other genre labels.

To solve all these problems, we lead classical latent dirichlet allocation (L DA) [1] model into our model and propose a supervised latent dirichlet allocation called Dynamic Labeled LDA (DLLDA) model to tag time-sync commented videos. Our model utilize bullet-screen comments and genre labels of videos to tag unlabeled videos with both genre labels and keywords. To deal with the omitted minor genre labels, our model considers both the temporal differences in each video and the similarity of same label among videos. By utilizing a supervised LDA model, we combine each genre label to a topic which guarantee the similarity of same label among videos. And by splitting bullet-screen comments into slides, each slide are allocate to a topic distribution which reflect the temporal differences in each video. Furthermore, video-specific word distributions are also introduced into our model to extract keywords for each video.

Fig. 1. An example for bullet-screen comments in time-sync commented videos

---

[1] https://www.imdb.com/title/tt5311514/.

[2] https://www.imdb.com/.

The results of experiments shows that our model performs better both in genre label classification and keyword extraction comparing with baseline methods, which validates the potentiality of our model in bullet-screen comments understanding.

2   Related Work

Traditional Video Tagging Techniques. Some traditional video tagging techniques [9,10,12] propose methods to tag whole videos mainly using identical scenes in visual. Some other works combine both visual and textual information, Xu et al. [14] and Chiu et al. [4] take advantages of both web-casting text and video shots to tag events in sport lives and Chakrabarti et al. [2] use tweets to label sport videos. However, these technics are limited in solving problem about big events videos. It is difficult for them to handle videos with plots such as movies and animations.

Time-Sync Commented Videos. Compared with works in traditional videos, the researches on time-sync commented videos are comparatively rare. Yoshii et al. [17] develop a music commentator to automatic generate music comments using time-sync commented videos. Chen et al. [3] uses both visual and textual features of time-sync commented videos to achieve personalized key frame recommendation. Wu et al. [13] and Yang et al. [16] use bullet-screen comments to tag videos by keywords. Similarly, Xu et al. [15] introduce a summarization model to extract key sentences for each slide. However, without supervised knowledge, their generated tags are extracted from user comments, which is insufficient to reflect the content of videos. Lv et al. [5] also introduce an approach to tag videos using bullet-screen comments. They propose a supervised method utilizing human labeled tags for each slide as training data. However, this kind of data is rare in natural extracted data and expensive to label by experts.

Supervised LDA Model. Since the LDA [1] model is an unsupervised model, to adapt supervised learning, a number of methods have been proposed. However, approaches such as Semi-LDA [11], Supervised Topic Model [6], MedLDA [18] are adaptations to solve single label classification problems. These methods have defects such as label independency in handling multi-label classification problem. Ramage et al. [7] propose a supervised LDA model for multi-label classification by corresponding labels with topics. Rubin et al. [8] also solve multi-label classification problem assuming that each topic correspond to a multinomial distribution over label-topics. However, as aforementioned, temporal differences are fairly important for time-sync commented videos. For ignoring this property, these approaches are not general enough in dealing with time-sync videos.

3   Problem Definition

In this section, we first introduce the basic properties of time-sync commented videos and bullet-screen comments. Then we will define the problem formally.

3.1 Properties of Bullet-Screen Comments

The first property is that bullet-screen comments often have temporal correlation in adjacent period of time, but in a long period of time, the comments reflect the content of the video and the correlation become weak. To model this temporal property, we sort bullet-screen comments in a video by playback time and split the comments into slides with same length in time. As videos with plots such as movies and animations are usually too dramatic to model the relations between scenes, we simply assume that slides in a video are independent.

Another property of bullet-screen comments is the aforementioned temporal differences in Sect. 1, which indicates that some contents of a video is not limited in the genre labels of the video. To solve this problem, we introduce a variable to control if the current slide is related to the genre labels of the video or not.

3.2 Formal Problem Definition

As mentioned in Sect. 3.1, we split the bullet-screen comments of time-sync commented videos into some isolated slides which have equal length. In our DLLDA model, bullet-screen comments in same slides will be treated as a document under the "bag-of-words" assumption. The notations to describe our model are showed in Table 1.

Table 1. Notations of DLLDA model

Notations	Interpretation
V	Set of Videos which contains $|V|$ videos
v_i	The ith video in V
C	Set of videos' bullet-screen comments
S_i	The list of $|S_i|$ slides of the bullet-screen comments of v_i
$s_{i,j}$	A set of $|s_{i,j}|$ words containing all the words in j^{th} slide of S_i
$w_{i,j,n}$	The nth word in $s_{i,j}$
$\mathcal{W}$	Vocabulary set which contains $|\mathcal{W}|$ words
$z_{i,j,n}$	The topic of word $w_{i,j,n}$
φ	The topic-word distribution
φ'_i	The video-specific topic-word distribution of video v_i
K	The total number of topics
t_k	The kth topic
$\theta_{i,j}$	The slide-topic distribution of slide $s_{i,j}$
$l_{i,j}$	The label-related selector of slide slide $s_{i,j}$
Λ_i	A one-hot binary vector to represent the genre labels of video v_i
η_i	The variable to control the proportion of label related slides in S_i
π	The hyperparameter to control the strength of labels
α	The Dirichlet prior of θ
β	The Dirichlet prior of φ and φ'_i
γ	The Binomial prior of η

The problems can be defined as following. By giving the video comment set C and label Λ for video set V as training set and video comment C' for video set V' as testing set, our model aims to find a label set Λ' and a keyword set D for video set V'.

4   DLLDA Model

In this section, we propose a supervised dynamic labeled LDA model to represent latent topics of labeled videos, classify genre labels and extract keywords from unlabeled videos.

4.1   The Training Model

To construct our training model, we first add two special labels. The "general" label represents general words in all comments and the "video-specific" label represent words specific for each video such as the names of characters. Since words in all slides may contain general words and video-specific words, these two labels are always equal to 1 for each slide.

Then each genre label are related to a latent topic which means there is a one-to-one relation between labels and topics φ, thus the supervised multi-label classification problem can be solved with a supervised LDA model. For the "video-specific" label mentioned above, we also introduce a video-specific topic φ' for each video, thus the keywords extraction problem is transformed to a video-specific topic learning problem. For simplification, we set the number of topics as k and the Kth label refers to the video-specific topic.

For each video in our training model, all the words of bullet-screen comments and the labels of the videos are regarded as observed variables and other variables are latent. For each word $w_{i,j,n}$ in a slide $s_{i,j}$, we assign a topic variable $z_{i,j,n}$. As mentioned in Sect. 3.1, a slide may in correlation with the labels of the video or not. In order to model this temporal difference between slides, for each slide $s_{i,j}$, we assign a binary variable $l_{i,j}$ to determine the topic in slide $s_{i,j}$ is in correlation with labels Λ_i of the video v_i or not. The variable $l_{i,j}$ is generated from distribution η_i. There is also a slide-topic distribution $\theta_{i,j}$ for each slide $s_{i,j}$, the prior of $\theta_{i,j}$ is described bellow.

To control the strength of relationship between label related slides and the genre labels, parameter π are introduced. If a slide $s_{i,j}$ is label related or in other words $l_{i,j} = 1$, then parameter π will determine the prior of $\theta_{i,j}$ of this slide: $\theta_{i,j} \sim Dirichlet(\Lambda_i \pi \alpha + (\mathbf{1} - \Lambda_i)(1 - \pi)\alpha)$ where $\mathbf{1}$ means a vector all elements are 1. If $l = 0$, the prior of θ is simply set as α, which means the prior of all topics are equal.

4.2   The Testing Model

For the testing model, we decide to adapt our DLLDA training model for testing task. With the consideration that some slides in a video may be noisy for learning its topics, e.g. the credits of movies or animations may only have the staff information and these slide will disturb the result of testing. In these case, simply averaging the topic distribution of all the slides in a video is not a good idea. To solve this problem, the bullet-screen comments of a video are regarded as a whole document to learn topic distribution of a video. In addition, we reserve the video-specific topic distribution to learn keywords in a video.

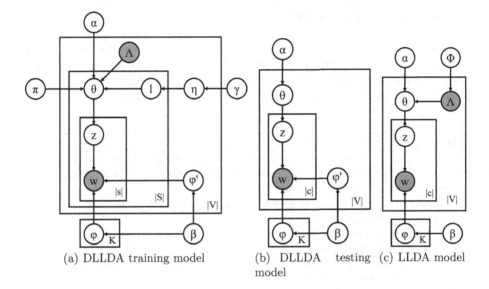

(a) DLLDA training model   (b) DLLDA testing model   (c) LLDA model

Fig. 2. Graphical model representations of DLLDA model and LLDA model

The probabilistic graphical model of our DLLDA model and traditional Labeled LDA model [7] is showed in Fig. 2.

4.3   The Generation Process

As the DLLDA model has been defined, it will be used for modeling the bullet-screen comments. The generation process of DLLDA model in training step is:

- For each topic t_k,
    - Draw $\varphi_k \sim Dirichlet(\beta)$
- For each video v_i,
    - Draw $\varphi'_i \sim Dirichlet(\beta)$
    - Draw $\eta \sim Binomial(\gamma)$
- For each slide $s_{i,j}$,
    - Draw $l_{i,j} \sim Bernoulli(\eta)$
    - If $l_{i,j} = 1$: draw $\theta_{i,j} \sim Dirichlet(\Lambda_i \pi \alpha + (1 - \Lambda_i)(1 - \pi)\alpha)$
    - If $l_{i,j} = 0$: draw $\theta_{i,j} \sim Dirichlet(\alpha)$
- For each word $w_{i,j,n}$,
    - Draw $z_{i,j,n} \sim Multi(\theta_{i,j})$

- If $z_{i,j,n} = t_k$: draw $w_{i,j,n} \sim Multi(\varphi'_i)$
- Else: draw $w_{i,j,n} \sim Multi(\varphi_{z_{i,j,n}})$

As for the testing step, generation process is similar to the traditional LDA model except the generation of φ' which is same as training process above.

5   Inference Algorithm

In this section, we introduce a collapsed Gibbs sampling algorithm to infer the latent variables in our DLLDA model with the generation process mentioned above. Gibbs sampling is a widely used Markov chain Monte Carlo (MCMC) algorithm for approximate inference in probabilistic graphical model.

Algorithm 1. DLLDA training algorithm

Input: Videos V, each video $v_i \in V$ refers to a list of slides S_i; each slide $s_{i,j} \in S_i$ contains $|s_{i,j}|$ words. Further more, each video also contains a set of label Λ_i.
Output: A topic-word distribution φ.
1: Initialize each variable $l_{i,j}$ in a slide $s_{i,j}$ and each topic $z_{i,j,n}$ of word $w_{i,j,n}$.
2: **for** #sampling iterations **do**
3:     **for** video v_i in V **do**
4:         Sample distribution η_i by (1)
5:         **for** slide $s_{i,j}$ in S_i **do**
6:             Sample $l_{i,j}$ by (2)
7:             **if** $l_{i,j} = 0$ **then**
8:                 Sample $\theta_{i,j}$ by (3)
9:             **else**
10:                Sample $\theta_{i,j}$ by (4)
11:            **for** word $w_{i,j,n}$ in $s_{i,j}$ **do**
12:                Sample $z_{i,j,n}$ by $\theta_{i,j}\varphi_{all}$
13:                **if** topic $z_{i,j,n} = t_K$ **then**
14:                    Sample $w_{i,j,n}$ by (5)
15:                **else**
16:                    Sample $w_{i,j,n}$ by (6)
17: **return** φ

At each iteration of our proposed Gibbs sampler and for each slide $s_{i,j}$, we first sample the variable η_i. As $\eta \sim Binomial(\gamma)$, the posterior probability of η_i can be represented as:

$$p(\eta_i|l_i,\gamma) = \frac{m_{i,l=1}^{\neg} + \gamma_1}{\sum_{h=0,1} m_{i,l=h}^{\neg} + \gamma_h}. \tag{1}$$

where $m_{i,l=h}^{\neg}$ represents the number of l which equals h in video v_i except the current one in slide $s_{i,j}$.

Then we sample the label $l_{i,j}$ for each slide $s_{i,j}$. For simplification, we define the prior of $\theta_{i,j}$ condition on $l_{i,j} = 1$ as $\lambda_i = \Lambda_i \pi \alpha + (1 - \Lambda_i)(1 - \pi)\alpha$, which denote that if a slide is related to the labels of the video, the strength of this correlation can be adjust by this formula. Thus the probability of $l_{i,j}$ can be represented as:

$$p(l_{i,j} = 1 | \theta_{i,j}, \eta_i) = \frac{Dir(\lambda_i)\eta_i}{Dir(\lambda_i)\eta_i + Dir(\alpha)(1 - \eta_i)}$$
$$= \frac{\prod_{k=1}^{K} \frac{\Gamma(\alpha_k)}{\Gamma(\lambda_{i,k})}\eta_i}{\prod_{k=1}^{K} \frac{\Gamma(\alpha_k)}{\Gamma(\lambda_{i,k})}\eta_i + \prod_{k=1}^{K} \theta_k^{\alpha_k - \lambda_{i,k}}(1 - \eta_i)}. \tag{2}$$

After inferencing the prior of the slide-topic distribution $\theta_{i,j}$, to estimate the posterior of distribution $\theta_{i,j}$, according to Sect. 3.1, each slide in a video is considered independent to other slides giving the parameter $l_{i,j}$. When the variable $l_{i,j} = 0$, we have:

$$\theta_{i,j,k} = \frac{m_{i,j,k}^{\neg} + \alpha}{\sum_{k=1}^{K}(m_{i,j,k}^{\neg} + \alpha)}. \tag{3}$$

When the variable $l_{i,j} = 1$, we have:

$$\theta_{i,j,k} = \frac{m_{i,j,k}^{\neg} + \lambda_{i,k}}{\sum_{k=1}^{K}(m_{i,j,k}^{\neg} + \lambda_{i,k})}. \tag{4}$$

where $m_{i,j,k}^{\neg}$ represents the number of words that topics are t_k in $s_{i,j}$, except the current word.

We also introduce a process to sample the video-specific topic distribution when the topic $z_{i,j,n}$ of a word $w_{i,j,n}$ is sampled as the Kth topic:

$$\varphi_{i,t}' = \frac{m_{k_i,t}^{\neg} + \beta}{\sum_{t=1}^{|\mathcal{W}|}(m_{k_i,t}^{\neg} + \beta)}. \tag{5}$$

where $m_{k_i,t}^{\neg}$ represents the number of words which topic is sampled as the video-specific topic in video v_i except the current word. And for other topics, the topic distribution is just same as traditional LDA model:

$$\varphi_{k,t} = \frac{m_{k,t}^{\neg} + \beta}{\sum_{t=1}^{|\mathcal{W}|}(m_{k,t}^{\neg} + \beta)}. \tag{6}$$

where $m_{k,t}^{\neg}$ represents the number of words which topic is sampled as the kth topic except the current word.

In each sample, $z_{i,j,k}$ can be drawn according to the probability proportional to $\theta_{i,j}\varphi_{all}$, where φ_{all} is the combination of video-specific topic distribution φ_i' and normal topic distribution φ.

6   Experiments

In this section, we firstly conduct experiments compared with several baseline models. Then we discuss the parameter sensitivity in Sect. 6.3, and show our learned results in the end of this section.

6.1   Dataset

Unfortunately, the data in English time-sync commented video website such as Viki[3] are not abundant enough for our experiments. We extract a real-world dataset from one of the most popular time-sync commented video website in China called bilibili[4], which focuses on animations and games. Our dataset contains over 500 animation series including their labels as well as the bullet-screen comments of each episode. In this paper, we simply regard one season of an animation as a video, which usually contains 13 or 26 episodes. As the bullet-screen comments are usually noisy, the raw data we extracted must be pre-processed.

First, we remove some of the uninformed genre label e.g. "Adapted from a comic", "UltraShort" and "Original" which have little information about content of videos. We also remove some video that have too few bullet-screen comments.

Next, we segment our bullet-screen comments data and remove stopwords, symbols and emoticons in comments and merge words only different in suffix with repeat characters. For instance, "www" and "wwwww" would be merged.

In the end of pre-processing part, we divide each bullet-screen comments of a video into slides. As mentioned above, the slides are length sensitive. According to our experience, we chose 1 min as the length of a slide.

After the pre-processing progress, we finally get 232 videos which correspond to 132688 min videos. We split videos into two datasets and for each dataset, we randomly select 80% of the videos as training set and the rest as testing set.

6.2   Experimental Setup

Before the experiment, we first discuss parameters for our model. We empirically set the Dirichlet prior $\alpha = 50/K$, $\beta = 0.01$ without optimization for both training and testing. And we set $\gamma_0 = 50$, $\gamma_1 = 100$ based on our observation of the videos. For the parameter π, we will explain the parameter learning in detail in Sect. 6.3.

To evaluate the performance of our model, we introduce three baseline models: Labeled-LDA model [7], Dependency-LDA model [8] and SW-IDF model [16]. In addition, the iterations of all the models are set to 100.

Labeled LDA model is a model to solve multi-label classification problem and generating topics for each label. In this model, the whole bullet-screen comments of a video are treated as an independent document.

---

[3] https://www.viki.com/.
[4] http://www.bilibili.com/.

Dependency LDA model is also used to solve multi-label classification problem. This model is a LDA-based method for multi-label document classification. The Dirichlet prior parameters $\alpha = 50/K$, $\beta = 0.01$ are same as our model and other parameters are set default.

SW-IDF model is an unsupervised keywords extraction algorithm using bullet-screen comment data. We use this baseline model for case study to compare keywords extracted by our video-specific topics.

Figure 3(a) shows variance of topics in training step of our model.

6.3 Parameter Sensitivity

In this section, we examine the optimization of the parameter π mentioned above.

It is obvious that if $\pi = 0.5$, the model will have no supervised information. To enforce supervised information, the value of π should be much greater than 0.5.

Thus we introduce a parameter tuning experiment to evaluate π. Figure 3(b) shows the performance of π and we can see that when π is set to 0.94, the model performs best.

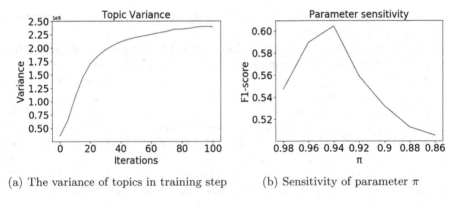

(a) The variance of topics in training step   (b) Sensitivity of parameter π

Fig. 3. Graphical model representations of DLLDA model and LLDA model

6.4 Experiment Results

To evaluate the multi-label task, we choose some widely used multi-label evaluation metrics. For the Jaccard index, Precision, Recall and F1-score, higher values refer to better performance and for the Hamming loss and One error, the lower values refer to better performance. The overall results of experiment are shown in Table 2 and "D1" and "D2" means the two datasets mentioned in Sect. 6.1.

We can see that our DLLDA model outperforms other models in nearly all of these metrics in two datasets. This result shows that the micro F1 score of two baseline models are close to each other and our model outperforms the two baselines over 20% in micro F1 score. Our model outperforms especially in precision metrics. This is because the large amount of slides are noisy in training

Table 2. Performance of three models in two datasets

	Jaccard index		Precision		Recall		F1-score		Hamming loss		One error	
	D1	D2	D1	D2	D1	D2	D1	D2	D1	D2	D1	D2
LLDA	0.288	0.337	0.382	0.458	0.539	**0.559**	0.447	0.504	0.300	0.346	0.186	0.179
D-LDA	0.292	0.280	0.382	0.383	0.553	0.511	0.452	0.438	0.300	0.318	0.196	0.206
DLLDA	**0.455**	**0.433**	**0.625**	**0.691**	**0.625**	0.537	**0.625**	**0.604**	**0.250**	**0.136**	**0.100**	**0.115**

data and our model consider the temporal differences which helps distinguish noise and improve the precision of learned labels. The one error metric shows that our method also performs better than baselines in ranking performance.

For comparison, we also use pair-t test to show the confidence of the improvement. We compare the results in our DLLDA model and D-LDA model of every animation in two datasets. The result p-values of two datasets are 0.002095 and 0.001097, which means there is a strong evidence that our model does work better than the baseline model on average.

6.5   Case Study

In this section, we show some results to illustrate the performance of our topic learning and keywords extraction.

Firstly, to compare the topics obtained by our DLLDA model and the baseline Labeled LDA model [7], we randomly choose three topics and demonstrate 10 most probable words from each topic[5] in Table 3. Since the relationship of labels and topics is one-to-one, we simply use label names to represent topics.

In general, both two model generate some meaningful words. Most of the words generated by our model have strong relationship with the topic. For the Labeled LDA model, many words such as "Joan of Arc", "Holy Grail" and "green hair" only have relationship with some animation labeled with these topics. In addition, we can notice that words such as "leading actor" and "2333"[6] are really general and have little information to reflect the content of topics.

We also compare the keywords extracted from our DLLDA model and the baseline SW-IDF [16] model. We randomly choose three animations in our test datasets and the result are showed in Table 4.

In the first sample, our DLLDA Model correctly extract the name of two main character, but for the SW-IDF model, the extracted names refer to characters from other irrelevant animations. In the second sample, some keywords extracted by our model have strong relationship with the animation. But for the baseline model, the most probable words wrongly focus on the taste of a kind of food which have absolutely no relation with the animation. In the third sample, two models get similar results and both model successfully extract some names of main characters.

---

[5] These words and topic names are manually translated to English by the authors.

[6] A Chinese internet slang means laughing.

Table 3. 10 most probable words of three topics generated from two models

Topic	DLLDA model	Labled LDA model
Topic "Battle"	king, **sword**, online, **powerful, chop, fight, falchion, physical strength**, infinite	**summon**, Glittering, king, my king, Joan of Arc, leading actor, **sword**, tutor, Holy Grail, world
Topic "Delicacy"	**eat, fish, delicious, drink**, China, Japan, **meat, meal, taste, hungry**	**eat**, eye protection, leading actor, 2333, **delicious**, drug, **meat**, lady, like, wife
Topic "Super Robot Wars"	**the earth, aircraft, cannon, human beings, universe, gundam, fly, war**, protagonist, **enemy**	queen, green hair, sing song, leading actor, uncle, sing, song, farewell, 2333, beautiful

Table 4. 10 most probable keywords extracted from three animations by two models

Animation	DLLDA model	SW-IDF model
From Me to You	panda, **like**, heroine, **Kurumi**, leading actor, husband, **cute, Sawako**, second heroine, **girl's heart**	husband, wife, panda, benefit, **heart**, girl, Takashi, Natsume, hope
Croisée in a Foreign Labyrinth	leading actor, **lolita, Yune**, elder sister, **Japan**, grandfather, **Kimono**, China, lily, **France**	eat, watery tender bean curd, sweat, salty, leading actor, **Japan, lolita**, lily, China, younger sister
Higurashi When They Cry	leading actor, **Shion, Mion, Rika**, perfect, crime, **Higurashi, Rena, breakdown**, air conditioner	**Shion**, leading actor, perfect, **Mion, crime**, like, **Rika**, 2333, **Higurashi, Ooishi**

7 Conclusions

In this paper, we proposed a supervised dynamic LDA model utilizing both genre labels and keywords to tag videos. We regard video labels as topics and introduce a supervised LDA model. To deal with the temporal differences in a video, we split each video into slides and introduce a selector to determine whether the slide is related to the labels of the video. Then we use a LDA based multi-label classification algorithm to label videos and extract keywords. The Experiments on large real-world datasets prove our effectiveness both in genre label classification and keyword extraction compared with baseline models.

Acknowledgments. This work is partially supported by National Key Research and Development Program of China.

References

1. Blei, D.M., Ng, A.Y., Jordan, M.I.: Latent Dirichlet allocation. J. Mach. Learn. Res. Arch. **3**, 993–1022 (2003)
2. Chakrabarti, D., Punera, K.: Event summarization using tweets. ICWSM **11**, 66–73 (2011)
3. Chen, X., Zhang, Y., Ai, Q., Xu, H., Yan, J., Qin, Z.: Personalized key frame recommendation. In: Proceedings of the 40th International ACM SIGIR Conference on Research and Development in Information Retrieval, pp. 315–324. ACM (2017)
4. Chiu, C.Y., Lin, P.C., Li, S.Y., Tsai, T.H., Tsai, Y.L.: Tagging webcast text in baseball videos by video segmentation and text alignment. IEEE Trans. Circuits Syst. Video Technol. **22**(7), 999–1013 (2012)
5. Lv, G., Xu, T., Chen, E., Liu, Q., Zheng, Y.: Reading the videos: temporal labeling for crowdsourced time-sync videos based on semantic embedding. In: AAAI, pp. 3000–3006 (2016)
6. Mcauliffe, J.D., Blei, D.M.: Supervised topic models. In: Advances in Neural Information Processing Systems, pp. 121–128 (2008)
7. Ramage, D., Hall, D., Nallapati, R., Manning, C.D.: Labeled LDA: a supervised topic model for credit attribution in multi-labeled corpora. In: Proceedings of the 2009 Conference on Empirical Methods in Natural Language Processing, vol. 1, pp. 248–256. Association for Computational Linguistics (2009)
8. Rubin, T.N., Chambers, A., Smyth, P., Steyvers, M.: Statistical topic models for multi-label document classification. Mach. Learn. **88**(1–2), 157–208 (2012)
9. Siersdorfer, S., San Pedro, J., Sanderson, M.: Automatic video tagging using content redundancy. In: Proceedings of the 32nd international ACM SIGIR Conference on Research and Development in Information Retrieval, pp. 395–402. ACM (2009)
10. Ulges, A., Schulze, C., Koch, M., Breuel, T.M.: Learning automatic concept detectors from online video. Comput. Vis. Image Underst. **114**(4), 429–438 (2010)
11. Wang, Y., Sabzmeydani, P., Mori, G.: Semi-latent Dirichlet allocation: a hierarchical model for human action recognition. In: Elgammal, A., Rosenhahn, B., Klette, R. (eds.) HuMo 2007. LNCS, vol. 4814, pp. 240–254. Springer, Heidelberg (2007). https://doi.org/10.1007/978-3-540-75703-0_17
12. Wang, Z., Yu, J., He, Y., Guan, T.: Affection arousal based highlight extraction for soccer video. Multimed. Tools Appl. **73**(1), 519–546 (2014)
13. Wu, B., Zhong, E., Tan, B., Horner, A., Yang, Q.: Crowdsourced time-sync video tagging using temporal and personalized topic modeling. In: 20th ACM SIGKDD International Conference on Knowledge Discovery and Data Mining, pp. 721–730. ACM (2014)
14. Xu, C., Wang, J., Wan, K., Li, Y., Duan, L.: Live sports event detection based on broadcast video and web-casting text. In: Proceedings of the 14th ACM International Conference on Multimedia, pp. 221–230. ACM (2006)
15. Xu, L., Zhang, C.: Bridging video content and comments: Synchronized video description with temporal summarization of crowdsourced time-sync comments. In: AAAI, pp. 1611–1617 (2017)
16. Yang, W., Ruan, N., Gao, W., Wang, K., Ran, W., Jia, W.: Crowdsourced time-sync video tagging using semantic association graph. In: 2017 IEEE International Conference on Multimedia and Expo (ICME), pp. 547–552. IEEE (2017)

17. Yoshii, K., Goto, M.: Musiccommentator: Generating comments synchronized with musical audio signals by a joint probabilistic model of acoustic and textual features. In: ICEC (2009)
18. Zhu, J., Ahmed, A., Xing, E.P.: Medlda: maximum margin supervised topic models for regression and classification. In: Proceedings of the 26th annual international conference on machine learning. pp. 1257–1264. ACM (2009)

Accurate Q-Learning

Zhihui Hu[1], Yubin Jiang[1], Xinghong Ling[1], and Quan Liu[1,2,3(✉)]

[1] School of Computer Science and Technology, Soochow University,
Suzhou 215006, Jiangsu, China
quanliu@suda.edu.cn
[2] Collaborative Innovation Center of Novel Software Technology
and Industrialization, Nanjing 210000, China
[3] Key Laboratory of Symbolic Computation and Knowledge Engineering
of Ministry of Education, Jilin University, Changchun 130012, China

Abstract. In order to solve the problem that Q-learning can suffer from large overestimations in some stochastic environments, we first propose a new form of Q-learning, which proves that it is equivalent to the incremental form and analyze the reasons why the convergence rate of Q-learning will be affected by positive bias. We generalize the new form for the purpose of easy adaptations. By using the current value instead of the bias term, we present an accurate Q-learning algorithm and show that the new algorithm converges to an optimal policy. Experimentally, the new algorithm can avoid the effect of positive bias and the convergence rate is faster than Q-learning and its variants on several MDP problems.

Keywords: Reinforcement learning · Q-learning · Positive bias
Accurate Q-learning

1 Introduction

Q-learning is a popular off-policy reinforcement learning algorithm proposed by Watkins [16], which is widely used to solve optimal control problems in Markov decision processes (MDPs) [4,17,18]. In finite state-action problems, it has been proved that Q-learning can converge to an optimal action-value function [6,9,11], but the convergence rate of it will be slow when the discount factor is close to one [15]. In some stochastic environments with highly random rewards, Q-learning will cause high statistical errors because of the max operator. The performance of Q-learning can suffer from the significant overestimation of action values [8,10]. This is proved under the multi-armed bandit problem [1,12]. The estimated expected return is overly optimistic which can hinder the performance.

There have been several papers appeared with proposed improvements. Bias corrected Q-learning [10] constructs the bias-correction term and retains the asymptotic convergence of Q-learning. This algorithm can only be used when the number of actions is not less than 5. Another method using a single action-value function estimate is weighted Q-learning [5], which is based on a weighted

© Springer Nature Switzerland AG 2018
L. Cheng et al. (Eds.): ICONIP 2018, LNCS 11303, pp. 560–570, 2018.
https://doi.org/10.1007/978-3-030-04182-3_49

average of the sample means. Since the weights are computed using Gaussian approximations for the distributions of the sample means, the calculation amount is huge. Double Q-learning [7] uses two estimators, one is to determine the maximum action and the other is to provide the estimate of its value. Double Q-learning has been shown to reduce bias but underestimate the action values in some cases. It inspired weighted double Q-learning [19], which is based on the construction of the weighted double estimator in order to strike a balance between the overestimation in the single estimator and the underestimation in the double estimator.

The main contribution of this paper is to propose the accurate Q-learning algorithm which can avoid the negative effects of positive bias by modifying one of the simplified Q-learning terms. We prove that the algorithm converges to the optimal solution in the limit. We demonstrate the benefits of our algorithm for several problems and compare our experiment results with Q-learning and its variants.

The rest of the paper is organized as follows: In Sect. 2, we give the notations used in this paper. In Sect. 3, we present the accurate Q-learning algorithm. We first propose a new form which is equivalent to Q-learning update rule in Sect. 3.1 and we generalize the new form in Sect. 3.2, then we construct an accurate Q-learning algorithm by using the current value instead of the bias term in Sect. 3.3. In Sect. 4 we give the experimental results and illustrate the policy quantity and convergence rate of the new algorithm. Finally, We conclude the paper and give the future directions of our work.

2   Background

A Markov decision process is defined as a five-tuple (S, A, P, R, γ), where S is a set of states and A is a set of actions, S and A are finite. $P : S \times A \times S' \to [0, 1]$ is the state transition distribution, where $P(s, a, s')$ is the probability of taking action a in state s will lead to state s'. $R : S \times A \to \mathbb{R}$ is a reward function, where $R(s, a)$ is the immediate reward obtained by taking action a in state s. $\gamma \in [0, 1)$ is a discount factor for the $\mathbb{E}[\sum_{t=1}^{\infty} \gamma^{t-1} R(s_t, a_t)]$ which represents the difference in importance degree between current rewards and long-term returns.

A policy $\pi : S \to A$ specifies the action that the agent will choose in state s. The action-value function of a policy π are represented by $Q^\pi : S \times A \to \mathbb{R}$. The main point of MDPs is to find an optimal policy π^* that maximizes $Q^\pi(s, a)$, which satisfies the Bellman optimality equation in MDPs [3]:

$$Q^*(s, a) = \mathbb{E}\{R(s, a) + \gamma \max_{a'} Q^*(s', a')\} \tag{1}$$

The key idea of Q-learning is to apply incremental estimations to Bellman optimization equation [2]. The update of Q-learning can be written as

$$Q_{t+1}(s, a) = (1 - \alpha_t(s, a))Q_t(s, a) + \alpha_t(s, a)(R_t(s, a) + \gamma \max_{a'} Q_t(s', a')) \tag{2}$$

where $\alpha_t(s, a) \in (0, 1]$ is the learning rate parameter associated with the state-action pair at time step t. Q_t is guaranteed to converge to Q^* if each state-action

562   Z. Hu et al.

pair can be visited infinite times and the learning rates are chosen under the following conditions [13,14]:

$$\sum_{t\geq 0}\alpha_t(s,a) = \infty, \quad \sum_{t\geq 0}\alpha_t^2(s,a) < \infty \tag{3}$$

3 Accurate Q-Learning

In this section, we introduce a new form of Q-learning, which proves that it is equivalent to the incremental form and analyzes the reasons for overestimation of action value. Then we generalize the new form and get the corresponding update rule. We propose a new algorithm called accurate Q-learning based on the new form and give its incremental form.

3.1 A New Form of Q-Learning

Let Q_n be the action-value function used in the (n+1)th update of state-action pair (s,a) and simplify $R_n(s,a)$ and $\alpha_n(s,a)$ as R_n and α_n, respectively. Define $\mathcal{M}$ as the maximum function on action-value functions that $\mathcal{M}(s) = \max_a Q(s,a)$. Then the update rule Eq. 2 may be written as

$$Q_{n+1}(s,a) = (1-\alpha_n)Q_n(s,a) + \alpha_n\big(R_n + \gamma\mathcal{M}_n(s')\big) \tag{4}$$

From the practical point of view, we obtain a new form of Q-learning by recurring the update rule according to Eq. 4.

Proposition 1. *For all $n \geq 0$, the update rule of Q-learning may be rewritten as*

$$Q_{n+1}(s,a) = H_0^n Q_0(s,a) + \sum_{i=0}^{n} H_{i+1}^n \alpha_i\big(R_i + \gamma\mathcal{M}_i(s')\big) \tag{5}$$

where H_i^j is defined as

$$H_i^j = (1-\alpha_i)(1-\alpha_{i+1})\ldots(1-\alpha_j), \quad \text{for } j \geq i \tag{6}$$

and $H_{j+1}^j := 1$.

Proof. For $n = 0$ we have:

$$Q_1(s,a) = (1-\alpha_0)Q_0(s,a) + \alpha_0\big(R_0 + \gamma\mathcal{M}_0(s')\big)$$

Now for any $k - 1 \geq 0$, let us assume that the equation Eq. 5 hold, which is

$$Q_k(s,a) = H_0^{k-1}Q_0(s,a) + \sum_{i=0}^{k-1} H_{i+1}^{k-1}\alpha_i\big(R_i + \gamma\mathcal{M}_i(s')\big)$$

Thus

$$Q_{k+1}(s,a) = (1-\alpha_k)Q_k(s,a) + \alpha_k\big(R_k + \gamma\mathcal{M}_k(s')\big)$$

$$= (1-\alpha_k)\Big[H_0^{k-1}Q_0(s,a) + \sum_{i=0}^{k-1} H_{i+1}^{k-1}\alpha_i(R_i + \gamma\mathcal{M}_i(s'))\Big]$$
$$+ \alpha_k\big(R_k + \gamma\mathcal{M}_k(s')\big)$$

$$= H_0^k Q_0(s,a) + \alpha_k\big(R_k + \gamma\mathcal{M}_k(s')\big) + \sum_{i=0}^{k-1} H_{i+1}^k\alpha_i\big(R_i + \gamma\mathcal{M}_i(s')\big)$$

$$= H_0^k Q_0(s,a) + \sum_{i=0}^{k} H_{i+1}^k\alpha_i\big(R_i + \gamma\mathcal{M}_i(s')\big)$$

In order to reduce the influence of initialization of the action values, we let $\alpha_0 = 1$, then $H_0^k Q_0(s,a) = 0$. When the rewards exist a considerable randomness, although $\sum_{i=0}^k H_{i+1}^k\alpha_i R_i$ is unbiased eventually, it is inaccurate in the early updates. Because its value is not equivalent to the expected value of the reward. $\gamma\sum_{i=0}^k H_{i+1}^k\alpha_i\mathcal{M}_i(s')$ may overestimate due to the inaccuracy of the value function. The unbiased estimate can be expressed as $\max_{a^*}\sum_{i=0}^k H_{i+1}^k\alpha_i Q_i(s',a^*)$ and we obtain the inequation:

$$\sum_{i=0}^{k} H_{i+1}^k\alpha_i\mathcal{M}_i(s') \geq \max_{a^*}\sum_{i=0}^{k} H_{i+1}^k\alpha_i Q_i(s',a^*) \tag{7}$$

The inequality is strict if $\exists i, \mathcal{M}_i(s') > Q_i(s',a^*)$. Since Q-learning use the max operator to determine the value of the next state, it always selects the maximum action value. In the early stage of training, the action value it chooses always has a positive bias compared to the expected value.

The positive bias can cause two main negative effects. One is the diffusion of positive bias in the value function. Due to using the maximum of the estimates as an estimate of the maximum of the true values, the overestimated action values are always selected so that the positive biases propagate throughout the value function. But this may not prevent Q-learning from converging to the optimal policy. The other effect is the slow convergence of Q-learning. With the continuous training, $\sum_{i=0}^k H_{i+1}^k\alpha_i R_i$ will converge to $\mathbb{E}[R]$. The bias of each action value in this trial will be corrected, but this correction process is unbearably slow. Due to the coefficient α_n, unbiased estimates do not immediately correct the biases, but rather require multiple visits and updates for each state action pair. This phenomenon is especially evident when we use decaying learning rates. With the number of times each state-action pair visited increases, α_n is gradually approaching 0, the correction of bias will be unacceptable slow. This results in very slow convergence to an optimal policy.

3.2   Generalization

Due to using $\sum_{i=0}^{k} H_{i+1}^{k} \alpha_i \mathcal{M}_i(s')$, the convergence rate of Q-learning is slow, so we intend to improve on this term. First, we propose a new form of generalization that facilitates the improvement of the estimator and makes it easy to obtain update rules. Let $E(s)$ represent the estimate of the state s. Then we can construct a general form as

$$Q_{n+1}(s,a) = H_0^n Q_0(s,a) + \gamma E_n(s') + \sum_{i=0}^{n} H_{i+1}^n \alpha_i R_i \tag{8}$$

Proposition 2. *Assume that $E_{-1} = 0$, then we have, for all $n \geq 0$, Eq. 8 is equivalent to the following incremental form:*

$$Q_{n+1}(s,a) = (1 - \alpha_n)\Big[Q_n(s,a) - \gamma E_{n-1}(s')\Big] + \gamma E_n(s') + \alpha_n R_n \tag{9}$$

Proof. For $n = 0$ we have:

$$Q_1(s,a) = (1 - \alpha_0)\Big[Q_0(s,a) - \gamma E_{-1}(s')\Big] + \gamma E_0(s') + \alpha_0 R_0$$

$$= H_0^0 Q_0(s,a) + \gamma E_0(s') + \sum_{i=0}^{0} H_{i+1}^0 \alpha_i R_i$$

Now for any $k - 1 \geq 0$, let us assume that Eq. 8 hold, which is

$$Q_k(s,a) = H_0^{k-1} Q_0(s,a) + \gamma E_{k-1}(s') + \sum_{i=0}^{k-1} H_{i+1}^{k-1} \alpha_i R_i$$

Thus

$$Q_{k+1}(s,a) = (1 - \alpha_k)\Big[Q_k(s,a) - \gamma E_{k-1}(s')\Big] + \gamma E_k(s') + \alpha_k R_k$$

$$= (1 - \alpha_k)\Big[H_0^{k-1} Q_0(s,a) + \sum_{i=0}^{k-1} H_{i+1}^{k-1} \alpha_i R_i\Big] + \gamma E_k(s') + \alpha_k R_k$$

$$= H_0^k Q_0(s,a) + \gamma E_k(s') + \alpha_k R_k + \sum_{i=0}^{k-1} H_{i+1}^k \alpha_i R_i$$

$$= H_0^k Q_0(s,a) + \gamma E_k(s') + \sum_{i=0}^{k} H_{i+1}^k \alpha_i R_i$$

For Q-learning, $E_n^Q(s) = \sum_{i=0}^{n} H_{i+1}^n \alpha_i \mathcal{M}_i(s) = \alpha_n \mathcal{M}_n(s) + (1 - \alpha_n)E_{n-1}(s)$. It is easy to see

$$Q_{n+1}(s,a) = H_0^n Q_0(s,a) + \gamma E_n^Q(s') + \sum_{i=0}^{n} H_{i+1}^n \alpha_i R_i$$

$$= H_0^n Q_0(s,a) + \gamma \sum_{i=0}^{n} H_{i+1}^n \alpha_i \mathcal{M}_i(s') + \sum_{i=0}^{n} H_{i+1}^n \alpha_i R_i$$

$$= (1 - \alpha_n)Q_n(s,a) + \alpha_n\big(R_n + \gamma \mathcal{M}_n(s')\big)$$

We can use Eq. 9 as the update rule of Q-learning, where $E_n(s') = E_n^Q(s')$. Comparing Eq. 8 with Eq. 9, it can be easily found that $H_0^n Q_0(s,a) + \sum_{i=0}^{n} H_{i+1}^n \alpha_i R_i = (1 - \alpha_n)\big[Q_n(s,a) - \gamma E_{n-1}(s')\big] + \alpha_n R_n$. Therefore, only the influence brought by the change of $E_n(s)$ needs to be considered when improving $E_n(s)$.

3.3   Accurate Q-Learning Algorithm

Because Q-learning uses $E_n^Q(s) = \sum_{i=0}^{n} H_{i+1}^n \alpha_i \mathcal{M}_i(s)$, the correction process of Q-learning is very slow and Q-learning suffers from prohibitively large overestimations. When the next state of the state-action pair (s,a) is determined and unique, i.e., $P(s,a,s') = 1$, we replace $\sum_{i=0}^{n} H_{i+1}^n \alpha_i \mathcal{M}_i(s)$ with $\sum_{i=0}^{n} H_{i+1}^n \alpha_i \mathcal{M}_n(s)$ to obtain a new estimator $E_n^A(s)$. We propose a modified version of Q-learning, that we call accurate Q-learning, which uses $E_n^A(s) = \sum_{i=0}^{n} H_{i+1}^n \alpha_i \mathcal{M}_n(s)$ to estimate the action values. Bringing $E_n^A(s)$ into Eq. 9, we can get the update rule of accurate Q-learning as

$$Q_{n+1}(s,a) = (1 - \alpha_n)\big(Q_n(s,a) - \gamma E_{n-1}^A(s')\big) + \gamma E_n^A(s') + \alpha_n R_n \qquad (10)$$

Algorithm 1. Accurate Q-learning

1: Initialize $F(s,a) = Q(s,a), \forall s \in S, a \in A$, arbitrarily
2: Initialize $H(s,a) \leftarrow 1, \forall s \in S, a \in A$
3: Initialize s
4: **loop**
5:     choose a from s using policy derived from Q
6:     Take action a ,observe r, s'
7:     $\Delta \leftarrow Q(s,a) - \gamma F(s,a)$
8:     $H(s,a) \leftarrow (1 - \alpha(s,a))H(s,a)$
9:     $a^* \leftarrow \arg\max_a Q(s',a)$
10:    $F(s,a) \leftarrow (1 - H(s,a))Q(s',a^*)$
11:    $Q(s,a) \leftarrow (1 - \alpha(s,a))\Delta + \gamma F(s,a) + \alpha(s,a)R$
12:    $s \leftarrow s'$

Accurate Q-learning algorithm, as shown in Algorithm 1, stores three functions: an action-value function Q and two auxiliary functions including H and F. Function H stores $H_0^n = (1 - \alpha_0) \ldots (1 - \alpha_n)$ and function F stores $E_n^A(s')$ for all state-action pairs.

Comparing the Q-learning update rule of Eq. 9 with the one for accurate Q-learning in Eq. 10, it can be found that the two estimators are different. $\sum_{i=0}^n H_{i+1}^n \alpha_i \mathcal{M}_i(s)$ includes all values that have been experienced before. It is a inaccurate value with a positive bias at the early stage of training because we use the maximum of the estimates as an estimate of the maximum of the expected values. The values will then be decreased in the next few steps, but the correction process is slow. In comparison, $\sum_{i=0}^n H_{i+1}^n \alpha_i \mathcal{M}_n(s)$ uses the maximum value of the current value function which is used to replace the maximum value of the previous value function. It allows the correction to propagate faster in the value function. This may be the reason why accurate Q-learning has a better performance than Q-learning.

In terms of computation-time complexity, accurate Q-learning is as same as that of Q-learning at each time step — $\mathcal{O}(1)$. While the space complexity is only slightly higher than Q-learning due to the two auxiliary terms that have to be accounted for. Accurate Q-learning needs $\mathcal{O}(3 \times |S| \times |A|)$ memory space while Q-learning only needs $\mathcal{O}(|S| \times |A|)$.

4   Experiments

This section shows the performance of Q-learning and its improved algorithms on roulette and grid world. The reward functions of these two experiments are highly random. Through roulette, which is a simple single-state MDP problem, we not only show that the performance of Q-learning can suffer from the significant positive bias but also present that the mean action values of these algorithms. Through grid world we show the empirical result of accurate Q-learning algorithm to analyze the estimation of state-action values and policy quality. In these two experiments, the discount factor was unified to 0.95. The policy for action selection of all algorithms was ϵ-greedy. The exploration parameter was $\epsilon(s) = 1/\sqrt{n(s)}$, where $n(s)$ is the number of times the state s has been visited.

4.1   Roulette

The game of Roulette is modeled as a MDP problem with one state and 171 actions, containing 170 different betting actions and one action corresponding to forfeit. Suppose that without considering the issue of funds, the player can choose randomly among 170 betting actions. If he bets \$1 each turn, he will get an expected payout of $\frac{1}{38}\$36 = \$0.947$ and an expected loss of $-\$0.053$. If the player chooses to forfeit, he will get \$0 and end the episode, so the optimal action is to forfeit and the optimal action value is \$0.

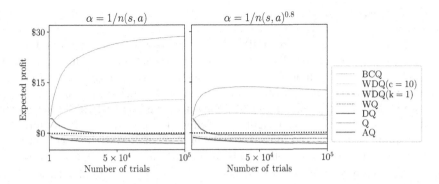

Fig. 1. The mean action values over all betting actions according to Q, DQ, AQ, WQ, WDQ(c = 10) and WDQ(k = 1) on the roulette. The learning method is step-by-step. These data are averaged over 10 experiments.

Figure 1 shows the mean action values over all betting actions created by Q-learning (Q), double Q-learning (DQ), accurate Q-learning (AQ), bias-corrected Q-learning (BCQ), weighted Q-learning (WQ), weighted double Q-learning (WDQ) with different values of parameter c = 10 and k = 1. The learning rate is either linear learning rate $1/(n(s,a)+1)$ or polynomial learning rate $1/(n(s,a)+1)^{0.8}$. We adopt the step-by-step update approach. Compared with the synchronous updates used in [7], which need each state to be visited at least once for each update, the asynchronous value iteration update is more meaningful in the actual application because it only updates the value of a single state at a time. During the whole learning process, the accurate Q-learning obtains the value closest to the true value $-0.053. The other algorithms seem to be more severely affected by the max operator. Moreover, after 100,000 trials, the bias of estimations still exist.

Table 1 shows the mean action values after 100,000 trials. Q-learning with a linear learning rate values all betting actions at almost $30 and it does not gradually decrease with the increase of iteration times. Even with a polynomial learning rate which is proven to be superior to a linear learning rate [8], the expected rewards still have a large bias more than $12. BCQ also has a clear positive bias on this problems. There is an not obvious underestimation in DQ. WDQC, WQ and AQ after the 100,000 step trials converge to a good value close to $0. It is worth mentioning that AQ is very close to the expected loss $-0.053.

Table 1. The mean action values over all betting actions on the roulette using asynchronous updates.

	Q	DQ	AQ	WDQ (c = 10)	WDQ (k = 1)	BCQ	WQ
Linear learning rate	28.62	−3.15	**−0.38**	−1.33	−2.60	9.904	−1.762
Polynomial learning rate	12.45	−3.34	**−0.62**	−1.80	−2.85	4.99	−1.868

4.2   Grid World

Consider a $n \times n$ grid world MDP problem where the cells correspond to the states of the environment. There are four possible actions (north, south, west, east) at each state. Each action deterministically causes the agent to move to an respective adjacent state, but keeps the current state unchanged when the agent will walk off the grid. The starting state is set in the lower left position and the goal state is set in the upper right. We define the reward for any actions at a non-terminal state is r and for any actions ending an episode from the terminal state is R_T. The optimal policy ends an episode after $2n - 1$ actions, so the optimal average reward per step is $\frac{\mathbb{E}[R_T]+2(n-1)\times\mathbb{E}[r]}{2n-1}$. We examined the action value in the starting state s_0. Note that the optimal value of this state would be $\max_a Q(s_0, a) = \mathbb{E}[R_T]\gamma^{2(n-1)} + \mathbb{E}[r]\sum_{i=0}^{2n-3}\gamma^i$. We use the linear learning rate $\alpha_n(s, a) = 1/(n(s, a) + 1)$ in this experiment. Each time a non-terminal state is visited, a reward of -8 or $+6$ will be awarded with equal probability, and a reward of $+5$ will be awarded to the terminal state and the episode will be ended immediately.

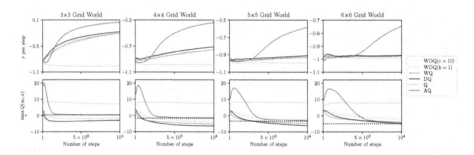

Fig. 2. The average rewards per time step and the maximal action value in the initial state s_0 according to Q, DQ, AQ, WQ, WDQ(c = 10) and WDQ(k = 1) on the grid world. These data are averaged over 1000 different sequences of runs.

We purposely add randomness to non-terminal state rewards in ordinary grid world, making the environment more challenging. The first row shows the average rewards and the second row is the maximum action value in the starting state. The reason for choosing the action value function using the state of s_0 is because this state is difficult to update. The black horizontal dotted lines in the second row of Fig. 2 represent the optimal average rewards per step on grid world problems with difference sizes from 3×3 to 6×6.

On all the questions shown in the Fig. 2, the maximal action value of Q-learning in the starting state has significant overestimation and mean action rewards is lower than others. The reason for this phenomenon is that if an action is overestimated, then this action will always be selected. Q-learning find that this action can not earn a expected high reward, then it will choose another action. This will lead to a similar dilemma, eventually causing unacceptable slow convergence of Q-learning. On these domains, the performances of WQ, WDQ(c = 10) and WDQ(k = 1) are similar by showing that the average rewards are very close to each other and the rates of convergence are slow. The performance of all algorithms suffer from the increase of the problem size. However, the AQ still maintains a very high reward value and a low error with the maximum action value of the s_0.

On each problems, there is a big increase in the stating state action value of AQ in the early stage of the training and affects the value of mean rewards. This is due to the $(1 - H_0^n)\mathcal{M}_n(s')$ in the update equation, using the max operator over the action values which is inaccurate in the early updates. Soon AQ can converge to a approximate true value and result in very high rewards.

5   Conclusion

This paper proposes a new form of Q-learning update equation which can make it more convenient to improve the algorithm. We use this new form to analyze the reasons for overestimation in Q-learning. In this new form, we find the biased term in Q-learning and use the current estimate to replace it. Using the new estimator we propose an accurate Q-learning algorithm that can converges to the optimal policy. The experimental results show that, compared with Q-learning and its variants, the new algorithm can avoid the performance suffer from positive bias in some MDP problems with highly random reward functions.

The future work is to use the function approximation to scale up the accurate Q-learning algorithm to solve the reinforcement learning problems in continuous state space. Another direction for future work is to use the general form to get other improved algorithms.

Acknowledgments. This work was funded by National Natural Science Foundation (61272005, 61303108, 61373094, 61502323, 61272005, 61303108, 61373094, 61472262). We would also like to thank the reviewers for their helpful comments. Natural Science Foundation of Jiangsu (BK2012616), High School Natural Foundation of Jiangsu (13KJB520020), Key Laboratory of Symbolic Computation and Knowledge.

References

1. Auer, P., Cesa-Bianchi, N., Fischer, P.: Finite-time analysis of the multiarmed bandit problem. Mach. Learn. **47**(2–3), 235–256 (2002)
2. Azar, M.G., Munos, R., Ghavamzadeh, M., Kappen, H.J.: Speedy Q-learning. In: Proceedings of the 24th International Conference on Neural Information Processing Systems, pp. 2411–2419. Curran Associates Inc. (2011)
3. Bellman, R.: Dynamic Programming. Courier Corporation, North Chelmsford (2013)
4. Bertsekas, D.P.: Stable optimal control and semicontractive dynamic programming. SIAM J. Control Optim. **56**(1), 231–252 (2018)
5. DEramo, C., Restelli, M., Nuara, A.: Estimating maximum expected value through gaussian approximation. In: ICML, pp. 1032–1040 (2016)

6. Even-Dar, E., Mansour, Y.: Learning rates for Q-learning. JMLR **5**, 1–25 (2003)
7. van Hasselt, H.P.: Double Q-learning. In: NIPS, pp. 2613–2621 (2010)
8. van Hasselt, H.P.: Insights in reinforcement learning: formal analysis and empirical evaluation of temporal-difference learning algorithms. Ph.D. thesis, Utrecht University, Netherlands (2011)
9. Kearns, M., Singh, S.: Finite-sample convergence rates for Q-learning and indirect algorithms. In: NIPS, pp. 996–1002 (1999)
10. Lee, D., Powell, W.B.: An intelligent battery controller using bias-corrected Q-learning. In: Proceedings of the Twenty-Sixth AAAI Conference on Artificial Intelligence, 22–26 July 2012, Toronto, Ontario, Canada, pp. 316–322 (2012)
11. Littman, M.L., Szepesvári, C.: A generalized reinforcement-learning model: convergence and applications. In: ICML, vol. 96, pp. 310–318 (1996)
12. Pandey, S., Chakrabarti, D., Agarwal, D.: Multi-armed bandit problems with dependent arms. In: Proceedings of the 24th International Conference on Machine Learning, pp. 721–728. ACM (2007)
13. Singh, S., Jaakkola, T., Littman, M.L., Szepesvári, C.: Convergence results for single-step on-policy reinforcement-learning algorithms. Mach. Learn. **38**(3), 287–308 (2000)
14. Sutton, R.S., Barto, A.G.: Reinforcement Learning: An introduction, vol. 1. MIT Press, Cambridge (1998)
15. Szepesvári, C.: The asymptotic convergence-rate of Q-learning. In: NIPS, pp. 1064–1070 (1997)
16. Watkins, C.J.: Learning from delayed rewards. Robot. Auton. Syst. **15**(4), 233–235 (1989)
17. Wiering, M., Van Otterlo, M.: Reinforcement Learning. Adaptation, Learning, and Optimization, vol. 12. Springer, Heidelberg (2012). https://doi.org/10.1007/978-3-642-27645-3
18. Yu, H., Bertsekas, D.P.: Q-learning and policy iteration algorithms for stochastic shortest path problems. Ann. Oper. Res. **208**(1), 95–132 (2013)
19. Zhang, Z., Pan, Z., Kochenderfer, M.J.: Weighted double Q-learning. In: IJCAI, pp. 3455–3461 (2017)

Monotonicity Extraction for Monotonic Bayesian Networks Parameter Learning

Jingzhuo Yang$^{(\boxtimes)}$, Yu Wang, and Qinghua Hu

School of Computer Science and Technology, Tianjin University, Tianjin, China
{yangjingzhuo,armstrong_wangyu,huqinghua}@tju.edu.cn

Abstract. Bayesian networks (BNs) parameter learning is a challenging task as it relies on a large amount of reliable and representative training data. Unfortunately, it is often difficult to obtain sufficient samples in many real-world applications. Monotonicity, as a class of prior information, widely exist in various practical tasks. This information is helpful for BN parameter learning. However, monotonicity is set by users traditionally. In this paper, we propose a data-dependent BN parameter learning method which can construct monotonicity constraints for BN parameters automatically. Firstly, we introduce rank mutual information (RMI) and Spearman rank correlation coefficient (RHO) to detect monotonicity among network nodes, and then construct monotonicity constraints for BN parameters. Finally, we transform the problem of parameter learning with monotonicity constraints into a convex Lagrange function and obtain the global optimum solution in polynomial time. Experimental results on real-world classification data and standard BNs show the effectiveness of our proposed algorithms with limited data.

Keywords: Bayesian Networks · Parameter learning
Monotonicity extraction · Monotonicity constraints

1 Introduction

Bayesian networks (BNs) are one of the most important probabilistic graph models proposed by Pearl in 1988 [15], which combine probability theory and graph theory for representing knowledge with uncertainty and efficient inference. During the last two decades, Bayesian networks have received increasing attention. BNs have been widely used in various fields and achieve a good performance, including fault detection [3,18], medical diagnosis [5,7], risk assessment [6,12], etc.

Parameter learning is considered as one of the most challenging tasks [4]. It estimates the conditional probability tables (CPTs) for network nodes to make a BN with predefined structure fit training samples. The performance of BN depends heavily on parameter learning, and the BN will infer unnatural results if the estimated parameters are inaccurate [1].

© Springer Nature Switzerland AG 2018
L. Cheng et al. (Eds.): ICONIP 2018, LNCS 11303, pp. 571–581, 2018.
https://doi.org/10.1007/978-3-030-04182-3_50

Traditional parameter learning shows good performance if training data are sufficient [4]. Unfortunately, it is often difficult to obtain a large number of labeled samples. Even in the big data era, the problem of data sparsity still exists in many real-world applications. In these tasks, data distribution satisfies the long-tailed distribution [8,17], in which a few categories occupy the majority of samples while most categories have few samples. As shown by previous researches [2,19], traditional parameter learning methods cannot perform well without sufficient training data.

To solve this problem, expert knowledge has been introduced as prior information for BN parameters [1,4,14]. This information help robustly and accurately estimate the parameters with limited data. Among expert knowledge, monotonicity constraints are a kind of prior knowledge to control the relationship between network nodes. Generally, monotonicity constraints are derived from the monotonic relationship of network nodes given by experts [1]. Several researchers have focused on combining monotonic relationship with BN parameter learning [1,4,9]. They introduce monotonicity constraints based on expert-defined knowledge and obtain improved performance compared with traditional methods. However, monotonicity constraints given by domain experts in some tasks are inconsistent with data, which reduce the performance of models. Moreover, it is difficult and costly to specify the monotonicity constraints of all BN parameters by domain experts, especially when the number of BN parameters is very large and the network structure of BN is complex [4].

To this end, we propose a data-dependent BN parameter learning method which can construct monotonicity constraints for parameters. Based on some monotonicity metrics, such as rank mutual information (RMI) [11] and Spearman rank correlation coefficient (RHO) [13], we detect the monotonicity relationship between network nodes, and then construct a set of monotonicity constraints for parameters of BN. We transform the problem of parameter learning with monotonicity constraints into a convex Lagrange function and obtain the globally optimum solution in polynomial time.

The rest of this paper is organized as follows. Section 2 introduces preliminary knowledge for this paper. Section 3 shows the details of the proposed parameter learning method. Experimental results on UCI datasets and publicly available BN repository are presented in Sect. 4. Finally, we give our conclusions for this work in the last section.

2   Preliminaries

2.1   Bayesian Networks and Its Parameter Learning

BN can be represented by $B = \langle G, \theta \rangle$, where $G = \langle V, E \rangle$ is a directed acyclic graph (DAG), where each node $V_i \in V$ corresponds with a stochastic variable X_i and the directed edge $E_i \in E$ captures the qualitative dependence relation between stochastic variables $X_1, \cdots, X_n$; θ is a set of conditional probability tables (CPTs)[15]. More specifically, each parameter $\theta_{ijk} = P(V_i = k \mid \Pi_i = j)$ represents the probability value of node V_i when its parent-nodes configuration

Π_i takes j-th value and its state is k-th value, where $i \in \{1, \cdots, n\}$ corresponds to all network nodes in a BN, $j \in \{1, \cdots, q_i\}$ corresponds to all the possible configurations of Π_i, and $k \in \{1, \cdots, r_i\}$ corresponds to all possible states of node V_i.

BN parameter learning is to estimate the CPTs for network nodes of BN where the network structure is known in advance. Given a training dataset D, parameter learning is to find the most probable CPTs that make the BN model fit the dataset D better. Classical maximum likelihood estimation (MLE) estimate the CPTs by maximizing the log-likelihood function

$$l\left(\theta \mid D\right) = \sum_{i=1}^{n} \sum_{j=1}^{q_i} \sum_{k=1}^{r_i} N_{ijk} \log \theta_{ijk}, \tag{1}$$

where N_{ijk} is the number of observation data in D corresponding to parameter θ_{ijk}. MAP(Maximum a Posteriori) method considers the prior information of the parameters by introducing a Dirichlet prior. MAP usually uses a flat prior $\alpha_{ijk} = 1$ or BDeu prior $\alpha_{ijk} = \frac{1}{r_i \cdot q_i}$ if no expert gives the hyperparameter α_{ijk} of the Dirichlet prior [19].

2.2 Monotonicity in Bayesian Networks

In 2004, Van Der Gaag et al. [10] introduced the definition of monotonicity in distribution for BN. A BN is said to be isotone in distribution for the node V_i and its parent node π_i if

$$k \leq k' \to P\left(V_i \leq c \mid \pi_i = k, \Pi_i \setminus \pi_i\right) \geq P\left(V_i \leq c \mid \pi_i = k', \Pi_i \setminus \pi_i\right), \tag{2}$$

where k, k' are two states of node π_i, and $c = 1, \cdots, r_i$; if

$$k \leq k' \to P\left(V_i \leq c \mid \pi_i = k, \Pi_i \setminus \pi_i\right) \leq P\left(V_i \leq c \mid \pi_i = k', \Pi_i \setminus \pi_i\right), \tag{3}$$

then the BN is said to be antitone in distribution for node V_i and its parent node π_i. Namely, BN is isotone in distribution for node pair $\langle \pi_i, V_i \rangle$, if it is more likely for the child node V_i to obtain higher-ordered values when a higher-ordered value is assigned to the parent node π_i.

2.3 Monotonicity Metrics

Several metrics have been proposed for measuring the monotonicity relationship between two stochastic variables. Rank mutual information (RMI) [11] and Spearman rank correlation coefficient (RHO) [13] are two widely used monotonicity metrics in monotonicity researches [16]. RMI is a robust metric for monotonicity, it combines the advantage of robustness of Shannon's entropy with the ability of dominance rough sets in extracting ordinal structures from dataset [11]. RHO is used to estimate the correlation between two ordered variables in statistics. The estimated value ρ can reach $+1$ or -1, if the two variables have a strong monotonicity relationship.

Given a training dataset D with N samples, the rank mutual information (RMI) between variables A and B is computed by

$$RMI\,(A,B) = -\frac{1}{N}\sum_{i=1}^{N}\log_2\frac{\left|[x_i]_A^{\leq}\right| \times \left|[x_i]_B^{\leq}\right|}{N \times \left|[x_i]_A^{\leq} \bigcap [x_i]_B^{\leq}\right|}, \tag{4}$$

where $[x_i]_v^{\leq} = \{x_j \in D \mid x_i \leq_v x_j\}$ is the ordinal relation between samples in terms of variable v and $\left|[x_i]_v^{\leq}\right|$ is the cardinality of set $[x_i]_v^{\leq}$. Spearman rank correlation coefficient (RHO) is a nonparametric estimation of the statistical dependence between two ordinal variables. The coefficient $\rho\,(A,B)$ of RHO between variables A and B is computed by

$$\rho\,(A,B) = 1 - \frac{6\sum_{i=1}^{N}(V(x_i,A) - V(x_i,B))^2}{N\,(N^2-1)}, \tag{5}$$

where $V(x_i,v)$ is the variable value of sample x_i on v.

3   The Proposed Method

3.1   Construction of Monotonicity Constraints

For network node V_i and one of its parent node $\pi_i \in \Pi_i$, rank mutual information (RMI) and Spearman rank correlation coefficient (RHO) can reflect the degree of monotonicity between their associated stochastic variables. If the value of $RMI\,(V_i,\pi_i)$ or $\rho\,(V_i,\pi_i)$ is larger than the given threshold ϵ (or smaller than $-\epsilon$), we can believe that parent node π_i has a positive (or negative) monotonic influence on node V_i, namely, V_i get higher state values be more (less) likely if π_i is a higher state value regardless of the configuration of other parents $\Pi_i \setminus \pi_i$. Without loss of generality, we assume that π_i has a positive monotonic influence on V_i in the rest of the paper, and negative monotonic influence can be defined analogously.

We use P_i, P_i' to denote two parent-nodes configurations of V_i, where $P = (\pi_i = k, \Pi_i \setminus \pi_i)$, $P_i' = (\pi_i = k', \Pi_i \setminus \pi_i)$, and $k \leq k'$. So the two parent-nodes configurations satisfy the partial order $P_i \preceq P_i'$, which means that parent-nodes configuration P_i' makes higher values for V_i more likely. But how to formulate the monotonic influence for parameters of node V_i? Intuitively, there have $\theta_{iP} \preceq \theta_{iP'}$, where θ_{ij} is a discrete probability distribution for all states in node V_i. In this paper, we follow the definition of monotonicity in Bayesian networks (isotone in distribution) given by Van Der Gaag in [10], and use first order stochastic dominance (FSD) to formulate the partial order relation between two parent-nodes configurations $P_i \preceq P_i'$ of V_i, the conditional probability distribution for V_i have

$$P\,(V_i \mid P_i) \preceq P\,(V_i \mid P_i'). \tag{6}$$

Equation (6) can be formulated by the cumulative distribution function as followed

$$P\left(V_i \leq c \mid P_i\right) \geq P\left(V_i \leq c \mid P_i'\right), \forall c \in \{1, \cdots, r_i\}, \tag{7}$$

namely,

$$\sum_{k=1}^{c} \theta_{iP_ik} \geq \sum_{k=1}^{c} \theta_{iP_i'k}, \forall c \in \{1, \cdots, r_i\}. \tag{8}$$

For each node in a BN, a set of inequality constraints as Eq. (8) can be constructed. Therefore, we have a set of monotonicity constraints Φ for BN parameters :

$$h_{iP_ic}(\theta) = \sum_{k=1}^{c} (\theta_{iP_i'k} - \theta_{iP_ik}) \leq 0, \forall c \in \{1, \cdots, r_i\}, \forall i \in \{1, \cdots, n\}, \forall P_i \preceq P_i'. \tag{9}$$

So, given the training data D, BN structure G and the threshold-value ϵ, we can obtain a set of monotonicity constraints Φ.

3.2 Parameter Learning with Monotonicity Constraints

The task of parameter learning with monotonicity constraints is to maximize the log-likelihood function $l(\theta \mid D)$ while also satisfying the constraints set Φ. In order to ensure the sum of all estimated parameters for each state in a parent-nodes configuration equal to one, we add the following constraints:

$$g_{ij}(\theta) = \sum_{k=1}^{r_i} \theta_{ijk} - 1 = 0, \forall i \in \{1, \cdots, n\}, \forall j \in \{1, \cdots, q_i\}. \tag{10}$$

Therefore, the parameter learning problem can be reformulated as:

$$\min_{\theta} - \sum_{i=1}^{n}\sum_{j=1}^{q_i}\sum_{k=1}^{r_i} N_{ijk} \log \theta_{ijk}$$

$$s.t \begin{cases} h_{iP_ic}(\theta) \leq 0, & \forall i \in \{1, \cdots, n\}, \forall c \in \{1, \cdots, r_i\}, \forall P_i \preceq P_i', \\ g_{ij}(\theta) = 0, & \forall i \in \{1, \cdots, n\}, \forall j \in \{1, \cdots, q_i\} \end{cases}. \tag{11}$$

All conditional constraints are linear functions of parameters and the objective function $l(\theta \mid D)$ is concave since a nonnegative sum of logarithms is concave. Hence, the Eq. (11) minimizes a convex function in a convex set, which means the objective function of the proposed method is a constrained convex optimization problem. As we know, constrained convex optimization problems have a number of attractive properties, such as any local optimum is global optimum and its optimal solution can be found in polynomial time [4]. This minimization problem of Eq. (11) can be solved by using Karush-Kuhn-Tucker (KKT) theorem. We introduce Lagrange multipliers λ_{ij} for equality constraints $g_{ij}(\theta)$ and μ_k for inequality constraints $h_{iP_ic}(\theta)$, then have the Lagrange function

$$L(\theta, \lambda_{ij}, \mu_k) = -l(\theta \mid D) - \sum_{ij} \lambda_{ij} g_{ij}(\theta) - \sum_{k} \mu_k h_{iP_ic}(\theta). \tag{12}$$

Therefore, the KKT conditions for Eq. (11) are

$$
\begin{cases}
\nabla_\theta L(\theta, \lambda_{ij}, \mu_k) = 0 \\
g_{ij}(\theta) = 0 \\
\mu_k \cdot h_{iP_ic}(\theta) = 0 \\
h_{iP_ic}(\theta) \leq 0 \\
\mu_k \geq 0
\end{cases}
\tag{13}
$$

The optimal parameters can be obtained through the KKT conditions. There are already several existing methods that have been proposed to solve this problem. In this paper, we employ the Mosek[1] software to solve the constrained convex optimization problem.

4   Experiments

To verify the performance of the proposed algorithms, we conduct a series of experiments with real-world tasks and standard BNs. We set the threshold $\epsilon = 0.05$, and all experimental results are the average of 20 times repeated operation. In addition, we use BDeu prior $\alpha_{ijk} = \frac{1}{r_i \cdot q_i}$ as the prior of parameters for the MAP method. In order to ensure the fairness to MLE method, we set a uniform distribution on all states when the parent-nodes configuration observation samples is zero.

4.1   Experiments on UCI Datasets

In this experiment, we collects four datasets from UCI ML repository[2], including Auto-mpg, Pima-Indian-Diabetes, Haberman, and Car. The network structure and domain knowledge about monotonic relationship between network nodes are shown in Fig. 1, defined by Altendorf in [1]. The continuous-valued attributes are discretized into five intervals by the equal-frequency discretization method in our experiments.

First, we test the performance of monotonicity metrics RMI and RHO in detecting monotonicity. We verify the monotonicity relationship between nodes Price, Doors, Safety and node Class in Car BN, respectively. The ground truth data distribution and metric values are shown in Fig. 2. From the figure we can see that the values of Class decrease with the increase of attribute Price, which means there is a monotonically decrement relationship between Price and Class. Obviously, Doors and Safety are monotonically increment with Class. The attribute Safety has larger impact on the Class than the attribute Doors. The two monotonicity metrics values accurately reflect these monotonicity information.

In order to compare the classification performance of the traditional parameter learning methods with the proposed parameter learning algorithms, we split the data into training set and test set randomly by 80% and 20%, respectively.

---

[1] http://www.mosek.com/.
[2] http://archive.ics.uci.edu/ml.

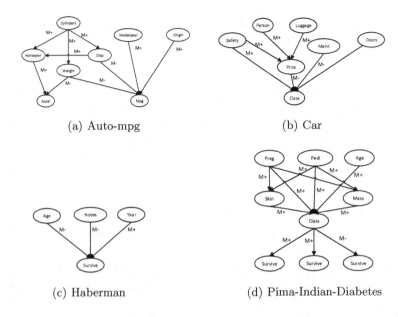

(a) Auto-mpg   (b) Car

(c) Haberman   (d) Pima-Indian-Diabetes

Fig. 1. Bayesian networks structure and domain knowledge of UCI datasets.

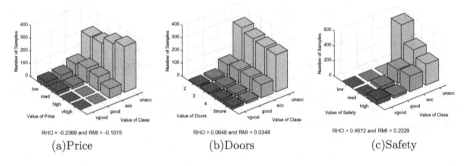

RHO = -0.2369 and RMI = -0.1075   RHO = 0.0646 and RMI = 0.0348   RHO = 0.4672 and RMI = 0.2229

(a)Price   (b)Doors   (c)Safety

Fig. 2. Monotonicity relationship between nodes Price, Doors, Safety and Class in Car dataset, respectively.

For testing the influence of the size of training sets, we use 10% training samples to 100% training samples to train the model, and then compute the test accuracy. The experimental results are shown in Fig. 3. From the figure we can see that the proposed algorithms significantly improve the prediction accuracy compared with the conventional methods MLE and MAP. In addition, both of our proposed methods can achieve comparable performance to the model (EXP-CML) which introduces domain knowledge given by experts. Specifically, our approach RHO-CML can achieve better performance than EXP-CML on Car, which shows that domain knowledge given by experts may be not correct in practical applications. For some complex tasks, experts do not necessarily know the exact relationship between the attributes. We should design some effective measurements to objectively compute the monotonicity.

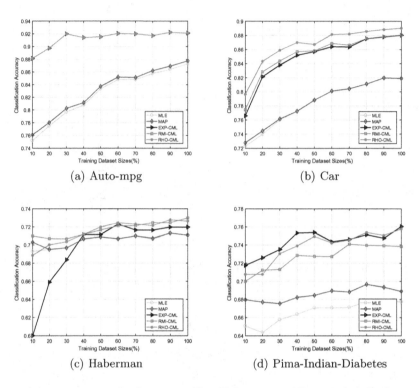

(a) Auto-mpg

(b) Car

(c) Haberman

(d) Pima-Indian-Diabetes

Fig. 3. Average accuracy for Auto-mpg, Car, Haberman and Pima-Indian-Diabetes BN under different sizes of training samples (curves EXP-CML, RMI-CML and RHO-CML are completely overlapping in Fig. 3(a)).

4.2   Experiments on Standard BNs

In this experiment, we compare our algorithms with traditional MLE and MAP methods on the standard BNs with some publicly available BN repository[3]. They are widely used to evaluate the performance BN parameter learning algorithms. These standard BNs include small networks(< 20 nodes) as well as very large networks ($100 - 1000$ nodes). For instance, Cancer only has 5 nodes, 4 edges and 10 parameters, while Munin1 has 186 nodes, 273 edges and 15622 parameters.

We apply the forwards sampling method to generate training samples from standard BN in this experiment, and then use the dataset to learning the parameters of the BN. To measure the difference between the estimated parameters $\hat{\theta}$ and actual parameters θ of the standard BN, we use the Kullback-Leibler (K-L)

---

[3] http://www.bnlearn.com/bnrepository/.

divergence between them. The smaller the value of K-L divergence is, the better the performance is. K-L divergence calculation formula as:

$$KL\left(\theta_{ij}, \hat{\theta}_{ij}\right) = \sum_{k=1}^{r_i} \theta_{ijk} log \frac{\theta_{ijk}}{\hat{\theta}_{ijk}}. \tag{14}$$

In order to ensure that the K-L divergence can be calculated if an estimated parameter is zero, we use a small number (1×10^{-7}) instead of it.

Table 1 shows the learning results of K-L divergence and standard deviation for MLE, MAP, RMI-CML, RHO-CML in fixed 100 training samples. The Average K-L row presents the average results over all BNs, and Average Rank row presents the average ranking over all BNs. The best results are presented in bold and statistically significant improvements of the best results over competitors are indicated by asterisks * (at 5% significance level). From the table, we can see clearly that our proposed algorithms RMI-CML, RHO-CML achieve good performance compared with others two conventional methods, and RHO-CML achieves the best performance in most BNs. 100 training samples are sufficient for parameter learning in small networks, so these methods achieve similar performance, such as Cancer and Weather BN. The performance of our algorithms improved obviously in large BNs, such as Hailfinder and Hepar2 BN.

In order to compare the learning performance of these methods on different sizes of training data, we compare the learning results of binary BN Andes and multivalued BN Water under different samples sizes ranging from 50 to 1000. The

Table 1. Learning results of K-L divergence and standard deviation for MLE, MAP, RMI-CML, RHO-CML on the standard BNs with 100 training samples.

BN	MLE	MAP	RMI-CML	RHO-CML
Andes	$0.253 \pm 0.019^*$	$0.066 \pm 0.003^*$	$0.061 \pm 0.003^*$	$\mathbf{0.053 \pm 0.005}$
Asia	$0.144 \pm 0.090^*$	$0.048 \pm 0.013^*$	$0.042 \pm 0.017^*$	$\mathbf{0.040 \pm 0.019}$
Cancer	$0.156 \pm 0.171^*$	$\mathbf{0.020 \pm 0.010}$	$\mathbf{0.020 \pm 0.010}$	$\mathbf{0.020 \pm 0.010}$
Child	$0.483 \pm 0.084^*$	$0.091 \pm 0.013^*$	$\mathbf{0.89 \pm 0.015}$	0.090 ± 0.010
Earthquake	$0.406 \pm 0.330^*$	0.066 ± 0.034	0.066 ± 0.034	$\mathbf{0.063 \pm 0.037}$
Hailfinder	$0.811 \pm 0.027^*$	$0.285 \pm 0.008^*$	$0.227 \pm 0.020^*$	$\mathbf{0.206 \pm 0.028}$
Hepar2	$0.490 \pm 0.052^*$	$0.170 \pm 0.009^*$	$0.169 \pm 0.010^*$	$\mathbf{0.151 \pm 0.018}$
Munin1	$0.569 \pm 0.016^*$	$0.423 \pm 0.005^*$	$\mathbf{0.409 \pm 0.009}$	$0.421 \pm 0.023^*$
Sachs	$0.476 \pm 0.093^*$	$0.149 \pm 0.018^*$	$\mathbf{0.111 \pm 0.015}$	$0.122 \pm 0.014^*$
Survey	$0.258 \pm 0.128^*$	$0.035 \pm 0.015^*$	$0.035 \pm 0.015^*$	$\mathbf{0.028 \pm 0.015}$
Water	$0.477 \pm 0.003^*$	$0.468 \pm 0.001^*$	$0.452 \pm 0.004^*$	$\mathbf{0.427 \pm 0.020}$
Weather	$0.149 \pm 0.186^*$	$\mathbf{0.020 \pm 0.025}$	$\mathbf{0.020 \pm 0.025}$	$\mathbf{0.020 \pm 0.025}$
Average K-L	0.893 ± 0.010	0.153 ± 0.128	0.142 ± 0.015	$\mathbf{0.137 \pm 0.019}$
Average rank	4.000	2.750	1.833	**1.417**

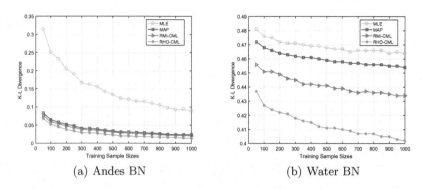

(a) Andes BN   (b) Water BN

Fig. 4. The learning results of MLE, MAP, RMI-CML and RHO-CML for Andes and Water BN under different sizes the training dataset.

experimental results shown in Fig. 4, which shows that the learning performance of the four methods are all improved as the samples increases. Moreover, the proposed algorithms always achieve the best performances which demonstrates that the proposed methods can effectively improve the learning performance by explore monotonicity constraints on parameters.

5   Conclusions

We propose a data-dependent BN parameter learning method which can extract monotonicity constraints between the network parameters automatically in this work. We use the rank mutual information or Spearman rank correlation coefficient to detect the monotonicity relationship between network nodes and then construct a set of monotonicity constraints for parameter of the BN. We transform this learning problem into a convex Lagrange function and obtain a global optimum solution. Experiments are conducted on the real-world classification data from the UCI ML repository and the standard Bayesian networks. With limited data, experimental results shows that our proposed algorithms achieve a significant improvement compared to traditional methods, and even surpass the method using expert-designed monotonicity constraints. In this work, we only consider the monotonicity relationship between network nodes without the degree of monotonicity. In the future, we will explore the influence of monotonicity degree on BN parameter learning methods with limited and incomplete data.

Acknowledgement. This work is partly supported by National Natural Science Foundation of China under Grants 61732011, 61432011 and U1435212.

References

1. Altendorf, E., Restificar, A.C., Dietterich, T.G.: Learning from sparse data by exploiting monotonicity constraints. In: Proceedings of the 21st Conference in Uncertainty in Artificial Intelligence, pp. 18–26 (2005)
2. Barber, D.: Bayesian Reasoning and Machine Learning. Cambridge University Press, New York (2012)
3. Cai, B., Huang, L., Xie, M.: Bayesian networks in fault diagnosis. IEEE Trans. Ind. Inform. **13**(5), 2227–2240 (2017)
4. de Campos, C.P., Tong, Y., Ji, Q.: Constrained maximum likelihood learning of Bayesian networks for facial action recognition. In: Forsyth, D., Torr, P., Zisserman, A. (eds.) ECCV 2008. LNCS, vol. 5304, pp. 168–181. Springer, Heidelberg (2008). https://doi.org/10.1007/978-3-540-88690-7_13
5. Constantinou, A.C., Fenton, N.E., et al.: From complex questionnaire and interviewing data to intelligent Bayesian network models for medical decision support. Artif. Intell. Med. **67**, 75–93 (2016)
6. Constantinou, A.C., Freestone, M., et al.: Causal inference for violence risk management and decision support in forensic psychiatry. Decis. Support Syst. **80**, 42–55 (2015)
7. Constantinou, A.C., Yet, B., et al.: Value of information analysis for interventional and counterfactual Bayesian networks in forensic medical sciences. Artif. Intell. Med. **66**, 41–52 (2016)
8. Farid, M.H., Ilyas, I.F., et al.: LONLIES: estimating property values for long tail entities. In: Proceedings of the 39th International ACM SIGIR Conference on Research and Development in Information Retrieval, pp. 1125–1128 (2016)
9. Feelders, A.J., van der Gaag, L.C.: Learning Bayesian network parameters with prior knowledge about context-specific qualitative influences. In: Proceedings of the 21st Conference in Uncertainty in Artificial Intelligence, pp. 193–200 (2005)
10. van der Gaag, L.C., Bodlaender, H.L., Feelders, A.J.: Monotonicity in Bayesian networks. In: Proceedings of the 20th Conference in Uncertainty in Artificial Intelligence, pp. 569–576 (2004)
11. Hu, Q., Guo, M., Yu, D., Liu, J.: Information entropy for ordinal classification. Sci. China Inf. Sci. **53**(6), 1188–1200 (2010)
12. Marvin, H.J.P., Bouzembrak, Y., Janssen, E.M., et al.: Application of Bayesian networks for hazard ranking of nanomaterials to support human health risk assessment. Nanotoxicology **11**(1), 123–133 (2017)
13. Myers, J.L., Well, A.D.: Research Design and Statistical Analysis (Second Edition ed.), 2nd edn. L. Erlbaum Associates, Mahwah (2010)
14. Niculescu, R.S.: Exploiting parameter domain knowledge for learning in Bayesian networks. Technical Report CMU-TR-05-147. Carnegie Mellon University (2005)
15. Pearl, J.: Probabilistic Reasoning in Intelligent Systems: Networks of Plausible Inference. Morgan Kaufmann, San Mateo (1988)
16. Pei, S., Hu, Q., Chen, C.: Multivariate decision trees with monotonicity constraints. Knowl. Based Syst. **112**, 14–25 (2016)
17. Powers, D.M.W.: Applications and explanations of Zipf's law. Adv. Neural Inf. Process. Syst. **5**(4), 595–599 (1998)
18. Wang, Z., Wang, Z., et al.: Fault detection and diagnosis of chillers using Bayesian network merged distance rejection and multi-source non-sensor information. Appl. Energy **188**, 200–214 (2017)
19. Zhou, Y., et al.: An empirical study of Bayesian network parameter learning with monotonic influence constraints. Decis. Support Syst. **87**, 69–79 (2016)

Dynamic Maintenance of Decision Rules for Decision Attribute Values' Changing

Yingyao Wang[1], Jianhua Dai[2(✉)], and Hong Shi[1]

[1] School of Computer Science and Technology,
Tianjin University, Tianjin 300350, China
`wangyingyao@tju.edu.cn`
[2] Hunan Provincial Key Laboratory of Intelligent Computing and Language
Information Processing, College of Information Science and Engineering,
Hunan Normal University, Changsha 410081, China
`jhdai@hunnu.edu.cn`

Abstract. Rule induction method based on rough set theory (RST) which can generate a minimal set of decision rules by using attribute reduction and approximations has received much attention recently. In real-life, the variation of objects, attributes and attributes' values affects reducts and approximations, e.g., the coarsening and refining of attributes' values. The goal of this paper is dynamic maintenance of decision rules for decision attribute values' coarsening and refining. Two incremental rough-set based methods are proposed to deal with this issue by updating assignment discernibility matrix dynamically without recomputing the reducts from the beginning, which increases the efficiency.

Keywords: Rough set theory · Incremental learning
Attribute reduction · Decision rule

1 Introduction

Rough set theory, originally proposed by Pawlak, provides the mathematical formulation of the concept of approximative (rough) equality of concepts in a given approximation space [9,13]. In real-life applications, the objects, attributes and attributes' values in the information system often vary with time, and rough-set based method for rule induction has received much attention as it can acquire a minimal set of rules from the decision system using attribute reduction and approximations [6].

Nowadays, many rough-set based methods have been widely used in machine learning [1,15,16], image processing [3,12] and data mining [2,4,5]. In rough set theory, the variation of objects, attributes and attributes' values affects reducts and approximations of a concept in information systems. Chen et al. [7] defined coarsening and refining of attribute values in information systems and proposed an incremental method for updating approximations of a concept. Liu et al. [10]

L. Cheng et al. (Eds.): ICONIP 2018, LNCS 11303, pp. 582–592, 2018.
https://doi.org/10.1007/978-3-030-04182-3_51

presented the strategies and mechanisms for incrementally learning knowledge in consistent information systems in which attributes' values change. Liu et al. [11] proposed new strategies of dynamically updating approximations in probabilistic rough sets. Li et al. [9] presented an incremental approach for updating approximations of dominance-based rough sets approach. Chan [4] proposed a method for updating approximations of a concept incrementally. The dynamic mechanisms for updating approximations in multigranulation rough sets while refining or coarsening attribute values was presented by Hu et al. [8]. Zheng et al. [19] developed an incremental knowledge acquisition algorithm combining rough set theory and rule tree. Two dynamic approaches in computing rough approximations for time-evolving information granule interval-valued ordered information system were presented by Yu et al. [17].

Chen et al. [6] proposed an incremental method for dynamic maintenance of decision rules for attribute values' coarsening and refining. However, they only considered about the variation of conditional attributes' values with time. So far, there are few studies on variation of decision attribute values. In this paper, we focus on rough-set based methods for updating decision rules of the decision system for decision attribute values' coarsening and refining, and propose two incremental methods which can avoid partial redundancy calculation and reduce computation consumption. They will save a lot of time costs when solving large-scale practical problems. The paper is organized as follows: In Sect. 2, the terms and definitions in rough set theory are introduced. In Sect. 3, the related propositions and algorithms for dynamic maintenance of decision rules for decision attribute values' coarsening and refining are presented, respectively. In Sect. 4, experimental evaluation under the datasets from UCI is given. In Sect. 5, we conclude the whole paper.

2   Preliminaries

In this Section, some basic concepts in rough set theory are introduced.

Definition 2.1 [18]. A quadruple $S = (U, A, V, f)$ is an information system, where U is a nonempty finite set of objects, called the universe. A is a nonempty set included conditional attributes C and decision attributes D. $V = \bigcup_{a \in A} V_a$, V_a is a domain of attribute a. $f : U \times A \to V$ is an information function, which gives values to every object on each attribute, $\forall a \in A, x \in U, f(x, a) \in V_a$.

Definition 2.2 [14]. The equivalence relation on $B \subseteq C$ is defined as follows:

$$R_B = \{(x, y) \in U \times U | \forall a \in B, f_a(x) = f_a(y)\} \tag{1}$$

The pair (U, R) is named as an approximation space. The equivalence relation R induces a partition of U, $U/R = \{E_1, E_2, ..., E_m\}$, $U/R_D = \{D_1, D_2, ..., D_k\}$. $[x]_B = \{y \in U | (x, y) \in R_B\}$ denotes the equivalence class with the object x.

Definition 2.3 [14]. If $D = \{d_1, d_2, ..., d_s\}$, then let $D_iC = \{f_{d_1}(x_k), ..., f_{d_s}(x_k)\}(x_k \in D_i, D_i \in U/R_D)$ denotes the characteristic value of decision class D_i. Let $\delta(Ei) = \{D_jC|E_i \cap D_j \neq \varnothing\}(E_i \in U/R, D_j \in U/R_D)$, which represents the generalized decision of an equivalence class E_i.

Definition 2.4 [18]. $(U, C \cup D, V, f)$ is a decision system, $U/R_C = \{E_1, E_2, ..., E_m\}, D^* = \{(E_i, E_j) : \delta_C(E_i) \neq \delta_C(E_j)\}$. $f_{a_k}(E_i)$ represents the attribute value of objects in E_i on attribute $a_k, (a_k \in A)$. We define the assignment discernibility attribute set between E_i and E_j, which is denoted by $D(E_i, E_j)$:

$$D(E_i, E_j) = \begin{cases} \{a_k \in C : f_{a_k}(E_i) \neq f_{a_k}(E_j)\} & (E_i, E_j) \in D^*; \\ \varnothing & (E_i, E_j) \notin D^*. \end{cases} \quad (2)$$

$M_D = (D(E_i, E_j) : i, j \leq m)$ is the Assignment Discernibility Matrix (ADM) of $(U, C \cup D, V, f)$.

Definition 2.5 [18]. $(U, C \cup D, V, f)$ is a decision system and $M_D = (D(E_i, E_j) : i, j \leq m)$ is the ADM. Then the assignment discernibility formal is $M = \wedge\{\vee\{a_k : a_k \in D(E_i, E_j)\} : i, j \leq m\}$. $M_{min} = \vee_{k=1}^{p}(\wedge_{s=1}^{q_k}a_s)$ is the minimal conjunctive formula of M. If $B_k = \{a_s : s = 1, 2, ..., q_k\}$, then $Red = \{B_k : k = 1, 2, ..., p\}$ is assignment reduct set of $(U, C \cup D, V, f)$.

Computing the decision rules with the ADM is a NP problem. Thus, several works have been done to deal with this issue. Chen et al. proposed a feasible method $(GRMDAS)$ to generate the reduct from ADM [6].

Definition 2.6 [6]. The minimal discernibility attribute set $(MDAS)$ is defined as $Att_{min} = \{Att_0, ..., Att_i, ..., Att_t\}$, where $\forall Att_i \in Att_{min}, \exists D(E_j, E_k)$, s.t., $Att_i \subseteq D(E_j, E_k)$, and $\forall D(E_j, E_k), \exists Att_i \in Att_{min}$, s.t., $Att_i \subseteq D(E_j, E_k)$. $\forall Att_i, Att_j \in Att_{min}(i \neq j), \neg \exists Att_i \subseteq Att_j$ or $Att_j \subseteq Att_i$. Att_i is called a minimal discernibility attribute.

Definition 2.7. Let $S = (U, C \cup d, V, f)$ be an information system. $f(x_i, d)$ and $f(x_k, d)$ are the value of x_i and $x_k(k \neq l)$ on the decision attribute d, respectively. $f(x_i, d) \neq f(x_k, d)$. Then, $U_d = \{x_{i'} \in U|f(x_{i'}, d) = f(x_i, d)\}$. On the one hand, let $f(x_{i'}, d) = f(x_k, d), \forall x_{i'} \in U_d$. Then we call the attribute value $f(x_i, d)$ is coarser than $f(x_k, d)$, and we call this case as Decision Attribute Values' Coarsening $(DAVC)$. On the other hand, Let $f(x_{j'}, d) = v$, where $\forall x_{j'} \in U_d$, $v \notin V_l$. Then we call the decision attribute value $f(x_j, d)$ on the object $x_{j'}$ is finer than v, called as Decision Attribute Values' Refining $(DAVR)$.

3 Incremental Method for Variation of Decision Attribute Values

In real-life applications, there are many scenarios where the class labels vary with time, i.e., in systems where the tag is address information, the tag of the system changes as the accuracy of the address in the new data changes. Thus, in this paper, we focus on the variation of decision attribute values.

3.1 An Incremental Method for Updating Decision Rules on DAVC (IMDAVC)

In a decision system, the decision attribute value v_1 on d is coarsened to $v_2(v_1 \neq v_2)$. Let $D_j^{v_1} \in U/R_d, \delta(D_j^{v_1}) = v_1$, Let $D_j^{v_2} \in U'/R_d, \delta(D_j^{v_{21}}) = v_2, E^{v_1} = \{E_i \in U/R | E_i \cap D_j^{v_1} \neq \varnothing\}, E^{v_2} = \{E_i \in U'/R | E_i \cap D_j^{v_2} \neq \varnothing\}$. Let $E^\wedge$ denotes the equivalence class E after coarsening, M_D denotes the ADM of the decision system before changed, and $M_D^\wedge$ denotes the ADM of the decision system after coarsening.

Proposition 3.1. M_D is updated only by replacing row i and field j with empty set, if $E_i, E_j \in U/R, \delta(E_i) \neq \delta(E_j), \delta(E_i^\wedge = E_j^\wedge)$

Proof: The elements of M_D cannot change from empty set to nonempty set because it's impossible for $\delta(E_i) = \delta(E_j)$ before coarsening and $\delta(E_i^\wedge) \neq \delta(E_j^\wedge)$ after coarsening.

For $M_D, D(E_i, E_j)$ will change, that is, $D(E_i^\wedge, E_j^\wedge) \neq D(E_i, E_j)$ only if $E_i, E_j \in E^{v_1} \cap E^{v_2}$.

Proposition 3.2. Update rules for $M_D^\wedge$.
IF $E_i \in E^{v_1} \cap E^{v_2}, E_j \in E^{v_1} \cap E^{v_2}$
  IF $\delta(E_i^\wedge) = \delta(E_j^\wedge)$, then $D(E_i^\wedge, E_j^\wedge) = \varnothing$

Proof: By Proposition 3.1, we can prove this proposition easily.

According to these propositions, an incremental algorithm for dynamic maintenance of decision rules for $DAVC$ is presented as Algorithm 1. For incrementally updating the assignment discernibility matrix, Algorithm 1 will be applied on $M_D^\wedge$ to get the decision rules after DAVC.

Algorithm 1. An Incremental Method for Updating Assignment Discernibility Matrix after DAVC(IMDAVC)

Input: M_D, E^{v_1}, E^{v_2}
Output: $M_D^\wedge$
1: $M_D^\wedge \leftarrow M_D$
2: **for** each E_i in $E^{v_1} \cap E^{v_2}$ **do**
3:    **for** each E_j in $E^{v_1} \cap E^{v_2}$ where $j \geq i$ **do**
4:       compute $\delta(E_i^\wedge)$ and $\delta(E_j^\wedge)$
5:       **if** $\delta(E_i^\wedge) = \delta(E_j^\wedge)$ **then**
6:          $D(E_i^\wedge, E_j^\wedge) = \varnothing$
7:       **end if**
8:    **end for**
9: **end for**

Example 3.1. A decision system is shown in the Table 1, where $U = \{x_1,$ $x_2, x_3, x_4, x_5, x_6, x_7, x_8\}$, $C = \{a_1, a_2, a_3\} = \{Height, Hair, Eyes\}$, $D = \{d\} = \{Nationality\}$, $U/R_C = \{E_1, E_2, E_3, E_4\} = \{\{x_1, x_4\}, \{x_2, x_6\}, \{x_3\}, \{x_5, x_7, x_8\}\}$.

By Definition 2.4, because $\delta(E_1) \neq \delta(E_2)$, so $D(E_1, E_2) = \{a_1, a_2, a_3\}$. Since $\delta(E_1) \neq \delta(E_3)$, then $D(E_1, E_3) = \{a_1\}$. As $\delta(E_1) \neq \delta(E_4)$, we have $D(E_1, E_4) = \{a_1, a_2\}$. Because $\delta(E_2) \neq \delta(E_3)$, so $D(E_2, E_3) = \{a_2, a_3\}$. Since $\delta(E_2) \neq \delta(E_4)$, then $D(E_2, E_4) = \{a_1, a_2, a_3\}$. As $\delta(E_3) \neq \delta(E_4)$, we have $D(E_3, E_4) = \{a_1, a_2\}$. Then we can get the assignment discernibility matrix M_D.

$$M_D = \begin{pmatrix} \varnothing & & & \\ a_1, a_2, a_3 & \varnothing & & \\ a_1 & a_2, a_3 & \varnothing & \\ a_1, a_2 & a_1, a_2, a_3 & a_1, a_2 & \varnothing \end{pmatrix}$$

Then we coarsen the decision attribute d of x_1, x_3, x_5, x_7 to $Europe$, Since $v_1 = \{UnitedKingdom, French\}$, $v_2 = \{Europe\}$, then $E^{v_1} = \{E_1, E_3, E_4\}$, $E^{v_2} = \{E_1, E_3, E_4\}$, $E^{v_1} \cap E^{v_2} = \{E_1, E_3, E_4\}$. By Proposition 3.2, $D(E_1, E_4) = \varnothing$, so we can get the $M_D^{\wedge}$.

$$M_D^{\wedge} = \begin{pmatrix} \varnothing & & & \\ a_1, a_2, a_3 & \varnothing & & \\ a_1 & a_2, a_3 & \varnothing & \\ \varnothing & a_1, a_2, a_3 & a_1, a_2 & \varnothing \end{pmatrix}$$

$Att_{min} = \{\{a_1\}, \{a_2, a_3\}\}$, then according to the GRMDAS method, we can get the reduct of decision system is $\{a_1, a_2\}$. In the end, we can get seven decision rules which were deduced from $M_D^{\wedge}$.

Table 1. Decision system 1

U	Height	Hair	Eyes	Nationality
x_1	tall	blond	blue	United Kingdom
x_2	medium	dark	hazel	China
x_3	medium	blond	blue	French
x_4	tall	blond	blue	Belgien
x_5	short	red	blue	United Kingdom
x_6	medium	dark	hazel	Singapore
x_7	short	red	blue	French
x_8	short	red	blue	Belgien

3.2 An Incremental Method for Updating Decision Rules on DAVR (IMDAVR)

In a decision system, $x_k(x_k \in E_l)$ is refined on attribute d and $x_k^\vee$ denotes the object x_k after refining. Let $E_i^\vee(1 \leq i \leq m, m = |U/R|)$ denotes the equivalence after refining. Let $M_D^\vee$ denotes the ADM of the decision system after refining.

Proposition 3.3. $\delta(E_l^\vee) \nsubseteq \delta(E_i^\vee)(i \neq l, 1 \leq i \leq m)$.

Proof: According to Definition 2.3, the value of x_k on attribute d after refining doesn't belong to the domain of d before refining.

Proposition 3.4. Update rules for $M_D^\vee$.

(1) if $D(E_i, E_l) \neq \varnothing$, then $D(E_i^\vee, E_l^\vee) = D(E_i, E_l)$

(2) if $D(E_i, E_l) = \varnothing$, then computed $D(E_i^\vee, E_l^\vee)$ according to the Definition 2.4

(3) otherwise, $D(E_i^\vee, E_j^\vee) = D(E_i, E_j)$

Proof: Since only $\delta(E_l)$ changed after refining, M_D is updated according to E_l. If $D(E_i, E_j) \neq \varnothing$, since $\delta(E_l^{lor}) \neq \delta(E_i^\vee)$, then $D(E_i^\vee, E_l^\vee) = D(E_i, E_l)$. If $\delta(E_l) = \delta(E_i)$, then we have to recompute $D(E_i^\vee, E_l^\vee)$.

These propositions show when to change the elements of M_D and when we need to recompute the assignment discernibility attribute set. And according to these propositions, an incremental algorithm for dynamic maintenance of decision rules for $DAVR$ is presented as Algorithm 2. For incrementally updating the assignment discernibility matrix, Algorithm 2 will be applied on $M_D^\wedge$ to get the decision rules after DAVR.

Example 3.2. A decision system is shown in Table 2, where $U = \{x_1, x_2, x_3, x_4, x_5, x_6, x_7, x_8\}$, $C = \{a_1, a_2, a_3\} = \{Height, Hair, Eyes\}$, $D = \{d\} = \{Nationality\}$, $U/R_C = \{E_1, E_2, E_3, E_4\} = \{\{x_1, x_4\}, \{x_2, x_6\}, \{x_3\}, \{x_5, x_7, x_8\}\}$.

By Definition 2.4, $\delta(E_1) = \delta(E_2)$, so $D(E_1, E_2) = \varnothing$. Since $\delta(E_1) \neq \delta(E_3)$, then $D(E_1, E_3) = \{a_1\}$. As $\delta(E_1) \neq \delta(E_4)$, we have $D(E_1, E_4) = \{a_1, a_2\}$. Because $\delta(E_2) \neq \delta(E_3)$, so $D(E_2, E_3) = \{a_2, a_3\}$. Since $\delta(E_2) \neq \delta(E_4)$, then $D(E_2, E_4) = \{a_1, a_2, a_3\}$. As $\delta(E_3) \neq \delta(E_4)$, we have $D(E_3, E_4) = \{a_1, a_2\}$. Then we can get the assignment discernibility matrix (ADM).

$$M_D = \begin{pmatrix} \varnothing & & & \\ \varnothing & \varnothing & & \\ a_1 & a_2, a_3 & \varnothing & \\ a_1, a_2 & a_1, a_2, a_3 & a_1, a_2 & \varnothing \end{pmatrix}$$

Now we refine the decision attribute d of x_2 and x_5 from *Asia* to *EastAsia*, then the index set for refined examples is $\{x_2, x_5\}$, by Proposition 3.4, we need

---

Algorithm 2. An Incremental Method for Updating Assignment Discernibility Matrix after DAVR(IMDAVR)

Input: Index set for refined examples:$\{x_1, x_2, ..., x_k\}$;
   Assignment Discernibility Matrix$(ADM) : M_D$.
Output: $M_D^{\vee}$
1: $M_D^{\vee} \leftarrow M_D$
2: **for** $i \leftarrow 1$ to k **do**
3:    compute E_l, that $x_i \in E_l$
4:    **for** $j \leftarrow 1$ to l **do**
5:       **if** $D(E_j, E_l) \leq \varnothing$ **then**
6:          $D(E_j^{\vee}, E_l^{\vee}) = D(E_j, E_l)$
7:       **else if** $D(E_j, E_l) = \varnothing$ **then**
8:          $D(E_j^{\vee}, E_l^{\vee}) = \varnothing$
9:          compute $D(E_j^{\vee}, E_l^{\vee})$
10:      **end if**
11:   **end for**
12:   **for** j$\leftarrow$ l+1 to $|M_D\{:, 1\}|$ **do**
13:      **if** $D(E_l, E_j) \leq \varnothing$ **then**
14:         $D(E_l^{\vee}, E_j^{\vee}) = D(E_l, E_j)$
15:      **else if** $D(E_l, E_j) = \varnothing$ **then**
16:         $D(E_l^{\vee}, E_j^{\vee}) = \varnothing$
17:         compute $D(E_l^{\vee}, E_j^{\vee})$
18:      **end if**
19:   **end for**
20: **end for**

---

recompute $D(E_1^{\vee}, E_2^{\vee})$. Since $D(E_1^{\vee}, E_2^{\vee}) = \{a_1, a_2, a_3\}$, we can get the new ADM.

$$M_D^{\vee} = \begin{pmatrix} \varnothing & & & \\ a_1, a_2, a_3 & \varnothing & & \\ a_1 & a_2, a_3 & \varnothing & \\ a_1, a_2 & a_1, a_2, a_3 & a_1, a_2 & \varnothing \end{pmatrix}$$

$Att_{min} = \{\{a_1\}, \{a_2, a_3\}\}$, then according to the GRMDAS method, we can get the reduct of DS is $\{a_1, a_2\}$. Finally, we can get seven decision rules which were deduced from $M_D^{\vee}$.

4   Complexity Analysis

Let $S = (U, C \cap D, V, f, G)(|U| = n, |C| = l, |D| = d, |U/R_C| = m)$ be an decision system, then the non-incremental approach includes five steps: calculating the EFM, calculating the ADM and MDAS, generating the reduct, and generating decision rules. In view of this, the computational complexity of non-incremental approach is $O(\frac{n^2}{2}l) + O(\frac{m^2}{2}l) + O(\frac{m^2}{2}l^2) + O(\frac{m^2}{2}l) + O(m)$. The computational complexity of incremental algorithms (IMDAVC, IMDAVR) proposed in this paper are analyzed as follows, respectively.

Table 2. Decision systems 2

U	Height	Hair	Eyes	Nationality
x_1	tall	blond	blue	Europe
x_2	medium	dark	hazel	Asia
x_3	medium	blond	blue	Europe
x_4	tall	blond	blue	Asia
x_5	short	red	blue	Asia
x_6	medium	dark	hazel	Europe
x_7	short	red	blue	Asia
x_8	short	red	blue	Asia

4.1 IMDAVC

In algorithm IMDAVC, there is no need to compute the EFM again, and we can get the computational complexity of updating ADM with E^{v_1} and E^{v_2}. Thus, the computational complexity of this algorithm is $O(\frac{|E^{v_1} \cap E^{v_2}|}{2}d) + O(\frac{m^2}{2}l^2) + O(\frac{m^2}{2}l) + O(m)$. In summary, the computational complexity of the incremental algorithm IMDAVC is lower than that of non-incremental approach.

4.2 IMDAVR

In algorithm IMDAVR, there is no need to compute the EFM either, then we can get the computational complexity of updating ADM with E^v. Thus, the computational complexity of this algorithm is $O(|E^v|l) + O(\frac{m^2}{2}l^2) + O(\frac{m^2}{2}l) + O(m)$. Based on the above analysis, we can know that the computational complexity of non-incremental approach is larger than that of incremental algorithm IMDAVR.

5 Experiments

Data sets from UCI[1] are selected to verify the effectiveness of the algorithms. The data sets are summarized in Table 3. In the experiment process, all continuous attributes have undergone corresponding discretization and all missing data are treated as fixed value for calculation. For non-incremental algorithm, the ADM is recalculated after the decision system changes. The number of the attribute values to be coarsened and refined were the same both for the non-incremental and incremental algorithms. The algorithms are developed in MATLAB 2016a on a computer with 3.2 GHz CPU, Intel core i5 and 8 GB of memory.

---

[1] http://archive.ics.uci.edu/ml/.

Table 3. Description of data sets

No.	Data set	Attribute	Sample	Class	Missing value
1	audiology	69	226	24	yes
2	soybean	35	683	19	yes
3	primary-tumor	17	339	22	yes
4	movement-libras	91	360	15	no
5	handwritten	256	1593	10	no
6	arrhythmia	279	452	16	yes

5.1   Performance of Algorithm 1 on DAVC

10% to 100% of each data set listed in Table 3, with step of 10 are used in the
experiments to compare the performance between non-incremental algorithm
and Algorithm 1 proposed in this paper. The results are shown in Fig. 1, in
which the x-axis represents the data sets which is from 10% to 100% of each
total data set and the y-axis represents the computation time. The attribute's
value to be coarsened are selected randomly. After the decision attribute value
is coarsened, the number of decision classes will decrease. From Fig. 1, we can
see the Algorithm 1 proposed in this paper is much faster than non-incremental
algorithm in most data sets. Since the decision attributes are randomly selected
for coarsening, the computation time fluctuates greatly.

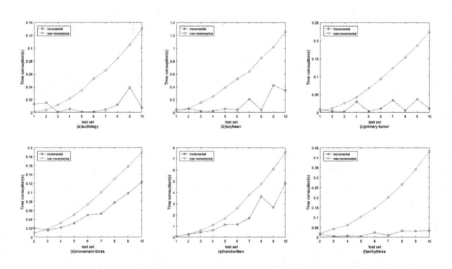

Fig. 1. The comparison between non-incremental algorithm and Algorithm 1.

5.2   Performance of Algorithm 2 on DAVR

10% to 100% of each data set listed in Table 3, with step of 10 are used in the experiments to compare the performance between non-incremental algorithm and Algorithm 2 proposed in this paper. The results are showed in Fig. 2, the x-axis represents the data sets which is from 10% to 100% of each total data set and the y-axis represents the computation time. The attribute's value to be refined is selected randomly. After the decision attribute value is refined, the number of decision classes will increase. From Fig. 2, we can see the Algorithm 2 proposed in this paper is much faster than non-incremental algorithm in every data sets.

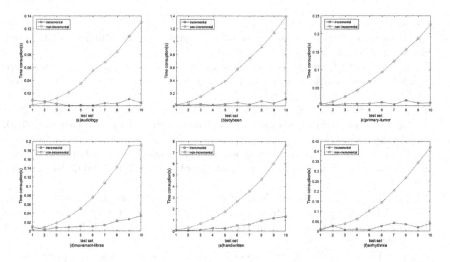

Fig. 2. The comparison between non-incremental algorithm and Algorithm 2.

6   Conclusion

In this paper, we have discussed the principles for dynamic maintenance of decision rules for decision attribute values' coarsening and refining. Then two incremental rough-set based methods are proposed to update the assignment discernibility matrix dynamically without unnecessary repetitive calculations. Compared with non-incremental method, the proposed algorithms are obviously efficient. At last, comparative experimental results verify the effectiveness of the algorithms proposed in this paper.

Acknowledgment. This work was partially supported by the National Natural Science Foundation of China (Nos. 61473259, 61502335, 61070074, 60703038) and the Hunan Provincial Science & Technology Program Project (2018TP1018, 2018RS3065).

592 Y. Wang et al.

References

1. Ananthanarayana, V., Murty, M.N., Subramanian, D.: Tree structure for efficient data mining using rough sets. Pattern Recognit. Lett. **24**(6), 851–862 (2003)
2. BŁaszczyński, J., Greco, S., SŁowiński, R.: Inductive discovery of laws using monotonic rules. Eng. Appl. Artif. Intell. **25**(2), 284–294 (2012)
3. Bouchet, A., Pastore, J., Ballarin, V.: Segmentation of medical images using fuzzy mathematical morphology. J. Comput. Sci. Technol. **7**(3), 256–262 (2007)
4. Chan, C.C.: A rough set approach to attribute generalization in data mining. Inf. Sci. **107**(1–4), 169–176 (1998)
5. Chen, H., Li, T., Luo, C., Horng, S.J., Wang, G.: A decision-theoretic rough set approach for dynamic data mining. IEEE Trans. Fuzzy Syst. **23**(6), 1958–1970 (2015)
6. Chen, H., Li, T., Luo, C., Horng, S., Wang, G.: A rough set-based method for updating decision rules on attribute values' coarsening and refining. IEEE Trans. Knowl. Data Eng. **26**(12), 2886–2899 (2014)
7. Chen, H., Li, T., Qiao, S., Da, R.: A rough set based dynamic maintenance approach for approximations in coarsening and refining attribute values. Int. J. Intell. Syst. **25**(10), 1005–1026 (2010)
8. Hu, C., Liu, S., Huang, X.: Dynamic updating approximations in multigranulation rough sets while refining or coarsening attribute values. Knowl.-Based Syst. **130**, 62–73 (2017)
9. Li, S., Li, T., Liu, D.: Dynamic maintenance of approximations in dominance-based rough set approach under the variation of the object set. Int. J. Intell. Syst. **28**(8), 729–751 (2013)
10. Liu, D., Li, T., Chen, H., Ji, X.: Approaches to knowledge incremental learning based on the changes of attribute values. In: Intelligent Decision Making Systems - International ISKE Conference on Intelligent Systems and Knowledge Engineering, pp. 94–99 (2010)
11. Liu, D., Li, T., Zhang, J.: Incremental updating approximations in probabilistic rough sets under the variation of attributes. Knowl.-Based Syst. **73**(81), 81–96 (2015)
12. Pal, S.K., Mitra, P.: Multispectral image segmentation using the rough-set-initialized EM algorithm. IEEE Trans. Geosci. Remote. Sens. **40**(11), 2495–2501 (2002)
13. Pawlak, Z.: Rough sets. Int. J. Comput. Inf. Sci. **11**(5), 341–356 (1982)
14. Pawlak, Z., Skowron, A.: Rudiments of rough sets. Inf. Sci. **177**(1), 3–27 (2007)
15. Qian, Y., Liang, J., Pedrycz, W., Dang, C.: An efficient accelerator for attribute reduction from incomplete data in rough set framework. Pattern Recognit. **44**(8), 1658–1670 (2011)
16. Swiniarski, R.W., Skowron, A.: Rough set methods in feature selection and recognition. Pattern Recognit. Lett. **24**(6), 833–849 (2003)
17. Yu, J., Chen, M., Xu, W.: Dynamic computing rough approximations approach to time-evolving information granule interval-valued ordered information system. Appl. Soft Comput. **60**, 18–29 (2017)
18. Zhang, W.X., Mi, J.S., Wu, W.Z.: Approaches to knowledge reductions in inconsistent systems. Int. J. Intell. Syst. **18**(9), 989–1000 (2003)
19. Zheng, Z., Wang, G., Wu, Y.: A rough set and rule tree based incremental knowledge acquisition algorithm. Fundam. Inform. **59**(2–3), 299–313 (2004)

Label Distribution Learning Based on Ensemble Neural Networks

Yansheng Zhai[1], Jianhua Dai[2(✉)], and Hong Shi[1]

[1] School of Computer Science and Technology,
Tianjin University, Tianjin 300350, China
yszhai@tju.edu.cn
[2] Hunan Provincial Key Laboratory of Intelligent Computing and Language
Information Processing, College of Information Science and Engineering,
Hunan Normal University, Changsha 410081, China
jhdai@hunnu.edu.cn

Abstract. Label distribution learning (LDL), as an extension of multi-label learning, is a new arising machine learning technique to deal with label ambiguity problems. The maximum entropy model is commonly used in label distribution learning. However, it does not consider the correlation between the labels and is not suitable for nonlinear relationships, and the prediction performance is also limited. In this paper, we propose a label distribution learning algorithm based on ensemble neural networks. The algorithm trains neural networks with preferences using training sets with different label sets to construct base learners, and combines the base learners with the weights, which is learned by the combined learner to obtain the final learning results. Experimental results show that the proposed algorithm is effective for label distribution data.

Keywords: Label distribution learning · Neural networks
Ensemble learning · Maximum entropy model

1 Introduction

At present, single-label learning and multi-label learning [14] are two kinds of machine learning paradigms to deal with the problem of label ambiguity. In single-label learning, each sample corresponds to a label. Obviously, single-label learning does not solve the case where a sample is related to multiple labels. Hence multi-label learning is presented. In multi-label learning, each sample is connected to a set of labels, which solves the problem that a sample has multiple labels. However, there are still some problems that can not be solved using multi-label learning. For example, in some image classification tasks, if two images are composed of the same elements, but the significance of each element in the image is different, then the two images tend to have different meanings. But the label set is the same in multi-label learning. In these application contexts, what people want to know is not just what elements are included in an image,

L. Cheng et al. (Eds.): ICONIP 2018, LNCS 11303, pp. 593–602, 2018.
https://doi.org/10.1007/978-3-030-04182-3_52

but more importantly, the difference in the significance of these elements. Thus, the set of labels for a sample not only indicates whether the sample has the labels but the significance of the labels for the samples, which is called the label distribution classification problem. To solve this problem, Geng [3] proposed the label distribution learning (LDL) method.

Label distribution learning supplies a method for solving the problems of label ambiguity. However, so far, the correlation between labels is not well utilized, and there is room for improving prediction performance. At the same time, ensemble learning can transform "weak learners" into "strong learners" through integration well. It will be a good way to improve the performance of the label distribution learning. Based on this idea, we use ensemble neural networks to predict the label distribution data, and set the base learners with preference according to the characteristics of the training sets, then obtain the final learning results by the learned weights. The main contributions of this paper are mainly the following three points,

(1) We use the combine learner to learn the correlation between labels and effectively combine the label distribution.
(2) We use multi-layer neural networks instead of linear combination in the maximum entropy model, which is more effective for nonlinear feature relationships.
(3) This is an attempt to solve the problem of label distribution learning by using ensemble learning ideas. Experiments show that the method is effective.

2   Related Work

So far, label distribution learning has been widely used to solve the label ambiguity problems [17]. According to the label distribution learning framework, k-means clustering method and least squares method were used to build the label distribution learning algorithm in [10], and a method of handling label enhancements containing only logical label data was proposed in [5]. Further, the label distribution learning was introduced into the problems of population count [15]. In this way, the discrete numerical label sets were transformed into continuous label distributions, and the adjacent samples can be used as the training samples to increase the number of training samples. Therefore, it solved the problems of estimating the number of people in public video surveillance effectively. In addition, in order to solve the problem of age estimation, it generated label distributions by single label datasets, and increased the original training samples, then put forward IIS-LDL [6] algorithm to further improve the age estimation performance. Based on that, the article [8] had optimized the labels of the age estimation, and used the correlation of the datasets, which increased the performance of age estimation. Furthermore, the label distribution learning algorithm was applied to facial expression recognition, and the emotion distribution learning algorithm (EDL) [19] was proposed. The researchers treated the emotion as a mixture of basic emotions, and expanded the single label into the label distribution to increase the categories of emotion recognition. The algorithm was effective in emotions classification [12], and it was a further application of the label distribution algorithms. In the recognition of the face image [18], it also achieved a good result. A deep label distribution learning method (DLDL), which achieved good predictive performance by using fewer training sets was proposed in [2], and it had a significant effect on age estimation and head attitude recognition.

3 Ensemble Neural Networks Framework

3.1 The Maximum Entropy Model

Let $X = R^m$ be the instance's features space and $Y = \{y_1, y_2, ..., y_L\}$ be the space of label distribution. The goal of LDL is to learn the mapping function between a set of instance features and a set of labels $(f = X \rightarrow Y)$ which can predict the label distributions for unseen instances. Given training set $S = \{(x_1, D_1), (x_2, D_2), ..., (x_n, D_n)\}$, where $x_i \in X$ is the instance and $D_i = \{d_i^1, d_i^2, ..., d_i^L\}$ is the label distribution associated with x_i. d_x^y represents the description of the instance x on the label y. We assume that the label set is complete, so $\sum_{j=1}^L d_i^j = 1$. The maximum entropy model is to learn a conditional probability mass function $P(x|y; \theta)$, where θ is the parameter vector. The goal of LDL is to find the θ that can predict a distribution similar to D_i given the instance x_i. The algorithm use the Kullback-Leibler to divergence define by

$$KL(P_a||P_b) = \sum_j P_a^j \ln \frac{P_a^j}{P_b^j} \tag{1}$$

where P_a^j and P_b^j are the j-th elements of the two distributions P_a and P_b, respectively. Accordingly, the best vector parameter θ^* is determined as fellows

$$\theta^* = argmin_\theta \sum_i KL(D_i||P_i) + \lambda_1||\theta||_F^2$$
$$= argmin_\theta \sum_i \sum_j (d_{x_i}^{y_j} \ln \frac{d_{x_i}^{y_j}}{p(y_j|x_i; \theta)}) + \lambda_1||\theta||_F^2 \tag{2}$$

Assuming that it fellow a maximum entropy [1] model,

$$p(y_k|x_i; \theta) = \frac{1}{Z_i} \exp(\sum_r \theta_{kr} x_i^r) \tag{3}$$

where $Z_i = \sum_k \exp(\sum_r \theta_{kr} x_i^r)$. x_i^r is the r-th feature of x_i, and θ_{kr} is an element in θ. Substituting Eq. 3 into Eq. 2 yields, it become that

$$T(\theta) = \sum_i \sum_j d_{x_i}^{y_j} \ln_{x_i}^{y_j} - \sum_i \sum_j d_{x_i}^{y_j} \sum_r \theta_{kr} x_i^r$$
$$+ \sum_i \ln \sum_k \exp(\sum_r \theta_{kr} x_i^r) + \lambda_1||\theta||_F^2 \tag{4}$$

To get the minimization of the function $T(\theta)$, it can use the limited-memory quasi-Newton method (L-BFGS) [13]. In the maximum entropy model, the values of the labels are predicted by θ_{kr} through linear combination of feature values. This method has limitations. If the relationship between the features and the labels does not conform to the linear relationship, a good prediction performance cannot be achieved, and the relationship between the labels is ignored. In order to solve this problem, this paper uses neural networks instead of $\theta_{kr}x_i^r$, because neural networks have a strong ability to fit and can predict nonlinear relationships. At the same time, the ensemble learning is used to predict the labeling relationship. The algorithm definition is introduced in the next section.

3.2   Construct Base Learners

We assume that the networks have n hidden layers. The first dimension of the neural networks is the number of features (m) and the dimensionality of the hidden layers are $H_1, H_2, ..., H_n, H_0 = m$. Output layer is $P^{(H_n \times L)}$, and $W_n^{(H_{n-1} \times H_n)}$ represents the weights matrix of the nth hidden layers of neural networks. $W_{output}^{(H_n \times L)}$ is the weights matrix of the output layer. $b_n^{(1 \times H_n)}$ indicates the offset values of the nth hidden layers. $b_{output}^{(1 \times L)}$ indicates the offset values of the output layer. $B_n^{(N \times H_n)} = \{b_n^1; b_n^2; ...; b_n^L\}$ indicates the matrix of the offset values. $B_{output}^{(N \times L)} = \{b_{output}^1; b_{output}^2; ...; b_{output}^L\}$ indicates the matrix of the output offset values. The activation function is the $relu$ activation function [7]. $I_j^{(N \times H_j)}$ represents the input values of the jth layer. N indicates the size of the training batch. The $relu$ is $f = \max(0, x)$.

The first layer of neural networks input:

$$I_0^{(N \times H_0)} = \{x_1; x_2; ...; x_N\} \tag{5}$$

The nth hidden layer of the neural network input ($n \geq 1$):

$$I_n^{(N \times H_n)} = relu(I_{n-1}W_n + B_n) \tag{6}$$

The output layer's output is:

$$Prd = softmax(I_n W_{output} + B_{output}) \tag{7}$$

Let V_i be the ith output cell of the output layer, then

$$Prd_i = \frac{\exp V_i}{\sum_j \exp(V_j)} \tag{8}$$

Since the goal of LDL is to make the predicted distribution be as similar to the true distribution as possible, the loss function should be able to measure the similarity of two distributions. In this paper we use K-L divergence [11] as the loss function, defined by

$$KL(P_a||P_b) = \sum_j P_a^j \ln \frac{P_a^j}{P_b^j} \tag{9}$$

where P_a^j and P_b^j are the jth elements of the two distribution P_a and P_b respectively. We assume that d_i^l represents the true distribution label element of the ith instance, Prd_i^l represents the predicted label distribution element of the ith instance. So the loss function of the algorithm is defined by

$$loss_{base} = (D||Prd) = \frac{\sum_{i=1}^{N} \sum_{l=1}^{L} (d_i^l \ln(\frac{d_i^l}{Prd_i^l}))}{N} \tag{10}$$

The combined learner is used to predict the relationship between labels, so we use the Pearson's correlation coefficient to measure the relationship between labels. p_i^k represents the k-th element of the prediction labels of the i-th sample. The cost function is defined as follows:

$$loss_{combine} = \sum_{k=1}^{L-1} \sum_{j=k+1}^{L} (\frac{\sum_{i=1}^{N} (p_i^k - \bar{p^k})(p_i^j - \bar{p^j})}{\sqrt{\sum_{i=1}^{N}(p_i^k - \bar{p^k})}\sqrt{\sum_{i=1}^{N}(p_i^j - \bar{p^j})}}$$
$$- \frac{\sum_{i=1}^{N} (d_i^k - \bar{d^k})(d_i^j - \bar{d^j})}{\sqrt{\sum_{i=1}^{N}(d_i^k - \bar{d^k})}\sqrt{\sum_{i=1}^{N}(d_i^j - \bar{d^j})}})^2 \tag{11}$$

In order to ensure that the base learners are "good and different", we designed the following base learners and combining strategies.

(1) Set the number of base learners according to the number of labels, It can make the base learners form the label "preference".
(2) Calculate the dominant labels of label distributions (label distribution values are larger). We set a threshold θ, and the labels whose values are greater than θ are the dominant labels.
(3) Divide the training set into subsets according to the dominant labels, and the training instances which have the same dominant labels are divided into the same set. In this way, the training instances with the same dominant labels will train the same base learners, so the learners can predict the dominant labels better.
(4) Use the gradient descent method to train each base learner. For each base learner, we divide the training subset into small batches for training. After training a round, randomly disrupt the whole training instances' order, and divide the training subset again, then continue to train until the model converges.
(5) The whole training sets are used to train the combined learner to describe the correlation between labels. They can gain the weights of the labels, and we use them to describe the relationship between labels. According to the weights, we combine the base learners by the following ways:
$P_n^{(1 \times L)} = \{p_n^1, p_n^2, ..., p_n^L\}$ indicates the label distribution predicted by the learner n for the instance. $Q^{(1 \times L)} = \{Q_1, Q_2, ..., Q_L\}$ indicates the weights of the label predicted by the combined learner. $Prediction_i$ represents the i-th element of the predicted distribution.

$$Prediction_i = \frac{\sum_j p_i^j Q_j}{\sum_k \sum_h p_k^h Q_h} \qquad (12)$$

The pseudo-code is shown in Algorithm 1.

Algorithm 1. The ENN-LDL algorithm

Data: training set $\{X, D\}$, parameters learn rate λ, basis learner number N,
          parameter θ
Result: the label distribution D_t

1 Train:
2 initialize θ, λ, N;
3 calculate the dominant labels according to θ;
4 divide the training set into $\{S_1, S_2, ..., S_N\}$ by the dominant labels;
5 $i = 1$
6 **while** $i \neq N$ **do**
7   training base learner i with S_i until *loss* convergence;
8   i++;

9 train the combine learner;
10 the prediction model R is obtained by combining the basis learner with Eq. 12;
11 Test:
12 return the label distribution D_t according to R;

---

4 Experiment

4.1 The Data Set of Label Distribution

The datasets[1] are based on the true label distribution datasets collected from yeast biochemical experiments. Each dataset corresponds to an experiment. Each instance in the data set represents a yeast gene, which contains 2465 yeast genes. An instance feature vector is a 24-dimensional phylogenetic profile vector. The details of the datasets are summarized in Table 1.

Table 1. The number of labels in datasets

Dataset	yeastdiau	yeastheat	yeastspo	yeastspo5	yeastcold	yeastdtt	yeastspoem	yeastcdc	yeastelu
Samples	2465	2465	2465	2465	2465	2465	2465	2465	2465
Features	24	24	24	24	24	24	24	24	24
Labels	7	6	6	3	4	4	2	14	15

---

[1] http://cse.seu.edu.cn/personalpage/xgeng/ldl/.

4.2   Evaluation Measures

In this paper, we selected three of the six evaluation indicators to evaluate the algorithm. The names and formulas are listed in Table 2. P_j and Q_j represent the j-th element of the true label distribution and the predicted label distribution respectively. For various evaluation measures [16], "↓" means smaller is better, "↑" means bigger is better.

Table 2. Three evaluation measures for LDL algorithms

Name	Chebyshev↓	Kullback-Leibler↓	Intersection↑
Formula	$d_1 = \max_j |P_j - Q_j|$	$d_4 = \sum_{j=1}^{c} P_j \ln \frac{P_j}{Q_j}$	$s_2 = \sum_{j=1}^{c} \min(P_j - Q_j)$

4.3   Experiment Setting

ENN-LDL proposed in this paper was compared with five algorithms, i.e., PT-SVM [4], AA-kNN [6], AA-BP [6], IIS-LLD [6] and BFGS-LLD [3]. The parameters setting of the compared algorithms were summarized as fellows. The number of the neighbors k in AA-kNN was set to 4. The number of hidden layer neurons for AA-BP were set to 60. For ENN-LDL, we set up the neural networks with three hidden layers. The numbers of neurons were 600, 800, 600 respectively. We used the Adam [9] optimization algorithm to optimize the model, and the parameter was set to $1e^{-4}$. To prevent the over-fitting problem from setting the drop layers in the hidden layers, the parameter was set to 0.8. The parameter θ was set to $\frac{1}{N}$. The program ran on the Tensorflow framework. The experimental results are shown in the following tables and figures.

Table 3. The result of experiment on *chebyshev*

Dataset	AA-BP	AA-kNN	PT-SVM	BFGS-LLD	IIS-LLD	ENN-LDL
yeastcold	0.0572 ± 0.0020	0.0555 ± 0.0022	0.0574 ± 0.0056	0.0512 ± 0.0018	0.0617 ± 0.0015	**0.0508 ± 0.0020**
yeastspo	0.0651 ± 0.0031	0.0644 ± 0.0024	0.0629 ± 0.0164	0.0584 ± 0.0038	0.0654 ± 0.0034	**0.0582 ± 0.0041**
yeastspo5	0.0957 ± 0.0065	0.0962 ± 0.0044	0.0953 ± 0.0048	0.0914 ± 0.0054	0.0970 ± 0.0052	**0.0904 ± 0.0052**
yeastdtt	0.0450 ± 0.0018	0.0393 ± 0.0016	0.0388 ± 0.0020	0.0361 ± 0.0013	0.0492 ± 0.0012	**0.0358 ± 0.0014**
yeastheat	0.0535 ± 0.0032	0.0451 ± 0.0012	0.0443 ± 0.0015	0.0423 ± 0.0009	0.0525 ± 0.0007	**0.0414 ± 0.0010**
yeastspoem	0.0934 ± 0.0047	0.0924 ± 0.0040	0.0917 ± 0.0057	0.0869 ± 0.0052	0.0928 ± 0.0042	**0.0862 ± 0.0044**
yeastdiau	0.0486 ± 0.0041	0.0393 ± 0.0010	0.0446 ± 0.0031	0.0370 ± 0.0015	0.0454 ± 0.0011	**0.0367 ± 0.0017**
yeastelu	0.0404 ± 0.0019	0.0177 ± 0.0005	0.0170 ± 0.0006	**0.0163 ± 0.0006**	0.0240 ± 0.0011	**0.0163 ± 0.0006**
yeastcdc	0.0409 ± 0.0020	0.0177 ± 0.0010	0.0171 ± 0.0010	0.0163 ± 0.0009	0.0233 ± 0.0010	**0.0162 ± 0.0009**

4.4   Experimental Results

Tables 3 and 4 show the experimental results of five algorithms based on evaluation indicators *chebyshev* and *intersection* on nine real label distribution datasets. Since the actual datasets are compared with the prediction effect of

Table 4. The result of experiment on *intersection*

Dataset	AA-BP	AA-kNN	PT-SVM	BFGS-LLD	IIS-LLD	ENN-LDL
yeastcold	0.9338 ± 0.0025	0.9335 ± 0.0032	0.9340 ± 0.0064	0.9407 ± 0.0023	0.9287 ± 0.0021	**0.9411 ± 0.0025**
yeastspo	0.9046 ± 0.0044	0.9075 ± 0.0036	0.9089 ± 0.0188	0.9154 ± 0.0055	0.9049 ± 0.0048	**0.9155 ± 0.0057**
yeastspo5	0.9043 ± 0.0065	0.9038 ± 0.0044	0.9047 ± 0.0048	0.9086 ± 0.0054	0.9030 ± 0.0052	**0.9096 ± 0.0052**
yeastdtt	0.9478 ± 0.0020	0.9547 ± 0.0017	0.9554 ± 0.0018	0.9583 ± 0.0013	0.9431 ± 0.0010	**0.9586 ± 0.0013**
yeastheat	0.9228 ± 0.0053	0.9356 ± 0.0018	0.9371 ± 0.0019	0.9401 ± 0.0012	0.9242 ± 0.0012	**0.9412 ± 0.0013**
yeastspoem	0.9066 ± 0.0047	0.9076 ± 0.0040	0.9083 ± 0.0057	0.9131 ± 0.0052	0.9072 ± 0.0042	**0.9138 ± 0.0044**
yeastdiau	0.9206 ± 0.0066	0.9367 ± 0.0040	0.9283 ± 0.0042	0.9403 ± 0.0028	0.9256 ± 0.0021	**0.9405 ± 0.0032**
yeastelu	0.8875 ± 0.0053	0.9546 ± 0.0011	0.9561 ± 0.0013	**0.9588 ± 0.0002**	0.9406 ± 0.0017	0.9586 ± 0.0013
yeastcdc	0.8792 ± 0.0066	0.9527 ± 0.0027	0.9553 ± 0.0022	**0.9573 ± 0.0027**	0.9396 ± 0.0022	**0.9573 ± 0.0029**

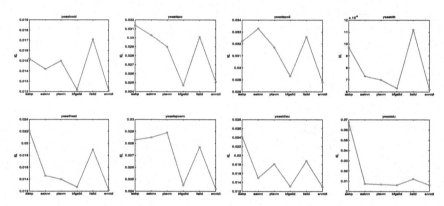

Fig. 1. KL divergence values of different algorithms on the datasets

the five algorithms by ten-fold cross validation, the experimental results are given in the format of "average ± standard deviation". Here, the mean values and standard deviation are obtained by statistics on ten experimental results. The best results of the compared algorithms were blacked.

By analyzing the experimental results shown in Tables 3 and 4, we can conclude that ENN-LDL has a good performance in datasets yeastcold, yeastspo5, yeastdtt, yeastheat, yeastdiau, yeastspoem, yeastcdc, while has the sub-optimum performance on dataset yeastelu. Figure 1 shows the variation of KL divergence of different algorithms on 8 data sets. We can see in the figure that the algorithm can effectively improve the prediction performance of label distribution learning. Experimental results indicate that, algorithm ENN-LDL perform better than compared algorithms.

In order to test the influence of parameters on the performance of the algorithms, we also conduct experiments on the parameter θ. The experimental parameters are set to $\frac{1}{2}, \frac{1}{3}, ..., \frac{1}{n}$. Due to page limitations, we only provide the experimental analysis results of yeastdtt data on six indicators shown in Fig. 2. From Fig. 2, we can conclude that when θ is larger, fewer samples are included in the dominant labels, and there are fewer training samples. This can easily lead to the lack of training data for the base learners and the phenomenon of "less fitting". When the θ value is small enough, it is easy to known that most labels intend to be classified as dominant labels, which leads to the lack of diversity of the base learners and reduces the generalization ability. The experimental results show that when θ is near the mean of the label distribution, the experimental results are the best.

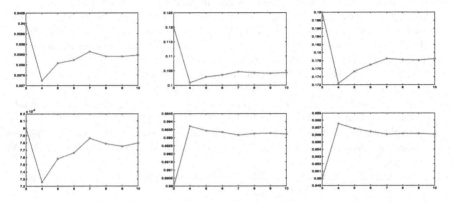

Fig. 2. Influence of θ with 6 indicators on dataset *yeastdtt*

5 Conclusion

As an extension of single-label learning and multi-label learning, label distribution learning can take more consideration of the label's information to deal with the label ambiguity problems. In order to improve the prediction performance of the LDL algorithm, and utilize the relationship between labels, we adopt the idea of ensemble learning and construct an label distribution learning model based on ensemble neural networks (ENN-LDL). Experimental results on multiple experimental datasets show that the proposed algorithm is suitable for label distributed learning framework, and it can achieve a good performance. In the future work, we will study relevant strategies to improve the proposed method.

Acknowledgment. This work was partially supported by the National Natural Science Foundation of China (Nos. 61473259, 61502335, 61070074, 60703038) and the Hunan Provincial Science & Technology Program Project (No. 2018TP1018).

References

1. Berger, A.L., Pietra, S.A.D., Pietra, V.J.D.: A maximum entropy approach to natural language processing. Comput. Linguist. **22**, 39–71 (1996)
2. Gao, B., Xing, C., Xie, C., Wu, J., Geng, X.: Deep label distribution learning with label ambiguity. IEEE Trans. Image Process. **26**(6), 2825–2838 (2017)
3. Geng, X.: Label distribution learning. IEEE Trans. Knowl. Data Eng. **28**(7), 1734–1748 (2014)

4. Geng, X., Xia, Y.: Head pose estimation based on multivariate label distribution. In: Computer Vision and Pattern Recognition, pp. 1837–1842 (2014)
5. Geng, X., Xu, N., Sao, R.: Label enhancement for label distribution learning. J. Comput. Res. Dev. **54**(6), 1171–1184 (2017)
6. Geng, X., Yin, C., Zhou, Z.: Facial age estimation by learning from label distributions. IEEE Trans. Pattern Anal. Mach. Intell. **35**(10), 2401–2412 (2013)
7. Glorot, X., Bordes, A., Bengio, Y.: Deep sparse rectifier neural networks. In: International Conference on Artificial Intelligence and Statistics, pp. 315–323 (2012)
8. He, Z., et al.: Data-dependent label distribution learning for age estimation. IEEE Trans. Image Process. **26**(8), 3846–3858 (2017)
9. Kingma, D.P., Ba, J.: Adam: a method for stochastic optimization. CoRR abs/1412.6980 (2014). http://arxiv.org/abs/1412.6980
10. Sao, D., Yang, W.Y., Zhao, H.: Application of k-means algorithm to realize label distribution learning. Trans. Intell. Syst. **12**(3), 325–332 (2017)
11. Sung-Hyuk, C.: Comprehensive survey on distance/similarity measures between probability density functions. Int. J. Math. Model. Methods Appl. Sci. **1**(4), 300–307 (2007)
12. Wang, Q., Geng, X.: Face estimation based on adaptive label distribution. Master's thesis, Southeast University, NanJing (2015)
13. Yuan, Y.X.: A modified bfgs algorithm for unconstrained optimization. IMA J. Numer. Anal. **11**(3), 325–332 (1991)
14. Zhang, M., Zhou, Z.: ML-KNN: a lazy learning approach to multi-label learning. Pattern Recogn. **40**(7), 2038–2048 (2007)
15. Zhang, Z., Wang, M., Geng, X.: Crowd counting in public video surveillance by label distribution learning. Neurocomputing **166**(1), 151–163 (2015)
16. Zhao, Q., Geng, X.: Study on several problems of label distribution learning. Master's thesis, Southeast University, NanJing (2016)
17. Zhao, Q., Geng, X.: Selection of objective function in label distribution learning. J. Front. Comput. Sci. Technol. **11**(5), 708–719 (2017)
18. Zhou, Y., Geng, X.: Application of label distribution in face image feature recognition. Master's thesis, Southeast University, NanJing (2016)
19. Zhou, Y., Xue, H., Geng, X.: Emotion distribution recognition from facial expressions. In: Proceedings of the 23rd ACM International Conference on Multimedia, pp. 1247–1250. ACM (2015)

Neural Information Processing in Hierarchical Prototypical Networks

Zilong Ji[1], Xiaolong Zou[2], Xiao Liu[1], Tiejun Huang[2], Yuanyuan Mi[3], and Si Wu[2(✉)]

[1] State Key Laboratory of Cognitive Neuroscience and Learning,
Beijing Normal University, Beijing 100875, China
[2] School of Electronics Engineering and Computer Science, IDG/McGovern Institute for Brain Research, Peking University, Beijing 100871, China
`siwu@pku.edu.cn`
[3] Institute for Neurointelligence, School of Medicine, Chongqing University, Chongqing, China

Abstract. Prototypical networks (PTNs), which classify unseen data points according to their distances to the prototypes of classes, are a promising model to solve the few-shot learning problem. Mimicking the characteristics of neural systems, the present study extends PTNs in two aspects. Firstly, we develop hierarchical prototypical networks (HPTNs), which construct prototypes at all layers and minimize the weighted classification errors of all layers. Applied to two benchmark datasets, we show that a HPTN has comparable, or slightly better, performances than a PTN. We also find that after training, the HPTN generates good prototype representations in the intermediate layers of the network. Secondly, we demonstrate that the classification operation via distance computation in a PTN can be replaced approximately by the attracting dynamics of the Hopfield model, indicating the potential realization of metric-learning in neural systems. We hope this study establishes a link between PTNs and neural information processing.

Keywords: Prototype · Few-shot learning · Metric-learning Hopfield model

1 Introduction

Recently, a deep neural network model, called prototypical networks (PTNs), was proposed to solve the few-shot learning problem [1]. The formulation of a PTN is based on metric-learning, as it classifies unseen data points according to their distances to the prototypes of different classes. Moreover, these prototypes, which are constructed by a set of support examples through an embedding function implemented by a deep network (Fig. 1A), can be optimized via end-to-end training. PTNs have displayed promising performances in benchmark

---

Z. Ji and X. Zou—Equal contribution.

© Springer Nature Switzerland AG 2018
L. Cheng et al. (Eds.): ICONIP 2018, LNCS 11303, pp. 603–611, 2018.
https://doi.org/10.1007/978-3-030-04182-3_53

datasets, and therefore have received large attention in the field. For instances, Oreshkin et al. explored different ways of defining the distance metric between data points [2], Sung et al. proposed an approach to learn a non-linear distance metric [3], Ren et al. augmented PTNs with the ability of using unlabeled examples to construct prototypes [4], and Liu et al. considered a transductive setting, which utilizes both the support and query sets to exploit the structure of class space across episodes [5].

The metric-learning idea in PTNs is very appealing to neural information processing, as the experimental data has indicated that neural systems categorize and recognize objects based on their semantic similarities [6]. The goal of this study is to investigate the potential link between PTNs and neural information processing. Motivated by the characteristics of neural systems, we extend the structure of PTNs in two aspects. Firstly, we know that a deep neural network mimics the layered architecture of the ventral visual pathway (Fig. 1B), and that in the visual system, object information are read-out at different layers, rather than only at the last layer as in a PTN. For instance, Matsumoto et al. [6] found that macaque monkeys can still assign stimuli to previously learned categories even after bilateral removal of the anterior inferior temporal cortex, and the latter is known to play an important role in object recognition. To mimic this property, we extend a PTN by considering that class prototypes are optimized at each layer of the network, which we call a hierarchical PTN (HPTN). Secondly, in a neural system, all computations are carried out via the dynamics of neural networks. The distance computation in a PTN is similar to the pattern overlap computation in the dynamics of an attractor network (Fig. 1C). It is valuable to explore whether a PTN can be realized by a biologically plausible attractor network. Overall, we hope this study will give us insight into understanding metric-learning in neural systems and shed light on developing advanced brain-inspired computational models.

The rest of the paper is organized as follows. Section 2 introduces the idea of HPTNs and their applications to two tasks on few-shot learning, Sect. 3 presents a link between PTNs and attractor neural networks, and Sect. 4 gives an overall conclusion.

2   Hierarchical Prototypical Networks

2.1   Prototypical Networks

To start, we briefly introduce the idea of PTNs. Following the fact that PTNs were originally proposed to solve the few-shot learning problem, we describe PTNs in the framework of few-shot learning. Specifically, we adopt the episodic learning paradigm proposed by Vinyals et al. [7], and call the condition that K examples for each of N classes are available in an episode to be N-way K-shot learning. Given the support set S, i.e., the training examples, a PTN computes the prototype of each class by the mean of the support data points belong to that class, which is written as,

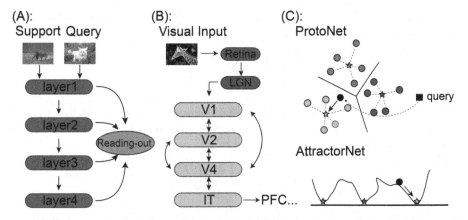

Fig. 1. (A). The architecture of a HPTN consists of four layers, each of which is formed by a Convolution-BatchNorm-ReLU-Pooling block. The classification probabilities at four layers are weighted properly to decide the final classification result. (B) The ventral visual pathway, along which object information are extracted and read-out layer by layer. (C) Upper panel: in a PTN, classification is done by computing the distances between the query data and prototypes; Lower panel: in an attractor neural network, classification is done through evolving the network state to one of its attractors.

$$c_m = \frac{1}{K} \sum_{(x_i, y_i) \in S_m} f_\phi(x_i), \tag{1}$$

where c_m denotes the prototype of class m and $S_m = \{(x_i, y_i)\}$, for $i = 1, \dots, K$, the corresponding support set. The function $f_\phi : R^D \to R^M$ represents the embedding function implemented by the deep network used, with ϕ the learnable parameters. When a new query data x is presented, the PTN calculates its probabilities belonging to N classes according to its distances to all prototypes, which is written as,

$$p_\phi(y = m|x) = \frac{\exp\{-d\,[f_\phi(x), c_m]\}}{\sum_n \exp\{-d\,[f_\phi(x), c_n]\}}, \tag{2}$$

where $d(\cdot)$ denotes the distance measure, and Euclidean distance is used in the present study. Through minimizing the negative log-probability $J(\phi) = -\log p_\phi(y = m|x)$ with respect to training examples, the PTN optimizes the network parameters ϕ, such that the network generates representations that are most suitable to classify data points based on their distance between each other.

2.2 Hierarchical Prototypical Networks

In contrast to a PTN, where prototypes are only explicitly optimized in the last layer of the network, in a HPTN, prototypes are defined at all layers and the network tries to minimize the weighted classification errors based on prototypes of all layers.

Suppose the network has L layers. Denote c_m^l the prototype of class m at layer l, which is calculated to be,

$$c_m^l = \frac{1}{K} \sum_{(x_i, y_i) \in S_m} f_{\phi^l}(x_i), \tag{3}$$

where $f_{\phi^l}(\cdot)$ represents the embedding function implemented by the first l layers of the network with ϕ^l is the corresponding parameters. Similarly, the class probabilities at layer l are calculated to be,

$$p_{\phi^l}(y = m|x) = \frac{\exp\left\{-\beta^l d\left[f_{\phi^l}(x), c_m^l\right]\right\}}{\sum_n \exp\left\{-\beta^l d\left[f_{\phi^l}(x), c_n^l\right]\right\}}. \tag{4}$$

Following the work of [2], we scale the distance metric by adding a learnable scalar β^l to each layer, which enables the HPTN to optimize the regime for each similarity metric $d\left[f_{\phi^l}(x), c_m^l\right]$.

The negative log-probability loss at layer l is given by $J_l(\phi^l) = -\log p_{\phi^l}(y = m|x)$. The overall loss function of the HPTN is defined to be

$$J(\phi) = \sum_{l=1}^{L} \alpha^{L-l} J_l(\phi^l), \tag{5}$$

where $0 \le \alpha \le 1$ is a discount factor. We assign larger weights to higher layers, since normally representations at higher layers better discriminate objects. When $\alpha = 0$, it returns to a PTN.

After training, the class probabilities for an unseen data are calculated by weighting the results from all layers, which are given by

$$p_\phi(y = m|x) = \frac{1}{Z} \sum_{l=1}^{L} \gamma^{L-l} p_{\phi^l}(y = m|x), \tag{6}$$

where Z is a normalization factor and γ the weighting factor. $\gamma = \alpha$ is used in the present study, but in practice γ can also be optimized via learning.

2.3   Experimental Results

In the experiments below, we choose the network architecture to be the same as that in [7], which consists of four stacked layers, with each layer containing a 3×3 convolution with 64 filters followed by batch normalization, a ReLU non-linearity, and 2×2 max-pooling. Prototypes are calculated at each layer after the ReLU operation.

Two datasets, Omniglot and miniImageNet, are used. The Omniglot dataset has 1623 classes from 50 alphabets [8]. There are 20 examples in each class and each of them was drawn by a different human subject. To reduce over-fitting, data augmentation was performed with rotations in multiples of 90 degrees. Following Vanyals et al. [7], all the gray scale images are resized to 28×28.

We use 1200 characters plus rotations (4800 characters in total) for episodic training and the rest plus rotations for testing. The miniImageNet dataset was originally proposed by Vinyals et al. [7], which is derived from the ILSVRC-12 dataset [9]. It consists of 100 classes and each class contains 600 colored images of size 84×84. In our experiments, we use the splits introduced by Ravi and Larochelle [10], which has a different set of 100 classes including 64 for training, 16 for validation, and 20 for test, and the validation dataset is only used for early-stopping. We find that it takes long time to calculate distances in shallow layers due to the large dimensionality of shallow features, we therefore scale down the original 84×84 images to 64×64, but the performance of PTNs is retained. Similar to [1], we set 60-way episodes for 1-shot and 5-shot classifications in the Omniglot task, and 30-way episodes for 1-shot classification and 20-way episodes for 5-shot classification in the miniImageNet task.

Table 1 presents the experiment results for the Ominglot task. Since the authors didn't release the detailed configuration, we could not reproduce the results in [1]. Therefore we implement PTNs by ourselves using the same hyper-parameters as in [1] but with more training epochs. We see that the overall performance of the HPTN is comparable to that of the PTN in [1] but better than our implementation of the PTN. The results for the miniImageNet are similar, see Table 2.

Table 1. Classification accuracies on Omniglot. In our implementation of PTN and HPTN, we initialize all the parameters with the same seed. Training epochs is set to be 200, episodic training classes is 60, learning rate is initialized to be 10^{-3}, and we cut the learning rate in half every 4000 episodes. No regularization was used other than batch normalization.

Model	5-way		20-way	
	1-shot	5-shot	1-shot	5-shot
Neural Statistician [11]	98.1	99.5	93.2	98.1
Matching Networks [7]	98.1	98.9	93.8	98.5
PTN [1]	98.8	**99.7**	**96.0**	**98.9**
PTN (Ours)	98.7	99.6	95.4	98.6
HPTN	**98.9**	**99.7**	95.6	98.8

Interestingly, we observe that the classification accuracies in the intermediate layers of the HPTN are significantly improved compared to that of the PTN (note that in the PTN, although we did not include classification errors of the intermediate layers in the loss function, we can still calculate their classification performances by computing the prototypes using the support set accordingly). As shown in Fig. 2, for the Omniglot task, the improvements of the HPTN are around 5% in layer 2 and 3% in layer 3 compared to the PTN (Fig. 2A); for the miniImageNet task, the improvements are around 3% in layer 2 and 5% in layer 3 (Fig. 2B). These improvements are intuitively understandable, since the HPTN

Table 2. Classification accuracies on miniImageNet. Parameters are initialized with the same seed. Training epochs are set to be 200. We use 30-way episodes for one-shot classification and 20-way episodes for 5-shot classification. Learning rate is initialized to be 10^{-3}, and we cut the learning rate in half every 4000 episodes. No regularization was used other than batch normalization.

Model	5-way	
	1-shot	5-shot
Matching Networks [7]	46.6	60.0
Meta-Learner LSTM [10]	43.4	60.6
PTN [1]	49.4	**68.2**
PTN (Ours)	48.6	65.5
HPTN	**49.7**	67.3

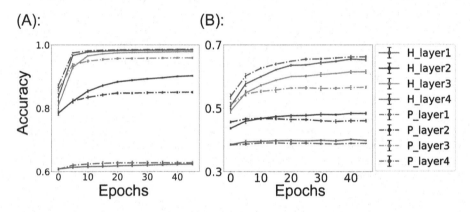

Fig. 2. Comparing classification accuracies at different layers of a HPTN and a PTN, denoted by solid and dashed lines, respectively. Different layers are colour-coded. The results are obtained by averaging over 1000 test episodes and reported with 95% confidence interval. A. The Omniglot dataset, 20-way 5-shot learning. B. The miniImageNet dataset, 5-way 5-shot learning. (Color figure online)

is trained to minimize also the classification errors in the intermediate layers of the network (Eq. (5)). This property is appealing to a neural system, as it means that the HPTN not only achieves an overall good classification performance, but also generates a set of good representations of increasing complexity along the network hierarchy. These hierarchical representations reflect that object features are extracted from simple to complex along the network hierarchy, and that the object information can be read-out from different layers depending on the requirement, achieving the so-called rough-to-fine object recognition.

3 Linking to Attractor Neural Networks

The appealing properties of PTNs/HPTNs are that, on one hand, they classify objects based on the distances/similarities between objects (metric-learning), and on the other hand, they provide an efficient way to optimize the representations of objects via end-to-end supervised training. Notably, the distance computation in a PTN/HPTN (in particular, when the Euclidean distance is used) is analogy to the pattern overlap computation in the attractor dynamics of a Hopfield network, and the latter is known to mimic information retrieval in neural systems [12]. Therefore, we explore whether the classification operation via distance computation in a PTN/HPTN can be replaced by the dynamics of an attractor network. For illustration, we only consider that the last step of classification by distance in a PTN is replaced by the dynamics of a Hopfield network, but extension to multiple layers in a HPTN is straightforward.

To construct a Hopfield model, we binarize the activities in the last layer of a PTN (± 1) and define the prototypes of all classes as the patterns the Hopfield network needs to memorize, which are denoted as μ_i, for $i = 1, \ldots, N$. We adopt two ways to construct the neuronal connections $\mathbf{W}$: one uses the conventional Hebb rule, which gives rise to $\mathbf{W} = \sum_i^N \mu_i^T \mu_i / N$, and the other uses the orthogonal Hebb rule, which orthogonalizes the memory patterns before applying the Hebb rule (for detail, see [13]). It is known that only the orthogonal Hebb rule ensures that the Hopfield model holds correlated memory patterns as its attractors ([14]). Once the Hopfield network is established, given a query pattern, the network evolves to an attractor (corresponding to a prototype) automatically, which outputs the class label of the input (see illustration in Fig. 1C).

Table 3 summarizes the experimental results, which show that: (1) when the representations are binarized (no attractor dynamics is applied yet), the performance of the PTN is slightly decreased compared to the case of using continuous representations; (2) when the Hopfield model with the conventional Hebb rule is applied, the performance of the network is degraded dramatically, in particular, in the difficult 20-way learning case; (3) when the Hopfield model with the orthogonal Hebb rule is applied, the performance of the network is still

Table 3. Classification accuracies of PTNs and a Hopfield model. Features are binarized according to mean values. The embedding network is trained using the Omniglot dataset. The hyper-parameters are the same as in Sect. 2.3.

Model	5-way		20-way	
	1-shot	5-shot	1-shot	5-shot
PTN	98.6	99.6	94.9	98.5
Binarized PTN	96.5	98.6	87.4	94.7
Hopfield with Hebb	89.1	92.9	25.9	27.3
Hopfield with Ortho_Hebb	94.4	97.9	78.2	90.6

decreased compared to the PTN, but the gap is not huge. Overall, our study indicates that a prototypical network can be approximately implemented by a biologically plausible attractor network.

4 Conclusions

In the present study, motivated by the characteristics of neural systems, we have extended PTNs in two aspects. First, we developed HPTNs to optimize representations for prototypes at all layers. Applied to benchmark datasets, the overall performances of HPTNs are comparable to, or slightly better than, that of PTNs. However, HPTNs generate much better representations for prototypes at the intermediate layers of the network. Secondly, we explored a potential link between PTNs and attractor neural networks by demonstrating that the classification operation via distance in PTNs can be realized approximately by the attracting dynamics of the Hopfield model. We hope that this study gives us insight into understanding metric-learning in neural systems (i.e., processing information based on semantic similarity) and sheds light on developing advanced brain-inspired computational models.

Acknowledgement. This work was supported by National Key Basic Research Program of China (2014CB846101), BMSTC (Beijing municipal science and technology commission) under grant No: Z161100000216143(SW), Z171100000117007(DHW & YYM); the National Natural Science Foundation of China (No: 31771146,11734004, YYM), Beijing Nova Program (No: Z181100006218118, YYM).

References

1. Snell, J., Swersky, K., Zemel, R.: Prototypical networks for few-shot learning. In: Advances in Neural Information Processing Systems, pp. 4080–4090 (2017)
2. Oreshkin, B.N., Lacoste, A., Rodriguez, P.: TADAM: task dependent adaptive metric for improved few-shot learning. arXiv preprint arXiv:1805.10123 (2018)
3. Sung, F., Yang, Y., Zhang, L., Xiang, T., Torr, P.H., Hospedales, T.M.: Learning to compare: relation network for few-shot learning. arXiv preprint arXiv:1711.06025 (2017)
4. Ren, M., et al.: Meta-learning for semi-supervised few-shot classification. arXiv preprint arXiv:1803.00676 (2018)
5. Liu, Y., Lee, J., Park, M., Kim, S., Yang, Y.: Transductive propagation network for few-shot learning. arXiv preprint arXiv:1805.10002 (2018)
6. Matsumoto, N., Eldridge, M.A., Saunders, R.C., Reoli, R., Richmond, B.J.: Mild perceptual categorization deficits follow bilateral removal of anterior inferior temporal cortex in rhesus monkeys. J. Neurosci. **36**(1), 43–53 (2016)
7. Vinyals, O., Blundell, C., Lillicrap, T., Wierstra, D.: Matching networks for one shot learning. In: Advances in Neural Information Processing Systems, pp. 3630–3638 (2016)
8. Lake, B.M., Salakhutdinov, R., Tenenbaum, J.B.: Human-level concept learning through probabilistic program induction. Science **350**(6266), 1332–1338 (2015)

9. Russakovsky, et al.: ImageNet large scale visual recognition challenge. Int. J. Comput. Vis. **115**(3), 211–252 (2015)
10. Ravi, S., Larochelle, H.: Optimization as a model for few-shot learning (2016)
11. Edwards, H., Storkey, A.: Towards a neural statistician. arXiv preprint arXiv:1606.02185 (2016)
12. Hopfield, J.J.: Neural networks and physical systems with emergent collective computational abilities. Proc. Natl. Acad. Sci. **79**(8), 2554–2558 (1982)
13. Srivastava, V., Sampath, S., Parker, D.J.: Overcoming catastrophic interference in connectionist networks using Gram-Schmidt orthogonalization. PloS One **9**(9), e105619 (2014)
14. Zou, X., Ji, Z., Liu, X., Mi, Y., Wong, K.M., Wu, S.: Learning a continuous attractor neural network from real images. In: Liu, D., Xie, S., Li, Y., Zhao, D., El-Alfy, E.S. (eds.) ICONIP 2017. LNCS, vol. 10637, pp. 622–631. Springer, Cham (2017). https://doi.org/10.1007/978-3-319-70093-9_66

Regularized Tensor Learning with Adaptive One-Class Support Vector Machines

Ali Anaissi[1]($\boxtimes$) (iD), Young Lee[2], and Mohamad Naji[3]

[1] The University of Sydney, Sydney, Australia
ali.anaissi@sydney.edu.au
[2] National University of Singapore, Singapore, Singapore
leey@comp.nus.edu.sg
[3] University of Technology Sydney, Ultimo, Australia
Mohamad.Naji@student.uts.edu.au

Abstract. The extraction of useful information from multi-sensors data requires fairly involved methodologies and algorithms. We propose an L_1 regularized tensor decomposition to decrease learning sensitivities, coupled with an adaptive one-class support vector machine (OCSVM) for anomaly detection purposes. This new framework yields sparse and smooth representations of the desired outcomes. An automatic parameter selection method based on the euclidean metric is also proposed to adaptively tune the kernel parameter inherent in OCSVM. These positive characteristics of tensor analysis allow us to fuse data from multiple sensors and further analyze them at the same time at which informative features are being extracted. This work is challenging because it is cross disciplinary; and thus it requires coherency to the specific domain applications fundamentals (such as structural health monitoring), on the one hand, and its diversity on machine learning techniques on the other. Compared to the state-of-the-art approaches for learning tensor and anomaly detection, our proposed methods work well on experiments and show better performance in terms of decomposition quality and stability of the extracted features.

Keywords: Tensor · One-class support vector machine
Online learning · Structural health monitoring · Anomaly detection

1 Introduction

Machine learning algorithms have been heavily employed in various application domains including bioinformatics, transportation and civil infrastructure [2,3,15]. A promising success has been reported in the literature. However, there are various application domains such as structural health monitoring (SHM) [18] that require development of new methodologies to analyze such excessive complex data. SHM is a continuous automated monitoring process of civil infrastructure condition using data obtained from several sensors mounted the structure

© Springer Nature Switzerland AG 2018
L. Cheng et al. (Eds.): ICONIP 2018, LNCS 11303, pp. 612–624, 2018.
https://doi.org/10.1007/978-3-030-04182-3_54

[10, 21]. The wealth of measured vibration responses values being generated by many sensors leads to complex high dimensional, multi-way and correlated data. This kind of multi-way form data raises many challenges to analyze and extract informative features that can be used to learn a classification model. Moreover, this type of data prohibits the use of a traditional decision-making classifier, such as random forest [6] or support vector machine [8] since only positive data samples (i.e. healthy samples) are accessible, and the instances from damage state, if not possible, are too difficult or costly to acquire.

These exceptional complexities led to the adoption of tensor analysis which allows the learning from such multi-way data in multiple modes at the same time and extracting damage sensitive features. These challenges also led to the development one-class classification models which can be constructed using one class data.

Several methods have been proposed in the literature for learning tensor known as tensor decomposition. However, two typical approaches are mostly used in the literature known as CANDECOMP/PARAFAC (CP) and Tucker decomposition [13]. This paper implements the tensor decomposition using CP approach since it has gained much popularity over tucker decomposition and it is the most widely used algorithm due to its ease of interpretation. Alternating Least Squares (ALS) technique is often used to solve this kind of tensor decomposition. However, it is well known that ALS can lead to sensitive solutions [9]. Thus, regularization methods are required to decrease this sensitivity.

This paper presents a novel algorithm to extract damage sensitive features form multi-way tensor data based on L_1 regularization to decrease ALS learning sensitivities, and adaptive one-class SVM (OCSVM) [19] to detect and assess damage severity.

This framework is extensively evaluated using laboratory-based and real-life structures datasets. The evaluation shows that our L_1 regularization method for learning tensor has the capability to extract damage sensitive features which were able to accurately detecting damage. It also reflects the fact that it has the potential to estimate the severity of damage in the specimen using the obtained decision values from the adaptive OCSVM. The contribution of this paper is as follows.

- Sensing multi-way data are fused in a tensor, from which L_1 regularization method for tensor decomposition was proposed to efficiently extract damage sensitive features.
- Damage detection and severity assessment are accomplished using adaptive OCSVM which has the capability to construct an optimal decision boundary without encountering the over-fitting nor the under-fitting problems.
- Experiments using data obtained from laboratory-based and real-life structures datasets show the effectiveness of our approach in damage identification and severity assessment.

The remainder of this paper is structured as follows. Section 2 reviews some related work. Section 3 describes our novel L_1 regularization method for learning tensor and adaptive OCSVM algorithms, while Sect. 4 presents our experimental results and evaluations. Finally, Sect. 5 discusses the contributions, future work and concludes this paper.

2   Related Work

Tensor analysis has been successfully applied in many application domains such as chemistry, social network analysis and computer vision [1,5,11], and produced significant results. For instance, [17] applied tensor analysis for damage detection and feature selection in SHM data, but without studying the capability of assessing the damage severity. [20] used the tensor analysis in online applications (e.g. computer network intrusion detection) and they worked on the problem of incrementally updating the component matrices in Tucker decomposition. However, few work reported in the literature to discuss the regularization learning of tensor decomposition to fuse data from multiple sensors and extract the features that has the ability to asses the severity of damage. [24] proposed a method to incrementally update the component matrices in CP decomposition over time. It adopts the alternating least square (ALS) method but without using any regularization parameters.

Successful applications of one-class support vector machine (OCSVM) for anomaly detection have been also reported in the literature. For instance, [23] designed a robust OCSVM to eliminate the influence of outliers to the learned boundary and used it to detect damage in a simulated structure. [14] also used OCSVM coupled with SVM-recursive feature elimination method for error detection. Further, the authors in [23] and [14] used OCSVM to detect damage in a rotating machinery and the results showed that the performance of the proposed method is superior to the state-of-the art methods. However, the work above focused on damage detection using data for each individual sensor which might help in detecting the damage but not in damage severity assessment in an unsupervised approach. They have also used the default setting of kernel parameter in OCSVM which may also over-fit or under-fit the model.

3   Methods

3.1   Tensor Analysis

Sensing data are usually collected from several networked sensors mounted the structure (e.g. a bridge) to measure the vibration signal over time. The data in SHM can be considered as a three-way tensor (feature × location × time) as described in Fig. 1. The feature in Fig. 1 is the information extracted from the raw signals in time domain. The sensors are represented in the location matrix, and time is data snapshots at different timestamps. Each slice along the time axis shown in Fig. 1 is a frontal slice representing all feature signals across all locations at a particular time.

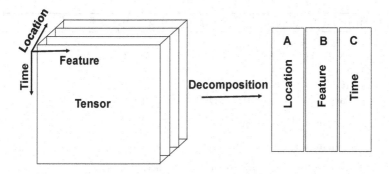

Fig. 1. Tensor data with three modes in SHM.

Given a tensor $X \in \Re^{I \times J \times K}$, CP method decomposes it into three matrices A, B and C as shown in Fig. 1. Matrix A represents the location mode, B represents feature mode and C represents time mode. In this sense, a tensor X can be written as

$$X \approx \sum_{r=1}^{R} \lambda_r \, A_r \circ B_r \circ C_r \equiv [\lambda; A, B, C] \qquad (1)$$

where "$\circ$" is a vector outer product. R is the latent element, A_r, B_r and C_r are r-th columns of component matrices $A \in \Re^{I \times R}$, $B \in \Re^{J \times R}$ and $C \in \Re^{K \times R}$, and λ is the weight used to normalize the columns of A, B, and C.

L_1 **Regularization for Learning Tensor:** The main goal of CP decomposition is to decrease the sum square error between the model and a given tensor X:

$$\min_{A,B,C} \|X - \sum_{r=1}^{R} \lambda_r \, A_r \circ B_r \circ C_r\|_f^2, \qquad (2)$$

where $\|X\|_f^2$ is the sum squares of X, and the subscript f is the Frobenius norm. It seems at first that the function presented in Eq. 2 is a non-convex problem since it aims to optimize the sum squares of three matrices. However, the problem can be reduced to a linear least squares problem by fixing two of the factor matrices, and solve only the third one. Following this approach, the ALS technique can be employed here which repeatedly solves each component matrix by locking all other components until it converges [16].

We remark that ALS can lead to sensitive solutions and it is not in general robust and hence motivates the need to incorporate the notion of penalty and regularization. The incorporation of regularization and penalization parameters into the L_1 norms make it possible to achieve smooth representations of the outcome and thus bypassing the perturbation surrounding the local minimum problem. The algorithm for CP decomposition using regularized ALS (RALS) is described in Algorithm 1. The L_1 penalty terms $\|X\|_{L_1} = \sum_. |x.|$ enforces the intensity of sparsity in X.

Algorithm 1. Regularized Alternating Least Squares for CP

Input: Tensor $X \in \Re^{I \times J \times K}$
Output: Matrices $A \in \Re^{I \times R}$, $B \in \Re^{J \times R}$, $C \in \Re^{K \times R}$, and λ
1: Initialize A, B, C
2: Repeat
3: $A = \arg\min\limits_{A} \frac{1}{2} \|X_{(1)} - A(C \odot B)^T\|^2 + \gamma_{X_A} \|X_{(1)}\|_{L_1}$
4: $B = \arg\min\limits_{B} \frac{1}{2} \|X_{(2)} - B(C \odot A)^T\|^2 + \gamma_{X_B} \|X_{(2)}\|_{L_1}$
5: $C = \arg\min\limits_{C} \frac{1}{2} \|X_{(3)} - C(B \odot A)^T\|^2 + \gamma_{X_C} \|X_{(3)}\|_{L_1}$
   ($X_{(i)}$ is the unfolded matrix of X in a current mode)
6: until converged

Incremental Tensor: Resolving the CP decomposition from scratch in online applications seems impractical in case of big training set of healthy samples. Therefore, there is an urgent need for incremental learning of tensor in online applications to update its components matrices when addition training data arrived. Similar to the RALS approach described in Algorithm 1 and as proposed by [24], we fix the two components A and B then update the temporal mode C, and sequentially update the non-temporal modes A and B, by fixing the other two.

Update Temporal Mode C:

$$C = \arg\min_C \frac{1}{2} \|X_{(1)} - C(B \odot A)^T\|^2 = \arg\min_C \frac{1}{2} \left\| \begin{bmatrix} X_{old(3)} - C_{old}(B \odot A)^T \\ X_{new(3)} - C_{new}(B \odot A)^T \end{bmatrix} \right\|^2$$

The new time mode C_{new} can be estimated by projecting the new arrived training sample $X_{new(3)}$ into the old matrices A and B. The new component C is then updated as follows

$$C = \begin{bmatrix} C_{old} \\ C_{new} \end{bmatrix} = \begin{bmatrix} C_{old} \\ X_{new(3)}((B \odot A)^T)^\dagger \end{bmatrix} \tag{3}$$

where $\dagger$ represents the pseudo-inverse of a matrix

Update Non-temporal Modes A and B: The optimization functions for A and B can be written as $\frac{1}{2} \|X_{(1)} - A(C \odot B)^T\|^2$ and $\frac{1}{2} \|X_{(2)} - C(B \odot A)^T\|^2$, respectively. The resultant derivatives of these two functions w.r.t A and B and setting them to zero are:

$$A = \frac{\overbrace{X_{(1)} - (C \odot B)}^{P}}{\underbrace{(C \odot B)^T(C \odot B)}_{Q}} \tag{4}$$

and

$$B = \overbrace{\frac{\overbrace{X_{(1)} - (C \odot A)}^{U}}{\underbrace{(C \odot A)^T (C \odot A)}_{V}}} \tag{5}$$

The computational time of $(C \odot B)$ and $(C \odot A)$ is costly since the resultant matrix size is very large. Therefore the simplified version of this equation can be estimated based on the old and new information of $X_{(i)_1^2}$ and C.

$$P = P_{old} + X_{new(1)}(C_{new} \odot B) \tag{6}$$

$$Q = Q_{old} + C_{new}^T C_{new} \circ B^T B \tag{7}$$

$$U = U_{old} + X_{new(2)}(C_{new} \odot A) \tag{8}$$

$$V = V_{old} + C_{new}^T C_{new} \circ A^T A \tag{9}$$

3.2 Adaptive One-Class Support Vector Machine Based Spatial Distance Algorithm

Given a set of training positive data samples $X = \{x_i\}_{i=1}^n$, where n is the number of samples, OCSVM uses a function ϕ to transform these positive samples into a high dimensional kernel space through the kernel $K(x_i, x_j) = \phi(x_i)^T \phi(x_j)$. Several kernel functions have been used in support vector machines such as Gaussian and polynomial kernels. These functions have a free critical kernel parameter γ which determines the width of the Gaussian kernel or the degree of the polynomial, respectively.

K-fold cross validation is often used at a training stage in order to tune the kernel parameter. However, in case of one class training, this technique is not possible because it selects the parameter that works only on the training class data and thus lack for generalization (over-fitting problem). Therefore, alternative approaches have been proposed for tuning this parameter for one class data [4,12,22]. These techniques only work to tune the Gaussian kernel parameter. This paper proposed a new algorithm called spatial distance (SD) for tuning all the possible type of kernel parameters based on inspecting the spatial locations of the edge and interior samples, and their distances to the enclosing surface of OCSVM. Following the objective function $f(\gamma_i)$ described in Eq. 10, the SD algorithm selects the optimal value of the kernel parameter $\hat{\gamma} = \underset{\gamma_i}{\mathrm{argmax}}(f(\gamma_i))$ which generates a hyperplane that is maximally distant from the interior samples but close to the edge samples. Note that we define the function $f(\cdot)$ as follows:

$$f(\gamma_i) := \underset{x_n \in \Omega_{IN}}{\mathcal{M}\{d_N(x_n)\}} - \underset{x_n \in \Omega_{ED}}{\mathcal{M}\{d_N(x_n)\}} \tag{10}$$

where Ω_{IN} and Ω_{ED}, respectively, represent the sets of interior and edge samples in the training positive data points identified using a hard margin linear support vector machine, $\mathcal{M}$ is the median value of the estimated normalized distances d_N. The d_N is distance for any sample to the hyperplane calculated using the following equation:

$$d_N(x_n) = \frac{d(x_n)}{d_\pi} \tag{11}$$

where d_π is the distance of a hyperplane to the origin described as $d_\pi = \frac{\rho}{\|w\|}$, and $d(x_n)$ is the distance of the sample x_n to the hyperplane obtained using the following equation:

$$d(x_n) = \frac{g(x_n)}{\|w\|} = \frac{\sum_{i=1}^{n} \alpha_i k(x_i, x_n) - \rho.}{\sqrt{\sum_{ij}^{n} \alpha_i \alpha_j K(x_i, x_j)}} \tag{12}$$

where w is a perpendicular vector to the decision boundary, α are the Lagrange multipliers and ρ known as the bias term learnt from OCSVM.

4   Experimental Results

This section presents two case studies to illustrate how our feature-based L_1 regularization for tensor decomposition and adaptive OCSVM based SD methods are capable to identify and estimate the structural damage severity. The core consistency diagnostic technique (CORCONDIA) technique described in [7] was used to determine the number of rank-one tensors R when it decomposed using CP method. The CORCONDIA suggests $R = 2$ for all experimented data sets.

In all experiments, the OCSVM uses the Gaussian kernel function defined in Eq. 13 to map the positive samples into high dimensional kernel space.

$$K(x_i, x_j) = \exp\left(-\frac{\|x_i - x_j\|^2}{2\gamma^2}\right). \tag{13}$$

The SD method were also used in all the experiments to adaptively tune the Gaussian kernel parameter. The accuracy values were obtained using the F-Score (FS), defined as $F\text{-}score = 2 \cdot \dfrac{\text{Precision} \times \text{Recall}}{\text{Precision} + \text{Recall}}$, $Precision = \dfrac{\text{TP}}{\text{TP} + \text{FP}}$ and $Recall = \dfrac{\text{TP}}{\text{TP} + \text{FN}}$, where the number of true positive is abbreviated by (TP), false positive (FP), and false negative (FN), respectively.

4.1   Case Study I: Sydney Bridge

Experiments Setup and Data Collection. Our main experiments were conducted using structural vibration based datasets acquired from a network of accelerometers mounted on the Sydney Bridge. The bridge has several joints each mounted by three sensors (left, right and middle). However, only one joint was used in this study because it is the only joint was known as a cracked joint. The data used in this study was collected over a period o nine months and it contains 7,625 events each has 300 features representing the frequencies of each event. The first month of the data (1800 samples) collected before the presence of any damage were used to derive the damage sensitive features and train the adaptive OCSVM. The remaining samples of the healthy data (before damage and after partial/full repair) were used for testing in addition to the damage samples.

The resultant matrices from the three sensors of the training dataset were fused in a tensor $X \in \Re^{1800 \times 300 \times 3}$, which was decomposed using RALS method (see Algorithm 1) into three matrices $A \in \Re^{3 \times 2}$, $B \in \Re^{300 \times 2}$, and $C \in \Re^{1800 \times 2}$. At the first of the experiment, we initialize the values of P, Q, U and V using Eqs. 6, 7, 8 and 9, respectively. The matrix C was used to construct the adaptive OCSVM model. For each arrive X_{new} datum, we used Eq. 3 to calculate C_{new}, and then we update the matrices A and B following Eqs. 4 and 5, respectively.

Results and Discussions. This section presents the classification performance of the OCSVM using the damage sensitive features extracted from the tensor using RALS algorithm. A 98% accuracy was achieved in the test dataset using the regularized tensor learning and the adaptive value of Gaussian kernel parameter in OCSVM. The damage samples were successfully detected while maintaining a low false alarm rates. On the other hand, 87% accuracy was achieved using the non-regularized tensor learning method as the OCSVM model classified many of the damage samples as healthy events. This is what we anticipated from the non-regularized tensor learning method which produced unstable features that cannot be used to accurately detect damage in the structure. Further and in order to show the ability of our regularized tensor learning approach in assessing the severity of the detected damage, we have calculated the decision values for the obtained classification results which shown in Figs. 2 and 3. The horizontal axis represents the date of the data instance and the vertical axis represents the decision values. 5,825 events were tested in this experiment; the average value for each ten events was used to report the health score of the structure. The first 418 green data points represent the healthy instances collected before the existence of the damage, the next 27 red data points represent the damaged instances. The following 75 and 65 samples (shown in orange and blue, respectively) refer to the healthy samples after partial and full damage repair, respectively. The average of all the decision values for each group were recorded and presented in Figs. 2 and 3. A black line was drawn to line the average values. As can be clearly seen from Fig. 2, the constructed OCSVM model using RALS features has the capability to identify the damage in the structure and to asses the condition of the joint when the mean value of the decision values clearly increased after repair. The non-regularized tensor learning, on the other hand, failed to detect all the damage samples and produced false alarms.

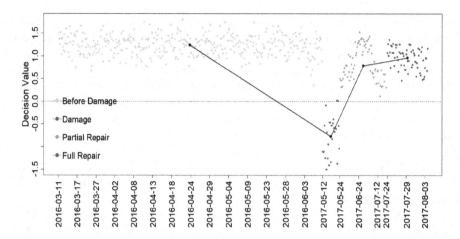

Fig. 2. Damage identification results applied on Sydney Bridge data using regularized tensor learning (Color figure online)

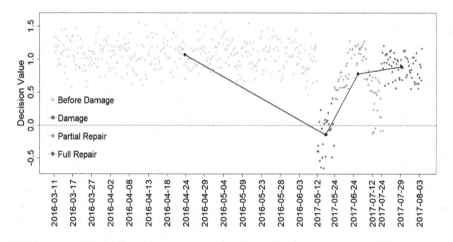

Fig. 3. Damage identification results applied on Sydney Bridge data using non-regularized tensor learning (Color figure online)

4.2 Case Study II: A Reinforced Concrete Jack Arch

Experiments Setup and Data Collection. The data in this case study contains 950 samples separated into two main groups, Healthy (190 samples) and Damaged (760 samples). Each sample has 8000 attributes representing the frequencies of each sample. The damaged cases were further sub-grouped into 4 different damaged cases (190 samples each) based on their severity.

We randomly chosen 80% of the positive data points (152 instances) to extract the damage sensitive feature by fusing the measured responses from the six sensors of the test positive samples into a tensor $X \in \Re^{152 \times 8000 \times 6}$ since only six sensors were used in these experiments. The remaining 20% of the healthy data was used for testing including the data collected from the four damage cases. Similar to the previous case study, we applied RALS algorithm to decompose the tensor X into three matrices A, B, and C which was used to construct the adaptive OCSVM. For each arrive X_{new} datum, we used Eq. 3 to calculate C_{new}, and Eqs. 4 and 5 to update the matrices A and B, respectively.

Results and Discussions. As can be seen in Fig. 4, although no information of damaged events has been employed to construct the OCSVM model and only data from the healthy events have been utilized for the purpose of training, the trained model can successfully predict the healthy and the damaged events using the damage sensitive features extracted from the tensor using RALS algorithm. Further, it can be clearly observed that by increasing the damage severity, the decision values were further decreased (i.e. the data were moving away from the positive data points). These results demonstrate the capability of the RALS method in detecting and assessing the evolution of damage in the structure based on the decision values. The model accuracy of the test data was 95.7% using RALS method for learning tensor. This suggests that the constructed adaptive model is well generalized on unseen samples and has the ability to detect healthy and damaged samples although the level of damage case 1 in this case study is considerably small. Moreover, the method has also shown the capability to identify the progression of the damage which illustrated in Fig. 4 by calculating and showing a solid lack line to connect the mean of all the decision values for each category.

A 81.5% accuracy, on the other hand, was achieved using the non-regularized ALS algorithm. As shown in Fig. 4, the non-regularized method produced less accurate damage sensitive features which yields OCSVM to miss the damage samples related to the small cracks in the structure i.e. damage cases 1 and 2.

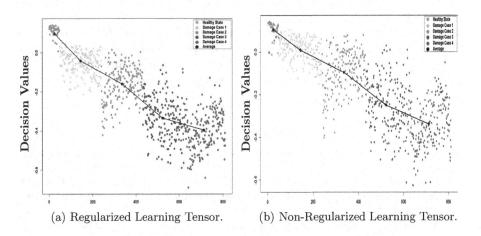

(a) Regularized Learning Tensor.   (b) Non-Regularized Learning Tensor.

Fig. 4. Damage identification results applied on lab specimen data.

5   Conclusions

This paper presents a novel and insightful L_1 regularized tensor learning coupled with an adaptive OCSVM for online damage identification and assessment. The online tensor analysis was used in this study as a data amalgamation technique to combine acceleration from multiple sensors and to extract desired features. The core consistency diagnostic technique method described in [7] was used to determine the number of rank-one tensors R in the CP [13] method. An adaptive OCSVM was then utilized to build a statistical model using the data from the healthy state of the structure. OCSVM approach is well suited this kind of applications when we lacking of damaged data instances. Gaussian kernel method was employed in OCSVM and an automatic parameter selection method called SD was proposed for tuning the Gaussian kernel parameter. Incremental tensor was then applied for each arrived new datum to adiabatically update the three tensor matrices obtained during the training stage without resolving the problem from scratch.

Two comprehensive case studies were investigated considering different damage scenarios including single and multiple damage states which were progressively increasing. It was demonstrated that our RALS method generated damage sensitive features which used by OCSVM to detect the presence of damage while maintaining a low false positive rate on data from the healthy structure. Our adaptive model was able to detect damage with a very small severity which was missed using the non-regularized tensor method. Furthermore, this paper illustrated how our RALS method generates damage sensitive features which can also reliably monitor the progress of damage in the structure by providing decreasing negative decision values.

These findings indicate that the application of unsupervised learning using online tensor to extract damage sensitive features along with the implementation of adaptive OCSVM can provide a robust method to detect and evaluate the progress of damage, which is of great importance during structural condition assessment.

We conclude that the availability of regularization in our framework leaves us the modelling freedom to maneuver and adapt to each domain specific application. Each application presents quite distinct specificity that require certain amount of improvised inventiveness fitted to that application.

References

1. Acar, E., Dunlavy, D.M., Kolda, T.G., Mørup, M.: Scalable tensor factorizations for incomplete data. Chemom. Intell. Lab. Syst. **106**(1), 41–56 (2011)
2. Anaissi, A., Goyal, M., Catchpoole, D.R., Braytee, A., Kennedy, P.J.: Ensemble feature learning of genomic data using support vector machine. PloS One **11**(6), e0157330 (2016)
3. Anaissi, A., Kennedy, P.J., Goyal, M., Catchpoole, D.R.: A balanced iterative random forest for gene selection from microarray data. BMC Bioinform. **14**(1), 261 (2013)

4. Anaissi, A., et al.: Adaptive one-class support vector machine for damage detection in structural health monitoring. In: Kim, J., Shim, K., Cao, L., Lee, J.-G., Lin, X., Moon, Y.-S. (eds.) PAKDD 2017. LNCS (LNAI), vol. 10234, pp. 42–57. Springer, Cham (2017). https://doi.org/10.1007/978-3-319-57454-7_4
5. Bader, B.W., Harshman, R.A., Kolda, T.G.: Temporal analysis of semantic graphs using ASALSAN. In: Seventh IEEE International Conference on Data Mining, ICDM 2007, pp. 33–42. IEEE (2007)
6. Breiman, L.: Random forests. Mach. Learn. **45**(1), 5–32 (2001)
7. Bro, R., Kiers, H.A.: A new efficient method for determining the number of components in parafac models. J. Chemom. **17**(5), 274–286 (2003)
8. Cortes, C., Vapnik, V.: Support vector machine. Mach. Learn. **20**(3), 273–297 (1995)
9. Eldén, L.: Perturbation theory for the least squares problem with linear equality constraints. SIAM J. Numer. Anal. **17**(3), 338–350 (1980)
10. Farrar, C.R., Worden, K.: Structural Health Monitoring: A Machine Learning Perspective. Wiley, Chichester (2012)
11. Ho, J.C., Ghosh, J., Sun, J.: Marble: high-throughput phenotyping from electronic health records via sparse nonnegative tensor factorization. In: Proceedings of the 20th ACM SIGKDD International Conference on Knowledge Discovery and Data Mining, pp. 115–124. ACM (2014)
12. Khazai, S., Homayouni, S., Safari, A., Mojaradi, B.: Anomaly detection in hyperspectral images based on an adaptive support vector method. IEEE Geosci. Remote. Sens. Lett. **8**(4), 646–650 (2011)
13. Kolda, T.G., Bader, B.W.: Tensor decompositions and applications. SIAM Rev. **51**(3), 455–500 (2009)
14. Mahadevan, S., Shah, S.L.: Fault detection and diagnosis in process data using one-class support vector machines. J. Process. Control **19**(10), 1627–1639 (2009)
15. Menon, A.K., Cai, C., Wang, W., Wen, T., Chen, F.: Fine-grained od estimation with automated zoning and sparsity regularisation. Transp. Res. Part B **80**, 150–172 (2015)
16. Papalexakis, E.E., Faloutsos, C., Sidiropoulos, N.D.: Tensors for data mining and data fusion: Models, applications, and scalable algorithms. ACM Trans. Intell. Syst. Technol. (TIST) **8**(2), 16 (2016)
17. Prada, M.A., Toivola, J., Kullaa, J., Hollmén, J.: Three-way analysis of structural health monitoring data. Neurocomputing **80**, 119–128 (2012)
18. Ricci, S.: Best achievable modal eigenvectors in structural damage detection. Exp. Mech. **40**(4), 425–429 (2000)
19. Schölkopf, B., Williamson, R.C., Smola, A.J., Shawe-Taylor, J., Platt, J.C., et al.: Support vector method for novelty detection. In: NIPS, vol. 12, pp. 582–588. Citeseer (1999)
20. Sun, J., Tao, D., Papadimitriou, S., Yu, P.S., Faloutsos, C.: Incremental tensor analysis: Theory and applications. ACM Trans. Knowl. Discov. Data (TKDD) **2**(3), 11 (2008)
21. Worden, K., Manson, G.: The application of machine learning to structural health monitoring. Philos. Trans. R. Soc. Lond. A Math. Phys. Eng. Sci. **365**(1851), 515–537 (2007)
22. Xiao, Y., Wang, H., Xu, W.: Parameter selection of gaussian kernel for one-class SVM. IEEE Trans. Cybern. **45**(5), 941–953 (2015)

23. Yin, S., Zhu, X., Jing, C.: Fault detection based on a robust one class support vector machine. Neurocomputing **145**, 263–268 (2014)
24. Zhou, S., Vinh, N.X., Bailey, J., Jia, Y., Davidson, I.: Accelerating online CP decompositions for higher order tensors. In: Proceedings of the 22nd ACM SIGKDD International Conference on Knowledge Discovery and Data Mining, pp. 1375–1384. ACM (2016)

Semi-supervised Multi-label Dimensionality Reduction via Low Rank Representation

Yezi Liu[✉]

School of Computing and Information, University of Pittsburgh,
Pittsburgh, PA 15260, USA
yel32@pitt.edu

Abstract. Multi-label dimensionality reduction is an appealing and challenging task in data mining and machine learning. Previous works on multi-label dimensionality reduction mainly conduct in an unsupervised or supervised way, and ignore abundant unlabeled samples. In addition, most of them emphasize on using pairwise correlations between samples, therefore, unable to utilize the high-order sample information to improve the performance. To address these challenges, we propose an approach called Semi-supervised Multi-label Dimensionality Reduction via Low Rank Representation (SMLD-LRR). SMLD-LRR first utilizes the low rank representation in the feature space of samples to calculate the low rank constrained coefficient matrix, then it adapts the coefficient matrix to capture the high-order structure of samples. Next, it uses low rank representation in the label space of labeled samples to explore the global correlations of labels. After that, SMLD-LRR further employs the learned high-order structure of samples to enforce the consistency between samples in the original space and the corresponding samples in the projected subspace by maximizing the dependence between them. Finally, these two high-order correlations and the dependence term are incorporated into the multi-label linear discriminant analysis for dimensionality reduction. Extensive experimental results on four multi-label datasets demonstrate that SMLD-LRR achieves better performance than other competitive methods across various evaluation criteria; it also can effectively exploit high-order label correlations to preserve sample structure in the projected subspace.

Keywords: Dimensionality reduction
Multi-label Linear Discriminant Analysis · Semi-supervised learning
Low Rank Representation

1 Introduction

Different from traditional supervised learning where each sample is associated with a single class label that denotes its semantic category. A sample, in many real-world applications, is often annotated with multiple labels. For example,

© Springer Nature Switzerland AG 2018
L. Cheng et al. (Eds.): ICONIP 2018, LNCS 11303, pp. 625–637, 2018.
https://doi.org/10.1007/978-3-030-04182-3_55

a news report can be tagged with multiple labels, such as economics, politics and culture. Multi-label learning is a paradigm proposed to address these scenarios, and has attracted much attention in data mining and machine learning research literature [1,2]. However, in multi-label learning, the dimensionality of data is usually very high. Directly working on such high-dimensional data is not only time consuming and computational unreliable [3], but may also degrade the classification performance due to the possible existence of redundant or noise features [4]. Dimensionality reduction, as a crucial preprocessing for many data mining (or machine learning) tasks on high-dimensional samples [4–8], can efficiently reduce the dimensionality of samples and boost the performance of the later analysis.

In the past few decades, many multi-label dimensionality reduction methods have been proposed in literature [9–12]. For example, Zhang et al. [11] introduced a supervised multi-label dimensionality reduction method called MDDM. MDDM learns a low-dimensional subspace by maximizing the feature-label dependence between the original features of samples and the associated labels of these samples under the Hilbert-Schmidt Independence Criterion [13]. Wang et al. [14] introduced the Multi-label Linear Discriminative Analysis (MLDA) by extending the classical Linear Discriminant Analysis (LDA) [15] via additionally utilizing label correlations. In addition, several classic techniques, such as Canonical Correlation Analysis (CCA) [9] is also extended to handle the multi-label problem [16]. These supervised methods are based on the sufficient labeled training data but it is always difficult to get the labeled data in reality [17]. Hence these methods limit their effectiveness by excluding a large amount of unlabeled samples, which can be used to promote the performance of multi-label dimensionality reduction [18].

To make full use of sufficient unlabeled samples and bypass the scarce labeled ones, a few semi-supervised multi-label dimensionality reduction methods have been proposed in recent years. For instance, Guo et al. [3] proposed a Semi-supervised Multi-label Dimensional Reduction method (SSMLDR). SSMLDR first tries to enlarge labeled samples by assigning pseudo labels to unlabeled samples via label propagation algorithm, it then combines all the labeled samples into MLDA to optimize the projective matrix. Nevertheless, the labels of labeled samples may be wrongly propagated to the unlabeled ones and thus degrades the performance of SSMLDR. Yuan et al. [19] proposed a method called Multi-label Linear Discriminant Analysis with Locality Consistency (MLDA-LC). MLDA-LC takes advantage of a kNN graph constructed on both labeled and unlabeled sample, and incorporates a smoothness term into the standard MLDA for dimensionality reduction. Yu et al. [20] proposed another approach Semi-supervised Multi-label Linear Discriminant Analysis (SMLDA). It also extends MLDA by maximizing the dependence between pairwise similarity derived from samples in the ambient space and the similarity from corresponding samples in the projected subspace. However, all of these aforementioned methods can not adequately exploit the global structure of samples. Moreover, they also can not explicitly utilize label correlations, which is very important in multi-label learning tasks [2].

In this paper, a novel multi-label dimensionality reduction approach called Semi-supervised Multi-label Dimensionality reduction via Low Rank Representation (SMLD-LRR) has been proposed. Unlike existing semi-supervised linear discriminant analysis methods that mainly focus on pairwise similarity of samples and can not use high-order label correlations, we take advantage of Low Rank Representation (LRR) [21] to capture high-order correlations in both feature and label spaces of samples. To be specific, SMLD-LRR first computes the coefficient matrix of samples (including both labeled and unlabeled samples) in the feature space by LRR and then adopts the matrix to learn the high-order structure of samples in feature space. In addition to that, SMLD-LRR explores the global relationships among labels by using LRR again in the label space of labeled samples. Next, SMLD-LRR utilizes the high-order similarity of samples to further enforce the consistency of samples in the original feature space and in the projected subspace by maximizing the dependence between them. Finally, SMLD-LRR fuses the feature-based and semantic-based correlations, as well as the dependence term into the MLDA framework. The empirical study on several publicly available datasets shows that SMLD-LRR not only can find more discriminant subspace than other related methods, but also can utilize label correlations to preserve sample similarity in the learned subspace.

The reminder of this paper is organized in 3 sections. We first elaborate on SMLD-LRR, then present the experiments, the analysis and conclusions.

2   Methodology

In this section, we briefly introduce the LRR, the capture of high-order structure of samples and the exploration of high-order label correlations, and the proposed Low Rank Representation based multi-label dimensionality reduction model. Before that, we introduce the notations that will be used in this paper. Let $\{\mathbf{x}_i, \mathbf{y}_i\}$ be a collection of samples, where $\mathbf{x}_i \in \mathbb{R}^D$ indicates the i-th sample. $\mathbf{y}_i \in \{0, 1\}^C$ is the label vector of the i-th sample. If sample i is annotated with label c, $\mathbf{y}_{ic} = 1$; otherwise, $\mathbf{y}_{ic} = 0$. Here we assume that among N samples, the first l samples are labeled and the left u samples are unlabeled, $N = l + u$. Our goal is to learn a discriminant subspace by leveraging both labeled and unlabeled samples.

2.1   Low Rank Representation for High-Order Sample Structure Exploration

Low rank representation (LRR) has been recently introduced to capture the global structure of data in multi-label data [21–24]. Since LRR is robust to noisy or redundant features, it has been regarded as an appropriate approach to explore the high-order structure and global mixture of the subspace structure of samples in many applications [25]. In this paper, we take advantage of LRR to explore the high-order structure of samples and use this structure to advance the performance of dimensionality reduction. Assuming $\mathbf{X} = [\mathbf{x}_1, \mathbf{x}_2, ..., \mathbf{x}_N] \in \mathbb{R}^{D \times N}$

is the feature space of both labeled and unlabeled samples. $\mathbf{Y} = [\mathbf{y}_1, \mathbf{y}_2, ..., \mathbf{y}_l] \in \mathbb{R}^{C \times l}$ denotes the label matrix for l labeled samples. Each sample therefore can be reconstructed as a linear combination of bases from a dictionary $\mathbf{A} = [\mathbf{a}_1, \mathbf{a}_2, \mathbf{a}_3, ..., \mathbf{a}_M] \in \mathbb{R}^{D \times M}$ as follow:

$$\mathbf{X} = \mathbf{A}\mathbf{Z}_1 \tag{1}$$

$\mathbf{Z}_1 \in \mathbb{R}^{N \times N}$ is the coefficient matrix and $\mathbf{Z}_1(\cdot, i) \in \mathbb{R}^N$ is the representation coefficient vector of sample $\mathbf{x}_i$ with respect to N samples. We also set $\mathbf{A} = \mathbf{X}$ for simplicity [22]. Note that entry in $\mathbf{Z}_1(i, j)$ is actually the contribution of $\mathbf{x}_j$ to the reconstruction of $\mathbf{x}_i$ based on $\mathbf{A}$. Based on this idea, the general LRR problem is to enforce $\mathbf{Z}_1$ to be low rank and solve the following optimization problem:

$$\begin{cases} \min_{\mathbf{Z}_1} rank(\mathbf{Z}_1) \\ \text{s.t. } \mathbf{X} = \mathbf{A}\mathbf{Z}_1, \mathbf{Z}_1 \geqslant 0 \end{cases} \tag{2}$$

Equation (2) is called low rank representation [21], where $\mathbf{X}$ is reconstructed by the low rank constrained matrix $\mathbf{Z}_1$. Solving Eq. (2) is a NP hard problem, Zhang et al. [26] suggests relax Eq. (2) as follows:

$$\begin{cases} \min_{\mathbf{Z}_1} \|\mathbf{Z}_1\|_* + \lambda\|\mathbf{E}\|_{2,1} \\ \text{s.t. } \mathbf{X} = \mathbf{A}\mathbf{Z}_1 + \mathbf{E}, \mathbf{Z}_1 \geqslant 0 \end{cases} \tag{3}$$

where $\|\mathbf{E}\|_{2,1} = \sum_{j=1}^{n} \sqrt{\sum_{i=1}^{m} (\mathbf{E_{ij}})^2}$ is a noise term and λ is a trade-off parameter. In this paper, we employ the Linearized Alternating Direction technique with Adaptive Penalty (LADMAP) [25] to accelerate the solution of LRR, where the time complexity is $O(rn^2)$ and r is the rank of $\mathbf{Z}_1$. As low rank representation jointly finds the low-ranked coefficient matrix $\mathbf{Z}_1$ for all samples, it is a natural encoding of the global structure among samples. For this reason, we adapt $\mathbf{Z}_1$ to measure the high-order structure of samples $\mathbf{W}$, where $\mathbf{W} = (\mathbf{Z}_1 + \mathbf{Z}_1^T)/2$.

2.2 Low Rank Representation for High-Order Label Correlation Exploitation

Classical supervised dimensionality reduction techniques often focus on single label problem, but the data in real-life usually have multiple labels simultaneously. For instance, considering two images, one is labeled with "cow", "prairie" and "green color" while the other is annotated with "deer", "forest" and "green color". Intuitively, the two images are similar from the feature aspect as they have similar backgrounds. From the aspect of semantic labels, however, these images are quite different because one describes a cow on the prairie and the other shows a deer in the forest. Therefore, only by computing the feature similarity between the samples can not differentiate some samples with polysemous labels but different concepts. To avoid this dilemma, it is necessary to build techniques that can additionally utilize semantic information for dimensionality reduction. Most previous multi-label dimensionality reduction methods mainly focus on pairwise label correlations; such as pairwise label correlation computed by cosine similarity [14, 20]. Recently, Wang et al. [27] suggests that the high-order semantic similarity between two multi-label samples can be derived from the labels associated with these two samples by Sparse Representation (SR) [28]. However, considering that SR [27] is optimized by separately treating each labeled samples; thus, it may be failed to capture the global semantic relationship between samples, which is very important as discussed above. Inspired by this, we try to use LRR to capture the high-order label correlations between labeled samples. Concretely, we first compute the low rank coefficients of samples based on label space $\mathbf{Y}$ by LRR as suggested in Eq. 3. Then we utilize the low rank coefficients again to define the semantic based high-order similarity of samples $\mathbf{S} \in \mathbb{R}^{N \times N}$, where $\mathbf{S} = (\mathbf{Z}_2 + \mathbf{Z}_2^T)/2$ and $\mathbf{Z}_2$ is the LRR coefficient matrix in regard to labeled samples.

2.3   The SMLD-LRR Method

In this subsection, we will show how to incorporate the feature-based and label-based high-order correlations into the standard MLDA framework so that it can not only leverage both labeled and unlabeled samples, but can also preserve sample similarity in the projected subspace. Let $\mathbf{P} \in \mathbb{R}^{D \times d}$ denotes the target projective matrix, which projects $\mathbf{x}$ into a d-dimensional discriminative subspace via $\mathbf{P}^T \mathbf{x}$. MLDA defines the within-class, between-class and the total-class scatter matrices as follows:

$$\mathbf{S}_b = \sum_{c=1}^{C} \mathbf{S}_b^c, \quad \mathbf{S}_b^c = \sum_{i=1}^{N} y_{ic}(\mathbf{m}_c - \mathbf{m})(\mathbf{m}_c - \mathbf{m})^T \tag{4}$$

$$\mathbf{S}_w = \sum_{c=1}^{C} \mathbf{S}_w^c, \quad \mathbf{S}_w^c = \sum_{i=1}^{N} y_{ic}(\mathbf{x}_i - \mathbf{m}_c)(\mathbf{x}_i - \mathbf{m}_c)^T \tag{5}$$

$$\mathbf{S}_t = \sum_{c=1}^{C} \mathbf{S}_t^c, \quad \mathbf{S}_t^c = \sum_{i=1}^{n} y_{ic}(\mathbf{x}_i - \mathbf{m})(\mathbf{x}_i - \mathbf{m})^T \tag{6}$$

Where $\mathbf{S}_b$, $\mathbf{S}_w$ and $\mathbf{S}_t$ are the corresponding between-class, within-class and the total class-wise scatter matrices for all the class labels, respectively. $\mathbf{m}_c$ denotes the centroid of the c-th class and $\mathbf{m}$ is the global centroid of labeled samples, which are defined as:

$$\mathbf{m}_c = \frac{\sum_{i=1}^{N} y_{ic}\mathbf{x}_i}{\sum_{c=1}^{C} y_{ic}}, \quad \mathbf{m} = \frac{\sum_{c=1}^{C} \sum_{i=1}^{N} y_{ic}\mathbf{x}_i}{\sum_{c=1}^{C} \sum_{i=1}^{N} y_{ic}} \tag{7}$$

In multi-label learning, the proper usage of label correlation usually can boost the performance [2]. First, MLDA defines the pairwise correlation between labels as follows:

$$\mathbf{M}(c1, c2) = \frac{\mathbf{Y}_{.c1}^T \mathbf{Y}_{.c2}}{\|\mathbf{Y}_{.c1}\| \|\mathbf{Y}_{.c2}\|} \tag{8}$$

where $\mathbf{Y}_{.c1}$ is the $c1$-th column of $\mathbf{Y}$ that includes all the member samples of this label. Then MLDA replaces the $\mathbf{Y}$ in Eq. (8) as $\tilde{\mathbf{Y}} = \mathbf{YM}$. Next, MLDA optimizes the optimal projective matrix $\mathbf{P}$ by solving the following optimization problem:

$$\max_{\mathbf{P}} \frac{tr(\mathbf{P}^T \mathbf{S}_b \mathbf{P})}{tr(\mathbf{P}^T \mathbf{S}_w \mathbf{P})} \quad \text{or} \quad \max_{\mathbf{P}} \frac{tr(\mathbf{P}^T \mathbf{S}_b \mathbf{P})}{tr(\mathbf{P}^T \mathbf{S}_t \mathbf{P})} \tag{9}$$

Where $tr(\cdot)$ represents the matrix trace operator. Eq. (9) is efficient when there are sufficient labeled samples. However, in real-life applications, it is rather difficult or impractical to collect sufficient labeled samples. Here, we extend MLDA to a semi-supervised way and construct a semi-supervised multi-label dimensionality reduction model like this:

$$\min_{\mathbf{P}} \frac{tr(\mathbf{P}^T \mathbf{S}_w \mathbf{P})}{tr(\mathbf{P}^T \mathbf{S}_b \mathbf{P}) + \alpha \Psi(\mathbf{W}) + \beta \Psi(\mathbf{S}) + \gamma \Psi(\mathbf{P})} \tag{10}$$

where the second term $\Psi(W)$ is the regularization on the global structure of labeled and unlabeled samples, the third term $\Psi(S)$ is to take advantage of the semantic relationship between labeled samples and the last term is to maximize the dependence between high-order similarity of samples in the original space and the corresponding samples in the projected subspace. α, β and γ are three trade-off parameters. The first term is defined as follows:

$$\Psi(\mathbf{W}) = \frac{1}{2} \sum_{i,j=1}^{N} ||\mathbf{P}^T \mathbf{x}_i - \mathbf{P}^T \mathbf{x}_j||^2 \mathbf{W}_{ij} = tr(\mathbf{P}^T \mathbf{X} \mathbf{L} \mathbf{X}^T \mathbf{P}) \tag{11}$$

where $\mathbf{W}_{ij}$ is the weight of the edge between samples $\mathbf{x}_i$ and $\mathbf{x}_j$. $\mathbf{D}$ is a diagonal matrix with $\mathbf{D}_{ii} = \sum_{j=1}^{N} \mathbf{W}_{ij}$; $\mathbf{L} = \mathbf{D} - \mathbf{W}$ is the graph Laplacian matrix [29]. The reason to maximize Eq. (11) is to preserve the local structure of samples in the projective subspace.

Similar to the assumption in Wang *et al.* [30], in this paper, we assume that the label of each sample can be reconstructed by the other samples, while the reconstructs coefficients are derived from the low rank representation matrix of sample vectors. Thus the linear reconstruction coefficients in the low rank matrix can be used to predict the labels of unlabeled samples, since the weight $\mathbf{W}_{ij}^2$ reflects the likelihood of sample $\mathbf{x}_i$ to have the same label as sample $\mathbf{x}_j$. Based on this label reconstruction assumption, the second term of Eq. (10) is defined as:

$$\Psi(\mathbf{S}) = \frac{1}{2} \sum_{i,j=1}^{N} ||\mathbf{P}^T \mathbf{x}_i - \sum_{j \neq i} \mathbf{S}_{ij} \mathbf{P}^T \mathbf{x}_j||^2 = tr(\mathbf{P}^T \mathbf{X} (\mathbf{I} - \mathbf{S})(\mathbf{I} - \mathbf{S})^T \mathbf{X}^T \mathbf{P})$$
$$= tr(\mathbf{P}^T \mathbf{X} \mathbf{M} \mathbf{X}^T \mathbf{P}) \tag{12}$$

where $\mathbf{S}_{ij}$ represents the similarity of label vector $\mathbf{y}_i$ and $\mathbf{y}_j$. $\mathbf{I} \in \mathbb{R}^{N \times N}$ is an identity matrix, and $\mathbf{M} = (\mathbf{I} - \mathbf{S})(\mathbf{I} - \mathbf{S})^T$. The motivation to use this term is to replenish possible missing features of labeled samples by taking advantage of the semantic based sample similarity. If a labeled sample has some missing features and its semantic neighbors have these features, this term can replenish missing features of that sample to some extent. In addition, by using this term, we can utilize the semantic information to induce the projective subspace learning, which not only helps to obtain a discriminative subspace but also may alleviate the widely spread semantic gap [31] between the low-level feature space and the semantic label space. Another advantage of this term is that by jointly work with the first term, the possible replenished features and semantic information further propagate to other unlabeled samples, and thus, SMLD-LRR can learn more discriminate features.

Note that the semantic similarity between samples always has a positive correlation with the feature similarity [11], the last term of Eq. (10) is defined as:

$$\Psi(\mathbf{P}) = HSIC(\mathbf{X}, \mathbf{Y}) = (N-1)^{-2} tr(\mathbf{KHQH}) \tag{13}$$

$\mathbf{K}, \mathbf{H}, \mathbf{Q} \in \mathbb{R}^{N \times N}$, $\mathbf{K}(i,j) = \mathcal{K}(\mathbf{x}_i, \mathbf{x}_j)$, $\mathbf{Q}(i,j) = \mathcal{Q}(\mathbf{y}_i, \mathbf{y}_j)$ and $\mathbf{H}(i,j) = \delta_{ij} - \frac{1}{N}$. $\mathcal{K}$ and $\mathcal{Q}$ are the kernel functions in the feature space and label space. This term measures the dependence between similarity of samples in the original space and the corresponding similarity in the projected subspace. In fact, this idea is already adopted by [20], which maximizes the dependence between pairwise samples and the corresponding samples in the learned subspace and to measure the dependence by using Hilbert-Schmidt Independence Criterion (HSIC) [13]. As discussed before, however, since multi-label high-dimensional data often contains noise or redundant features, directly calculating the pairwise similarity of samples in such space may be not correct and thus compromise the performance of dimensionality reduction. To bypass the risk, we maximize the dependence between high-order similarity of samples in the original space and in the projected subspace as follows:

$$HSIC(\mathbf{X}, \mathbf{Y}, \mathbf{P}) = \frac{tr(\mathbf{KHX}^T \mathbf{PP}^T \mathbf{XH})}{(N-1)^2} \tag{14}$$

Here, we replace $\mathbf{Q}$ with $\mathbf{X}^T \mathbf{PP}^T \mathbf{X}$ and $\mathbf{K} = \mathbf{W}$. It is worth to note that one advantage of adding this term is that it can further preserve local structure of samples in the projected subspace.

Based Eq. (11)–(14), we can rewrite Eq. (10) as:

$$\min_{\mathbf{P}} \frac{tr(\mathbf{P}^T \mathbf{S}_w \mathbf{P})}{tr(\mathbf{P}^T \mathbf{S}_b \mathbf{P}) + \alpha tr(\mathbf{P}^T \mathbf{XLX}^T \mathbf{P}) + \beta tr(\mathbf{P}^T \mathbf{XMX}^T \mathbf{P}) + \gamma tr(\mathbf{KHHX}^T \mathbf{PP}^T \mathbf{X})} \tag{15}$$

This equation is a generalized Rayleigh quotient problem, and the optimal $\mathbf{P}$ is composed of eigenvectors corresponding to the smallest d eigenvalues of $\mathbf{S}_w \mathbf{P} = \lambda(\mathbf{S}_b + \alpha \mathbf{XLX}^T + \beta \mathbf{XMX}^T + \gamma \mathbf{X}^T \mathbf{KHHX})\mathbf{P}$. Since the rank of $\mathbf{S}_w$ is much smaller than $D(N)$ and can be larger than $C-1$, the target dimensionality d can be larger than $C-1$.

3 Experimental Results

3.1 Experimental Setup

Datasets. The four popular multi-label datasets used in the experiments (Yeast, Scene, Corel5k and Reference) are summarized in Table 1. These datasets are obtained from Mulan[1]. For each dataset, we randomly sample 20% of the data for training, and use the remaining 80% data for testing (unlabeled data).

Table 1. Statistic of the experimental datasets. N is the number of instances, D is the dimensionality of instances, C is the number of distinct labels of instances, LC (Label Cardinality) is the average number of labels for every instance.

Dataset	N	D	C	LC	Domain
Yeast	2417	103	14	4.24	Gene
Scene	2407	294	6	2.158	Image
Corel5k	4395	1000	260	3.61	Image
Reference	7929	26397	15	1.15	Text

Comparing Methods. We compare SMLD-LRR against five related multi-label dimensionality reduction approaches: MDDM [11], MLDA [14], MLDA-LC [19], SMLDA [20] and SSMLDR [3]. The first two are supervised methods and the last three are semi-supervised methods. These methods have been introduced in the Introduction Section. In the experiment, we estimate the performance of different methods by first project the high-dimensional samples into a subspace, and then adopt the popular ML-kNN [32] to classify unlabeled samples in the respective subspace projected by a comparing method. In addition, we introduce SMLD-LRR_Nc and SMLD-LRR_pair to investigate the benefit of using label correlations and high-order structure of samples. SMLD-LRR_Nc excludes label information, namely $\beta = 0$. SMLD-LRR_pair substitutes the high-order structure of samples with pairwise similarity calculating by cosine similarity.

In the experiments, unless other specified, we set the parameters of the comparing methods according to what the author suggested in the original papers or codes. For our method, we selected the parameters α, β and γ from $\{10^{-6}, 10^{-5}, ...10^{1}\}$. We find that SMLD-LRR yields relatively stable performance with α, γ around 10^{-1} and β around 10^{-2}, and therefore we use these values. All the experiments are repeated ten times, and both the average and standard deviation are reported.

---

[1] Available at http://mulan.sourceforge.net/datasets-mlc.html.

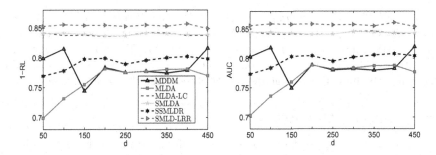

Fig. 1. $1 - RL$ and AUC results of all methods on the Corel5k dataset with different dimensions.

Table 2. Results on all datasets. **boldface** indicates the best (or comparable best) results which under a pairwise t-test at 95% significance level.

Dataset	Method	MicF	MacF	1 − RL	AP	AUC
Corel5k	MDDM	0.100 ± 0.002	0.005 ± 0.001	0.786 ± 0.003	0.137 ± 0.003	0.791 ± 0.003
	MLDA	0.113 ± 0.002	0.032 ± 0.001	0.781 ± 0.002	0.157 ± 0.002	0.787 ± 0.002
	MLDA-LC	0.280 ± 0.002	0.079 ± 0.001	0.848 ± 0.001	0.317 ± 0.001	**0.854 ± 0.000**
	SMLDA	0.280 ± 0.002	0.079 ± 0.001	0.849 ± 0.001	0.319 ± 0.002	0.853 ± 0.001
	SSMLDR	0.177 ± 0.002	0.033 ± 0.002	0.788 ± 0.001	0.213 ± 0.002	0.794 ± 0.001
	SMLDA-LRR_Nc	0.256 ± 0.004	0.069 ± 0.002	0.817 ± 0.001	0.319 ± 0.012	0.822 ± 0.002
	SMLDA-LRR_pair	0.253 ± 0.001	0.068 ± 0.003	0.823 ± 0.001	0.318 ± 0.002	0.827 ± 0.000
	SMLD-LRR	**0.300 ± 0.002**	**0.095 ± 0.002**	**0.851 ± 0.001**	**0.338 ± 0.001**	**0.856 ± 0.001**
Scene	MDDM	0.460 ± 0.002	0.480 ± 0.004	0.745 ± 0.003	0.660 ± 0.003	0.804 ± 0.003
	MLDA	0.427 ± 0.005	0.441 ± 0.005	0.714 ± 0.005	0.625 ± 0.004	0.776 ± 0.006
	MLDA-LC	0.479 ± 0.006	0.495 ± 0.009	0.766 ± 0.009	0.683 ± 0.005	0.819 ± 0.009
	SMLDA	0.494 ± 0.006	0.517 ± 0.006	0.781 ± 0.005	0.696 ± 0.004	0.832 ± 0.006
	SSMLDR	0.478 ± 0.004	0.513 ± 0.004	0.752 ± 0.003	0.675 ± 0.003	0.805 ± 0.003
	SMLDA-LRR_Nc	0.558 ± 0.002	0.596 ± 0.003	0.822 ± 0.001	0.786 ± 0.003	0.819 ± 0.001
	SMLDA-LRR_pair	0.561 ± 0.002	0.611 ± 0.006	0.848 ± 0.002	0.798 ± 0.005	0.826 ± 0.002
	SMLD-LRR	**0.601 ± 0.001**	**0.638 ± 0.003**	**0.894 ± 0.001**	**0.826 ± 0.002**	**0.921 ± 0.001**
Yeast	MDDM	0.599 ± 0.001	0.379 ± 0.002	0.786 ± 0.001	0.704 ± 0.001	0.799 ± 0.001
	MLDA	0.608 ± 0.001	0.403 ± 0.003	0.790 ± 0.001	0.714 ± 0.001	0.803 ± 0.001
	MLDA-LC	0.609 ± 0.002	0.405 ± 0.002	0.794 ± 0.001	0.716 ± 0.001	0.806 ± 0.001
	SMLDA	0.610 ± 0.001	0.403 ± 0.003	0.793 ± 0.001	0.717 ± 0.001	0.805 ± 0.001
	SSMLDR	0.610 ± 0.002	0.365 ± 0.001	0.794 ± 0.001	0.717 ± 0.001	0.807 ± 0.001
	SMLDA-LRR_Nc	0.606 ± 0.001	0.358 ± 0.001	0.781 ± 0.000	0.701 ± 0.000	0.790 ± 0.002
	SMLDA-LRR_pair	0.604 ± 0.002	0.336 ± 0.003	0.780 ± 0.003	0.714 ± 0.002	0.782 ± 0.004
	SMLD-LRR	**0.628 ± 0.002**	**0.410 ± 0.003**	**0.808 ± 0.001**	**0.734 ± 0.001**	**0.818 ± 0.001**
Reference	MDDM	0.326 ± 0.003	0.049 ± 0.004	0.690 ± 0.001	0.336 ± 0.004	0.737 ± 0.001
	MLDA	0.326 ± 0.003	0.049 ± 0.004	0.690 ± 0.001	0.336 ± 0.004	0.737 ± 0.001
	MLDA-LC	0.344 ± 0.019	0.053 ± 0.009	0.700 ± 0.005	0.348 ± 0.023	0.748 ± 0.004
	SMLDA	**0.354 ± 0.001**	0.053 ± 0.003	**0.714 ± 0.001**	**0.352 ± 0.002**	**0.757 ± 0.001**
	SSMLDR	0.332 ± 0.011	0.044 ± 0.006	0.686 ± 0.004	0.336 ± 0.015	0.732 ± 0.003
	SMLDA-LRR_Nc	0.179 ± 0.002	0.029 ± 0.001	0.595 ± 0.000	0.283 ± 0.002	0.657 ± 0.000
	SMLDA-LRR_pair	0.214 ± 0.000	0.022 ± 0.000	0.596 ± 0.000	0.220 ± 0.000	0.687 ± 0.000
	SMLD-LRR	**0.354 ± 0.003**	**0.058 ± 0.000**	0.713 ± 0.001	0.351 ± 0.003	**0.759 ± 0.002**

Evaluation Metrics. Five widely used multi-label evaluation metrics are adopted for performance comparisons, i.e., *Micro Average F1* (MicF), *Macro*

Average F1 (MacF), *Ranking Loss* (RL), *Average Precision* (AP), and *Area Under the Curve* (AUC). A formal definition of the first three metrics can be found in [1]. The adapted *AUC* is suggested in [33]. To maintain consistency with other evaluation metrics, in our experiments, we report $1 - RL$ instead of RL. Thus, the higher the value of $1 - RL$, the better the performance is.

3.2   Results on All Datasets

In this section, we conduct experiments on different types of datasets to investigate the performance of SMLD-LRR for multi-label dimensionality reduction on Table 2. In this table, the target dimensionality (d) is fixed to $C - 1$, *i.e.*, 13, 5, 259 and 14 for Yeast, Scene, Corel5k and Reference correspondingly.

From Table 2, it can be seen that SMLD-LRR outperforms the other methods, and the semi-supervised approaches are generally perform better than the supervised ones. These results show the benefit of leveraging both labeled and unlabeled samples. MLDA-LC, SSMLDR and SMLDA are semi-supervised approaches, but SMLDA outperforms the former two in most cases. The possible cause is that SMLDA introduces a dependence term to maximize the dependence between the original feature space and the projected one. Both SMLDA and SMLD-LRR introduce the dependence term, but SMLDA loses to SMLD-LRR in many cases. The possible reason is that SMLD-LRR additionally utilizes label information of labeled samples to preserve the local structure of samples. In addition, we also studied the performance of directly applying ML-kNN in the original feature space. The results are much lower than SMLD-LRR so the corresponding results are not included in this paper for page limitation. These comparisons show the effectiveness of the proposed method.

SMLD-LRR_Nc is a degenerate case of SMLD-LRR, which obtained by excluding label information. SMLD-LRR almost performs better than SMLD-LRR_Nc. This result demonstrate our motivation to exploit label information. SMLD-LRR_*pair* is obtained by using pairwise similarity of samples from SMLD-LRR, and is almost outperformed by SMLD-LRR. This fact validates the effectiveness of the proposed method in capturing high-order structure of samples.

Figure 1 demonstrates the results of comparing methods on different dimensions on Corel5k; similar results were obtained for the other datasets as well. We again see that SMLD-LRR obtains better low-dimensional subspaces than others. These results further show the effectiveness of the proposed method.

3.3   Parameter Analysis

In this section, we test the sensitivity of SMLD-LRR w.r.t. α, β and γ. Here, we first fix $\alpha = 10^{-1}$ and $\beta = 10^{-2}$, and run SMLD-LRR with γ from 10^{-6} to 10^{1}. Then we fix α and γ to $= 10^{-1}$, and run SMLD-LRR with $\beta = 10^{-6}$ to 10^{1}. Finally, we fix $\beta = 10^{-1}$ and $\gamma = 10^{-1}$, and run SMLD-LRR with α from 10^{-6} to 10^{1}. Due to the limit page space, we only report $1 - RL$ result on Corel5k in Fig. 2. From this figure, we can see that SMLD-LRR achieves relatively stable and good performance when $\alpha \approx 10^{-1}$, $\beta \approx 10^{-2}$ and $\gamma \approx 10^{-1}$. We also observe

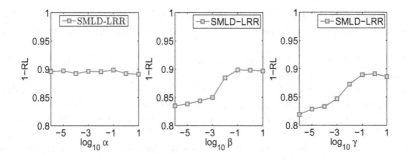

Fig. 2. Parameter analysis w.r.t. α, β and γ on Scene.

that when $\gamma = 10^{-6}$, SMLD-LRR has the smallest $1 - RL$. This result confirms the contribution of preserving geometric structure of samples using high-order label correlations.

4   Conclusions and Feature Work

In this paper, we take advantage of low-rank representation to capture high-order correlations of samples and introduce a Semi-supervised Multi-label Dimensionality Reduction approach based on Low Rank Representation (SMLD-LRR). Unlike existing methods that often require sufficient labeled samples and only utilize pairwise similarity of samples in the feature space, we explore the high-order structure of samples in both feature and label spaces, and utilize these high-order similarity of samples to leverage the labeled and unlabeled samples as well as preserve the sample structure in the projected subspace. The experimental results show that SMLD-LRR outperforms other competitive methods. In addition, it also can effectively preserve the structure of samples in the projected subspace by using high-order label correlations. In the future, we want to define a more robust multi-label dimensionality reduction classifier under missing labels or feature scenarios.

References

1. Zhang, M., Zhou, Z.: A review on multi-label learning algorithms. IEEE Trans. Knowl. Data Eng. **26**(8), 1819–1837 (2014)
2. Wu, X.Z., Zhou, Z.H.: A unified view of multi-label performance measures. In: ICML, pp. 3780–3788. IMIS, Sydney (2017)
3. Guo, B., Hou, C., Nie, F., Yi, D.: Semi-supervised multi-label dimensionality reduction. In: IEEE ICDM, pp. 919–924. IEEE, Barcelona (2016)
4. Nie, F., Xu, D., Li, X., Xiang, S.: Semisupervised dimensionality reduction and classification through virtual label regression. SMC Man Cybern. Part B (Cybern.) **41**(3), 675–685 (2011)
5. Jolliffe, I.: Principal Component Analysis. In: International Encyclopedia of Statistical Science, pp. 1094–1096. Springer, Heidelberg (2011)

6. Belkin, M., Niyogi, P., Sindhwani, V.: Manifold regularization: a geometric framework for learning from labeled and unlabeled examples. J. Mach. Learn. Res. **7**, 2399–2434 (2006)

7. Roweis, S.T., Saul, L.K.: Nonlinear dimensionality reduction by locally linear embedding. Science **290**(5500), 2323–2326 (2000)

8. Tenenbaum, J.B., De Silva, V., Langford, J.C.: A global geometric framework for nonlinear dimensionality reduction. Science **290**(5500), 2319–2323 (2000)

9. Hotelling, H.: Relations between two sets of variates. Biometrika **28**(3–4), 321–377 (1936)

10. Ji, S., Tang, L., Yu, S., Ye, J.: A shared-subspace learning framework for multi-label classification. TKDD **4**(2), 8 (2010)

11. Zhang, Y., Zhou, Z.: Multilabel dimensionality reduction via dependence maximization. TKDD **4**(3), 14 (2010)

12. Zhang, Z., Chow, T.W.: Robust linearly optimized discriminant analysis. Neurocomputing **79**, 140–157 (2012)

13. Gretton, A., Bousquet, O., Smola, A., Schölkopf, B.: Measuring statistical dependence with Hilbert-Schmidt norms. In: Jain, S., Simon, H.U., Tomita, E. (eds.) ALT 2005. LNCS (LNAI), vol. 3734, pp. 63–77. Springer, Heidelberg (2005). https://doi.org/10.1007/11564089_7

14. Wang, H., Ding, C., Huang, H.: Multi-label linear discriminant analysis. In: Daniilidis, K., Maragos, P., Paragios, N. (eds.) ECCV 2010. LNCS, vol. 6316, pp. 126–139. Springer, Heidelberg (2010). https://doi.org/10.1007/978-3-642-15567-3_10

15. Fisher, R.A.: The use of multiple measurements in taxonomic problems. Ann. Hum. Genet. **7**(2), 179–188 (1936)

16. Sun, L., Ji, S., Yu, S., Ye, J.: On the equivalence between canonical correlation analysis and orthonormalized partial least squares. In: IJCAI, Padadena, pp. 1230–1235 (2009)

17. Yu, G., Zhang, G., Domeniconi, C., Yu, Z., You, J.: Semi-supervised classification based on random subspace dimensionality reduction. Pattern Recognit. **45**(3), 1119–1135 (2012)

18. Wu, H., Prasad, S.: Semi-supervised dimensionality reduction of hyperspectral imagery using pseudo-labels. Pattern Recognit. **74**, 212–224 (2018)

19. Yuan, Y., Zhao, K., Lu, H.: Multi-label linear discriminant analysis with locality consistency. In: Loo, C.K., Yap, K.S., Wong, K.W., Teoh, A., Huang, K. (eds.) ICONIP 2014. LNCS, vol. 8835, pp. 386–394. Springer, Cham (2014). https://doi.org/10.1007/978-3-319-12640-1_47

20. Yu, Y., Yu, G., Chen, X., Ren, Y.: Semi-supervised multi-label linear discriminant analysis. In: Liu, D., Xie, S., Li, Y., Zhao, D., El-Alfy, E.S. (eds.) ICONIP 2017. LNCS, vol. 10634, pp. 688–698. Springer, Cham (2017). https://doi.org/10.1007/978-3-319-70087-8_71

21. Liu, G., Lin, Z., Yan, S., Sun, J., Yu, Y., Ma, Y.: Robust recovery of subspace structures by low-rank representation. TPAMI **35**(1), 171–184 (2013)

22. Yang, S., Wang, X., Wang, M., Han, Y., Jiao, L.: Semi-supervised low-rank representation graph for pattern recognition. IEEE Trans. Image Process. **7**(2), 131–136 (2013)

23. Zhuang, L., Wang, J., Lin, Z., Yang, A.Y., Ma, Y., Yu, N.: Locality-preserving low-rank representation for graph construction from nonlinear manifolds. Neurocomputing **175**, 715–722 (2016)

24. Wen, J., Zhang, B., Xu, Y., Yang, J., Han, N.: Adaptive weighted nonnegative low-rank representation. Pattern Recognit. **81**, 326–340 (2018)

25. Lin, Z., Liu, R., Su, Z.: Linearized alternating direction method with adaptive penalty for low-rank representation. In: NIPS, pp. 612–620. MIT Press, Spain (2011)
26. Zhang, H., Lin, Z., Zhang, C.: A counterexample for the validity of using nuclear norm as a convex surrogate of rank. In: Blockeel, H., Kersting, K., Nijssen, S., Železný, F. (eds.) ECML PKDD 2013. LNCS (LNAI), vol. 8189, pp. 226–241. Springer, Heidelberg (2013). https://doi.org/10.1007/978-3-642-40991-2_15
27. Wang, C., Yan, S., Zhang, L., Zhang, H.: Multi-label sparse coding for automatic image annotation. In: IEEE CVPR, pp. 1643–1650. IEEE, Miami Beach (2009)
28. Wright, J., Ma, Y., Mairal, J., Sapiro, G., Huang, T.S., Yan, S.: Sparse representation for computer vision and pattern recognition. Proc. IEEE **98**(6), 1031–1044 (2010)
29. Chung, F.R.: Spectral Graph Theory. American Mathematical Soc (No. 92) (1997)
30. Wang, F., Zhang, C.: Label propagation through linear neighborhoods. TKDE **20**(1), 55–67 (2007)
31. Datta, R., Joshi, D., Li, J., Wang, J.Z.: Image retrieval: ideas, influences, and trends of the new age. ACM Comput. Surv. (CSUR) **40**(2), 5 (2008)
32. Zhang, M., Zhou, Z.: ML-kNN: a lazy learning approach to multi-label learning. Pattern Recognit. **40**(7), 2038–2048 (2007)
33. Bucak, S.S., Jin, R., Jain, A.K.: Multi-label learning with incomplete class assignments. In: CVPR, pp. 2801–2808. IEEE, Colorado, Colorado Springs (2011)

Visualization Method of Viewpoints Latent in a Dataset

Hideaki Ishibashi[1,2P] (✉)

[1] Department of Life Science and Systems Engineering, Kyushu Institute of
Technology, Kitakyushu, Japan
o899004h@mail.kyutech.jp,ishibashi.hideaki@ism.ac.jp
[2] Research Center for Statistical Machine Learning, The Institute of Statistical
Mathematics, Tachikawa, Japan

Abstract. The purpose of this study is to propose a paradigm visualizing the viewpoints from datasets observed by multiple viewpoints, which is referred to as Latent Viewpoint Visualization (LVV). Since LVV visualizes similarity/dissimilarity among the viewpoints, it has many applications such as the authors' perspective from news articles and the psychological measurements' aspect from psychological surveys. In this study, we propose the concept of LVV and develop a preliminary algorithm. Furthermore, we experimentally show what kind of information can be visualized by LVV using several datasets.

Keywords: Latent viewpoint visualization · Multi-view learning
Data integration · Grassmann manifold

1 Introduction

In news media, each media reports an event from its own viewpoint and often gives a biased perspective. This is called media bias problem which has being addressed by several researchers [1–3]. A similar problem is also occurring on the internet. A search system on the Internet provides information adjusting to a viewpoint of each user. However, this often causes a filter bubble problem, i.e., the search system only shows the information desired by the user [4]. Such bias of viewpoints becomes an important issue in various fields such as recommendation system [5–7], news aggregator [1], statistics [8], cognitive science [9].

The purpose of this study is to propose a paradigm for visualizing similarity/dissimilarity among the viewpoints from a dataset observed by multiple viewpoints. In this study, we refer to the paradigm as "Latent Viewpoint Visualization (LVV)".

Various applications can be considered for LVV. For example, when news articles reported by several media is obtained, LVV shows that similarity between the media's perspective about each event. As a result, LVV can recommend different perspectives to us. Those perspectives could lead us to fairly judge the news sources and alleviate the media-bias.

© Springer Nature Switzerland AG 2018
L. Cheng et al. (Eds.): ICONIP 2018, LNCS 11303, pp. 638–647, 2018.
https://doi.org/10.1007/978-3-030-04182-3_56

Another example is to visualize psychological measures' aspect from psychological surveys. In psychology, a psychological state of subjects is observed using various psychological measures whose aspects are different from each other. Then, by visualizing the viewpoint using LVV, it becomes easier to select a psychological measure suitable for the purpose, and to compare experimental results using different psychological measures.

In order to realize the above visualization, we must address following two issues: (i) integrating the datasets observed from various viewpoints, and (ii) estimating the latent viewpoints from the integrated dataset.

The structure of this paper is as follows. The concept of LVV is formulated in Sect. 2, and related works are introduced in Sect. 3. In Sect. 4, a preliminary algorithm to instantiate LVV is proposed. Through the application of LVV to two-real datasets, we consider what kind of information can be visualized by LVV. Sections 5 and 6 are discussion and conclusion, respectively.

2 Problem Formulation

Suppose that N samples are observed from I different viewpoints. The observed dataset by the i-th viewpoint is denoted by $\mathbf{X}_i \in \mathbb{R}^{D_i \times N}$, which consists of N data $\mathbf{x}_n^i \in \mathbb{R}^{D_i}$. Here, we assume that $\sum_{n=1}^{N} \mathbf{x}_n^i = \mathbf{0}$ holds.

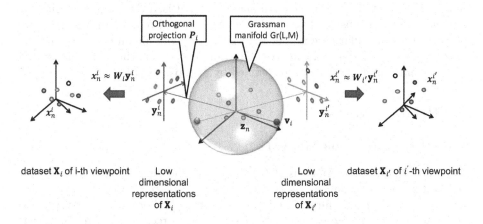

Fig. 1. The concept of the Latent Viewpoint Visualization.

Figure 1 shows the concept of LVV. When $\{\mathbf{X}_1, \mathbf{X}_2, \cdots, \mathbf{X}_I\}$ are given, our aim is to integrate all of the datasets and to estimate the latent viewpoints $\{\mathbf{v}_1, \mathbf{v}_2, \cdots, \mathbf{v}_I\}$. To this end, we adopt a 3d reconstruction method which estimates three-dimensional coordinates of objects and camera-viewpoints from images taken by different cameras [10], i.e., the latent viewpoints are modeled as camera-viewpoints. In this work, we reconstruct M-dimensional coordinates and viewpoints from L-dimensional corresponding points among viewpoints. Let

$\mathbf{Y}_i$ and $\mathbf{Z}$ be the L-dimensional corresponding points of i-th viewpoints and M-dimensional points. Then, the following relationship holds between $\mathbf{Y}_i$ and $\mathbf{Z}$.

$$\mathbf{Y}_i \approx \mathbf{P}_i \mathbf{Z}, \tag{1}$$

where $\mathbf{P}_i \in \mathbb{R}^{L \times M}$ is an orthogonal projection matrix which satisfies $\mathbf{P}_i \mathbf{P}_i^{\mathrm{T}} = \mathbf{I}$. Furthermore, we assume that $\mathbf{X}_i$ is generated by mapping $\mathbf{Y}_i$ through linear transformation $\mathbf{W}_i$. Here, in order to guarantee that the solution uniquely exists, we assume that $\mathbf{W}_i^{\mathrm{T}} \mathbf{W}_i = \mathbf{I}$ holds. Hence, the following equation holds.

$$\mathbf{X}_i \approx \mathbf{W}_i \mathbf{P}_i \mathbf{Z}. \tag{2}$$

In this study, the latent viewpoint $\mathbf{v}_i$ is defined as a parameter controlling a projection direction of $\mathbf{P}_i$. For example, when $L = 2$ and $M = 3$, the i-th latent viewpoint $\mathbf{v}_i$ becomes a normal vector of a subspace spanned by $\{\mathbf{p}_1^i, \mathbf{p}_2^i\}$, where $\mathbf{p}_l^i$ is the l-th row vector of $\mathbf{P}_i$. Then, $\mathbf{v}_i$ can be represented by cross product of $\mathbf{p}_1^i$ and $\mathbf{p}_2^i$. Generally, $\mathbf{v}_i$ is represented by a point on a Grassmann manifold which corresponds to $\mathbf{P}_i$. Grassmann manifold $Gr(L, M)$ is the Riemannian manifold formed by a set of L-dimensional subspaces in an M-dimensional space, and defined as follows:

$$Gr(L, M) \triangleq St(L, M) / \mathcal{O}(L), \tag{3}$$

where $St(L, M) = \{\mathbf{P} \in \mathbb{R}^{L \times M} | \mathbf{P}\mathbf{P}^{\mathrm{T}} = \mathbf{I}\}$, which is called a Stiefel manifold, and $\mathcal{O}(L)$ is a set of $L \times L$ orthogonal matrices. That is, a Grassmann manifold is defined as a set of subspaces spanned by L orthonormal vectors. In this study, we consider the particular case $L = M - 1$. Then, the Grassmann manifold and the L-sphere in the M-dimensional space are isomorphic, i.e., the $\mathbf{v}_i$ is represented as a normal vector of a subspace spanned by $\{\mathbf{p}_1^i, \mathbf{p}_2^i, \cdots, \mathbf{p}_L^i\}$.

3   Related Works

3.1   Data Integration Methods

Canonical Correlation Analysis (CCA) is one of the most basic algorithms for integrating multiple datasets [11]. CCA integrates two datasets by expressing the datasets in a low dimensional space so that the correlation between two projected vectors in these datasets is maximized. In recent years, research that integrates data among viewpoints like CCA is called multi-view learning [12]. In multi-view learning, various methods have been proposed [12–14]. The aim of the multi-view learning is to integrate datasets observed from different viewpoints. Therefore, the purpose of the multi-view learning and our proposed framework, which aims to analyze the viewpoints, is different.

3d reconstruction methods used in computer vision can also be regarded as a data integrating method, since they estimate 3d-object from multiple camera-images. The early work in 3d reconstruction is a factorization method which estimates 3d-coordinates and camera-angles from points on camera-images associated with each other among cameras [10]. In computer vision, various 3d reconstruction methods have been developed such as a method assuming the perspective projection [15], a method for reconstructing a three-dimensional object from a large number of independent images [16]. However, these methods heavily relies on the assumption the space to be three dimensional. On the other hand, the factorization method does not consider three-dimensional constraints when integrating data. Therefore, in this study, we develop the LVV based on the factorization method.

3.2 Viewpoint Estimation Methods

In the literature of the viewpoint estimation, meta-visualization is one of the related works. The meta-visualization visualizes a set of visualizations, each of which visualizes the high dimensional dataset in a two-dimensional space [17]. The LVV also visualizes the set of visualizations, each of which is visualized by two-dimensional space by a 3d reconstruction method. Therefore, the meta-visualization can also be seen as a viewpoint estimation method. The difference between LVV and meta-visualization lies in the visualization method. Meta-visualization displays the similarity of each visualization result on a two-dimensional plane, while LVV visualizes observation targets and viewpoints in a three-dimensional space.

4 Algorithm of the Latent Viewpoint Visualization

LVV is formulated as a constrained minimization problem to estimate the integrated data $\mathbf{Z}$ and the latent viewpoints $\{\mathbf{v}_i\}_{i=1}^I$:

$$F = \sum_{i=1}^I \|\mathbf{X}_i - \mathbf{W}_i\mathbf{P}_i\mathbf{Z}\|_F^2, \qquad s.t. \mathbf{W}_i^T\mathbf{W}_i = \mathbf{I}, \quad \mathbf{P}_i\mathbf{P}_i^T = \mathbf{I}. \tag{4}$$

Here, $\|\cdot\|_F$ is the Frobenius norm. LVV is composed of two procedures: data integration and viewpoint estimation.

Data Integration. Firstly, $\mathbf{X}_i$ is decomposed into $\mathbf{W}_i$ and $\mathbf{Y}_i$ by Singular Value Decomposition (SVD), i.e., $\mathbf{X}_i = \mathbf{U}_i\boldsymbol{\Sigma}_i\mathbf{V}_i^T$. Then, $\mathbf{W}_i$ and $\mathbf{Y}_i$ are estimated as follows:

$$\mathbf{W}_i = \mathbf{U}_i, \tag{5}$$

$$\mathbf{Y}_i = \boldsymbol{\Sigma}_i\mathbf{V}_i^T. \tag{6}$$

Next, in order to decompose $\mathbf{Y}_i$ into $\mathbf{P}_i$ and $\mathbf{Z}$, the following objective function is minimized with the constrain $\mathbf{P}_i\mathbf{P}_i^T = \mathbf{I}$:

$$F = \sum_{i=1}^I \|\mathbf{Y}_i - \mathbf{P}_i\mathbf{Z}\|_F^2 \tag{7}$$

$$= \|\mathbf{Y} - \mathbf{P}\mathbf{Z}\|_F^2, \tag{8}$$

where $\mathbf{Y}$ and $\mathbf{P}$ are matrices constructed by concatenating $\mathbf{Y}_i$ and $\mathbf{P}_i$ in the row direction, respectively. Then, Eq. (8) is minimized by decomposing $\mathbf{Y}$ using SVD. When $\mathbf{Y}$ is decomposed into $\mathbf{U}\boldsymbol{\Sigma}\mathbf{V}^T$, $\tilde{\mathbf{P}}$ and $\tilde{\mathbf{Z}}$ are estimated as follows:

$$\tilde{\mathbf{P}} = \mathbf{U}\boldsymbol{\Sigma}^{\frac{1}{2}}, \tag{9}$$

$$\tilde{\mathbf{Z}} = \boldsymbol{\Sigma}^{\frac{1}{2}}\mathbf{V}^{\mathrm{T}}. \tag{10}$$

Note that $\tilde{\mathbf{P}}$ does not satisfy $\tilde{\mathbf{P}}_i\tilde{\mathbf{P}}_i^{\mathrm{T}} = \mathbf{I}$. Furthermore, $\tilde{\mathbf{P}}$ and $\tilde{\mathbf{Z}}$ have equivalent solutions with respect to non-singular matrix $\mathbf{Q}$, i.e., $\mathbf{PZ} = \tilde{\mathbf{P}}\mathbf{Q}\mathbf{Q}^{-1}\tilde{\mathbf{Z}}$. Thus, in order to satisfy $\tilde{\mathbf{P}}_i\mathbf{Q}\mathbf{Q}^{\mathrm{T}}\tilde{\mathbf{P}}_i^{\mathrm{T}} = \mathbf{I}$, we estimate $\mathbf{Q}$ by minimizing the following objective function:

$$E = \sum_{i=1}^{I} \|\tilde{\mathbf{P}}^i\mathbf{Q}\mathbf{Q}^T\tilde{\mathbf{P}}^i - \mathbf{I}\|_F^2. \tag{11}$$

$\mathbf{Q}$ is estimated by using the method of Morita et al. [18].

Viewpoint Estimation. In this procedure, we estimates $\mathbf{v}_i$ from $\mathbf{P}_i$. In this research, we consider the case of $L = M - 1$. In this case, the viewpoint $\mathbf{v}_i$ is the normal vector for the subspace spanned by $\{\mathbf{p}_1^i, \mathbf{p}_2^i, \cdots, \mathbf{p}_L^i\}$. Then, $\mathbf{v}_i = (v_{i1}, v_{i2}, \cdots, v_{iM})^{\mathrm{T}}$ can be obtained as follows:

$$v_{im} = \det(\mathbf{e}_m, \mathbf{p}_1^i, \mathbf{p}_2^i, \cdots, \mathbf{p}_L^i). \tag{12}$$

Here, $\mathbf{e}_m$ is a standard basis. In the case of $L = 2$ and $M = 3$, Eq. (12) means the cross product of $\mathbf{p}_1^i$ and $\mathbf{p}_2^i$.

5   Experimental Results

In order to examine that the LVV can visualize many kinds of information, we apply the proposed algorithm to two real datasets.

5.1   Psychological Measures' Aspect Visualization

The first dataset is the psychological surveys dataset which measures psychological states of football players by multiple psychological scales. The dataset consists of 17 psychological scales surveyed for 439 football players in the Japanese university league [19]. These psychological scales are categorized into 10 self-assessments and 7 coach-assessments. The self-assessments categorized are composed of three types: (1) one motivation (M), (2) eight self-management skill (SM_1–SM_8) and (3) one football skill (F). Likewise, the coach-assessments are also composed of three types: (1) one coaching acceptance (CA), (2) two Performance and Maintenance (PM_1–PM_2) and (3) four Perceived Coaching Effectiveness (PCE_1–PCE_4). Each psychological survey is composed of about 4 to 13 queries of Likert scale. In the study, each query is normalized so that the average zero, variance 1.

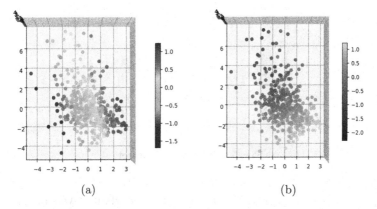

Fig. 2. The psychological state of the football players integrating the psychological scales. (a) The color means each player's average score of the self-assessments. (b) The color means each player's average score of the coach-assessments. (Color figure online)

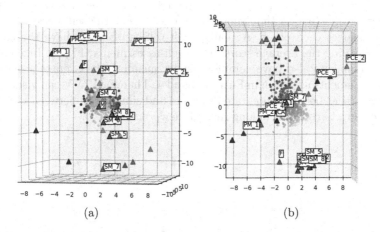

Fig. 3. The latent viewpoints of each psychological measure. The color of the psychological measures means category of the measures: *red* self-assessments and *blue* coach-assessments. (Color figure online)

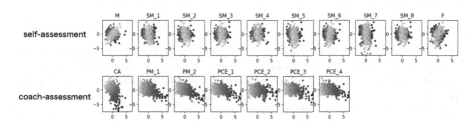

Fig. 4. The psychological state of the football players observed from each psychological scale. The color has the same meaning as in the Fig. 2. (Color figure online)

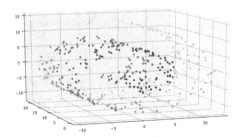

Fig. 5. The original dataset.

Figure 2 shows the psychological states of the football players integrating the multiple psychological scales. The result indicates that the psychological states have two features: whether the average score of the self-assessment is high and whether the average score of the coach-assessment is high.

The latent viewpoints of the psychological measures for the integrated datasets are shown in Fig. 3. As shown in Fig. 3(a), the self-assessments are distributed on a line on the sphere. This means that the self-assessments have common aspect each other. Actually, each self-assessments have the common feature meaning the average score of the self-assessments (Fig. 4). Likewise, the coach-assessments also have common aspect meaning the average score of the coach-assessments as shown in Figs. 3(b) and 4. Therefore, LVV can visualize not only similarities between integrated samples and similarities between viewpoints but also aspects common among several viewpoints.

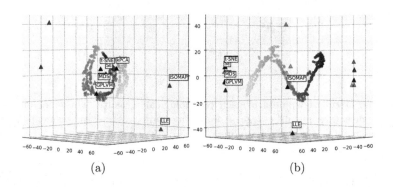

(a)   (b)

Fig. 6. The viewpoint of each dimensionality reduction method. Each point means a sample-point integrated the results of dimensionality reductions. Each triangle is a latent viewpoint of a dimensionality reduction. (a) and (b) are the results viewing the visualization result in three dimensions at different angles. 2d relationship of data viewed each dimensionality reduction is shown by Fig. 7.

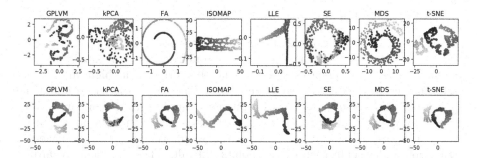

Fig. 7. The relationships between the data observed by each dimensionality reduction method's perspective. The upper row is the visualization results of each dimensionality reduction methods, while the lower row is the reconstruction results by LVV.

5.2 Dimension Reduction Techniques' Perspective Visualization

When we apply various dimensionality reduction methods to a high-dimensional dataset, each method maps the dataset to low-dimension space by their own perspective. Therefore, we visualize the perspectives of the methods by using a set of visualization methods for the original data shown in Fig. 5. We use the eight methods: GPLVM, kernel PCA (kPCA), Factor Analysis (FA), ISOMAP, LLE, Spectral Embedding (SE), Multi-Dimensional Scaling (MDS) and t-SNE.

Figure 6 shows that each dimensionality reductions are clustered into two types: the first cluster consists of the kPCA, GPLVM, FA,SE, MDS and t-SNE; the second cluster consists of the ISOMAP and LLE. As shown in Fig. 7, the first cluster expresses the shape of the roll as it is, while the second cluster reduces the dimension along the shape of the manifold. In this way, the LVV visualizes the similarity/dissimilarity between the visualization methods.

6 Discussion

6.1 Data Integration

In the proposed method, we use a 3d reconstruction method to the dimension reduced data. However, this approach is not necessarily appropriate, since features truncated by the dimension reduction methods are not taken into account by the viewpoint. Therefore, an integration method in the estimation of the latent point of view should be developed.

There is an interesting issue of the proposed method. It is able to estimate common features among viewpoints as the multi view learning. Especially, unlike multi-view learning which estimates features common to all viewpoints, the proposed method can estimate common features among some of the viewpoints as shown in Fig. 3. Since there are $2^I - 1$ patterns of common features for I viewpoints, it is difficult to automatically estimate the common features to some of the viewpoints in multi-view learning. Therefore, the LVV approach is also interesting as a multi-view learning.

6.2 Viewpoint Estimation

In this paper, we demonstrate that LVV visualizes various information. However, due to this visualization, the algorithm is compromised in flexibility despite the fact that the viewpoint estimation is possible even in the high dimensional case. Furthermore, although we assumed $L = M - 1$ in order to estimate the viewpoints, this is also a strong constraint, since there are $L-1$ common features necessarily between any two viewpoints. However, since the viewpoint can no longer be expressed as a normal vector, the assumption $L \neq M - 1$ make the estimation and visualization of viewpoints difficult. These problems regarding visualization should be addressed in our future work.

In this paper, we did not quantitative evaluate validity of the visualization results. It is non-trivial to quantitative evaluate the results. Therefore, to set the evaluation criteria is also remained task.

7 Conclusion

In this paper, we proposed a learning paradigm to visualize the latent viewpoints from dataset observed by multiple viewpoints. We developed a preliminary algorithm to instantiate the paradigm and applied it to the aspect analysis of psychological scales and perspective analysis of dimensionality reduction methods. Since the algorithm proposed in this paper is preliminary, it is necessary to develop a better algorithm.

Acknowledgement. The author is grateful to Prof. Tetsuo Furukawa from the Kyushu Institute of Technology for advising about many topics treated in this paper. The author thanks Assoc. Prof. Hideitsu Hino from the Institute of Statistical Mathematics for his valuable comments. A part of this work was supported by KAKENHI (17H01793) and JST CREST (JPMJCR1761). This work was partially supported by Dr. Takashi Ohkubo.

References

1. Park, S., Kang, S., Chung, S., Song, J.: Newscube: delivering multiple aspects of news to mitigate media bias. In: Proceedings of the SIGCHI Conference on Human Factors in Computing Systems, CHI 2009, pp. 443–452. ACM (2009). http://doi.acm.org/10.1145/1518701.1518772
2. Groseclose, T., Milyo, J.: A measure of media bias*. Q. J. Econ. **120**(4), 1191–1237 (2005). https://doi.org/10.1162/003355305775097542
3. Lampe, C., Garrett, R.K.: It's all news to me: the effect of instruments on ratings provision. In: 2007 40th Annual Hawaii International Conference on System Sciences (HICSS 2007), pp. 180b–180b (2007)
4. Pariser, E.: The Filter Bubble: What the Internet Is Hiding from You. Penguin Press, New York (2011)
5. Kamishima, T., Akaho, S., Asoh, H., Sakuma, J.: Enhancement of the neutrality in recommendation. In: Proceedings of the 2nd Workshop on Human Decision Making in Recommender Systems, vol. 893, pp. 8–14 (2012)

6. Fukuchi, K., Sakuma, J., Kamishima, T.: Prediction with model-based neutrality. In: Blockeel, H., Kersting, K., Nijssen, S., Železný, F. (eds.) ECML PKDD 2013, Part II. LNCS (LNAI), vol. 8189, pp. 499–514. Springer, Heidelberg (2013). https://doi.org/10.1007/978-3-642-40991-2_32
7. Pedreshi, D., Ruggieri, S., Turini, F.: Discrimination-aware data mining. In: Proceedings of the 14th ACM SIGKDD International Conference on Knowledge Discovery and Data Mining, KDD 2008, pp. 560–568. ACM (2008). http://doi.acm.org/10.1145/1401890.1401959
8. Field, A.P., Raphael, G.: How to do a meta-analysis. Br. J. Math. Stat. Psychol. 63(3), 665–694 (2010). https://onlinelibrary.wiley.com/doi/abs/10.1348/000711010X502733
9. Mohanani, R., Salman, I., Turhan, B., Marín, P.R., Ralph, P.: Cognitive biases in software engineering: A systematic mapping and quasi-literature review. CoRR abs/1707.03869 (2017). http://arxiv.org/abs/1707.03869
10. Tomasi, C., Kanade, T.: Shape and motion from image streams under orthography: a factorization method. Int. J. Comput. Vis. 9(2), 137–154 (1992). https://doi.org/10.1007/BF00129684
11. Hotelling, H.: Relations between two sets of variates. Biometrika 28(3/4), 321–377 (1936). http://www.jstor.org/stable/2333955
12. Xu, C., Tao, D., Xu, C.: A Survey on Multi-view Learning. ArXiv e-prints (2013)
13. Sun, S.: A survey of multi-view machine learning. Neural Comput. Appl. 23(7), 2031–2038 (2013). https://doi.org/10.1007/s00521-013-1362-6
14. Zhao, J., Xie, X., Xu, X., Sun, S.: Multi-view learning overview. Inf. Fusion 38(C), 43–54 (2017). https://doi.org/10.1016/j.inffus.2017.02.007
15. Triggs, B.: Factorization methods for projective structure and motion. In: Proceedings CVPR IEEE Computer Society Conference on Computer Vision and Pattern Recognition, pp. 845–851 (1996)
16. Snavely, N., Seitz, S.M., Szeliski, R.: Photo tourism: exploring photo collections in 3D. ACM Trans. Graph. 25(3), 835–846 (2006). https://doi.org/10.1145/1141911.1141964
17. Peltonen, J., Lin, Z.: Information retrieval approach to meta-visualization. Mach. Learn. 99(2), 189–229 (2015). https://doi.org/10.1007/s10994-014-5464-x
18. Morita, T., Kanade, T.: A sequential factorization method for recovering shape and motion from image streams. IEEE Trans. Pattern Anal. Mach. Intell. 19(8), 858–867 (1997)
19. Shinriki, R., Hagiwara, G., Isogai, H.: Effective leadership behavior of football coaches: a study using pm theory. Jpn. J. Sport. Ind. (In Jpn.) 26(2), 203–216 (2016)

Localized Multiple Sources Self-Organizing Map

Ludovic Platon[1,2]([✉]), Farida Zehraoui[1], and Fariza Tahi[1]

[1] IBISC, Univ. Evry, Université Paris-Saclay, Evry, France
ludovic.platon@univ-evry.fr
[2] IPS2, CNRS, INRA, Université Paris-Sud, Université d'Evry, Université
Paris-Saclay, Gif sur Yvette, France

Abstract. We present in this paper a new approach based on unsupervised self organizing maps called MSSOM. This approach combines multiple heterogeneous data sources and learns the weights of each source at the level of clusters instead of learning the same source weights for the whole space. This allows to improve the performances of our model especially in applications where a local feature selection is important. We evaluate our method using several artificial and real datasets and show competitive results compared to the state-of-art.

Keywords: Unsupervised learning · Dimensionality reduction
Neural network · Kernel methods · Self-Organizing Map

1 Introduction

In many application areas, data of interest can be described using multiple heterogeneous sources. Each source can contain a partial information about the objects. With multiple information sources simultaneously available, it is a challenging task how to conduct integrated exploratory analysis. For example, on biological studies about genes, one objective can be the extraction of the sequences information but also their expression profiles and their related epigenetics markers. The main exploratory analysis approach of data is clustering. Clustering finds subgroups of objects that are similar. It can help to guide the analysts to understand the data, when the goal of the analysis is well defined and captured by the clustering model. The data sources can be of different types: numerical or complex (trees, graphs, sequences, etc.). The complex types can be represented using similarity or dissimilarity matrices. When dealing with heterogeneous mixed (numerical and complex) data sources, existing approaches transform all the data sources to one type (numerical vectors or similarity (dissimilarity) matrices) and apply adapted clustering algorithms. Few clustering algorithms address the problem of learning the weights associated to the sources [19,22]. In addition, most of the clustering algorithms learn the same source combination for all the clusters.

© Springer Nature Switzerland AG 2018
L. Cheng et al. (Eds.): ICONIP 2018, LNCS 11303, pp. 648–659, 2018.
https://doi.org/10.1007/978-3-030-04182-3_57

In this paper, we propose to address the clustering from heterogeneous mixed sources by representing the complex data sources using kernels and by keeping the numerical data sources without any transformation. In addition, we propose to learn the source weights at the level of a cluster. Instead of learning the same combination for the whole space, we learn a different kernel combination for each cluster. This makes sense in several applications since a cluster can represent a group or a class of objects that share the same characteristics (data sources) and different groups can have different characteristics. For this purpose, we use self-organizing maps (SOM) [7], which are among the most used connectionist models for data clustering and visualization.

The paper is organized as follows: we first address the clustering problem using multiple heterogeneous sources and present the related works to multiple sources SOMs. Then we introduce our new multiple sources SOM algorithm called MSSOM. We show the efficiency of our algorithm by presenting results on artificial and real data. We conclude by giving some perspectives of this work.

2   State of Art

The basic problem of exploiting multiple information sources for unsupervised learning approaches has been extensively studied in the literature [13, 21, 22]. This problem is often known as multi-view clustering which has been successfully applied in many applications. The information sources can represent heterogeneous data that are of different types: digital, texts, graphs, trees, categories, etc. Complex data can be represented using dissimilarity or similarity (kernel) matrices. A common approach to cluster these heterogeneous data is to convert all data types to the same type: numerical, dissimilarity or kernel matrices.

Classical clustering algorithms concatenate (or combine) all multiple sources into a single one [21] before performing clustering. This is not appropriate in many real world applications because the data sources may not have the same importance and some noisy sources can deteriorate the clustering results. To overcome such limitations, other approaches are proposed in the literature. These approaches can be classified into three groups as proposed in [17]: co-training, multiple kernel clustering, and subspace learning. Co-training approach [1, 2, 20] alternately maximizes the mutual agreement on two distinct sources of the unlabelled data. Multiple kernel clustering approach [19, 22] proposes to learn an optimal linear or non-linear combination of kernels for clustering. Subspace approach [3] aims to obtain a latent subspace with low dimension shared by multiple sources by assuming that the input sources are generated from this latent subspace.

The SOM clustering models comprise an important unsupervised class of competitive neural models. The output neurons of the SOM are arranged in a specific geometrical form. Each neuron unit r in the SOM is associated to a prototype vector $p_r = [p_r^1, p_r^2, ..., p_r^m]^T \in R^m$ with the same dimension as the input vector $x = [x_1, x_2, ..., x_m]^T \in R^m$. Through an unsupervised learning process, the output neurons become tuned and organized after several presentations of the data. The learning algorithm that leads to a self-organization can be summarized in two steps. A winning or best-matching unit of the map, denoted $BMU(x)$, is found by using a distance or similarity measure (Euclidean distance, scalar product, etc.) between the input and the weight vectors:

$$BMU(x) = \arg\min_{r \in A} d(x, p_r) \tag{1}$$

where A is the set of neurons and $d()$ is a distance measure. Then the winner and its neighbours in the map have their weights $p_r(t)$ updated towards the current input x:

$$p_r(t+1) = p_r(t) + \alpha(t)h_r(t)[x - p_r(t)] \tag{2}$$

where $\alpha_h(t)$ is the learning rate and $h_r(t)$ is the neighbourhood function. In order to improve the convergence speed, a batch version of SOM is proposed in the literature [7]. The difference with the standard SOM is that the update step is performed after the BMU selection of each batch. The prototype vector associated to each neuron will represent the mean of the closest inputs: for all $p \in P$, $p = \frac{\sum_{i=1}^{U} N_i h_{i,p} \overline{x_i}}{\sum_{i=1}^{U} N_i h_{i,p}}$ where $\overline{x_i}$ and N_i are respectively the mean and the number of the inputs associated to prototype i.

Some approaches are proposed in the literature in order to deal with heterogeneous data sources using SOM. The most common approach is to combine the different sources before using the SOM. For numerical data sources, an augmented vector is formed by concatenating the vectors associated to each source. This vector is presented as input to classical SOM [15]. For complex data sources, each source is represented using a kernel function. The resulting kernels are combined using a fixed rule without any parameters (e.g., summation or multiplication of the kernels) and then presented to a kernel SOM.

To our knowledge, only one approach learns the combination parameters of the data sources. It is called Multiple Kernel SOM [12]. Each data source is represented by a kernel function and the algorithm learns the optimal linear combination of the kernels. A convex combination of the kernels is defined by:

$$K(x_i, x_j) = \sum_d \alpha_d K_d(x_i, x_j) \tag{3}$$

where $\alpha_d \in [0,1]$ and $\sum_d \alpha_d = 1$. Following the general framework of kernel SOM, the prototypes can be written as a convex combination of the input data in a Hilbert space (called the feature space) H:

$$p_u = \sum_{u=1}^{U} \gamma_{ui} \phi(x_i) \tag{4}$$

where the application $\phi : G \rightarrow H$ is symmetric and positive. The squared distance between x_i and a prototype p_u is then computed by:

$$\| \phi(x_i) - p_u \|^2 = k(x_i, x_i) - 2 \sum_{j=1}^{n} \gamma_{uj} k(x_i, x_j) + \sum_{s,l=1}^{n} \gamma_{us} \gamma_{ul} k(x_l, x_s) \tag{5}$$

The multiple kernel SOM optimizes the weights in order to minimize the following energy function:

$$\varepsilon((\gamma_{ui})_{ui},(\alpha_d)_d) = \sum_{i=1}^{n}\sum_{u=1}^{U} h_{BMU_{i,u}}(t) \parallel \phi(x_i)^\alpha - P_u^\alpha \parallel_\alpha^2 \qquad (6)$$

where $\phi(x_i)^\alpha$, P_u^α and $\parallel . \parallel_\alpha$ are used to emphasize that these quantities depend on α. The multiple kernel SOM learns the same kernel combination for each cluster. We propose to combine the different sources inside the algorithm and to represent the importance of each source locally for each cluster.

3 Method

In this paper we present a new multiple data sources SOM algorithm called MSSOM. Our algorithm can handle complex and numerical data. Complex (non numerical) data are represented using kernels and numerical data are given as input to the algorithm without any transformation.

3.1 MSSOM Algorithm

We propose a three layer architecture (see Fig. 1). The first layer represents the input layer. Let be a dataset $X = \{x_1, x_2, ..., x_n\}$ where x_i is described using d sources $S = \{S^1, S^2, ..., S^d\}$. These sources can be represented by numerical vectorial data or by complex data. Let be $K = \{K^1, K^2, ..., K^l\}$ the kernels representing the complex sources and $V = \{V^1, V^2, ..., V^{d-l}\}$ the vectors representing the numerical sources where K^i represents the complex features of the source i and V^i the numerical features of the source $i + l$.

The second layer is composed of SOMs computed from each source. In order to obtain good performances, we use the batch version of SOM [7]. For the numerical data we use the batch SOM defined in [7]. For the complex data we propose a batch version of the Bagged Kernel SOM [10]. The Bagged Kernel SOM [10] approach overcomes the computation cost due to kernels by using a small part of the kernels, called a bag, to compute the map. The resulting prototypes represent a sparse combination of the kernel values (they are composed of the representative values).

The third layer combines the outputs of the SOMs presented in the second layer to train the single final map. The output of each SOM represents a similarity between the input data and the prototypes in the maps. We define a Gaussian similarity measure $s^i(x_j, p_u^i)$ between x_j, the jth element of X, and p_u^i, the uth prototype of the ith map. We define a function $\theta_i : X \to \mathbb{R}^{U^i}$, returning the similarity between an element x_j in X and all the prototypes of the SOM for the source i such that:

$$\theta_i(x_j) = [s^i(x_j, p_1^i), s^i(x_j, p_2^i), ..., s^i(x_j, p_{U^i}^i)]$$
$$for\ i = 1, ..., d$$

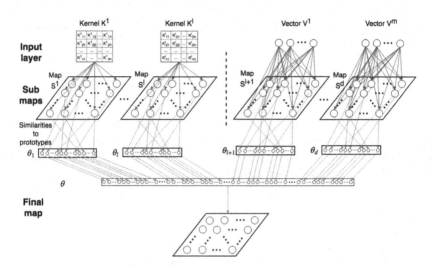

Fig. 1. Multiple sources SOM architecture

Let be $\theta : X \to \mathbb{R}^{U^{total}}$ (with $U^{total} = \sum_{i=1}^{d} U^i$) a function returning a vector composed of the similarity between an element x_j in X and all the prototypes of all the maps (see Fig. 1) defined by:

$$\theta(x_j) = [\theta_1(x_j), \theta_2(x_j), ..., \theta_d(x_j)] \tag{7}$$

With these definitions we can create a map S^{final} using the function $\theta(x)$ as an input. This map is defined by the U^{final} prototypes.

Let be W the three dimensional $(U^{final} \times d \times U^j)_j$ structure containing the weights vectors of the prototypes such that $W = [w_1, w_2, ..., w_{U^{final}}]$ with $w_u = [w_{u,1}, w_{u,2}, ..., w_{u,d}]$ and $w_{u,i} = [w_{u,i}^1, w_{u,i}^2, ..., w_{u,i}^{U^i}]$. We define a two dimensional $(U^{final} \times d)$ matrix α containing the prototypes local source weights such that $\alpha = [\alpha_1, \alpha_2, ..., \alpha_{U^{final}}]$ with $\alpha_u = [\alpha_{u,1}, \alpha_{u,2}, ..., \alpha_{u,d}]$.

The BMU denoted $BMU(x)$ of the final map S^{final} is found by using the dot product of the input and the weight vectors:

$$BMU(x) = arg \max_{u \in U^{final}} \sum_{i=1}^{d} \alpha_{u,i} \left(\theta_i(x) \cdot w_{u,i} \right) \tag{8}$$

Let $H(x_i, u)$ a neighborhood function that returns the "structural" similarity between the BMU of the input x_i and the prototype u of the map S^{final}:

$$H(x_i, u) = \exp \left(\frac{-d'(BMU(x_i), u)}{(1 - \frac{t}{T}) * \sigma'} \right)$$

where $d'(BMU(x_i), u)$ is the Hamming distance between the BMU and the prototype u, T is the maximal number of iterations and σ' is the radius of the map. We define a new objective function f that we maximize, in order to create the final map B^{final} by learning the local source weights for each cluster (represented by the map prototypes), as follows:

$$f = \sum_{j=1}^{N} \sum_{u=1}^{U^{final}} \sum_{i=1}^{d} H(x_j, u)\alpha_{u,i}\left(w_{u,i}.\theta_i(x_j)\right) \tag{9}$$

We update alternatively the final prototype and the source weights using the gradient descent method as follows:

$$w_{u,i}(t+1) = w_{u,i}(t) + \mu(t)\Delta w_{u,i}$$
$$\alpha_{u,i}(t+1) = \alpha_{u,i}(t) + \mu(t)\Delta\alpha_{u,i}$$

where $\mu(t) = 1 - \frac{t}{T}$ and T is the maximal number of iterations. The gradients of W and α are computed as follows:

$$\Delta w_{u,i} = H(x_j, u)\alpha_{u,i}\theta_i(x_j)$$
$$\Delta\alpha_{u,j} = H(x_i, u)\left(w_{u,j}.\theta_j(x_i)\right)$$

The vectors $w_{u,i}$ and $\alpha_{u,i}$ are then normalized by fixing the Euclidean norm of the prototype weights and source weights vectors to 1.

The originality of our approach lies in including the local source weights optimization in the SOM learning algorithm. The convergence of our approach depends of the convergence of final map (the third layer). As in the multiple kernel SOM [12], the cost induced by the learning of the α parameter is moderate. In practice, we multiplied by five the number of iterations necessary to train the classical SOM in order to insure the convergence of our learning algorithm.

3.2 Discussion

As presented before, we combine the information of the different sources using their respective SOM inside the algorithm and create a final SOM with their outputs. We associate to each cluster obtained in the final SOM a prototype representing the local weights of the sources associated to the different SOMs. Compared to the only one SOM algorithm in the literature which learns a global source weights for all the clusters (MKSOM [12]), our algorithm learns local weights of the data sources for each cluster. Localized multiple kernel learning approach was recently applied to k-means clustering [5,9,16]. Lei et al. [8] proposed an approach that can be formulated as a convex optimization problem over a given cluster structure. Instead of giving the same kernel weights for all input instances, the kernel combination is sample specific. Different kernel weights are calculated at the sample level rather than at the group level. In spite of the performance improvements proposed by these methods, the learning of a very large number of parameters leads to very expensive computation. Moreover, it is difficult to interpret the obtained results. In our approach, we associate a different sources combination to each cluster in order to distinguish it from the other clusters. This modelling makes sense in several application domains. Different categories of objects do not necessarily share the same importance of data sources. To our knowledge two approaches have been proposed in this sense. In [18], Yang et al. incorporate the notion of group in the MKL (Multiple Kernel Learning) framework using support vector machines for objects categorization. In [11], Mu and Zou use the graph embedding framework to tune the kernel weights that vary at the cluster level. They introduce in their work a non-uniform MKL. Finally, our method allows to learn mixed source weights at the cluster level in a self organizing map, which performs clustering as well as data visualization. This is very useful since it allows to interpret and explain the clustering results.

4   Experimentation

4.1   Datasets

We evaluate our approach called MSOM on six datasets (see Table 1). We generate two artificial datasets in order to show the interest of our approach and we use four datasets obtained from the UCI database [4] for the evaluation.

Table 1. Overview of the considered datasets.

Dataset	#Instances	#Attributes	#Classes
Artificial 1	800	4	4 balanced classes
Artificial 2	800	4	4 balanced classes
Dermatology	366	34	6 classes (1:112, 2:61, 3:72, 4:49, 5:52, 6:20)
E.coli	336	7	4 classes (1:143, 2:116, 3:52, 4:25)
Iris	150	4	3 balanced classes
WDBC	685	10	2 classes (1:458, 2:241)

The first artificial dataset is composed of four clusters of 4-dimensional points. For each dimension there are two clusters that can be separated from the other ones and so this dimension is informative for these clusters. The separated clusters are represented by Gaussian distributions with a standard deviation of 0.1 with two different centers (-5 and 5). The other non-separated clusters (noise) are represented by uniform distributions between -4 and 4. For the second artificial dataset, we use also four clusters containing 4-dimensional points. In this dataset, each cluster has one informative dimension. As for the first generated dataset, the separated clusters are represented by Gaussian distributions (with a standard deviation of 0.1 and a center of 1). The non-separated ones are obtained using a uniform distribution between -5 and 0. In Table 2 we show the informative dimensions for each cluster for both the datasets.

Table 2. Clusters informative dimension for the first and second artificial datasets where N stands for noise and I stands for informative.

Cluster id	Informative dimension							
	Artificial 1 dataset				Artificial 2 dataset			
	0	1	2	3	0	1	2	3
0	I	I	N	N	I	N	N	N
1	I	N	I	N	N	I	N	N
2	N	I	N	I	N	N	I	N
3	N	N	I	I	N	N	N	I

4.2 Protocol

To show the ability of our approach MSSOM to handle the mixed datasets, we used three representations of attributes (sources) in the considered datasets:

1. Datasets containing only numerical data: Each source is represented by a numerical attribute.
2. Datasets containing only kernels: We represent each attribute of the dataset by a Gaussian kernel defined by: $K(x, y) = \exp(-\gamma \times \|x - y\|^2)$. In order to select the parameter γ, we follow a simple rule defined in [6]. The value of parameter γ is computed as follows: $\gamma = \frac{\sqrt{2*U}}{dist_{max}}$ where U is the number of neurons and $dist_{max}$ represents the maximal distance between the instances.
3. Mixed datasets containing kernels and numerical data: In this case, we represent, for each dataset, some attributes using numerical values and others using kernels.

In the following, we will note by MSSOM_num the variant of our approach where only numerical attributes are used, MSSOM_kernel the one where only kernels are used and MSSOM_mix the one that uses mixed data sources.

We compare the results obtained by the different variants of our approach to the SOM [7] which uses the numerical attributes and to MKSOM [12] which uses the kernels. The three methods use two common parameters which are the grid dimension and the maximal number of iterations. For all the datasets we use a 3×3 grid and the maximal number of iterations is equal to five times the number of instances. We used small grids in order to highlight the clustering capability of our approach. However, our method works well with higher map sizes and produces the same kind of results as classical SOM. The other parameters of MKSOM are selected by doing a grid search. MSSOM and SOM methods use the same neighbourhood function.

To measure the clustering performance, we use the Normalized Mutual Information (NMI) and the purity measures. The NMI measures how the classes are distributed in the clusters. A high NMI means that the clusters tend to contain data of one class and the class tends to be in one cluster. The purity measures the distribution of classes in the clusters. A high purity means that the clusters tend to contain only one class of data. These measures are defined by:

$$NMI = \frac{2 \times \sum_{i=1}^{U} \sum_{j=1}^{C} N_{ij} \times \log\left(\frac{N}{N_i \times N_j}\right)}{\sum_{i=1}^{U} N_i \times \log\left(\frac{N_i}{N}\right) + \sum_{j=1}^{C} N_j \times \log\left(\frac{N_j}{N}\right)}, \; purity = \frac{1}{N} \sum_{i=1}^{U} \max_j \left(N_{ij}\right) \text{ where}$$

N is the number of instances, N_i is the number of instances associated to the cluster i, N_j is the number of the instances in class j and $N_{i,j}$ is the number of instances of class j associated to the cluster i.

4.3 Results

Figure 2 shows the performance of our methods against two state-of-art methods. On the generated datasets as well as on the real world datasets, the different variants of our method give the best results.

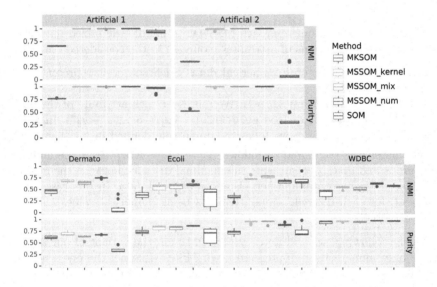

Fig. 2. Performance of the different methods on the artificial and real datasets.

As shown in Fig. 2, our method gives an NMI and a purity close to 1 for both artificial datasets. The performance of MSSOM is not sensitive to the different attributes (sources) representations. This means that all the variants of our MSSOM are able to well separate the clusters. These good results can be explained by the ability of our method to associate an adequate source combination for each cluster instead of one combination for the whole space.

We can see that all the variants of MSSOM give good results on the real datasets compared to the SOM state-of-art methods. For example, on the Iris dataset, our method gives an NMI greater than 0.7 and a purity greater than 0.9 (for the three variants) when the classical SOM shows an NMI of 0.7 and a purity lower than 0.8 and the MKSOM shows an NMI lower than 0.4 and a

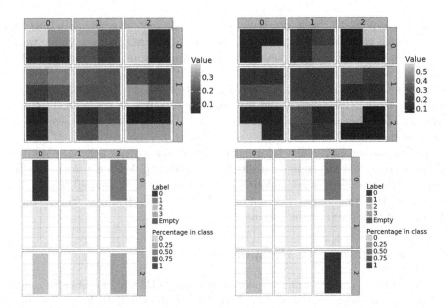

Fig. 3. Sources weights (top) and label repartition (bottom) of the final map for the first artificial dataset (left) and the second artificial dataset (right) using the MSSOM_kernel.

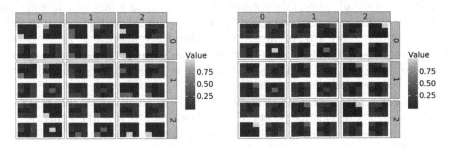

Fig. 4. Visualization of the final map weights of MSSOM on the kernel version of the first (left) and second (right) artificial datasets.

purity lower than 0.8. This confirms that learning the local source combinations for each cluster allows to improve the clustering results.

Figure 3 shows the source weights for each unit (neuron) in the final map for the first and second artificial datasets, as well as the distribution of the examples among the map units (neurons). We can see that the units 0: $(0,0)$, 2: $(0,2)$, 6: $(2,0)$ and 8: $(2,2)$ represent well the four classes. We can also see clearly that for the four units representing the classes, the informative sources have the greatest weights. For example, on the first artificial dataset, the unit 0 which contains the inputs of class 0 has sources weights close to 0.4 for the first and the second sources and close to 0.1 for the other sources. The sources with the biggest weight values correspond to the informative ones. On the second artificial dataset, the unit 2 which contains the inputs of class 1 has approximatively a source weight of 0.56 for the second source which is the informative one. The weight values of the other sources are around 0.1. The MKSOM method gives the same kernel weights for all the clusters. It gives the greatest weight for the fourth source in the first artificial dataset, and for the second source in the second artificial dataset. The other weight values are close to 0. This does not make sense on our artificial datasets.

In Fig. 4 we propose a new informative visualization of our approach MSSOM, where we represent the source weights in the final map. Each neuron in the final map is represented by a set of four subplots (one for each source). For a given neuron in the final map, each subplot allows to identify for each cluster its representative neuron in the sub maps.

5   Conclusions and Perspectives

In this paper we present a new approach based on SOM which is able to combine heterogeneous (numerical and complex) data sources and to learn the source weights locally for each cluster. We show, using artificial datasets, that our method can select informative sources for each cluster. We show that our method gives good results and is competitive to other SOM approaches. Moreover, our approach can handle hybrid data sources by representing in an efficient way the numerical and complex datasets. Experimental results show that our approach is not sensitive to the kind of inputs we use (kernels, numerical or mixed data). In all the cases, we obtain approximatively the same performances. Keeping the numerical data without transforming them to kernels reduces the time and memory complexities of our method.

In [14] we proposed a new supervised SOM approach that uses rejection options in order to identify potential new classes and reduce the ambiguity of prediction. One of our future work is to propose a supervised variant of MSSOM which will be inspired from [14]. This method will be used in a bioinformatics project which purpose is to identify and classify non-coding RNAs. For this purpose we will use heterogeneous data such as the sequences and the secondary structures of the RNAs, as well as epigenomics markers. By using the rejection option with the information of the different sources, we will be able to identify new classes of non-coding RNAs and to characterize them.

References

1. Bickel, S., Scheffer, T.: Multi-view clustering. In: ICDM, vol. 4, pp. 19–26 (2004)
2. Blum, A., Mitchell, T.: Combining labeled and unlabeled data with co-training. In: Proceedings of the Eleventh Annual Conference on Computational Learning Theory, pp. 92–100. ACM (1998)
3. Chaudhuri, K., Kakade, S.M., Livescu, K., Sridharan, K.: Multi-view clustering via canonical correlation analysis. In: Proceedings of the 26th Annual International Conference on Machine Learning, pp. 129–136. ACM (2009)

4. Dheeru, D., Taniskidou, E.K.: UCI machine learning repository (2017)
5. Gönen, M., Margolin, A.A.: Localized data fusion for kernel k-means clustering with application to cancer biology. In: Advances in Neural Information Processing Systems, pp. 1305–1313 (2014)
6. Haykin, S.: Neural Networks: A Comprehensive Foundation. Macmillan College Publishing Company, New York (1994)
7. Kohonen, T.: Self-Organizing Maps. Springer Series in Information Sciences, 3rd edn. Springer-Verlag, Heidelberg (2001). https://doi.org/10.1007/978-3-642-56927-2
8. Lei, Y., Binder, A., Kloft, M.: Localized multiple kernel learning a convex approach (2016)
9. Li, M., Liu, X., Wang, L., Dou, Y., Yin, J., Zhu, E.: Multiple kernel clustering with local kernel alignment maximization (2016)
10. Mariette, J., Olteanu, M., Boelaert, J., Villa-Vialaneix, N.: Bagged kernel SOM. In: Villmann, T., Schleif, F.-M., Kaden, M., Lange, M. (eds.) Advances in Self-Organizing Maps and Learning Vector Quantization. AISC, vol. 295, pp. 45–54. Springer, Cham (2014). https://doi.org/10.1007/978-3-319-07695-9_4
11. Yadong, M., Zhou, B.: Non-uniform multiple kernel learning with cluster-based gating functions. Neurocomputing 74(7), 1095–1101 (2011)
12. Olteanu, M., Villa-Vialaneix, N., Cierco-Ayrolles, C.: Multiple kernel self-organizing maps, p. 83 (2013)
13. Pelckmans, K., Vooren, S.V., Coessens, B., Suykens, J.A., Moor, B.D.: Mutual spectral clustering: microarray experiments versus text corpus. In: Proceedings of the Workshop on Probabilistic Modeling and Machine Learning in Structural and Systems Biology, pp. 55–58 (2006)
14. Platon, L., Zehraoui, F., Tahi, F.: Self-organizing maps with supervised layer. In: 2017 12th International Workshop on Self-Organizing Maps and Learning Vector Quantization, Clustering and Data Visualization (WSOM), pp. 1–8. IEEE (2017)
15. Wan, W., Fraser, D.: A multiple self-organizing map scheme for remote sensing classification. In: Kittler, J., Roli, F. (eds.) MCS 2000. LNCS, vol. 1857, pp. 300–309. Springer, Heidelberg (2000). https://doi.org/10.1007/3-540-45014-9_29
16. Wang, Q., Dou, Y., Liu, X., Xia, F., Lv, Q., Yang, K.: Local kernel alignment based multi-view clustering using extreme learning machine. Neurocomputing 275, 1099–1111 (2018)
17. Xu, C., Tao, D., Xu, C.: A survey on multi-view learning. arXiv preprint arXiv:1304.5634 (2013)
18. Yang, J., Li, Y., Tian, Y., Duan, L., Gao, W.: Group-sensitive multiple kernel learning for object categorization. In: 2009 IEEE 12th International Conference on Computer Vision, pp. 436–443. IEEE (2009)
19. Zhao, B., Kwok, J.T., Zhang, C.: Multiple kernel clustering. In: SDM, pp. 638–649 (2009)
20. Zhao, X., Evans, N., Dugelay, J.-L.: A subspace co-training framework for multi-view clustering. Pattern Recognit. Lett. 41, 73–82 (2014)
21. Zhou, D., Burges, C.J.C.: Spectral clustering and transductive learning with multiple views. In: ICML 2007, Proceedings of the 24th International Conference Machine Learning, New York, NY, USA, pp. 1159–1166. ACM (2007)
22. Zhuang, J., Wang, J., Hoi, S.C.H., Lan, X.: Unsupervised multiple kernel learning. J. Mach. Learn. Res. 20, 129–144 (2011). Proceedings Track

Heterogeneous Dyadic Multi-task Learning with Implicit Feedback

Simon Moura[1]([✉]), Amir Asarbaev[1,4], Massih-Reza Amini[1],
and Yury Maximov[2,3]

[1] Univ. Grenoble Alps, CNRS, Grenoble INP - LIG, Grenoble, France
{Simon.Moura,Amir.Asarbaev,Massih-Reza.Amini}@univ-grenoble-alpes.fr,
amir.asarbaev@grenoble-inp.org
[2] Skolkovo Institute of Science and Technology, Moscow, Russia
y.maximov@skoltech.ru
[3] Theoretical Division T-5 and CNLS Los Alamos National Laboratory,
Los Alamos, USA
[4] Moscow Institute of Physics and Technology, Dolgoprudny, Russia

Abstract. In this paper we present a framework for learning models
for Recommender Systems (RS) in the case where there are multiple
implicit feedback associated to items. Based on a set of features, representing the dyads of users and items extracted from an implicit feedback
collection, we propose a stochastic gradient descent algorithm that learn
jointly classification, ranking and embeddings for users and items. Our
experimental results on a subset of the collection used in the RecSys 2016
challenge for job recommendation show the effectiveness of our approach
with respect to single task approaches and paves the way for future work
in jointly learning models for multiple implicit feedback for RS.

Keywords: Recommendation systems · Multiple implicit feedback
Dyadic prediction · Muti-task learning

1 Introduction

The aim of Recommender Systems (RS) is to present products to users by adapting the displayed offers to their taste. Recently, there was a surge of interest in
the design of efficient RS especially after the NetFlix challenge [1]; and also
because of many new problems such as the study of an accurate and scalable
RS presents. As most of the users interactions are now provided in the form
of clicks, an active line of research on RS is to learn models based on implicit
feedback. Although there is no evidence of the actual value of such feedback,
as a positive (respectively negative) feedback on an item does not necessarily
represents a user preference (respectively dislike), almost all approaches assume
that positive feedback conveys relevant information for the problem at hand.

In this work, we consider the case where there are multiple implicit feedbacks
for each items and propose a multi-target learning algorithm that enhance prediction over each of these feedbacks. We cast the problem as a dyadic prediction

© Springer Nature Switzerland AG 2018
L. Cheng et al. (Eds.): ICONIP 2018, LNCS 11303, pp. 660–672, 2018.
https://doi.org/10.1007/978-3-030-04182-3_58

problem, where the aim is to predict multiple outputs for observations that are constituted by pairs of examples formed by users and items. A classical approach when dealing with multiple outputs is to divide the problems into *simpler* sub-prediction problems and deal with them separately without considering their relationships. However, the intuition and the consensus is that, when some tasks are interdependent and potentially heterogeneous (that is, for instance, when some tasks deal with classification while others deal with ranking or regression), the learner will benefit from learning them jointly by taking into account the shared information. Following this intuition, multi-task (MTL) approaches[1] [2,3] consider the first example of the dyad, an *observation*, and the second example, a *task*, and propose to solve the general prediction problem by taking advantage of the correlation between the tasks [4–7]. Most of these approaches learn a different model for each task using the feature representation of observation and model the dependencies in the objective function using a shared regularization term that enforces correlated tasks to have close models [4,5,8]. However, by simultaneously taking into account both instances of a dyad, one can expect to do better by building a single model and by using all the available information at once.

Contributions. Although, dyadic prediction problems are common [9], the case where they are associated with multiple and heterogeneous outputs is rare and still not fully studied mainly due to the lack of data collections. In this paper, we propose a generic method to extract a meaningful representation from multiple implicit feedback data for RS. Based on this method, we provide a dataset built over the RecSys 2016 competition for job recommendation. The competition consisted in proposing job offers to users that would be of their interest, with the particularity that users may have simultaneously *clicked, bookmarked, replied,* and *removed* specific offers that they have been proposed. We adapted the method of [10] that was initially designed for explicit feedback single task learning to the case of multiple implicit feedbacks. To evaluate the user-offer dyadic representations and analyze the usefulness of taking into account the relationship between the tasks (different implicit feedback), we propose a MTL stochastic gradient descent (SGD) algorithm that combines classification and ranking predictors and the learning of user and item embeddings. We show that the combination of tasks allows to considerably enhance the prediction of most of the tasks, particularly the predictions for the clicks. Empirical comparisons of the MTL approach with single task approach that considers each of the implicit feedback independently shows the efficiency of the proposed strategy.

Organization of the Paper. In Sect. 2 we briefly present the RecSys 2016 challenge dataset and the features that were extracted from the implicit feedback. In Sect. 3 we describe the MTL learning framework considered throughout this paper. In Sect. 4, we describe the gradient descent for jointly learning multiple heterogeneous tasks and embeddings. In Sect. 5 we present experimental results that evaluate the effect of taking into account the dependency among the

---

[1] https://www.ngdata.com/icml-2013-tutorial-multi-target-prediction/.

outputs. Finally, in Sect. 6, we discuss the outcome of this study and give some pointers for further research.

2 Feature Extraction Based on Implicit Feedback

In this Section, we briefly describe the RecSys 2016 challenge and present the steps we followed to extract the dataset as well as the proposed learning strategy for combining the outputs[2].

The RecSys 2016 challenge[3], hosted by the XING social network platform, was defined as a ranking problem of clicks for job offers. For this competition, four main files were made available, each containing information on users and jobs offers. The offers that were displayed to the users are called *impressions*. Each offer displayed could be interacted in four different manners, by clicking, bookmarking, replying or deleting the offer.

In this collection, users provided implicit feedback and may have different interactions with the same offer. In particular, a user could have clicked and also replied to the same offer. We used both of these information to extract characteristics for the pairs of users and items. We did not consider the information concerning deleted offers as it does not help to decide whether a user is willing to interact positively with an offer or not, which is the primary goal.

The statistics of the original interactions and impressions are summarized in the left column of Table 1. In this table, the *sparsity* represents the user-offer pairs for which there is no interaction at all. Figure 1 shows the number of users that have positive interactions with respects to the clicks, bookmarks and replies. Table 2 shows the number of offers displayed to users, as well as the number of suggestions that clicked, bookmarked and replied. As expected, the vast majority of users make very few interactions while few users interact a lot. It comes out that the offers that are bookmarked and replied, represent less than 5% of those that are displayed to the users, making both associated tasks challenging.

To extract relevant features, we relied on the approach described in [10], which is a method that uses neighbors preferences and retrieves statistics about the closest users interactions (regarding a predefined similarity). The main idea is to compute statistics that describe the net preference of a user for a given offer. The dyadic representation for a user-offer pair is hence composed of statistics summarized in 15 features to which is added 2 biases.

Due to a significant amount of data available and the substantial sparsity, we subsampled the dataset to keep only the users for which we had enough information to extract meaningful statistics. We decided to keep users with more than 30 interactions and offers which had been interacted with at least 30 times. Following this process, we obtained **819.226** dyadic user-offer pairs that we randomly split into training (30% of the original dataset) and test (70% of the original dataset) collections. The right column of Table 1 contains statistics describing the dataset that we obtain after using the subsampling method described above.

---

[2] We make available the extracted dataset as well as the codes for research purpose.
[3] https://recsys.acm.org/recsys16/challenge/.

Table 1. Statistics over the *original* RecSys 2016 (left) and DAEMON (right) datasets. In the latter we only consider users which had at least 30 interactions and the offers which had been interacted at least 30 times. (*Med.* stands for median, *Av.* for average).

Table 2. Number of offers *displayed, clicked, bookmarked* and *replied*. *All* is the sum of the three latter. The dataset is highly imbalanced and contains only few *Bookmarked* and *Replied* interactions.

	RecSys'16	DAEMON
# Users	**770.859**	**5.949**
# Offers	**1.002.161**	**25.184**
Av. nb interactions/user	7	137.71
Med. nb interactions/user	3	120
Av. nb users/offer	5	32.53
Med. nb users/offer	2	12
Av. nb displayed/user	76	115.04
Av. nb clicked/user	7	42.19
Av. nb bookmarked/user	0.27	1.38
Av. nb replied/user	0.42	3.40
Sparsity	99.991%	99.45%

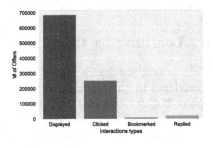

The rational behind the subsampling of the original dataset is twofold. First, the users who did not interact enough are not relevant to learn a predictive model as we lack information about them. Secondly, the method we use to extract features [10] heavily relies on the similarity between users, which is less likely to be high in a sparse dataset.

To compute the statistics of for pairs (*user, item*), as proposed in [10], we associated to clicks, bookmarks and replies respectively the weights 1, 2 and 3, emphasizing the fact the replying to an offer shows more interest than simply clicking or bookmarking it.

Below we detail the exact steps followed to create representations for multiple implicit feedbacks:

1. Create a matrix (IM) of implicit interactions of users and offers summing over all positive interactions:

$$IM_{(u,o)} = \sum_{i \in \mathcal{I}} i * \mathbb{1}_{(u,o)_i = 1} \,,$$

where $(u, o)_i = 1$ means that user u interacted positively with offer o, $\mathcal{I}$ represents the set of all possible interactions and $i = 1, 2$ or 3 for respectively clicked, bookmarked and replied interactions;

2. Evaluate users similarities (e.g. using cosine distance) using IM. Keep top K closest users for each users;

3. For each dyad (u, o) for which we have a ground truth (at least one positive interaction), compute the *win/tie/loss* vectors based on similar users than u that interacted with offer o as described in [10].
4. Finally, the features associated to the dyad (u, o) is obtained by considering 5 statistics over each of the *win/tie/loss* vectors, namely: the mean, the standard deviation, the max, the min and the normalized number of respectively wins, ties or losses.

3   Learning by Combining the Outputs

This section provides a formal definition of the multi-target learning framework considered in this article. We suppose that dyads are represented in an input space $\mathcal{X} \subseteq \mathbb{R}^d$ and that the output space $\mathcal{Y} = \mathcal{Y}_1 \times \ldots \times \mathcal{Y}_T$ is a product of T different output spaces corresponding each to an interaction. Further, we assume that the pairs and their associated output $(\mathbf{x}, \boldsymbol{y}) \in \mathcal{X} \times \mathcal{Y}$ are generated i.i.d with respect to a fixed yet unknown joint probability distribution $\mathcal{D} = (\mathcal{D}_1, \ldots, \mathcal{D}_T)$. The aim of learning is to find a prediction function in some predefined function set $\mathcal{H} = \{\boldsymbol{h} : \mathcal{X} \to \mathcal{Y}\}$, that minimizes the expected risk

$$\mathcal{L}(\boldsymbol{h}) = \mathbb{E}_{(\mathbf{x}, \boldsymbol{y}) \sim \mathcal{D}}[\boldsymbol{\ell}(\boldsymbol{h}(\mathbf{x}), \boldsymbol{y})], \tag{1}$$

where $\boldsymbol{\ell}(\boldsymbol{h}(\mathbf{x}), \boldsymbol{y})$ is an instantaneous loss measuring the discrepancy of the predictions over different outputs $\boldsymbol{h}(\mathbf{x}) = (\boldsymbol{h}_1(\mathbf{x}), \ldots, \boldsymbol{h}_T(\mathbf{x})) \in \mathcal{Y}$ of observation $\mathbf{x}$ and its desired output $\boldsymbol{y} = (\boldsymbol{y}_1, \ldots, \boldsymbol{y}_T) \in \mathcal{Y}$. Following the Empirical Risk Minimization principle, we achieve this aim by minimizing an empirical loss over a training set $\mathcal{S} = (\mathbf{x}_i, \boldsymbol{y}_i)_{i=1}^m$ of size m, where examples are be generated i.i.d with respect to the same probability distribution $\mathcal{D}$.

In the case where we consider multiple tasks, the general formulation of the empirical loss function on a train set $\mathcal{S}$ can be written as follow:

$$\mathcal{L}_m(\boldsymbol{h}, \mathcal{S}) = \frac{1}{m} \sum_{i=1}^m \frac{1}{T} \sum_{j=1}^T \ell_j(\boldsymbol{h}_j(\mathbf{x}_i), \boldsymbol{y}_i) + \lambda \Omega(\boldsymbol{h}), \tag{2}$$

where ℓ_j is the loss for task j, $\boldsymbol{h}_j$ is the hypothesis function for tasks j and $\Omega(\boldsymbol{h})$ is a regularization term on the parameters of the models. In this work we aim at minimizing multiple loss functions over all models $\{\boldsymbol{h}_j\}_{j \in \{1, \ldots, T\}}$.

We considered two different setups to achieve this goal:

- Single task learning (STL) where the goal is to learn each prediction function $(h_j)_{1 \leq j \leq T}$ independently of the others by minimizing the associated empirical loss;
- Multi-target learning (MTL) where the goal is to learn all prediction functions jointly by taking into account dependencies between their outputs. While the classical way of binding models is to use a shared regularization across the tasks, in our approach, the multi-task is carried out by a shared representation, that is also learned during the same training phase.

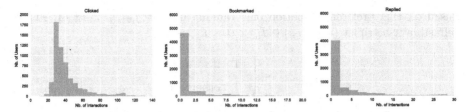

Fig. 1. The number of users with respect to the number of interactions. From left to right, for *Clicked*, *Bookmarked* and *Replied* tasks.

4 SGD for Multi-target Heterogeneous Dyadic Learning

Based on the dataset extracted following the steps described in Sect. 2, we defined a set of relevant tasks for RS. As the main source of revenue for many online website rely on the number of clicks, the main goal remains to improve the predictions for the clicks. In this sense, we define the following tasks:

1. A *classical* learning to rank task in RS for clicking interactions, where the goal is to provide a ranking list of offers on which a user is willing to click. As explained earlier, the main source of revenues of many website is based on the number of clicks they register. In this scenario, it makes sense to "specialize" the models for the predictions of clicks. We will discuss more about this question and how we do that in practice in the definition of the fourth task.
2. Two binary classification tasks, where the goal is to predict if a user is willing to interact positively with specific offers in terms of bookmark and reply interactions.
3. A representation learning task, where the goal is to learn a meaningful representation $\mathcal{U}$ for users and $\mathcal{I}$ for items. This tasks is used as a binding between of the different models.

Based on the definition of the four tasks described above, we can formally re-write the loss functions for the tasks at hand as:

$$
\mathcal{L}_m(\boldsymbol{h}, \mathcal{S}) = \mathcal{L}_{rank}^{click}(\boldsymbol{h}_1, \mathcal{S}) + \mathcal{L}_{class}^{book}(\boldsymbol{h}_2, \mathcal{S}) + \mathcal{L}_{class}^{reply}(\boldsymbol{h}_3, \mathcal{S}) + \mathcal{L}_{rank}^{emb}(\mathbb{U}, \mathbb{I}, \mathcal{S}) +
$$
$$
\lambda \left(||\boldsymbol{h}_1||_2^2 + ||\boldsymbol{h}_2||_2^2 + ||\boldsymbol{h}_3||_2^2 + ||\mathbb{U}||_2^2 + ||\mathbb{I}||_2^2 \right). \tag{3}
$$

We define each tasks as the average of all logistic losses evaluated for each example in the sample. In the case of pairwise ranking we note as $\mathcal{U} \subseteq N$ (resp. $\mathcal{I} \subseteq N$) the set of indexes over users (resp. the set of indexes over items). Furthermore, for each user $u \in \mathcal{U}$, we consider two subsets of offers, the offers interacted negatively $\mathcal{I}_u^- \subset \mathcal{I}$ (the user did not click) and the offers interacted positively $\mathcal{I}_u^+ \subset \mathcal{I}$ (the user clicked) such that:

- $\mathcal{I}_u^- \neq \emptyset$ and $\mathcal{I}_u^+ \neq \emptyset$.
- For any pair of offers $(i^+, i^-) \in \mathcal{I}_u^+ \times \mathcal{I}_u^-$, $i^+ \underset{u}{\succ} i^-$ mean that user u has a preference for item i^+ over item i^-.

Based on this preference relation, the ranking output $y_{i^+,u,i^-} \in \{-1,+1\}$ is defined over a triplet $(i^+, u, i^-) \in \mathcal{I}_u^- \times U \times \mathcal{I}_u^-$ as:

$$y_{i^+,u,i^-} = \begin{cases} +1 & \text{if } i^+ \underset{u}{\succ} i^- \\ -1 & \text{otherwise.} \end{cases} \tag{4}$$

In the case of the ranking problem for the clicks, h_1 is of the form

$$h_1(\mathbf{x}) = \langle \mathbf{w}_1, \mathbf{x} \rangle,$$

and we say that the model $\mathbf{w}_1$ is making an error on a prediction when it ranks higher a negative example over a positive example. Thus, we can compute the error made on one triplet $(i^+, u, i^-) \in \mathcal{I}_u^- \times U \times \mathcal{I}_u^-$ as

$$\mathbb{1}_{\langle \mathbf{w}_1,(u,i^+)\rangle < \langle \mathbf{w}_1,(u,i^-)\rangle} = \mathbb{1}_{\langle \mathbf{w}_1,(u,i^+)-(u,i^-)\rangle < 0}. \tag{5}$$

The loss function of Eq. 5 is hard to optimize but can be approximated by a smooth surrogate function, such as the logistic loss function. We can re-write the loss for the ranking problem over a training set $\mathcal{S}$ as follow

$$\mathcal{L}_{rank}^{click}(\mathbf{w}_1, \mathcal{S}) = \frac{1}{|\mathcal{U}|} \sum_{u \in \mathcal{U}} \frac{1}{|\mathcal{I}_u^+||\mathcal{I}_u^-|} \sum_{(i^+,i^-) \in \mathcal{I}_u^+ \cup \mathcal{I}_u^-} \mathbb{1}_{\langle \mathbf{w}_1,(u,i^+)\rangle < \langle \mathbf{w}_1,(u,i^-)\rangle}$$

$$\approx \frac{1}{|\mathcal{U}|} \sum_{u \in \mathcal{U}} \frac{1}{|\mathcal{I}_u^+||\mathcal{I}_u^-|} \sum_{(i^+,i^-) \in \mathcal{I}_u^+ \cup \mathcal{I}_u^-} \log(1 + e^{-y_{i^+,u,i^-} \langle \mathbf{w}_1,(u,i^+)-(u,i^-)\rangle}) + \lambda_1 ||\mathbf{w}_1||_2^2.$$

Remark 1. Note that, for a fixed user, it is possible to have a different polarity of interactions with respect to clicks, bookmarks and replies. In other words, as we select pairs of offers (i^+, i^-) regarding the clicks polarity, it is possible to have i^+ and i^- both positives, or negatives, for bookmarks or replies.

In terms of binary classification with outputs $y_i \in \{-1, +1\}$, a model h_2 makes an error when its prediction differs from the ground truth y. We can write the zero-one loss for classification as

$$\mathbb{1}_{y_i h_2(\mathbf{x}) < 0}, \text{ where } h_2(\mathbf{x}) = \langle \mathbf{w}_2, \mathbf{x} \rangle.$$

Similar to the case of ranking described above, we use a logistic loss in order to approximate the zero-one loss for the two classification tasks at hand. We also average the loss over the two pairs. Note that the polarity can be different for the bookmarks and replies tasks, we will note $y_{i^+}^b$ (respectively $y_{i^-}^b$) the polarity for the positives (respectively negatives) interactions for the bookmark task for user $u \in \mathcal{U}$ and offers $(i^+, i^-) \in \mathcal{I}^2$ and $y_{i^+}^r$ (respectively $y_{i^-}^r$) the polarity for the positives (respectively negatives) interactions for the reply task:

$$\mathcal{L}_{class}^{book}(\mathbf{w}_2, \mathcal{S})$$

$$= \frac{1}{|\mathcal{U}|} \sum_{u \in \mathcal{U}} \frac{1}{|\mathcal{I}_u^+||\mathcal{I}_u^-|} \sum_{(i^+,i^-) \in \mathcal{I}_u^+ \cup \mathcal{I}_u^-} \left[\frac{1}{2} \mathbb{1}_{y_{i+}^b \langle \mathbf{w}_2, (u,i^+) \rangle < 0} + \frac{1}{2} \mathbb{1}_{y_{i-}^b \langle \mathbf{w}_2, (u,i^-) \rangle < 0} \right]$$

$$\approx \frac{1}{|\mathcal{U}|} \sum_{u \in \mathcal{U}} \frac{1}{|\mathcal{I}_u^+||\mathcal{I}_u^-|} \sum_{(i^+,i^-) \in \mathcal{I}_u^+ \cup \mathcal{I}_u^-} \left[\frac{1}{2} \log(1 + e^{-y_{i+}^b \langle \mathbf{w}_2, (u,i^+) \rangle}) \right.$$

$$\left. + \frac{1}{2} \log(1 + e^{-y_{i-}^b \langle \mathbf{w}_2, (u,i^-) \rangle}) \right] + \lambda_2 ||\mathbf{w}_2||_2^2,$$

$$\mathcal{L}_{class}^{reply}(\mathbf{w}_3, \mathcal{S})$$

$$= \frac{1}{|\mathcal{U}|} \sum_{u \in \mathcal{U}} \frac{1}{|\mathcal{I}_u^+||\mathcal{I}_u^-|} \sum_{(i^+,i^-) \in \mathcal{I}_u^+ \cup \mathcal{I}_u^-} \left[\frac{1}{2} \mathbb{1}_{y_{i+}^r \langle \mathbf{w}_2, (u,i^+) \rangle < 0} + \frac{1}{2} \mathbb{1}_{y_{i-}^r \langle \mathbf{w}_3, (u,i^-) \rangle < 0} \right]$$

$$\approx \frac{1}{|\mathcal{U}|} \sum_{u \in \mathcal{U}} \frac{1}{|\mathcal{I}_u^+||\mathcal{I}_u^-|} \sum_{(i^+,i^-) \in \mathcal{I}_u^+ \cup \mathcal{I}_u^-} \left[\frac{1}{2} \log(1 + e^{-y_{i+}^r \langle \mathbf{w}_3, (u,i^+) \rangle}) \right.$$

$$\left. + \frac{1}{2} \log(1 + e^{-y_{i-}^r \langle \mathbf{w}_3, (u,i^-) \rangle}) \right] + \lambda_3 ||\mathbf{w}_3||_2^2.$$

Finally, the representation learning task aims at automatically learning an embedding for users $\mathbb{U}$ and an embedding for items $\mathbb{I}$. We insist on two important points here. The first one is that the embedding representations are trained jointly with the other tasks and are used also as inputs in the other tasks. Secondly, in a sense, the embeddings are *specialized* for the click task, as we select pairs with different polarity for the clicks. We consider that the embedding is provides a mistake if

$$\mathbb{1}_{\langle \mathbb{U}_u, \mathbb{I}_{i+} \rangle < \langle \mathbb{U}_u, \mathbb{I}_{i-} \rangle} = \mathbb{1}_{\langle \mathbb{U}_u, \mathbb{I}_{i+} - \mathbb{I}_{i-} \rangle < 0},$$

where $\mathbb{U}_u \in \mathbb{R}^d$ denotes the vector representation for user u and $\mathbb{I}_i \in \mathbb{R}^d$ denotes the vector representation for item i. Again, we use the logistic loss as an approximation to this error and we have

$$\mathcal{L}_{rank}^{emb}(\mathbb{U}, \mathbb{I}, \mathcal{S}) = \frac{1}{|\mathcal{U}|} \sum_{u \in \mathcal{U}} \frac{1}{|\mathcal{I}_u^+||\mathcal{I}_u^-|} \sum_{(i^+,i^-) \in \mathcal{I}_u^+ \cup \mathcal{I}_u^-} \mathbb{1}_{\langle \mathbb{U}_u, \mathbb{I}_{i+} - \mathbb{I}_{i-} \rangle < 0}$$

$$\approx \frac{1}{|\mathcal{U}|)} \sum_{u \in \mathcal{U}} \frac{1}{|\mathcal{I}_u^+||\mathcal{I}_u^-|} \sum_{(i^+,i^-) \in \mathcal{I}_u^+ \cup \mathcal{I}_u^-} \log(1 + e^{-\langle \mathbb{U}_u, (\mathbb{I}_{i+} - \mathbb{I}_{i-}) \rangle})$$

$$+ \lambda_4 (||\mathbb{U}_u||_2^2 + ||\mathbb{I}_{i+}||_2^2 + ||\mathbb{I}_{i-}||_2^2).$$

4.1 SGD for Multi-output and Heterogeneous Tasks

As a result of the multiple loss functions defined above, we can write our heterogeneous multi-target learning problem as a constrained convex optimization where the goal is to minimize the general loss functions defined over the sum of all losses defined above and its regularization parameters:

$$\mathcal{L}(\mathcal{F}, S) + \lambda\Omega(\mathbf{w}_1, \mathbf{w}_2, \mathbf{w}_3, u, i^+, i^-) \to \min. \tag{6}$$

Algorithm 1 describe the SGD strategy we propose to optimize the parameters of our models.

The algorithm works as follow. First, we provide the hyper-parameters values: two stopping criteria parameters, ϵ and the maximum number of epochs #$epochs$. We also provide the learning rate η, the number of iteration per epoch #$iters$ and the $\lambda_1, \lambda_2, \lambda_3$ and λ_4 parameters that control the regularization terms for respectively the clicks, bookmarks, replies and learning of the embeddings. Then, we randomly initialize all the weights of the models $\mathbf{w}_1, \mathbf{w}_2$ and $\mathbf{w}_3$ and the embeddings for the users $\mathbb{U}$ and for the offers $\mathbb{I}$.

Then, the idea is the following, at each step of the algorithm, the learning is carried out by computing the gradient of each tasks simultaneously and updating the weights of each models, and the weights of the representation $\mathbb{U}$ and $\mathbb{I}$. As the embeddings are initialized randomly, the first iterations of the algorithm mainly relies on the handcrafted features. After few iterations, as quality the

---

Algorithm 1. Multi-target learning based on SGD algorithm

1: **Inputs:**

    2: $(extracted, context, \mathbb{U}, \mathbb{I}) \in S$ #Training set
    3: η (Learning rate), $\lambda_1, \ldots, \lambda_4$ (Regularization)
    4: ϵ # Stopping criterion
    5: nb_epochs # Maximum number of epochs
    6: nb_iters # Number of iterations per epochs

7: **Initialize:**

    8: Randomly intialized: $W^{(0)} = \{\mathbf{w}_1^{(0)}, \mathbf{w}_2^{(0)}, \mathbf{w}_3^{(0)}, u^{(0)}, i^{+(0)}, i^{-(0)}\}$
    9: $epoch = 0$
    10: $global\_loss\_old \leftarrow 0$

11: $global\_loss\_new \leftarrow \mathcal{L}(\mathcal{F}, S) + \lambda\Omega(\mathbf{w}_1, \mathbf{w}_2, \mathbf{w}_3, u, i^+, i^-)$

12: **while** $global\_loss\_new - global\_loss\_old > \epsilon$ and $epoch < nb\_epochs$ **do**

13:     $global\_loss\_old \leftarrow global\_loss\_new$

14:     $local\_loss \leftarrow 0$

15:     **for** $t = 1...nb\_iters$ **do**

16:         **randomly choose:** user u, positive offer i^+ and negative offer i^-

17:         for the click interaction

18:         $W^{(t)} \leftarrow W^{(t-1)} - \eta \cdot [\nabla\mathcal{L}(\mathcal{F}, S) + \lambda\Omega(\mathbf{w}_1, \mathbf{w}_2, \mathbf{w}_3, u, i^+, i^-)]$

19:         $local\_loss^{(t)} \leftarrow local\_loss^{(t-1)} + [\ell_{rank}^{click}(u, i^+, i^-)] + [\ell_{rank}^{emb}(\mathbb{U}_u, \mathbb{I}_{i+}, \mathbb{I}_{i-})]$
                $+[\frac{1}{2}\ell_{class}^{book}(u, i^+) + \frac{1}{2}\ell_{class}^{book}(u, i^-)] + [\frac{1}{2}\ell_{class}^{reply}(u, i^+) + \frac{1}{2}\ell_{class}^{reply}(u, i^-)]$
                $+\lambda_1||\mathbf{w}_1||_2^2 + \lambda_2||\mathbf{w}_2||_2^2 + \lambda_3||\mathbf{w}_3||_2^2 + \lambda_4(||\mathbb{U}_u||_2^2 + ||\mathbb{I}_{i+}||_2^2 + ||\mathbb{I}_{i-}||_2^2)$

20:     **end for**

21:     $global\_loss\_new \leftarrow \frac{local\_loss}{nb\_iters}$

22:     $epoch = epoch + 1$

23: **end while**

---

embeddings get better, the models rely on both, the handcrafted and the embeddings features. Finally, after updating models and embeddings, we compute the new general loss and compare it to the old one to evaluate if we need to continue the descent.

Remark 2. Here we used l_2 regularization for all loss functions and bind the task using the embedding. The use of l_2 regularization is a decision that can be debated, however, it is not the focus of this study. Here the goal is to evaluate the impact of learning jointly a representation and different models.

Remark 3. In MTL, when one want to optimize over multiple tasks jointly, the question of the stopping criteria arise. For the SGD algorithm at hand, we defined two different stopping criteria: (i) ϵ is used to measure the losses difference between two epochs. If the difference is lower than a given threshold, say $\epsilon = 10^{-3}$, then we stop the descent and keep the weights. In our MTL scenario, we set ϵ to be a shared criteria accross all tasks. It means that some tasks might continue to update, even if they reached a point where the difference in losses between 2 epochs is lower than ϵ. In other words, the task that takes the slowest task to converge is going to set the number of iteration of the whole optimization. (ii) *max_iters* is used as a safeguard, in the case where the ϵ does not stop the algorithm and defines the maximum number of iteration.

5   Experiments

In order to evaluate the proposed framework and the quality of the feature extraction method, we conducted a series of experiments aimed at showing the benefit of the jointly learning the representations and the models with SGD strategy. We release the source code[4] and the dataset for research purposes. The aim of the experiments are twofold: (a) to empirically assess if the tasks are effectively interdependent; (b) to evaluate the benefits of learning the tasks jointly.

Setup and Evaluation Measures. In both frameworks, MTL and STL, we used ℓ_2 regularized Logistic Regression implemented using SGD algorithm in order to minimize the loss functions for the classification and pairwise ranking tasks. The tuning of hyper-parameters has been made by cross validation over the F1-measure of 3 hyper-parameters: the regularization parameters λ in the range $[10^{-1}, 10^{-4}]$, the class weights in the set $\{1, 3, 5, 7, 9\}$ and the decision threshold probability in $\{0.3, 0.35, 0.4, 0.45, 0.5\}$.

We evaluated the ranking results using two metrics. First, we used the area under the ROC curve (AUC) averaged over all users, and the Mean Average Precision at k (MAP@k). In the case of highly imbalanced datasets, a classical classification metric is the F1-measure that is defined as the harmonic mean of precision and recall.

---

[4] https://github.com/asarbaev/Multi-Target-learning.

		Single task				Multitask			
		F_1	AUC	MAP@1	MAP@5	F_1	AUC	MAP@1	MAP@5
Ranking	Clicking	$0.51^{\downarrow}$	$0.62^{\downarrow}$	$0.659^{\downarrow}$	$0.537^{\downarrow}$	**0.54**	**0.65**	**0.708**	**0.584**
	Embedding	0.42	$0.52^{\downarrow}$	**0.427**	**0.318**	**0.46**	**0.54**	0.421	0.314
Classification	Bookmarked	$0.03^{\downarrow}$	$0.51^{\downarrow}$	$0.008^{\downarrow}$	$0.005^{\downarrow}$	**0.06**	**0.54**	**0.026**	**0.015**
	Replied	**0.17**	$0.55^{\downarrow}$	$0.062^{\downarrow}$	$0.038^{\downarrow}$	0.16	**0.58**	**0.083**	**0.05**

Fig. 2. Results on classification (F1-measure) and pairwise ranking (AUC) averaged over users for the single-task and the multi-task/stacking strategies on the DAEMON dataset. The best results are shown in bold, and a $\downarrow$ indicates a result that is statistically significantly worse than the best, according to a Wilcoxon rank sum test with $p < 0.01$.

Implementation Details and Running Time. In our implementation, the feature space is a n-dimensional space, $\mathbf{x} \in \mathbb{R}^n$. n consists in 15 extracted statistical features (see Sect. 2), 16 contextual features directly extracted from the original dataset, 30 features for the user embedding and 30 features for the offers embedding. Thus, $(w_1, w_2, w_3) \in \mathbb{R}^{91}$, $\mathbb{U} \in \mathbb{R}^{\#users \times 30}$ and $\mathbb{I} \in \mathbb{R}^{\#offers \times 30}$. The stopping criteria is set to $\epsilon = 10^{-3}$, the learning rate $\eta = 10^{-6}$ and $\#max_iters = 100$. We set the number of iterations for each epochs to be the product between the number of unique users, the minimum number of positive interacted offers and the minimum number of negative interacted offers as

$$\#iters = \#unique\_users \times \#min\_nb\_pos \times \#min\_nb\_neg,$$

where $\#unique\_users$ is the number of unique users in dataset, $\#min\_nb\_pos$ (respectively $\#min\_nb\_neg$) represents lowest number of positives (resp. negatives) clicks interactions that a user can have.

We implemented the SGD described in Algorithm 1 in Python for both, the single task learning and MTL models. The computations for 1 epoch takes about 10 min on a single 3.2 GHz core, that is about 6.5 h for the training of all models for MTL when considering a learning rate of $\eta = 10^{-6}$ and a stopping criteria set to $\epsilon = 10^{-3}$.

Models Performance Analysis. Figure 2 presents the results for all the tasks and both approaches, heterogeneous multi-target learning and single task learning. We use a bold font to highlight the highest performance rates and a $\downarrow$ to show that a performance is significantly worse than the best result, according to a Wilcoxon rank sum test used at a p-value threshold of 0.01 [11]. First, and without any surprise, the predictions performance are better for the balanced tasks, in terms of positives and negatives examples. In both setups, the predictions of clicks provides better results than any other tasks, up to 0.65 in terms of AUC and 0.708 in terms of $MAP@1$. In most cases, we observe that multi-task ranking approach achieves statistically significant improvements compared to the single-task learning approach. While quite close regarding $MAP@k$ measure, multi-task improves AUC measure consistently for all the tasks, apart from the embedding learning tasks. Intuitively, the learning of the embedding in the single task learning case is specialized for the clicks task while in the multitask case it bias all tasks. Regarding F1-measure, the multi-target learning approach is also better in all tasks but in the case of replies.

6 Conclusions and Future Work

In this paper, we consider the problem of MTL dyadic prediction in a recommendation systems setting. Based on the collection of the RecSys 2016 challenge for a job recommendation, we propose a method to extract meaningful dyads representation based on multiple implicit feedbacks and extracted DAEMON a dataset for this task. We also propose and implement a SGD algorithm that learns jointly over multiple heterogeneous tasks. To the best of our knowledge, this work is the first to learn an embedding jointly with multiple heterogeneous tasks using a SGD approach. We show that this algorithm allows a substantial improvement over the single-target learning case. These results also bring evidence that the proposed dataset is of interest for the problem at hand. An interesting starting point for future work would be to extend the proposed heterogeneous SGD algorithm with a shared regularization.

References

1. Bennett, J., Lanning, S.: The Netflix prize. In: KDD Cup and Workshop 2007, p. 35 (2007)
2. Ben-David, S., Schuller, R.: Exploiting task relatedness for multiple task learning. In: Schölkopf, B., Warmuth, M.K. (eds.) COLT-Kernel 2003. LNCS (LNAI), vol. 2777, pp. 567–580. Springer, Heidelberg (2003). https://doi.org/10.1007/978-3-540-45167-9_41
3. Caruana, R.: Multitask Learning. Mach. Learn. **28**(1), 41–75 (1997)
4. Yang, X., Seyoung, K., Xing, E.P.: Heterogeneous multi-task learning with joint sparsity constraints. In: Advances in Neural Information Processing Systems 22, Vancouver, pp. 2151–2159 (2009)
5. Sculley, D.: Combined regression and ranking. In: 16th ACM SIGKDD International Conference on Knowledge Discovery and Data Mining, Washington, pp. 979–988 (2010)
6. Kumar, A., Daume, H.: Learning task grouping and overlap in multi-task learning. In: 29th International Conference on Machine Learning, New York, pp. 1383–1390 (2012)
7. Chapelle, O., Shivaswamy, P., Vadrevu, S., Weinberger, K., Zhang, Y., Tseng, B.: Multi-task learning for boosting with application to web search ranking. In: 16th ACM SIGKDD International Conference on Knowledge Discovery and Data Mining, Washington, pp. 1189–1198 (2010)
8. Evgeniou, T., Pontil, M.: Regularized multi-task learning. In: 10th ACM SIGKDD International Conference on Knowledge Discovery and Data Mining, New York, pp. 109–117 (2004)
9. Stock, M., Pahikkala, T., Airola, A., De Baets, B., Waegeman, W.: Efficient Pairwise Learning using Kernel Ridge Regression: an Exact Two-Step Method. Technical report (2016)

10. Volkovs, M., Zemel, R.S.: Collaborative ranking with 17 parameters. In: Advances in Neural Information Processing Systems 25, Lake Tahoe, pp. 2294–2302 (2012)
11. Lehmann, E.L., Romano, J.P.: Testing Statistical Hypotheses. Springer Texts in Statistics. Springer, New York (2005). https://doi.org/10.1007/0-387-27605-X

Parallel Cooperative Ensemble Learning by Adaptive Data Weighting and Error-Correcting Output Codes

Shota Utsumi[1(✉)] and Keisuke Kameyama[2]

[1] Graduate School of Systems and Information Engineering, University of Tsukuba,
1–1–1, Tennoudai, Tsukuba-shi, Ibaraki, Japan
utsumi@adapt.cs.tsukuba.ac.jp
[2] Faculty of Engineering, Information and Systems, University of Tsukuba, 1–1–1
Tennoudai, Tsukuba-shi, Ibaraki, Japan
keisuke.kameyama@cs.tsukuba.ac.jp

Abstract. AdaBoost uses the weights assigned to samples to make the latest weak hypothesis adapt to classification mistakes of existing weak hypotheses. However, AdaBoost is very sensitive to the outliers and the existing hypotheses cannot be further trained to cooperate with the newer one. We proposed a new algorithm which prepares all weak hypotheses from the beginning of the training and trains all of them in parallel. Thus, the weak hypotheses are able to cooperate with each other during training. Also, we changed the function which update the weights of the samples to suppress the effects of the weights of outliers. We compared the performances of the new algorithm on several error-correcting output codes and weak hypothesis types. It was found that the proposed PCEL improves the accuracies of multi-class classification task in most datasets.

Keywords: Ensemble learning · AdaBoost · Multi-class classification
Error-correcting output codes

1 Introduction

Ensemble learning is one of the machine learning schemes which uses multiple classifiers called "weak hypotheses". Among them, AdaBoost is the most popular. AdaBoost is composed of weak hypotheses that are trained using weight-assigned samples of which weights are calculated observing the misclassifications of previous weak hypotheses. However, there are cases when the performance do not improve due to the outliers. The weak hypotheses which have already been trained are fixed, thus they cannot cooperate with the newly-added weak hypotheses.

In this paper, we propose an ensemble method which trains all weak hypotheses in parallel and allow them to cooperate with each other via the weights

© Springer Nature Switzerland AG 2018
L. Cheng et al. (Eds.): ICONIP 2018, LNCS 11303, pp. 673–683, 2018.
https://doi.org/10.1007/978-3-030-04182-3_59

assigned to the data. We aim to improve the performance of multi-class classification using AdaBoost.OC, which employs error-correcting output codes [8].

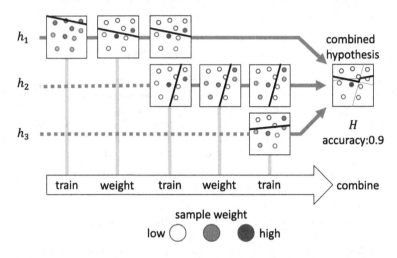

Fig. 1. Serial training and weight assignment in conventional AdaBoost.

2 Related Work

2.1 AdaBoost.OC

AdaBoost [3] is one of the boosting algorithms which trains several classifiers named "weak hypotheses" in a serial order. When training a new weak hypothesis, the importances of the samples which were misclassified by the previous weak hypotheses are increased, and a new weak hypothesis is made concentrating to such samples which were difficult to classify. The final combined hypothesis is a weighted vote of the weak hypotheses (Fig. 1). AdaBoost is an algorithm for two-class classification task, whose weak hypotheses only need to be slightly better than a random guess. However, when AdaBoost classifies a multi-class classification task, it is more difficult for the weak hypotheses to achieve error rates of $\frac{1}{2}$ or less. Hence, AdaBoost.M2 and AdaBoost.OC [8] were proposed.

In AdaBoost.M2, weak hypotheses output a set of labels which is likely to be correct instead of a single label corresponding to the input sample. A combined hypothesis outputs a label which is most common in the output sets of all weak hypotheses.

AdaBoost.OC, proposed by Schapire [8], makes weak hypotheses to output a binary code. A weak hypothesis outputs a single bit of error-correcting output code [2] which is set in advance for converting a multi-class classification task to multiple two-class classification tasks. A combined hypothesis outputs a label whose assigned output code is closest to the actual outputs of weak hypotheses according to the weighted Hamming distance.

Algorithm 1 shows the training procedure of AdaBoost.OC. Table 1 shows the variables used.

In Algorithm 1, first, all weights of samples of weak hypotheses $\tilde{D}_t(i, l)$ are initialized (line 1). Then, a new weak hypothesis $\tilde{h}_t(\mathbf{x})$ is trained under $\tilde{D}_t(i, l)$ (line 4). The error rate $\tilde{\epsilon}_t$ (line 4), the weight of a weak hypothesis α_t (line 5), and weights of samples $\tilde{D}_t(i, l)$ of the weak hypothesis are updated based upon all other outputs (line 6) as,

$$\tilde{D}_{t+1}(i, l) = \frac{1}{Z_t} \tilde{D}_t(i, l) \cdot \exp\left(\alpha_t \left([y_i \notin \tilde{h}_t(\mathbf{x}_i)] + [l \in \tilde{h}_t(\mathbf{x}_i)]\right)\right). \quad (1)$$

The parameters of the existing hypotheses are frozen, and a new hypotheses is added. This process is repeated until a stopping criterion is met, or for predetermined iterations T. Finally, the combined hypothesis $H(\mathbf{x})$ is organized from all weak hypotheses (line 8).

Table 1. List of variables used in AdaBoost.OC

Variables	Role
K	Number of classes
M	Number of samples
$\{(\mathbf{x}_i, y_i)\}_{i=1}^{M}$	Set of samples and their corresponding labels
$\tilde{D}_t(i, l)$	The weight of sample $\mathbf{x}_i$ when $l = \tilde{h}_t(\mathbf{x}_i)$
Z_t	A normalization factor which makes sum of $\tilde{D}_{t+1}$ to 1
$l \in \{-1, +1\}$	The label of 2-class classification tasks
$[\pi]$	The notation defined to be 1 if π is *true* holds and 0 otherwise
$\tilde{h}_t(\mathbf{x})$	t-th weak hypothesis for sample $\mathbf{x}$
T	Maximum boosting iteration
$\tilde{\epsilon}_t$	Error rate under $\tilde{D}_t$
α_t	Weight of weak hypothesis $\tilde{h}_t(\mathbf{x})$ in $H(\mathbf{x})$
$H(\mathbf{x})$	The combined hypothesis

Figure 1 shows the process of AdaBoost(.OC). This example uses the linear classifier as the weak hypotheses. The training data are distributed in two-dimensional space and are for two classes. First, weak hypothesis h_1 is trained, and the weights of the samples are updated based upon its outputs. Then, h_1 is frozen and a new weak hypothesis h_2 is trained under the weights of the samples. After training, h_2 is frozen and the weights of the samples are updated. Further, another weak hypothesis h_3 is trained under the updated weights of the samples. Finally, the combined hypothesis is built from the three weak hypotheses h_1, h_2, and h_3.

Algorithm 1. Adaboost.OC

Require:
  $\{(\mathbf{x}_1, y_1), \ldots, (\mathbf{x}_M, y_M)\}, \mathbf{x}_i \in X, y_i \in Y$
Provides:
  combined hypothesis $H(\mathbf{x})$
1: Initialize $\tilde{D}_1(i, l) = \frac{[l \neq y_i]}{M(K-1)}$
2: **for** $t = 1, 2, \ldots, T$ **do**
3:    Train weak hypothesis $\tilde{h}_t$ using weight $\tilde{D}_t$
4:    Let $\tilde{\epsilon}_t = \frac{1}{2} \sum_{i=1}^{m} \sum_{l \in Y} \tilde{D}_t(i, l) \cdot \left([y_i \notin \tilde{h}_t(\mathbf{x}_i)] + [l \in \tilde{h}_t(\mathbf{x}_i)] \right)$
5:    Let $\alpha_t = \frac{1}{2} \ln \left(\frac{1 - \tilde{\epsilon}_t}{\tilde{\epsilon}_t} \right)$
6:    Update $\tilde{D}_{t+1}(i, l) = \frac{\tilde{D}_t(i, l) \cdot \exp\left(\alpha_t \left([y_i \notin \tilde{h}_t(\mathbf{x}_i)] + [l \in \tilde{h}_t(\mathbf{x}_i)] \right) \right)}{Z_t}$
7: **end for**
8: Organize combined hypothesis $H(\mathbf{x}) = \underset{l \in Y}{\mathrm{argmax}} \sum_{t=1}^{T} \alpha_t [l \in \tilde{h}_t(\mathbf{x})]$

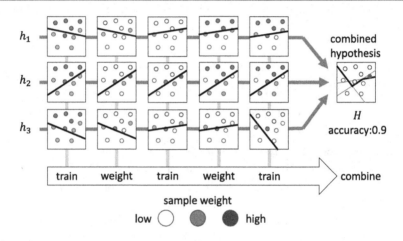

Fig. 2. Parallel training and mutual weight assignment in PCEL.

2.2 Aim of This Work

In conventional AdaBoost(.OC), the older weak hypotheses cannot change to cooperate with the newer one. One way to introduce cooperation among hypotheses is to train them in parallel, allowing them to communicate via weights assigned to training data. Although there exist other algorithms focusing on parallelization of AdaBoost such as AdaBoost.PL [7] and MULTBOOST [4], AdaBoost.PL focuses on speedup of AdaBoost and MULTBOOST focuses on privacy preservation.

AdaBoost, including the AdaBoost.OC, uses an exponential function for calculating the weights of the samples. This enables weak hypotheses to adjust to the weights of the samples which is hard to be classified by the others. However, when a sample is an outlier, its weight becomes very high because many weak hypotheses misclassifies this particular sample. This can lead to a weak hypotheses which is harmful to the combined hypotheses.

In this paper, we propose a new algorithm which prepares all weak hypotheses from the beginning of training and trains all of them in parallel. This algorithm lets the weak hypotheses to be able to cooperate with each other during training. Also, we change the function to update the weights of the samples to improve the handling of the outliers.

3   Parallel Cooperative Ensemble Learning (PCEL)

To achieve cooperation among all weak hypotheses, we propose a method that trains all weak hypotheses in parallel without freezing. However, the intervention from other hypotheses should decrease when the error rate of a weak hypothesis is low. Also, to control the sensitivity to the samples which are misclassified by many weak hypotheses, we change the function for calculating the weights to saturate at a modest value.

3.1   Parallel Training for Cooperation

In the proposed algorithm, all weak hypotheses are trained in parallel to cooperate with each other. Each hypothesis focuses on the training samples which are misclassified by other weak hypotheses. The weights of samples for a weak hypothesis are calculated according to the performances of the other weak hypotheses, and this is done for all weak hypotheses in parallel. Samples that are difficult to learn are assigned large weights by multiple hypotheses. It will be solved by other hypotheses, achieving cooperation among them.

3.2   Sample Weight Assignment Function

The effect of the outliers can be made smaller by changing the function for calculating the weights of samples. AdaBoost uses an exponential function for calculating the weights. The weights of samples (D) are determined exponentially by the weighted sum of outputs of others. Thus, the trained weak hypothesis becomes too sensitive to the samples which is misclassified by many others. In order to solve this issue, the proposed algorithm uses a modestly-saturating increasing function such as the sigmoid function.

In addition, the deviation of the weights is affected by the number of weak hypotheses. For example, if the performances of all weak hypotheses are the same, the number of weak hypotheses which misclassify one data is in proportion to their number. Thus, the samples which tend to be misclassified by weak hypotheses such as the outliers will have very high weights. In this work, the function which determines the weights of the samples is stretched in proportion to the number of hypotheses T to keep validations of the weights of the data constant regardless of T. Equation (2) shows the stretched sigmoid function used in this work and Eq. (3) shows the proposed weighting function W_s.

$$s(x) = \frac{1}{1 + \exp(-\frac{x}{T})} \tag{2}$$

$$W_s(u,t) = \frac{s\left(\sum_{v=1}^{U}[v \neq u] \cdot \frac{\alpha_v^t}{T}\left([y_i \notin \tilde{h}_v^t(\mathbf{x}_i)] + [l \in \tilde{h}_v^t(\mathbf{x}_i)]\right)\right)}{Z_u^t} \tag{3}$$

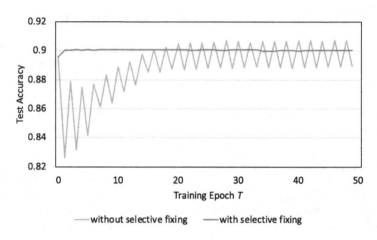

Fig. 3. Fluctuation of test accuracies in PCEL without hypothesis fixing. (Color figure online)

3.3   Fixing Weak Hypotheses According to Their Accuracy

In the proposed method, all weak hypotheses are trained to mutually make up for the classification error of the other weak hypotheses via weights of the samples. However, the blue plot in Fig. 3 shows the transition of the test accuracies through the training epochs. There, fluctuation of accuracies is observed. A weak hypothesis is trained with certain weights of data, which is calculated by the other weak hypotheses in one epoch. However, the weights of the data may be changed and can differ from the previous epoch. Then, the target of a weak hypothesis can be changed from the previous one. Figure 3 shows the state in which the optimum of weak hypotheses are changed frequently. If one hypotheses is performing well, the necessity of retraining it should be low. In such cases, weak hypotheses ought to be fixed.

In this work, we introduce a momentum γ_u in updating the weights of data. This momentum is calculated linearly proportional to the loss $\tilde{\epsilon}_u^t$. Equation (4) shows the modified calculation of the weights of data. By increasing γ_u as $\tilde{\epsilon}_u^t$ decreases, it is possible to fix weak hypotheses which has high accuracy rate. Here, γ_u is determined according to Eq. (5).

$$\tilde{D}_u^{t+1}(i,l) = \frac{\gamma_u \tilde{D}_u^t(i,l) + (1-\gamma_u)W_s(u,t)}{Z_u^t} \tag{4}$$

$$\gamma_u = -2|\tilde{\epsilon}_u - 0.5| + 1 \tag{5}$$

3.4 PCEL

The proposed algorithm is called Parallel Cooperative Ensemble Learning (PCEL). Algorithm 2 shows the algorithm of PCEL. Table 2 shows the variables used. The u-th weak hypotheses in iteration t is $\tilde{h}_u^t(\mathbf{x})$. First, all weights of the training samples for hypotheses are initialized (line 1). Then, all weak hypotheses $\tilde{h}_u^t(\mathbf{x})$ learns under weights $D_u^t(i,l)$ (line 4). Then the error rate $\tilde{\epsilon}_u^t$ (line 5), and the weight of weak hypothesis α_u^t are updated (line 6). Next, the weights of samples $D_u^t(i,l)$ of weak hypothesis are updated using all other outputs (line 9). This loop will be run in parallel. Finally, the combined hypothesis $H(\mathbf{x})$ is organized from all weak hypotheses $\tilde{h}_u^T(\mathbf{x})$ (line 11).

Table 2. List of variables used in PCEL

Variables	Role
$\tilde{h}_u^t(\mathbf{x})$	u-th weak hypothesis in iteration t
$\tilde{D}_u^t(i,l)$	Weight of samples of $\tilde{h}_u^t(\mathbf{x})$
Z_u^t	A normalization factor which makes sum of $\tilde{D}_{u+1}^t$ to 1
U	Maximum boosting iteration
T	Number of weak hypotheses
$\tilde{\epsilon}_u^t$	Error rate under $\tilde{D}_u^t$
α_u^t	Weight of a weak hypothesis $\tilde{h}_u^t(\mathbf{x})$
$H_U(\mathbf{x})$	Combined hypothesis

Figure 2 illustrates the process of PCEL in contrast with Fig. 1. There are three weak hypotheses h_1, h_2, and h_3 from the beginning. Then, these weak hypotheses are trained, and weights of the samples for each weak hypothesis h_t are updated based upon the outputs of all weak hypotheses except h_t. Then, the weak hypotheses are retrained under their updated weights of the samples. Finally, the combined hypotheses is built from three weak hypotheses h_1, h_2, and h_3.

4 Experimental Results

This section shows the comparison between AdaBoost.OC and PCEL in the performances of multi-class classification tasks.

Algorithm 2. PCEL

Require:
$\quad \{(\mathbf{x}_1, y_1), \ldots, (\mathbf{x}_M, y_M)\}, \mathbf{x}_i \in X, y_i \in Y$

Provides:
$\quad$ combined hypothesis $H(\mathbf{x})$

1: Initialize $D_u^t(i,l) = \frac{[l \neq y_i]}{M(K-1)} \quad (t \in \{1, 2, \ldots, T\})$

2: **for** $t = 1, 2, \ldots T$ **do**

3: $\quad$ **for** $u = 1, 2, \ldots, U$ **do**

4: $\quad\quad$ Train weak hypothesis $\tilde{h}_u^t$ using weight $\tilde{D}_u^t$

5: $\quad\quad$ Let $\tilde{\epsilon}_u^t = \frac{1}{2} \sum_{i=1}^{M} \sum_{l \in Y} \tilde{D}_u^t(i,l) \cdot \left([y_i \notin \tilde{h}_u^t(\mathbf{x}_i)] + [l \in \tilde{h}_u^t(\mathbf{x}_i)] \right)$

6: $\quad\quad$ Let $\alpha_u^t = \ln\left(\frac{1 - \tilde{\epsilon}_u^t}{\tilde{\epsilon}_u^t} \right)$

7: $\quad\quad$ Update $\gamma_u = -2|\tilde{\epsilon}_u - 0.5| + 1$

8: $\quad$ **end for**

9: $\quad$ Update $\tilde{D}_u^{t+1}(i,l) = \frac{\gamma_u \tilde{D}_u^t(i,l) + (1-\gamma_u)s\left(\sum_{v=1}^{U} [v \neq u] \cdot \frac{\alpha_v^t}{T} \left([y_i \notin \tilde{h}_v^t(\mathbf{x}_i)] + [l \in \tilde{h}_v^t(\mathbf{x}_i)] \right) \right)}{Z_u^t}$

10: **end for**

11: Organize combined hypothesis $H(\mathbf{x}) = \underset{l \in Y}{\text{argmax}} \sum_{u=1}^{U} \alpha_u^T [l \in \tilde{h}_u^T(\mathbf{x})]$

Table 3. Datasets used in the experiments

Dataset	Instance	Feature	Class	ECF-ECOC Length
DNA [5]	2000	180	3	3
Iris	150	4	3	3
Vehicle	846	18	4	4
Vowel	528	10	11	12

4.1 The Datasets

Table 3 shows the datasets which were used in the experiments. Iris, Vehicle and Vowel datasets were obtained from the UCI Machine Learning Repository [1].

4.2 Experimental Conditions

The Experiments were performed 50 times each and the average test accuracies for AdaBoost.OC and PCEL were recorded.

Weak Hypotheses. In the experiments, we used both simple perceptron and Multi Layer Perceptron (MLP) for weak hypotheses. They used hyperbolic tangent function as activation functions, and were trained by RMSProp [9]. A MLP has 10 units in the hidden layer. The weak hypotheses are generated as many as the length of ECOC.

Output Codes. The experiments used two types of output codes, which are 10 bits random output codes (Exp. 1) and ECOC generated by Error-Correcting Factorization (ECF) [6] (Exp. 2). ECF generates ECOC considering the inter-class distances from the dataset of each pair of classes. When there are two classes that are more difficult to distinguish than the other pairs, ECF assigns an ECOC which has a larger Hamming distance for this pair. The lengths of ECF-ECOC are shown in Table 3.

4.3   Results

Tables 4 and 5 show the results of the experiments.

Accuracies. In most cases, the accuracies of PCEL surpasses those of AdaBoost.OC. It reflects the benefits of cooperation among weak hypotheses in PCEL, where weak hypotheses compensated for the misclassified data mutually. However, in some datasets, the accuracy of PCEL were lower.

Table 4. Experiment 1 (Random output code): mean classification accuracies and variances.

Weak hypothesis	Dataset	AdaBoost.OC	PCEL
Perceptron	DNA	$0.887 \ (6.5 \times 10^{-5})$	$\mathbf{0.893} \ (1.4 \times 10^{-4})$
	Iris	$0.955 \ (3.1 \times 10^{-4})$	$\mathbf{0.973} \ (0)$
	Vehicle	$0.700 \ (8.2 \times 10^{-4})$	$\mathbf{0.751} \ (3.4 \times 10^{-4})$
	Vowel	$0.320 \ (2.2 \times 10^{-3})$	$\mathbf{0.405} \ (3.7 \times 10^{-3})$
MLP	DNA	$\mathbf{0.904} \ (1.2 \times 10^{-4})$	$0.900 \ (8.1 \times 10^{-5})$
	Iris	$\mathbf{0.972} \ (2.6 \times 10^{-5})$	$0.959 \ (1.1 \times 10^{-4})$
	Vehicle	$0.782 \ (3.2 \times 10^{-4})$	$\mathbf{0.805} \ (2.1 \times 10^{-4})$
	Vowel	$0.869 \ (1.4 \times 10^{-3})$	$\mathbf{0.946} \ (8.8 \times 10^{-4})$

Variance. In most cases, the accuracy variances of PCEL were lower than those of AdaBoost.OC. This shows the strength of PCEL against outlier samples, that can cause AdaBoost.OC to be unsuccessful when many of the frozen hypotheses fail to classify them.

Weak Hypotheses. In most cases, the accuracies of methods which used MLPs for the weak hypotheses surpasses those of simple perceptrons. It reflects the benefits of the performances of the weak hypotheses. In Vowel dataset, it is remarkable.

The choice of weak hypotheses did not affect variances.

Table 5. Experiment 2 (ECF-ECOC): mean classification accuracies and variances.

Weak hypothesis	Dataset	AdaBoost.OC	PCEL
Perceptron	DNA	**0.885** (4.3×10^{-5})	0.870 (2.6×10^{-8})
	Iris	0.933 (1.7×10^{-3})	**0.973** (0)
	Vehicle	0.605 (1.9×10^{-4})	**0.739** (6.3×10^{-6})
	Vowel	0.455 (3.3×10^{-4})	**0.487** (3.8×10^{-5})
MLP	DNA	**0.912** (2.0×10^{-5})	0.889 (3.2×10^{-5})
	Iris	**0.973** (0)	0.946 (9.2×10^{-5})
	Vehicle	0.782 (9.2×10^{-4})	**0.786** (1.6×10^{-4})
	Vowel	0.923 (2.8×10^{-4})	**0.975** (5.1×10^{-5})

Output Codes. The accuracies were slightly higher when random codes were used (Exp. 1) than ECF-ECOC (Exp. 2), except for the Vowel dataset. This may be explained by the lengths of the output codes. ECF-ECOC (Table 3) were much shorter than random codes (10 bits) except for Vowel. As shorter code means less number of weak hypotheses to cooperate, this might have caused the lower performance. The suitable lengths of ECOC for use in PCEL needs to be investigated further. As the codes were unchanged throughout the trials in ECF-ECOC, but varied in random codes, the accuracy variance for ECF-ECOC were lower.

5   Conclusion

In this paper, we proposed an ensemble learning algorithm based on Adaboost, which trains the weak hypotheses in parallel. This algorithm is named PCEL. All weak hypotheses train in parallel, and they cooperate via the weights assigned to the training data. Besides parallelism, two significant changes have been made from AdaBoost.OC. First is the choice of the function for calculating the weights of data. The exponential function used in AdaBoost.OC made the training too sensitive to outliers. Because of that, we changed the function to the sigmoid function. Second, we added a rule which fixes weak hypotheses according to their accuracies. Experiments showed that PCEL gave higher recognition performance when compared with AdaBoost.OC in most cases. However, some assigned codes brought down the performance of the other weak hypotheses and the accuracies of a combined hypothesis became lower. In the future, we plan to propose a method which adjusts the optimal output codes during training.

References

1. Dheeru, D., Karra Taniskidou, E.: UCI machine learning repository (2017). http://archive.ics.uci.edu/ml
2. Dietterich, T.G., Bakiri., G.: Solving multiclass learning problems via error-correcting output codes. J. Artif. Intell. Res. **2**(1), 263–286 (1995)
3. Freund, Y., Schapire., R.E.: A decision-theoretic generalization of on-line learning and an application to boosting. J. Comput. Syst. Sci. **55**, 119–139 (1997)
4. Gambs, S., Kégl, B., Aïmeur, E.: Privacy-preserving boosting. Data Mining Knowl. Discov. **14**(1), 131–170 (2007)
5. Hsu, C.W., Lin, C.J.: A comparison of methods for multiclass support vector machines. IEEE Trans. Neural Networks **13**(2), 415–425 (2002)
6. Martin, M.U.B., Pujol, O., l. Torre, F.D., Escalera, S.: Error-correcting factorization. IEEE Trans. Pattern Anal. Mach. Intell. **40**(10), 2388–2401 (2018)
7. Palit, I., Reddy., C.K.: Scalable and parallel boosting with mapreduce. IEEE Trans. Knowl. Data Eng. **24**(10), 1904–1916 (2012)
8. Schapire, R.E.: Using output codes to boost multiclass learning problems. In: 14th International Conference on Machine Learning, pp. 313–321. Morgan Kaufmann Publishers Inc., San Francisco (1997)
9. Tieleman, T., Hinton, G.: Lecture 6.5–RmsProp: Divide the gradient by a running average of its recent magnitude. COURSERA: Neural Networks for Machine Learning (2012)

Author Index

Printed in the United States
By Bookmasters